China Semiconductor Technology International Conference 2011 (CSTIC 2011)

Editors:

H. Wu
Semiconductor Manufacturing International
Corporation
Shanghai, China

C. Claeys
imec
Leuven, Belgium

Y. Kuo
Texas A&M University
College Station, Texas, USA

K. Lai
IBM, Thomas J. Watson Research Center
Yorktown Heights, New York, USA

A. Philipossian
The University of Arizona
Tucson, Arizona, USA

T. Jiang
Maxim Integrated Products Inc.
Sunnydale, California, USA

S. Xiaoping
imec
Leuven, Belgium

Q. Lin
IBM, Thomas J. Watson Research Center
Yorktown Heights, New York, USA

D. Huang
Praxair
Danbury, Connecticut, USA

R. Huang
Peking University
Beijing, China

Y. Zhang
Yorktown Heights, New York, USA

R. Liu
Fudan University
Shanghia, China

P. Song
IBM, Thomas J. Watson Research Center
Yorktown Heights, New York, USA

Published by
The Electrochemical Society

65 South Main Street, Building D
Pennington, NJ 08534-2839, USA

tel   609 737 1902
fax   609 737 2743
www.electrochem.org

ecstransactions ™

Vol. 34 No. 1

Published by:

The Electrochemical Society
65 South Main Street
Pennington, New Jersey 08534-2839, USA

Telephone 609.737.1902
Fax 609.737.2743
e-mail: ecs@electrochem.org
Web: www.electrochem.org

ISSN 1938-6737 (online)
ISSN 1938-5862 (print)

ISBN 978-1-56677-884-1 (CD-ROM)
ISBN 978-1-60768-234-9 (PDF)
ISBN 978-1-60768-235-6 (Soft-cover)

Printed in the United States of America.

Preface

On behalf of the conference committee and organizers, we would like to express our sincere thanks to all attendees for their active participation and significant contribution to one of the largest semiconductor conferences in China, CSTIC-2011 (China International Semiconductor Technology Conference). The 2011 conference will be the tenth international conference since ISTC was first initiated in 2001 and it has been constantly well organized, jointly, by The Electrochemical Society and SEMI, co-organized by CHTEC, and sponsored by IEEE, CEMIA, MRS and CSE.

Our mission is to provide a forum for world experts to discuss technologies, address the growing needs associated with silicon technology, and exchange their discoveries and solutions for current issues of high interest. We encourage collaboration, open discussion, and critical reviews at this conference. Furthermore, we hope that this conference will also provide collaborative opportunities for those who are interested in the semiconductor industry in Asia, particularly in China.

The conference committee consists of more than 100 world-class experts from high profile companies including, but not limited to Intel, IBM, SMIC, IMEC, Maxim, Applied Materials, TEL, Praxair, MEMC, Cabot Microelectronics, Dow, and also local companies and universities such as Anji, Gritek, Peking University, Fudan University and Zhejiang University. Guest speakers came from companies and institutions like those mentioned above, as well as renowned universities such as UC Berkeley, Stanford, Yale, Peking University, Fudan University, and other top universities around the world. We are proud of this prestigious volunteer team and feel honored by the distinguished speakers. With their contributions, we believe that we have prepared a world class international conference.

The conference offers more than 390 high quality presentations and posters in ten symposia to cover most of the aspects of semiconductor and solar technology. It is well known that the silicon technology development trend is primarily dependent of innovative materials application. In this volume, there are more than 50% presentations are new material and process related technologies, while the rest are related to device and process integration. As a part of low-carbon society, we have enhanced symposium X "Silicon Materials for Electronic and Photovoltaic Applications". The conference started with a plenary session before all the symposia, covering a critical review and high-level summary of the technology.

We have achieved a record number of paper submissions this year from all around the world. We've seen strong participation from the USA, Europe, areas in Asia, and other countries, and especially growing paper submission from China. The following table will provide you with a brief summary of the paper distribution in each symposium:

Table 1 – Conference Symposia and Papers

Symposium	Total Presentation
Plenary Session	4
I - Design and Device Engineering	55
II - Lithography and Patterning	35
III - Dry &Wet Etch and Cleaning	38
IV - Thin Film Technology	42
V - CMP and Post-CMP Cleaning	40
VI - Materials and Process Integration for Device and Interconnection	32
VII - Packaging and Assembly	33
VIII - Metrology, Reliability and Testing	42
IX - Emerging Semiconductor Technologies	35
X - Silicon Materials for Electronic and Photovoltaic Applications	39
Total	395

There is little doubt that our conference would have been as successful without active participation from the audience and its readers. We appreciate your support and trust that this issue of *ECS Transactions* will be a worthwhile addition and reference to your collection.

Han-Ming Wu (SMIC), Conference Chairman
Allen Lu, President of SEMI China
Roque J. Calvo, Executive Director of ECS
CSTIC Committee
March 13, 2011, Shanghai, China

ECS Transactions, Volume 34, Issue 1
China Semiconductor Technology International Conference 2011 (CSTIC 2011)

Table of Contents

Chapter 1
Design and Device Engineering

Chapter 2
Lithography and Patterning

Chapter 3
Dry and Wet Etch and Cleaning

Chapter 4
Thin Film Technology

xiii

Chapter 5
CMP and Post-CMP Cleaning

Chapter 6
Materials and Process Integration for Device and Interconnection

Chapter 7
Packaging and Assembly

Chapter 8
Metrology, Reliability and Testing

Chapter 9
Emerging Semiconductor Technologies

Chapter 10
Silicon Materials for Electronic and Photovoltaic Applications

Author Index

Facts about ECS

The Electrochemical Society (ECS) is an international, nonprofit, scientific, educational organization founded for the advancement of the theory and practice of electrochemistry, electrothermics, electronics, and allied subjects. The Society was founded in Philadelphia in 1902 and incorporated in 1930. There are currently over 7,000 scientists and engineers from more than 70 countries who hold individual membership; the Society is also supported by more than 100 corporations through Corporate Memberships.

The technical activities of the Society are carried on by Divisions. Sections of the Society have been organized in a number of cities and regions. Major international meetings of the Society are held in the spring and fall of each year. At these meetings, the Divisions and Groups hold general sessions and sponsor symposia on specialized subjects.

The Society has an active publications program that includes the following.

Journal of The Electrochemical Society — JES is the peer-reviewed leader in the field of electrochemical and solid-state science and technology. Articles are posted online as soon as they become available for publication. This archival journal is also available in a paper edition, published monthly following electronic publication.

Electrochemical and Solid-State Letters — ESL is the first and only rapid-publication electronic journal covering the same technical areas as JES. Articles are posted online as soon as they become available for publication. This peer-reviewed, archival journal is also available in a paper edition, published monthly following electronic publication. It is a joint publication of ECS and the IEEE Electron Devices Society.

Interface — *Interface* is ECS's quarterly news magazine. It provides a forum for the lively exchange of ideas and news among members of ECS and the international scientific community at large. Published online (with free access to all) and in paper, issues highlight special features on the state of electrochemical and solid-state science and technology. The paper edition is automatically sent to all ECS members.

Meeting Abstracts (formerly Extended Abstracts) — Abstracts of the technical papers presented at the spring and fall meetings of the Society are published on CD-ROM.

ECS Transactions — This online database provides access to full-text articles presented at ECS and ECS-sponsored meetings. Content is available through individual articles, or as collections of articles representing entire symposia.

Monograph Volumes — The Society sponsors the publication of hardbound monograph volumes, which provide authoritative accounts of specific topics in electrochemistry, solid-state science, and related disciplines.

For more information on these and other Society activities, visit the ECS website:

www.electrochem.org

CHAPTER 1

DESIGN AND DEVICE ENGINEERING

ECS Transactions, 34 (1) 3-8 (2011)
10.1149/1.3567551 ©The Electrochemical Society

Embedded Non-Volatile Memory Technologies

Danny Shum
Automotive ATM, Infineon Technologies Taiwan, Taipei 11503, Taiwan

The embedded Nonvolatile Memories (eNVM) technologies offer in-system re-programmability along with low voltage (LV), low power (LP), noise reduction, more reliable, and reduced system cost while maintain high speed, high density system performance over other NVM that including stand-alone Flash memory. The extent of market opportunity is driven by product cost. Manufacturability is the primary concern, rather than mere cell-size scalability. The design simplicity with integration compatibility of the logic core process to the additional NVM flow is a key factor when choosing memory cell. In this paper we will review different NVM concepts including cell design, from ROM to Flash, and to non-Si based memories and their applications in support of the diverse and emerging consumer market.

Introduction

eNVM typically consist of ROM (Read Only Memory), EPROM (Electrical Programmable ROM), EEPROM (Electrical Erasable PROM), and Flash EEPROM of Si-based memories, and non-Si-based emerging memories such as FeRAM, MRAM or PCRAM. Referred to as Si-based charge-trapping devices, its principle is to shift the transistor threshold voltage by adding electrons (-Q, high Vt state), remove charges (neutral state), or adding holes (+Q, low Vt state) from the storage node FG as illustrated in Fig.1. Non-Si-based memory cell does not depend on charge, i.e. Ferroelectric is based on polarization of crystal that creates hysteresis, as illustrated in Fig.8. eNVM offers broad areas of applications for MCU (microcontroller) in Automotive, Mobility, and Security markets with key advantages such as dense board designs with reduced number of parts, reduced system costs, reduced noise, higher system speed due to fast code access, in-system on-board re-programmability of code and data storage, lower power dissipation, improved reliability, and real-time control application. Fig.2 shows an example of 32b MCU products.

Requirements

General products requirements consist of faster access time to match the on-chip high performance core logic. Majorities are random read access, and no access degradation due to variable on-chip memory complexities. Variable modules are organized x8, x16, and x32 with tolerable memory module overhead. Most products used on-chip state machine since memory module does not depend on MCU. For Mobility market the eNVM must be LV/LP capability. Most chips require high precision high voltage (HV) regulators and efficient HV generators compatible with host-logic base process without disrupting the logic design rules and electrical parameters. Products in consumer market must be compatible to process simplicity with competitive cost advantage. Redundancy is only in high end MCU with dense memories. Reliability concern must be resolved with design-in test such as error code correction (ECC).

NVM Cell Concepts

Fig.3a illustrates the simplest concept of ROM which consist of a one-transistor cell, program either by channel Vt implant or some sort of fuse solutions. The content is stored or programmed at the IC manufacturing plant; hence, it does not offer re-programmability. But it offers smallest cell size of $8F^2$ (2Fx4F, where F is the minimum feature size). FG approach is to place a floating gate above the transistor channel area to offer flexibility (Fig.3b). Typical cell size is greater than $12F^2$ (3Fx4F) in which the lateral dimension needs extra 1F in bitline to add capacitive-coupling between control gate CG-to-FG. Any charge stored on the FG node shifts the transistor threshold. Obvious conflicting requirements exist on this type of cell concept: a reliable charge transport mechanism through the gate dielectrics, and reliable charge storage in the same dielectrics. EPROM was first introduced by Frohman-Bentchkowsky and mass-produced in 1971 as FAMOS (Floating gate Avalanche-injection MOS) [1]. Its charge transport employs channel hot-electron (CHE) injection from avalanche plasma in the drain region underneath the gate for programming (electron stored in FG, high Vt state, Fig.4b) from neutral state (Fig.4a). Endurance is limited to few cycles due to CHE degradation in tunnel oxide. EPROM is a step above ROM to offer flexibility to change charge, but not in-system. Typical UV exposure time takes 20min with dielectric thickness over 30nm to prevent hot electron damage. Next is in-system re-programming Electrical Erasability PROM by Fowler-Nordheim (FN) tunneling [2], the mechanism that allows electron transport at high field while retent a high barrier at low field for charge retention (Fig.4c).

Floating Gate (FG) Memories

FG cell technology falls into two categories: planar and non-planar. In planar case, the electric injection field is slightly above the Tunnel oxide (Tox) field. Examples are Flash EPROM, EEPROM in which electrons transport by FN tunneling through the thin Tox ~100Å oxide. In non-planar case, known as textured poly device, the electric injection field is much greater than the Tox field. The e-field is enhanced by the surface topography on the cathode, which bends the oxide conduction band for electron tunneling at low field. Examples are poly-poly tunneling of SST cell (Fig.5a) [3]. Typical oxide thickness is in the range of 200-350Å. Flash EEPROM (byte programmability and block/sector erase) can be classified by programming: CHE injection and FN tunneling. In CHE injection the cell is known as SIMOS (Stacked gate Injection MOS). The smallest cell is the 1T structure (Fig.4b) such as ETOX^{TM} [4] (EPROM Tunnel OXide) with asymmetric S/D junction. The abrupt drain junction is optimized for CHE programming while the graded source region is intended for source-erase type [5] with FN tunneling. This type of cell is usually small but often requires complicated 2-step erase algorithm [6] to control Vt distribution with added circuitry. The split-gate (SG) 1.5T structure (SCSG of Fig.5b) is a relax cell in series with an enhancement transistor to offer design simplicity for over-erase Vt control without complicated algorithm or erase scheme [7]. The series transistor ensures a high immunity to drain turn-on, and put a limit on the minimum L_{eff} of the NVM transistor. It also improves source-drain punch-through (PT) problems. Triple-poly structure with Source-Side Injection (SSI, Fig.5c) is designed to reduce programming current by 10x to few μA for LP applications [8]. In FN tunneling the cell is referred to as FLOTOX (FLOating gate Thin OXide, Fig.5d), where the tunneling area is a small thin oxide region over the drain. A variation to incorporate thin oxide over the entire FG transistor channel area creates alternative memory cells with improved performance. This type of NVM cells is called FETMOS (FG Electron Tunneling MOS). The device structure is 1T cell and is further divided into different memory organizations such as NAND (Fig.6a) [9], NOR (Fig.6b), AND (Fig.6c) [10], and other combinations for high density, LP applications. 2T cell (Fig.6d) with a select transistor for byte-addressable function [11] is excellent choice for embedded applications without complicated algorithm or overhead peripheral circuitry for Vt control and still offer LP compatibility with FN tunneling W/E operations.

Non-Floating Gate Memories

MNOS (metal-nitride-oxide-silicon) is the first non-FG device that stored charges in discrete traps in the nitride layer [12]. SONOS (Silicon-Oxide-Nitride-Oxide, Fig.7) is the improved MNOS device to inhibit gate injection and to block charges injected from the top oxide-nitride interface, resulting in a higher trapping efficiency and more reliable memory cell. The device can be programmed by FN tunneling [13] or CHE injection [NROM, 14] or SSI [SPIN, 15], and erased by hot hole injection. The key advantage is the FG elimination; thus reduce process complexity and compatible to most single polysilicon CMOS manufacture. The cell size is reduced to $8F^2$ by elimination of 1F from the gate coupling regions.

Emerging memories (FeRAM, MRAM, PCRAM)

Non-Si-based memories that offer alternate NVM solution whose operation is based on resistive effect and reading the stored data by sensing a change in resistance value. Key advantages over conventional Si-based NVM are in read/write speed in ns and extended rewrite endurance cycles to 10^{12}-10^{15} ranges with low program voltages. Some may offer opportunity to build up "universal memories" of NVM + SRAM + DRAM within the same memory, ideally for system-on-chip (SoC) applications.

FeRAM (Ferro-electric RAM) operation is based on non-linear dielectric crystals. Lattice displacements of atoms induce a polarization of the crystal under an external E-field and to remain polarized after the field removal. The polarization can be reversed by applying an opposite polarity field. Thus, a nonvolatile capacitor is obtained in which stored information is based on polarization state rather than on stored charge [16]. If a read voltage is applied to the capacitor of polarity opposite to the previous write voltage, the polarization state will switch to cause electrical hysteresis, giving rise to a large displacement current that sensed by Sense Amplifier (Figs.8-9). Top two of today's most important

Ferroelectrics materials are Strontium Bismuth Tantalate (SBT) and Lead Zirconate Titanate (PZT).

MRAM is composed of a transistor and two magnetic layers (a fixed-orientation layer and a free magnetic layer) separated by thin tunnel barrier < 2nm (Fig.8). Data is written when the free layer is reversed by changing the magnetic orientation induced by currents through BL (bitline) and WL (wordline). High ~1mA of current per cell is required for the cell switching [17]. Newly developed Spin-transfer-torque STT-MRAM [18] uses the magnetization of the spinning electron to change the magnetization of the MRAM bit cell by momentum transfer. A stream of spin-polarized electrons flow through the free layer, caused the free layer to polarize either spin up or down pending flow direction; thus eliminate the need to run current through the BL, reduce power by 100x and potential to smaller cell size.

Phase Change occurs within a chalcogenide layer that has a meta-stable glassy (amorphous, high R) structural state and a stable polycrystalline (low R) structural state (Fig.8). PCM exploits electrical differences between these two states to store data. The speed at which it is allowed to cool determines whether it will crystallize (conductive) or remain amorphous (nonconductive). Switching occurs by applying current pulses to melt the PCM materials which change the layer phase [19].

Process Issues

A double-polysilicon stacked gate CMOS process is needed to fabricate NVMs. The first poly is used as FG with the second poly as CG. A high quality thin Tox is formed between FG and Si with junctions overlaps the gate edge to reduce substrate current. Post-gate poly process can significantly affect the quality of the thin Tox due to the post-annealing temperature. Local oxide thinning and oxide surface roughness at the poly/oxide interface are too influenced by post-gate temperature treatment. They are the direct consequence of the grain growth of the poly gate and viscous flow of the oxide, which are enhanced with rising annealing temperature and time. In addition, poly gate doping level and species play a major role in local variation of tunneling current. Stacked O-N-O interpoly dielectrics are formed between FG and CG to insulate FG from electron leakage or to block holes injection from CG. Typical thickness is in the scaling range of 120-150Å [20]. A high selective gate etch is another key process development to achieve robust manufacturability. The double-polysilicon self-aligned (SA) stacked-gate structure requires a high precision anisotropic dry etching process, new to most wafer manufacture [21]. It etches through the CG-ONO-FG stacked structure and stop at thin Tox to avoid Si damage from etching. Well formation is critical that requires MeV implanter to form N-band region that isolate array P-well from the P-type substrate. For 0.25µm and below, shallow trench isolation (STI) process or modified scheme with more rounded corner is common practice among the industry. HV module is formed with dual gate oxide (DGO), conventional resist patterning and wet etch, with gate oxide thickness in the range of 125-250Å, pending on the cell type. Fig.10 shows the cross-section along the WL and the BL directions [22] that best illustrate many process issues such as gate oxide thinning around STI or FG corner, STI oxide thinning through the stacked gate cell, general oxide qualities, poly doping, and others [23]. Emerging memories present contamination issues when introduce new materials to conventional CMOS manufacturing, different electrodes in Pt, Ir, Ru and their barriers that are critical to integration, polycrystalline thin film and its properties that are a function of morphology, stochiometry, and thickness. The process must be compatible with Cu metallization, ILD, forming gas and other BEOL processing steps.

Challenges of Embedded NVMs

Several process approaches have been discussed with the main stream is either stand-alone approach for low cost and simplifies process modules development, or logic based approach for high performance and simplifies design reuse. The challenge for logic based is the integration of HV transistors (DGO) and Flash cells (Tox and ONO dielectrics) and the additional thermal budget that pleases the NVM reliability but nightmare for logic transistor design of maintaining the same transistors characteristics. Logic based process must use modular approach for compatibility of core logic and embedded memory process [22]. The challenge for stand-alone is how to convert the single gate oxide, buried P-channel transistors, and polycide process to high performance DGO, dual work function salicide process for the high speed peripheral circuitry that is required by the system design. Other issues are modeling of aging device effects, Spice simulations, general CAD and analog circuit support

Reliability Issues

The issues are wide open that include topics such as Tox defect density and burn-in screening, ONO trapping density, cell design, endurance, retention, stress induced (dielectrics) leakage current (SILC) through trap center after endurance, charge induced dielectric breakdown and others. Tox inline monitor is to exclude the extrinsic defect density after deposition in addition to wafer level reliability (WLR) at the end of process. WLR includes traditional TDDB on special test structures, in addition to endurance and retention tests on actual product macros through random sample selection by number of wafers and number of lots per week. Recent improvement of Tox with nitridation of the preoxide to form N_2O oxynitride can improve endurance characteristics well beyond 10^5 cycle's range [24]. Data retention is lost of stored information due to FG charge leakage. Typically below 1V charge loss requires less than 1 electron per day of leakage. Endurance is hard bit fails due to repeat operations of write-read-erase. Program disturbs as in Fig.6b cause adjacent cell high leakage along the same BL, resulting in read error if initial Vt-state is low, short channel effect, punch-through, or create hot hole injection.

BEOL processing issues of multi-level metallization (MLM), Cu wire, HDP oxide, low K dielectric, and how in dealing with the cell endurance have just begun. Electrostatic charging of the WL, boron density of BPSG and hydrogen content of IMD all has reliability effects on the cell retention. Hydrogen killer effects extend into Ferroelectrics materials as well

Most semiconductor components are tested by voltage and temperature acceleration assuming some level of activation. NVM has the same reliability problems and follows the same acceleration. The "classical" burn-in screening is by high temperature acceleration such as 150°C to exclude the hard bit fails. Soft fails are the abnormal "moving bits" that screen with lower temperature such as below 80°C [22]. Other issues include erratic bits, endurance fails, and program disturbs. The work-around solutions to aid process reliability with soft fails are the implementation of design-in tests such as ECC, recovery algorithms or on-chip failure corrections.

Embedded NVM Module Design and Test Challenge

Several key areas of design must enter into considerations. Array architecture is the number one priority that include program or erase disturbs, high performance and low stand-by power, erase block size variations, unit block design flexible for different array size implementations, and chip area optimization. Several analog blocks of designs are the basis for memory: sense amplifier and reference circuits, WL and BL drivers and decoders, global control and glue logic for bus interface, charge pumps and DC power system, high voltage circuits and level shifters [25].

Issues are cost concern that embedded NVM must be tested along with high speed logics using a high speed tester that is more expensive and less efficient for NVM than a specialized memory tester. Test cost for embedded NVM can be a significant portion of the device cost. The test time could be reduced through the write time reduction such as design for multiple-word or page mode programming, or simultaneous program of multiple NVM modules on-chip. A highly reliable process and a robust design can lead to significant tests required. Use of on-board CPU and RAM to perform self-test and other tasks can reduce test code development [26].

Conclusion

We have shown that embedded NVM offer performance, cost advantages and more robust system solutions over stand-alone NVM. MCU with eNVMs will gain market share while stand-alone will have difficulty to match performance. There are different NVM concepts (ROM...Flash) established. The applications define the most practicable one. The extent of market opportunity is driven by product cost. The parts are expected to operate at -40°C and extend to 125°C, suitable for a wide range of automotive, Mobility, and Security products. For most applications, manufacturability for eNVM is the primary concern, not cell size, followed by ease of design solutions and test methodologies. Emerging memories is extremely interesting; however, they also have higher new materials costs.

Acknowledgments

The author wishes to thank all Infineon NVM colleagues who have contributed to this and previous publications.   Special thanks to Georg Tempel and Ronald Kakoschke for many technical discussions; also to Robert Allinger, Robert Strenz, and Alexander Rahm for management supports.

References

[1] D.Frohman-Bentchkowsky, *Appl. Phys. Lett.*, vol.18, p.332 (1971)
[2] M.Lenzlinger & E.Snow, J. Appl. Phys., vol.40, p.278 (1969)
[3] S.Kianian, et al, Symp. VLSI Tech., p.71 (1994)
[4] N.N.Kynett et al, Proc. ISSCC Dig. Tech. Pap., pp.132-133 (1988)
[5] S.Mukherjee, et al, IEDM Tech. Dig., p.616 (1985)
[6] S.Yamada, et al, IEDM Tech. Dig, p.307 (1991); K.Oyama, et al, IEDM Tech. Dig, p.607 (1992), D.Shum et al, IEDM Tech. Dig, p.41 (1994)
[7] C.Kuo, et al, Symp. VLSI Circuit, p.87 (1991)
[8] A.T.Wu, et al, IEDM Tech. Dig., p.584 (1986)
[9] R.Shirota, et al, Symp. VLSI Tech., p.33 (1988); F.Masuoka, et al., IEDM Tech. Dig., p.552 (1988)
[10]H.Kume, et al, IEDM Tech. Dig., p.991 (1992)
[11]G.Yaron, et al, J. Sol. St. Circ., v.SC-17, no.5, p.833 (1982)
[12]H.Wegener, et al, IEDM Tech. Dig., (1967)
[13]C.Chao, et al, Sol. St. Elec. v.30, p.307 (1987); E.Suzuki, et al, IEEE TED, v.ED-30, p.122 (1983)
[14]B.Eitan et al, SSDM, p.522 (1999)
[15]W.Chen, et al, Symp. VLSI Tech., p.63 (1997)
[16]J.Evans, et al, J. Sol St. Circ, v.SC-23, p.1171 (1982)
[17]G.Mueller et al, p.567, Y.Asao et al, p.571; C.Hung et al, p.575 IEDM Tech. Dig., (2004)
[18]C.Lin et al, IEDM Tech. Dig., p.279 (2009)
[19]S.Lai, et al, IEDM Tech. Dig., p.803 (2001); M.Gill et al, ISSCC p.202 (2002)
[20]D.Shum, et al, IEEE TED, (1995)
[21]G.Gerosa, et al., IEDM Tech. Dig., p.631 (1985)
[22]G.Tempel, System-On-A-Chip, Short Course IEDM 1999
[23]D.Shum, et al, IEDM Tech. Dig., p.665 (1997)
[24]B.Maiti, et al, IEEE Proc. IRPS, (1996)
[25]C.Kuo, Nonvolatile RAM Technology and Applications, Short Course IEDM 1995
[26]Nonvolatile Semiconductor Memory Technology, W.D. Brown & J.E. Brewer; IEEE Press 1998

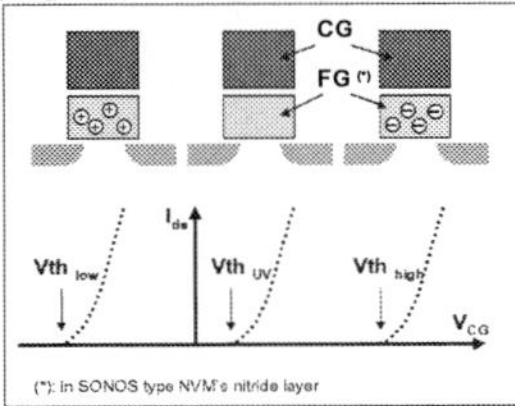

Fig.1 Principle functions of FG cell operations.

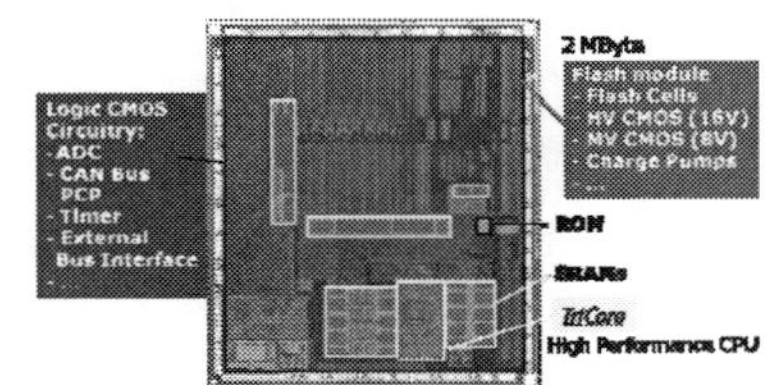

Fig.2 Infineon TC1796 32b-MCU products with 2MB Program and 4KB Data flash at 150MHz.

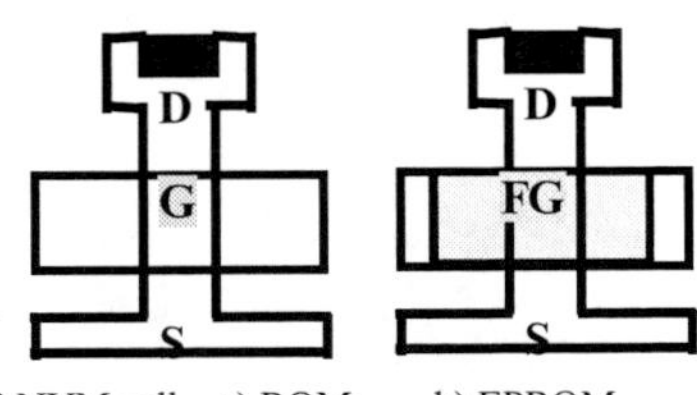

Fig.3 NVM cell   a) ROM;   b) EPROM.

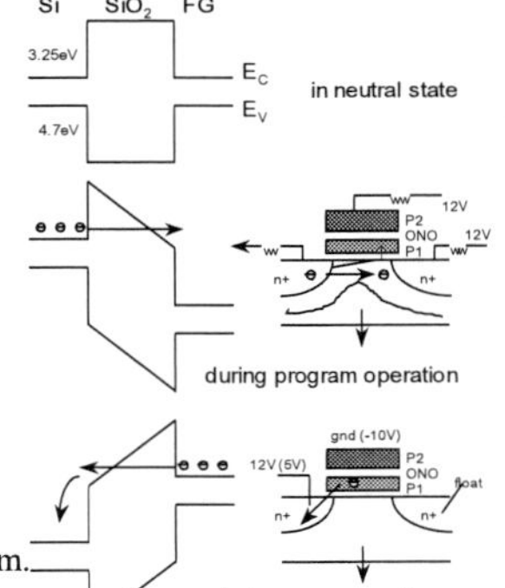

Fig.4 EPROM Si/SiO$_2$ Energy Band Diagram and its transport mechanism.
   a) neutral state; b) program operation; c) erase operation.

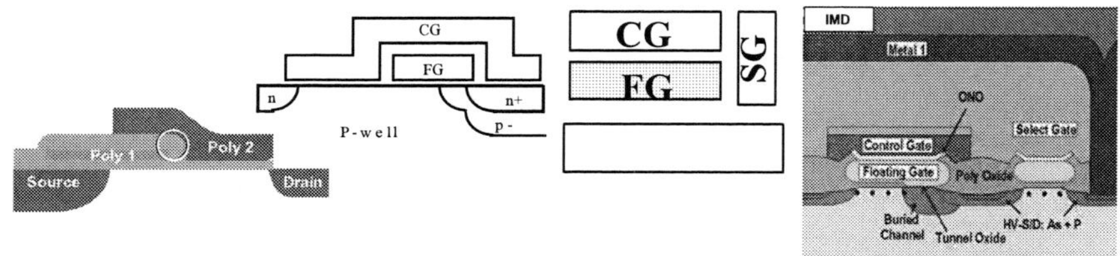

Fig.5 a) SST cell: electron b) Motorola 1.5T SCSG 5c) Triple poly SSI cell   5d) FLOTOX EEPROM.
tunneling through P1-P2 tip

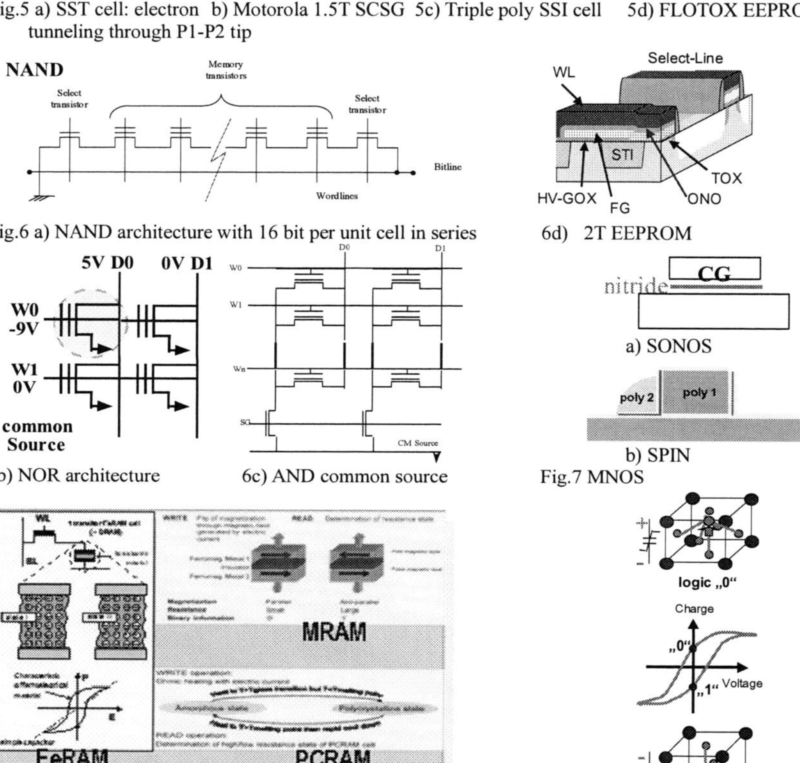

Fig.6 a) NAND architecture with 16 bit per unit cell in series   6d) 2T EEPROM

a) SONOS

b) SPIN

6b) NOR architecture   6c) AND common source   Fig.7 MNOS

Fig.8) Principle functions of cell operations of non-Si-based NVM (emerging memory)

Fig.9) Principle of SBT capacitor

Fig.10 Cell cross-section   a) along WL   10b) along BL

ECS Transactions, 34 (1) 9-15 (2011)
10.1149/1.3567552 ©The Electrochemical Society

A Novel High Programming Efficiency and Highly Scalable Flash Memory Cell Based on Tunneling FET (TFET)

Shiqiang Qin[a], Poren Tang[a], Yimao Cai[a], Qianqian Huang[a], Yu Tang[a], Ru Huang[a]*

[a] Institute of Microelectronics, Peking University, Beijing 100871, China

A novel flash memory cell based on Tunneling Field Effect Transistor (TFET) is proposed and investigated in this paper. Based on the TFET structure, the proposed novel flash memory cell shows high programming efficiency, low power consumption, and good punch-through immunity.

Unlike traditional NOR Flash cell which adopts the channel hot carrier (CHE) injection near drain side for programming operation, the TFET flash memory cell utilizes hot electrons injection introduced by band-to-band tunneling near source side for programming, which can achieve 3 orders higher programming efficiency due to greatly alleviated competition between the vertical electric field and lateral electric field. Moreover, the programming leakage current is 2 orders lower due to the high punch-through immunity for TFET structure. Our results show that the programming leakage current of TFET flash memory can be tolerably low while the gate length continues scaling down, although flash cell has relatively stringent requirements for leakage current. The obtained results and theoretical analysis demonstrate that the newly proposed flash cell can be a potential candidate for low power, and high density NOR-type flash applications.

1. Introduction

As technology node advances, non-volatile memory cells continue aggressively scaling down towards high density, low power to meet the requirement of fast-growing wireless communication market [1]. One of them, NOR-type flash memory is widely used in code storage area due to its fast random access speed. However, for the conventional floating gate flash memory, the injection efficiency (IE) (=I_g/I_d) is very low [2], resulting in high power consumption. Even worse, the unselected cell, which share the same bit-line with the cell programming has high leakage current, causing additional power dissipation and may even lead to programming failure [3]. In addition, drain-to-source punch-through (PT) effect [4] seriously challenges the conventional flash scaling down.

To overcome these problems, some new structures and programming methods have been reported, such as the source side injection (SSI) [5] and FBD bias scheme [6]. To a certain extent, these solutions improve the programming efficiency and alleviate the punch through (PT) effect, but it's hard to solve these problems at the same time, and also increase the complexity and fabrication cost.

In this paper, a novel flash memory cell based on TFET is proposed for the first time. Through two-dimensional simulations, we have demonstrated that, compared with the MOSFET based flash memory, the TFET based flash memory has higher programming efficiency and better scalability. These characteristics make it favorable in mobile application with low power consumption and high memory density.

2. Device Structure and Simulation Method

Fig.1 shows the cross-sectional view of the TFET based flash memory cell and the MOSFET based flash memory cell. The structure of memory based on TFET is similar to the conventional flash memory except employing the TFET as the underlying part rather than MOSFET. The TFET flash memory, in this work, adopts underlapping structure to avoid P-type working mode. In order to investigate the memory performance of this novel memory structure, numerical simulations have been performed with device simulator TCAD Sentaurus 10.0. Both the novel flash memory cell and conventional flash cells are simulated and compared. In our device simulations, hydrodynamic transport, Poissons' equation, and continuity equation are solved self-consistently. SRH recombination model, Auger recombination model and BTB tunneling generation model are adopted. Lucky-electron model with electron temperature calculated by hydrodynamic transport is used for hot channel electron injection during programming, while FN tunneling model is used during erasing. In our simulations, the gate length of device is $0.15\,\mu$ m (except specially mentioned), Gauss S/D doping profile is used. The thickness of tunnel and block oxide is 8nm and 13nm, respectively.

3. Results and Discussion

<u>3.1 Program and Erase Characteristics</u>

The TFET based flash memory works in N-TFET mode, that is the N+ region works as drain, and the P+ region works as source [7]. The reason for this is because that in N-TFET mode, electrons rather than holes will be induced near the source side and flow through the channel, which can be used for programming. Fig.2 shows the program efficiency (I_{fg}/I_d) versus the floating gate voltage. It can be found, in comparison with the MOSFET based flash memory, that the TFET based cell shows higher programming efficiency (about 2-3 orders) than that of the conventional one. To further study the physical mechanism of high programming efficiency, the electric field distribution is investigated when programming. Fig.3 shows distribution of the lateral electric field along the channel when programming. From it, we can see that the peak of lateral electric field is greatly enhanced and moves to source side, comparing with the conventional cell. The TFET based flash cells alleviate competition between the lateral electric field and vertical electric field for the MOSFET based flash memory cells, and thus high programming efficiency is achieved.

Fig.4 shows the programming and erasing characteristics of TFET flash, using CHE injection and FN tunneling respectively.

<u>3.2 Unselected Cells Current Characteristics</u>

To surmount the Si/SiO_2 barrier, a high drain voltage has to be applied to obtain enough elector energy. The high voltage will cause the sub-threshold current flowing through the unselected cells, which share the same bit-line of the one to be programmed during programming. When the flash memory cell continues scaling down, the total sub-threshold current can't be neglected, which leads to large extra power consumption and even result in programming failure [3]. Fig.5 shows drain current versus gate voltage with the drain biased at 4V of the two flash structures. The gate length L_g of these two cells is $0.09\,\mu$ m. From Fig.5, we find that 6 orders reduction of sub-threshold current of flash cell based on TFET is achieved at high drain bias. The significantly decreased sub-threshold current means that the flash cell based on TFET has low unexpected power consumption by the unselected cells and better programming failure immunity.

3.3 Punch-Through (PT) Characteristics

As the NOR flash memories continually scaling down, the cells face a serious challenge of punch-through (PT) effect, which seriously challenges the memory advances for high density application. In this section, the PT characteristics of the TFET based flash cell is studied and is compared with conventional cell.

Fig.6 shows the punch-through voltage V_{pt} versus gate length L_g for flash memory based on TFET and conventional cells. Here we defined the V_{pt} as the drain voltage biased when the drain current of the cell is 10^{-7} A/μm with the gate grounded and the V_{pt} of the MOSFET based flash cells is from our previous work [8]. The Fig.6 clearly shows that V_{pt} of the TFET based flash cell is much larger than the conventional one, for the same gate length L_g of the short channel cells. As an example, when L_g is 0.09 μ m, V_{pt} of the normal flash memory cell is about 1.2V, while it is about 6.8V for the flash memory cell based on TFET. So it is hard for the normal flash memory cell to operate when L_g shrinks to 0.09 μ m, since V_d is not high enough to accelerate electrons to reach the floating gate. Therefore, with the alleviated punch-through effect, the flash memory cell based on TFET have better scalability, and can be regard as a promising candidate structure for high density NOR-type flash memory application.

4. Conclusion

For the first time, a novel flash memory based on TFET is proposed and studied through simulation. The simulation results show that this type flash memory exhibits higher programming efficiency due to preferable electric field distribution for injection. And the current of unselected cells is much smaller. These features demonstrate that the TFET based flash has a good application potential in low power flash memory field. In addition, the novel structure also shows better scalability due to the improvement of PT characteristics. Thus, the proposed flash cell can be considered as a promising flash memory structure for the low-power and high density application.

(a)

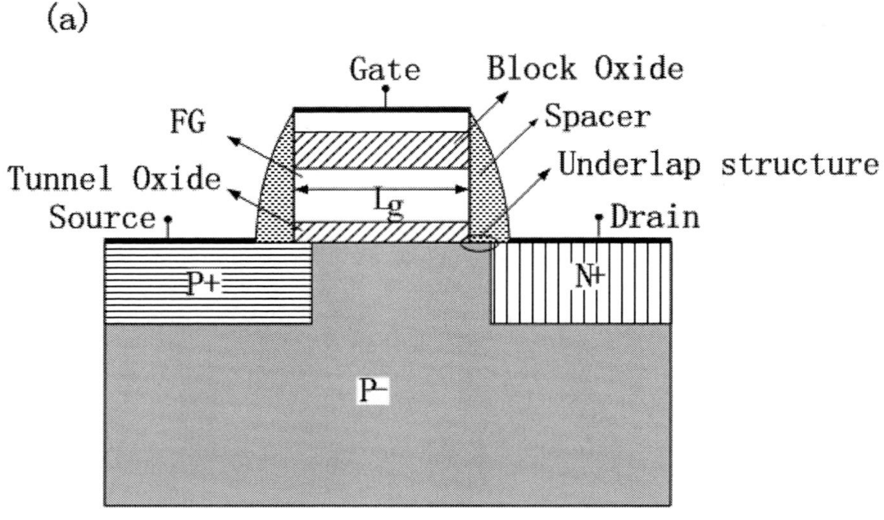

(b)

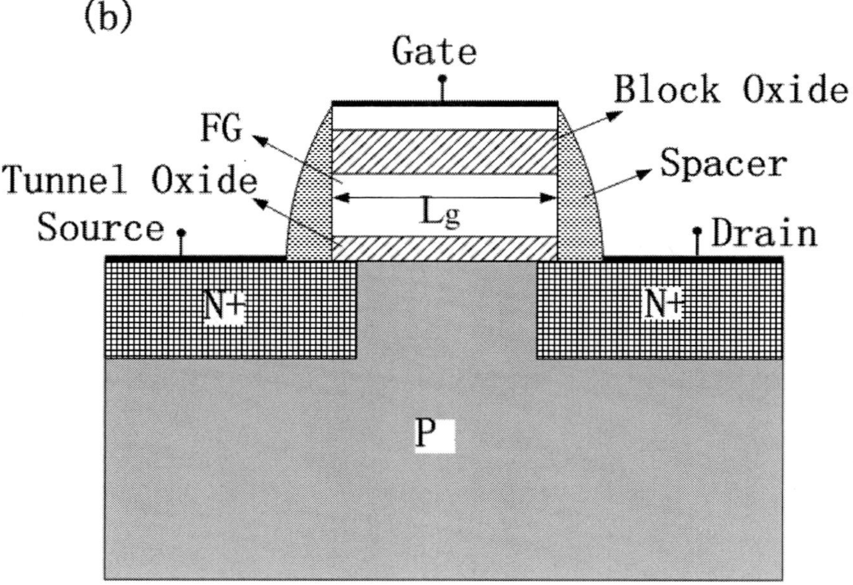

Figure 1. Schematic cross-sectional view of (a) cell based on TFET (b) conventional cell.

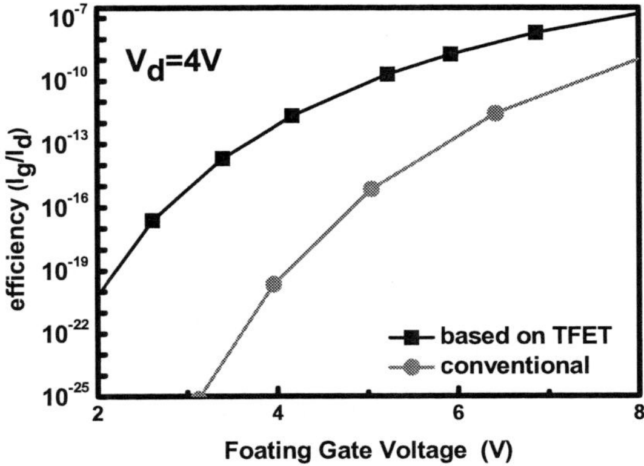

Figure 2. Injection efficiency as a function of gate voltage, when V_d=4.0V.

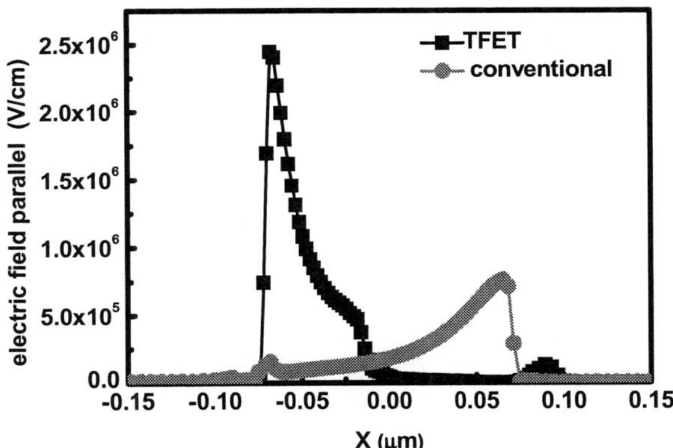

Figure 3. electric field distribution along the channel.

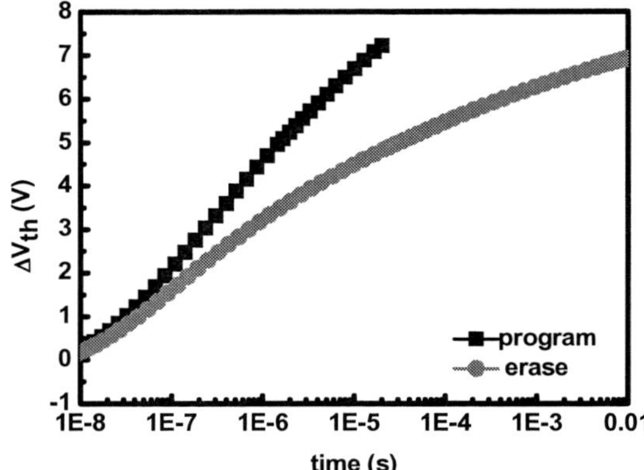

Figure 4. Program & erase characteristics of the cell based on TFET (bias condition: for program, V_g=12V, V_d=4V, V_s=0V; for erase, V_g=-10V, V_d=7V, V_s=7V)

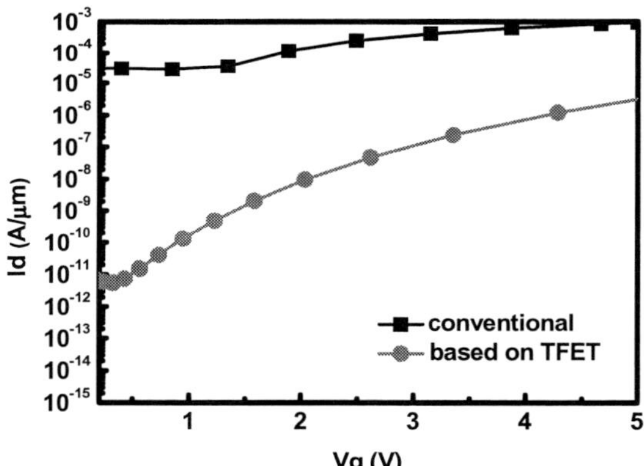

Figure 5. Unexpected leakage current (L_g=0. 09 μ m)

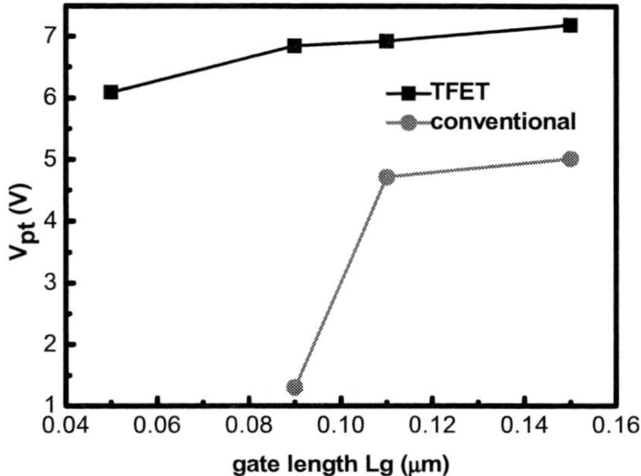

Figure 6. Punch-through voltage for different L_g.

Acknowledgments

This work was supported by the National Natural Science Foundation of China (Grant No. 61006062).

References

1. M. Wei, R. Banerijee et al., VLSI Symposium Technical Digest,p.226,Kyoto(2007).
2. S. Tanaka and M. Ishikawa, *IEEE Trans. Electron Devices*,**28**,10(1981).
3. K. Prall, W.I. Kinney, and J. Macro, *IEEE Trans. Electron Device*,p.2463,**ED-34**(1987).
4. Chankwang Park, Sangpil Sim, Jungin Han, and etl., *VLSI Technology,*P.238,Tampa(2004).
5. A.T. Wu, T.Y. Chan, P.K. Ko,and C. Hu,*IEDM Tech.Dig.*,p.584,(1986).
6. Wen-Jer Tsai,T.F Huang and etl., *IEDM Tech.Dig.*, p.1,San Francisco(2008)
7. P.-F. Wang, K. Hilsenbeck,Th. Nirschl, and etl.,Solid-State Electronics,**48**(2004)
8. Poren Tang, Dake Wu, Ru Huang, *ICSICT*,p.846,Beijing(2008)

ECS Transactions, 34 (1) 17-24 (2011)
10.1149/1.3567553 ©The Electrochemical Society

Design of a 1-T image sensor by simulation

Xin-Yan Liu, Song-Gan Zang, Xi Lin, Cheng-Wei Cao, Peng-Fei Wang, and David Wei Zhang

State Key Laboratory of ASIC and System, Dept. of Microelectronics, Fudan University, Shanghai, China

A 1T silicon-based image sensor has been investigated in this paper. This device can fully accomplish the functions of the conventional 3T CMOS image sensor. In this paper, the basic performances including the dynamic range, read margin (signal-to-noise ratio) and information storage of the 1-T image sensor have been discussed. The applied voltages have been studied in detail, including the reset operation, read operation and exposure operation. The advantage of this device is the minimized cell size. The structure with a photo-diode integrated into the 1-T image sensor may give a higher density in the same chip area while the performance is not degraded. With more pixels in the same area, the fill factor will be raised to a large extent, and the resolution can be further improved.

Introduction

CMOS image sensor became popular since the in-pixel amplification circuit was introduced in the unit cell. The advantage of a CMOS image sensor is the high compatibility to the conventional CMOS fabrication process. CMOS image sensors can be classified into two categories: Passive Pixel Sensor (PPS) and Active Pixel Sensor (APS). The first commercially used MOS image sensor was a PPS which was composed of a photodetector and a switching transistor. Because of its simple structure, PPS has a relatively large FF (Fill Factor). But PPS has a big SNR challenge that makes PPS hard to meet the high performance requirement. APS is based on PPS. It is compatible to the conventional CMOS fabrication processes. However, the structure of APS is relatively complex. Conventional active pixel sensors consist of at least a photodetector and three transistors: a reset transistor, a source follower (amplifier) and a switching transistor (1). For the better control of the signal charge, a transfer gate (TX) can be introduced, forming a 4-transistor configuration. Now the performance of the 4T-APS is comparable with CCDs. The 3T or 4T APS has a tradeoff between FF and SNR (2). Further, the device matching, which results in the fixed pattern noise (FPN), has been more and more severe. In this paper, we propose a novel APS with 1T-configuration. The device working principle will be introduced and the performance will be investigated by TCAD simulation tools.

Device Structure and Working Principle

The proposed device structure is shown in Figure 1. A p-type silicon bulk is used as the substrate. The 1T-APS is composed of a vertical pnp junction as the photo-generation region and a horizontal N-MOSFET. A novel structure called floating junction gate is used in this device (3). It can be seen as a quasi-floating gate, which is used to store the signal charge, and convert the signal charge to potential signal. With the change of the potential in the floating junction gate, the threshold voltage of the N-MOS transistor will be changed as well. Compared to a conventional 3T-APS, the floating gate NMOS transistor in our device works as a reset transistor, a source follower (amplifier) and a switching transistor in different operation. As appropriate voltages are applied on it, this device can fully accomplish the function of a conventional 3T-APS. In our simulation, the channel length of this NMOS device is 260nm. The photo-detection region is 1 μm^2. The working principle of the device is described as the following: Firstly, the device is reset to a certain level and the V_{th} is getting higher. Secondly, the photodetector (PD) is exposed under a monochromatic light with a fixed intensity and the signal charge is generated and transferred into the floating junction gate. The V_{th} is getting lower. Importantly, the storing principle of our device is somewhat different from flash. Several MIM and MOS capacitances are involved in. So the read operation is quasi-destructive and it is particularly discussed in the next section. Thirdly, the gate is positively biased and the read-out drain current is sensed. Finally, the first step is repeated and the device recovers to the original state.

Simulation Operation, Results and Discussion

In this section, the voltage settings for every operation are given. Some electrical parameters are particularly studied and discussed, including transient simulation figure, dynamic range, read margin (signal-noise ratio). We use a monochromatic light with the certain intensity to illuminate the device. We use light of the intensity of 0.08 W as an example.

The transient simulation is shown in Figure 2. The time arrangement is as follows: 7 ms for reset, 20 ms for exposure, 2 ms for standby and 1 ms for readout. The voltage settings for every operation are shown in table I. Standby operation can be ignored when the device is in the practical application. Because of the similarity to the PPS, the proposed device has a quasi-destructive read operation. So the readout time is limited, and we assumed it as 1 ms. The read voltages are as follows: V_{drain}=2V, V_{cgate}=3V and V_{source}=0V. Figure 3 shows the dynamic range of the proposed device. Considering the need of at least 24 frames in a second, we do not use the full capacity. Figure 3(a) is the floating gate potential changes of 20 ms exposure and figure 3(b) is the potential changes of 46 ms exposure using the full capacity. We can see the potential change in figure 3(a) is up to 1.284V. When we use the full capacity of the proposed device, the potential change of the floating junction gate will be up to 2.363V, as shown in figure 3(b).

Figure 4 shows the read margin of the drain current between exposure of 0.08W monochromatic light and dark condition. A big window in potential in the floating junction gate results in a large read-out current margin of 8 orders of magnitude. That indicates a good signal-to-noise ratio.

Because of the quasi-destructive read, the retention of the signal is important. From figure 2, we see the standby before and after readout operation are at the same level. That means our operation is very legal and the loss of the signal charges is under control. Most importantly, the proposed device can accomplish the functions of 3T-APS using only a single composite-transistor. Further investigation in the potentials of the floating junction gate also confirms the operation feasibility, as can be seen in figure 5.

The simulation is also performed under the conditions of exposure light intensities of 0.03W and 0.06W. The transient simulation figure is shown in figure 6. The operation sequences of the proposed device are as following: reset (0-7 ms), exposure (7-27 ms), standby (27-29 ms), readout (29-30 ms), standby (30-32 ms), reset (32-39 ms). Figure 7 shows read margins of four different conditions. Because of the low drain current of the dark condition, this device can be used in different light intensity cases.

<u>Summary</u>

A simple FJG typed 1T-APS has been investigated. Because of the multiple usages of the single transistor, the cell size can be greatly decreased. This design will bring a large fill factor. Simulations show that this 1T-APS has a quasi-destructive read. However, the loss of the signal charges is fully under control and it can entirely accomplish the functions of 3T-APS. Because the operation speed meet the requirement of at least 24 frames per second, the device can be used in both digital still camera and video camera. The dynamic range and read margins meet most requirements of camera applications. If the applied voltages are changed, we can realize different resolution usages as well.

Acknowledgments

This research work is supported by the Program for Professor of Special Appointment (Eastern Scholar) at Shanghai Institutions of Higher Learning.

References

1. J. Nakamura, in *Basics of Image Sensors/2005*, J. Nakamura, editor, Image Sensors and Signal Processing for Digital Still Cameras. CRC Press, Boca Raton (2005).
2. J. Ohta, *Smart CMOS Image Sensors and Applications*, p. 11, CRC press, Boca Raton (2007).
3. PF. Wang, *Solid-State Electron*, **54**, 985 (2010).

Table I. voltage settings

Operation	Drain voltage (V)	Control gate voltage (V)	Source voltage (V)
reset	0	3	0
exposure	1.2	0	0
standby	1.2	0	0
readout	2	3	0

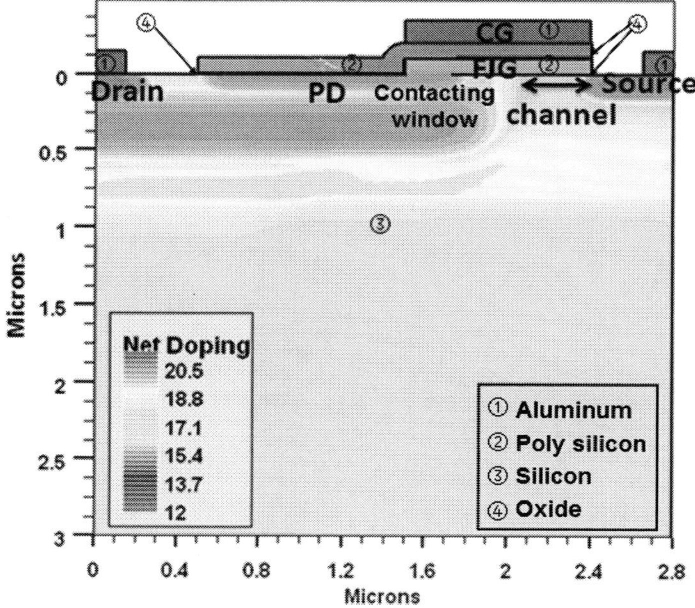

Figure 1 Proposed 1T-APS structure using p typed Si as substrate and shallow BF$_2$ implant with n typed drain as a photodetector. The channel length is 260nm. Anyi-punch-through implant is applied to suppress DIBL. The collection of photo-generated carriers by the drain-substrate junction flows to the substrate and do not contribute.

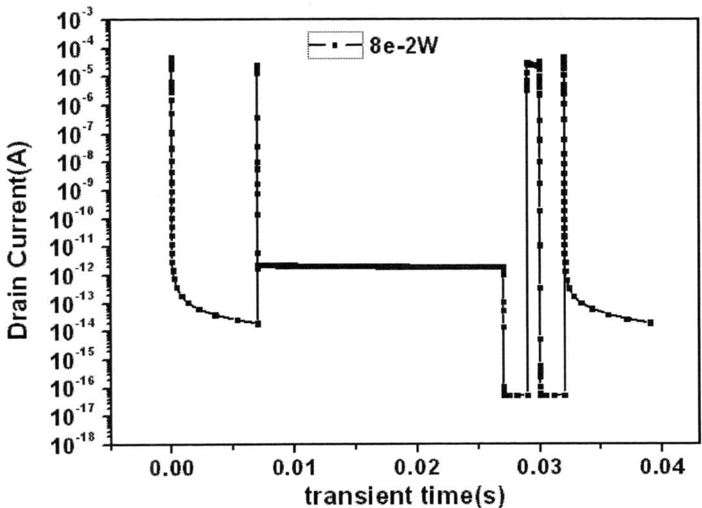

Figure 2 Transient simulation of the condition of 8e-2W illuminance. The operation sequences are as following: reset (0-7 ms), exposure (7-27 ms), standby (27-29 ms), readout (29-30 ms), standby (30-32 ms), reset (32-39 ms).

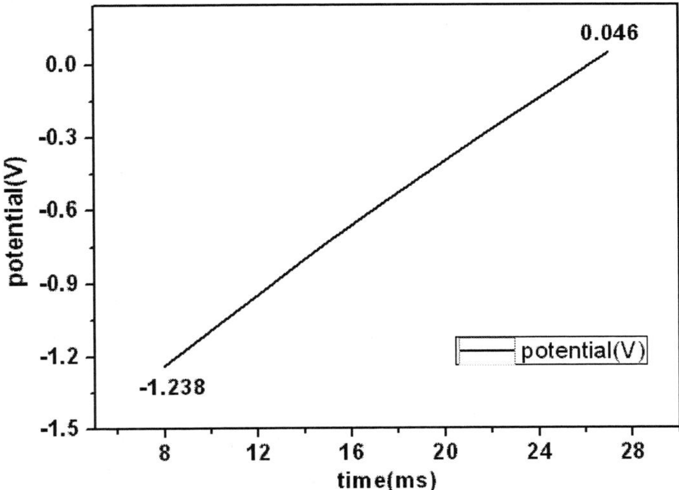

Figure 3(a) Potential changes of 20ms exposure in the practical condition. The potential change of the floating junction gate is up to 1.284V.

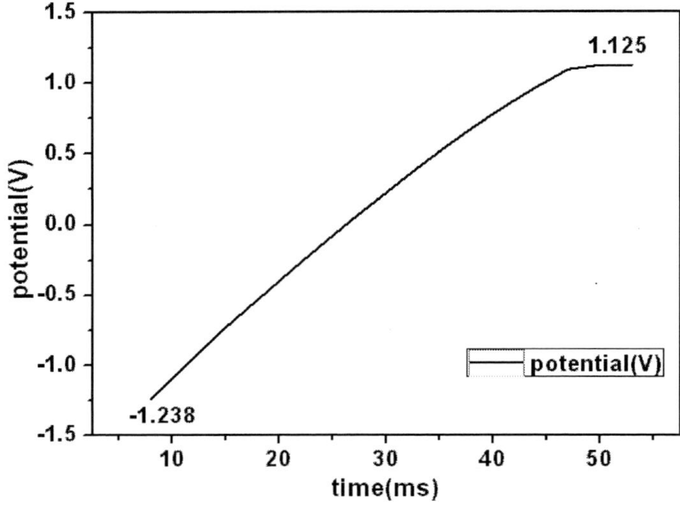

Figure 3(b) Potential changes of 46ms exposure using the full capacity in the ideal condition. The potential change of the floating junction gate is up to 2.363V.

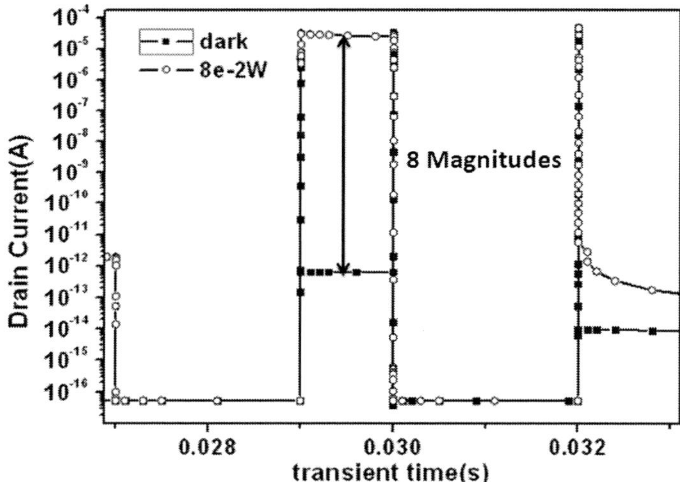

Figure 4 Read-out current margin between the conditions of the 0.08W illuminance and dark. An 8 orders of magnitude window can be seen.

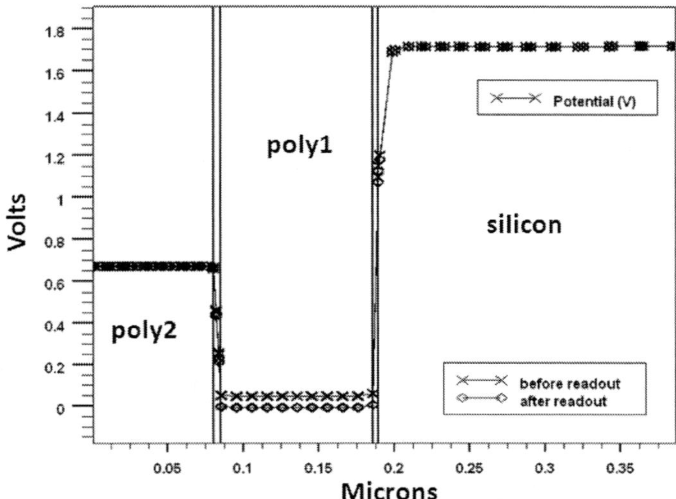

Figure 5 Potential changes in the floating junction gate before and after readout. The loss of the signal charges results in 0.05V changes in potential. Compared with the 1.284V window, the loss is under control.

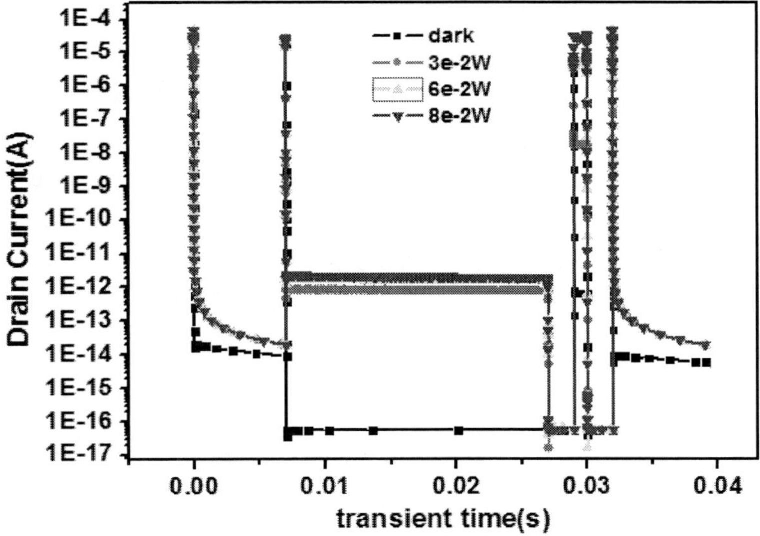

Figure 6 Comprehensive transient simulations of dark, 3e-2W, 6e-2W and 8e-2W. The operation sequences are as following: reset (0-7 ms), exposure (7-27 ms), standby (27-29 ms), readout (29-30 ms), standby (30-32 ms), reset (32-39 ms).

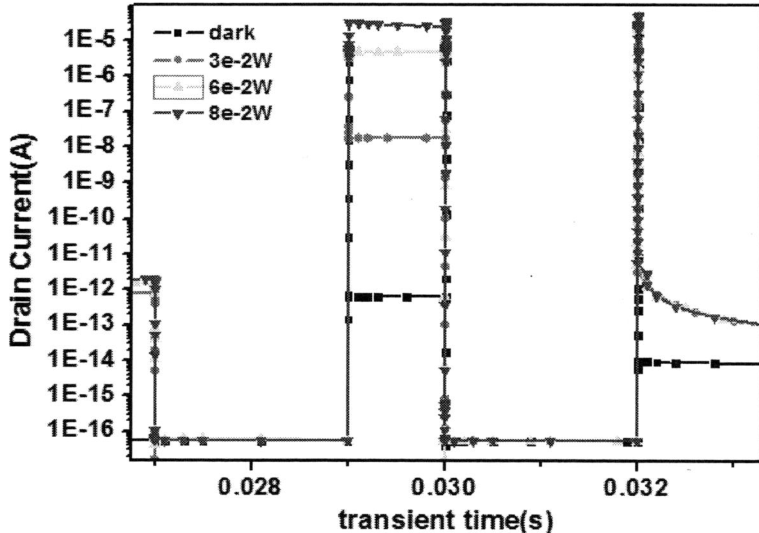

Figure 7 Read-out current margins of four different conditions: dark, 3e-2W, 6e-2W and 8e-2W. We can see at least 4 orders of magnitude between the conditions of 3e-2W and dark.

ECS Transactions, 34 (1) 25-30 (2011)
10.1149/1.3567554 ©The Electrochemical Society

Deposition of ZnO films by sputtering and its resistive switching properties

Fang Wang [1,2], Kailiang Zhang [1]*, Baohe Yang [1,2], Lanlan Wang [1], and
Kai Song [1]

1. School of Electronics Information Engineering, Tianjin Key Laboratory of Film
Electronic & Communication Devices, Tianjin University of Technology, Tianjin,

China, 300384 , *corresponding author , kailiang_zhang@163.com

2. Tianjin University, Tianjin, China, 300072

In this paper, ZnO thin films were deposited on the
$Cu/Ti/SiO_2/Si$ substrates by radio frequency (RF) magnetron
sputtering at different ratio of O_2/Ar with 7:3, 6:4, 5:5, 4:6, 3:7.
The effects of different ratio of O_2/Ar on crystallinity and
resistive switching behaviors of film were studied. XRD
indicated that ZnO films are highly c-axis oriented, the crystal
size is smallest at 60% oxygen ratio. AFM indicated the film
has most uniform morphology at the same sample. Morphology
and the resistive switching characteristics of this film were
studied in situ by CSAFM. The results showed that the film
was smooth and dense. The current image showed stable and
superior electrical switching under different scan. This is
proved by SPA (B1500A), which reveal the resistive switching
characteristics of same sample.

Introduction

Zinc oxide (ZnO) is a polar group II–VI semiconductor material with a direct band
gap of 3.37 eV and a large exciton binding energy of 60 meV, and high temperature
(melting point of 1975°C) and chemical stability. Zinc oxide films are employed to
several applications such as varistors (1), gas sensors (2), SAW devices (3),
transparent electrodes etc. The various applications of ZnO are due to the specific
chemical, surface and microstructure properties of this material. The microstructure
and physical properties of ZnO can be modified by introducing changes into the
procedure of its deposition. In recent time, with the rapid development of resistance
random access memory (RRAM), the resistive switching behaviors of zinc oxide
films has attracted an extensive attention (1) while the research is still in the initial
stage.

In this paper, the ZnO thin films are deposited on the $Cu/Ti/SiO_2/Si$ substrates by
radio frequency (RF) magnetron sputtering at different ratio of O_2/Ar with 7:3, 6:4,
5:5, 4:6, 3:7. The effects of different ratio of O_2/Ar on crystallinity and resistive
switching behaviors of ZnO thin films are studied.

Experiment

Devices were fabricated as a conventional metal–insulator–metal structure. ZnO thin films were grown on obtained Cu/Ti/SiO$_2$/Si substrates by radio frequency (RF) magnetron sputtering at different ratio of O$_2$/Ar. The deposition chamber was initially evacuated to a base pressure of 2×10^{-4} Pa using a turbo molecular pump and then filled with argon and oxygen ambient at flow velocity of 40sccm ,while the deposition chamber pressure: 1.0Pa , base temperature : 250°C and RF power:100W were keeping for 30min. The phase, morphology, and electrical properties were analyzed by X-ray diffraction analyses (XRD, D/MAX III-C) ,current sense atomic force microscope(CSAFM, Agilent 5500) and semiconductor parameter analyzer (SPA, Agilent B1500) respectively.

Results and Discussion

Microstructure and Surface morphology

The Microstructure and Surface morphology of ZnO films were analyzed by X-ray diffraction analyses (XRD), atomic force microscope(AFM).

X-ray diffraction. Fig 1. shows the typical XRD spectra of the fabricated ZnO films. Fig 1.(a)Shows the XRD patterns of ZnO films. All the films grown in the range of 7:3, 6:4, 5:5, 4:6, 3:7 were found to have a high intensity diffraction peaks at scattering angles (2θ) of 34.4° which correspond to the reflection from (002) crystal plane, and indicate all the films have high c-axis orientation. No other ZnO film peaks were found in the XRD spectra.

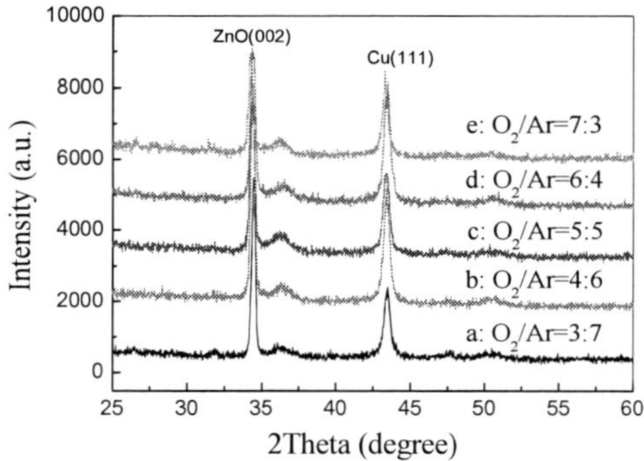

(a) effect of ratio of oxygen on the peak intensity

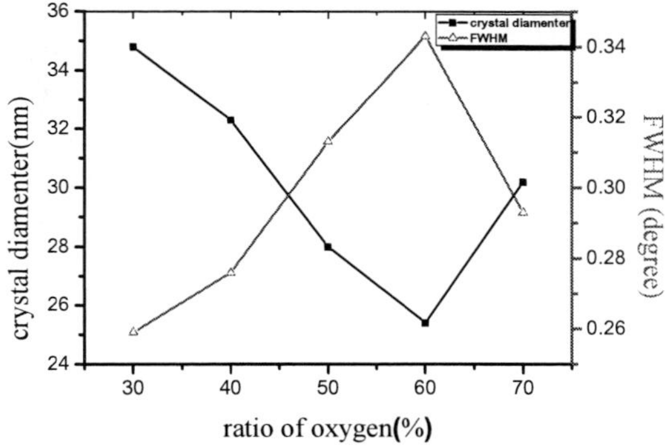

(b) effect of ratio of oxygen on the crystal size and FWHM of ZnO films

Figure 1. XRD spectra of the ZnO/Cu/Ti/SiO$_2$/Si

The effects of ratio of O$_2$/Ar on the full-width at half Maximum (FWHM) values and crystal size of XRD spectra of the ZnO films are shown in Fig 1.(b): FWHM values of the films increase with ratio of oxygen increases until ratio of oxygen 60%, then decrease at 70%,the value is biggest at ratio of oxygen 60% ; crystal size of the film decrease with ratio of oxygen increases until ratio of oxygen 60%, then increase at 70%, crystal diameter is smallest at ratio of oxygen 60%.

AFM. Fig. 2 show AFM images of ZnO thin films at different ratio of oxygen. The scanning area of AFM was 1μm×1μm, the surface RMS (root mean square) roughness of images are (a):8.15nm; (b):5.51nm; (c):5.94nm; (d):8.19nm; (e):9.25nm respectively. However, from fig. 2, we can see the morphologies of (a), (b), (c) samples consisted of bigger grains and unevener film surface than d, e samples, and the most uniform morphology is obtained from d film's surface with 60% oxygen ratio.

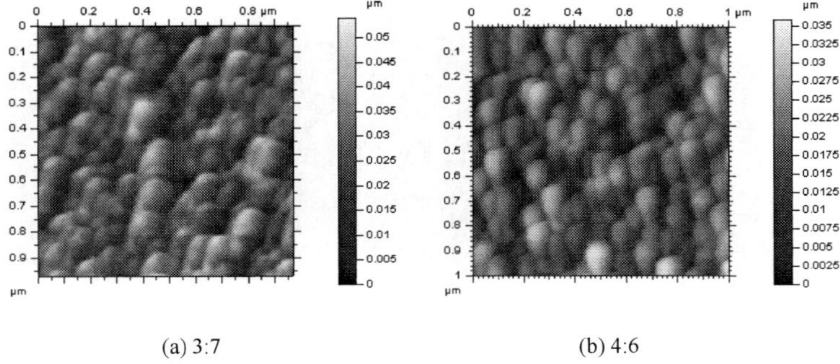

(a) 3:7 (b) 4:6

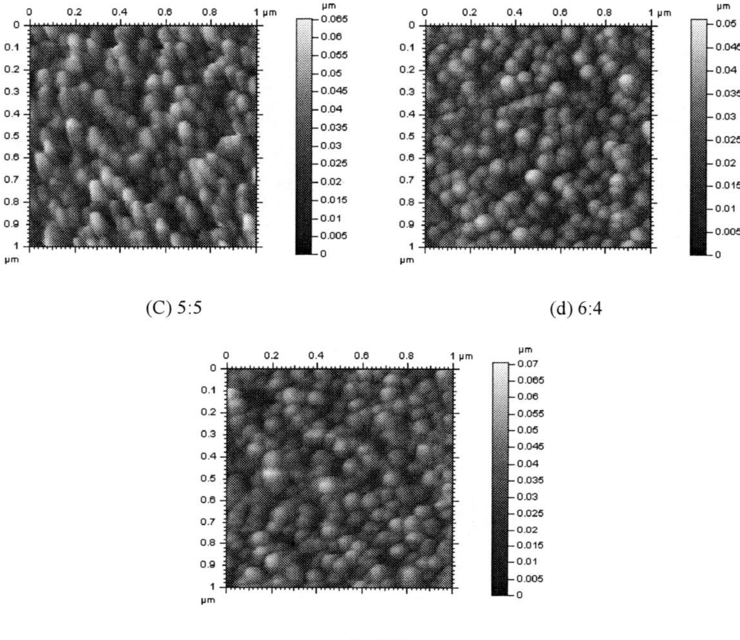

(C) 5:5   (d) 6:4

(e) 7:3

Figure 2. AFM images of ZnO thin films

<u>Electrical properties</u>

Based on XRD and AFM results, the sample of 60% oxygen ratio with smaller particle size and uniform particle has been selected to research electrical properties.

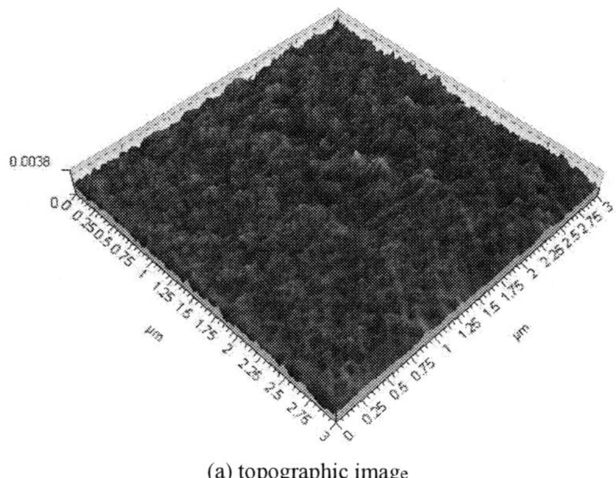

(a) topographic image

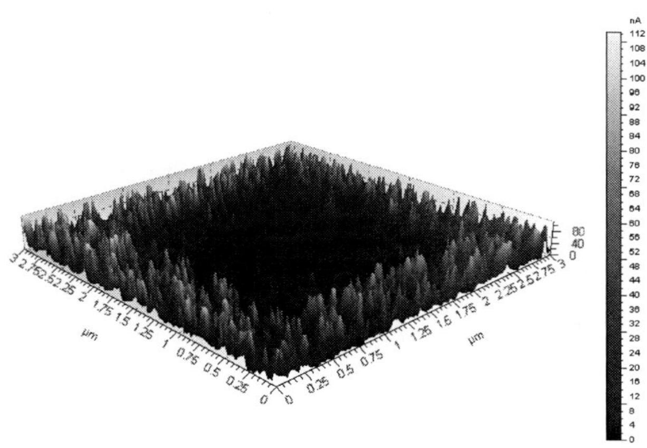

(b) current image of ZnO/Cu film

Figure 3. CSAFM images of ZnO thin film with 60% ratio in oxygen

CSAFM. Fig.3 is CSAFM images of ZnO thin film with 60% ratio in oxygen , Fig.3 (a) : topographic image and Fig.3 (b) :current image of ZnO/Cu film are observed Simultaneously: first, select 2μm × 2μm range from sample , scan with 5V bias, then in the same center select 3μm × 3μm range , scan with 5V bias, at last obtain Fig.3. from Fig.3 we observe that there are no current signal in the center area of 2μm × 2μm, while in the other edge of the area can observe a strong current signal, which shows the different scan ways result in change of resistance. It indicated that the film have Performance with variable resistance.

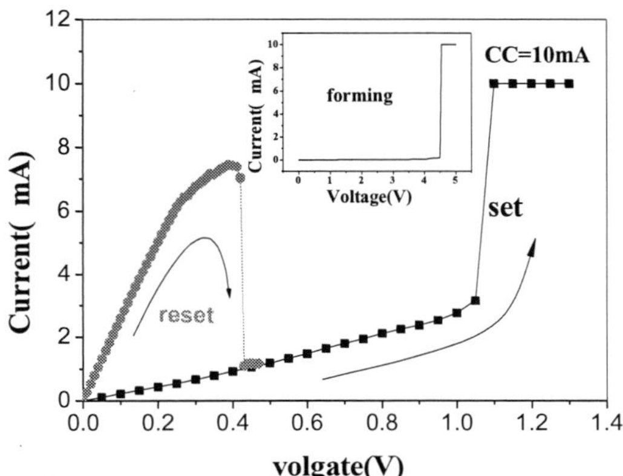

Figure 4. Typical I–V curves of Pt/ZnO/Cu structure. The inset displays forming process under a high electrical stress.

SPA. To confirm the switching characteristics of 60% ZnO film, the typical current–voltage (I-V) curves are obtained by SPA, as shown in Fig 4. indicating "Forming", "SET" and "RESET" processes. A electrical stress with a compliance current of 10mA is necessary to initiate the switching behavior of the Au/ZnO/Cu structure, which is known as the forming process (the forming voltage), as illustrated in the inset of Fig. 4. After the forming process, the device is in the LRS. Then, by sweeping the voltage from zero to a certain value (the reset Voltage), current suddenly drops and the resistance of the device switches to the HRS. While sweeping again, an abrupt jump of current appears at a specific voltage (the set voltage), which is always larger than the reset voltage. The resistance switches to the LRS again. It is noticed that compliance current was required to protect the sample from a permanent breakdown during the set process. The switching characteristics of the sample 60% oxygen ratio are shown in Fig. 4. The forming voltage value for the sample is 4.5v, the reset voltage value is 0.42v and the set voltage value is 1.05v.

Conclusion

The effects of different ratios of O_2/Ar on crystallinity and resistive switching behaviors of ZnO thin films were studied. The phase and electrical properties were analyzed by X-ray diffraction analyses (XRD), atomic force microscope (AFM) current sense atomic force microscope (CSAFM) and semiconductor parameter analyzer (SPA). XRD indicated that all the ZnO films are highly c-axis oriented, while the crystal size is smallest at the ratio of O_2/Ar 6:4. And AFM indicate the film has most uniform morphology at the oxygen ratio 60%. Morphology and the resistive switching characteristics of these films were studied in situ by CSAFM. the results show that the grains were evenly distributed, morphology of films were Smooth and dense, the current image showed stable and superior electrical switching of the films at the 60% oxygen ratio under different scan. This is proved by SPA (B1500A), which reveal the resistive switching characteristics of the same sample.

Acknowledgments

We acknowledge the financial support provided by the National Natural Science Foundation of China under Grant No 60806030, and Tianjin Natural Science Foundation under Grant No 08JCYBJC14600, No 10SYSYJC27700 and Tianjin Science and Technology Developmental Funds of Universities and Colleges under Grant No ZD200709.

Reference

1. Hui Lu, Yuele Wang and Xian Lin, *Materials Letters*,63(27),2321(2009)
2. S Roy and S Basu, *Bull. Mater. Sci*, **25**(6,), 513(2002)
3. X. Y. Du, Y. Q. Fu, S. C. Tan, *Appl. Phys. Lett.* **93**, 094105 (2008)
4. L M KUKREJA, A K DAS and P MISRA, *Bull. Mater. Sci.,* **32**(3), 247(2009)

ECS Transactions, 34 (1) 31-36 (2011)
10.1149/1.3567555 ©The Electrochemical Society

Leakage Engineering Enabling PDSOI Ring Oscillators Operating in Sub-100pA/µm I_{off} Regime

Zhibin Ren, [1]Jin Cai, [2]Robert R. Robison, Basanth Jagannathan, [1]Dae-Gyu Park, [1]Tak H. Ning

IBM Semiconductor Research & Development Center, Hopewell Junction, NY 12533
[1]IBM Research Division, T. J. Watson Research Center, Yorktown Heights, NY 10598
[2]IBM Semiconductor Research & Development Center, Essex Junction VT 05452
Phone: 914-945-3965; E-mail: zhibinr@us.ibm.com.

Abstract

This work presents hardware demonstration of low power operation of PDSOI CMOS (I_{off} down to ~10pA/µm at V_{dd}=0.9V, 45nm node) transistors, exhibiting successful operation of low leakage ring oscillators (channel and junction leakage ~30pA/invertor stage at V_{dd}=0.9V, 25C with 101 stages, pFET width=1.2µm, nFET width=0.8µm in each stage). The work highlights device and process feasibility study (through an asymmetric source/drain design) of enabling a PDSOI CMOS based low power application platform.

Results

PDSOI (partially depleted) MOSFETs have been widely used in high performance, and RF chips [1,2,3]. However, it is not straightforward to achieve low leakage operation with such devices. The limitations are 1) junction leakage and 2) floating body effect. In order to reduce channel leakage, a high dose halo implant is typically needed to increase threshold voltage (V_t) of the transistor. The higher halo dose, in general, induces higher junction leakage, which adds to the device total leakage (limitation 1). More than that, the increased junction leakage results in a forward biased SOI body with respect to the device source (floating body effect-FBE), which in turn lowers threshold voltage, leading to increased channel leakage (limitation 2). In low power SOI CMOS design, the key challenge is to achieve required high V_t without inducing much penalty of junction leakage and FBE.

The leakage is illustrated in nFET IVs in Figure 1. The channel current increases as FBE reduces the thermal emission barrier height, the junction leakage may dominate the current at the off-state, when the drain-side junction is highly reverse biased. These two effects are further illustrated in Figure 2, where nI_{off} (V_g=0V, V_d=0.9V) taken on floating body mode is plotted against source-side I_{off} of the same device operating on body grounded mode. The source current of a body-grounded FET contains no junction leakage and FBE enhanced channel current, both of them can be seen clearly when the device switches to the floating body mode.

Simply increasing channel doping (e.g. halo dose) will hit a limit of decreasing off-current. In Fig. 3, nI_{off} is plotted vs. threshold voltage. As I_{off} approaching 100pA/µm level, the use of higher halo dose fails to decrease off-current. The current is essentially modulated through SCEs improvement at longer channel lengths in each group.

Basically, increasing V_t would simply degrade the drive current, the off-current saturates at the junction leakage floor.

We chose a source-side-only-halo approach [4,5,6] to address the leakage issues. High halo doses are used in the source-side only to set high threshold voltages, the drain side junction leakage remains low, FBE is also minimized. Effectiveness of such solution is assessed in experiments. Figure 4 shows the schematic structure of an asymmetric source/drain PDSOI MOSFET design. Reduced diode current at the reverse-bias condition (drain side of a MOSFET), and unbalanced forward and reserve diode currents in favor of reduced FBE are shown in Fig. 5. The two elements are essential in leakage control in PDSOI transistors.

Upon implementing such source-side-only-halo design, the floating body induced DIBL reduces by ~ 50mV as shown in Fig. 6. The FBE improvement is also shown in Fig. 7, where the off-current of asymmetric devices becomes closer to the leakage of body-grounded FETs. Comparison of IVs between a symmetrical design nFET and an asymmetrical design is shown in Fig. 8, at V_{dd} of 0.9V, I_{off} reduces by ~ 1 decade to below 10pA/μm in an asymmetrical nFET.

Figure 9 plots drive current vs. off-state leakage for 2 different designs. It shows 1) asymmetric design achieves leakage of ~10pA/μm level, while that of the symmetric design stays around 100pA/μm, 2) asymmetric nFET drive current also over-performs the symmetric counterpart, due to reduced DIBL.

PDSOI pFETs suffer much less junction leakage issues as compared to nFETs. Figure 10 plots I_{dsat}-I_{off} for pFETs. pFET I_{off} can easily be engineered to below 100pA/μm level in conventional symmetrical source/drain designs. In this experiment, a blanket tensile liner was used to enhance nFET performance and <100> channel orientation was chosen for its weak response to tensile liner on pFETs as shown in the chart [7]. A separate compressive liner can be applied on pFETs for higher performance at higher process cost.

The asymmetric nFET design was tested in a 101-stage ring oscillator circuit. Each active stage contains 2 gate fingers, sharing a drain contact. Additional masks are required to separate asymmetrical implants for top device and bottom device as shown in Fig. 11.

Low leakage RO operation was successfully demonstrated. The result is compared against that of ring oscillators using a conventional PDSOI CMOS design in Fig. 12. These are 101-stage ROs on 45nm ground rule. Physical gate length measured ~ 40-45nm, nFET width is 0.8um, pFET width is 1.2um in each stage. Each stage also drives 2 stages of gate capacitance load with the same device dimensions of the active stage. It is shown that nFET junction current dominates the leakage of the symmetric ROs, saturating at ~100pA/μm level at V_{dd}=0.9V and room temperature. The leakage level reduces by a factor of ~3 in the asymmetric ROs. Figure 13 shows RO delay vs. dynamics current in the active mode. At the same operating speed, the asymmetric ROs consume similar amount of power to the symmetric ROs, suggesting the new design does not introduce extra parasitic components.

Discussion

As the size of CMOS continues to scale, SOI transistors might become more attractive building blocks for VLSI chips because of its superior device isolation capability. At low power supply voltages, it would be relatively easy to obtain low off-state leakage in SOI FETs. As shown in Fig. 14, both junction current and FBE reduces with V_{dd}, more or less exponentially and linearly, respectively. It might be possible that a conventional PDSOI design would meet leakage requirements for low power applications at reduced V_{dd}. FDSOI (Fully-depleted) is well-known for its excellent scalability [8,9]. It is also an ideal device candidate for low power circuits. In an extremely thin (ET) SOI MOSFET, SCEs control can be achieved without use of doped channels. The FD and undoped body not only minimize floating body effect, but also suppress junction leakage, solving the 2 key issues addressed in this paper. Figure 15 shows example IVs of low leakage ETSOI CMOS. Looking forward, there seems to be more room available to enable low power SOI CMOS as device scaling continues.

Summary

This experiment demonstrated that low power logic operation ($\sim$30pA/μm off-state source-drain current in ROs) can be achieved in PDSOI CMOS using an asymmetrical halo-extension scheme. Combined with high performance and RF modules, the low power feature would enable SOI CMOS a powerful platform for portable device applications.

Acknowledgements

This work was performed at IBM SRDC, East Fishkill, NY. We wish to thank Haizhou Yin, Nivo Rovedo, Noah Zamdmer, Jun Yuan, Mike West, David Harame, G. Shahidi and T.C. Chen for technical assistance and management support.

References:
[1] S. Narasimha et al., et al., IEDM, (2006).
[2] S. Lee et al., et al., VLSI Technology, p. 54, (2007).
[3] D. Wendel et al., ISSCC, p. 102, (2010).
[4] Samuel K.H. Fung, et al., IEDM, p. 1035, (2007).
[5] J. Cai et al., IEEE SOI Conf. p. 15, (2008).
[6] H.M. Nayfeh et al., IEEE, TED, p. 3097, (2009).
[7] J. Yuan et al., ICSICT, Beijing, China, p. 1130, (2008).
[8] A. Majumdar et al., IEEE, TED, p. 2270, (2009).
[9] K. Cheng, et al., IEDM, p. 49, (2009).

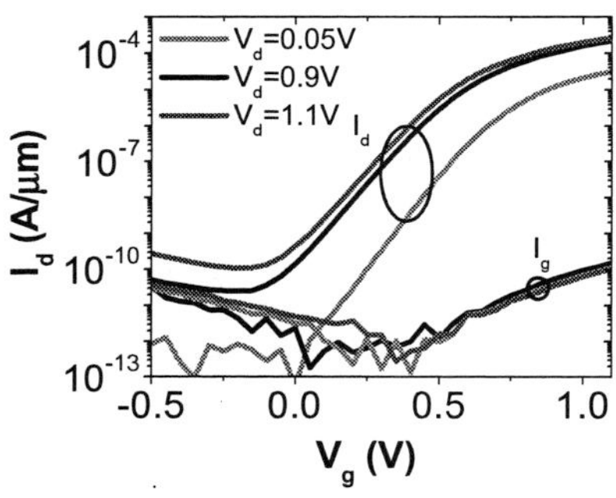

Fig. 1 Drain current and gate leakage vs. gate bias in a typical PDSOI nFET: the thermal emission leakage is enhanced by DIBL and SOI floating body effect, and the junction leakage sets the minimum I_{off} floor.

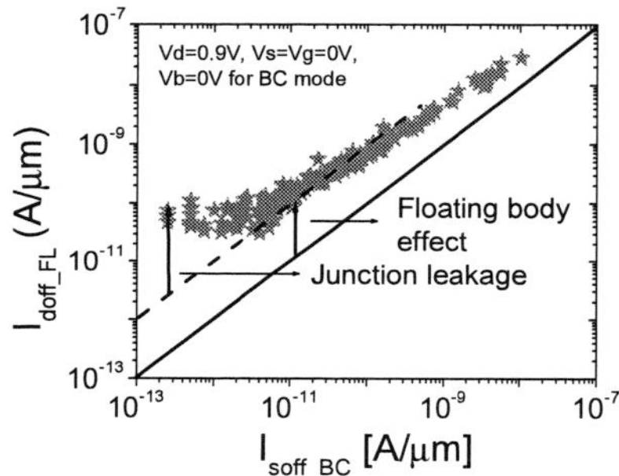

Fig. 2 For the same devices, the floating body mode I_{doff} is plotted against the body contacted mode I_{soff}, showing the floating body effect induced I_{off} increase, and junction leakage induced I_{off} increase.

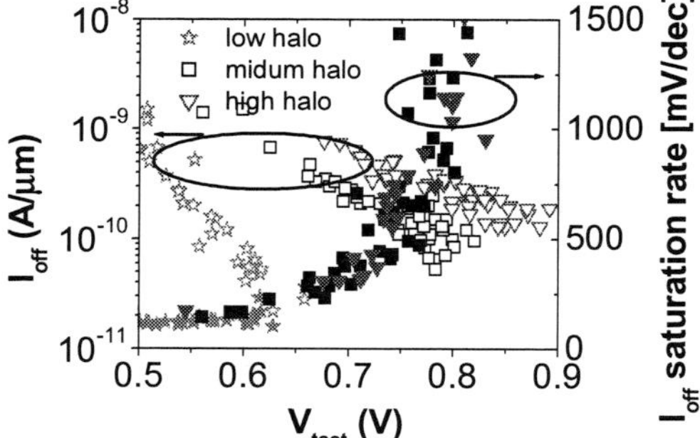

Fig. 3 I_{off} vs. V_{tsat} and I_{off} saturation rate (defined as inverse of I_{off}-V_g slope at V_g=0V) vs. V_{tsat}. As I_{off} decreases to the junction leakage limited level, it becomes V_{tsat} insensitive. Increasing V_{tsat} would simply degrade the drive current.

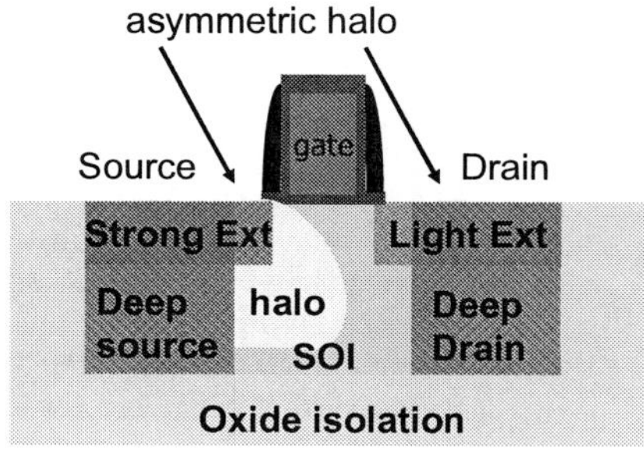

Fig. 4 Asymmetric source side halo enables a source forward leakage dominant operation, reducing floating body effect and drain side junction leakage.

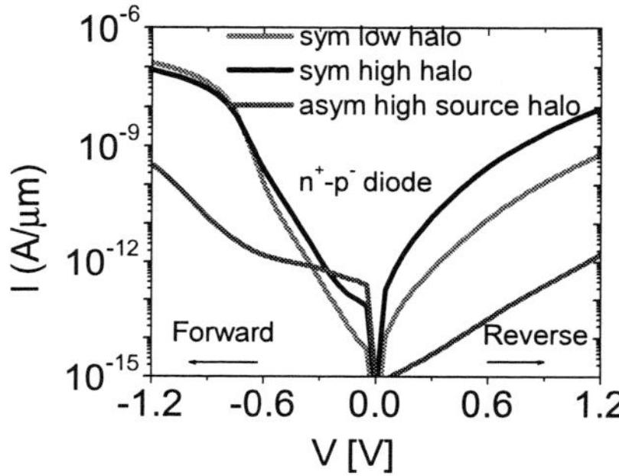

Fig. 5 Source side only halo and low dose symmetric halo reduce reverse junction leakage. For the asymmetric case, forward current measured from source side junction and reverse current measured from drain side junction.

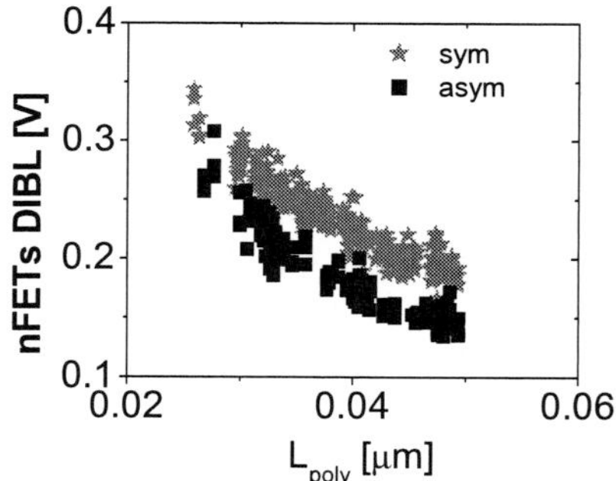

Fig. 6 The source side only halo technique helps pin the body potential to the source potential, reducing floating body DIBL and SOI history effects, V_{dd}=0.9V.

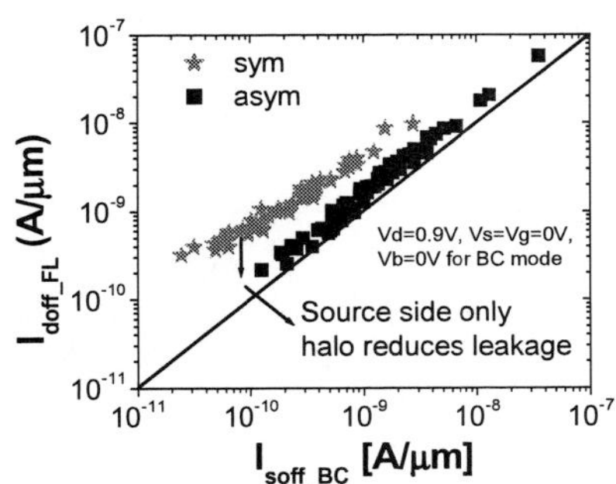

Fig. 7 Using the source side only halo technique, both floating body leakage and junction leakage reduce

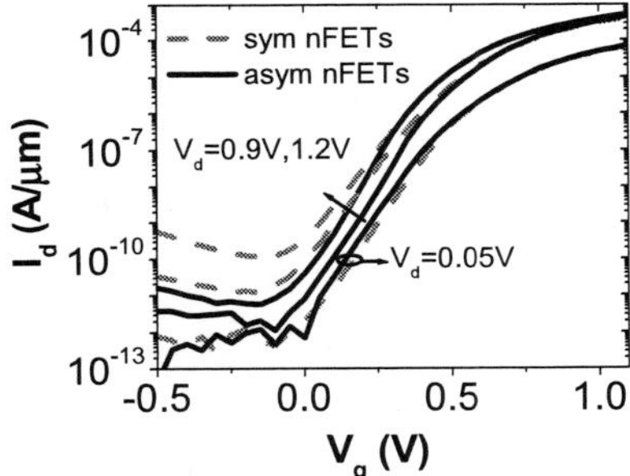

Fig. 8 Comparison of I_d vs. V_g curves for nFET with symmetric halo and source side only halo. The latter shows much reduced Vt and junction leakage sensitivity to drain biases.

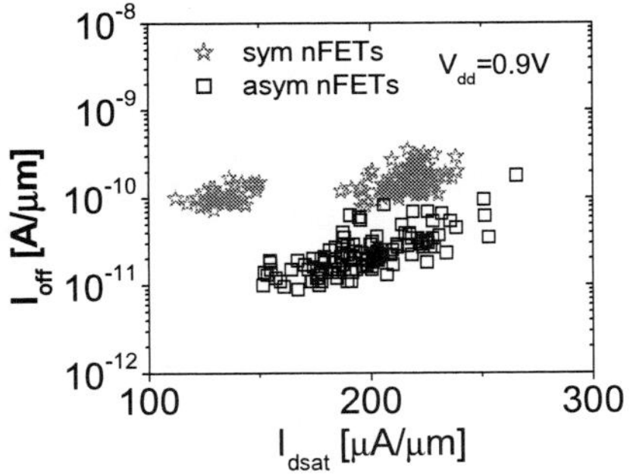

Fig.9 Drive current (I_{dsat}) vs. I_{off} for nFETs with symmetric halo and nFETs with source side only halo. The latter achieves much lower leakage, V_{dd}=0.9V

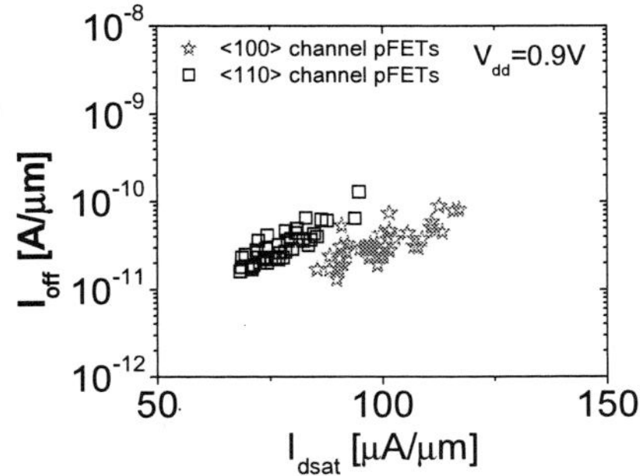

Fig. 10 Drive current (I_{dsat}) vs. I_{off} for symmetric pFETs built with a simplified process: with no eSiGe stressors and compressive liners. The <100> channel orientation is chosen for enhanced performance at a lower cost.

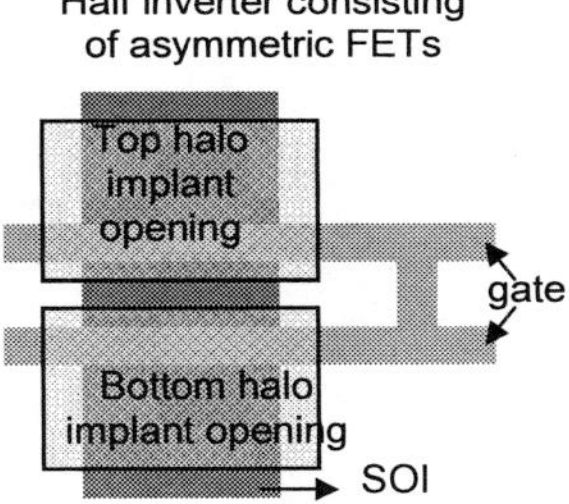

Fig. 11 A schematic top-down view of half inverter consisting of asymmetric devices. The source side halo for the top device and bottom device are implanted through two separate masks.

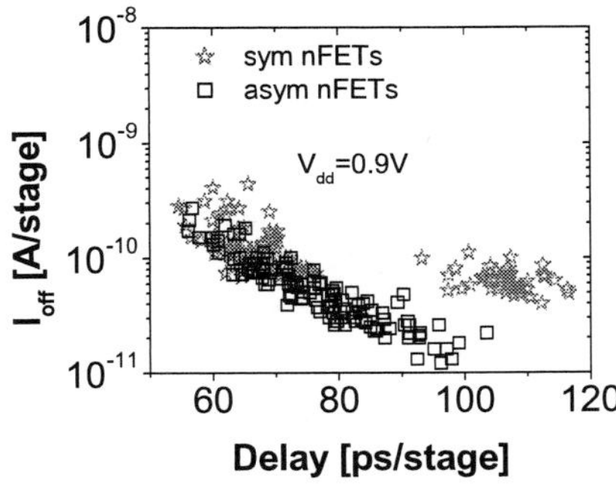

Fig.12. Delay of 101-stage ring oscillators vs. device leakage (excluding gate leakage). The circuits consisting of asymmetric devices achieve low standby leakage down to ~30pA/stage.

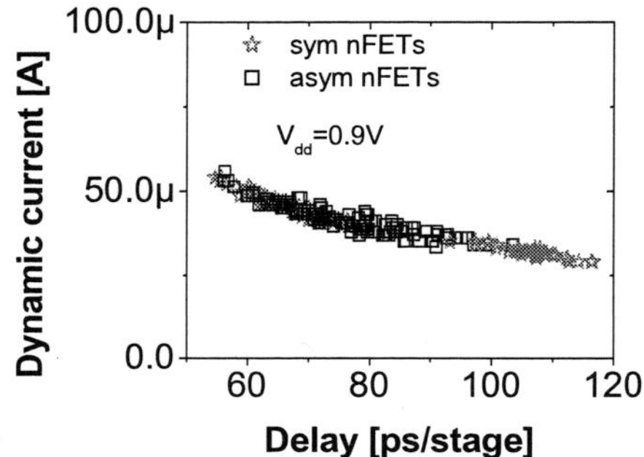

Fig.13 Power consumption vs. performance for ring oscillators (101 stages).

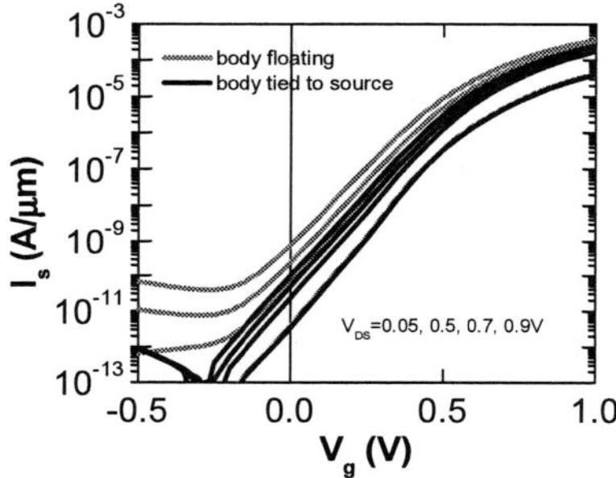

Fig.14 Source-side current of a PDSOI nFET (using symmetrical halos) operating at different V_{dd}. Comparing body floating mode current vs. grounded body mode current. Both junction leakage and FB-leakage reduces at low V_{dd}.

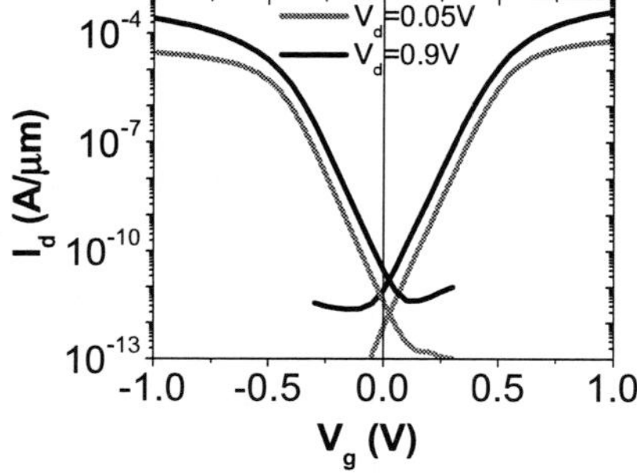

Fig.15 IVs of SOI CMOS with an extremely thin undoped SOI body, showing low leakage and good SCEs control. Channel length is ~ 35nm. A good device candidate for low power applications.

ECS Transactions, 34 (1) 37-42 (2011)
10.1149/1.3567556 ©The Electrochemical Society

Ultra-thin Body and BOX (UTBB) Device for Aggressive Scaling of CMOS Technology

Q. Liu[a], A. Yagishita[b], A. Kumar[c], N. Loubet[a], T. Yamamoto[d], P. Kulkarni[c], F. Monsieur[a], A. Khakifirooz[c], S. Ponoth[c], K. Cheng[c], B. Haran[c], M. Vinet[e], J. Cai[c], P. Khare[a], S. Monfray[a], F. Boeuf[a], S. Mehta[c], J. Kuss[c], E. Leobandung[c], M. Hane[d], H. Bu[c], K. Ishimaru[b], T. Skotnicki[a], W. Kleemeier[a], M. Takayanagi[b], T. Hook[c], M. Khare[c], S. Luning[f], B. Doris[c], R. Sampson[a]

[a]STMicroelectronics, [b]Toshiba, [c]IBM, [d]Renesas, [e]CEA-LETI, [f]GLOBALFOUNDRIES, Albany NanoTech, New York 12203, USA

We report fully-depleted UTBB devices with a gate length (L_G) of 25nm and BOX thickness (T_{BOX}) at 25nm, fabricated with a 22nm technology ground rule, featuring conventional gate first high-K/metal and raised source/drain (RSD) process. Competitive drive currents are achieved. Effective V_t modulation is observed with back bias (V_{bb}) and different ground plane (GP) polarity. Excellent local V_t variation is demonstrated. A functional UTBB $0.108\mu m^2$ 6-T SRAM is reported with V_{dd} down to 0.4V. We also present simulation studies that suggest UTBB device has superior scalability compared with thick BOX ETSOI device, thanks to better short channel effect (SCE) control.

Introduction

Fully depleted SOI (FDSOI) device is a viable option for the ever aggressive scaling of CMOS technology, thanks to the excellent short channel control, the elimination of random dopant fluctuation (RDF) effect, and its compatibility with mainstream planar CMOS manufacturing [1~5]. Compared with thick buried oxide (BOX) ETSOI device, thinner BOX UTBB device provides superior scalability by introducing a ground plane (GP) beneath the BOX region which further reduces electric field coupling between source and drain. Other than setting the V_t by the metal gate work function, the UTBB device allows the back bias to modulate the V_t and potentially simplify the gate stack while providing a multi-V_t solution. A dynamic power management scheme can also be realized by adjusting back bias depending on circuit block in action or idle. The use of un-doped channel in FDSOI provides excellent local V_t variability, which is essential for SRAM stability and can lead to lower supply voltage and lower power.

<u>Device Integration</u>

UTBB devices are fabricated at 22nm technology ground rules with a contacted gate pitch (CPP) of 100nm. Figure 1 shows a simplified integration flow. STI isolation is followed by GP implantation and annealing. Arsenic is used for N-type GP, while Indium is used for P-type GP. Hafnium-oxide based high-k metal gate is formed and patterned by a litho-etch-litho-etch (LE^2) step. A thin offset spacer is then formed and Si RSD is epitaxially grown. A high energy implantation is implemented to penetrate the RSD EPI and form extension. After S/D spacer formation and S/D implantation, RTA and laser annealing steps are used to diffuse and activate the dopant. Silicide, contacts and metallization are subsequently formed to complete the fabrication. Figure 2 shows the cross-section TEM picture of the final device, featuring channel thickness (T_{si}) ~6nm, L_G=25nm and spacer ~12nm. The challenges in UTBB integration include the preservation of the thin Si to insure RSD EPI growth and the reduction of the extension resistance. Both gate etch and offset spacer etch processes are optimized to render minimum Si loss. Only 1nm Si was consumed before RSD EPI, shown also in figure 2.

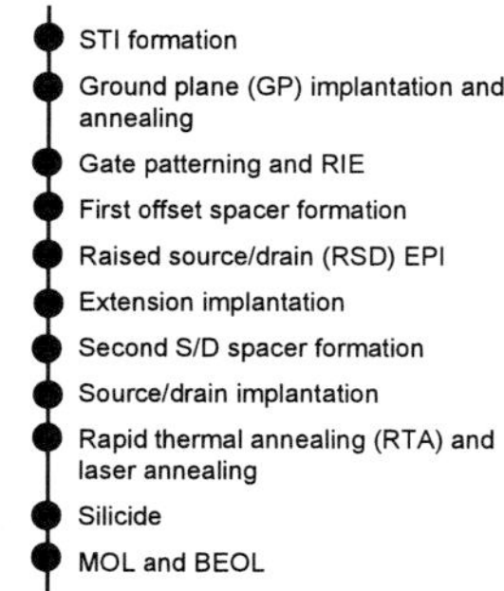

Figure 1 A simplified UTBB integration flow.

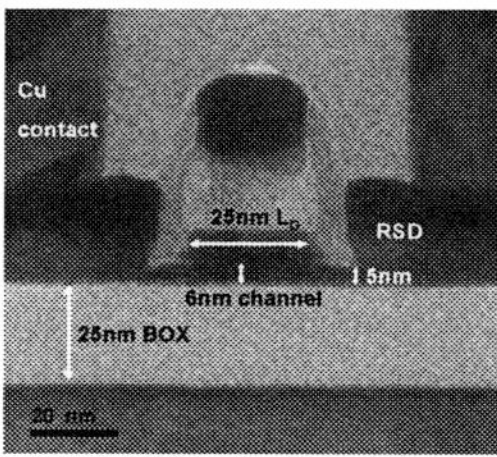

Figure 2 TEM cross-section of UTBB device, featuring T_{si}=6nm and L_G=25nm. The silicon loss, from optimized gate and offset spacer etch process, is only 1nm.

<u>Device Characteristics</u>

Figure 3 shows I_d/V_G curves of UTBB NFET and PFET. Excellent SCE control is achieved with drain induced barrier lowering (DIBL) at 60mV/V (NFET) and 80mV/V (PFET), sub-threshold swing (SS) at 80mV/dec for both N/PFET. At V_{dd}=0.9V, competitive drive currents (I_{on}) of 470/480 µA/µm, with off-state currents (I_{off}) at 2/1.5 nA/µm, for N/PFET, are achieved. The NFET I_{on}/I_{off} characteristics with respect to GP polarity and V_{bb} are shown in figure 4a. GP doping concentration is 5E18cm^{-3}. No I_{on}/I_{off} degradation is observed. A V_t shift of 80mV is obtained when V_{bb} changes by 0.9V on both N/P-type GP, with good V_t rolloff down to L_G 21nm (figure 4b). Further improvement on NFET performance is achieved by adopting a multi-step implantation (MSI) at S/D, which is essential to lower link-up resistance between extension and S/D region. Figure 5 shows that, by the implantation and RTA optimization, at constant

I_{off}=1nA/µm, a 28% improvement on I_{on} is obtained, while at I_{off}=10nA/µm, a 37% improvement on I_{on} is achieved.

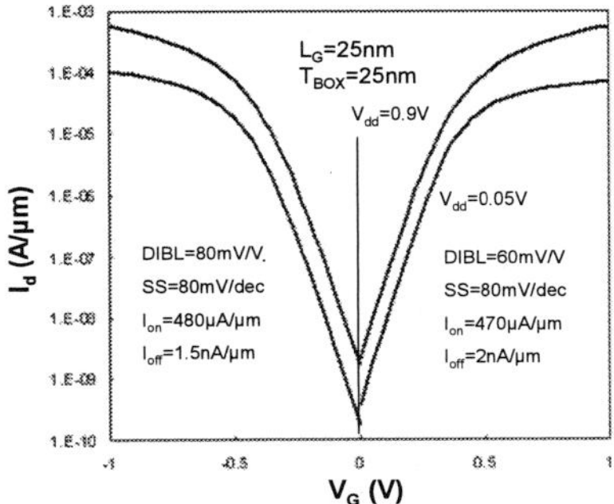

Figure 3 I_d/V_G characteristics of UTBB N/PFET, showing excellent SCE control and competitive drive current.

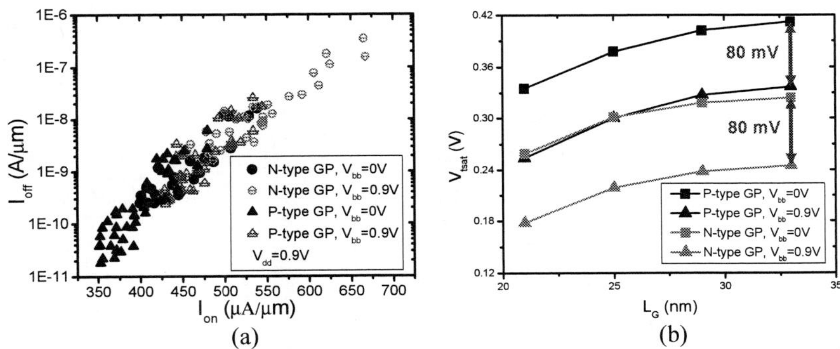

Figure 4 (a) UTBB NFET I_{on}/I_{off} showing no degradation from GP polarity and V_{bb}. (b) A V_t shift of 80mV is observed, with good V_t rolloff down to L_G 21nm.

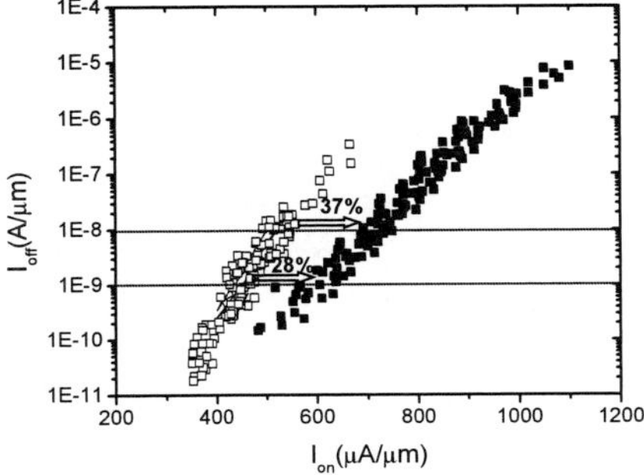

Figure 5 With optimized S/D implantation and RTA, I_{on} improvements of 28% and 37% are achieved at I_{off} 1nA/µm and 10nA/µm, respectively.

Thanks to the absence of RDF in the channel region, low A_{Vt} of 1.32 mV•µm and 1.27 mV•µm were measured with/without GP (Figure 6). The SRAM pair transistors, in the measurement, have L_G ranging from 25nm to 37nm, and width from 50nm to 250nm. The small difference on A_{Vt}, between with and without GP, shows that the GP RDF has a negligible impact on local V_t variability when T_{BOX} is at 25nm.

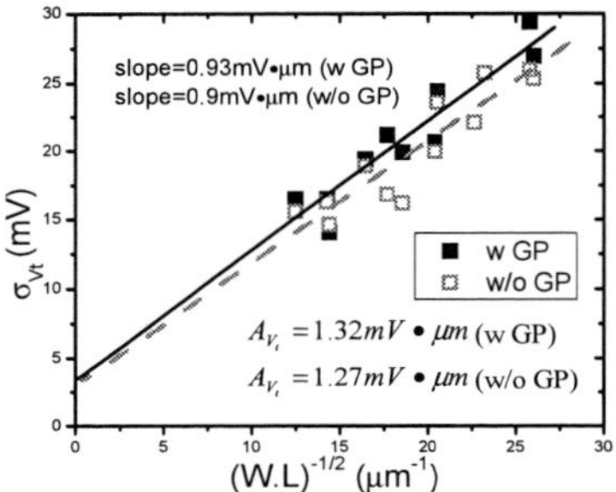

Figure 6 Measured A_{Vt} with/without GP, showing negligible GP RDF impact when T_{BOX}=25nm.

Furthermore, a UTBB 0.108µm^2 (0.2µm×0.54µm) 6-T SRAM is fabricated and presented. The topdown SEM is shown in figure 7. The CPP is 100nm. The butterfly curves are shown in figure 8. The SRAM remains functional with V_{dd} down to 0.4V.

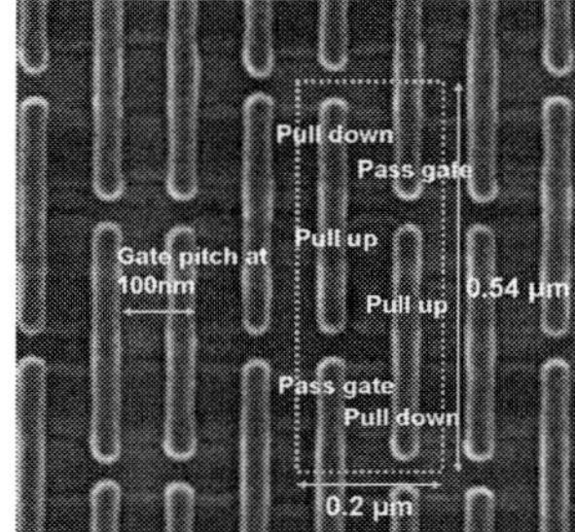

Figure 7 Topdown SEM of 0.108µm^2 6-T SRAM after gate formation and RSD EPI.

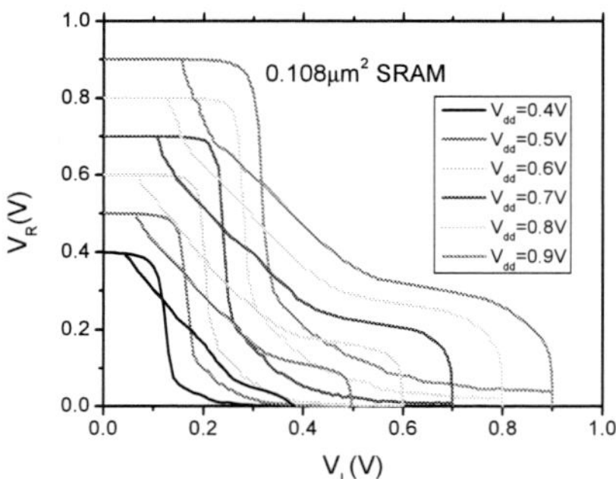

Figure 8 Butterfly curves of 0.108µm^2 SRAM with V_{dd} from 0.9V down to 0.4V.

Simulations and Discussion

The TCAD simulations of SCE with BOX thickness from 140nm down to 10nm, featuring L_G at 20/25nm and inversion gate dielectric (T_{inv}) at 1nm, are shown in figure 9. More than 30mV/V DIBL reduction is obtained for L_G at 25nm, when scaling T_{BOX} from 140nm to 10nm. With L_G at 20nm, 40mV/V DIBL reduction is achieved. SS keeps almost constant with T_{BOX} down to 25nm. SS shows a limited degradation (<3mV/dec) due to the stronger capacitance coupling from the back gate when reducing T_{BOX} down to 10nm. Thin BOX shows excellent gate length scaling capability, other than further T_{inv} reduction. It was also reported that to keep the constant device electrostatics, both T_{inv} and T_{Si} can be relaxed by scaling of T_{BOX} [3]. The simulations demonstrate the superior scalability of UTBB vs. thick-BOX ETSOI. Furthermore, the local V_t variability from the GP RDF effect is simulated (figure 10). In agreement with experimental data, at T_{BOX}=25nm, the effect on $\Delta\sigma_{Vt}$ is negligible. For a 10nm T_{BOX}, the $\Delta\sigma_{Vt}$ shows an increased impact from GP RDF due to its closer proximity to the channel region. However, this increased V_t variation can be mitigated by a factor of 2 by applying V_{bb}=-0.9V, due to improved DIBL.

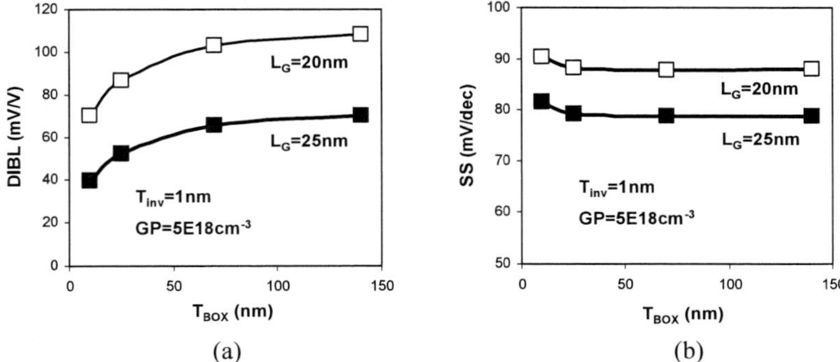

(a)　　　　　　　　　　　　　　　　(b)

Figure 9 TCAD simulations show (a) much improved DIBL (b) similar SS with T_{BOX} scaling from 140nm to 10nm.

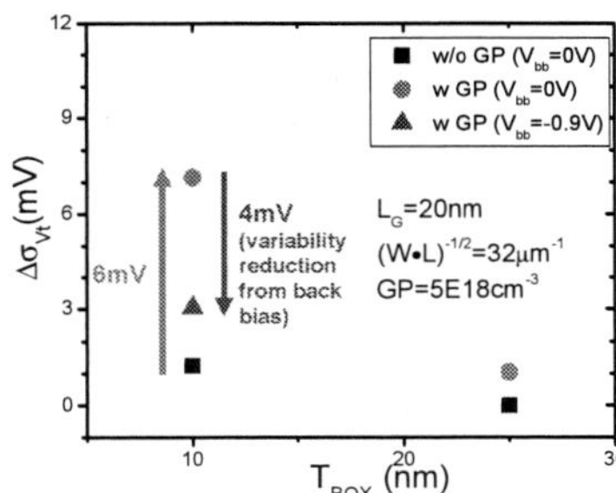

Figure 10 TCAD simulations of $\Delta\sigma_{Vt}$ response to BOX thickness, GP and V_{bb}, showing GP RDF effect on $\Delta\sigma_{Vt}$ and the recovery by V_{bb}.

Conclusions

UTBB devices with L_G=25nm at CPP 100nm have been demonstrated. Significant V_t modulation by GP polarity and V_{bb} was observed. Respectable NFET drive current of 740µA/µm at I_{off} 10nA/µm was achieved by optimizing S/D conditions. UTBB scalability and local V_t variation were also studied. It is concluded that with the excellent device electrostatics, UTBB is an attractive option for the aggressive scaling of CMOS technology.

Acknowledgments

The authors thank G. Shahidi for extensive discussions and managerial support. We also thank SOITEC for continued support. This work is performed by the research alliance teams at various IBM facilities.

References

1. K. Cheng, A. Khakifirooz, P. Kulkarni, S. Ponoth, J. Kuss, D. Shahrjerdi, L.F. Edge, A. Kimball, S. Kanakasabapathy, K. Xiu, S. Schmitz, A. Reznicek, T. Adam, H. He, N. Loubet, S. Holmes, S. Mehta, D. Yang, A. Upham, S.-C. Seo, J.L. Herman, R. Johnson, Y. Zhu, P. Jamison, B.S. Haran, Z. Zhu, L.H. Vanamurth, S. Fan, D. Horak, H. Bu, P.J. Oldiges, D.K. Sadana, P.Kozlowski, D. McHerron, J. O'Neill, and B. Doris, *IEDM Tech. Dig.*, **49** (2009).
2. C. Fenouillet-Beranger, P. Perreau, L. Pham-Nguyen, S. Denorme, F. Andrieu, L. Tosti, L. Brevard, O. Weber, S. Barnola, T. Salvetat, X. Garros, M. Casse, C. Lerous, J.P. Noel, O. Thomas, B. Le-Grantiet, F. Baron, M. Gatefait, Y. Campidelli, F. Abbate, C. Perrot, C. de-Butter, R. Beneyton, L. Pinzelli, F. Leverd, P. Gouraud, M. Gros-Jean, A. Bajolet, C. Mezzomo, C. Leyris, S. Haendler, D. Noblet, R. Pantel, A. Margain, C. Borowiak, E. Josse, N. Planes, D. Delprat, F. Boedt, K. Bourdelle., B.Y. Nguyen, F. Boeuf, O. Faynot and T. Skotnicki, *IEDM Tech. Dig.*, **667** (2009).
3. Q. Liu, A. Yagishita, N. Loubet, A. Khakifirooz, P. Kulkarni, T. Yamamoto, K. Cheng, M. Fujiwara, J. Cai, D. Dorman, S. Mehta, P. Khare, K. Yako, Y. Zhu, S. Mignot, S. Kanakasabapathy, S. Monfray, F. Boeuf, C. Koburger, H. Sunamura, S. Ponoth, A. Reznicek, B. Haran, A. Upham, R. Johnson, L. F. Edge, J. Kuss, T. Levin, N. Berliner, E. Leobandung, T. Skotnicki, M. Hane, H. Bu, K. Ishimaru, W. Kleemeier, M. Takayanagi, B. Doris and R. Sampson, *Symp. VLSI Tech.*, **61** (2010).
4. J.-L. Huguenin, S. Monfray, G. Bidal, S. Denorme, P. Perreau, S. Barnola, M.-P. Samson, C. Arvet, K. Benotmane, N. Loubet, Q. Liu, Y. Campidelli, F. Leverd, F. Abbate, L. Clement, C. Borowiak, A. Cros, A. Bajolet, S. Handler, D. Marin-Cudraz, T. Benoist, P. Galy, C. Fenouillet-Beranger, O. Faynot, G. Ghibaudo, F. Boeuf and T. Skotnicki, *Symp. VLSI Tech.*, **59** (2010).
5. O. Faynot, F. Andrieu, O. Weber, C. Fenouillet-Beranger, P. Perreau, J. Mazurier, T. Benoist, O. Rozeau, T. Poiroux, M. Vinet, L. Grenouillet, J-P. Noel, N. Posseme, S. Barnola, F. Martin, C. Lapeyre, M. Casse, X. Garros, M-A. Jaud, O. Thomas, G. Cibrario, L. Tosti, L. Brevard, C. Tabone, P. Gaud, S. Barraud, T. ernst and S. Deleonibus, *IEDM Tech. Dig.*, **50** (2010).

ECS Transactions, 34 (1) 43-48 (2011)
10.1149/1.3567557 ©The Electrochemical Society

Simulations of FDSOI CMOS with Sharing Contact between Source/Drain and Back Gate

Miao XU, Qingqing LIANG, Huilong ZHU[*], Haizhou YIN, Zhijiong LUO, Dapeng CHEN, Tianchun YE

IC Advanced Process R&D Center,
Institute of Microelectronics Chinese Academy of Sciences, Beijing 10029, CHINA
[*]Email: zhuhuilong@ime.ac.cn

In this paper, a new ultra-thin fully-depleted SOI CMOS structure with sharing contact between source/drain and back gate is presented to save area and increase threshold voltage tuning capability. TCAD simulations are used to investigate the back-gate effect on the ultra-thin SOI CMOS. A new process flow to make the fully-depleted SOI CMOS structures is also proposed.

Introduction

SOI CMOS technology has proven to be compatible with bulk CMOS in many ways, ranging from circuit design, layout to wafer processing. Ultra-Thin Fully-Depleted (FD) SOI transistors have gained extensive attention for their improved scaling potential over the Partially-Depleted (PD) SOI and bulk Si counterparts. It was demonstrated [1] that the FD-SOI devices exhibit the following advantages over bulk and PD-SOI MOSFETs: (1) sharper sub-threshold slope, (2) reduced parasitic-capacitance, (3) reduced short-channel-effect (SCE) and random-doping-fluctuation (RDF).

Fully depleted UTSOI CMOS is attractive for continued CMOS scaling due to the superior short channel control and reduced threshold voltage variation enabled by the undoped channel which provides the additional benefits of smaller random-doping-fluctuation (RDF). The new device structure combines the benefits of short channel control because some of the charge that would otherwise be needed to support the channel depletion-region-charge is instead produced by charge in the substrate under the BOX [1]. The induced charge in the substrate will have the effect of trying to establish an inversion layer along the interface between the Si surface layer and the BOX (back-gate). A coupling exit between the potentials on the top-gate and the back-gate vary with bias applied to the top-gate.

It has proposed that a fully depleted SOI CMOS with double BOX and dual STI for electrically isolating the back gate from the ultra-thin FD-SOI MOSFET [2]. The effectiveness of back gate bias was illustrated by leakage and performance compensation and necessary threshold voltage (Vt) adjustment. However, it has area penalty since extra contacts and additional wirings are needed for the back gate. To save area and increase Vt tuning capability, new fully-depleted SOI CMOS structures with sharing contact between source/drain and back gates are presented in this paper.

Device Design and Fabrication

As shown in Fig.1, an asymmetric back gate shared source/drain contact is implemented to control threshold voltage of an UT-SOI MOSFET with an ultra-thin BOX (BOX_1) between SOI and the back gate. The device design parameters for the structure are studied by TCAD design. The thickness of the thin BOX_1 layer under the SOI layer (Tsi~10nm) is chosen to be 20nm for good Vt controllability and low parasitic capacitance penalty. The bottom thick BOX_2 enables dielectric isolation of the complementary-doped polysilicon back gate.

The back gate is connected to source or drain to save area and used to adjust the Vt of the UT-SOI MOSFET. A process flow for fabricating the UT-SOI MOSFET is also proposed which is shown in Fig.2. The contact hole above the asymmetry ground plane extends into STI so that the asymmetry ground plane is exposed.

Heavy doping in the back gate is required to minimize depletion effect and maintain Vt under control.

Results and Discussion

Effective Vt control by asymmetric back gate shared source/drain contact and complement doping is demonstrated by the Id-Vg characteristics for ultra thin FD-SOI CMOS devices with Lgate=30nm, shown in Fig.3. The solid and dash lines represent the asymmetric back gate shared contacted with source and drain respectively, in which the back gate is doped with the same type dopant as the source/drain. The dot lines denote the asymmetric back gate shared contact with source and complement dope to the source. This clearly shows good threshold voltage controllability with asymmetric back gate. A large adjustable range of the UT-SOI MOSFET Vt, ~0.308V, is achieved selectively connecting the back gate to source or drain [1] and adjusting dopant type or concentration level [2]. The Vt roll-off characteristics vs. asymmetric back gates are shown in Fig.4. Low-Vt (LVT), Regular-Vt (RVT) and High-Vt (HVT) can respectively be achieved by selectively connecting the back gate to drain or source and complement doping type. Figure 5 shows the DIBL vs. gate length for FDSOI PMOS with asymmetric back gates. The effects of doping concentration in the back gate on the control of Vt are investigated as well. The impacts of back gate doping concentration and SOI thickness on Vt controllability are illustrated in Fig.6. It turns out that high back gate doping and thick SOI improves Vt controllability. The solid symbols represent PMOS with Tsi=10nm and the open symbols represent the Tsi=7nm. The adjustable range of the UT-SOI PMOS Vt, ~0.240V, when Tsi=7nm, which is less than 0.308V when Tsi=10nm. The thicker Tsi gets higher Vt controllability when changing N-type to P-type back-gate is mainly caused by sub-threshold degradation. Figure 7 shows the impacts of the thickness of thin BOX layer under SOI on Vt control. And thin BOX improves the Vt controllability.

Summary

We present a new UT-SOI MOS structure and process flow enabling back-gate asymmetrically tied to source or drain. Compared with the previous work, the new device structure combining the benefits of short channel control, and Vt turnability is attractive for an ultimate planar CMOS technology and smaller area penalty from the back-gate contact window. Therefore, it is a promising candidate for the sub-22nm CMOS technology.

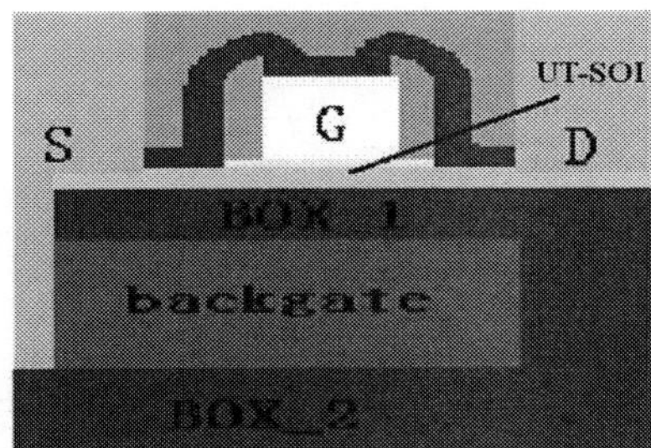

Figure 1. Structures of ultra-thin SOI MOSFET with asymmetric back gate and shared source/drain contact.

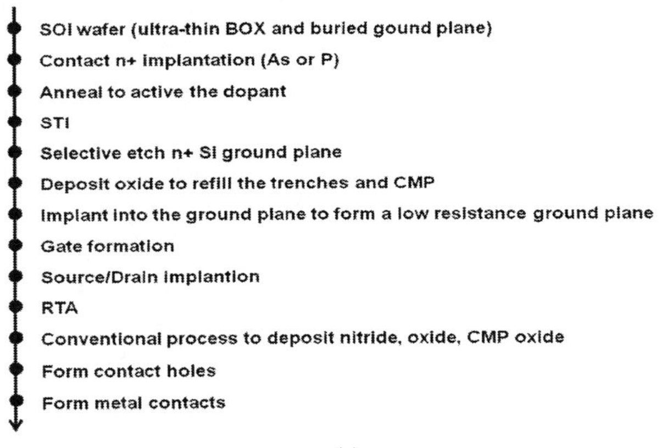

(a)

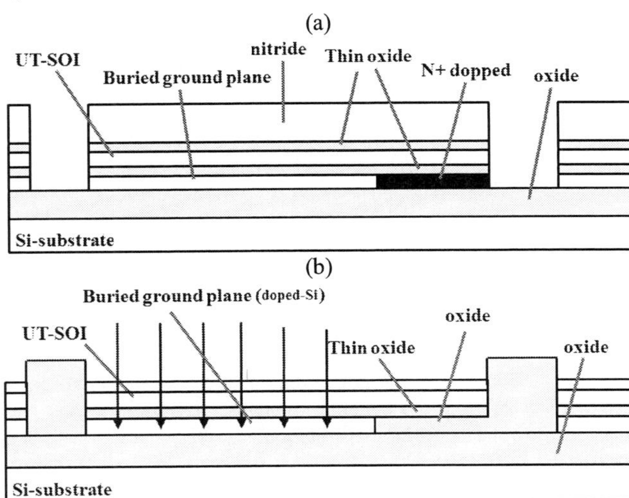

(b)

(c)

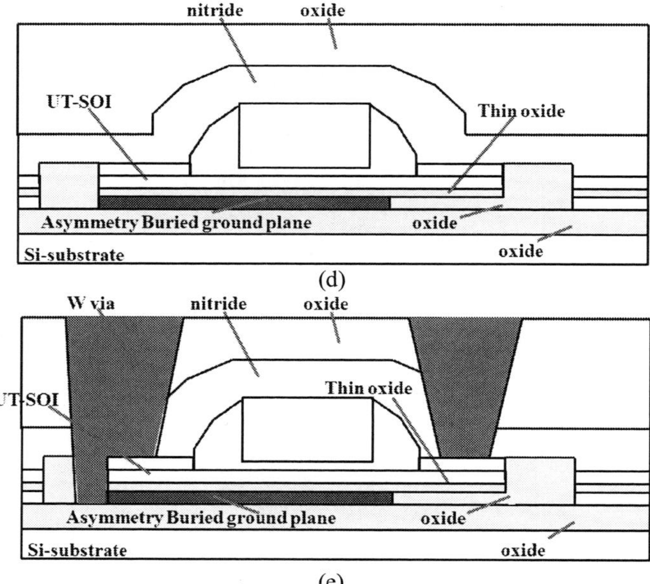

Figure 2. (a) A process flow for making an ultra-thin FD-SOI MOSFET device with sharing a contact between SD and back gate. (b) After contact implantation follow conventional process to form STI. (c) Conduct an implantation into the ground plane to form a low resistance ground plane. (d) Follow conventional process to form a MOSFET. (e) The contact hole above the asymmetry ground plane extends into STI.

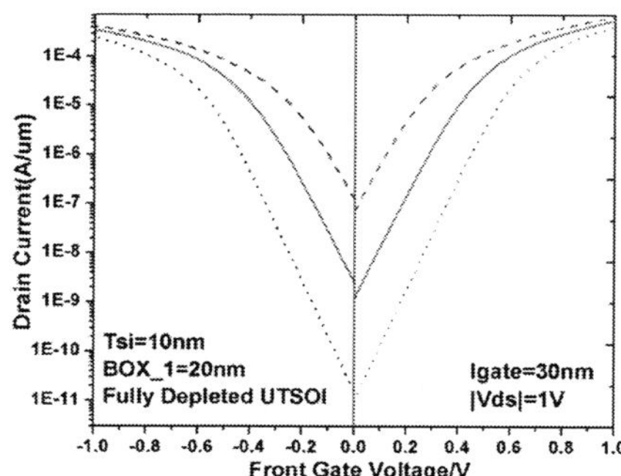

Figure 3. Id-Vg characteristics vs. asymmetric back gate for FD-SOI CMOS with Lgate=30nm. The solid and dash lines respectively represent the asymmetric back gate shared contacted with source and drain, while the back gate is the same type doped with the source/drain. The dot lines represent the asymmetric back gate shared contacted with source and complement dope to the source.

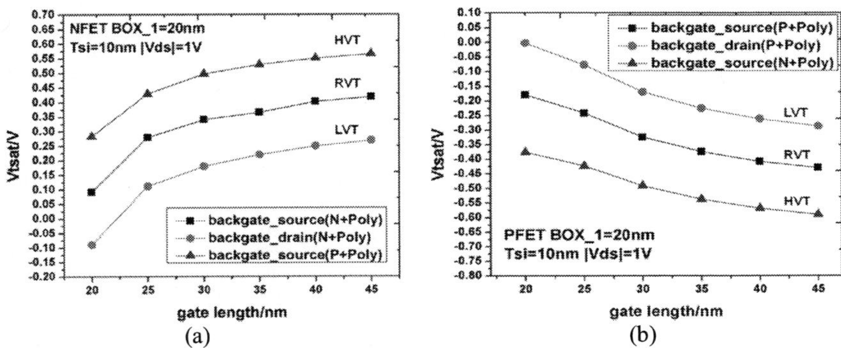

(a)   (b)

Figure 4. Vt roll-off characteristics vs. asymmetric back gate for ultra-thin FD-SOI MOSFETs with BOX_1=20nm and Tsi=10nm. Low-Vt (LVT), Regular-Vt (RVT) and High-Vt (HVT) can be achieved by selectively connecting the back gate to source or drain and adjusting doping type.

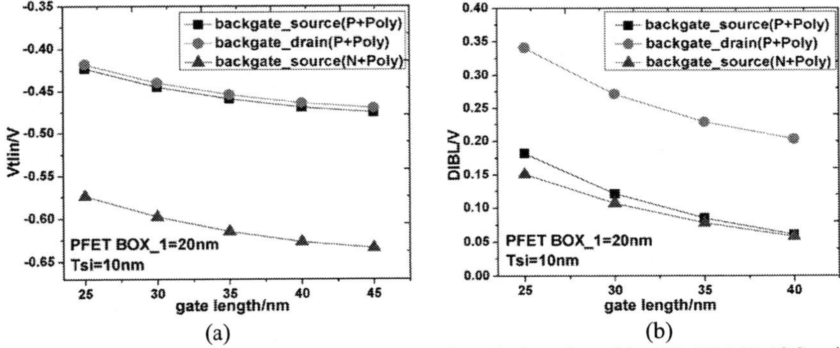

(a)   (b)

Figure 5. Simulated Vtlin and DIBL vs. gate length for ultra-thin FD-SOI PMOS with asymmetric back gates.

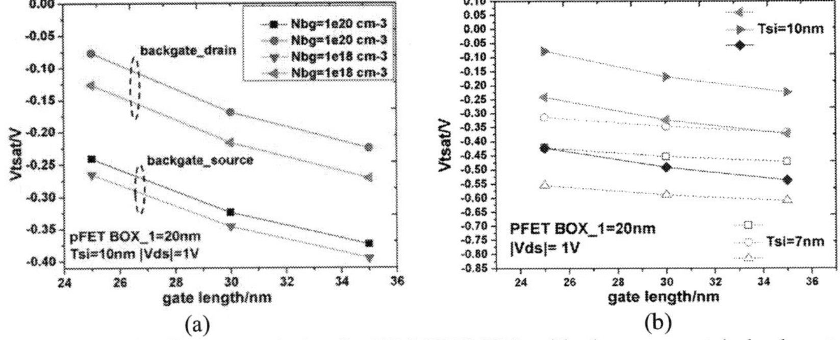

(a)   (b)

Figure 6. Vt roll-off characteristics for FDSOI PMOS with the asymmetric back gates shared to source & drain, while Nbg (back gate doping) =10^{20} cm^{-3} & 10^{18} cm^{-3} (a) and Tsi=10nm & 7nm (b). The Square & LeftTri symbols represent the PMOS with the asymmetric back gates shared contacted with source while the RightTri & Circle symbols represent the asymmetric back gates shared contacted with drain. The Diamond & UpTriangle symbols represent the asymmetric back gate shared contacted with source and complement dope to the source. High back gate doping improves Vt controllability.

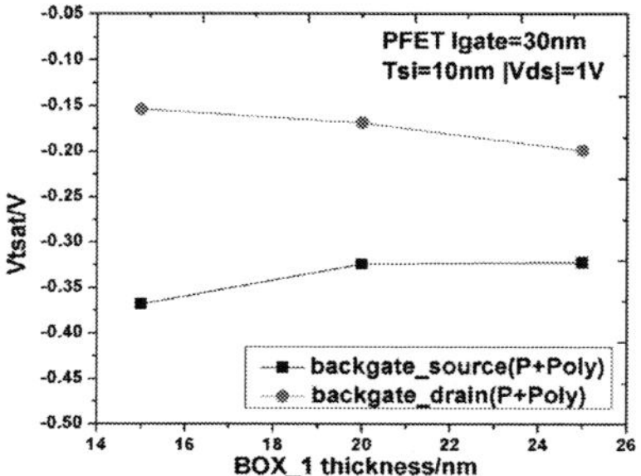

Figure 7. Vtsat vs. the thickness of the thin BOX under SOI for FD-SOI PMOS with asymmetric back gates shared to source/drain, with Lgate=30nm, Tsi=10nm.

Acknowledgments

This work is sponsored by China National S&T Major Project 02 and CAS funding for China Recruitment Program of Global Experts.

References

1. S.Wolf, *Silicon Processing for the VLSI Era*, 4(2002) p.522-524.
2. M. Khater, J. Cai et al, VLSI Technology Digest of Technical Papers, p.43 (2010).

ECS Transactions, 34 (1) 49-54 (2011)
10.1149/1.3567558 ©The Electrochemical Society

Scaling MOSFETs with Self-aligned Super-Steep-Retrograded Halo (3SRH)

Binneng WU, Weiping XIAO, Huilong ZHU [a], Qingqing LIANG, Hao WU,
Haizhou YIN, Zhijiong LUO, Hongyu YU [b], Dapeng CHEN, Tianchun YE
Institute of Microelectronics, Chinese Academy of Sciences, Beijing 100029, PR China
[b]Nanyang Technological University, School of Electrical & Electronic Engineering
[a] Email: zhuhuilong@ime.ac.cn

In order to take the advantages of the replacement gate process, improve the control of SCE, and enhance the performance of CMOSFETs, a novel method of self-aligned super-steep-retrograded halo (3SRH) implantation is designed and presented in this paper. With process and device simulations, it's demonstrated that 3SRH can be used to enhance MOSFET performance.

1. Introduction

With aggressively scaling MOSFET gate length, the control of short-channel-effect (SCE) of the MOSFET becomes more and more difficult. Halo implantations are usually used to improve SCE. The shorter gate length, the heavier halo dose is needed. However, the heavy halo dose or high halo dopant concentration reduces channel carrier mobility, increases extension resistance, and then causes large degradation of MOSFET performance.

Recently, CMOSFETs with replacement gate have been implemented in the state of the art HKMG (high-K metal gate) 32nm technology node [1]. In order to take the advantages of the replacement gate process, improve the control of SCE, and enhance the performance of CMOSFETs, a novel method of self-aligned super-steep-retrograded halo (3SRH) implantation is designed and presented in this paper. Process and device simulations are conducted to demonstrate the benefits of device performance and scaling due to 3SRH.

2. Device Design and Process Flow

Conventional halo implantation creates higher dopant concentration in the channel of short-channel MOSFET than that of long-channel MOSFET, which causes reverse SCE (RSCE) and reduce carrier mobility in the channel. To reduce extension resistance and form abrupt doping profile, Vt adjustment ion implantation is conducted after dummy gate being removed to form a self-aligned super-steep-retrograded island (3SRI). For nMOSFET, p-type dopants of boron and indium can be used for 3SRI. Due to indium's low solid solubility and possible implant damage to channel, the peak concentration of indium is kept at about or less than $2.0 \times 10^{18} \mathrm{cm}^{-3}$. For pMOSFET, n-type dopants of arsenic and antimony can be used for 3SRI.

For short channel devices, good control of SCE is obtained with 3SRI. However, the threshold voltage (Vt) for long channel devices with 3SRI is too high. Inspired by reverse halo [2] application, reverse halo implantation (RHI) is applied to lower Vt for long channel devices without covering short channel devices by lithographic mask, which will be explained by the process flow introduced later. It is seen from Fig. 1 that RHI can effectively reduce Vt in long channels. The combination of the 3SRI and the RHI results in a 3SRH profile, which improves CMOSFET performance and scaling capability. The process flow is showed in Fig. 2. The RHI is conducted vertically and good for further scaling without shadowing effect. We call reverse halo implantation because that short channel MOSFET receives less halo dose in the channel than that in long channel MOSFET. Species used for the implantation are listed in TABLE I. The design parameters are described in next section. In addition, the process steps forming 3SRH is compatible with HKMG replacement gate flow.

A simplified process flow for the formation of 3SRH:

- Conventional replacement process flow without silicidation
- Dummy gate etching
- Self-aligned super-steep-retrograded island implantation
- Reverse halo implantation
- Laser/flash anneal
- Formation of a replacement gate
- Silicidation

3. Device Characteristics

Process and device simulations are carried out to investigate how 3SRH improves the performance and scaling of nMOSFETs and pMOSFETs. A comparison of the characteristics between un-strained bulk CMOSFETs with conventional halo and with 3SRH is made.

To make apple-to-apple comparison, the values of the Vt_sat and overlap capacitance of CMOSFETs with 3SRH are fitted to that with conventional halo. Vt_sat roll-off curves for control CMOSFETs and for CMOSFETs with 3SRH are plotted in Fig.3. It was shown in [3] that, due to DIBL, an effective drive current, Ieff, needs to be used to calculate the actual delay of an inverter instead of Ion since the actual switching current could be significantly lower than the Ion. The value of Ieff is an averaged value of I_{high} and I_{low}. The value of I_{high} is equal to the drive current at Vgs=Vdd and Vds=Vdd/2. The value of I_{low} is given by the drive current at Vgs=Vdd/2 and Vds=Vdd. Fig. 4 shows that Ieff improvement at Ioff=1×10^{-7} A/um is ~24% for nMOSFET with 3SRH and ~12% for pMOSFET with 3SRH. And Fig. 5 shows that Ion improvement at Ioff=1×10^{-7} A/um is ~20% for nMOSFET with 3SRH and ~11% for pMOSFET with 3SRH. Our simulations show that the improvement of performance is mainly due to the reduction of external resistance, Ron, for nMOSFET and scaling of gate length for pMOSFET, which can be seen from Fig. 6(a) and Fig. 3(b), respectively. The overlap capacitances of control devices and 3SRH devices are almost the same, namely about 0.36×10^{-15}F/um for nMOSFET and 0.3×10^{-15}F/um for pMOSFET. The EOT of both nMOSFET and pMOSFET is 1.12nm. The work function is 4.2eV for nMOSFET and 5.0eV for pMOSFET.

4. Discussion

Implantations with different dopants are conducted for nMOSFETs and pMOSFETs. For nMOSFET, BF_2 and In are used to form 3SRI and As to form RHI. For pMOSFET, As and Sb are used to form 3SRI and BF_2 and In to form RHI. Those dopants are implanted for several times with different dose and energy to optimize doping profiles. The abruptness of the 3SRI is about 12nm/decade for nMOSFET and 3.5nm/decade for pMOSFET, which can be seen from Fig. 7. To match control device Vt roll-off or reduce Vt for long channel MOSFET, reverse halo implantation (RHI) is used as shown in Figs. 2(b) and 2(c). The performance gain for nMOSFET with 3SRI is mainly due to Ron reduction as shown in Fig. 6(a). It is attributed to 3SRI implantation shown in Fig. 2(a) is blocked by the oxide and nitride spacer and, unlike conventional halo implantation, 3SRI profile does not overlap with extension doping. It is particularly important for nMOSFET due to boron fast diffusion during SD anneal and boron pile-up occurs on the tip of the extension of the nMOSFET if conventional halo implantation is used. The boron pile-up can cause external resistance increases since it compensates the extension n-type doping. However, the performance gain for pMOSFET with 3SRI is from scaling of gate length instead of Ron (ref. Fig. 6(b)). The Vt roll-off curves in Fig. 3(b) show that the pMOSFETs with 3SRH has better scaling capability than that with conventional halo. For Ioff at 1×10^{-7} A/um, indeed, the gate length of the pMOSFET with 3SRH is 30nm, while the gate length of the pMOSFET with conventional halo is 36nm. It indicates that the gate length scaling contributes the performance enhancement of the pMOSFET with 3SRI.

5. Conclusion

A novel method of self-aligned super-steep-retrograded halo (3SRH) implantation is designed and presented. Our simulations show that 3SRH profiles can be used to obtain good control of short-channel-effects and to enhance performances for both n- and p-MOSFET. With 3SRH Ieff at Ioff=1×10^{-7} A/um is improved by ~24% for nMOSFET and ~12% for pMOSFET. The process of 3SRH is compatible with current replacement gate flow for manufacturing HKMG CMOSFETs.

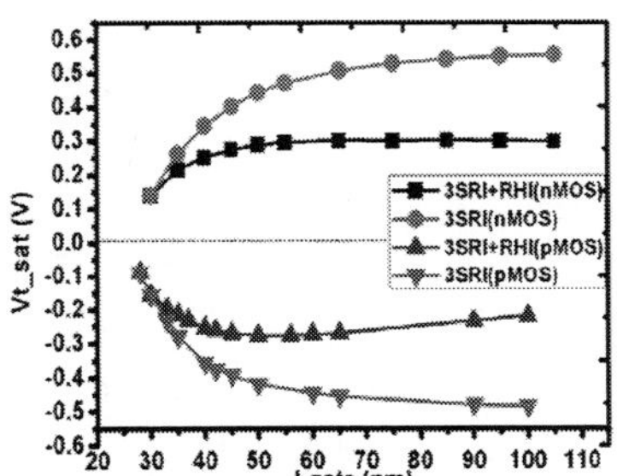

Fig.1. A comparison of Vt_sat between 3SRI and 3SRI+RHI for nMOSFET and pMOSFET.

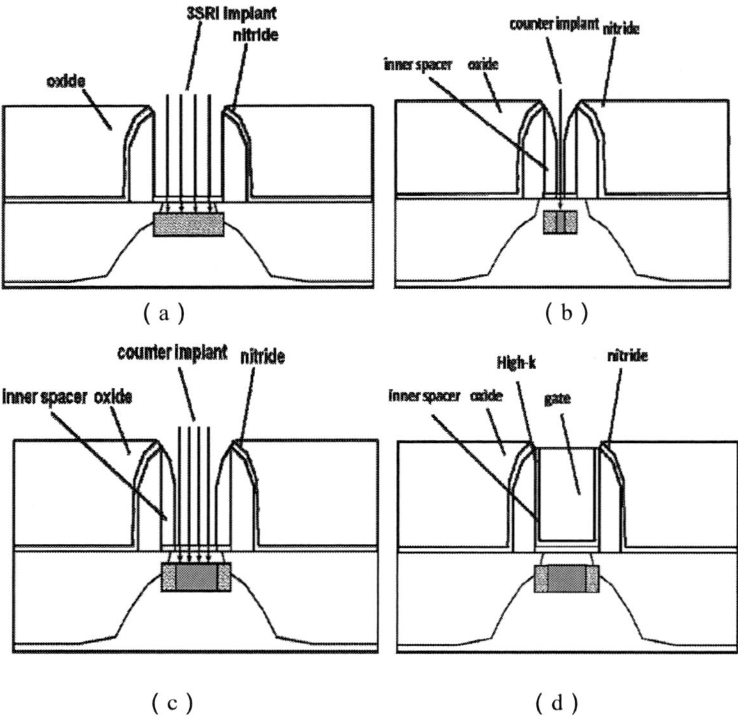

Fig. 2. (a) Formation of a self–aligned super-steep-retrograded island (3SRI). (b) Reverse halo implantation for a short channel device. (c) Reverse halo implantation for a long channel device. (d) Removal of the inner spacer and formation of a replacement gate.

TABLE I. Dopants used for self-aligned super-steep-retrograded halo (3SRH).

Device	3SRI	RHI
nMOS	BF_2、 Indium	Arsenic
pMOS	Arsenic、 Antimony	Indium、 BF_2

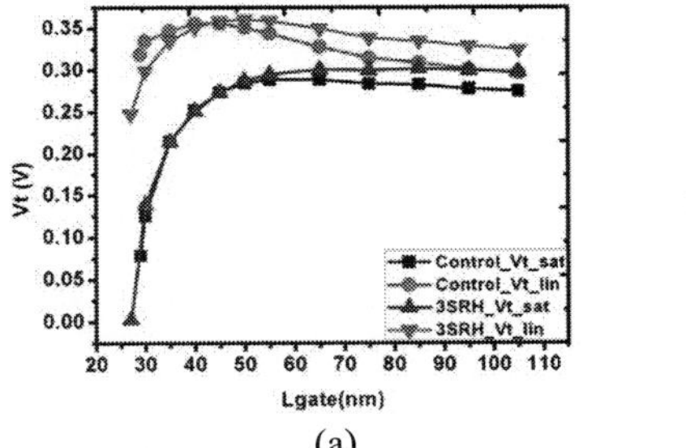

(a)

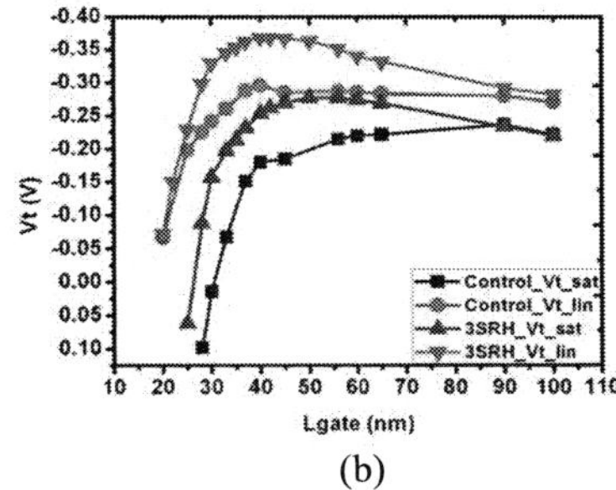

(b)

Fig. 3. Vt_sat roll-off curves for control MOSFETs and for the MOSFETs with self-aligned super-steep-retrograded halo (3SRH): (a) for nMOSFET and (b) for pMOSFET.

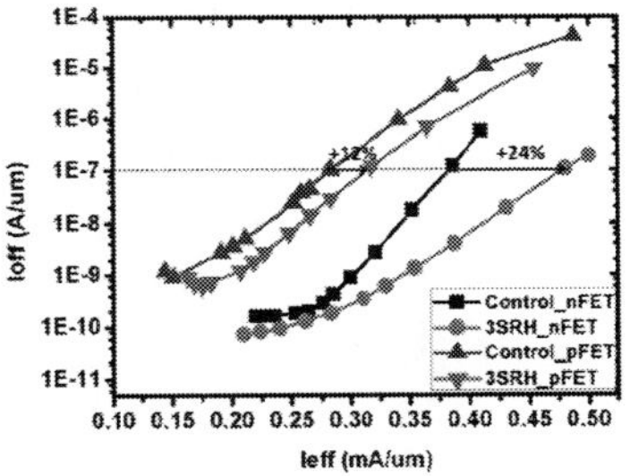

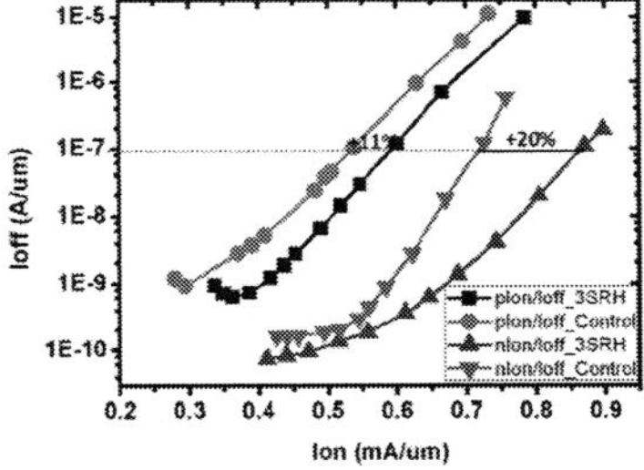

Fig. 4. A comparison of Ieff/Ioff between control devices and the devices with self-aligned super-steep-retrograded halo (3SRH). With 3SRH, the Ieff @ Ioff=1x10^{-7} A/um improvement is ~24% for nMOSFET and ~12% for pMOSFET.

Fig. 5. A comparison of Ion/Ioff between control devices and the devices with self-aligned super-steep-retrograded halo (3SRH). The Ion @ Ioff=1x10^{-7} A/um improvement is ~20% for the nMOSFET with 3SRH and ~11% for the pMOSFET with 3SRH.

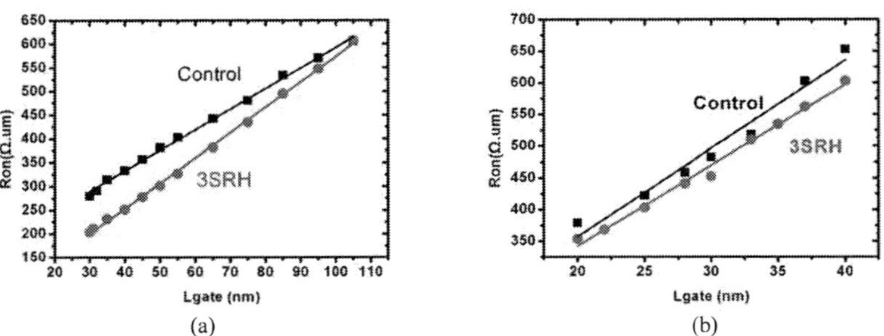

Fig. 6. A comparison of Ron between control devices and the devices with 3SRH (a) for nMOSFET and (b) for pMOSFET.

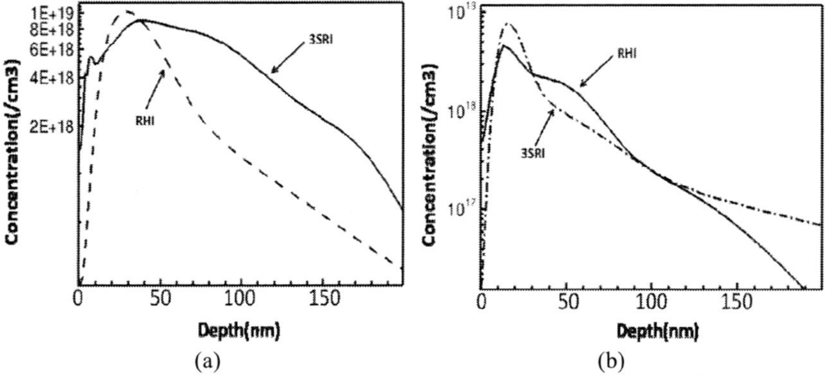

Fig.7. 3SRH vertical doping profiles in the channel (a) for nMOSFET and (b) for pMOSFET.

Acknowledgments

This work is sponsored by China National S&T Major Project 02 and CAS funding for China Recruitment Program of Global Experts (千人计划).

References

1. P. Packan *et al.*, IEDM Tech. Dig., pp659-662 (2009)
2. H. Zhu *et al.*, EDL, Vol. 28, pp168-170 (2007)
3. M. H. Na *et al.*, IEDM Tech. Dig., pp121–124 (2002)

ECS Transactions, 34 (1) 55-60 (2011)
10.1149/1.3567559 ©The Electrochemical Society

Electrostatic Discharge (ESD) Protection Challenges of Gate-All-Around Nanowire Field-Effect Transistors

W. Liu[a], J. J. Liou[a], N. Singh[b], G. Q. Lo[b], J. Chung[c] and Y. H. Jeong[c]

[a] School of EECS, University of Central Florida, Orlando, USA, liou@eecs.ucf.edu
[b] Institute of Microelectronics, A*STAR (Agency for Science, Technology and Research), Singapore
[c] Pohang University of Science and Technology, Pohang, Korea

Electrostatic discharge (ESD) performance, including trigger voltage, failure current, on-resistance and leakage current, of the gate-all-around silicon nanowire field-effect transistor depends on nanowire diameter, gate length and nanowire numbers. The device with best ESD robustness has medium gate length, small diameter and a large number of nanowires with multi-finger-multi-channel layout.

I. Introduction

The Si nanowire field-effect transistor has been regarded as a promising alternative CMOS technology in the beyond Moore Era due to its superior electrostatic channel control, high on/off current ratio, better noise performance, and ease of fabrication in the existing Si processing (1-4). However, the sub-10nm dimension, silicon-on-insulator (SOI), and silicided diffusion of Si nanowire (NW) FETs also lead to localized high current density and high electric field, making them highly susceptible to the electrostatic discharge (ESD) stress. Moreover, tens of thousands of paralleled NWs may be needed to achieve a satisfactory on-current level, and unevenly distributed currents among these NWs can further degrade the device's ESD tolerance and raise challenges to the commercialization of the nanowire-based electronics.

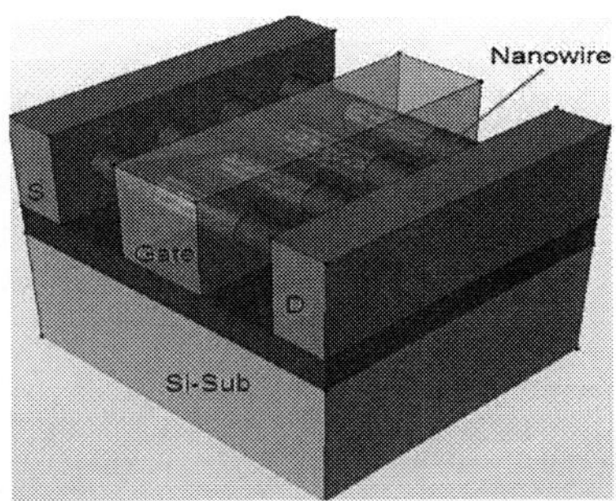

Fig.1 Three dimensional schematic of multi-channel gate-all-around NWFET (not to scale).

In an effort to determine the safe operating area and layout parameter dependency of nanowire-based ESD devices, we characterized gate-all-around silicon nanowire transistors (NWFET) fabricated using a CMOS compatible process by the Institute of Microelectronics, Singapore. The detailed process can be found in (4). As shown in Fig. 1, multiple NWs (1~1000) are formed between silicon drain and source on top of SOI wafer. The diameters of NWs can be trimmed down to 5~10nm with 5nm SiO2 surrounding the NWs. The variable parameters of the structure include gate length Lg, NW diameter D and NW numbers N. Human Body Model (HBM) ESD stresses with 100 ns pulse width and 10 ns rise time were applied to the NWFETs by using the transmission line pulsing (TLP) tester (Barth 4002).

II. Gate Length and Nanowire Diameter Dependency

N- and P-type NWFETs with NW diameters of 10 nm and 5 nm operating in the bipolar mode (gate and source connected) were characterized as a function of gate length. Available data were extracted from TLP I-V curves and plotted in Fig.2, where trigger voltage Vt1 is the voltage when the devices start to turn on and failure current It2 is the current before the devices are catastrophically damaged. The failure current density Jt2 and normalized on resistance Ron' are calculated by the following equations:

$$Jt2 = It2 / (N \times D) \tag{1}$$

$$Ron' = Ron / (N \times D) \tag{2}$$

$$Ron = (\text{slope of the TLP I-V curve after device turning on})^{-1} \tag{3}$$

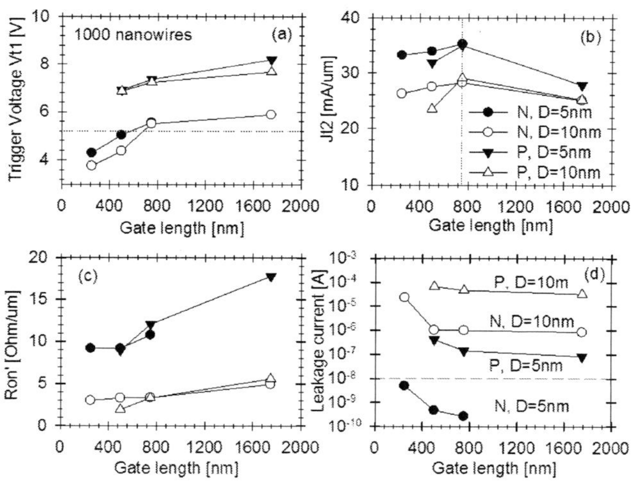

Fig.2 (a) Trigger voltage, (b) failure current density, (c) normalized on resistance and (d) DC leakage current as a function of gate length for N- and P-type NWFETs with NW diameters of 5nm and 10nm.

According to Fig.2 (a), the trigger voltage increases with increasing gate length, N-type devices triggers at lower voltage than P-type devices, and NWFETs with wider NWs have lower Vt1 than those with narrower NWs. This can be attributed to the fact that the mobility of electrons is higher than that of holes, and longer channel length, as well as narrower NW cross-section result in a lower avalanche multiplication factor. Longer Lg also means wider base width of the parasitic BJT, which in turn requires higher current to trigger the device and thus a higher trigger voltage.

In Fig. 2 (b), failure current density Jt2 steps up with increasing gate length initially but drops when the gate length exceeds a certain value. Longer Lg benefits the uniform-distribution of currents among multiple channels, makes it harder for filamentation to occur between drain and source, and improves heat dissipation by larger volume under gate (5-8), thus improves It2. Meanwhile, loner Lg leads to higher voltage drop along the channel and more energy dissipation. This explains the maximum Jt2 obtained at Lg=750nm.

The normalized on resistance Ron' as a function of gate length is shown in fig. 2(c). The overall trend follows R = (L×ρ) /Area. The leakage current $I_{leakage}$ (measured at Vds=1V for N-type and Vds=-1V for p-type) decreases with increasing Lg (fig. 2(d)). Among the devices measured, only N-type narrow (D=5nm) NWFETs shows applicable potential for their low level leakage current.

Good candidacies for ESD protection application should have high It2, small Ron and low leakage current, which can be represented by the following Figure of Merit (FOM):

$$FOM = It2 / (Ron \times I_{leakage})$$
[4]

Inserting the measurement data into the above equation, we can plot the FOM against Lg. As indicated in Fig. 3, the n-type medium Lg narrow NWFETs have the highest FOMs, followed by P-tye narrow and n-type wide devices.

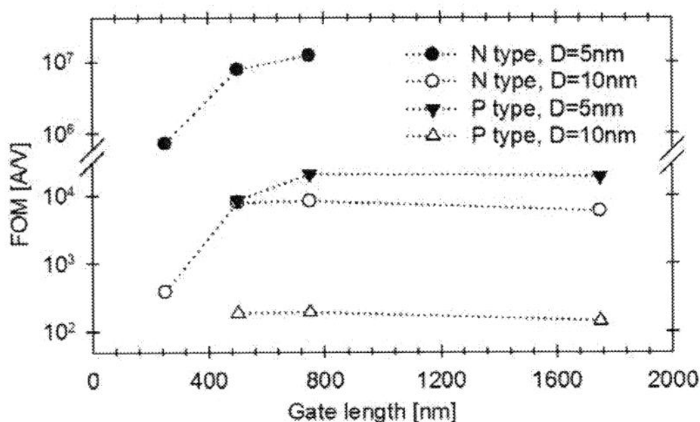

Fig. 3 Figure of merit as a function of gate length.

III. Nanowire number and layout dependency

Multi-channel NWFETs were fabricated to realized high on-state current and their simplified layout view is shown in Fig. 4(a). In order to study the ESD performance scalability with the number of NWs, N-type devices with a 5nm diameter and 500nm Lg, 100, 500 and 1000 NWs were characterized in Fig. 5. The results suggest that It2 of these devices scales almost linearly with the number of NW, at the cost of increasing leakage current when the NW number goes higher. Nevertheless, better FOM can be achieved with increased channel numbers.

The area efficiency of NWFET is not comparable to the sub-65nm planar MOS devices due to the relatively large NW spacing (S=400nm) and single-finger layout (9). Besides reducing the value of S, implementing multi-finger-multi-channel layout (Fig. 4 (b)) is capable of reducing 21% of effective layout area and facilitating a uniform turn-on among the multiple NWs, for the NWFET with 100 NWs, 750nm Lg and 5nm diameter.

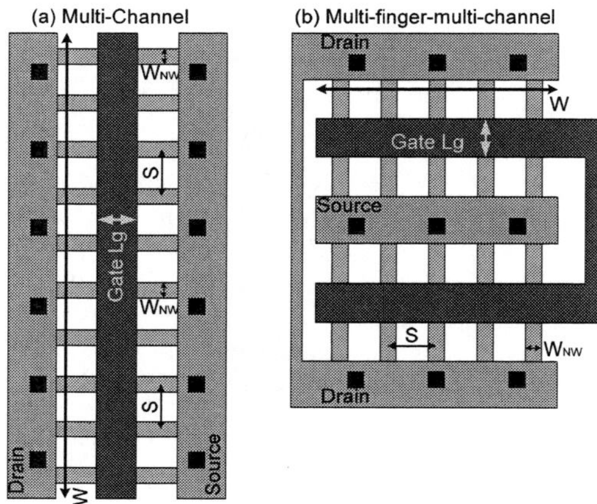

Fig. 4 (a) Simplified layout view of multi-channel NW FET and (b) proposed layout view of multi-finger-multi-channel NWFET.

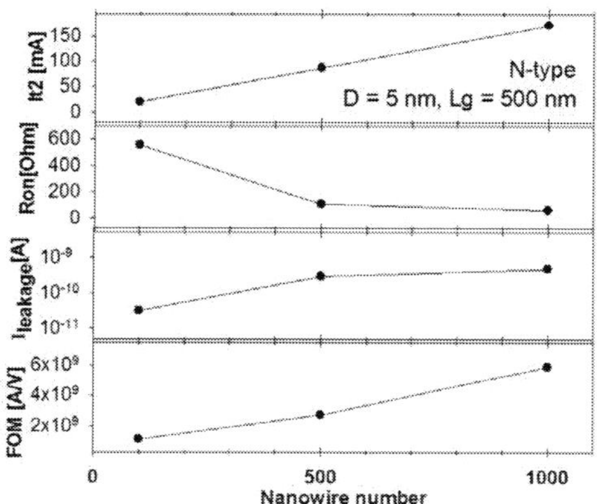

Fig. 5 Multi-channel NWFETs' ESD performance dependency on NW numbers.

IV. Safe Operation Area

Summarizing the measurement data, the NWFET safe operation area and ESD design window can be defined in Fig. 6. The DC operation voltage of NWFET is below 1.5V. The gate oxide breakdown voltage for CMOS technology can be approximated by the following equation (10):

$$V_{bd} = t_{ox} \times 1V / nm \qquad [5]$$

Given that t_{ox}=5nm, the gate oxide breakdown voltage of NWFET is about 5V. The MOS-diode mode NWFET can be turned on at 1V and several devices will be connected in series to achieve higher trigger voltage. The GGNMOS mode NWFET does not show snapback TLP I-V curve, which is a major mechanism of conventional GGNMOS under ESD stress. Further improvements are required to enhance the failure level and constrain the range of trigger voltage and drain-source breakdown voltage to 1.5V ~ 5V.

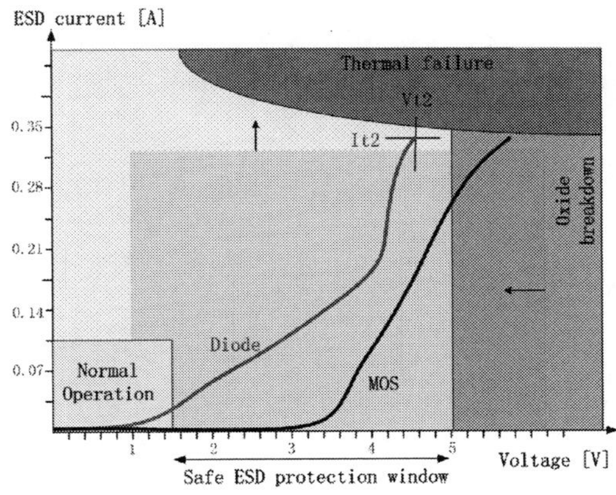

Fig. 6 Safe operation area and ESD design window for NWFETs.

V. Conclusions

Aimed to understand the ESD robustness of gate-all-around Si nanowire FETs, the layout parameter dependency was investigated experimentally and figure of merit was proposed in this work. It was found that adjusting the gate length, nanowire diameter, and nanowire number is an efficient and simple way to design the Si NWFET with optimal ESD performance for meeting the ESD design constrains and targets. The use of multi-finger and multi-channel layout topologies can further enhance the ESD device's area efficiency.

References

1. H. Iwai, *International Workshop on Junction Technology,* p.1-2, (2008).
2. 2007 International Roadmap for Semiconductors (ITRS). http://public.itrs.net.
3. N. Singh et al, *IEEE Electron Device Letters,* volume: 27, issue: 5, p. 383 – 386 (2006).
4. Y. Jiang et al, *International Electron Devices Meeting*, p. 1-4 (2008).

5. A. Griffoni, et al, *Proc. EOS/ESD Symp.* (2009).
6. C. Russ, *Microelectronics Reliability,* 48, p. 1403-1411(2008).
7. D. Tremouilles, et al, *Proc. EOS/ESD Symp.* (2007).
8. Karlheinz Bock, et al, *IEEE Transaction on CPMT*, vol. 21, no. 4 (1998).
9. W. Liu, et al, *IEEE Electron Device Letters*, vol. 31, no. 9, p. 915 – 917 (2010).
10. P. Leroux and N. Steyaert, *LNA-ESD Co-Design for Fully Integrated CMOS Wireless Receivers*, p.110, Springer Dordrecht, Berlin, Heidelberg, New York (2005).

ECS Transactions, 34 (1) 61-66 (2011)
10.1149/1.3567560 ©The Electrochemical Society

Characterization of Random Telegraph Signal Effects for 0.18um Technology

Yuanqian Ji, Sutter Dai, Minxia Wei, Xiangdang Lu, Sunny Zhang, Daniel Xu
Technology Development, Grace Semiconductor Manufacturing Corporation,
1399 Zuchongzhi Road, Shanghai, China 201203

Abstract

The RTS noise of MOSFETs (both P-type and N-type) for Grace 0.18uM CMOS technology was extensively investigated as a function of device size (W*L) and biasing conditions. The RTS noise was observed as the drain current of MOSFET drops randomly.

The RTS effect appears more severe when gate voltage is 0.2V lower than the threshold voltage at gate voltage side. At channel area side, it will be easier to find RTS effect when device channel area is smaller than $0.2um^2$. Analysis of $\Delta Id/Id$ versus channel area and gate voltage is also given. N channel MOSFET shows clear trend of $\Delta Id/Id$ versus gate bias for different channel area devices. Taking NMOS 1.2/0.18 as an example, when Vd is ~0.1V, $\Delta Id/Id$ decreases from 0.27% to 0.068% when gate voltage increases from Vth-0.2 to Vth+0.4. At P channel devices side, although a little bit fluctuation, we still can get the similar trend as NMOS. Also take PMOS 1.2/0.18 as an example, at |Vd|=0.1V, $\Delta Id/Id$ decreases from 0.21% to 0.025% when |Vg| increases from |Vth|-0.2 to |Vth|+0.4. The peak value is 0.24%, which occurs when |Vg| is |Vth|-0.1 and |Vth|+0.1.

Finally, we also captured the RTS data in frequency domain, indicating RTS noise contribution is important when frequency is below 10Hz, for our 0.18um process technology.

Introduction

The random signal telegraph (RTS) has been reported as one of the root cause for noise level at low frequencies [1], [2].

RTS has posted challenges for high precision analog design, especially at advanced technology node, such as 65nm or beyond. Though, the root cause of RTS has yet to be fully clarified. But most believe it is related interface traps [1], [3].

So far, most reports are focused in very advanced technology node. Indeed, RTS is worse at the gate area gets smaller [4], [5], [6]. However, to this date no detailed investigation has been reported on 0.18uM technology node for RTS effects.

In this work, we used 0.18uM CMOS transistors to investigate the RTS. Though, the transistor size is much bigger than advanced node transistor size, the RTS effect indeed exits. By varying transistor biasing conditions, for same transistor can switch from no RTS to showing RTS effects. [Figure1 vs. Figure 2]

Moreover, we only found very few dice showing RTS among very large number of tested dice, under most favorable biasing conditions for RTS. This suggests the at 0.18uM technology, for Grace

0.18uM CMOS process technology, the RTS only exits in very rare occasions.

Furthermore, we also tried to measure the RTS ($1/f^2$) noise in frequency domain, that measurement was proven to be a big challenge. The data was finally obtained, showing at 10Hz, RTS noise going down to traditional 1/f noise level. [Figure3]

Experimental

Three dimensions were selected for the RTS verification for NMOS, 1.2/0.18, 10/0.18 and 10/0.5. $\Delta Id/Id$ clearly decreases when gate voltage increases when drain voltage is 0.1V and 0.5V separately.

It has been recognized that there are two kinds of trap response for RTS, acceptor-type and donor-type. Assuming electrons as charge carriers, acceptor type traps are neutral when empty and negatively charged when filled. Donor-type traps are neutral when filled and positively charged when empty [4]. Plots corresponding to two kinds of response have been collected as shown in Figure 1.

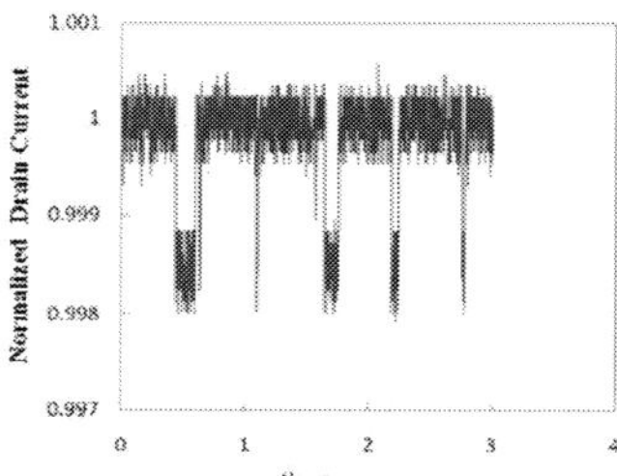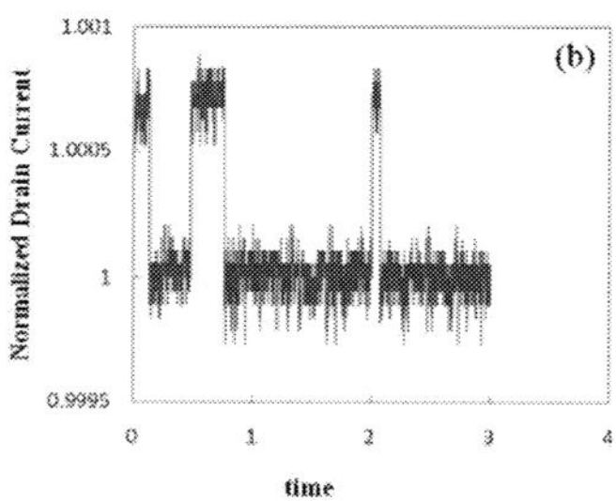

Figure 1. RTS plot from N-channel MOSFET for (a) Acceptor-type two-level impulse of normalized drain current caused by RTS effect for N-channel MOSFET 1.2/0.18 when Vd=0.1V and Vg=Vth+0.2V. The trap is negatively charged; (b) Donor-type two-level impulse. This data is captured from N-channel MOSFET 1.2/0.18 and the bias condition is Vd=0.1V and Vg=Vth+0.4V. The trap is positively charged.
(Note: Drain current is normalizeded to the baseline current)

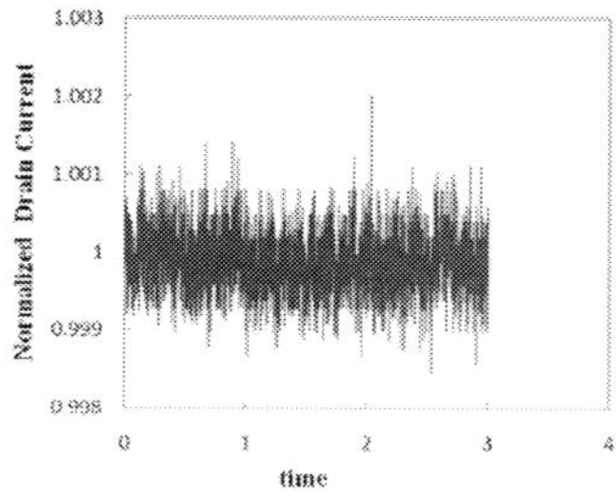

Figure 2. Normalized drain current versus time for the same device as shown in figure 1, bias condiction is Vd=0.1V, Vg=Vth. No RTS effect shown at this bias, very similar to large number of our other measurements.

RTS ($1/f^2$) noise is also founded when verifying noise feature for the same technology, as shown in Figure 3.

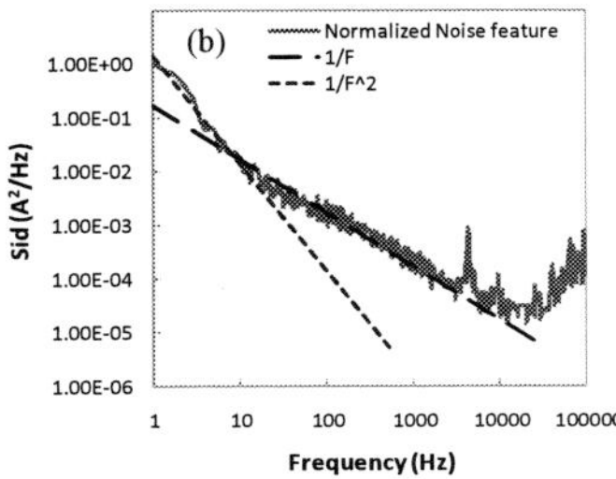

Figure 3. Noise feature for 0.18um technology with device channel area smaller than $1um^2$, RTS ($1/f^2$) noise were captured when frequency is lower than 10Hz. The device size is 2/0.18 in this particular measurement.

Table I-III and Figure3 give the summary of ΔId/Id versus gate voltage for different devices. RTS effect for 1.2/0.18 can be gathered at all bias condition. While for the larger devices 10/0.18 and 10/0.5, RTS effect does not exist at several bias conditions though the large sampling size. Data also shows 0.1V drain voltage is better for RTS effect monitoring on N channel MOSFET.

TABLE I. ΔId/Id for NMOS 1.2/0.18

|Vg|-|Vt0|	-0.2	-0.1	0	0.1	0.2	0.3	0.4
Vd=0.1V	0.27%	0.24%	0.16%	0.12%	0.14%	0.09%	0.07%
Vd=0.5V	0.32%	0.31%	0.22%	0.15%	0.19%	0.04%	0.03%

TABLE II. ΔId/Id for NMOS 10/0.18

|Vg|-|Vt0|	-0.2	-0.1	0	0.1	0.2	0.3	0.4
Vd=0.1V	0.52%	0.19%	N/A	N/A	0.06%	0.04%	0.02%
Vd=0.5V	1.20%	0.37%	0.16%	N/A	0.06%	0.05%	N/A

TABLE III. ΔId/Id for NMOS 10/0.5

|Vg|-|Vt0|	-0.2	-0.1	0	0.1	0.2	0.3	0.4
Vd=0.1V	0.48%	0.16%	0.21%	0.15%	0.04%	0.03%	0.03%
Vd=0.5V	1.20%	0.07%	0.15%	N/A	0.06%	0.04%	N/A

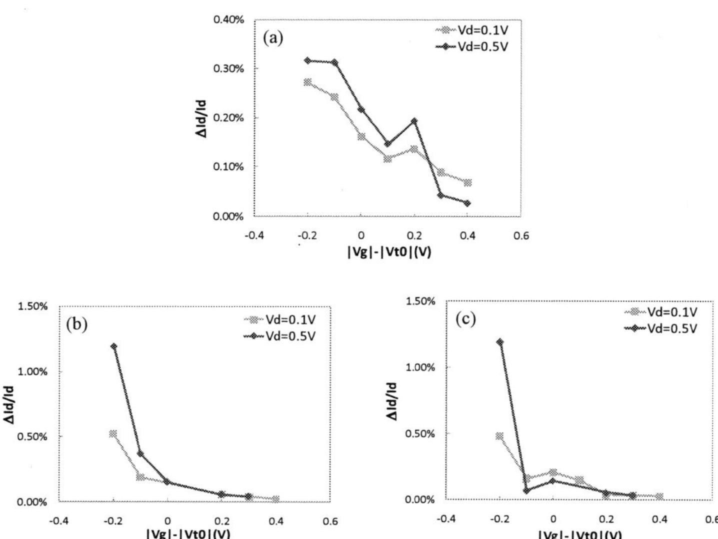

Figure 4.   RTS Effect versus gate voltage for: (a) NMOS 1.2/0.18; (b) NMOS 10/0.18; (c) NMOS 10/0.5.

We can also find out when channel width increases, $\Delta Id/Id$ decrease except when gate voltage is 0.2V lower than threshold voltage. Percentage of current variation of NMOS 10/0.5 is close to NMOS 10/0.18 though channel area has grown a lot.

Same dimensions were selected for the RTS verification for PMOS, 1.2/0.18, 10/0.18 and 10/0.5. Although a little bit fluctuation, trend of $\Delta Id/Id$ is similar with NMOS. The percentage of $\Delta Id/Id$ is much lower than that of NMOS with same size. As shown in Figure 4, plots corresponding to acceptor type traps and donor-type traps have also been collected.

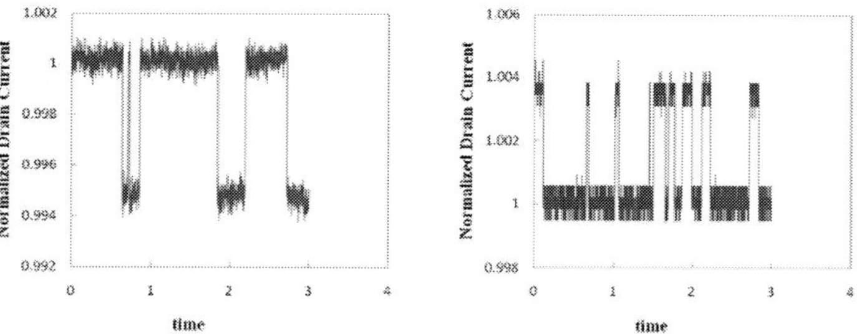

Figure 5. RTS plot from P-channel MOSFET for (a) Acceptor-type two-level impulse of normalized drain current caused by RTS effect for P-channel MOSFET 1.2/0.18 when $|Vd|=0.5V$ and $|Vg|=|Vth|+0.1V$. The trap is negatively charged; (b) Donor-type plot for PMOS. This data is

captured from 1.2/0.18, bias condition is |Vd|=0.5V and |Vg|=|Vth|+0.2V. The trap is positively charged.

Table IV-VI and Figure 3 gives the summary for P-channel MOSFET.

TABLE IV.   ΔId/Id for PMOS 1.2/0.18

|Vg|-|Vt0|	-0.2	-0.1	0	0.1	0.2	0.3	0.4
|Vd|=0.1V	0.207%	0.238%	0.150%	0.240%	0.027%	0.171%	0.025%
|Vd|=0.5V	0.346%	0.308%	0.207%	0.544%	0.360%	0.199%	0.127%

TABLE V.   ΔId/Id for PMOS 10/0.18

|Vg|-|Vt0|	-0.2	-0.1	0	0.1	0.2	0.3	0.4
|Vd|=0.5V	0.115%	0.122%	0.115%	0.074%	N/A	0.017%	0.032%

TABLE VI.   ΔId/Id for PMOS 10/0.5

|Vg|-|Vt0|	-0.2	-0.1	0	0.1	0.2	0.3	0.4
|Vd|=0.5V	N/A	0.151%	0.114%	0.070%	N/A	0.017%	0.011%

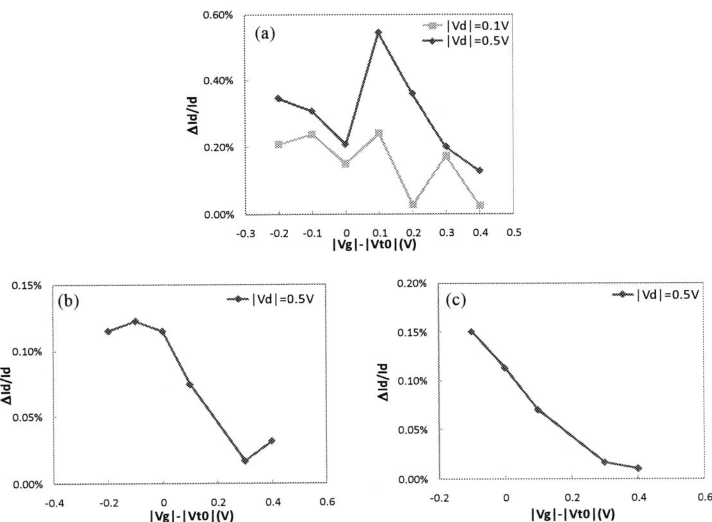

Figure 6.   RTS Effect versus gate voltage for: (a) PMOS 1.2/0.18; (b) PMOS 10/0.18; (c) PMOS 10/0.5.

For PMOS 1.2/0.18, as shown in Table IV and Figure5 (a), RTS effect can be found on each wafer at all bias condition. While for PMOS 10/0.18 and 10/0.5, we are not able to collect RTS data when drain voltage is -0.1V. When drain voltage is -0.5V, we can get similar trend with other devices. Different from NMOS, ΔId/Id at each bias of PMOS 10/0.18 is much lower than that of PMOS 1.2/0.18. No big difference for device 10/0.18 and 10/0.5, which shows result as that of NMOS.

Conclusion and Discussions

The RTS effect appears more severe when drain voltage is 0.1V for N-channel MOS and -0.5V for P-channel MOS. Gate voltage between Vth+0.2V and Vth-0.2V is recommended for easier RTS observation. At channel area side, it will be easier to find RTS effect when device channel area is less than $0.2uM^2$ in our case. Generally speaking, it is easier for small N-channel devices to get RTS effect data than large sized devices.

Due to the nature of RTS is randomness, it poses big challenges for high precision analog design. Special design techniques such as chopping stabilization are needed to combat this problem. Recently, we even heard at very advanced technology node, digital chips exhibit issues due to RTS. We believe advanced screening techniques are also needed to resolve these issues in addition to advanced design techniques. This topic, however, is far beyond the scope of this paper.

Acknowledgement

The authors wish to acknowledge their co-workers in process team, who tirelessly supported this work by providing large number of processed samples.

References

1. J.P. Campbell, L.C. Yu, K.P. Cheung, J. Qin1,3, J.S. Suehle, A. Oates, K. Sheng, *Large Random Telegraph Noise in Sub-Threshold Operation of Nano-Scale nMOSFETs*, 2009 IEEE
2. O. Roux, B. Dierickx, E. Simoen, C. Claeys, G. Ghibaudo and J. Brini, *Investigation of Drain Current RTS Noise in Small Area Silicon MOS Transistors*, Microelectronic Engineering, 15 (1991)
3. C. Leyris, S. Pilorget, M. Marin, M. Minondo, H. Jaouen, *Random telegraph signal noise SPICE modeling for circuit simulators*, IEEE, (2007).
4. Nicola Zanolla, Domagoj Šiprak, Marc Tiebout, Peter Baumgartner, Enrico Sangiorgi, and Claudio Fiegna, *Reduction of RTS Noise in Small-Area MOSFETs Under Switched Bias Conditions and Forward Substrate Bias*, IEEE TRANSACTIONS ON ELECTRON DEVICES, **VOL. 57, NO. 5**, MAY 2010
5. B. Razavi, "A study of phase noise in CMOS oscillators," IEEE J. Solid-State Circuits, vol. 31, no. 3, pp. 331–343, Mar. 1996.
6. M. J. Uren, D. J. Day, and M. J. Kirton, "1/f and random telegraph noise in silicon metal–oxide–semiconductor field-effect transistors," Appl. Phys. Lett., vol. 47, no. 11, pp. 1195–1197, Dec. 1985.

ECS Transactions, 34 (1) 67-73 (2011)
10.1149/1.3567561 ©The Electrochemical Society

Effect of AlGaN barrier thickness on the noise of AlGaN/GaN High electron mobility transistors

R. Yahyazadeh[a], Z. Hashempour[a]

[a]Department of Physics, Islamic Azad University of Khoy branch 58135/175, Iran

An analytical- numerical model for the drain source and gate source currents of AlGaN/GaN based HEMTs has been developed that is capable to predict accurately the effects of AlGaN barrier thickness on the cut off frequency and noise in different gate source and drain source biases. Salient features of the model are incorporated of fully and partially occupied sub-bands in the interface quantum well, combined with a self-consistent solution of the Schrödinger and Poisson equations. In addition current in the barrier of AlGaN, traps density in AlGaN are also take in to account. The calculated model results are in very good agreement with existing experimental data for HEMTs device.

1. INTRODUCTION

AlGaN/GaN high electron mobility transistors with high power and high-frequency characteristics are necessary for application in micrometers-wave length range. These advantages of AlGaN/GaN hetrostructure has been achieved because of a large critical breakdown electric field, a larger conduction band discontinuity between GaN and AlGaN and the presence of polarization fields that allows a large two-dimensional electron gas to be confined[1,2] High electron mobility transistors fabricated in the AlGaN/GaN semiconductor system can generate large amounts of RF power because of a unique   combination of material characteristics. The high breakdown fields ($3x10^6$ V/cm) resulting from the wide band gaps of GaN and AlGaN enable the use of much higher drain biases than can typically be used in the AlGaAs/GaAs system [3,4].   The noise of these transistors was previously calculated without including traps and depletion layer effects [5]. In the present work, a new analytical-numerical model for drain source and gate source currents presented.   That is capable of determining parallel conduction in AlGaN layer for gate bias. This is achieved by (i) using a self consistent solution to the Schrödinger and Poisson equations in order to obtain the 2DEG density, Fermi level (E_{FI}) specified relative to the bottom of triangular well, and band bending of GaN (E_B), (ii) take into account the current in AlGaN barrier, I_{AlGaN}, (non-depleted region current), (iii) take into account the two dimensional electron gas channel temperature, (iv) using a more accurate model for mobility, (v) using the electron traps effects in the surface, interface and buffer layers, and (vi) incorporating a simple , realistic model for the ungated area in the AlGaN/GaN HEMTs device.   That is capable of determining effect of AlGaN barrier thickness on the transconductanc, gate current, cut off frequency and noise. Close agreement with the experimental data confirms the validity of the present model.

2.MODEL DESCRIPTION

In order to obtain the transport parameter of AlGaN/GaN HEMTs, the basic structure has been considered as Fig.1,Where the x-direction is along the two dimensional electron gas (2DEG) channel, the z-direction is along the growth direction, and the regions I and II represents the ungated channel portions of the HEMT and the region II represents the gated due of the devices. The traps within the AlGaN barrier can be assumed as the barrier capacitance, C_{AlGaN}, which is related to depletion distance of d_1 and d_2 [6].To calculate the total drain current, the both 2DEG channel and AlGaN barrier currents have been calculated. The expression of device current can be obtained as [6,7]

$$I_{total} = \begin{cases} I_{AlGaN} + I_{2DEG} & for \quad d_2 \langle barrier \quad thickness \\ I_{2DEG} & for \quad d_2 \geq barrier \quad thickness \end{cases} \quad (1)$$

Where d_2 are the depletion layer thickness has been used [6] .

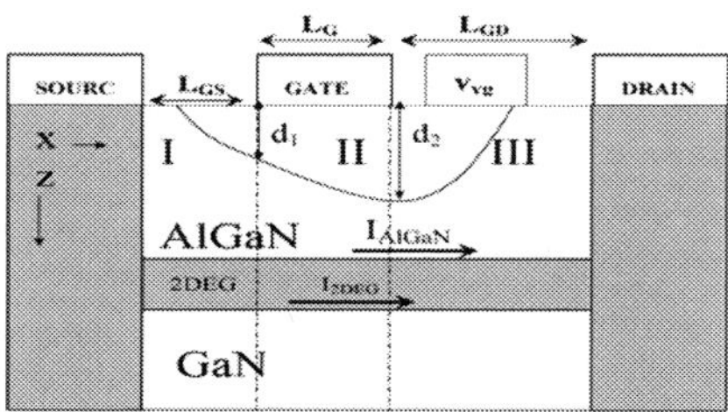

Fig. 1. Schematic HEMT structure used for model calculation

In the model, it has been assumed that the voltage varies linearly with x in the ungated regions, so the voltage is given by the following relation [7]:

$$V_{GS}(x) = \begin{cases} I: & \left(\dfrac{V_{GS}}{L_{SG}}\right)x \\ II: & V_{GS} \\ III: & \left(\dfrac{V_{DS}-V_{GS}}{L_{GD}}\right)(x - L_G - L_{SG}) + V_{GS} \end{cases} \quad (2)$$

where x is the distance along the channel from the source, V_{GS}, and V_{DS} are the applied gate and drain voltage with respect to the grounded source, so that L_{SG}, L_G and L_{GD} are the length of the ungated distance between source and gate, the gate length and the ungated distance between gate and drain, respectively.

To calculate the exact electrical properties in HEMTs, one need to include the electron traps effects such as interface and surface traps, and GaN barrier trap in calculation. As a result of lattice mismatch between AlGaN and GaN, and possibly compositional non-uniformity caused by alloy clustering, there could be considerable amount of trap states at the AlGaN/GaN interface. The electrical behavior of the interface trap states can be modeled as a capacitive (C_S) and a conductive (G_S), where G_S is negligible at the model biases. The insulating interfacial layer is only a few monolayers thick so that the metal can easily communicated electrons with the trap states at the semiconductor surface [8].

Capacitance of band bending for GaN is related the bulk field E_B [9 and is given by: $C_{GaN} = \varepsilon_{GaN} q \left(dE_B / du\right)$, since the buffer, GaN, is un-doped, the capacitance, C_{GaN}, is quite small.

The traps within the AlGaN barrier can be assumed as the barrier capacitance, C_{AlGaN}, which is related to depletion distance of d_1 and d_2 [9, 10, and 11].So that in this model total mobility can be obtained as [12]:

$$\mu_{total} = \begin{cases} \dfrac{\mu_{2DEG} N_S + \mu_{AlGaN} N_{AlGaN}}{N_S + N_{AlGaN}} & for \quad d_2 \langle barrier \quad thickness \\ \mu_{2DEG} & for \quad d_2 \geq barrier \quad thickness \end{cases} \qquad (3)$$

So that N_{AlGaN} is the density of electron real space transfer from the 2D (2DEG channel) to 3D state (in AlGaN barrier).It should be mentioned that even at room temperature, the mobility of 3D-electron gas is around 50% smaller the mobility of 2D-electrons.

In order to obtain accurate values for mobility of 2DEG channel, the occupancy of the various sub-bands, the intrasub-band and intersub-band coupling coefficients (H_{mn}), and the sheet carrier concentration for the 2DEG in AlGaN/GaN heterostructures; both the Schrödinger and Poisson equations must be solved self-consistently. This has been achieved by solving Schrödinger's equation and simultaneously taking into account the electrostatic potential obtained from Poisson's equation, as well as the image and exchange-correlation potentials using Numerov's numerical method. In the self-consistent calculation, the nonlinear formulism of the polarization–induced field as a function of Al mole fraction in $Al_mGa_{1-m}N$/GaN heterostructures has been assumed, as well as taking in to account all fully and partially–occupied sub-bands within the interface 2DEG potential well [13,14]. Using such an approach, it is possible to calculate the 2D-electron mobility taking into account the combined contributions from each of the individual electron scattering mechanisms. The cut off frequency (f_T) and shot noise (F_{min}) are given by the following relation [15, 16]

$$f_T = \frac{g'_m}{2\pi(C_{GS} + C_{GD}) + [(R_S + R_D)[g'_m(C_{GS} + C_{GD}) + g'_m C_{GI}]} \qquad (4)$$

$$g'_m = \frac{g_m}{1 + g_m R_S}$$

$$F_{min} = 1 + \frac{R_{in}}{R_{opt}} + b \frac{R_{opt}^2 + X_{opt}^2}{R_{opt}}$$

$$+ \frac{a}{R_{opt}} \left| R_{in} + R_{OPT} + j\left(X_{opt} - \frac{1}{\omega C} \right) \right|$$

$$a = g'_m \Gamma \left(\frac{\omega}{\omega_T} \right)^2 , \quad b = \frac{q I_{GT}}{2 K_B T_e} , \quad \omega_T = 2\pi f_T , \quad C'_{GS} = \frac{C_{GS}}{1 + g_m R_S}$$

Where R_S and R_D are the source and drain contact resistance respectively, g_m is the transconductance, I_{GT} is the total gate current, R_{opt} and X_{opt} are the optimum source impedance, ω is the angular frequency, R_{in} is the sum of gate and charging resistances, Γ is electric field dependent quantity, q is the electron charge, K_B is the Boltzmann's constant, T_e is the electron temperature in 2DEG.So that the gate-drain C_{GD} and gate-source C_{GS} depletion layer capacitances of AlGaN/GaN HEMTs, has been used [15] . The total gate current ($I_{GT} = I_{SM} + I_{MS} + I_S + I_V$) contain the tunneling gate current, is composed of two components: from the semiconductors, AlGaN and GaN, to the metal I_{SM} and from the metal to the semiconductors I_{MS} [16]. Note that each of the components, I_{SM} and Jms , is further divided into a tunneling component through the barrier, I_T , and a thermally emitted component, over the barrier, I_{TT} .To calculate the tunneling component, the two-step tunneling via trap, which is called the trap assisted tunneling (TAT) model, has been used. To calculate the surface current I_S , and the lateral tunneling current I_V AlGaN barrier   analytical relations has been used [17, 18].

3. RESULTS AND DISCUSSION

To assess the validity of this combined analytical–numerical model for charge transport, a comparative study has been undertaken comparing theoretically obtained $I-V$ curves, total gate current, cut off frequency and noise with experimental results from Refs. 16, 19and 20for AlGaN/GaN based HEMT.

Fig. 2 show the depletion layer in AlGaN barrier along x–direction for AlGaN/GaN based HEMT. For different barrier thickness (d_{AlGaN}) with increasing applied gate voltage, the depletion layer width (d_2) is decreasing so that for $d_{AlGaN} = 400 A^\circ$ and from V_{GS}= -3V, the 3D channel opening within the AlGaN barrier, i.e. depletion layer width(d_2) is thinner than the AlGaN barrier thickness($d_{AlGaN} \rangle d_2$) and the electrons within non-depleted 3D channel in AlGaN barrier can cause a conductivity between source and drain that is total drain source current is quall to ($I_{ds} = I_{2DEG} + I_{AlGaN}$) Thus, to calculate exact current-voltage(I-V)curve in AlGaN/GaN based HEMTs for positive applied gate bias and higher drain voltage, one needs to include the effects of all electron trap states and non-depleted 3D electron channel within the AlGaN barrier. For $d_{AlGaN} = 200 A^\circ$ depletion layer width is larger than the AlGaN barrier   thickness ($d_{AlGaN} \langle d_2$) and total current is quall to ($I_{ds} = I_{2DEG}$).Fig. 3 shows the characteristics drain source current versus drain source voltage for the $Al_{0.4}Ga_{0.6}N$/GaN HEMTs in different AlGaN barrier thickness. In this Fig all electron trap states, depletion layer effect and temperature effects (in conduction band discontinuity, total mobility, and threshold voltage and sheet carrier density) are included. As this Fig. demonstrates, this effects cause a small NDC (Negative Differential Conductivity) in I-V characteristics. As shown from this Fig. (including the non-depleted 3D channel for higher gate bias) a larger negative definitional conductivity happens because of wide non-depleted 3D channel opening in AlGaN barrier since even at room temperature, the mobility of 2D-electron gas larger than the mobility of 3D-electrons that is mobility and sheet carrier density decrease and it can cause a NDC in I-V characteristics. Since the sheet carrier concentrations and threshold voltage of two dimensional electron gases as a function of AlGaN barrier thickness. As a result, the absolute of threshold voltage (voltage that to require for formation of two dimensional quantum well in interface AlGaN/GaN) decrease with the decrease of AlGaN barrier thickness and cause decrease of depth and Fermi level in quantum well. With decreasing in depth of quantum well, decrease the discontinuity in band gaps and occupancy of the various sub-bands (sheet carrier concentrations). The result shows that change in sheet carrier concentrations 0.3×10^{12} cm^{-2} and total current $I_{ds} = 0.5 A / mm$ of device with increase in AlGaN barrier thickness from 2 nm to 6 nm in$V_{gs} = -1V$. Total gate current or gate leakages is important parameter for survey of shot noise , which includes the tunneling and thermionic currents under the gate region I_{GT} .As shown in Fig. 4 with decreasing in barrier thickness, decrease the gate current . In these devices we should mention that the vertical tunneling current through the main gate area is the most dominant leakage mechanism. Fig.5 shows the variation of cut-off frequency as a function of gate length with different AlGaN barrier thickness for $Al_{0.4}Ga_{0.6}N/GaN$. It is seen that with decreasing in barrier thickness, decrease the cut-off frequency. As seen in figure, cut off frequency falls with increases in gate length. This is because as channel length is increased the electron transit time through the channel also increases resulting in the reduction of the cut of frequency. As results with decreasing in barrier thickness, decrease minimum noise (shot noise) there are good agreement between the model and experimental data.

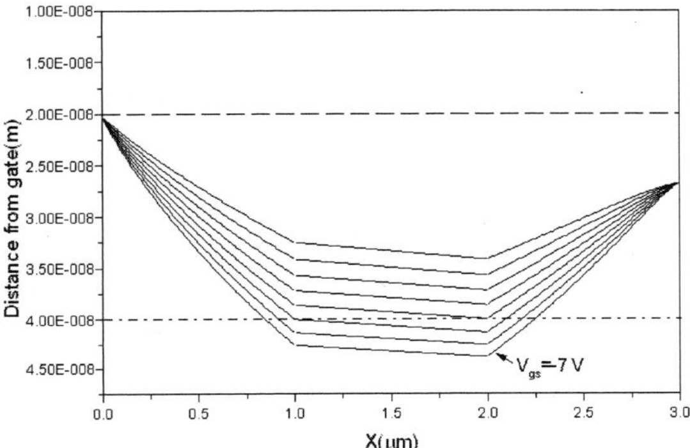

Fig. 2. Depletion layer profile for AlGaN barrier in $Al_{0.4}Ga_{0.6}N/GaN$ based HEMT, the gate voltage, V_{GS}, is stepped in 1 V from -7 V to +0V ,The horizontal lines is the barrier thickness from the gate.

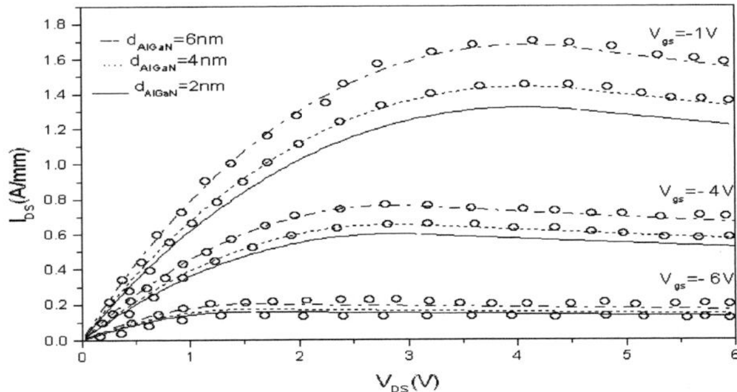

Fig.3. Drain source current versus drain source voltage for the $Al_{0.4}Ga_{0.6}N/GaN$ .The dots represent Exp. Data from Ref. [20].

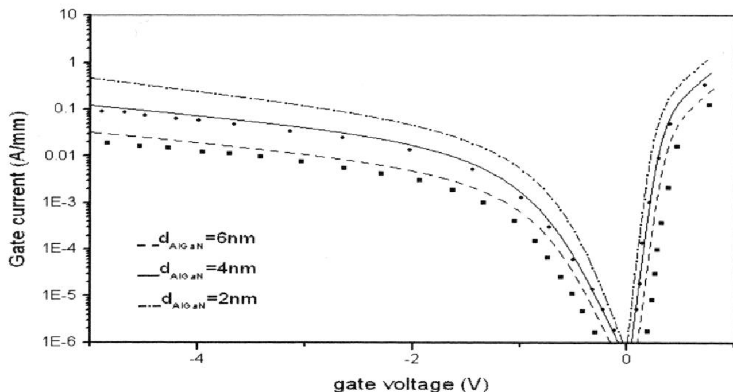

Fig.4 . The total gate current verse gate voltage for the $Al_{0.4}Ga_{0.6}N/GaN$ HEMTs in different AlGaN barrier thickness. The dots represent Exp. Data from Ref. [20].

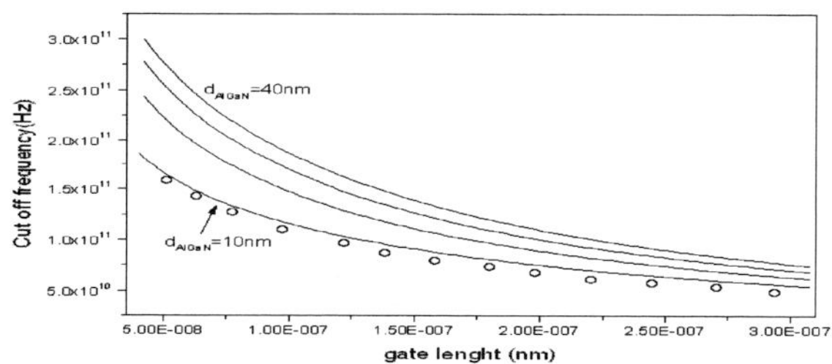

Fig.5. shows the variation of cut off frequency versus gate length HEMTs in different AlGaN barrier thickness for the $Al_{0.4}Ga_{0.6}N/GaN$ .The dots represent Exp. Data from Ref. [20].

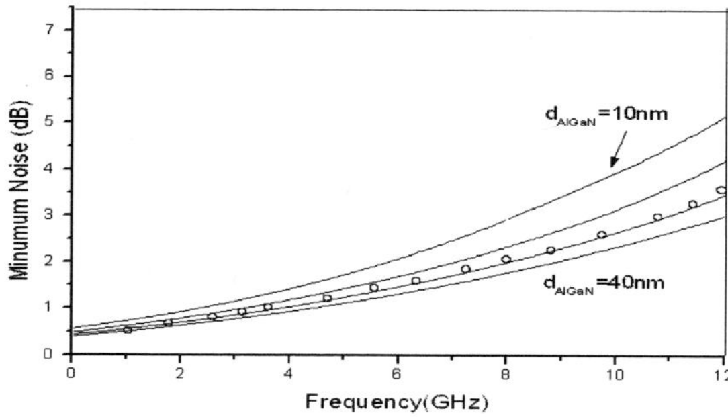

Fig. 6. Minimum noise (shot noise) versus frequency for the $Al_{0.4}Ga_{0.6}N/GaN$ based HEMT, d is stepped in 10 nm from10 nm to 40 nm; the dots represent Exp. Data from Ref. [20].

Refrenses:

1. W. Lu, V. Kumar, E. L. Pinter, and I. Adesida *IEEE Trans. Electron Devices*, **50** (2003).
2. J.Kuzmik, S.Bychikhin, M.Neuburger, A.Dadgar, A.Krost, E.Kohn, and D.Pogany, *IEEE Trans.Electron Devices*, **52** (2005)
3. M.K.Chattopadhyay, S.Tokekar, Solid – *State Electronic* **50** (2007)
4. H. Rohdin, N. Moll, A. M. Bratkovsky, and C. Y. Su, *Phys. Rev.* B **59** (1999).
5. G. Koley, V. Tilak, Ho-Young Cha, L. F. Eastman, and M. G. Spencer, *IEEE Electron Device Letters*, **8** (2002).
6. R.Yahyazadeh, A.Asgari, and M.Kalafi, *Physica E* **33**(2006).
7. Z.Hashempour, A.Asgari, S.Nikipar, M.R.Abolhasani, M.Kalafi, *Phisica E* **41** (2009).
8.H. Rohdin, N. Moll, A. M. Bratkovsky, and C. Y. Su, *Phys. Rev. B*, **59** (1999).
9. P. Roblin, H. Rahdin, "High-speed heterostructure devices from device concepts to circuit modeling", Cambridge university press, 277 (2002)
10E. J. Miler, X. Z. Dang, H. H. Wieder, P. M. Asbeck, and E. T. Yu, J. Appl. Phys. 87, 8070 (2000).
11.G. Koley, V. Tilak, Ho-Young Cha, L. F. Eastman, and M. G. Spencer, *IEEE Electron Device Letters*, **8** (2002).
[12]J. Deng, R. Gaska, M. S. Shur, M. A. Khan, J. W. Yang, *Materials Research Society Symposium-Proceedings* , V**595**(2000).

13. A. Asgari, M. Kalafi, *Material Science and Engineering* C **26** (2006).
14. E. J. Miller, X. Z. Dang, H. H. Wieder, P. M. Asbeck, and E. T. Yu, *J. Appl. Phys.* **87** (2000).
15. R.Yahayzadeh, Z.hashempour, M.Kalafi, M.R.Aboulhasani, Journal *of Material Science and engineering* 3(12) (2009)
16. C.Sanabria, A.Chakraborty, H.Xu, M.J.Rodwall, U.K.Mishra, R.A.York, *Electron Device Letter,* **27**(1) (2006)
17. A.J.Sierakowaskliand L.F.Estman, *Journal of Applied Physics*, **86**(6) (1999)
18. A.Asgari, M.Karamad, M.Kalafi, Superlattice *and Microstructures* **40** (2006)
19. M.Higashiwaki and Toshiaki Matsui, *Japanese Journal of Applied physics* **44**(16), (2005)
 20. M.Haghashiwalki, T.Mimura and T.Matsui, *Applied Physics Express* **1** (2008)

ECS Transactions, 34 (1) 75-80 (2011)
10.1149/1.3567562 ©The Electrochemical Society

Extraction and Analysis of Substrate Parameters in On-Chip Spiral Inductor Model

Xi Li[a], Zheng Ren[b], Dawei Chen[a], Yanling Shi[a]

[a] Department of E.E., East China Normal University, Shanghai 200241, China
[b] Shanghai Integrated Circuits Research & Development Center, Shanghai 201210, China

Inductors on high resistivity substrate, such as quartz or high resistivity silicon(HR-Si), can get better Q factor than those on low resistivity substrate, which makes it necessary to research the substrate influence on inductors. In this paper, a lumped element model for on-chip inductor that includes skin and proximity effects, substrate electric and magnetic losses at high-frequency range is presented. The proposed model has been verified with square spiral inductors with various geometrical configurations on quartz, SOI and high-resistivity silicon substrate. It shows good agreement with the simulation and tested results. This model can be easily used in circuit simulations and optimization design.

Introduction

On-chip spiral inductor is a vital element in RF IC design. It is widely used in filters, low-noise amplifiers and voltage-controlled oscillators due to its low cost and process integration. Q factor is one of the most important parameters that evaluates inductors' performance. Reducing substrate electric and magnetic losses is an important way to increase the Q factor especially in high frequency. Inductors on different substrate, such as SOI (1) and quartz substrates (2) have been reported.

While most spiral inductors models that have been reported are based on Si substrate (3)–(5). Model of spiral inductor on different substrate is seldom investigated in previous studies. Therefore, research of inductor model on different substrate is necessary.
Considering skin and proximity effects, substrate electric and magnetic losses, a lumped element model for on-chip inductor which emphasizes the influence of substrate is presented in this paper.

To verify the accuracy of the proposed model, square spiral inductors with various geometrical configurations on high-resistivity silicon, quartz and SOI substrate have been designed by HFSS and inductors on HR-Si substrate have been fabricated. The proposed model shows good consistency with the simulated and measured results under self-resonance frequency. The trends of extracted parameters show good agreement with physical analysis.

Proposed Inductor Model

One-$\prod$ conventional inductor model (6) is the foundation of inductor model, which has been shown in Fig.1. However, it cannot present performance of the inductor in high frequency accurately because of skin and proximity effects, substrate electric and magnetic losses. So combinations of RL and RC should be added to the model to show good agreement in wide-band frequency.

Due to skin and proximity effects, current does not flow uniformly in metal lines and resistance increases with increasing frequency. When the current flowing in wiring becomes less uniform with increasing frequency, inductance decreases (7, 8). So a combination of resistance Ro and inductor Lo in parallel has been added in the proposed model. It expresses the increase in series resistance and the decrease in series inductance due to the skin and proximity effects (9).

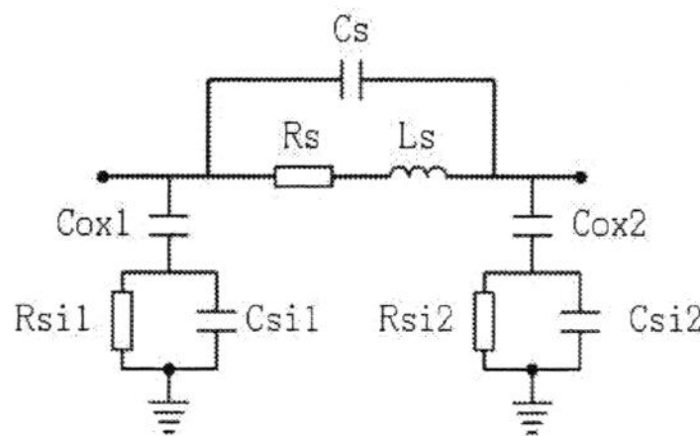

Fig.1.One-∏conventional inductor model

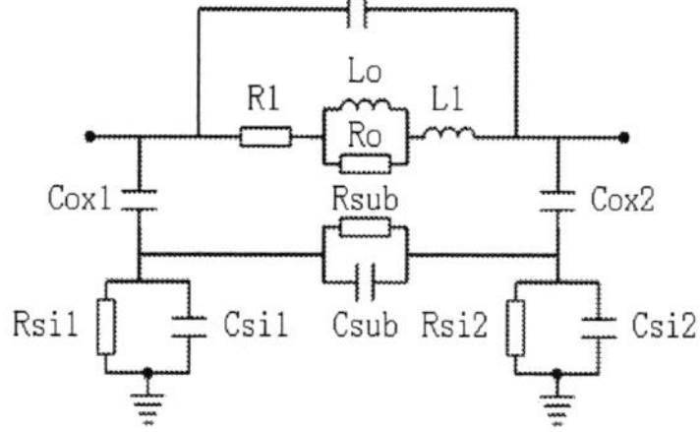

Fig.2. Proposed equivalent circuit model of the spiral inductor on different substrates

It has been reported that R_{sub} and C_{sub} in substrate can represent the lateral substrate coupling (10). A parallel combination of R_{sub} and C_{sub} is placed in the substrate, similar to the equivalent-circuit model for substrate coupling (11), and on-chip interconnects (12). Considering skin and proximity effects, substrate electric and magnetic losses, the proposed model has been shown in Fig.2. Series resistance $R_{(\omega)}$ and series inductance $L_{(\omega)}$ of equivalent circuit are extracted as

$$R_{(\omega)} = R_1 + R_o \frac{\omega^2 L_0^2}{R_o^2 + \omega^2 L_o^2} \qquad [1]$$

$$L_{(\omega)} = L_1 + L_o \frac{R_0^2}{R_o^2 + \omega^2 L_o^2} \qquad [2]$$

When $\omega \rightarrow 0$, $R_{DC}=R_1$, $L_{DC}=L_1+L_0$; when $\omega \rightarrow \infty$, $R_{HF}=R_1+R_0$, $L_{HF}=L_1$. R_{DC} and L_{DC} are series resistance and inductance in low frequency. R_{HF} and L_{HF} are series resistance and inductance in high frequency. So the proposed model in this paper can represent the frequency characteristics of spiral inductors. The series resistance increases and the series inductance decreases with increasing frequency.

Model Parameters Extraction and Discussion

To verify the accuracy of the proposed model, square spiral inductors with various geometrical configurations on high-resistivity silicon, quartz and SOI substrate have been designed by HFSS. Then inductors on HR-Si substrate ($\rho=10^3\Omega\cdot$cm) have been fabricated in the following processes. First, Ti/Au metals approximate 0.6µm are electroplated and patterned to form the underpass of the inductors. Secondly, a PECVD SiO_2 layer about 0.8µm is deposited for isolation. Subsequently, 1.5µm thick Ti/Au layer is electroplated for patterning spiral coil of inductor.

The physical dimensions of these inductors are summarized in Table 1. The parameters of the inductor include number of turns (N), outer opening diameter (Dout), metal width (W), conductor inter-turn space (S). Q1, Q4, Q7 are the inductors on HR-Si($\rho=10^3\Omega\cdot$cm). Q2, Q5, Q8 are the inductors on quartz. And Q3, Q6, Q9 are the inductors on SOI substrate. The SOI substrate adopted is realized by SIMOX procedure from high-resistivity silicon with more than $100\Omega\cdot cm$ resistivity. It has a $0.37\mu m$ thick and $0.21\mu m$ deep buried oxide layer.

TABLE 1. Various Physical Dimensions of The Inductors

Sample Number	d_{out} (μm)	N	$w_n + s_n$ (μm)	W (μm)	S (μm)	Substrate Type
Q1						HR-Si
Q2		3.5	30	15	15	quartz
Q3						SOI
Q4						HR-Si
Q5	400	3.5	40	20	20	quartz
Q6						SOI
Q7						HR-Si
Q8		4.5	30	15	15	quartz
Q9						SOI

Measurements are carried out at frequencies ranging from 100MHz to 10GHz by E8363B network analyser and Cascade on-wafer probe. Accurate measurements for the inductors alone can be obtained, by measuring S parameters of both the device under test (DUT), probe pads and ground planes (PAD), and subtracting the effects of PAD from DUT.

TABLE 2. Model Parameters Extracted from Simulation Data

Sample Number	L_1 (nH)	L_0 (nH)	R_1 (Ω)	R_0 (Ω)	C_{ox1} (fF)	C_{ox2} (fF)	R_{si1} (Ω)	R_{si2} (Ω)	C_{si1} (fF)	C_{si2} (fF)	C_s (fF)	R_{sub} (Ω)	C_{sub} (fF)
Q1	5.83	0.28	5.11	4.27	126.5	128.3	887.1	887.1	183.1	183.1	30.1	753.9	9.77
Q2	7.57	0.55	1.90	2.50	73.5	75.2	/	/	151.3	151.3	18.9	1186.4	8.37
Q3	6.27	0.35	1.25	2.52	105.5	107.4	993.2	993.2	282.8	282.8	24.4	1083.2	2.48
Q4	6.13	0.26	5.41	4.17	148.5	149.8	874.3	874.3	173.2	173.2	36.1	736.8	20.8
Q5	5.89	0.43	2.48	2.43	61.7	63.4	/	/	154.8	154.8	22.8	1153.4	12.19
Q6	5.29	0.32	1.23	2.68	103.7	105.4	964.4	964.4	294.5	294.5	26.1	1051.7	3.22
Q7	4.43	0.23	4.11	3.86	138.6	139.7	897.8	897.8	173.1	173.1	32.1	683.5	35.7
Q8	7.87	0.65	1.69	2.60	73.5	74.6	/	/	162.9	162.9	24.9	1132.5	11.3
Q9	7.37	0.39	1.38	2.77	96.6	98.8	982.1	982.1	310.5	310.5	29.4	1035.1	7.21

After the simulation data of the inductors on three different substrates are gained from HFSS and the measured data of samples on HR-Si are obtained from fabrication, each parameter in equivalent circuit model can be extracted using gradient algorithm in ADS. R_{si} and C_{si} in two branches should be extracted in the same value. Considering the asymmetry of the two ports, C_{ox1} and C_{ox2} should be extracted in the different value. The optimization of Q factor, the equivalent series resistance R_s and the equivalent series inductance L_s is added to the aim function. For the spiral inductor always works under the self-resonance frequency, the equivalent circuit model should meet the practical requirement under the self-resonance frequency. The equivalent circuit parameters for the proposed model are extracted in Table 2. Parameters on SOI and quartz As shown in Table 2, R_{sub} values of quartz substrate, SOI substrate and HR-Si substrate increase orderly. C_{sub} values of SOI substrate, quartz substrate and HR-Si substrate increase orderly. The Rsi value of SOI substrate is higher than the high- resistivity silicon substrate. The Capacitance C_{si} of SOI is greater than that of the quartz substrate and the HR-Si substrate.

Fig.3 shows an example for the comparison of Q factor calculated from the simulation results and the proposed model. The Q factor of the proposed model matches inductor Q8 on quartz and inductor Q9 on SOI well.

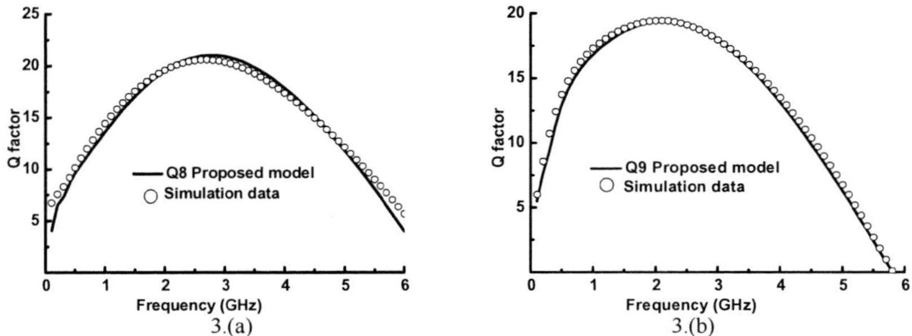

Fig.3 Comparison of Q factor between HFSS simulation and proposed model in ADS of Q8 and Q9

Fig.4 shows the simulated and the modelled S-parameters of inductors with different number of turns on SOI substrate. The S parameters of Q3 are in Fig.4a and the S parameters of Q9 are in Fig.4b. As seen from the figure, S-parameters of the proposed model match the simulated data well.

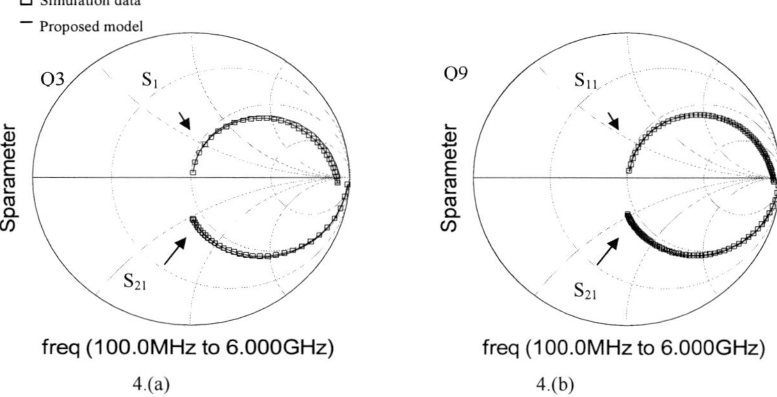

Fig.4 Q3 and Q9 S parameters comparison between HFSS simulation and proposed model results

Fig.5 shows the comparison of Q factor of inductors with different number of turns calculated from the proposed model and simulated data on SOI substrate. Q3 is 3.5 turns and Q9 is 4.5 turns. It is found that the proposed model shows good agreement with the SOI substrate structure of different number of turns.

Fig.6 gives the Q factor comparison for the experimental results and the proposed model of Sample Q7 on HR-Si. The proposed model also shows very good agreement with experimental results.

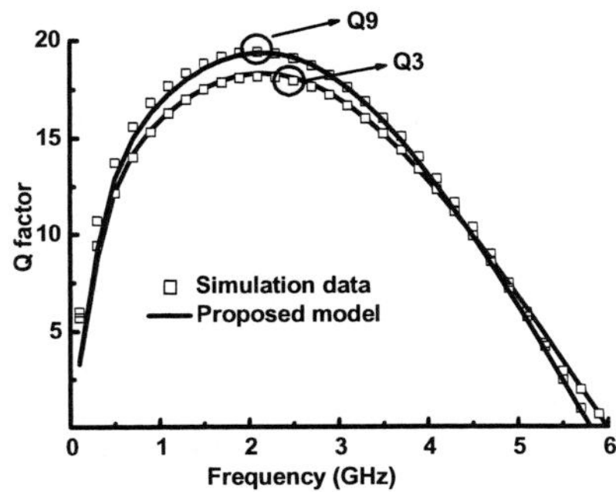

Fig.5 Comparison of Q factor between different number of turns of Q3 and Q9

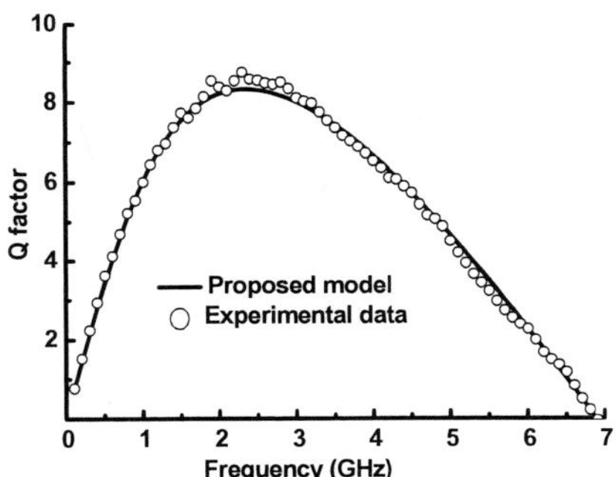

Fig.6 Comparison of Q factor between experimental results and the proposed model of Sample Q7

Conclusions

A lumped element model has been proposed to accurately present the influence of the different substrate on spiral inductor, which considers skin and proximity effects, substrate electric and magnetic losses in high frequency. The measured data from various geometrical configurations demonstrate the validity of the proposed model on high-resistivity silicon substrate. The simulated data validate the proposed model on quartz substrate and SOI substrate. In conclusion, the proposed model can be applied to the spiral inductor on different substrates. It is helpful to RF IC design.

Acknowledgment

This work was supported by Foundation of Shanghai Science &Technology Committee (075007033, 04QMX1419) and Natural Science Foundation of China (No. 60676047 , 60606010).

References

1. A. Tekin, E. Zencir, D. Huang, W. Liu, and N. S. Dogan, "A 700-MHz VCO using high-Q silicon on insulator (SOI) inductors," in Radio and Wireless Symposium, 2006, p.427-429.
2. K. L. Scott, T. Hirano, H. Yang, H. Singh, R. T. Howe, and A. M. Niknejad, "High-performance inductors using capillary based fluidic self-assembly," Journal of Microelectromechanical Systems, Vol. 13, pp.300-309, Apr. 2004.
3. F. Rotella, B. K. Bhattacharya, V. Blaschke, M. Matloubian, A. Brotman, Cheng Yuhua, R. Divecha, D. Howard, K. Lampaert, P. Miliozzi, M. Racanelli, P. Singh, and P. J. Zampardi, "A broad-band lumped element analytic model incorporating skin effect and substrate loss for inductors and inductor like components for silicon technology performance assessment and RFIC design," IEEE Transactions on Electron Devices, Vol. 52, pp.1429-1441, July. 2005.
4. K. Y. Lee, S. Mohammadi, P. K. Bhattacharya, and L. P. B. Katehi, "A Wideband Compact Model for Integrated Inductors," IEEE Microwave and Wireless Components Letters, Vol. 16, pp.490-492, Sept. 2006.
5. A. Goni, J. Del Pino, B. Gonzalez, and A. Hernandez, "An Analytical Model of Electric Substrate Losses for Planar Spiral Inductors on Silicon," IEEE Transactions on Electron Devices, Vol. 54, pp.546-553, Mar. 2007.
6. C. P. Yue, and S. S. Wong, "Physical modeling of spiral inductors on silicon," IEEE Transactions on Electron Devices, Vol. 47, pp.560-568, Mar. 2000.
7. W. B. Kuhn, and N. M. Ibrahim, "Analysis of Current Crowding Effects in Multitrun Spiral Inductors," IEEE Transactions on Microwave Theory and Techniques, Vol. 49, pp.31-38, Jan. 2001.
8. B. L. Ooi, D. X. Xu, P. S. Kooi, and F. J. Lin, "An Improved Prediction of Series Resistance in Spiral Inductor Modeling With Eddy-Current Effect," IEEE Transaction on Microwave Theory and Techniques, vol. 50, pp.2202-2206, Sept. 2002.
9. A. C. Watson, D. Melendy, P. Francis, H. Kyuwoon, and A. Weisshaar, "A comprehensive compact-modeling methodology for spiral inductors in silicon-based RFICs," IEEE Transactions on Microwave Theory and Techniques, pp.849-857, Mar. 2004.
10. A. Joonho Gil, and Hyungcheol Shin, "A Simple Wide-Band On-Chip Inductor Model for Silicon-Based RF ICs," IEEE Transaction on Microwave Theory and Techniques, Vol. 51, pp.2023-2028, Sept. 2003.
11. W. Jin, Y. Eo, J. I. Shim, W. R. Eisenstadt, M. Y. Park, and H. K. Yu, "Silicon substrate coupling noise modeling, analysis, and experimental verification for mixed signal integrated circuit design," in IEEE MTT-S Int. Microwave Symp, 2001, pp. 1727–1730.
12. J. Zheng, Y. C. Hahm, V. K. Tripathi, and A. Weisshaar, "CAD-oriented equivalent-circuit modeling of on-chip interconnects on lossy silicon substrate," IEEE Trans. Microwave Theory Tech., vol. 48, pp.1443-1451, Sept. 2000.

ECS Transactions, 34 (1) 81-86 (2011)
10.1149/1.3567563 ©The Electrochemical Society

Opportunities and challenges of FinFET as a device structure candidate for 14nm node CMOS Technology

T. Yamashita[a], V.S. Basker[a], T. Standaert[a], C-C Yeh[a], J. Faltermeier[a], T. Yamamoto[b], C-H Lin[a], A. Bryant[a], K. Maitra[c], P. Kulkarni[a], S. Kanakasabapathy[a], H. Sunamura[b], J. Wang[a], H. Jagannathan[a], A. Inada[b], J. Cho[c], R. Miller[c], B. Doris[a], V. Paruchuri[a], H. Bu[a], M. Khare[a], James O'Neill[a] and E. Leobandung[a]

[a]IBM Research, [b]RENESAS Electronics, [c]GLOBALFOUNDRIES·
IBM Research at Albany Nanotech, 257 Fuller Road, Albany, NY 12203 USA
E-mail: tyamash@us.ibm.com

FinFET is a promising device candidate for 14nm node CMOS technology. We have developed FinFET device showing superior short channel control at 25nm gate length. This FinFET device featuring gate first high-k/metal gate and merged Epi source/drain process. Key process improvements to resolve the FinFET unique challenges are presented. High drive currents have been obtained for both nFET and pFET. All these results show FinFET is the most promising candidate for 14nm node CMOS technology.

Introduction

FinFET is a promising device candidate for 14nm node CMOS technology. As the gate length scaling on conventional planar transistor is approaching the physical limit due to constraints in both equivalent oxide thickness scaling and junction scaling, fully depleted device such as FinFET can enable the aggressive gate length scaling and possible Vdd scaling for 14nm node due to its superior electrostatics. We have demonstrated the excellent FinFET electrostatics and performance on 22nm node gate pitch. A $0.063um^2$ cell size SRAM with static noise margin (SNM) ~150mV at Vdd=0.8V was also demonstrated. On the other hand, as a 3D device structure, FinFET has many challenges in both processes integration and device design. In process integration, Fin definition, gate patterning, and spacer formation in the 3D structure are challenging. Non-uniform vertical dopant profile along the gate sidewall can cause non-uniform gate-to-SD overlap and need to be resolved.

FinFET Process and Device Integration

Table 1 Show the critical dimensions of FinFET device fabricated on 22nm design rules. In order to improve the short channel effect (SCE) at the gate length of 25nm, Fin width (D_{fin}) has to be aggressively scaled to 10nm. Figure 1 show the schematic process flow of Fin formation using a novel SIT process to form the ultra-thin Fin of D_{fin}=10nm at Fin Pitch=40nm. Since the dummy fins are removed after Fin formation, there is no fat end fins seen in the conventional SIT process as shown in figure 2 (a). A good gate profile is achieved as shown in figure 2 (b). Spacer formation is extremely challenging in such high aspect ratio in the 3D structure. An advanced spacer RIE process has been developed to remove the SiN on Fin sides in the source/drain region while without causing Si erosion on Fin top as shown in Figure 3. Figure 4 shows 50ohm-um external resistance reduction is achieved by removing the SiN on Fin sides completely.

Fin Width (nm)	10
Fin Height (nm)	30
Fin Pitch (nm)	40
Gate Length (nm)	25
Gate Pitch (nm)	100
Contact Pitch (nm)	100

Table 1   Critical dimensions for FinFET Devices

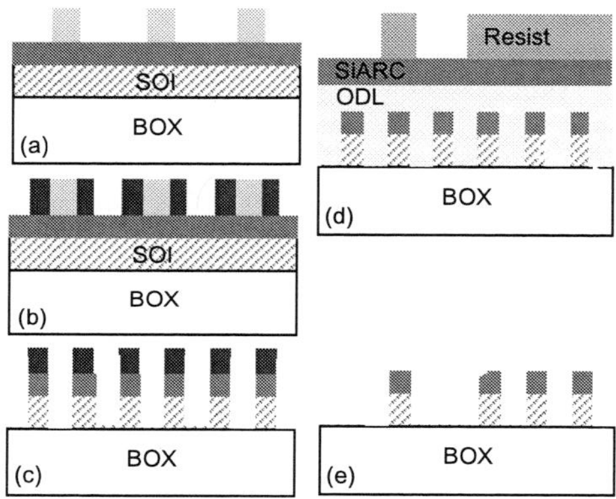

Figure 1   A novel SIT process for Fin formation with Fin Pitch=40nm. (a) a-Si mandrel patterning, (b) Si₃N₄ Spacer formation, (c) a-Si mandrel removal and SiO₂/Si RIE, (d) Litho for dummy fin removal, (e) dummy fin removal

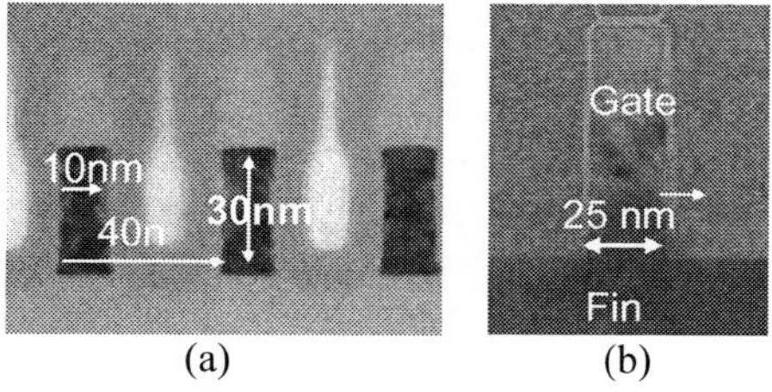

Figure 2   (a) Fins with good vertical profiles are formed. Fin width is10nm, Fin height is 30nm and SIT and Fin Pitch is40nm. (b) Gate profile of 25nm

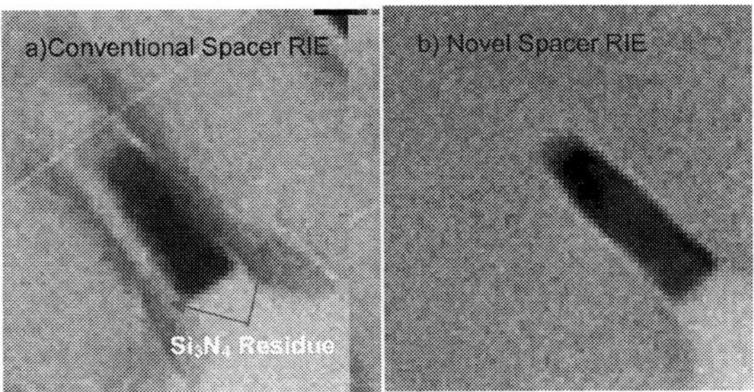

Figure 3   A novel spacer RIE process has been developed to remove SiN completely on Fin sides in source/drain region

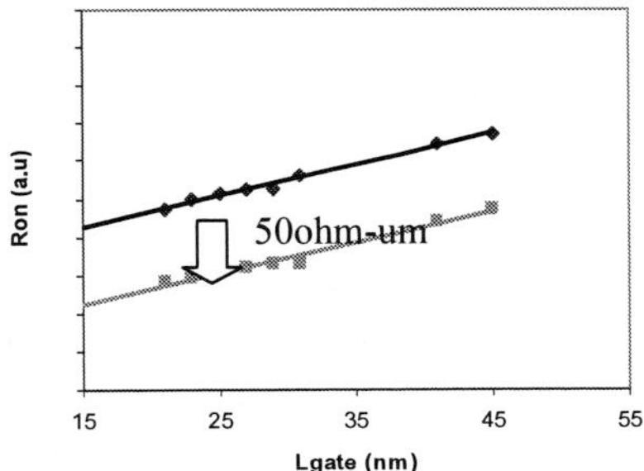

Figure 4   Significant external resistance reduction on PFET with removing the SiN completely by using the novel spacer RIE process.

Extension Doping

Extension doping is anther unique challenge for FinFET. Figure 4 show TCAD simulated activated Boron profile from the fin top to the fin bottom using the conventional angled implantation. 20 degree tilted BF2 implant was used in this study. With such non-uniform dopant profile, gate-to-SD overlap can vary within the width direction. It will cause Cov variation and external resistance variation. A conformal doping technique using *In-Situ* Boron doped SiGe Epi on PFET has been developed as shown in Figure 5. 30% Rext reduction can be seen on Figure 6.

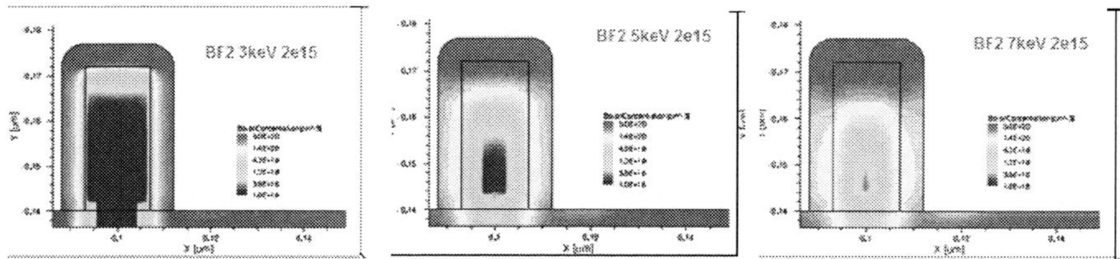

Figure 5   TCAD simulated Boron profiles with different implant energies. No uniform Boron profile can be obtained.

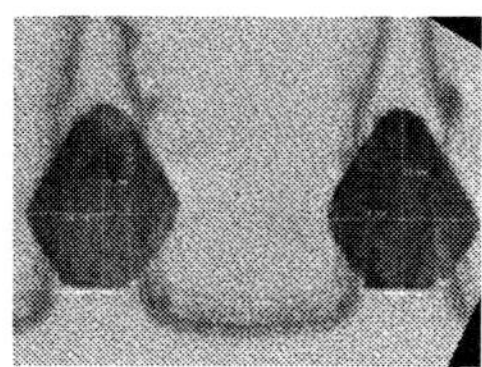

Figure 6   Conformal doping technique for FinFET using *in-situ* doped epitaxial film growth on fin sidewalls

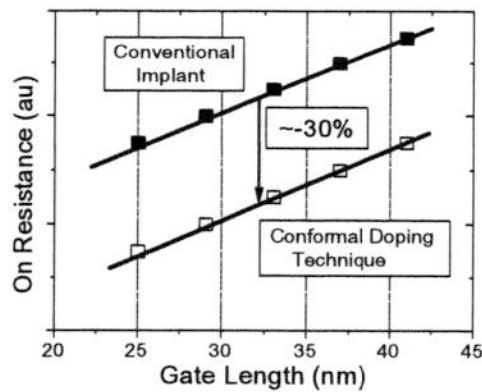

Figure 6   Resistance reduction obtained by conformal doping as compared to conventional implantation technique. 30% improvement is shown

FinFET Device and SRAM Characteristics

With the <100> surface orientation starting wafer, nFET channel is on <110> conduction plane and it was concerned on electron mobility degradation. In fact, as shown in Figure 7, electron mobility only degraded 20% although it is still much higher than the electron mobility in doped channel. Hole mobility gain ~100%. Net mobility benefit is significant and results high drive current at short channel devices.

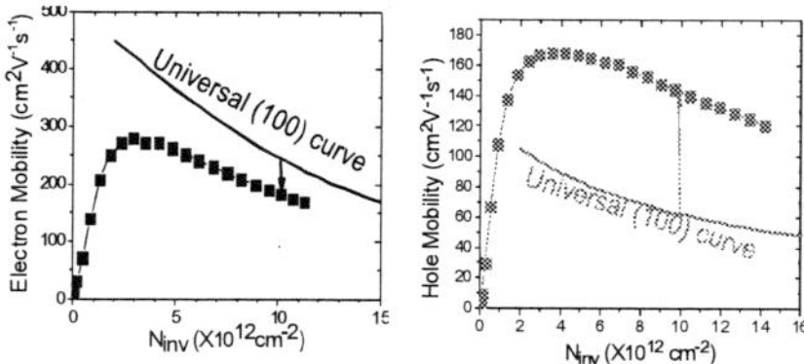

Figure 7   Mobility of electrons and holes in undoped (110) channel of FinFETs as compared to universal (100) mobilities. Significant improvement is shown for hole mobility, while the electron mobility degradation is <20%

Figure 8 shows (a) I_d/V_G curves and (b) I_{on}-I_{off} of FinFET NFET and PFET. Excellent electrostatics demonstrated at 25nm gate length with drain induced barrier lowering (DIBL) at 70mV/V (NFET) and 90mV/V (PFET) and substhrethold slope of 75mV/dec(NFET) and 85mV/dec(PFET). High drive current are achieved of Idsat=800uA/um for PFET and Idsat =1200uA/um for NFET at Ioff=100nA/um and Vdd=1V.

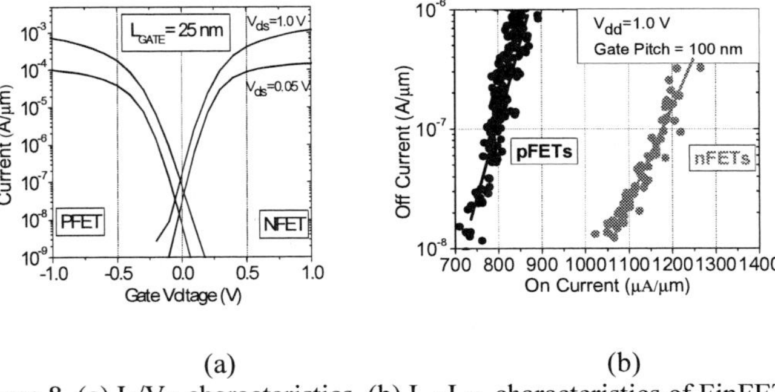

(a)   (b)

Figure 8   (a) I_d/V_G characteristics, (b) I_{on}-I_{off} characteristics of FinFET N/PFET

A $0.063\mu m^2$ 6-T SRAM is fabricated using SIT process. The topdown SEM image is shown in figure 9(a). The CPP is 100nm. The butterfly curves are shown in figure 9(b). The SRAM remains functional with V_{dd} down to 0.4V.

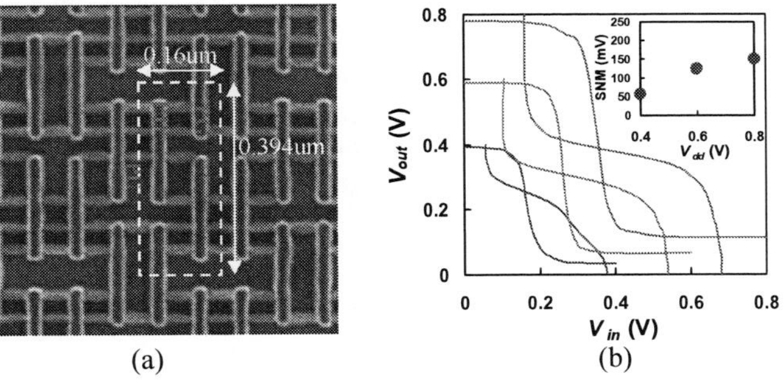

(a)   (b)

Figure 9   (a) Topdown SEM of $0.063\mu m^2$ 6-T SRAM fabricated using SIT process and merged source/drain epi

Conclusions

FinFET devices with excellent electrostatics and high drive current on gate length scaled down to 25nm under 22nm node design rules have been demonstrated. A novel SIT process has been developed for Fin formation and significant device variation reduction by eliminating the D_{fin} variation using this novel SIT process was observed. An advanced spacer RIE process to remove SiN completely from Fin sides in source/drain region is critical for external resistance reduction. Conformal extension doping technique by using *In-Situ* Boron doped SiGe for PFET results 30% external resistance reduction. We have achieved the high device drive current for both NFET and PFET contributed by resolving the key FinFET unique process integration challenges as described above. A 0.063um^2 SRAM cell with excellent statics noise margin (SNM) at low Vdd has been demonstrated. All these data help us to conclude that FinFET is a promising candidate for 14nm node CMOS and beyond.

Acknowledgments

This work was performed by the Research Alliance Teams at various IBM Research and Development Facilities. The authors would like to thank W. Haensch, E. Nowak and R. Divakaruni for the useful technical   discussions. We also thank the operations teams at Albany Nanotech and IBM East Fishkill for device fabrication.

References

1. W. Haensch, E. Nowak, R. Dennard, P. Solomon, A. Bryant, O.Dokumaci, A. Kumar, X. Wang, J. Johnson and M. Fischetti, *IBM J. Res. Dev., vol.* **50**, *p. 339(2006)*

2. M. Guillorn, J. Chang, A. Bryant, N. Fuller, O. Dokumaci, X. Wang, J. Newbury, K. Babich, J. Ott, B. Haran, R. Yu, C. Laovie, D. Klaus, Y.Zhang, E. Sikorski, W. Graham, B. To, M. Lofaro, J. Tornello, D. Koli, B. Yang, A. Pyzyna, D. Neumeyer, M. Khater, A. Yagishita, H. Kawasaki and W. Haensch, *Symp. VLSI Tech.* p. **12**(2008)

3. V. S. Basker, T. Standaert, H. Kawasaki, C.-C. Yeh, K. Maitra, T. Yamashita, J. Faltermeier, H. Adhikari. H. Jagannathan, J. Wang, H. Sunamura†, S. Kanakasabapathy, S. Schmitz, J. Cummings, A. Inada, C. -H. Lina, P. Kulkarni, Y. Zhua, J. Kuss, T. Yamamoto†, A. Kumara, J. Wahl, A. Yagishita, L. F. Edge, R. H. Kim, E. Mclellan, S. J. Holmes, R. C. Johnson, T. Levin, J. Demarest, M. Hane, M. Takayanagi, M. Colburn, V. K. Paruchuri, R. J. Miller, H. Bu, B. Doris, D. McHerron, E. Leobandung and J. O'Neill, *Symp. VLSI Tech.* p. **19**(2009)

4. H. Kawasaki, V.S. Basker, T. Yamashita, C-H. Lin, Y. Zhu, J. Faltermeier, S. Schmitz, S. Kanakasabapathy, H. Adhikari, H. Jagannathan, A. Kumar, K. Maitra, J. Wang, C-C. Yeh, C. Wang, M. Khater, M. Guillorn, N. Fuller, J. Chang, L. Chang, R. Muralidhar, A. Yagishita, R. Miller, Q. Ouyang, Z. Zhang, V.K. Paruchuri, H. Bu, B. Doris, M. Takayanagi, W. Haensch, D. McHerron, J. O'Neill and K. Ishimaru, *IEDM Tech. Dig.*, **289** (2009).

ECS Transactions, 34 (1) 87-92 (2011)
10.1149/1.3567564 ©The Electrochemical Society

Structural effects of channel cross-section on the gate capacitance of Silicon nanowire field-effect transistors

S. Sato[a], K. Kakushima[b], P. Ahmet[a], K. Ohmori[c], K. Natori[a], K. Yamada[c], and H. Iwai[a]

[a] Frontier Research Center, Tokyo Institute of Technology, Yokohama, 226-8502 Japan.
[b] Interdisciplinary Graduate School of Science and Engineering,
Tokyo Institute of Technology, Yokohama, 226-8502 Japan.
[c] Graduate School of Pure and Applied Sciences, University of Tsukuba, 1-1-1 Tennodai,
Ibaraki, 305-8573 Japan.

We investigated the gate capacitance of the silicon nanowire (SiNW) field-effect transistor (FET) using T-CAD software. The smaller equivalent oxide thickness of the SiNW FETs was obtained compared with the planar FET that had the same oxide thickness down to sub-1 nm. Gate capacitance of the SiNW FET substantially increased with thinner physical oxide thickness compared with planar FET. SiNW structure with thinner gate oxide thickness is effective for improvement of off-characteristics.

Introduction

Silicon nanowire (SiNW) field-effect metal-oxide-silicon field-effect transistor (MOSFET) is a promising candidate at the scaling limit of complementally MOS (CMOS) technology because of the immunity to the short channel effect (SCE) due to its surrounded gate structure with the narrow SiNW channel [1,2]. An increase of a gate capacitance leads to the suppression of the SCE and the reduction of off-current (I_{OFF}). We can expect that the normalized gate capacitance of SiNW FET is larger than the gate capacitance of the planar-type FET because of the surrounded gate structure. Using the cylindrical conductor model, the gate capacitance can be calculated as follows;

$$C_{NW} = \frac{\varepsilon_0 \varepsilon_{ox}}{r_{Si} \cdot \ln\left(1 + \frac{t_{ox}}{r_{Si}}\right)} \quad (1); \quad C_{planar} = \frac{\varepsilon_0 \varepsilon_{ox}}{t_{ox}} \quad (2)$$

, where ε_0 is the permittivity in vacuum, ε_{ox} is relative permittivity of dielectric, r_{si} is the radius of the cylinder, t_{ox} is the oxide thickness. The more r_{Si} decreases, the larger C_{NW} becomes compared with C_{planar}. Many researchers have calculated and evaluated the gate capacitance of SiNW FET [3, 4]. Recently electrical characteristics of tri-gate and SiNW FETs have been reported considering effects of channel cross-sectional shapes, especially corners of the channel cross-sections [5-7]. Therefore quantitative investigation on the gate capacitance considering the corners is necessary. In this work, we investigated the gate capacitance of the SiNW FETs using technology computer-aided design software (T-CAD) considering the corners of the channel cross-sections.

Computer Simulation Scheme

We used Taurus-MEDICI for the design of cross-sectional shapes of the SiNW channels. We used Taurus-DAVINCI for calculation of the electrical characteristics of the SiNW nFETs; drain current, inversion charge distribution, and magnitude of electric field with three-dimensional structures. Quantum-mechanical effects were introduced using the modified local density approximation (MLDA) method [8]. For the calculation of the gate capacitance, we used the quasi-static calculation scheme as follows:

$$C_{gate} = \frac{dQ_{gate}}{dV_{gate}} \qquad (3)$$

Inversion charge density of the SiNW FET was calculated with the normalization of Q_{gate} by the peripheral length of the SiNW channel. We set the doping concentration of the source/drain region to 10^{20} cm^{-3}. We used typical physical parameters of n-doped polycrystalline silicon (poly-Si) for nFETs and p-doped poly-Si for pFETs for the gate electrode. We set doping concentration in SiNW channel 10^{15} cm^{-3}. We adopted semi gate-around structure [9] as a gate structure of SiNW FET used in this work because of its readiness for a fabrication. We calculated equivalent oxide thickness of SiNW FET using CVC software by North Carolina State University [10].

Results and discussions

Figure 1 (a) shows a contour plot of inversion carrier distribution and **figure 1 (b)** shows a contour plot of the magnitude of electric field of SiNW FET at the drain bias voltage of 50 mV and gate overdrive voltage of 1.0 V at the center of the channel between source and drain. We can see higher electric field around corners than the electric field along flat surface. The higher electric field at corners leads to higher inversion carrier density around the corners, which is equal to a reduction of the equivalent oxide thickness around corners.

Figure 2 (a) shows inversion charge density of SiNW nFETs with rectangular cross-section. The channel width w_{NW} and the channel height h_{NW} were 19 and 12 nm, respectively, with different corner radii. We examined effects of corner radii on the gate capacitance. **Figure 2 (a)** suggests that the inversion charge density is almost consistent with different corner radii. We evaluated the peak inversion carrier density around the corners, and as the corner radius increased the peak carrier density decreased as shown in **figure 2 (b)**. On the other hand, an area in the SiNW channel that has inversion carrier density more than 3.2×10^{19} cm^{-2} increased as the corner radius increased as shown in **figure 2 (b)**. We speculate that a trade-off relationship between the peak inversion carrier density and the area of high inversion carrier region resulted in the consistent inversion charge density with different corner radii. We can also observe this trade-off relationship in **figure 2 (d), (e), and (f)**.

Next, we investigated inversion charge density and equivalent oxide thickness of SiNW FET with rectangular cross-section as a function of channel width w_{NW}. **Figure 3 (a)** shows inversion charge density at the overdrive voltage of 1.0 V. As w_{NW} decrease, inversion charge density at both low and high drain bias voltage increased. Equivalent oxide thickness (EOT) of SiNW nFETs are shown in **figure 3 (b)**. Higher EOT was obtained with smaller w_{NW}. We speculated the higher EOT with smaller w_{NW} was due to high inversion carrier region around the corners of channel cross-section.

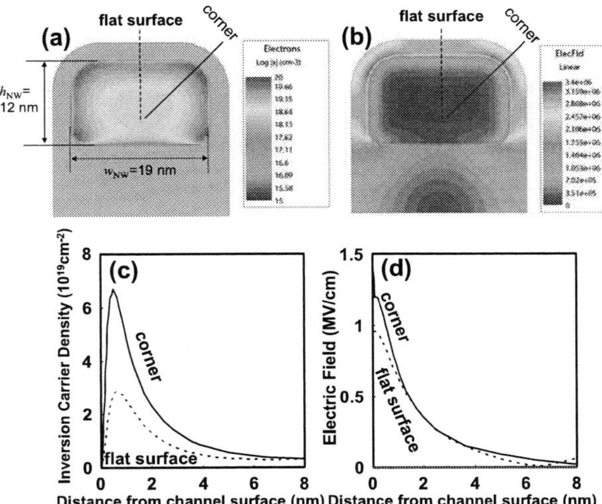

Figure 1. Contour plots of (a) inversion carrier density and (b) magnitude of electric field at gate overdrive voltage of 1.0 V and drain voltage of 50 mV. One dimensional plots of (c) inversion carrier density and (d) electric field at the corners and along flat-surface are also shown.

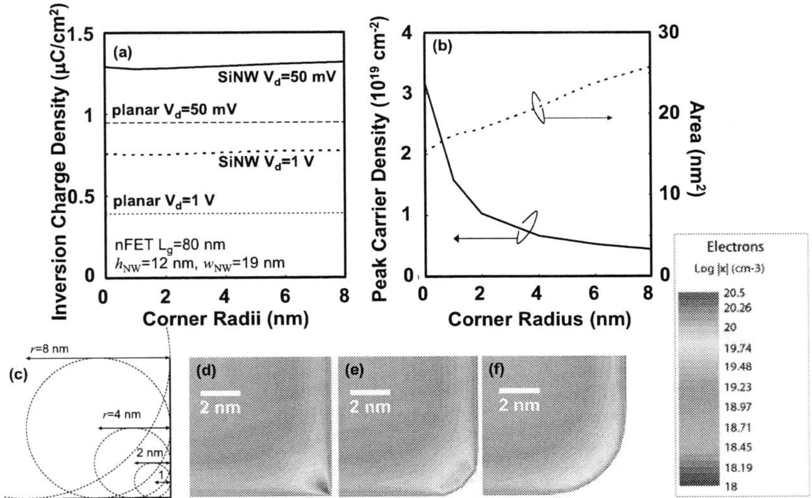

Figure 2 (a) Inversion charge density as a function of corner radius. (b) Peak inversion carrier density and area with inversion carrier density over 3.2×10^{19} cm^{-2}. (c) Schematic illustration of corner radius (r). Contour plot of inversion carrier density with corner radii of (d) 0, (e) 2 and (f) 4 nm at the upper-corner of the SiNW channel.

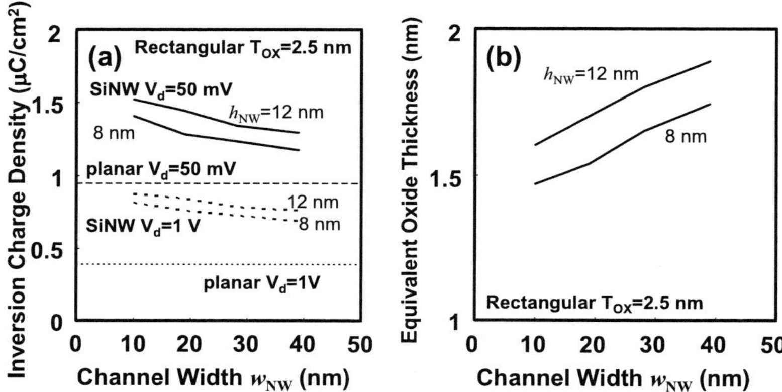

Figure 3. (a) Inversion charge density of SiNW nFETs with rectangular cross-sections (h_{NW}=8 and 12 nm) and the upper-corner radius of 4 nm. (b) Equivalent oxide thickness of SiNW nFETs with rectangular cross-sections (h_{NW}=8 and 12 nm) and the upper-corner radius of 4 nm.

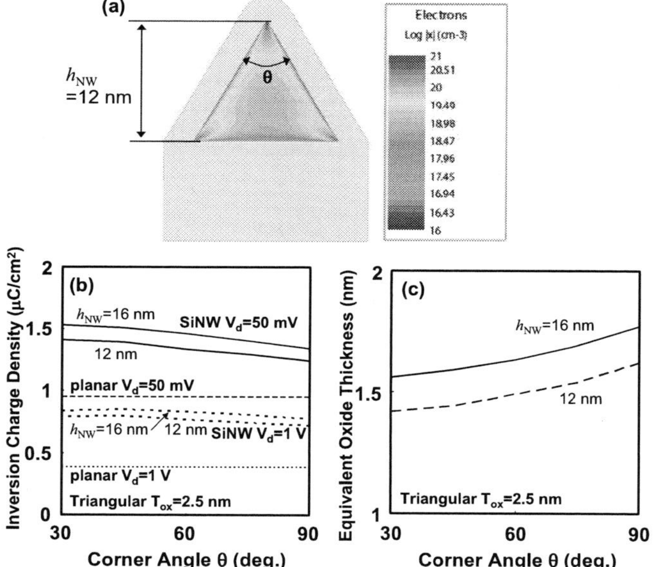

Figure 4. (a) Inversion carrier distribution and (b) inversion charge density of SiNW nFETs with triangular cross-sections. (b) Equivalent oxide thickness of SiNW nFETs with triangular cross-sections.

Figure 4 (a) shows inversion carrier distribution and **figure 4 (b)** shows inversion carrier density of SiNW nFETs with triangular channel cross-sectional shapes. Corner angle θ is the upper corner angle of triangular cross-section. Channel height h_{NW} is 12 and 16 nm. As the corner angle decreased, inversion charge density increased and equivalent oxide thickness decreased. We speculate that a higher peak inversion carrier density with a smaller corner angle θ is the main reason. Although lower corner angles of triangular cross-section decrease as the upper corner angle increases, contributions of the upper corner are larger than the contributions of lower corners because of the gate semi-around structure as shown in **figure 4 (a)**.

Next we investigated dependence of the equivalent oxide thickness of SiNW nFETs on gate oxide thickness. **Figure 5 (a)** shows equivalent oxide thickness as a function of physical gate oxide thickness of SiNW nFETs with rectangular cross-sections (w_{NW}=10 and 19 nm, h_{NW}=12 nm) and triangular cross-section (h_{NW}=12 nm, θ=45°). We can observe that structural effects of the SiNW FET are valid in thinner gate oxide region. **Figure 5 (b)** shows gate capacitance of SiNW FET. The gate capacitance of SiNW FET with triangular cross section is obviously larger than other structure. As the physical oxide thickness decrease, the gate capacitance of the SiNW FET substantially increased up to 81 % with triangular cross-section.

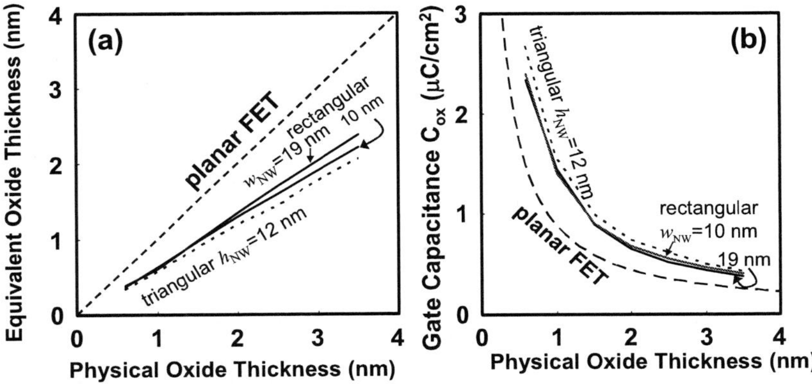

Figure 5. (a) Equivalent oxide thickness of SiNW nFETs as a function of physical oxide thicknesses. SiNW FETs with rectangular cross-sections (h_{NW}=12 nm, w_{NW}=10 and 19 nm, r=4 nm) and triangular cross-section (h_{NW}=12 nm and θ=45°) were evaluated. (b) Gate capacitance of SiNW nFETs as a function of gate oxide thickness.

Finally, we characterized transfer characteristics of the SiNW FETs with the channel width w_{NW} of 14 nm and channel height h_{NW} of 10 nm and the gate oxide thickness of 3, 2 and 1.5 nm as shown in **figure 6**. Absolute values of the drain current were shown. Detailed fabrication processes are described in our previous report [5]. We can observe that as the gate oxide thickness decreased the subthreshold slope of the SiNW pFETs improved. Drain current of the SiNW pFET with the oxide thickness of 1.5 nm at the gate voltage above about 0.5 V was larger than the drain current of SiNW pFETs with oxide thicknesses of 2 nm and 3 nm. This was due to thinner gate oxide thickness of 1.5 nm, which resulted in larger gate leakage current from gate to source and drain, and not due to the gate induced drain leakage. We experimentally demonstrated that the reduction of gate oxide thickness resulted in the improvement of off-characteristics.

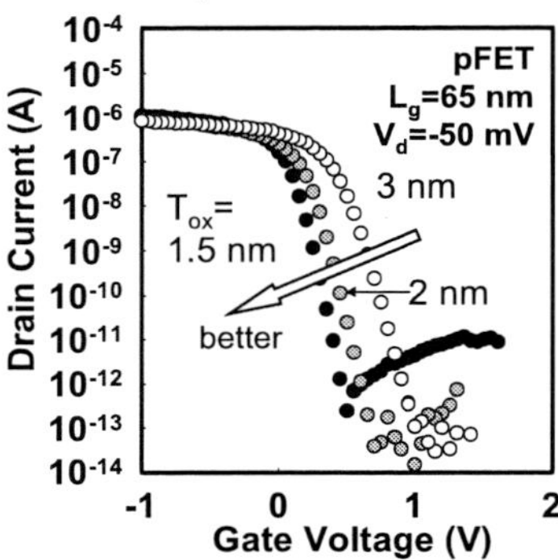

Figure 6. Transfer characteristics of SiNW pFETs with channel width w_{NW} of 14 nm and channel height h_{NW} of 10 nm with the gate oxide thickness of 3, 2, and 1.5 nm.

Conclusion

In this work, the gate capacitance of the SiNW FETs was investigated using T-CAD software. The equivalent oxide thickness of the SiNW FETs was smaller than the planar FET that had the same oxide thickness down to sub-1 nm region. Gate capacitance of the SiNW FET was substantially increased in thin oxide thickness region compared with planar FET. It was experimentally demonstrated that as the gate oxide thickness decreased off-characteristics improved. SiNW FET with thinner gate oxide thickness is effective for the increase of gate capacitance and the improvement of off-characteristics.

Acknowledgments

The authors thank all members of ASKA II line and the researchers in the front-end program in the R&D department 1, Selete, Tsukuba for device fabrication and fruitful discussions. This work was supported by the program 'Development of Nanoelectronic Device Technology' of the New Energy and Industrial Technology Development Organization (NEDO).

References

1. J.-P. Colinge, *Solid-State Electron.*, **48**, 897 (2004).
2. J.-T. Park *et al.*, *IEEE Trans. Electron Devices*, **49**, 2222 (2002).
3. O. Wunnicke. *Appl. Phys. Lett.*, **89**, 083102 (2006).
4. D. R. Khanal *et al.*, *Nano Lett.*, **7**, 2778 (2007).
5. S. Sato *et al.*, *Microelectronics Reliability.* doi:10.1016/j.microrel.2010.12.007.
6. B. Doyle *et al.*, *Proc. Symp. VLSI Technology.*, 133 (2003)
7. K. E. Moselund *et al.*, *Proc. ESSDERC,* 359 (2006).
8. M. G. Ancona and H. F. Tiersten, *Phys. Rev. B*, **35**, 7959 (1987).
9. S. Sato *et al.*, *Solid-State Electron.* **54**, 925 (2010).
10. J. R. Hauser *et al.*, *Proc. AIP Conf.*, 235 (1998).

ECS Transactions, 34 (1) 93-98 (2011)
10.1149/1.3567565 ©The Electrochemical Society

Process Impact and Design Optimization on the Soft Yield of 25nm FinFET SRAM Cells

Meng Li, Qingqing Liang, Huilong Zhu, Huicai Zhong, Dapeng Chen, and Tianchun Ye

Institute of Microelectronics of Chinese Academy of Sciences, Beijing 100029, China

A simulation methodology to correlate process variations to the soft yield of 6-T FinFET SRAM cells is demonstrated. By using the TCAD tool which is calibrated to the recently published 25nm FinFET technology, process impacts on the device electrical characteristics are simulated and stored in behavioral models. Circuit simulations can then be carried out to calculate the read/write noise margin. Employing the linear function approximation of the noise margin with respect to the six individual process variation parameters in sigma, one can analytically solve the yield when noise margin drops to zero. The calculation results show that instead of Vt variation, parasitic resistance fluctuation dominates fails at high bias voltages. A simple design optimization guideline is also demonstrated to reduce Vmin of the given technology (to 0.5V in this work). This methodology can serve as useful guidelines for both process development and device design.

Introduction

Due to excellent control of short channel effects (SCE), FinFETs are regarded as promising candidate device structure in 22nm technology node and beyond (1~3). In such aggressively scaled devices, variations become prominent in limiting the performance and degrading the yield of a design. SRAM yield is a well-recognized benchmark for evaluating variation in different technologies (4) (5). It is therefore necessary to analyze the possible SRAM yield bumpers and obtain rules-of-thumb for device design on FinFET process technology.

In this work, for the first time, we present a systematic approach to simulate, model, and optimize 6-T SRAM yield based on the state-of-the-art FinFET technology (3). With this approach, both process-level diagnosis and device-level optimization are conducted for an exploratory yield evaluation on emerging FinFET technology.

Simulation and Optimization Technique

Sentaurus process and device simulation used in this study are calibrated to the reported 0.094 μm^2 FinFET SRAM (3). Fig. 1 shows a bird-eye view of the simulated FinFET SRAM cell. Double-fins are used in pull-down NFETs to achieve β-ratio equals 2. The centered fin width, height, and gate length are 12nm, 30nm, and 25nm, respectively.

Behavioral statistical model (6) is used to capture process-induced variations. Six key process-induced variables are studied, including: gate length edge roughness (LER), Vt fluctuation (VTF), gate dielectric thickness (Tox), fin height (Fh), parasitic resistance (Rpar), and mobility (Mob) variations. Compared to planar device, FinFET exhibits improved Vt fluctuation due to un-doped channel (2). However, Rpar of FinFETs is generally worse than that of planar devices, which dominates yield fails at high operation points as shown later.

Fig. 2 shows the simulation flow of this work. The process impact on electrical characteristics were simulated and stored in behavioral models. These models then serve as lookup tables in the circuit simulation to calculate read/write static noise margin (RSNM/WSNM) of the 6-T cell. The advantages of using behavioral model are: it greatly simplifies model extraction and reduces circuit simulation time.

Fig. 3-4 show calculated left/right nodal voltages correlation under read/write mode (e.g. butterfly curves), and extracted RSNM and WSNM on the left side of the cell. RSNM and WSNM as functions of the six variables of the six FETs are extracted (as shown in Fig. 5-6). Here one can turn on/off different process-induced variations to determine the dominant contributions.

The soft yield is then estimated as the overall sigma when RSNM/WSNM drops to 0 (7) (e.g. sigma=4.89 is equivalent to 1 fail in 1e6 bits, sigma=6.11 is equivalent to 1 fail in 1e9 bits). Even though Monte-Carlo simulation is a reliable approach for yield calculation, it becomes prohibitive for design optimization when dealing with Mega-bit memory. To obtain faster first-order predictions, RSNM and WSNM are approximated to linear functions of the six process-induced variations (in sigma) and yield can be analytically solved. This linear approximation was validated by the experimental data plotted in Figure. 7 of Reference (8), which shows approximately linear dependency of SNM with respect to the variation up to ±4 sigma. The linear trend can be also observed from Fig. 5 and Fig. 6 in this paper. Using this algorithm, we first extracted Schmoo chart (e.g. yield vs. bit and word line voltages), then calculated yield as functions of design parameters (i.e. gate length and N-P Vth offset) at a fixed bias point to find the optimum design. The sigma values for the six variables used in our calculation are summarized in TABLE I, which is reasonable compared with the experimental data in Reference (2).

Results and Discussions

Fig. 7 shows read/write yield as functions of the bias voltages, with only one of the six process variations on and all variations on, respectively. It is clear that Rpar fluctuation dominates the yield fails at high voltage operation, whereas VTF and Tox dominate yield fails at low voltage operation.

Fig. 8 shows the yield Schmoo chart (e.g. sweeping bit and word line voltages), write fail dominates when bit line voltage is higher than word line voltage, whereas read fail dominates vice versa. The trend is consistent to reported data (9). Fig. 9 shows the yield contours on different bias conditions. Observe that at this device design, yield at 0.5V is less than 3.29 sigma, meaning that there is more than 1 fail in 1K-bit memory. Therefore, this cell design is not feasible for 0.5V operation. However, one may find a better device design such as different Lgate and N-P Vt offset to maximize the yield.

Fig. 10 shows the yield as functions of N-P Vt offset and gate length design, respectively. Here stronger NFET (i.e. 100mV lower Vth) is preferred to achieve yield>5 sigma. In addition, longer Lgate helps a little as well. Fig. 11 shows yield contours of different design point. Consistent to Fig. 7, write fail dominates at 0.5V operation, and weaker PU will make PG easier to drain the node voltage down. One can also find that 18nm is the minimum gate length to achieve Mega-bit cell yield (i.e. sigma>4.89) in this technology.

Conclusion

Systematic study on the impact of process-induced variation and design optimization on FinFET SRAM yield is conducted. Applying this analysis, we find that: Rpar dominates yield fails at high bias voltages, whereas VTF and Tox dominate yield fails at low bias voltages; weaker PU helps to improve write and overall yield in low bias range. For Mega-bit SRAM design, Vmin≤0.5V can be achieved as Lgate>18nm and N-P Vt offset ~-0.1V.

Acknowledgments

This work has been funded by "National Sciences and Technology Major Project 02".

References

1. S. O'uchi, T.Nakagawa, T. Matsukawa, *et al*, in *IEEE International SOI Conference*, p. 1, Foster City, CA (2009).
2. T. Matsukawa, S. O'uchi, K. Endo, *et al*, in *Symposium on VLSI Technology*, p. 118, Honolulu, HI (2009).
3. V.S. Basker, T. Standaert, H. Kawasaki, *et al*, in *Symposium on VLSI Technology*, p. 19, Honolulu, HI (2010).
4. F. J. List, in *European Solid-State Circuits Conference*, p. 16, Twelfth, European (1986).
5. A. Bhavnagarwala, S. Kosonocky, C. Radens, *et al*, in *International Electron Devices Meeting*, p. 659, Washington, DC (2005).
6. Q. Liang, J. Johnson, J. Walko, *et al*, in *International Semiconductor Device Research Symposium*, p. 12, College Park, MD (2005).
7. K. Agarwal, and S. Nassif, in *Design Automation Conference*, p. 57, San Francisco, CA (2006).
8. X. Song, M. Suzuki, T. Saraya, *et al*, in *IEEE International Electron Devices Meeting*, p. 62, San Francisco, CA (2010).
9. B. Greene, Q. Liang, K. Amarnath, *et al*, in *Symposium on VLSI Technology*, p. 140, Honolulu, HI (2009).

TABLE I. Comparison of Sigma Values for the Six Process-Induced Variables Used in This Paper and That in Ref (2).

Here	LER (nm)	VTF (mV)	Tox (nm)	Fh (nm)	Rpar (Ω)	Mob (%)
This paper	0.835	23	0.067	2	200	6.67
Ref (2)	0.8	21.4	0.04	NA	100~400	NA

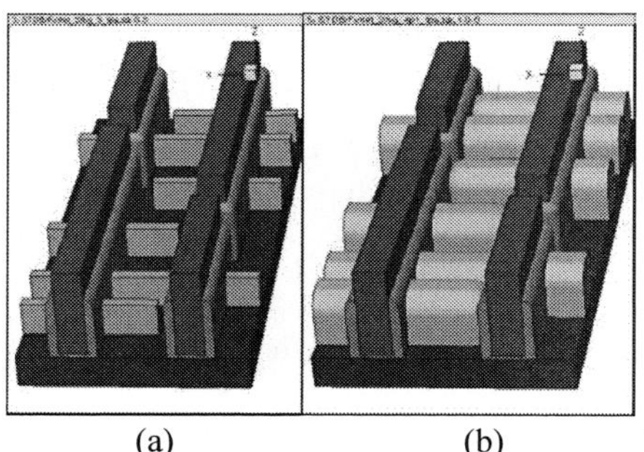

(a)   (b)

Figure 1. Bird-eye view of 6-T FinFET SRAM cell structure used in TCAD simulation: (a) before S/D epi, (b) after S/D epi. The centered fin width, height and gate length are 12nm, 30nm, and 25 nm, respectively.

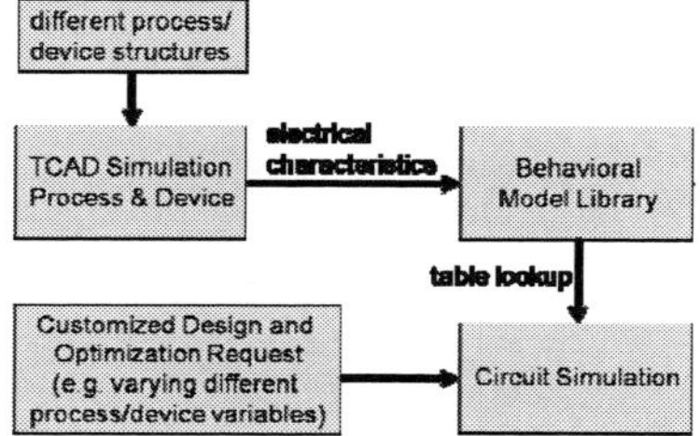

Figure 2. Illustration of the simulation flow used in this work.

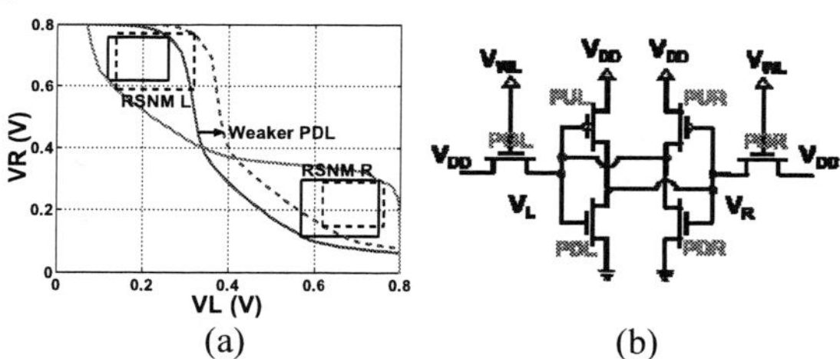

(a)   (b)

Figure 3. (a) Left and right bit node voltages (VL and VR) trends in read mode, and extracted left/right read-static-noise margin (RSNM), (b) circuit structure and bias conditions when the cell is in read mode.

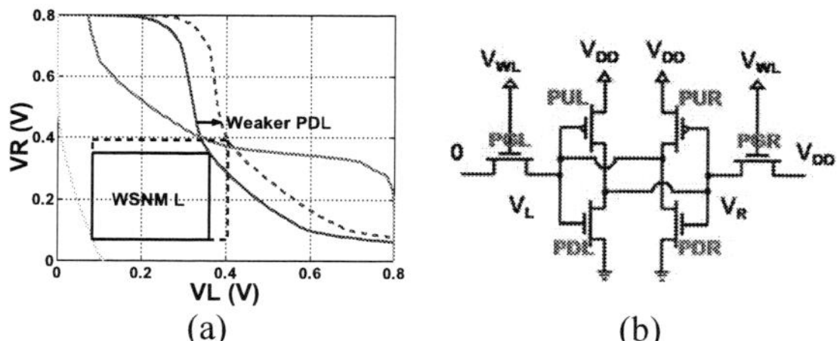

(a)   (b)

Figure 4. (a) Left and right bit node voltages (VL and VR) trends in read mode and write mode (green curve), and extracted left/right write-static-noise margin (WSNM), (b) circuit structure and bias conditions when the cell is in write mode (left node writes to "0").

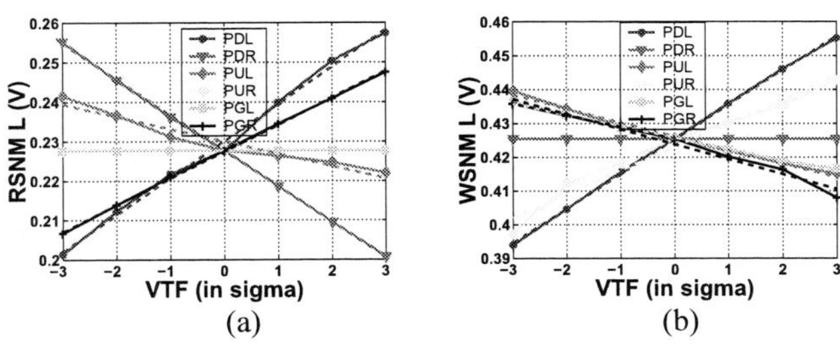

(a)   (b)

Figure 5. Read (a) and write (b) static noise margin on the left side (RSNM L and WSNM L) as a function of Vt fluctuation (VTF) on each device of the 6-T cell. VDD and VWL are both 0.8V. Negative sigma represents a lower Vt (i.e. stronger FET). Dashed lines are linear fits of the trend.

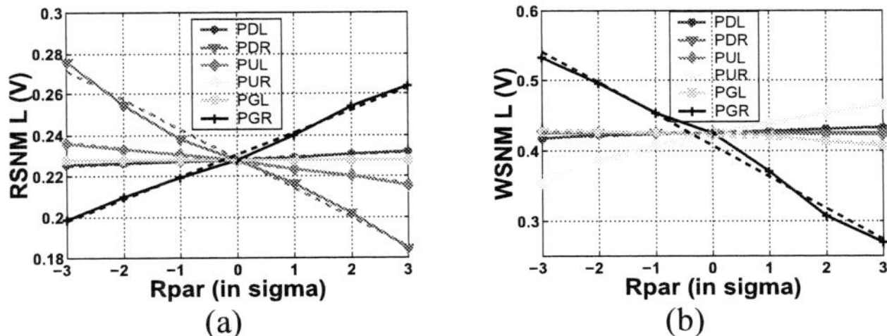

(a)   (b)

Figure 6. Read (a) and write static (b) noise margin on the left side (RSNM L and WSNM L) as a function of parasitic resistance fluctuation (Rpar) on each device of the 6-T cell. VDD and VWL are both 0.8V. Negative sigma represents a lower Rpar (i.e. faster FET). Dashed lines are linear fits of the trends.

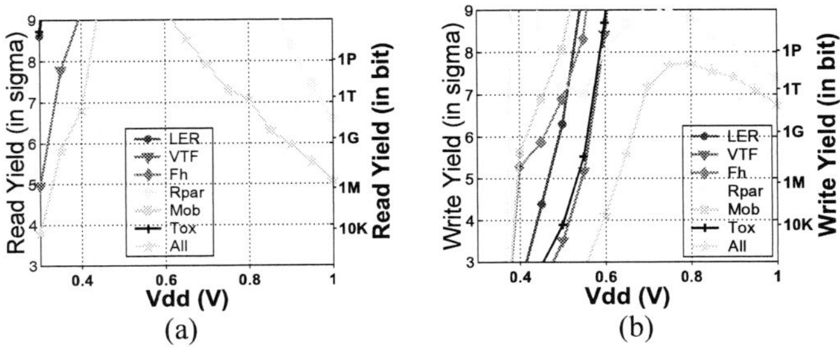

(a)   (b)

Figure 7. Read yield (a) and write yield (b) as functions of bias voltages, VDD and VWL are sweeping simultaneously from 0.3 to 1V, simulated with one of each process-induced variation turned on or all variations turned on.

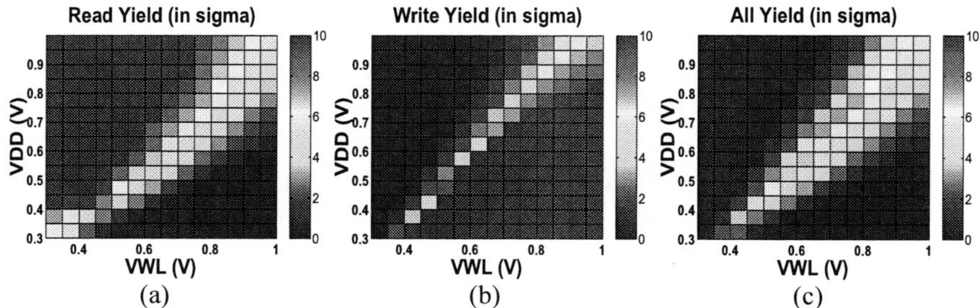

Figure 8. Read (a), write (b) and combined (c) yield Schmoo chart, where bit line voltage (VDD) and word line voltage (VWL) are the sweeping variables.

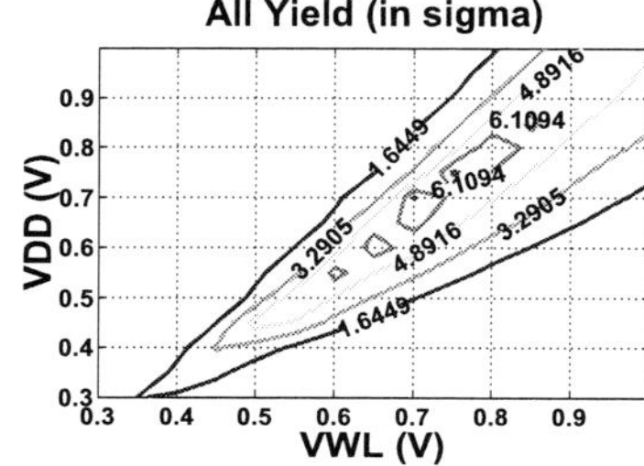

Figure 9. Yield contours of different bit line voltage (VDD) and word line voltage (VWL) bias points.

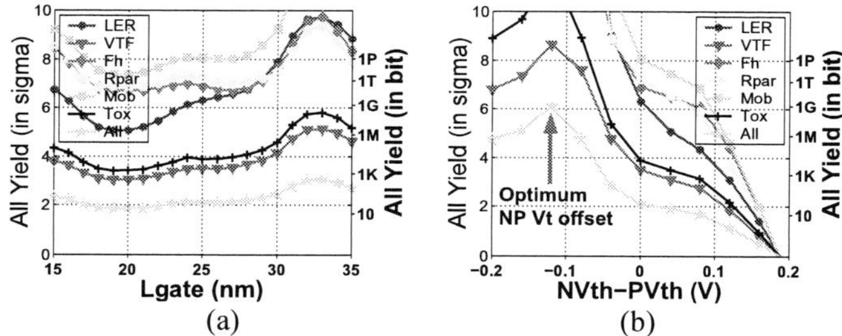

Figure 10. Yield as functions of different N-P Vt offset design , where Lgate is centered at 25nm (a); and different Lgate design, where N-P Vt is balanced (b), with VDD and VWL biasing at 0.5V.

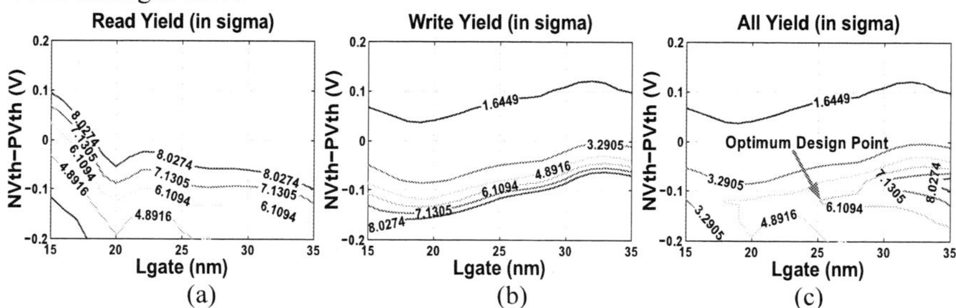

Figure 11. Read (a), write (b) and combined (c) soft yield contours of different N-P Vt offset and Lgate design, with VDD and VWL biasing at 0.5V.

ECS Transactions, 34 (1) 99-102 (2011)
10.1149/1.3567566 ©The Electrochemical Society

TiN/W/La$_2$O$_3$/Si High-k Gate Stack for EOT below 0.5nm

P. Ahmet[a], D. Kitayama[a], T. Kaneda[a], T. Suzuki[a], T. Koyanagi[a], M. Kouda[a],
M. Mamatrishat[a], T. Kawanago[a], K. Kakushima[b], K. Tsutsui[b], A. Nishiyama[b],
N. Sugii[b], K. Natori[a], T. Hattori[a], and H. Iwai[a]

[a] Frontier Research Center, Tokyo Institute of Technology, Yokohama 226-8502, Japan
[b] Interdisciplinary Graduate School of Science and Engineering, Tokyo Institute of
Technology, Yokohama 226-8502, Japan

Electrical properties of TiN/W/La$_2$O$_3$ high-k gate stack were
studied by fabricating MOS capacitors. Obtained results showed
that a W layer inserted at the interface between TiN and La$_2$O$_3$ is
the key factor in suppression of the equivalent oxide thickness
(EOT) increment during the annealing process. An EOT of 0.43nm
was achieved with a 3nm W inserted layer after annealed at 800°C
in a forming gas ambient. Our results show that TiN/W/La$_2$O$_3$ gate
stack is one of the promising candidates for realizing high-k gate
stack with EOT of 0.5nm and beyond.

Introduction

As the result of continues scaling in CMOS technology, required equivalent oxide
thickness (EOT) in MOSFETs is expected to down to a thickness of less than 0.5nm in
the near future. For such a small EOT, a directly contacted of high-k/Si structure is
required because of a SiO$_x$ based interfacial layer at the interface between high-k and Si
will limits the EOT value obtainable with the high-k gate dielectrics. However, realizing
a high quality interface between directly contacted high-k dielectric and Si substrate bring
about a great challenge, because of the interface between gate dielectric and Si substrate
has always been SiO$_2$/Si or SiON/Si in the past. Recently, several types of high-k gate
stacks without any SiO$_x$ based interfacial layer at the interface between high-k and Si
were investigated [1-4], however, further reduction of EOT to beyond 0.5nm with a high
quality interface is still the major obstacle in the future high-k gate stack technology.

In this paper, we report our approaches in realizing EOT of 0.5nm and below by
using La$_2$O$_3$ as the high-k dielectric and employing TiN/W as the gate electrode. An EOT
of 0.43nm has been achieved with TiN/W/La$_2$O$_3$ gate stack after annealed at 800°C by
optimizing the thickness of the W layer. Our results show that a proper gate electrode
structure is one of the most important factors for realizing EOT below 0.5nm in La$_2$O$_3$
high-k gate stack.

Experimental

La$_2$O$_3$ thin films were deposited by e-beam evaporation in an ultra-high vacuum
chamber on HF-last n-Si substrates. After the deposition of La$_2$O$_3$ thin films, the
substrates were transferred into a sputter chamber without breaking the ultra-high
vacuum. Tungsten (W) and TiN gate electrodes were deposited by RF sputtering and

reactive RF sputtering, respectively. After patterned by lithography and reactive ion etching (RIE), post-metallization annealing was carried out in a forming gas ambient (H_2:N_2 = 3:97%) at various temperatures. Back side contacts were formed by Al deposition. The samples were evaluated by electrical measurements and the EOT values of the samples were calculated by NCSU CVC program [5].

Results and Discussion

A schematic illustration of the structure of deposited samples and a cross sectional TEM image of the fabricated gate stack after annealed at 800°C for 2 seconds in a forming gas ambient are shown in Figure 1. No any SiO_x like interfacial layer was observed at the interface between La_2O_3 high-k dielectric and the Si substrate after such a temperature thermal processing.

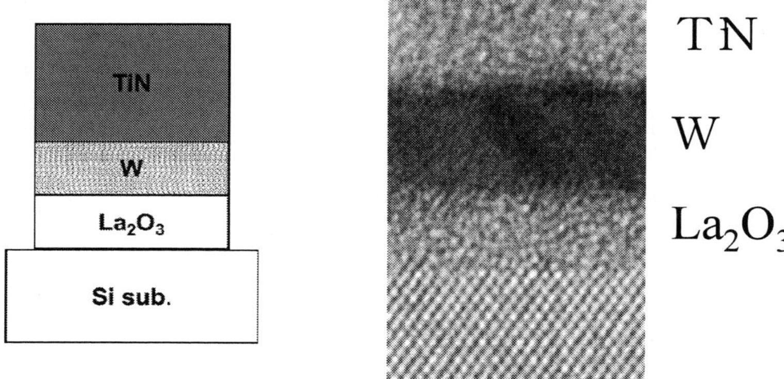

Figure 1. Schematic illustration of the structure of deposited samples and a cross sectional TEM image of the gate stack after annealed at 800°C for 2 seconds in a forming gas ambient.

Obtained lowest EOT from TiN(45nm)/W/La_2O_3(3nm)/n-Si MOS capacitors with different thicknesses of W layer are shown in Figure 2. These samples were annealed at 700°C in a forming gas ambient for 30 minutes. A minimum EOT of 0.57 nm was obtained with a 3nm W layer. The samples annealed at 800°C in a forming gas ambient for 2seconds also a give similar result where a minimum EOT of 0.43 nm was obtained with a 3nm W layer. Obtained EOTs without a W layer insertion were 2.3 nm for 700°C in a forming gas ambient for 30minutes and 1.5 nm for 800°C in a forming gas ambient for 2seconds (not shown). The huge difference in obtained EOT between the capacitors with and without a W layer insertion indicating that the W layer inserted at the interface between La_2O_3 high-k dielectric and TiN electrode plays the crucial role in suppression of the EOT increment during annealing process. Figure 2 also shows a low EOT can be obtained by optimizing the inserted W layer thickness.

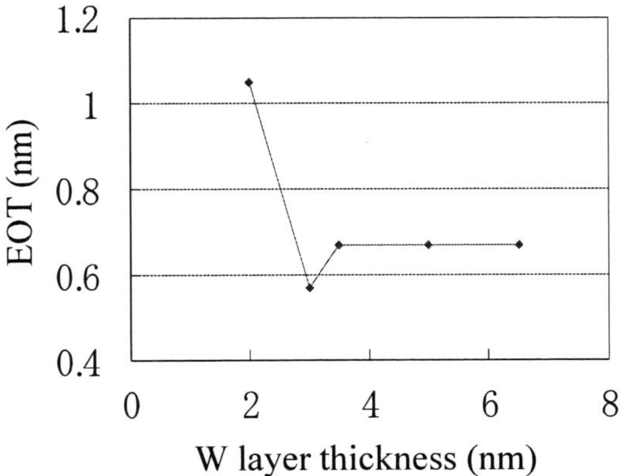

Figure 2. Obtained lowest EOT from TiN(45nm)/W/La$_2$O$_3$(3nm)/n-Si MOS capacitors with different W layer thicknesses. (Samples were annealed at 700°C in a forming gas ambient for 30min)

Figure 3 show the measured Capacitance-Voltage (C-V) curves from TiN(45nm)/W(3nm)/ La$_2$O$_3$(3nm)/n-Si MOS capacitors annealed at 500°C and for at 700°C for 30 minutes, and at 800°C for 2 seconds in an forming gas ambient. The C-V characteristics at higher voltages also show that the gate dielectric film quality was better in the samples annealed at 700°C for 30 minutes compare to that of the samples annealed at 500°C for 30 minutes. No significant reduction in k-value of the dielectric film was observed during the annealing process performed at 700°C for 30 minutes compare to the annealing process performed at 500°C for 30 minutes. Lowest EOT of 0.43nm was obtained when the samples annealed at 800°C for 2 seconds, however, the hysteresis was not as negligibly-small as those observed with the samples annealed at 500°C and 700°C for 30 minutes.

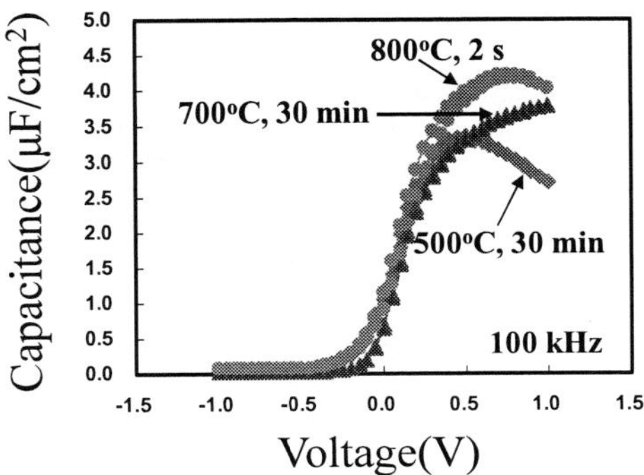

Figure 3. C-V characteristics of TiN(45nm)/W(3nm)/La$_2$O$_3$(3nm)/n-Si MOS capacitor annealed in an forming gas ambient at 500°C for 30minuts, 700°C for 30minuts and 800°C for 2 seconds (with hysteresis).

The results above show that the thermal stability properties of La_2O_3 and Si interface can be significantly improved by the TiN/W gate electrode. The results also show that there is a great possibility to overcome the thermal stability problems of high-k materials on Si by developing a proper gate electrode structure.

Summary

The importance of the gate electrode structure in realizing EOT of 0.5nm and below was demonstrated. Our results show that a W layer inserted at the interface between TiN and La_2O_3 is the key factor in suppression of the EOT increment during high temperature annealing process. An EOT of 0.43nm was achieved with TiN/W(3nm) /La_2O_3(3nm)/n-Si gate stack after annealed at 800°C in a forming gas ambient. An EOT of 0.57 nm was also obtained with TiN(45nm)/W(3nm)/La_2O_3(3nm)/n-Si MOS capacitor even after an annealing process was performed at 700°C for as long as 30 minutes. Our results also showed the possibility to overcome the thermal stability problems in extremely scaled high-k gate stack with EOT of 0.5nm and beyond by developing a proper gate electrode structure.

Acknowledgments

This work was supported by NEDO.

References

[1]K. Choi, H. Jagannathan, C. Choi, L. Edge, T. Ando, M. Frank, P. Jamison, M. Wang, E. Cartier, S. Zafar, J. Bruley, A. Kerber, B. Linder, A. Callegari, Q. Yang, S. Brown, J. Stathis, J. Iacoponi, V. Paruchuri, V. Narayanan, *VLSI Tech. Dig.*, pp.138-139(2009).
[2]J. Huang, D. Heh, P. Sivasubramani, P. D. Kirsch, G. Bersuker, D. C. Gilmer, M.A. Quevedo-Lopez, M. M. Hussain, P. Majhi, P. Lysaght, H. Park, N. Goel, C. Young, C.S. Park, C. Park, M. Cruz, V. Diaz,P. Y. Hung, J. Price, H.-H. Tseng and R. Jammy, *VLSI Tech. Dig.*, p.34 (2009).
[3]K. Kakushima, K. Tachi, J. Song, S. Sato, H. Nohira, E. Ikenaga, P. Ahmet, K. Tsutsui, N. Sugii, T. Hattori, and H. Iwai, *Journal of Applied Physics*, **106**, 124903 (2009)
[4]K. Kakushima, P. Ahmet, and H. Iwai, *ECS Transactions,* **25 (7)**, 171 (2009)
[5]J. R. Hauser, K. Ahmed, AIP Conf. Proc., **449**, p.235 (1998).

ECS Transactions, 34 (1) 103-106 (2011)
10.1149/1.3567567 ©The Electrochemical Society

PMOS SOURCE/DRAIN EXTENSION DOPANT SPECIES EFFECT ON DEVICE AND SRAM PERFORMANCE

Jinhua Liu[a], June Zhou, William Wang, Rita Guo, Leo Zhang, ZhaoXu_Shen, Brisk_Wang, Allan Zhou, Humbert Hao, Jimmy Cui, Jay Ning

[a] Logic Technology R&D Center, SMIC (BJ) Beijing 100176, PRC
Jinhua_liu@smics.com

Source/drain extension implantation species are critical for both ultra shallow junction formation and device electric performance. Phosphorus (Ph) and Arsenic (As) are two species quite widely used for PMOS pocket implantation. Arsenic is easy to generate amorphization layer, which is beneficial for shallow junction formation. Moreover, Arsenic has better TED (transient enhanced diffusion) characteristics hence causes less implantation induced device variation. It is shown in the study that Arsenic improves 30% device Idsat variation compared to Ph implantation. SRAM Vccmin value may be improved around 4% too. While it is found that, compared to Ph implantation, Arsenic critically increases junction leakage current, which may account for more than half part of the total off current for high threshold voltage (Vth) device with gate length less than 50nm. The large junction leakage current also leads to 5% device performance degradation and SRAM stand by leakage current (Istby) 15% increasing. Device reliability study exhibits that PMOS negative bias temperature instability (NBTI) performance is deteriorated by Arsenic implantation. It is believed that the NBTI degradation attributes to Arsenic implantation induced defects.

INTRODUCTION

Source/drain extension implantation aims to achieve ultra shallow junction with low series resistance for continuous device scaling and device performance improving. Beyond the shallow junction and low parasitic series resistance, another requirement for source/drain extension implantation is reducing device off leakage current. Source/drain extension implantation species play an important role in device junction leakage current. For devices with high threshold voltage, junction leakage current may occupy nearly half of the total off current. Souce/drain extension species may determine the device off leakage current level for those devices with high threshold voltage. Phosphorus (Ph) and Arsenic (As) are two dopant species generally used for PMOS pocket implantation. As has better TED (transient enhanced diffusion) effect than ph, hence is easy to form shallow junction and behaves less device variation (1,2).

In the paper, a hybrid PMOS pocket implantation (As+Ph) and single pocket (Ph only) are comprehensively studied with a 55nm CMOS technology. Device performance and off leakage current level are compared. Implantation species impact to PMOS NBTI (negative bias temperature instability) features is evaluated. Moreover, 16M SRAM

Vccmin properties and stand by leakage current are characterized also for a complete understanding of PMOS pocket implantation species effects.

EXPERIMENT

Fig.1 briefly shows the fabrication process flow of the paper. Device goes through the STI isolation, gate stack, spacer, and source/drain implantation. Equivalent gate oxide thickness is around 21Å. After poly pre-doping and gate stack etch, source/drain extension implantation is performed. BF2 is adopted for LDD implantation. Hybrid pocket (As +Ph) and single pocket (Ph only) implantation are then followed. Spike RTA annealing is carried out for dopant activation at 1050°C after deep source/drain implantation. NiSi is used as silicide in the process.

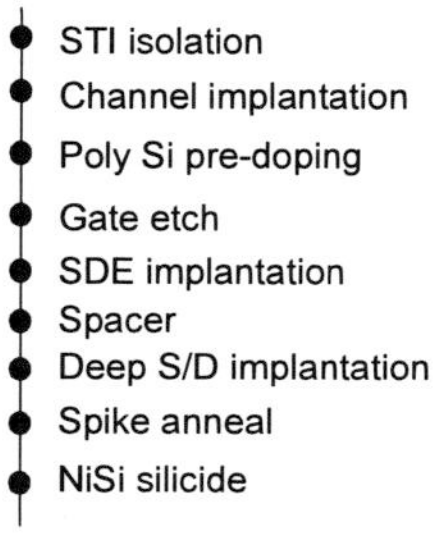

- STI isolation
- Channel implantation
- Poly Si pre-doping
- Gate etch
- SDE implantation
- Spacer
- Deep S/D implantation
- Spike anneal
- NiSi silicide

Fig.1 Process flow in the work

RESULTS AND DISCUSSION

<u>3.1. Device performance</u>

Fig.2 shows PMOS Ion-Ioff characteristics at Vdd =1.0V for process with hybrid pocket and single pocket implantation. It is clear that the device with single pocket displays better device performance than hybrid pocket. The performance difference does not owe to Vth as shown in Fig.3, the two devices' Vth are very similar. The device performance difference is answered in Fig.4, where junction leakage is displayed. Single pocket device shows much smaller junction leakage current than hybrid pocket device. As the devices' threshold voltage is quite high, junction leakage current plays an important role in the total off leakage current. In long channel device, junction leakage dominates the total off current. In short device, junction leakage current occupies nearly 40% of total off current. It is the reduced junction leakage current that leads to the improved device performance.

The junction leakage current difference may be explained like this. It is well known that As implantation has better TED features than Ph, which is helpful for shallow junction formation and results in a steeper doping profile meanwhile. The steep doping profile tends to increase the junction leakage current. Another reason for the large junction leakage current is the residual EOR (error of range) defects created by the As implantation. Despite the high temperature annealing in the following process, the defects could not be thoroughly removed (3).

Hybrid pocket implantation also impacts PMOS NBTI performance. Device Idsat variation under constant gate bias stress is monitored for NBTI performance evaluation. As shown in Fig.5, hybrid pocket exhibits higher Idsat shift than single pocket under same gate bias stress condition, which indicates a worse NBTI characteristics. It is speculated that interface states is induced between gate oxide and silicon surface during the tilt angle As pocket implantation. The interface states are not fully recovered during the subsequent process and deteriorate NBTI characteristics in the end (4).

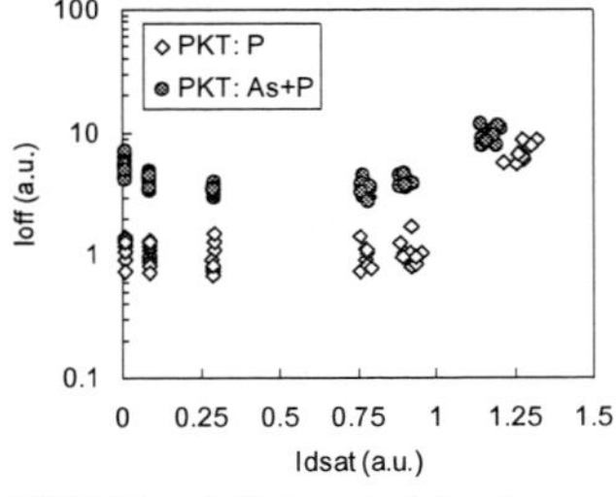

Fig.2 PMOS Ion-Ioff characteristics for process with hybrid pocket and single pocket

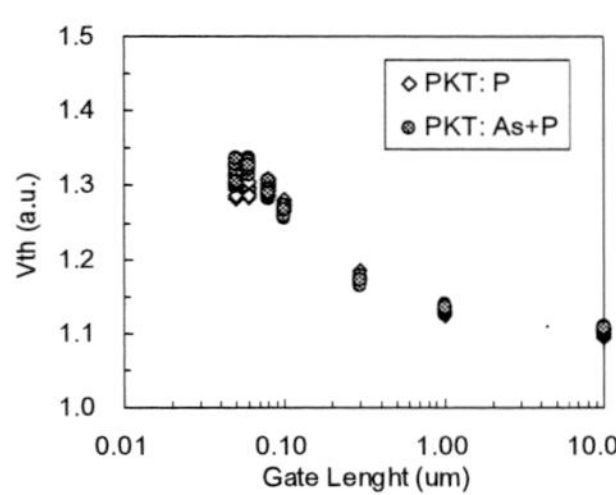

Fig.3 PMOS Vth roll-off characteristics along gate length for process with hybrid pocket and single pocket

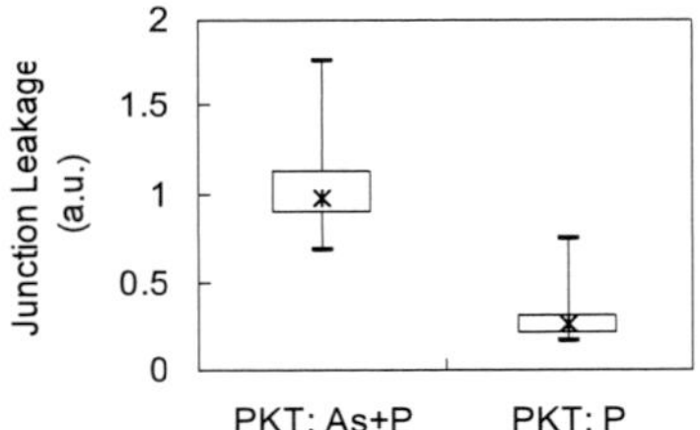

Fig.4 Junction leakage current of process with hybrid pocket and single pocket

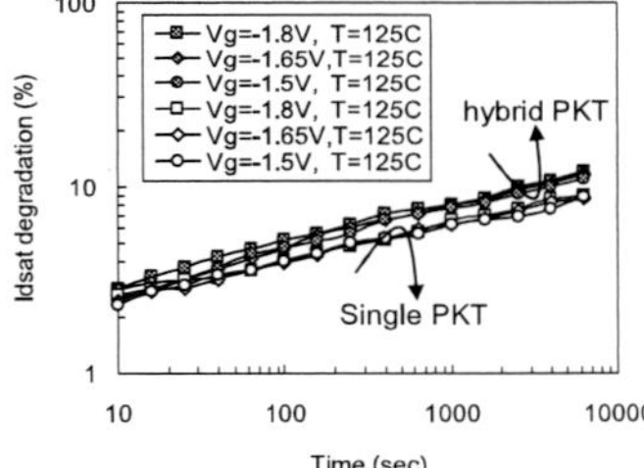

Fig.5 PMOS Idsat degradation under constant gate stress for process with hybrid pocket and single pocket

3.2. SRAM Characteristics

16M High density SRAM array has been characterized for a complete study of PMOS pocket implantation species effects. Vccmin distribution is given in Fig.6. Vccmin is defined as the minimum voltage SRAM needs to perform the required read/write function. Vccmin spread level is generally adopted to reflect the process and device variation. It is shown in Fig.6 that Vccmin median value is very similar for hybrid and single pocket processes, except the spread of the single pocket is larger than hybrid pocket. The result indicates that single pocket implantation process behaves larger variation. PU Idsat variation of the two processes proves the point as shown in Fig.7.

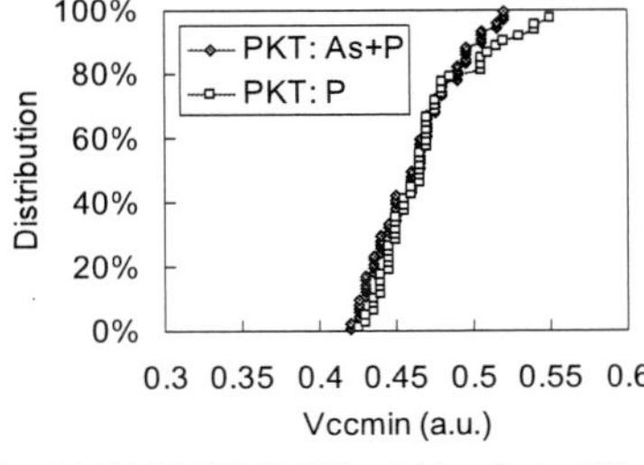

Fig.6 16M SRAM (0.525um2 bit cell size) Vccmin for process with hybrid pocket and single pocket

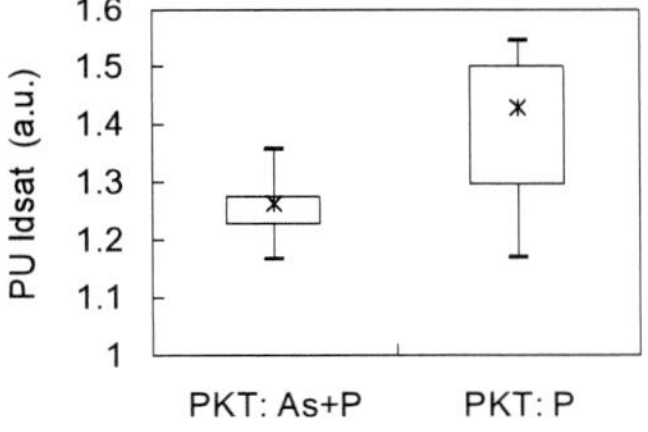

Fig.7 PU transistor Idsat variation for process with hybrid pocket and single pocket

Though hybrid pocket implantation shows better Vccmin distribution and smaller device variation, SRAM stand by leakage current of the process is larger. The result is consistent with device performance discussed above. It is the high junction leakage current caused by the hybrid pocket implantation that leads to the high SRAM stand by current. Thus from SRAM standy by leakage current point of view, hybrid pocket implantation is not advised for PMOS device tuning.

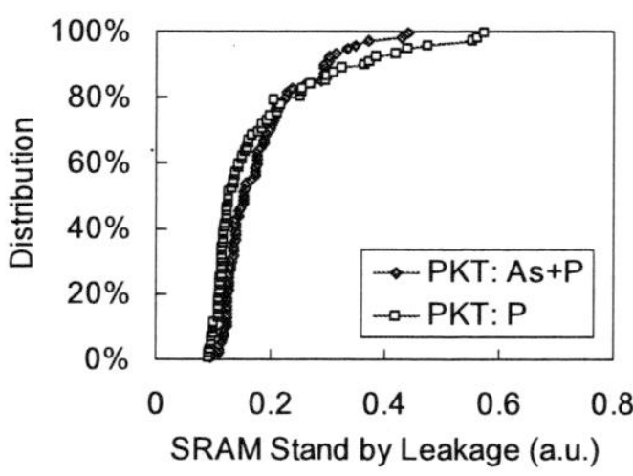

Fig.8 16M SRAM stand by leakage current for process with hybrid pocket and single pocket

CONCLUSION

PMOS hybrid and single pocket implantation process are examined. The experiments data shows that hybrid pocket implantation degrades PMOS device performance due to high junction leakage current of the process. The reason of the high junction leakage current is steep doping profile and EOR defects induced by the As implantation. Hybrid pocket process also shows degraded NBTI performance. However, it is demonstrated that hybrid pocket process has better SRAM Vccmin performance and smaller device variation compared to single pocket implantation process. Combining the process complexity and limited advantages hybrid pocket introduces, single pocket implantation is preferred in general CMOS process.

References

1. c.c. Wang, Simulation of Semiconductor Processes and Devices, 2003, pp 255- 258.
2. R.W.Mann et al, IBM J. Res.&Dev. Vol.47 No.5/6 2003, pp.553-563.
3. K. Roy et al, Proceedings of the IEEE, Vol.91, No.2, FEBRUARY 2003, pp.305-327.
4. John Borland, Junction Technology, 2006. IWJT '06, Page(s): 4 - 9

ECS Transactions, 34 (1) 107-111 (2011)
10.1149/1.3567568 ©The Electrochemical Society

A Novel Tunnel Oxide Based Tunnel FET

Hefei Wang, Zhijiong Luo, Haizhou Yin, Huilong Zhu, Jia Liu, Zhengyong Zhu

Institute of Microelectronics of Chinese Academy of Sciences, Beijing 100029,
P.R.China
Email: luozhijiong@ime.ac.cn

In this work, we propose a novel tunnel oxide based Tunnel FET. This novel Tunnel FET can achieve high drive currents, low SS, low off currents and many other superb device characteristics. From device simulation, this novel device can obtain drive currents 100x higher than that of a conventional Tunnel FET. This novel device also can realize SS as low as 30mV/dec. Moreover, this novel device has potential to significantly reduce off currents, and suppress the ambipolar behavior existing in conventional Tunnel FETs. In addition, the novel device can obtain the ability to scale down without performance degradation due to the unique scalability of tunnel oxide. These excellent device characteristics make the novel Tunnel FET an ideal device structure for future LSTP applications.

Introduction

The increasing power consumption has become a huge problem for continued CMOS scaling. One big contributor for the power problem is the difficulty of reducing the supply voltage, which is mainly due to the non-scalability of the subthreshold swing. Among various novel devices that have been proposed to address this issue, Tunnel FET based on carrier injection via tunneling, is considered the most promising alternative, because it offers the possibility to lower subthreshold swing, thus supply voltage as well.

However, so far, the experimental results have shown that the drive currents of Tunnel FETs are unacceptably low (1). The main reason is the difficulty of fabricating the critical ideal abrupt junctions in conventional Tunnel FETs. Additionally, Tunnel FETs suffer from severe ambipolar characteristic (2), due to which the devices can't be properly turned off.

In order to address these shortcomings, we propose a novel tunnel oxide based Tunnel FET employing an ultra-thin tunnel oxide layer between source and channel to block dopant diffusion and to enhance drive currents. Moreover, the novel device can successfully suppress the ambipolar characteristic, and realize low off currents and low SS.

Device structure and parameters

Fig. 1 shows the schematics of one conventional Tunnel FET and two novel tunnel oxide based Tunnel FETs, one has a vertical tunnel oxide layer, and the other has a titled one. For the conventional Tunnel FET shown in Fig. 1 (a), the source and drain are doped into different polarity, a PN junction is formed on the source side of channel. For the novel Tunnel FETs, a thin tunnel oxide layer is inserted between the source and channel, to block dopant diffusion and to enhance drive currents. The thickness of this ultra-thin tunnel oxide layer is usually less than

2nm for the purpose of increasing tunneling currents. Si_3N_4 is the preferred material for this critical insulator layer, due to its outstanding dopant diffusion blocking ability (3).

The process flow to fabricate the novel structure of the Tunnel FET with vertical tunnel oxide layer is shown in Fig. 2 (a) to Fig. 2 (d). The silicon on the source side is first etched away, while the drain side and gate are protected. And then, the source is refilled with conducting materials after growing a layer of tunnel oxide in the recessed source. Fig. 3 shows the final cross section of the novel Tunnel FET after process simulation. For the Tunnel FET with tilted tunnel oxide layer shown in Fig. 1 (c), after the source is etched away, another step of etching, which has high selective ratio over different crystal orientations, is implemented to shape the channel on the source side. Followed by the same methods to grow tunnel oxide layer and refill the source as mentioned above. Benefiting from this tilted tunnel oxide layer, we can further increase the drive currents and reduce performance degradation caused by current sinking problem (4).

Process and device simulations are carried out on Synopsys Sentaurus simulator. The typical Tunnel FETs used in this study are p-channel devices. Table 1 lists detailed device parameters for these transistors.

Simulation results and discussion

The novel tunnel oxide based Tunnel FETs are extensively studied, and compared with conventional Tunnel FETs through simulation.

In order to verify that the novel PIN structure has the potential for drive currents improvement, we compared the I-V curves of a reverse biased PN diode along with a PIN diode sandwiching a 1nm Si_3N_4 layer between the same P type and N type regions. Diffusionless laser

anneal simulation is used for the junction anneal, in order to achieve abrupt dopant profile. As shown in Fig. 4, when voltage is larger than 0.2V, the tunneling currents through PIN diode are significantly higher than that through PN diode. The larger tunneling currents indicate that PIN diode has much potential to increase drive currents.

Diffusionless laser anneal simulation is used for the conventional Tunnel FET fabrication process as well. Fig. 5 shows the simulated doping contours in the source region of the conventional Tunnel FET, it can be seen that, laser anneal can realize dopant profile less than 1nm/dec in the lateral direction.

Fig. 6 exhibits the transfer characteristics of the conventional Tunnel FET and two novel Tunnel FETs. V_T is defined by constant current method. Clearly, I_{ON} of the conventional Tunnel FET saturates at merely $1x10^{-8}$A/um. Nevertheless, the novel Tunnel FETs realize more than two orders of magnitude drive currents enhancement. Fig. 7 shows the output characteristics of two novel Tunnel FETs. It can be seen that, Tunnel FET using tilted tunnel oxide layer can obtain more drive currents enhancement, which can be explained by the reduction of barrier width. Coupling between the gate and source will deplete the source surface close to the gate edge. Therefore, carriers in the source side are forced to tunnel through a certain distance into the body, which makes the barrier wider. Since the tilted tunnel oxide layer can make the carriers pass through at the direction perpendicular to its surface, the tunnel distance can be consequently reduced.

Also, the off currents of novel Tunnel FETs are reduced by approximately 100x, to less than $1x10^{-17}$A/um. It can be explained by the formation of an extra barrier between source and channel in the PIN structure based novel Tunnel FETs. Normally in the off stage, currents derive from the diffusion of minority carriers. In the novel Tunnel FETs, these carriers will encounter a potential barrier formed by the tunnel oxide layer, instead of a single potential well, which is the case in the conventional PN junction based Tunnel FETs. As a result, fewer carriers will contribute to the off currents.

Low SS is the most favorable merit of Tunnel FETs, which offers the possibility to lower the supply voltage in future technology nodes. In Fig. 8, it can be seen that the novel devices can achieve SS as low as 30mV/dec, which is comparable to that of conventional Tunnel FETs.

One major drawback of conventional Tunnel FETs comes from their ambipolar behavior, which can increase the off currents and lead to increasing power density. Some solutions, such as adding gate drain underlap (2) and introducing hetero gate dielectric (5), are proposed. But these methods will either decrease integration density or add fabrication complexity. Our novel devices can successfully suppress the unwanted ambipolar behavior due to the presence of the tunnel oxide layer, which can cut off the current path when gate bias is reversed. As shown in Fig. 9, when a positive gate voltage is applied on these p-channel devices, the conventional Tunnel FETs show high leakage currents as expected. However, the currents of the novel Tunnel FETs show no increase, and keep negligible low across the whole range.

We also studied the scaling ability of the novel Tunnel FETs by means of tunnel oxide scaling. As can be seen in Fig. 10, when the tunnel oxide thickness is reduced to 8 angstrom and 4 angstrom from 10 angstrom, the drive currents can be doubled and tripled, respectively. This unique scaling tool can ensure the device outstanding characteristics in further scaling down device sizes.

Conclusion

Our work presents a novel tunnel oxide based Tunnel FET structure. This novel Tunnel FET shows many desired device characteristics such as high drive currents, low SS and low off currents. From device simulation, this novel device can obtain drive currents 100x higher than that of a typical conventional Tunnel FET. This novel device also can realize SS as low as 30mV/dec. Furthermore, this novel device can significantly reduce off currents and suppress the ambipolar behavior existing in conventional Tunnel FETs. In addition, tunnel oxide scaling in the novel Tunnel FET can ensure the devices excellent performance in further scaling down device sizes. These excellent device characteristics make the novel Tunnel FET very promising for future LSTP applications.

Acknowledgement

Authors would like to thank Dr. Qingqing Liang for useful discussion. This work is sponsored by the Hundreds People Project of Chinese Academy of Sciences, and by 02 National S&T Major Project.

Reference

1. C. Sandow, J. Knoch, C. Urban, Q.-T. Zhao, S. Mantl, Solid-State Electronics 53 (2009), p. 1126–1129.
2. Anne S. Verhulst, William G. Vandenberghe, Karen Maex, and Guido Groeseneken, APPLIED PHYSICS LETTERS 91, 053102 (2007).
3. Y. Wu and G. Lucovsky, IEEE International Reliability Physics Symposium (1998), p. 70.
4. Costin Anghel, Prathyusha Chilagani, Amara Amara, and Andrei Vladimirescu, APPLIED PHYSICS LETTERS 96, 122104 (2010).
5. Woo Young Choi, Woojun Lee. IEEE TRANSACTIONS ON ELECTRON DEVICES, VOL. 57, NO. 9 (2010).

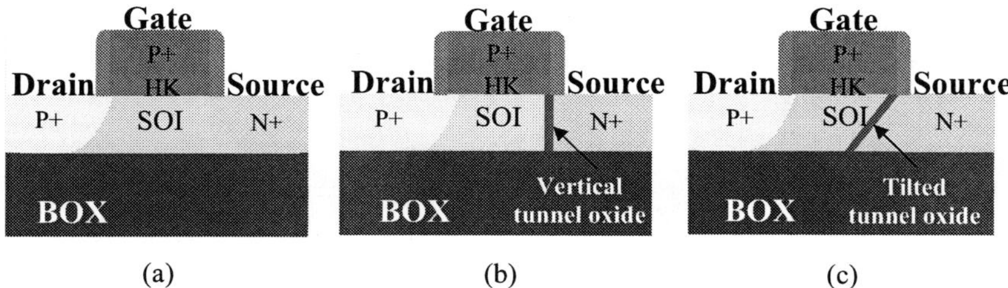

Fig. 1 Device structures of (a) a conventional Tunnel FET, (b) a novel Tunnel FET with a vertical tunnel oxide layer, and (c) a novel Tunnel FET with a tilted tunnel oxide layer. For the conventional Tunnel FET, the tunnel junction is formed on the source side of channel, while the novel Tunnel FETs feature a thin tunnel oxide layer between source and channel.

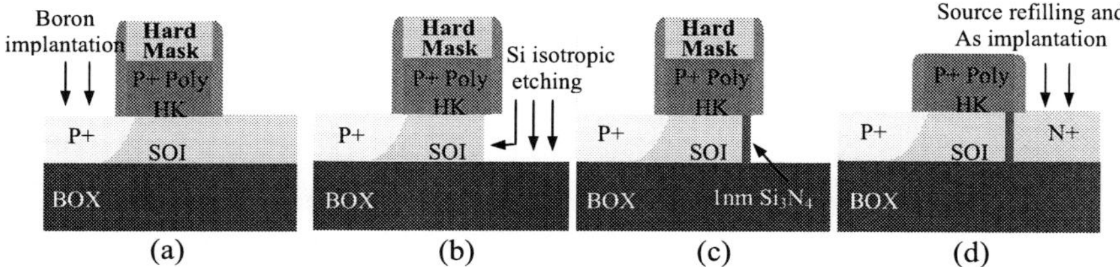

Fig. 2 Process flow to fabricate the Tunnel FET with vertical tunnel oxide layer.

TABLE 1. **Device parameters of the conventional Tunnel FET and novel Tunnel FETs**

Parameter	Value
Gate length	100nm
T_{ox} (HfO_2)	2.5nm
T_{SOI}	50nm
T_{BOX}	200nm
Gate doping concentration	Boron $1\times10^{20}(cm^{-3})$
Drain doping concentration	Boron $1\times10^{20}(cm^{-3})$
Source doping concentration	Arsenic $1\times10^{20}(cm^{-3})$
Channel doping concentration	Boron $1\times10^{15}(cm^{-3})$
Tunnel oxide thickness (Si_3N_4)	1nm

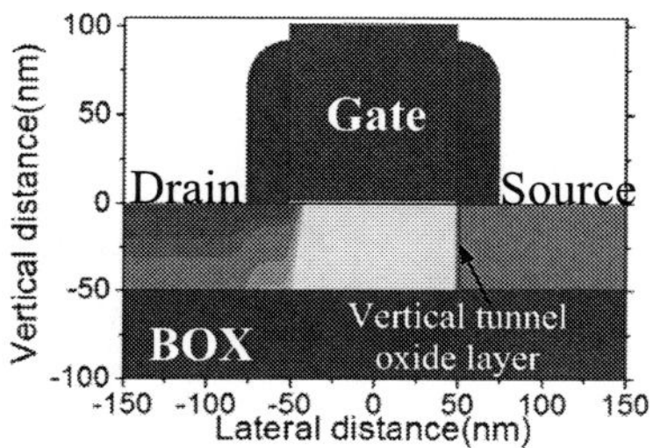

Fig. 3 Final cross section of the novel Tunnel FET after process simulation.

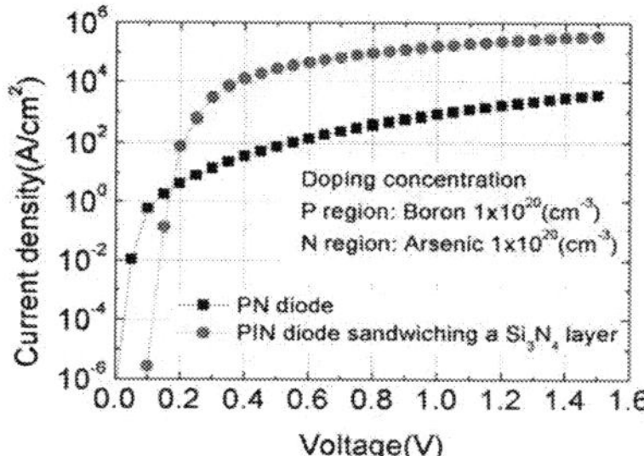

Fig. 4 Tunneling current density of a conventional PN diode and a PIN diode sandwiching a 1nm thick Si_3N_4 layer. Current density is significantly improved in the PIN diode.

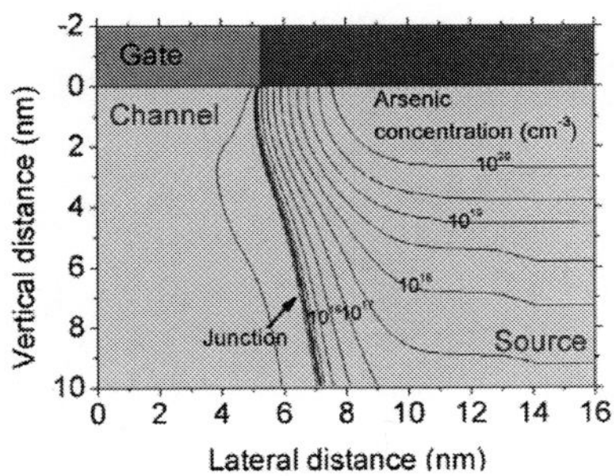

Fig. 5 Arsenic doping concentration contours in the source region of the conventional Tunnel FET. Dopant profile is less than 1nm/dec in the lateral direction after using laser anneal.

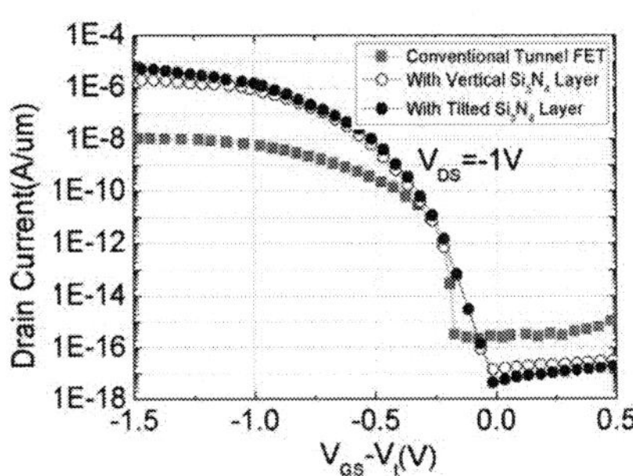

Fig. 6 Transfer characteristics of novel Tunnel FETs and a conventional Tunnel FET. V_T is defined by constant current method. More than 100x drive currents enhancement can be obtained in the novel Tunnel FETs.

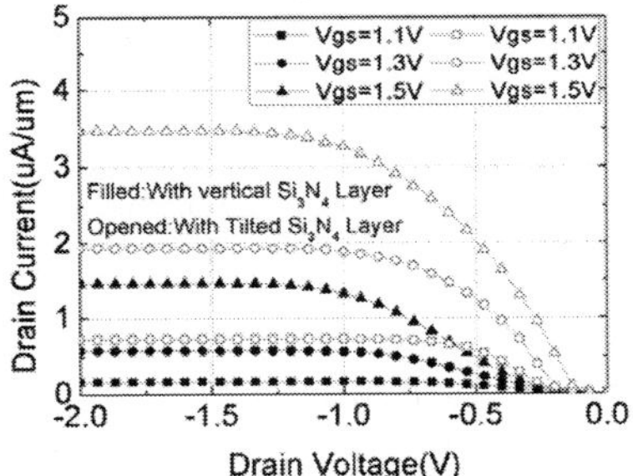

Fig. 7 Output characteristics of two novel Tunnel FETs. Tunnel FET using tilted tunnel oxide layer can achieve more drive currents enhancement.

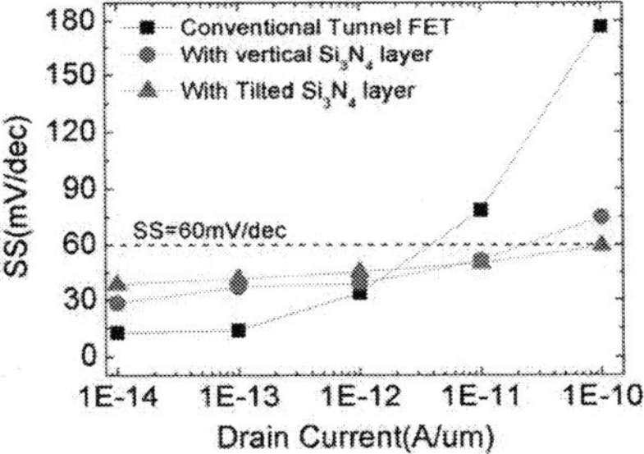

Fig. 8 SS of conventional Tunnel FETs and two novel Tunnel FETs. SS as low as 30mV/dec can be obtained in the novel Tunnel FETs.

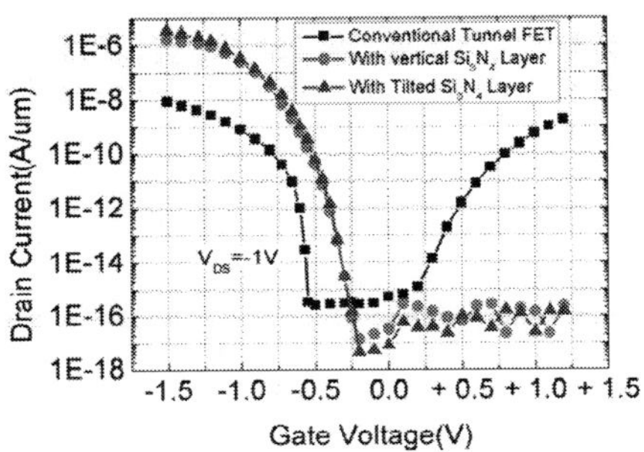

Fig. 9 I_D-V_G curves with V_G ranging from +1.5V to -1.5V. In the positive gate voltage region, the conventional Tunnel FET shows high leakage currents due to severe ambipolar behavior.

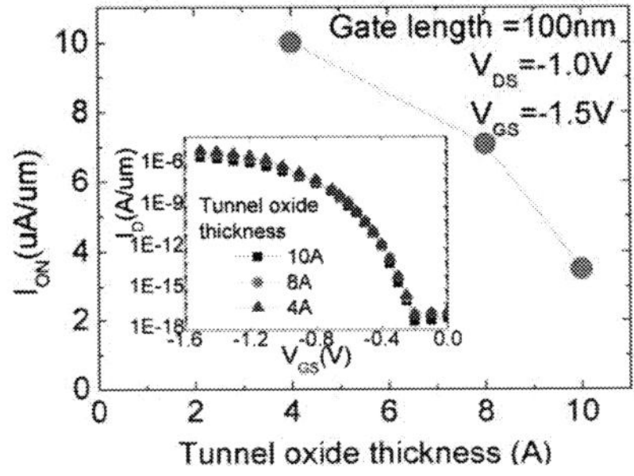

Fig. 10 I_{ON} of novel Tunnel FETs with tilted tunnel oxide layer vs. different tunnel oxide thicknesses. I_{ON} increase as tunnel oxide thickness decreasing. Inset is the I_D-V_G curves of Tunnel FET with different tunnel oxide thicknesses.

ECS Transactions, 34 (1) 113-117 (2011)
10.1149/1.3567569 ©The Electrochemical Society

STI CMP: Exploration of a Colloidal Silica Based Slurry System

Peter Song [1,a], Daisy Yao [1,b], Jack Daw sun [1c]

[1] Anji Microelectronics (Shanghai) Company Shanghai 201203, China

Abstract

With the size shrinking of IC transistors, more stringent polishing performance in shallow trench isolation (STI) CMP processes is requested in advanced fabrication processes. Up to date, most available STI slurries are ceria based. Technical challenges, such as scratches, rate instability, and handling, are the challenges to overcome. In addition, lower CoO is also desired. In this paper, colloidal silica based STI slurries are explored as an alternative. Different types of colloidal silica abrasives and chemicals are examined to understand their effects on HDP oxide removal rate and polishing selectivity of silicon nitride to silicon oxide. Planarization efficiency and slurry colloidal stability are also investigated. The results show that an abrasive type can make great difference on the removal rates of silicon nitride and silicon oxide. Certain proprietary chemicals can also greatly increase the selectivity of silicon oxide to silicon nitride without significant oxide losses in trenches. The capability of step height repairing can be tuned by carefully selected additives. The possible mechanism is also proposed. In addition, slurry handling can be improved by special designed dispersion agents. The preliminary data from the studies indicate that it is feasible to develop a silica based STI slurry to replace the traditional ceria based STI slurry. It also shows the potential to offer additional performance and cost benefits.

Key words: STI, selectivity, planarization efficiency

Instruction

STI isolation technology is adopted in the fabrication of advanced integration circuit for its lower leakage current and junction capacitance compared with LOCOS processes. Also, the less area of field oxide can avoid the bird beak issue at the edges of oxide layers occurred in previous isolation and then form a smaller isolated zone to allow the higher device density[1]. With the development of a deep submicron process, the technical requirements on STI planarization become more stringent in rate and selectivity tuning ability, defectivity control, within-die uniformity minimization, and reduction of total CoO. Traditional STI CMP slurries use ceria based abrasives which have disadvantages such as micro scratches, polishing rate variation, storage instability, handling complexity, and cost pressure, although it has higher HDP oxide removal rates and good selectivity to silicon nitride, and higher planarization efficiency[2,3]. Many other metal oxide abrasives were also studied in STI polishing [4,5], but no commercialized applications were reported. In this paper, colloidal silica based slurries were investigated for higher polishing performance and lower cost. Different types of colloidal silica particles and special designed chemical additives were studied on their effects on HDP silicon oxide/nitride removal rates. The effects of proprietary chemicals on the planarization efficiency are also optimized. And a possible mechanism of selectivity variation and step height repairing were also proposed. The results showed that the polishing rate of HDP oxide and selectivity to oxide/nitride can be adjusted to match the current baseline. The step height repairing capability can be greatly improved by applying certain surfactants.

Also, the storage stability is better than ceria based slurry. Therefore, it is potentially possible to realize the commercial application on STI CMP.

Experimental

Wafer polishing for formulation screening was on rotational polishing tool, Logitech LP50/PM5. The film thickness of HDP oxide and silica nitride was measured on Nano Spec 6100. Post CMP clean was done on an auto-cleaner using DIW. Step heights were measured on a scanning AFM X-300 。

HDP silicon oxide (high density plasma oxide with 15K thickness) and silicon nitride wafer with 2K thickness were used for removal rate collection and selectivity evaluation. The patterned wafers were from customers with a common film stack.

The polishing conditions are as follow:

Polishing pad: IC1000 pad   Down-force: 4.0psi

Polishing head rotation speed: 70rpm   Platen rotation speed: 90rpm

Flow rate: 100ml/min   Polishing time: 1-2min.

Colloidal silica and ceria abrasive samples and chemical additives are available on market.

Results and Discussions

Abrasive effects on the HDP oxide removal rate and selectivity to silica nitride

HDP oxide is a very stable silica compound with higher hardness. A high solid loading usually can speed up removal rates with increasing mechanical force. However, abrasive shape, mean size, and size distribution shows significantly different performance in rate enhancement. As a example, several kinds of particles in different size and size distribution were used for polishing HDP oxide, Fig.1 showing their removal rates of HDP oxide with various solid loadings. The results indicates that larger particles with wider distribution result in a rapid increase on HDP silicon oxide removal rates, and smaller particles yield relatively lower removal rate. As shown in Fig.2, a possible reason is that a larger mean size and wide distribution generates larger contact areas caused by inter particles space filling and stronger deformation caused by down-force and shearing strength, which can take out more materials an give a higher removal rate.

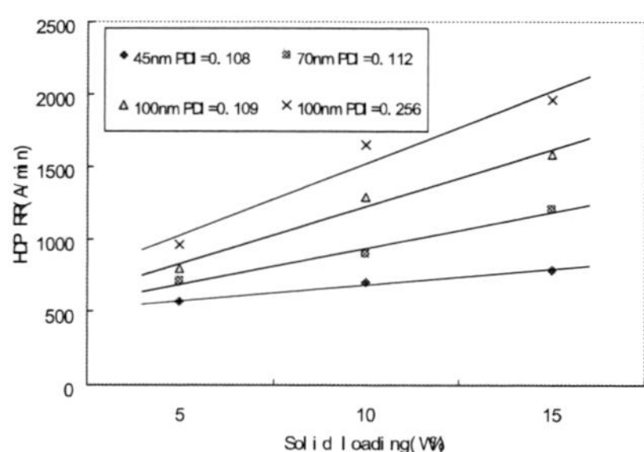

Fig.1   HDP oxide and silica nitride RR under different mean size and solid loading.

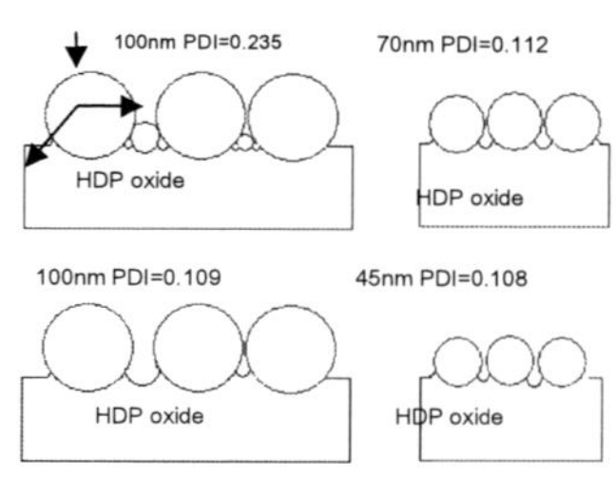

Fig.2 The model of interaction between particle to surface

Effects of surfactants on the selectivity of HDP oxide to silicon nitride.

Since silicon nitride has a similar hydrophilic surface to silicon dioxide and similar interaction of hydroxyl group between film surfaces and particle surfaces. Both removal rates tend to behave similiar. Therefore, in order to meet the high selectivity of HDP oxide to silicon nitride, we select several series of ionic surfactants to suppress nitride rates and keep HDP oxide rates stable. The following graph presents some polishing results, which indicates that certain structures in surfactants can greatly decrease silicon nitride rates but minimally affect HDP oxide rates. The selectivity is close to 20:1 which is higher than the typical ceria based slurry. Due to the weaker ionic properties of OH-Si-N in comparison with OH-Si-O, we deduce that there are more ionic groups of surfactants adsorbed by conjugation on the surface of silica nitrides than on the surface of oxides. As a result, less particle hydroxyl groups react with those of Si_3N_4 surface, compared those with oxide. The removal rates of Si_3N_4 are decreased to a very low level.

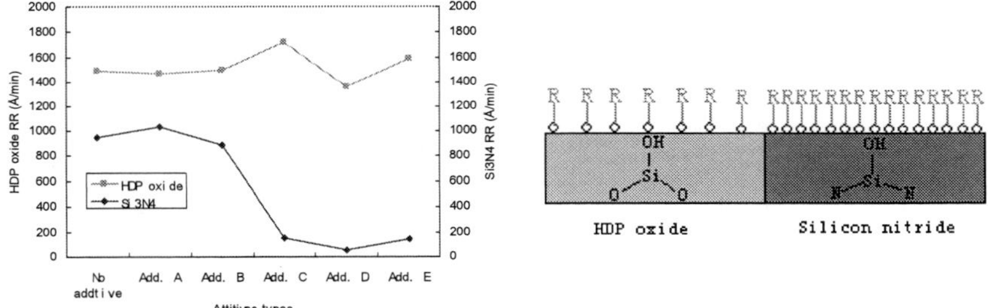

Fig.3 HDP oxide and silica nitride removal rate using different surfactants and the possible adsorption model

Effects of surfactant to step height correction ability

Since the step height and topography with concave and convex areas exist on a patterned wafer surface before polishing, local pressure distribution and temperature effects caused by friction during polishing are usually different than those on a blank wafer. Therefore, the actual material removal rate and selectivity on patterned wafer often vary at different degrees even the polishing conditions remain constant. The column map in Fig.5 shows that the step height variation before and after polishing in the presence of different chemical additives. It is found that even with the high HDP oxide removal rates and high selectivity of oxide/nitride, it only shows a small topography correction after patterned wafer polishing with additives A or B. However, chemical C clearly shows difference than others. With applying chemical C, the step height and dishing values decrease significantly, and the data show a higher planarization efficiency. Based on adsorption data of chemical C on HDP oxide surface measured by EDX analysis, we propose a model to demonstrate the planarization mechanism in Fig. 6. During the polishing, due to a lower surface tension, the surfactant molecules with different electrical charges adsorb or saturate on blocks or into gap areas to keep the surface from removing, which results in a 10-20% rate decrease, higher mechanical force at the top zone compared with concave area will yield the significant removal rate difference than the concave area. As a result, the step height and dishing are corrected to meet the performance requirements. The Fig.7 shows the profiles before and after polishing. The step height variation percentage indicate that the surface is almost planarized after polishing.

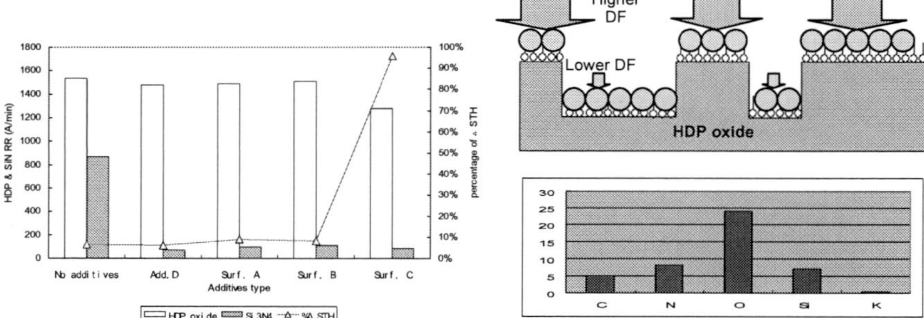

Fig4: Removal rate and selectivity of oxide/nitride and planarization efficiency under different chemicals

Fig. 5 The mechanism model of step height correction and surface EDX analysis results

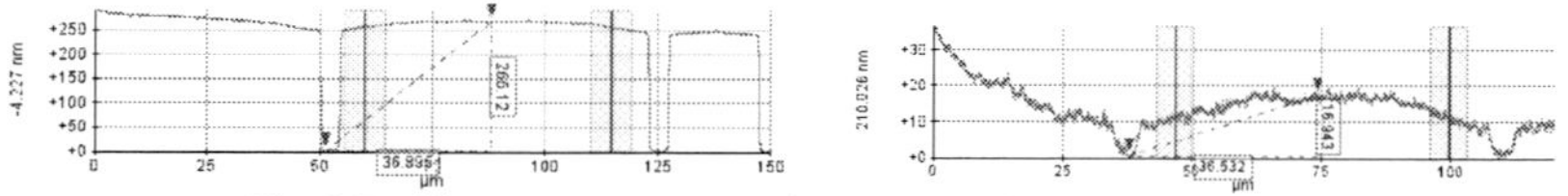

Fig.6 Profiles on measurement site before and after STI polishing

Polishing stability comparison with ceria based slurry

The variation of polishing rates along with time is a key factor for manufacturing viability. Precipitation of ceria based slurry during storage is well known. The soft aggregation can be recovered by agitation before online usage. In Fig. 8. we made a comparison between an optimized colloidal silica based slurry and a ceria based slurry in trend of polishing rate variation along with the storage time. The results shows that colloidal silica slurry is stable on HDP oxides. Due to the sedimentation of ceria, the slurry requires stirring before the usage, and a rate drop of about 30% was observed after 60 days storage.

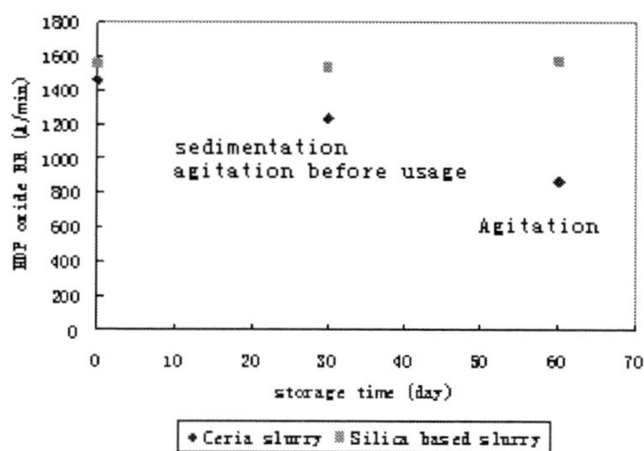

Fig. 7 Comparison of HDP oxide and slurry stability of ceria based and silica based slurry

Conclusion

1) Colloidal silica based STI slurries with a large particle (100nm) with a wide mean size distribution can yield high HDP oxide removal rates.
2) Certain surfactants can significantly increase the polishing selectivity of HDP oxide to silicon nitride..
3) Certain proprietary chemical additives can enhance the step height correction capability and obtain a high planarization efficiency
4) Removal rate variation trend of a colloidal silica based slurry could potentially show a better stability than a ceria based slurry.
5) The silica based colloidal silica slurry has a potential to realize the industrial application for its improved performance and cost of ownership.

Acknowledgments

The authors would like to acknowledge the contributions from R&D group at Anji Microelectronics for their technical supports and collaborations.

Reference

1. James D. Plummer, Michael D. Deal, Peter B. Griffin, Silicon VLSI Technology (fundamentals, Practice and Modeling) 2005, P47
2. David Merricks, Brian Suntora, Evolution and Revolution of Cerium Oxide slurries in CMP , ISTC, 2008
3. Dae Hyeong Kim, Gyeonggi-Do (KR), Slurry for CMP and method of polishing substrate using same, US 7364600 B2
4. Ramanthan Srinivasan, Slurry for Chemical Mechanical Polishing of Silicon Dioxide, US 6627107 B2.
5. Venigalla; Rajasekhar (Wappingers Falls, NY), Hannah; James W. (Ossining, NY), Ceria-based polish processes, and ceria-based slurries, US 7056192, 2006

ECS Transactions, 34 (1) 119-124 (2011)
10.1149/1.3567570 ©The Electrochemical Society

Linearity Improvement on MIM Capacitors

T.P. Chu[a], P. Yang[b], Evie Kho[b], Y.K. Ang[a], and S.H. Tia[a]

[a] Process Characterization; [b] Process Development, X-FAB Sarawak Sdn. Bhd.

A study has been carried out to improve metal-insulator-metal (MIM) capacitor's capacitance density and linearity performance. The scopes of the study included single MiM and stack MIM structures. Different dielectric schemes were evaluated with their corresponding capacitance density, breakdown voltages and linearity coefficient to voltage and temperature variation etc. characterised. Trade-off of capacitance density and linearity observed with respect to the dielectrics thickness. In this paper, a special layout design and a new integration concept from stack MIM device will be introduced, with this, we can achieve both high capacitance density as well as ultra low voltage coefficients of capacitance for the desirbale good analog performance on MIM passive devices.

Introduction

In the area of mixed signal analog application devices, precise performance of passive devices required across technologies. The importance for good quality of passive components commonly used for filters, mixers, dividers, converters ...etc, is as equally crucial as the characteristics of active devices in amplifiers, oscillators...etc. MIM (metal-insulator-metal) capacitors are widely used as compared to PIP (poly-insulator-poly) or MOS devices (varactors), as the latter exhibit depletion effects, the impact of associated parasitic capacitance cause undesirable capacitance variations with voltage bias fluctuations. Inherent advantages with MIM using metal plates provides depletion-free, high-conductance electrodes suitable for high speed applications at low cost. Quality of MIM depends on how high capacitance (storing electrical energy in terms of charges) without occupying too much area, i.e. unit capacitance density, and the stability of the MIM maintaining precise values independent of operating conditions, shifts in capacitance with different operating frequencies, known as dispersion, also lead to high distortion in analog signals, whereas linearity determine the changes in capacitance with bias voltage.

As such, for high performance analog circuits, desirable characteristics on MIM possess low voltage and thermal linearity, non-dispersive characteristics, low leakage, high capacitance density as well as with high breakdown field strength. In this study, different MIM capacitor schemes – oxide, nitrides, oxynitrides and its combinations .etc.) will be evaluated on their capacitance properties and linearity performance. In addition, the dependence of dielectics' thickness, layout architecture, capacitor area, stack effect will be discussed; a special design cross-coupling MIM primitive device with ultra low voltage coefficients ($< 1ppm/V^2$) as well as high capacitance density (upto ~ 5 fF/μm^2) will be demonstrated at the later session of this paper.

Results and Discussion

Linearity Characterization

Capacitance of both MIM and stacked MIM are measured using Agilent LCR meter 4284, by applying DC voltage sweep from -5V to 5V in step of 0.1V, AC signal amplitude configured to 50mV and 100 kHz; Capacitor voltage linearity is sensitive to the material properties of the plate and dielectrics composition. The normalized capacitance can be modeled as follows:

$$\frac{dC}{C_O} = \frac{C(V) - C_O}{C_O} = \alpha V^2 + \beta V \qquad [1]$$

where C_o is the capacitance at zero voltage and V is the voltage applied between the MIM electrodes. α and β are the quadratic and linear coefficients of the MIM capacitor in ppm/V^2 and ppm/V respectively

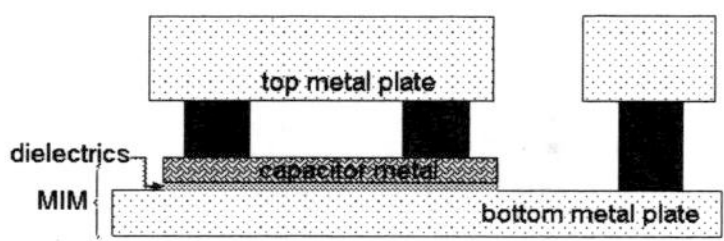

Figure 1. Formation of the MIM capacitor by a PECVD dielectric layer deposition on top of a bottom metal layer, while top plate is another thinner capacitor metal patterned using one additional masking step.

So in order to make an ideal MIM capacitor; the target will be to achieve high C_o as well as small α and β.

Normalized C-V curves on various dielectrics

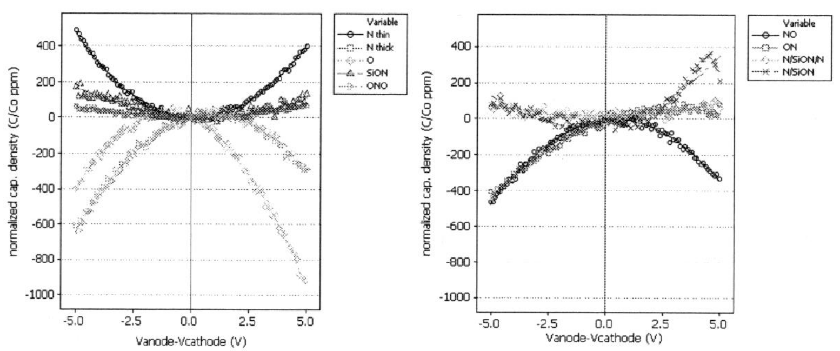

Figure 2. Normalized C-V (capacitance vs. bias voltage) curves from various dielectrics. Single MIM layer was measured, with ON , NO, ONO, N/SiON, N/SiON/N etc., refer to deposition layer by layer in sequence of processing individual layers of dielectrics - "oxide", "nitride" … etc.

<u>Trade off relationship on linearity vs. capacitance density:</u> Figure 2 shows the characterization linearity curves of normalized capacitance – bias voltage curves, as refer to eq. (1), the corresponding quadratic and linear coefficients can be extracted from each of the different dielectric curves. Figure 3 and 4 illustrated the characteristics of the capacitor's linearity variation with its corresponding capacitance density. Among all the different dielectrics with various compositions, a trade off relationship between the voltage coefficients and capacitance density does exist. (In general, trade off also had seen between capacitance and breakdown voltages). Nitride, SiON or the combination show alike properties with the curves concave upwards (corresponds to. +ve α), whereas ONO, ON, NO follow more to oxide kind of materials, with the curves concave downwards, (i.e. with opposite sign, –ve α). Oxide and Nitride shows opposite signs voltage linearity, while their combinations perform in between of the two materials. The degree of the linearity in fact is dependent on the respective dielectric thickness; with the thicker the dielectrics, the better the linearity, and thus the lower the capacitance. As such, the relations from linearity as well as capacitance can be modeled with dielectrics thickness accordingly.

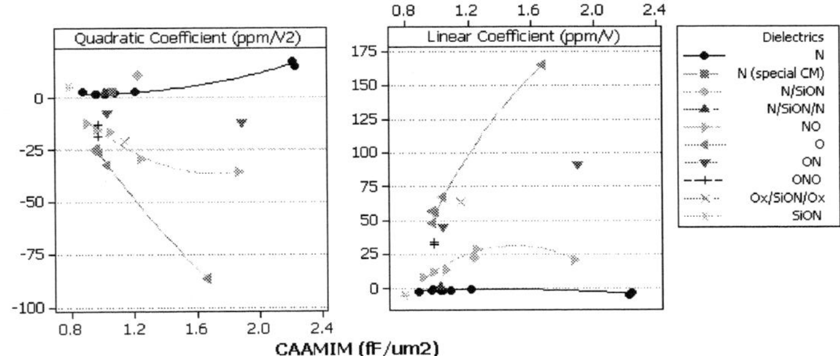

Figure 3 & 4. Quadratic coefficient α and linear coefficient β, of Voltage linearity of capacitance vs. different dielectrics in single layer MIM

<u>Optimization of stack MIM achieving both good capacitance density and linearity:</u> Quadratic coefficient, α determines the curvature of the C-V curves; whereas linear coefficient β governs how the curve shift with reference to the 0 bias x-axis, the smaller for both coefficients, the better linearity performance, little dependence of capacitance among voltage, i.e. can maintain precise values no matter under what operating conditions of the devices.

Reduction of the linear term of VCC observed by means of the stack structure in the DMIM (double MIM); the two terminals of the resultant stack MIM are connected in a way with one terminal linked the top plate and bottom plate for upper and lower individual MIM layer respectively, and vise versa for the other terminal, this results in a feedback loop canceling out the effect of linear term of voltage coefficient of capacitance, β, reduced almost close to 0ppm/V.

In general, to have a simple interpretation, varying the voltage from one terminal +ve to –ve equivalent to the other terminal –ve to +ve, graphical illustration is changing the C-V curve with respective to y-axis, (or mirror image to y-axis plane).

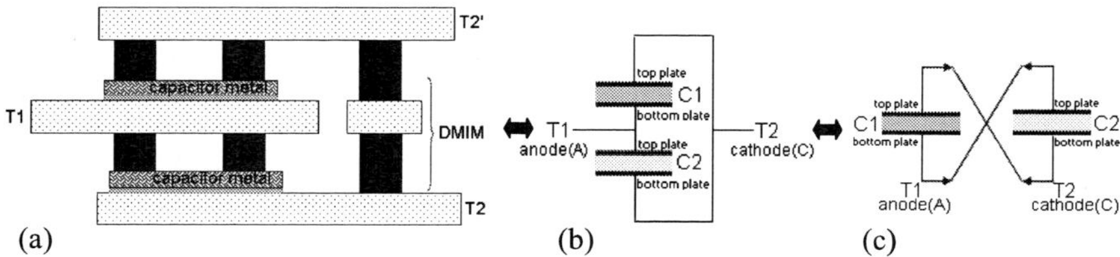

Figure 5. (a) DMIM primitive device layout x-section profile; (b) the equivalent circuitry of a DMIM and (c) Cross-coupling layout and its equivalent circuitry of 2 single MIM with identical dielectrics connected top to bottom, and bottom to top plates each other.

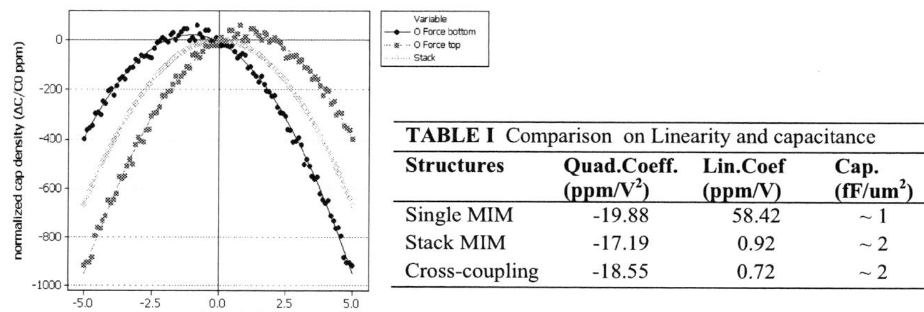

TABLE I Comparison on Linearity and capacitance

Structures	Quad.Coeff. (ppm/V^2)	Lin.Coef (ppm/V)	Cap. (fF/um^2)
Single MIM	-19.88	58.42	~ 1
Stack MIM	-17.19	0.92	~ 2
Cross-coupling	-18.55	0.72	~ 2

Figure 6. sample C-V curves are shown for the oxide used as the MIM dielectric. O Force bottom/top refers to anode at CM bottom/top plate respectively. Stack corresponds to DMIM connected as in Figure 5.

Table I. Comparison on Voltage coefficients and capacitance density for single MIM, stack MIM and cross-coupling of 2 individual MIM capacitors connected to each other from identical dielectrics, examples given here actual measurement data on oxide.

<u>Formulation of resultant capacitance and voltage coefficients for a stack MIM:</u> In order to achieve high capacitance density together with good linearity, it is possible to stack 2 different dielectrics together with opposite performance of voltage coefficients, the resultant capacitance and corresponding voltage coefficients can be modeled by the equations as follows. Connections of the two single MIM are in parallel as in Figure 5.

With different dielectrics in stack MIM, we have the respective normalized capacitance:

$$\frac{dC_i}{C_{iO}} = \frac{C_i(V) - C_{iO}}{C_{iO}} = \alpha_i V^2 + \beta_i V, i = 1, 2 \qquad [2]$$

Total Capacitance:

$$C_T = C_1 + C_2 \qquad [3]$$

Breakdown voltage:

$$BV_T = \min\left(BV_1, BV_2 \right) \qquad [4]$$

And also Linearity for the stack MIM from individual layer:

$$\alpha_T = \frac{C_{1O}\alpha_1 + C_{2O}\alpha_2}{C_{1O} + C_{2O}} \quad ; \quad \beta_T = \frac{C_{1O}\beta_1 + C_{2O}\beta_2}{C_{1O} + C_{2O}} \qquad [5],\,[6]$$

Based on the above equations[2]-[6], with the relations as established from single layer MIM from figures 3&4; the resultant electrical and linearity properties can be modeled from the individual dielectrics materials with their respective thickness for each single layer.

<u>Optimization conditions to achieve both good capacitance density and linearity:</u>
Based on the characteristics relation, the resultant stack MIM consist of dielectrics from opposite voltage coefficients can be possibly optimized with ultra low linearity, while having high capacitance density.

Take an example as below in Table II using nitride and oxide as dielectrics respectively for the individual MIM layers in the stack structures:

TABLE II. Resultant Stack MIM's key parameters from the integrated structures of dielectric1 and 2.

Dielectric 1					Dielectric 2					Stack MIM				
$R^{*}t_1$	C_1	BV_1	α_1	β_1	$R^{*}t_2$	C_2	BV_2	α_2	β_2	C_S	BV_S	α_S	β_S	β_{CS}
1.9	1.7	30	11	-4	1.6	0.8	46	-23	9	2.5	30	-0.2	0.1	~0
1.5	2.2	21	17	-5	1.5	0.9	44	-23	29	3.0	21	5.7	5	~0
1.5	2.2	21	17	-5	1.36	1.0	40	-27	57	3.1	21	3.5	15	~0
1.0	3.3	10	28	-6	1.0	1.4	30	-60	129	4.6	10	1.9	33	~0
0.9	3.6	8	30	-7	0.92	1.5	28	-72	146	**5.1**	8	**0.2**	38	**~0**
0.8	4.1	6	33	-7	0.87	1.6	27	-81	156	**5.7**	6	**0.9**	38	**~0**

*Thickness ratio R(s) with respect to the individual reference dielectric thickness t_1 for dielectric1 and t_2 for dielectric 2. (i.e. the values of R times t_1 and t_2 correspondingly.). $t_1 \sim 20nm$; $t_2 \sim 30nm$.

By means of the stack approach of two opposite VCC dielectrics; both high cap >5fF/um^2 and VCC < 1ppm/V^2. (with the last 2 optimized conditions in the above table II) is achievable. Therefore, we can achieve both high capacitance density as well as low VCC at the same time as demonstrated here.

As we have showed in the previous session, with the layout optimization using cross coupling interconnection of 2 DMIMs, we can further reduce the linear terms to almost 0 ppm/V.

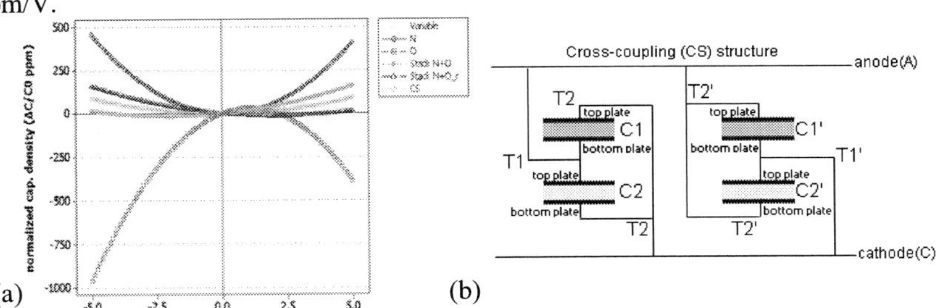

Figure 7. (a) Normalized C-V curves of an example with the optimized thickness of oxide and nitride: individual single MIM layers, resultant DMIM (stack of oxide and nitride) –

forward and reverse bias on the two terminals of the respective capacitor plates and connected in cross-coupling (CS) layout. Linear term of voltage coefficients can be further cancelled out from cross-coupling layout; reduced to almost 0ppm/V. (with the capacitance density remain unchanged); i.e. a nearly straight horizontal line resides across the Vanode-Vcathode x-axis. (b) Equivalent circuitry of a cross-coupling stack MIM.

Frequency dependence study has also been studied on individual MIM. Linearity level maintained relatively constant over 50kHz to 1MHz, no significant change in both the quadratic and linear voltage coefficients of capacitance, regardless their corresponding capacitance density gradually increase with decreasing frequency. The new MIM structure demonstrated here also follows.

Conclusions and Outlook

MIM linearity performance has been characterized among different dielectrics scheme composition. Trade off relationship observed for capacitance density and voltage coefficients. This leads to the difficulties to achieve both high capacitance density as well as low VCC concurrently. This paper demonstrated the use of stack MIM integration with the special layout structures and dielectrics material thickness optimization on different materials, with capacitance density doubled, as well as reducing the voltage coefficients down to almost 0 (a very good linearity performance) for analog applications. The stated stack MIM integration approach for linearity improvement can be extended to use with high-K dielectrics as well in advanced technology.

Acknowledgments

The authors would like to thank the operation group - etch, thin film modules and manufacturing with their kind support and great contributions for wafers processing and valuable technical advice.

References

1. Jeffrey A. et. al. "Analog Characteristics of Metal- Insulator-Metal Capacitors Using PECVD Nitride Dielectrics" in *IEEE ELECTRON DEVICE LETTERS*, VOL.**23**. NO.5, May 2001
2. S.Van Huylenbroeck, et.al. "Investigation of PECVD Dielectrics for Nondispersive Metal-Insulator-Metal Capacitors", *IEEE ELECTRON DEVICE LETTERS*, VOL.**23**. NO.4, April 2002.
3. C.H.Ng, et. al., "Characterization and Comparison of PECVD Silicon Nitride and Silicon Oxynitride Dielectric for MIM Capacitors", in *IEEE ELECTRON DEVICE LETTERS*, VOL. **24**, NO. 8, Aug. 2003.
4. S.J.Kim, et.al. "Improvement of Voltage Linearity in High-κ MIM Capacitors Using HfO_2-SiO_2 stacked Dielectric", in *IEEE ELECTRON DEVICE LETTERS*, VOL. **25**, NO. 8, Aug. 2004.
5. C.Besset, et.al. "Stability of Capacitance Voltage Linearity for High-κ MIM capacitor", in *43rd Annual International Reliability Physics Symposium*, San Jose, 2005.

ECS Transactions, 34 (1) 125-136 (2011)
10.1149/1.3567571 ©The Electrochemical Society

Modeling of electron transport in III- Nitride Compound Semiconductors for Low Field and Low Temperature Applications

Souradeep Chakrabarti, Shyamasree Gupta Chatterjee, Debasish Chattopadhyay and Somenath Chatterjee*

Department of Electronics and Communication Engineering, Techno India, Salt Lake, Kolkata, India

* Corresponding author: somenath@gmail.com

The doping concentration, doping compensation and temperature dependency of electron mobility for GaN, InN, AlN have been calculated using the relaxation time approximation method considering parabolic nature of energy bands. The change in position of Fermi energy with the temperature as well as electron concentrations has been plotted to observe the level of degeneracy for different III-nitride samples. According to the change of Fermi level we have implemented different statistics for mobility extraction modeling at low temperature and low electric field. We compare our simulated results with experimental data and find reasonable correlations, but with evidence that structural imperfection and compensation play crucial roles in our studied samples. Compensation ratio e.g. 0.23 for GaN has been significant in charge neutrality condition. We observe that after 150K the theoretical mobility values deviate from the experimental data. The reasons for the discrepancy are discussed.

Introduction

Wider band gap III-nitride semiconductors such as Gallium nitride [GaN, band gap: 3.4 eV], Indium nitride [InN, band gap: 1.9 eV], and Aluminum nitride [AlN, band gap: 6.2 eV] (1) with commensurately higher breakdown voltage [e.g. $3x10^6$ V/cm for GaN, 4 times larger than in either silicon or GaAs] are enabling in much higher maximum output power delivery, used in amplifiers capable of delivering between 10 and 20 W power at frequencies approaching the X-Band for emerging commercial and military applications. Additionally, they have higher electron saturation drift velocities implying potentially higher frequency performance, lower dielectric constant with a corresponding lower capacitance, strong bond strength [e.g. GaN: 2.3 eV/bond] and higher thermal conductivities improving tolerance to temperature increases. For these properties, III-nitrides make them promising materials used for a number of optoelectronic and electronic devices including laser diodes (2), light-emitting diodes (3), photo detectors (4), metal semiconductor field-effect transistors [MESFETs] (5), high electron mobility transistors [HEMTs] (6) and hetero junction bipolar transistors [HBTs] (7).

Recently, the wurtzite crystal phase of III-nitrides have been focused mainly due to the sapphire substrates which tends to transfer their hexagonal symmetry to the nitride films grown on them, that causes the higher energy band gap than zinc blende (cubic) structure. However, growth on sapphire substrates creates the large lattice mismatch incites dislocation and grain boundary contributions to relatively poor electron mobilities (8). Mobility is considered to be the *figure of merit* for materials used for electronic devices (1). Determination of fundamental material parameters and understanding of electron scattering mechanisms demand accurate comparison between experiment and theory (2). Given the growing importance of III-nitride semiconductors, proposed reliable mobility models are timely for device simulation. Although lots of experimental mobility data in the technical literatures and theoretical results on the carrier mobility have been published by several research groups, however, there is a lack of reliable mobility models suited for device simulation.

In the present work, numerical modeling of the electron mobility at low field and low temperature for III-nitride semiconductors such as GaN, InN, and AlN are carried out using the relaxation time approximation method for Elastic scattering processes including acoustic phonon scattering (with the two modes deformation potential and piezoelectric), neutral impurity scattering, dislocation scattering and ionized impurity scattering.

Model

In this work, we have reported an analytical model for the temperature and concentration-dependent electron mobility of GaN, InN and AlN. In this model, scatterings by ionized impurities, acoustic phonons, neutral impurity and dislocations have been considered to analyze the experimental behavior of mobility for III nitride devices.

The concentration of electron varies with the change of temperature as well as with the change of doping concentration. The presence of a doping impurity creates a bound level $[E_d]$ near the conduction band edge. We note that carrier densities of the electrons will be redistributed, but their numbers will be conserved and will satisfy the following equality resulting from charge neutrality condition.

To calculate densities of electrons at finite temperatures in doped semiconductors, $n + n_d = N_d$ and for compensated semiconductors, $n = N_d^+ - N_a^-$, where, n is the total free electrons in the conduction band and n_d is the electrons bound to the donors

$$\frac{n_d}{(n+n_d)} = \frac{1}{\left(1 + \dfrac{N_c}{2N_d} e^{-(E_c - E_d)/kT}\right)} \qquad [1]$$

$$N_c = 2\left(\frac{2\pi m_e kT}{h^2}\right)^{\frac{2}{3}} \qquad [2]$$

The factor 1/2 essentially arises from the fact that there are two states an electron can occupy at a donor site corresponding to the two spin-states. N_c is the density of states at Energy level "E". Again, according to the reference (9), we can

$$E_d = E_c - 13.6 \frac{m^*}{m_o}\left(\frac{\in_0}{\in}\right)^2 eV \qquad [3]$$

E_c = Conduction band energy, $\in_o$ = Permittivity of vacuum, $\in$ = Permittivity of donor, m^* = effective mass and m_o = rest mass of electron

The Fermi level has been calculated using the following equations to determine the degeneracy of samples on the basis of electron concentration and temperature. Degeneracy is very important factor as it is used to imply different statistics (Maxwell – Boltzmann or Fermi – Dirac) at different temperature. Generalized M-B statistics are used throughout because the samples we have used to compare our results are highly Non-degenerate.

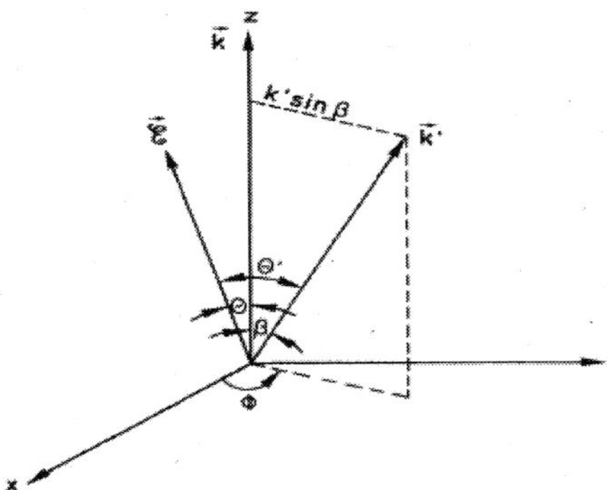

Figure 1: spherical coordinate system with the k direction as the polar axis.

We shall now introduce a spherical coordinate system with the k direction as the polar axis. In this system let (k', β, Φ) be the spherical coordinates of k', the azimuthal angle Φ being measured from the k-ε plane figure 1. One may then write (10)

$$n = \int \frac{2}{(2\pi)^3} f d\hat{k} \qquad [4]$$

$$k = \frac{\sqrt{2m^* E}}{\hbar} \qquad [5]$$

$$dk' = k'^2 dk' \sin \beta d\beta d\phi \qquad\qquad [6]$$

From this we derived

$$n = \int_{E_c}^{\infty} \frac{\left(2 m^* k_\beta T\right)^{3/2}}{2 \pi^2 \hbar^3} F_{1/2}(\eta) \qquad\qquad [7]$$

Here, the electron mobility considering various elastic scattering mechanisms is determined by solving Boltzmann transport equation using the relaxation time approximation method. The parameter for characterizing the various scattering mechanisms is the relaxation time τ, which determines the rate of change in electron momentum as it moves through the semiconductor crystal. Mobility is related to the scattering time by

$$\mu = \frac{e \langle \tau \rangle}{m^*} \qquad\qquad [8]$$

where τ is the average relaxation time over the electron energies and μ is the mobility, and m is the effective mass of electron. In the following sections, the expressions of relaxation time and mobility caused by different scattering mechanisms have been given.

Electron Mobility:

Phonon scattering
Deformation Potential Scattering
The acoustic mode lattice vibration induced changes in lattice spacing, which change the bandgap from point to point. Since the crystal is 'deformed' at these points, the potential is called the deformation potential. The corresponding scattering relaxation time can be written as

$$\tau(E) = \frac{\pi d h^4 V_s^2}{\sqrt{2} E_1^2 m^{*3/2} k_B T} E^{-\frac{1}{2}} \qquad\qquad [9]$$

Where, $C_l = dV_s^2$; acoustic longitudinal elastic constant, V_s= average velocity of sound in GaN, d= mass density of GaN, E_1= Acoustic deformation potential

Piezoelectric Scattering
Electrons can suffer scattering with piezoelectric mode of acoustic lattice vibrations. In this scattering mechanism. The relaxation time due to piezoelectric potential scattering is given by

$$\tau(E) = \frac{2\sqrt{2}\,\pi\hbar^2 \in E^{\frac{1}{2}}}{e^2 p^2 m^{*\,1/2} k_B T} \qquad [10]$$

Where,

$$p = \left[\frac{h_{pz}^2}{dV_s^2 \in} \right] \qquad \text{is the piezoelectric coupling coefficient,}$$

hpz=piezoelectric constant

Impurity Scattering

Ionized Impurity Scattering:

The amount of scattering due to electrostatic forces between the carrier and the ionized impurity depends on the interaction time and the number of impurities. Larger impurity concentrations result in a lower mobility (8). The relaxation time due to scattering of ionized impurities is given by

$$\frac{1}{\tau(E)} = \frac{NZ^2 e^4}{16\sqrt{2}\pi \in^2 m^{*\frac{1}{2}}} E^{-\frac{3}{2}} \left(\ln(1+b) - \frac{b}{1+b} \right) \qquad [11]$$

Where $b = \dfrac{4k^2}{\beta_s^2} = \dfrac{8m^* E}{\hbar^2 \beta_s^2}$

Where β_s is the inverse screening length $\beta_s^2 = \dfrac{ne^2}{\in k_B T} \dfrac{F_{-1/2}(\eta)}{F_{1/2}(\eta)}$

N=ionized impurity which is for a sample with dislocation density and un-compensation is

$N = n + f\left(\dfrac{N_s}{d}\right)$, Ns= dislocation density, d= c-lattice constant of a hexagonal GaN lattice, f=fraction of filled traps

For compensated $N = N_d^+ + N_a^-$, where N_d and N_a are donor and acceptor concentration

Neutral impurity scattering:

When an electron passes close to neutral atom, momentum can be transferred through a process in which the free electron exchanges with a bound electron on the atom. The relaxation time can be written as

$$\tau\left(E\right) = \frac{m^{*}}{20\ N_{n}\ \hbar\ a_{o}} \qquad [12]$$

Where, a_0 is the effective Bohr radius of donor, and Nn is the concentration of neutral impurities.

Dislocation Scattering

The high density of dislocations and native defects induced by nitrogen vacancies in GaN, dislocation scattering and scattering through nitrogen vacancies has also been considered as a possible scattering mechanism. Dislocation scattering is due to the fact that acceptor centers are introduced along the dislocation line, which capture electrons from the conduction band in an n-type semiconductor. The dislocation lines become negatively charged and a space-charge region is formed around it, which scatters electrons traveling across the dislocations, thus reducing the mobility. The relaxation time is

$$\tau\left(E\right) = \frac{8\left(\in\in_{0}\right)^{2}a^{2}m^{*2}}{Ne^{4}f^{2}L_{D}}\left(V_{t}^{2} + \hbar^{2}\Big/4\,m^{*2}L_{D}^{2}\right)^{3/2} \qquad [13]$$

Where V_t is the component of V perpendicular to dislocation line, a is the distance between imperfection centers along the dislocation line, and f is their occupation probability.

Analysis and Discussion

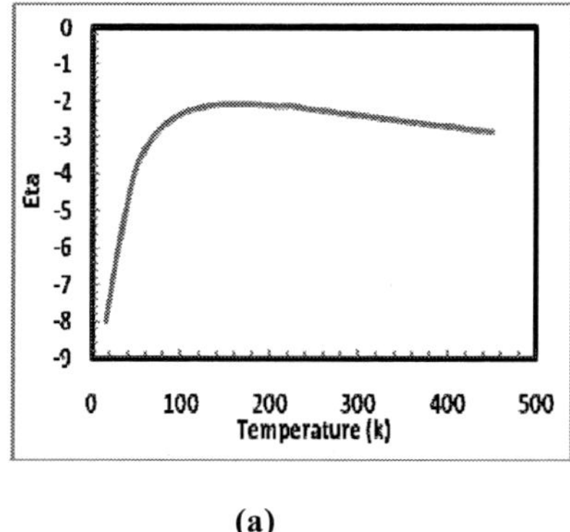

(a)

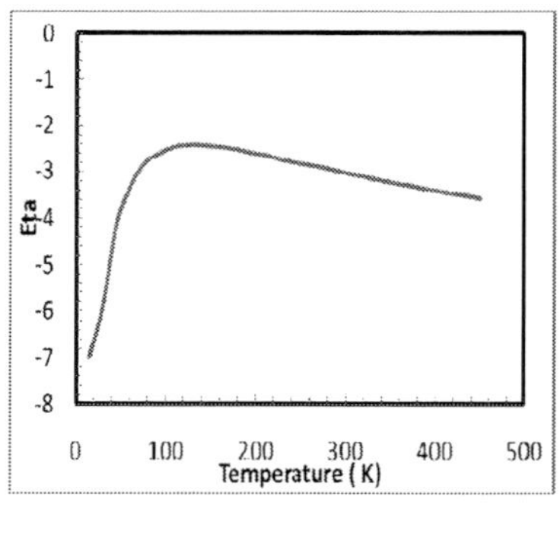

(b)

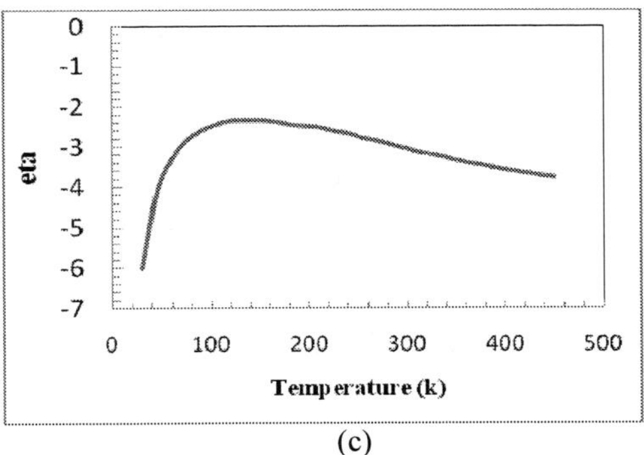

(c)

Figure 2:   The variation of Eta with temperature for (a) uncompensated InN, (b) uncompensated GaN and (c) compensated GaN

The fig 2 shows the change in position of Fermi energy ** with the temperature as well as electron concentrations to observe the level of degeneracy for bulk- GaN sample from (11) and bulk- InN sample from (1) . In fig. 2(b), the GaN sample is unintentionally doped and uncompensated n-type and grown by MOCVD technique and in fig. 2(b), the GaN sample is compensated. Fermi level has been calculated using equation 8 to determine the level of degeneracy of the samples. It is prominent from figure that the samples we have used to compare our theoretical model are highly non-degenerate, therefore Generalized M-B statistics was used throughout.

[** Fermi level= η $k_B T$ +Ec , where Ec is conduction band energy level.]

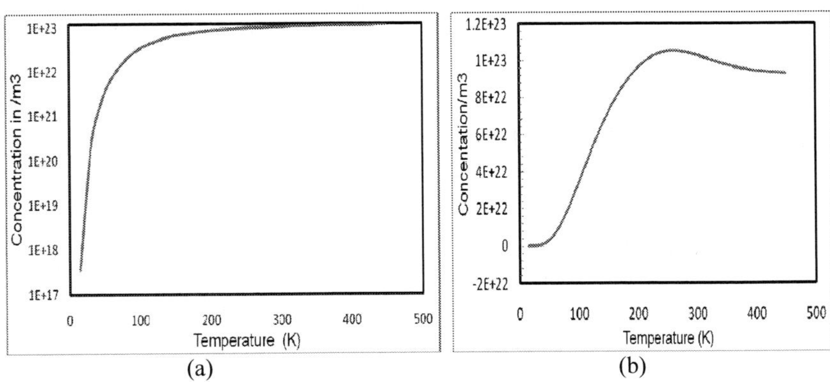

(a)   (b)

Figure 3:   The change of electron concentration at different temperature for   (a) uncompensated and (b) compensated GaN sample.

It is clear from figure 3 (a) that in the case of the uncompensated GaN sample electron concentrations increase with temperature and above 100 K it gradually saturates (12) and for compensated GaN sample electron concentration starts to decrease after approx. 250 K as shown in figure 3(b) because hole activation energy is higher, number of hole concentration increases at higher temperature.

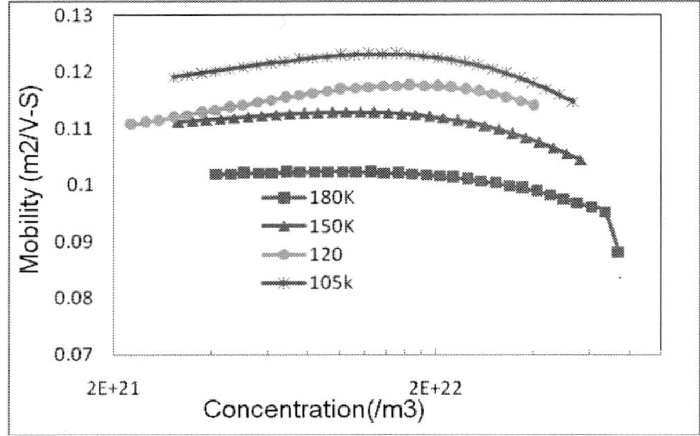

Figure 4: The calculated electron drift mobility as a function of carrier concentrations @ 105 K , 120K,150K and 180K

Figure 4 shows the calculated electron drift mobility as a function of carrier concentrations at 105K, 120K, 150K and 180K. As temperature increases the peak value of mobility decreases. It can be observed that the nature of our simulated curves at different temperatures resemble with the curves using Monte-Carlo Simulation reported at Chin et al. (13).

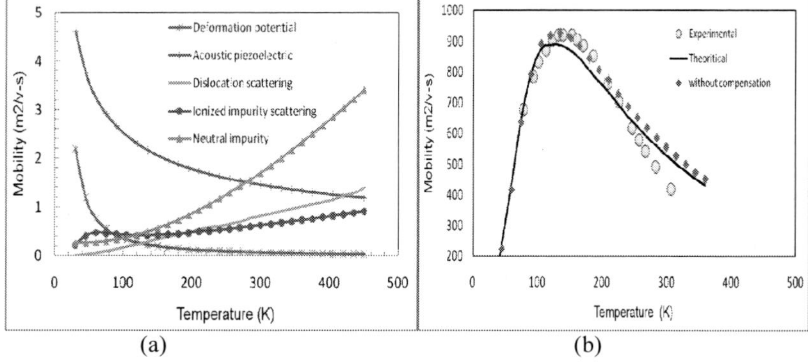

Figure 5: (a) Mobility for different scatterings with temperatures (b) Mobility variations at different temperatures for compensated and uncompensated for bulk- GaN sample

Figure 5(a) represents the variation of electron mobility with temperature for various types of scattering mechanism such as elastic process of acoustic phonon deformation potential scattering, acoustic piezoelectric scattering ,ionized impurity scattering, neutral impurity scattering and dislocation scattering individually using the parameters of uncompensated n-type GaN.

TABLE I. Sample parameters (11)

sample	Concentration at 300 K (cm^{-3})	Activation energy
RK-120c	$1.03*10^{17}$	13.5 eV

The measured values, shown in Table I, for the MOCVD sample (Ref 11) is used for theoretical calculations. Here, the dislocation density, estimated by our model is $9x10^{8}$cm^{-2} and average longitudinal elastic constant $C11$ is $1.85x10^{10}$ N/m^{2} respectively. In n-type GaN the activation energy of the donors is quite large, which keeps a large number of donors neutral at low temperatures from 15K to 60K. Between 60K to 100K larger impurities concentrations result in a lower mobility and Beyond 100K almost all impurity atoms are ionized and Drift velocity of carriers become dominant (14).

From figure 5 (a), it is seen that the electron mobility due to scattering by dislocations should increase monotonically with net carrier concentration, satisfies with ref. (15)

It is found that fraction of the filled traps (*f*) increases with the increase of carrier concentration and dislocation density. For a constant dislocation density, *f* increases with T to a maximum, then decreases. This behavior can be attributed to the competition between two opposing effects. First, the increase in temperature increases the number of free electrons, which leads to increasing *f*. Second, the increase in temperature provides energy for the electrons to escape from the traps, and this reduces *f*, it agrees well with our theoretical modeling.

From figure 5 (b) that GaN exhibits maximum mobilities in the temperature range (120-180) K depending on the electron density. This behavior is related to the interplay of acoustic phonon scattering at low carrier concentrations and ionized impurity scattering at higher carrier concentrations. Also, the mobility is affected over the whole temperature range from 40 to 400 K, due to the presence of the interfacial layer. This interfacial layer causes lattice mismatch at the interface. This mismatch generates highly dislocated and defective region to relieve the strain (12).

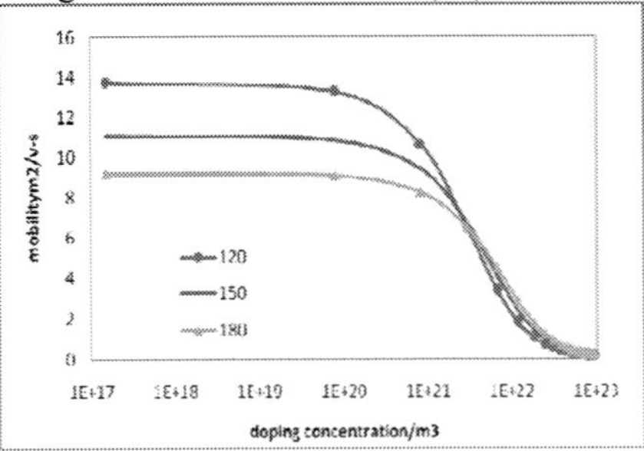

Figure 6: Calculated electron drift mobility of bulk-InN sample as a function of carrier concentrations in different temperatures

This thin interface layer acts as a parallel degenerate conduction channel to the conduction in the bulk GaN layer. Our modeling suited well with the experimental data at low temperature i.e. upto 120K for compensated and upto 220K for uncompensated GaN, but there is a deviation between our theoretical value with experimental data at higher temperature. This deviation can be well justified by the effect of polar optical phonon scattering and plasmon scattering, which is the reason of sharper fall of the experimental data after (200 – 250) K.

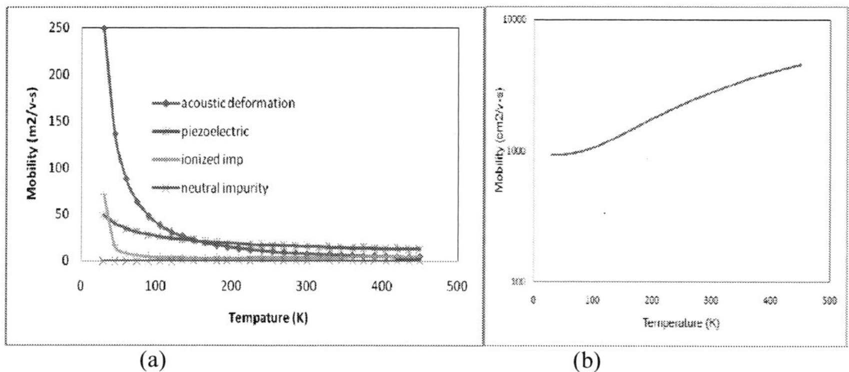

(a)   (b)

Figure 7: (a) Mobility for different scatterings at different temperatures (b) Mobility variation at different temperatures uncompensated - for bulk-InN sample.

It is evident from the figure 5 (b) that ionization scattering is dominant in the temperature range between 120 to 275K by comparing the mobility of compensated and uncompensated GaN sample. As in compensated sample, the number of ionized impurity concentration is much higher than that of uncompensated but after approx. 275 K all the dopents are fully ionized, so the effect of ionized impurity scattering is less dominant and the effect of compensation will be very less.

As shown in Figure 6, the predicted maximum low-field, low temperature mobility in InN (at 120K) is approx. 14 m^2/V-s. Keep in mind that m* is inversely proportional to mobility, the mobilities predicted in this work seem to be realistic (16).

Figure 7(a) represents the variation of electron mobility with temperature for various types of scattering mechanism such as elastic process of acoustic phonon deformation potential scattering, acoustic piezoelectric scattering ,ionized impurity scattering, neutral impurity scattering individually using the parameters of uncompensated *n*-type InN. From the figure, it is prominent that impact of neutral impurity scattering is very dominant from low (~ 30K) to high (450K) temperature because the effective mass of InN is much less than GaN and AlN , so the density of states at a particular temperature is less which directly affects ionization phenomena. The effect of acoustic phonon deformation potential scattering is minimum at low temperature. Figure 7(b) shows the mobility of carriers in 3D bulk InN for the donor concentration of 1.1x 10^{17}/ cm^3. Our calculated low-field mobility values for InN at a temperature of 300 K are approx. 3000 cm2/Vs corresponding to effective masses of 0.11 m0, which is well suited with ref (17). It has been observed that after 150K the theoretical mobility value has been deviating from the experimental data because of polar optical phonon scattering, plasmon scattering, and interfacial layer charge effect. A large disparity of the atomic radii of In and N is an additional contributing factor to the difficulty of obtaining InN of good quality. Because of all these factors, the electron mobilities obtained from various films have varied very widely. Electrical properties vary also substantially with the choice of growth techniques (18).

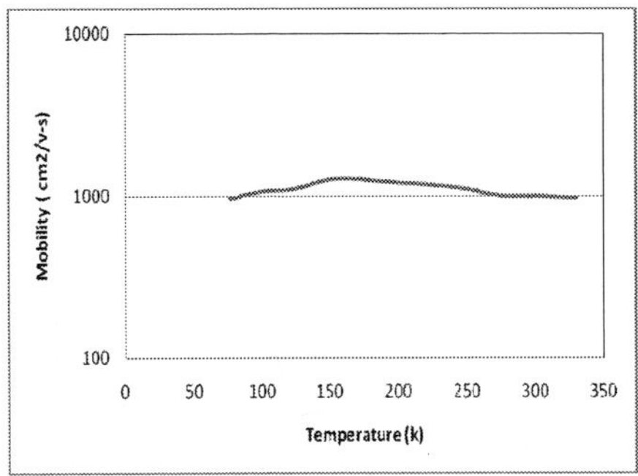

Figure 8: Mobility variation at different temperatures for uncompensated bulk AlN.

Figure 8 describes the low field electron mobility of AlN. Here we have considered all the elastic scattering effects including impurity scattering but in reference (13) only phonon limited scattering had been considered for mobility profile with temperature.

Conclusions

In conclusion, a sufficiently accurate model for the electron mobility at low temperature and low electric field of GaN, InN and AlN has been developed. In this work, we studied the dependence of three dimensional mobility for GaN, AlN, and InN on various parameters such as temperature, doping concentration etc to compare these materials with each other. The electron mobilities have been calculated using Relaxation Time Approximation method considering elastic process of acoustic phonon deformation potential scattering, acoustic piezoelectric scattering and ionized impurity scattering, neutral impurity scattering and dislocation scattering. It is obvious from our work that, i) Unlike InN, the 3D mobility values of GaN and AlN increase until they reach their maximum value between the temperatures 150–200 K and decreases again, ii) bulk InN has much better characteristics than GaN and AlN with regard to low-field 3D mobility even when the traditionally accepted parameters are used.

References

1) Z. Yarar, B. Ozdemir, and M. Ozdemir, Journal of Electronic Materials, Vol. 36, No. 10, 2007
2) S. Nakamura, M. Senoh, S. Nagahama, N. Iwasa, T. Yamada, T. Matsushita, H. Kiyoku, and Y. Sugimoto, Jpn. J. Appl. Phys., Part 2 **35**, L217 (1996)
3) D. Steigerwald, S. Rudaz, H. Liu, R. S. Kern, W. Götz, and R. Fletcher, JOM **49**, 18 (1997)
4) J. M. Van Hove, R. Hickman, J. J. Klassseen, P. P. Chow, and P. P. Ruden, Appl. Phys. Lett. **70**, 2282 (1997)
5) P. Kelin, J. Freitas, Jr., and S. Binai, Appl. Phys. Lett. **75**, 4016 (1999)
6) S. Sheppard, K. Doverspike, W. Pribble, S. Allen, J. Palmour, L. Kehias, and T. Jenkins, IEEE Electron Device Lett. **20**, 161 (1999)

7) L. McCarthy, P. Kozodoy, M. Rodwell, S. DenBaars, and U. Mishra, IEEE Electron Device Lett. **20**, 277 (1999)

8) Frank Schwierz, An electron mobility model for wurtzite GaN, Solid-State Electronics 49 , 889–895, (2005).

9) U. K. Mishra, and J. Singh, Semiconductor Device Physics and Design, Springer, (2008)

10) D. Chattopadhya and H. J. Queisser, Rev. Mod. Phys. **53**, 745 (1981)

11) I. M. Abdel-Motaleb, R. Y. Korotkov, J. Appl. Phys., **97**, 093715 (2005)

12) S. Dhar and S. Ghosh, J. Appl. Phys., 86, (1999)

13) V. W. L. Chin, T. L. Tansley, and T. Osotchan- J. Appl. Phys. 75 , ll (1994)

14) K. Alfaramawi, Journal of King Saud University - Science (2010)

15) H. Ng, D. Doppalapudi, T. Moustakas, N. Weimann, and L. Eastman, Appl. Phys. Lett. **73**, 821(1998)

16) F.Schwierz, V. M. Polyakov, Proc. Solid-State and Integrated Circuit Technology, Shanghai, China, (2006)

17) T. L. Tansley, C. P. Foley, J. S. Blakemore (Ed.): Proc. 3rd Int. Conf. on Semiinsulating III–V Materials, Springs, London , (1985)

18) Fehlberg, T.B. ; Umana-Membreno, G.A. ; Nener, B.D. ; Parish, G. ; Gallinat, C.S. ; Koblmuller, G. ; Bernardis, S. ; Speck, J.S. : Proc. Optoelectronic and Microelectronic Materials and Devices, Western Australia, (2006).

ECS Transactions, 34 (1) 137-142 (2011)
10.1149/1.3567572 ©The Electrochemical Society

Fast Flexible Electronics Based on Printable Thin Mono-Crystalline Silicon
(Invited)

Zhenqiang Ma[a,c], Kan Zhang[a], Jung-Hun Seo[a], Han Zhou[a], Lei Sun[a], Hao-Chih Yuan[a],
Guoxuan Qin[a], Huiqing Pang[a] and Weidong Zhou[b]

[a] University of Wisconsin-Madison, Department of Electrical and Computer Engineering
1415 Engineering Drive, Madison, WI 53706, USA
[b]University of Texas-Arlington, Department of Electrical Engineering, Arlington, TX
76019, USA
[c] Email: mazq@engr.wisc.edu

Traditional flexible electronics employ amorphous Si, poly Si and organic materials, but these materials cannot be used for fast (radio-frequency, RF) flexible electronics due to their low carrier mobilities. Instead, we employ monocrystalline Si as the active materials. These high quality materials were released from silicon-on-insulator by undercutting the buried oxide layer. They have equivalent mobility values as their bulk counterpart, yet with high mechanical flexibility. We realized the first RF flexible thin-film transistors by performing pre-release doping and changing the transistor fabrication process to gate-after-source/drain to avoid high temperature process on plastic substrate and achieve low source/drain contact resistance. We further increased the device speed with reduced source/drain access resistance through careful device structure design. By realizing smaller feature size (1 μm) with local gate alignment and higher fidelity membrane registration/transfer technique, 12 GHz flexible thin-film transistors were demonstrated. Strained channel and nanolithography are projected to further increase device speed.

Introduction

For the past decade, there has been an enthusiastic pursuit for high performance flexible electronics due to their unique advantages of being light weight, bendable, and mountable to uneven surfaces (1-3). Amorphous Si, poly Si and organic materials are traditional materials for flexible electronics, and have demonstrated their dominance in low-speed applications such as flexible displays, electronic tags, and low-cost integrated circuits (4). But their low carrier mobilities limit their applications in higher frequency domain, where rigid materials, such as monocrystalline Si (5-7) and III-V materials (8, 9), dominate. High speed flexible electronics can be used in personal Wi-Fi devices, wearable radios, radio frequency (RF) identification devices, biomedical telemetry devices, foldable phased-array antennas, large-area radars for remote sensing, surveillance, etc (10).

High-speed flexible electronics imposes specific requirements on both active materials and fabrication techniques to achieve comparable performance with its rigid counterparts. Active materials are required to possess acceptable carrier mobilities for high-speed operation, be readily integratable on flexible substrates, and be bendable without much sacrifice of performance. Within such criteria, monocrystalline Si nanomembrane (NM) released from silicon-on-insulator (SOI) emerges as one of the best choices due to its high carrier mobilities, commercial availability and mature fabrication techniques. Specific considerations must also be taken for fabrication techniques so that they are compatible with flexible substrates which usually have much worse physical and chemical properties than rigid semiconductor materials. For example, high temperature process like thermal annealing and corrosive solutions like Piranha must be avoided on most flexible substrates. In this paper, we will summarize our current effort in developing the process flow to fabricate flexible RF metal-oxide-semiconductor field transistors (MOSFETs) based on monocrystalline Si NMs on low-temperature plastic substrates. Our years of effort has led to the recent demonstration of a record 12 GHz flexible N-type MOSFET on polyethylene terephthalate (PET) substrates. Further research on continuing to increase the speed of thin film transistors (TFTs) will be projected.

Fabrication and Results

In this section, we illustrate the design of fabrication process flow in detail and explain how specific considerations and compromises are made to yield the finished devices with improved performance.

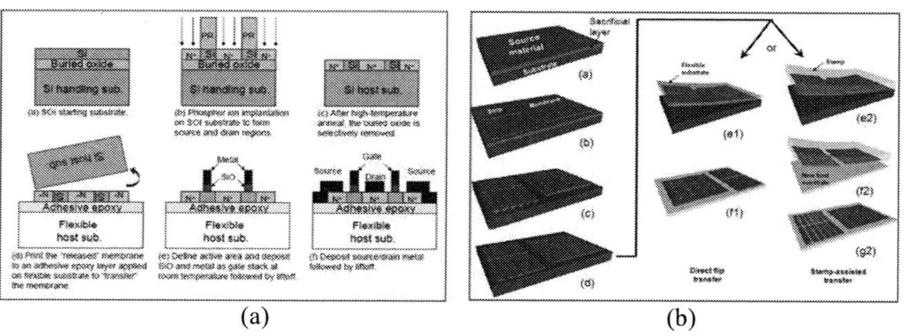

(a)   (b)

Figure 1. (a) Illustration of SiNM doping, release and transfer, and gate-after-source/drain device fabrication processes. (b) Comparison of two SiNM transfer methods: stamp-assisted transfer and direct flip transfer.

Monocrystalline Si NMs can be created from SOI material. The top Si on SOI is patterned into sheet or strips to expose the buried oxide (BOX) underneath by photolithography and reactive-ion etching (RIE). Then the patterned SOI is immersed in aqueous HF solution to fully remove the BOX and, if the BOX is thin enough, the top Si membrane will fall down and register on the Si handling substrate by weak van der Waals force (11). This free-standing Si NM is now ready for transfer onto flexible substrates. There are two transfer methods as shown in Fig. 1 (b). The first method (a, b, c, d, e2, f2, g2) utilizes a soft stamp to pick up the NM from the source substrate and transfer it onto a flexible destination substrate, which is PET in this case, with assistance of adhesives (12). This method preserves the NM orientation, i.e. the top side of the NM on the source substrate is still the top side on the flexible substrate. But it suffers from more complicated manipulation of NM (picking up and placing down) compared to the second method, which may be detrimental for yield if not very skillful. The second method employs a direct flip transfer from the source substrate directly onto the flexible substrate with adhesive, as illustrated in Fig. 1(b) (a, b, c d, e1 and f1). This method finishes the transfer with only one-time manipulation, but obviously the NM orientation is inversed, which requires more sophisticated design to achieve required doping profile (13), if doping is needed. Despite of the advantages and disadvantages of each method explained above, the flip transfer technique is adopted in our work because high-speed operation of TFTs requires short channel length and stringent alignment. Statistical study shows that the flip transfer method achieves higher fidelity NM registration and thus eases the alignment requirement (7).

Based on the choice of the flip transfer method, the fabrication process flow was designed taking both doping profile and properties of PET substrate into account. The flip transfer method uses bottom side of the top Si on SOI as the active material, thus requires higher ion implantation dose and energy to form adequate doping concentration on the bottom side. PET substrate also places constraint on thermal process temperature (below 170°C Vicat softening point) (14), thus forbidding annealing process. With these considerations, a process flow consisting of pre-release doping and gate-after-source/drain was developed to meet all the requirements. Conventional gate self-aligned procedure cannot be adopted so that the required doping profile can be formed before the membrane transfer, which we refer to as pre-release doping. The process started with commercial SOI material (from SOITEC) and ion implantation was performed selectively through photolithography pattern. Subsequent furnace annealing was performed to activate the dopants. Then the top Si NM was released from the handling substrate by undercutting the BOX, and flip transferred onto a PET substrate spin coated with SU-8 as adhesive. Further photolithography and RIE process was conducted to pattern active regions and gate dielectric (SiO) and metal contact (Ti/Au) were fabricated by photoresist lift-off procedure to conclude the process (15).

With this process flow, the first RF flexible NMOS TFT was demonstrated. The SOI material has 200 nm Si top membrane on 200 nm BOX. Phosphorous ion implantation has dose of 4×10^{15} cm^{-2} and energy of 40 keV. The furnace annealing was performed at 850°C for 40 min in N_2 ambience. The gate dielectric is 200 nm SiO and the metal contact consists of 40 nm Ti and 300 nm Au. Fig. 2 (a) shows the device structure and (b) shows the RF performance with current gain and power gain. Figure of merits (FOMs) including cut-off frequency (f_T) of 1.9 GHz and maximum oscillation frequency (f_{max}) of 3.1 GHz were demonstrated and they validate the feasibility of this process flow (5). Note that this device has (nominal) 2µm channel length and its channel does not overlap with source/drain regions but leaving 1µm gaps, causing high source/drain access resistance. Although it is the first multi-GHz TFT, their device parameters are a bit far from satisfactory as a practical RF MOSFETs, so further efforts were engaged to refine the structure. Some device parameters were improved in a following design, in which SiO was reduced down to 100 nm, gate length was shortened to 1.5 µm and gate overlapped on source/drain regions with 0.5 µm on each side, which reduced access resistance but induced more capacitance between gate and source/drain. Fig. 2 (c) shows the refined device structure and (d) shows the FOMs of f_T 2.04 GHz and f_{max} 7.8 GHz, which verify that the overlapping gate is actually beneficial (6).

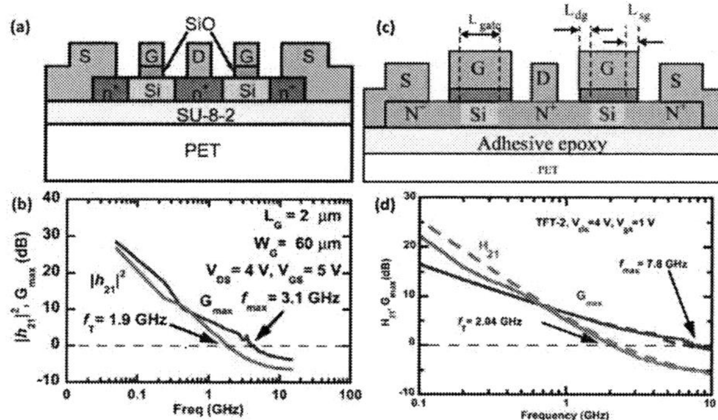

Figure 2. (a) Schematic illustration of RF flexible NMOS TFTs. Gate does not overlap with source/drain regions. (b) Curves of current gain and power gain vs. frequency, in which f_T of 1.9 GHz and f_{max} of 3.1 GHz were derived. (c) Illustration of the improved structure in which gate overlaps with source/drain regions. (d) Curves of current gain and power gain vs. frequency. Higher f_T of 2.04 GHz and f_{max} of 7.8 GHz were achieved.

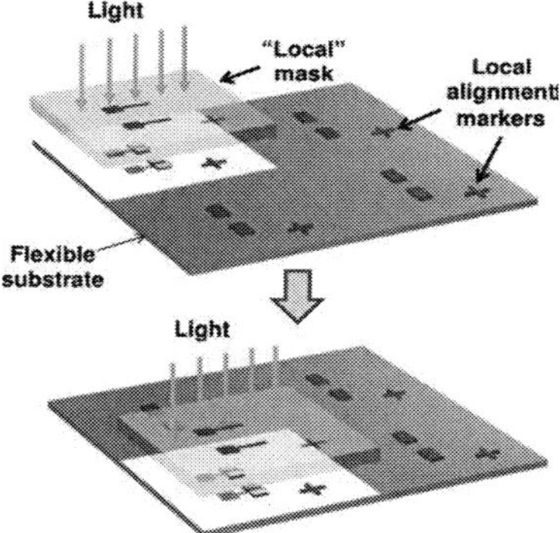

Figure 3. Schematic illustration of local alignment scheme. Devices in different regions on the flexible substrate are separately and repeatedly aligned to realize accurate alignment for the gate stack photolithography step.

Although channel scaling down proves very successful of improving device speed on rigid substrate, further scaling down beyond 1.5 µm on PET substrate turns out cumbersome due to thermal expansion of PET. Like other plastic material, PET has no definite melting point and it softens and expands as temperature increases. Although high temperature process is eliminated on PET, photoresist baking at 115°C is still indispensable. Statistical study proves that alignment for 1 µm channel length can only be feasible within 2 mm span. This limited alignment region requires a local scheme, which is conceptually similar with using a stepper for lithography, as shown in Fig. 3 (7). Each time only a small region was aligned and exposed for 1 µm pattern, and this procedure repeated until the sample was all exposed. Although this scheme increases fabrication time, it is still cost effective comparing to e-beam lithography for sub-micron pattern.

Using the local alignment scheme, device structure was further refined and channel length of 1 μm was achieved while other parameters were unchanged. The device structure is illustrated in Fig. 4 (a). Optical microscope images in Fig. 4 (b) and (c) proves the acceptable alignment accuracy. With this sole scaling down, the FOMs were significantly improved and f_T of 3.8 GHz and f_{max} of 12 GHz were achieved (7). The optical image shown in Fig. 4 (d) shows the sample in bending condition. As mentioned at the beginning, bendability is an essential merit of flexible electronics that allows conformal attaching to uneven surface without much performance sacrifice. Fig. 4 (f) shows the measurement results of the NMOS TFT under bending conditions and proves the acceptable performance variation (7).

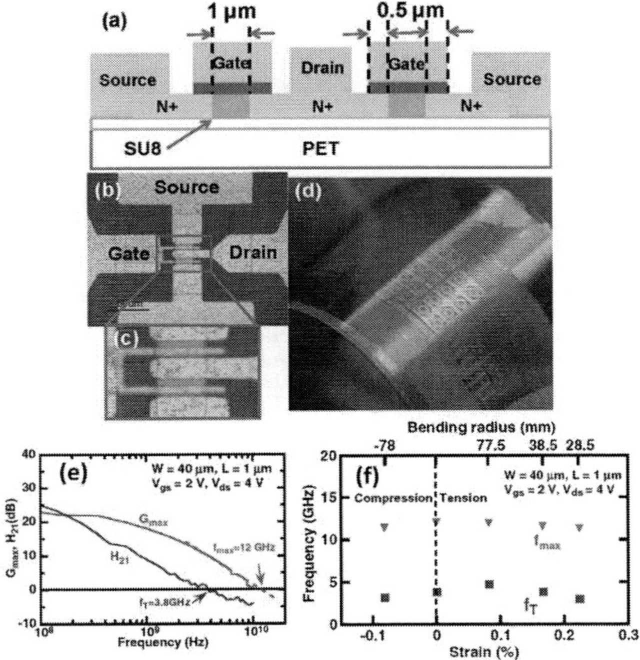

Figure 4. (a) Schematic illustration of the refined device structure. (b) and (c) Optical microscope images of the finished NMOS TFT, proving the acceptable alignment accuracy. (d) An optical image of a bended sample. (e) Curves of current gain and power gain vs. frequency, showing f_T of 3.8 GHz and f_{max} of 12 GHz. (f) Variations of f_T and f_{max} under different bending radii.

To further improve device performance, several research directions can be projected. One option is channel engineering (15), i.e. using strained channel. Another option is channel length further scaling down, which requires novel nanolithography methods can be applied to SiNMs .

Conclusion

Progress of fabricating RF flexible NMOS TFTs has been summarized. With the procedure of pre-release doping and gate-after-source/drain, the first RF flexible NMOS TFT was demonstrated. With reduced source/drain access resistance and channel length scaling down, device speed was further improved. Local alignment scheme proves essential for 1 μm channel length, and 12 GHz record NMOS TFT was fabricated with this scheme.

Acknowledgments

The work was supported partly by AFOSR MURI grant under FA9550-08-1-0337, NSF MRSEC under grant DMR0520527 and a PECASE grant under FA9550-091-0482. The program manager of the MURI and PECASE grants is Dr. Gernot Pomrenke.

References

1. J. A. Rogers, T. Someya and H. Yonggang, *Science*, **327**, 1603 (2010).
2. S.-I. Park, Y. Xiong, R.-H. Kim, P. Elvikis, M. Meitl, D.-H. Kim, J. Wu, J. Yoon, Y. Chang-Jae, Z. Liu, Y. Huang, K.-C. Hwang, P. Ferreira, L. Xiuling, K. Choquette and J. A. Rogers, *Science*, **325**, 977 (2009).
3. D.-H. Kim, J. Xiao, J. Song, Y. Huang and J. A. Rogers, *Advanced Materials*, **22**, 2108 (2010).
4. Y. Sun and J. A. Rogers, *Advanced Materials*, **19**, 1897 (2007).
5. H.-C. Yuan and Z. Ma, *Applied Physics Letters*, **89**, 212105 (2006).
6. H.-C. Yuan, G. K. Celler and Z. Ma, *Journal of Applied Physics*, **102**, 034501 (2007).
7. L. Sun, G. Qin, J.-H. Seo, G. K. Celler, W. Zhou and Z. Ma, *Small*, **6**, 2553 (2010).
8. Y. Sun, E. Menard, J. A. Rogers, H.-S. Kim, S. Kim, G. Chen, I. Adesida, R. Dettmer, R. Cortez and A. Tewksbury, *Applied Physics Letters*, **88**, 183509 (2006).
9. J. Yoon, S. Jo, I. S. Chun, I. Jung, H.-S. Kim, M. Meitl, E. Menard, X. Li, J. J. Coleman, U. Paik and J. A. Rogers, *Nature*, **465**, 329 (2010).
10. Z. Ma and L. Sun, in *IEEE 10th Annual Wireless and Microwave Technology Conference*, Clearwater, FL, United states (2009).
11. G. M. Cohen, P. M. Mooney, V. K. Paruchuri and H. J. Hovel, *Applied Physics Letters*, **86**, 251902 (2005).
12. E. Menard, R. G. Nuzzo and J. A. Rogers, *Applied Physics Letters*, **86**, 093507 (2005).
13. H.-C. Yuan, Z. Ma, M. M. Roberts, D. E. Savage and M. G. Lagally, *Journal of Applied Physics*, **100**, 013708 (2006).
14. G. Qin, H.-C. Yuan, G. K. Celler, W. Zhou and Z. Ma, *Journal of Physics D: Applied Physics*, **42**, 234006 (9 pp.) (2009).
15. H.-C. Yuan, M. M. Kelly, D. E. Savage, M. G. Lagally, G. K. Celler and Z. Ma, *IEEE Transactions on Electron Devices*, **55**, 810 (2008).

ECS Transactions, 34 (1) 143-148 (2011)
10.1149/1.3567573 ©The Electrochemical Society

HHNEC 0.18um BCD Technology for high density power integration

Zhang Shuai (张帅)[a,†], Qian Wensheng (钱文生)[a], Dong Ke (董科)[a]

(a Shanghai Hua Hong NEC Electronics Company, Limited)

A new 0.18um Bipolar CMOS DMOS (BCD) integrated platform is presented. This technology provides both N type and P type power LDMOS with capability of voltage up to 40V applications; these power LDMOS devices have competitive specific on-resistance; the parasitical miller capacitance and SOA of power LDMOS devices are optimized by careful device and layout design; it also provides HVCMOS with voltage up to 40V analog application; the bipolar devices in this technology are also optimized for power switching as standard bipolar process offers. The 1.8V/5.0V CMOS devices are performance compatible with industry main stream 1.8V/5.0V CMOS technology. This platform is modularized with various options; it can be separated into 1.8V/5.0V mixed 40V HVCMOS process, 5.0V Vgs 40V Vds 0.18um backend BCD process and 1.8V/5.0V Vgs 40V Vds 0.18um BCD process, which customers can easily choose for their product needs without any process tuning.

Introduction

BCD technology is one kind of most famous power integration technology which was invented by ST in early 1980s. It integrates high performance bipolar devices for high accuracy analog circuit, cmos devices for high speed and low leakage digital circuit, LDMOS for high efficient power switching output. Other passive devices such as high sheet resistance poly resistor, low temperature middle sheet resistance poly resistor, high density capacitance are also integrated for designer's more choices; some BCD technologies also provide several of diodes for different application, such as clamp, protection and even power switching; HHNEC 0.18um BCD technology provides special buried zener diode with ultra-low walk-out effect performance.

HHNEC's 0.18um BCD technology is one of the most advanced high density power integration technology. This technology is developed by HHNEC's technology development group with the fund aid of National 02 Special Project. There are many inventions relevant process integration scheme, device structure, device layout and process module. More than 100 patents were applied to protect these inventions. The device models and PDK are formal released.

HHNEC's 0.18um BCD technology provides 1.8V CMOS for core and standard library and 5.0V CMOS for I/O. These standard 1.8V/5.0V CMOS devices are matched with industry main stream 1.8V/5.0V CMOS devices in electrical parameters even models exactly, including small devices and narrow devices. Customer can port their complicated digital circuit to this platform directly from other main stream CMOS foundry, save much cost and development time for

---

† Corresponding author. Email: zhangshuai@hhnec.com

customers. This platform also provides 0.18um technology point standard cell and I/O library for digital circuit design. This platform provides many analog IPs for customer's product design, such as band-gap voltage circuit, AC/DC circuit, start-up circuit, over temperature protection circuit, over voltage protection circuit and oscillator circuit; this platform also provides OTP/MTP IPs for customer's application without additional mask layer. The most valuable work of this platform is providing standard power device cells for different application voltages up to 40V. Customer's layout engineers can use this cell to build up power devices as they want easily and reliably.

HHNEC's 0.18um BCD technology was designed by modularized method. Most of devices can be options without any effects on other devices; customers can freely choose base on their product design and cost considerations. Customer can skip barrier layer and EPI layers for some drivers application, then the platform becomes to be a standard 1.8V/5.0V mixed signal process; Customer can even skip 1.8V devices relevant process steps if they only use 5.0V CMOS and other high voltage devices, then the platform will be very cost effective for some DC/DC application and Class-D audio application. For the completed 0.18um BCD platform, heavy doped N type bury layer to decrease vertical parasitical PNP effect and heavy doped N sink to decrease lateral parasitical PNP effect during power device switching and bipolar NPN device work in saturation region.

Basic Structure of main devices

Main device types are shown in following table one.

Main Device	Name	Description
Bipolar	VNPN	low Vcesat
	LPNP	
	Zener	Low temperature coefficient and walk-out effect
CMOS	1.8V~5.0V LVNMOS	Industry main stream 1.8V/5.0V CMOS
	1.8V~5.0V LVPMOS	Industry main stream 1.8V/5.0V CMOS
	20V MVNMOS	Asymetric and symetric, iso and non iso
	20V MVPMOS	Asymetric and symetric
	40V HVNMOS	Asymetric and symetric, iso and non iso
	40V HVPMOS	Asymetric and symetric
DMOS	12V LDMOS	Self-alignment channel
	20V LDMOS	Self-alignment channel
	40V LDMOS	Self-alignment channel
Passive devices	HRPS	high sheet poly resistance
	MRPS	Low Tc poly resistance
	MIM CAP	Low Vc and Tc capcitance
Others	5V SBD	Co silicide schottky barrier diode
	20V SBD	Co silicide schottky barrier diode
	40V SBD	Co silicide schottky barrier diode
	PN junction diodes	

Table 1, main devices in HHNEC BCD180 technology

The basic structures of the key devices in 0.18um BCD technology are shown in following. We design the structures base on the different application field of the different devices for customer products. The devices are designed for better performance for power integration applications.

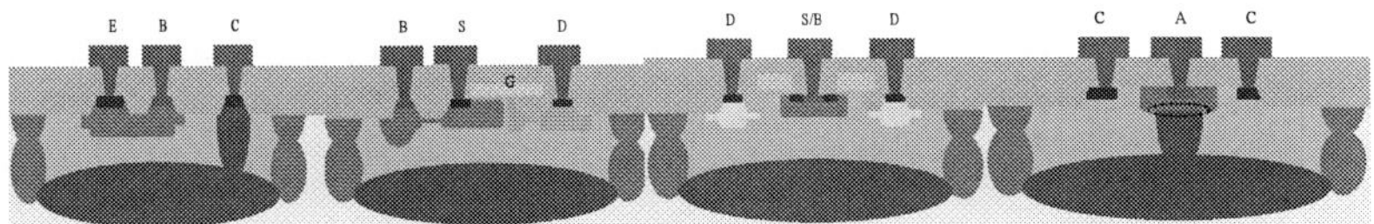

Figure 1: the structure of key devices: NPN, HVNMOS, HVNLDMOS and Zener

This vertical NPN devices have all of the conventional NPN's benefits, special PBASE designed for enough current gain, enough BVceo and enough high early voltage; The NPN in HHNEC's 0.18um BCD technology has performances to match all accurate analog application of designers. This device can work under saturation region safely.

The high voltage NMOS and PMOS devices are designed for high voltage analog application, such as current mirror, current source and other circuits. These devices can work under saturation region with high Vds and high Idsat up to 40V level in HHNEC's 0.18um BCD technology. Linearity is considered for these kinds of devices. 20V devices are provided with shrink device sizes.

The power LDMOS is the key of BCD technology. HHNEC's 0.18um BCD technology provides self-aligned channel power LDMOS up to 40V application including N type and P type. These self-alignment channels have good uniformity for large size power devices, independent to layer CD variation and misalignment. There are standard power cells of this kind of self-alignment LDMOS, the cells are designed with careful consideration of not only specific on resistance minimization and also miller capacitance minimization, customer can use this power cell to build up a large size power device easily. HHNEC BCD180 also provides one special layer to surround the power devices to decrease the current gain of parasitical lateral PNP of power device, thus can obviously decrease the hole current flows to isolation and P type substrate with the aid of NBL together. (Hole current to substrate will induce substrate de-bias failure.)

The invented buried Zener's break down voltage has low temperature coefficient, the break down point is in the body of silicon, not on the surface, the breakdown voltage along time is very stable, that means the break down voltage walk-out effect is very small.

Process integration

The process integration for these devices is base on optimized device design and layer sharing. Only four mask layers are added to the standard 1.8V/5.0V logic CMOS process to build up the baseline of BCD180 platform, other four mask layers are added for high performance power NLDMOS, PLDMOS, high sheet poly resistance and middle low Tc poly resistance together and heavy doped layer for low parasitical lateral PNP gain. This process integration scheme is the minimum one as we already know. The main process steps are shown as the following.

NBL implant and drive in→PBL implant and drive in→PEI growth→Deep Nwell implant and drive in→Deep NSINK implant and drive in→PBase implant and drive in→LVNwell and LV Pwell implant→ Vt adjust implant→Dual gate oxide growth→Gate Poly deposition→NBODY and PBODY implant→Gate poly pattern formation→LV and MV LDD implant→Poly resistor depostion→NPlus and PPlus implant→Silicde Block→Contact→Backend Metallization

1.8V/5.0V CD180 technology can be separated from the 0.18um BCD technology if remove NBL, PBL and PEPI. This CD180 is fully comparable to industry standard 1.8V/5.0V mixed 40V HVCMOS process. Backend is supported up to six metal layers.

The 1.8V CMOS can also be flexible removed from HHNEC 0.18um BCD technology if customers don't use small logic devices. Thus five mask layers will be saved for cost effective.

Silicon results and discussion

The silicon result of HHNEC BCD180 technology is shown in the table 2.

Main Device	Name	Performance
Bipolar	Low Vcesat VNPN	Bvceo=24V, Beta=90, Vcesat=90mv@Beta=10, Ic=1mA
	LPNP	Bvceo=-50V, Beta=100
	5.7V Zener	BV=5.7V, Tc1=100ppm/C
CMOS	1.8V~5.0V LVNMOS	Main stream 1.8V/5.0V compatible
	1.8V~5.0V LVPMOS	Main stream 1.8V/5.0V compatible
	20V MVNMOS	BVdss=28V, Vth=0.93V, Idsat=300μA/μm
	20V MVPMOS	BVdss=-33V, Vth=-0.80V, Idsat=-160μA/μm
	40V HVNMOS	BVdss=52V, Vth=0.93V, Idsat=320μA/μm
	40V HVPMOS	BVdss=-54V, Vth=-0.80V, Idsat=-160μA/μm
DMOS	12V LDMOS	BVdss=21V, Rsp=14mohm.mm2
	24V LDMOS	BVdss=33V, Rsp=23mohm.mm2
	40V LDMOS	BVdss=44V, Rsp=42mohm.mm2
Passive devices	HRPS	Rs=1.0kohm/[]
	MRPS	Rs=240ohm/[] , Tc=100ppm/C
	MIM CAP	~2.9fF/μm2

Table 2 device key performance of HHNEC BCD180

12V NPN I-V curs are shown as following.

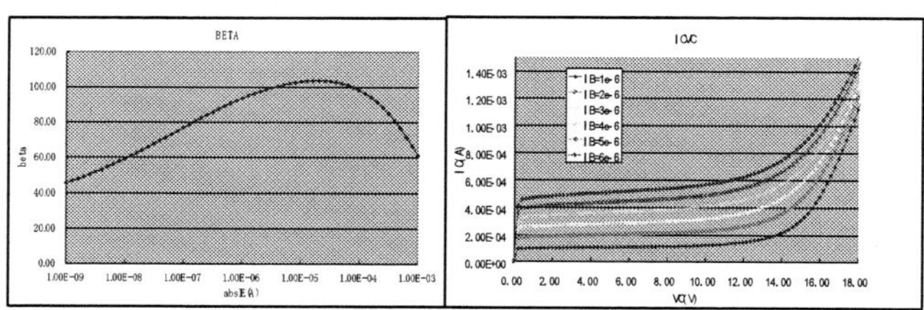

Figure 2, 12V NPN I-V curves

40V LPNP I-V curves are shown as following.

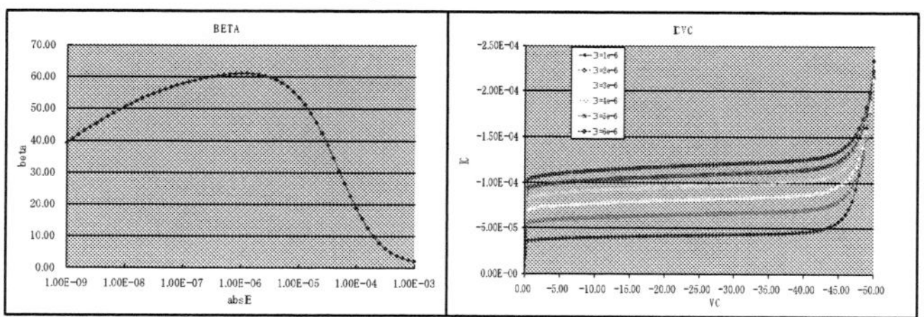

Figure 3, 40V LPNP I-V curves

All bipolar devices I-V curves are normal, performance are OK.

40V HVNMOS and 40V HVPMOS I-V curves are shown in following figure. All I-V curves are normal and device performances are OK.

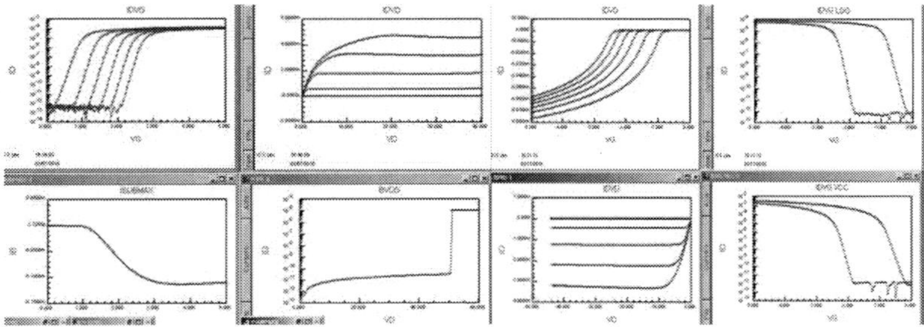

Figure 4, 40V HVCMOS I-V curves

40V NLDMOS and 40V PLDMOS I-V curves are shown in following. All curves are normal and device performances are OK.

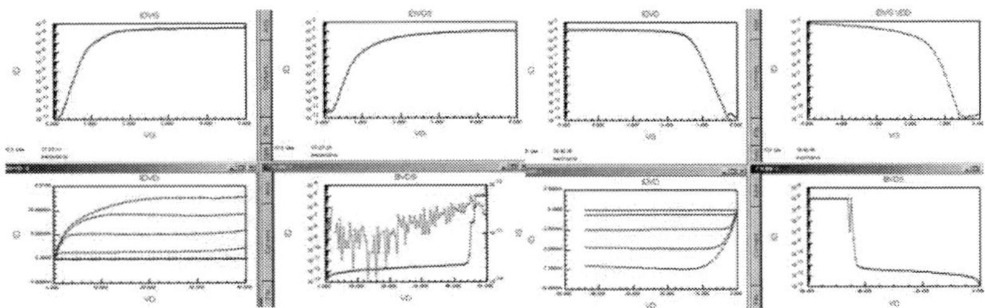

Figure 5, 40V LDMOS I-V curves

Conclusions

HHNEC 0.18um BCD technology provides self-alignment channel power LDMOS for power switching, the devices are designed for FOM optimized, not only Rsp minimum. The technology also provide heavy doped NSINK to decrease parasitical lateral PNP, with the aid of NBL for decreasing the parasitical vertical PNP gain, the total bipolar gain is very small during high current switching; The technology is modularized which 1.8V devices, power switching devices and high sheet poly resistance are optional base on customer's product design requirements. The CMOS devices are close compatible with industry standard 1.8V/5.0V CMOS technology which is very easy for customer logic porting in this technology. The mask layers of the technology are minimized with layer sharing technology to ensure this technology is cost effective.

Acknowledgements

This technology is developed with the fund aid of National 02 special project. Thanks for the aid of the development fund to the committee. Also thanks to the team of the 0.18um BCD project.

References

[1] T. Uhlig et al., "A18 – a novel 0.18um smart power SOC IC technology for automotive

applications," Proc. of ISPSD, pp. 237-240, 2007

[2] Rakesh Vaid and Naresh Padha, "A Novel CMOS Compatible LDMOS for Automotive Applications", *J. of Active and Passive Electronic Devices,* Vol. 3, pp. 101–108

[3] Damiano Riccardi et al., "BCD8 from 7V to 70V: a new 0.18umtechnology platform to address the evaluation of applications towards smart power Ics with high logic contents," Proc. of ISPSD, pp. 73-76, 2007

[4] Il-Yong Park et al., "BD180 – a new 0.18um BCD (bipolar-CMOSDMOS) technology from 7V to 60V," Proc. of ISPSD, pp. 64-67, 2008

[5] F. De Pestel et al., "Deep trench isolation for a 50V 0.35um based smart power technology," Proceeding of the European Solid-State Devices Research Conference (ESSDERC), pp. 191-194, 2003

ECS Transactions, 34 (1) 149-153 (2011)
10.1149/1.3567574 ©The Electrochemical Society

Performance Improvement of Si-NC Memory Device by Using a Novel Junction Assisted Programming Scheme

Dandan Jiang[a, b], Zongliang Huo[a], Manhong Zhang[a], Qin Wang[a], Jing Liu[a], Zhaoan Yu[a], Xiaonan Yang[a], Yong Wang[a], Bo Zhang[c], Junning Chen[b], Ming Liu[a*]

[a] Institute of Microelectronics, Chinese Academy of Sciences, Beijing 100029, China
[b] School of Electronic Sci. and Tech., Anhui University, Hefei 230039, China
[c] Grace Semiconductor Manufacturing Corporation, Shanghai 201203, China

*Corresponding author (E-mail: liuming@ime.ac.cn, Tel: 86-10-82995578, Fax: 86-10-82995583)

Abstract: A junction assisted programming scheme is proposed to enlarge the memory window of Si nanocrystal (Si-NC) memory devices. During this programming scheme, with source, substrate and gate grounded, a negative bias applied to the drain (a forward bias to drain-substrate pn junction for the present NMOS device) in the first programming step; then in the second programming step gate and drain voltages are switched to a conventional channel hot electron (CHE) program setup. Compared to the conventional CHE programming scheme, this new programming scheme is shown to give a much larger memory window. Meanwhile, high program speed, good endurance (over 104 cycling times) and better data retention have also been achieved.

1. Introduction

Silicon nano-crystal (Si-NC) memory device has attracted extensive attention from both academy and industry due to the potential advantages in scalability, low lost and full compatibility with CMOS processes [1-3]. Its potential superior data retention characteristics make it very useful in automobile electronics as an embedded memory component [4].

Usually, a Si-NC memory is erased by Fowler-Nordheim (FN) method and programmed by either FN or CHE method. Due to a limited surface coverage rate of Si-NCs and a low intrinsic coupling ratio in such kind of devices, Si-NC memory device has a small memory window under FN program/erase (P/E) operation. Conventional CHE programming scheme can achieve a fast programming speed. However, due to the very narrow CHE injection region, hot electrons can only be injected into very few Si-NCs near the drain junction. Therefore, even using a CHE program, a Si-NC memory still does not have a large enough memory window. In addition, the narrow injection region also brings up a severe local damage to the tunneling oxide causing reliability issues. To meet highly reliable embedded application, searching better operation method becomes necessary.

In this paper, a new p-n junction assisted programming method is proposed and demonstrated. This scheme consists of two steps. In this first step, with gate, source and substrate grounded, a forward bias pulse is applied to the drain-substrate p-n junction and causes non-equilibrium electrons to inject into the junction space charge region. Then in the second step, the drain and gate bias are switched to the conventional CHE programming setup. The wide distributed non-equilibrium electrons are accelerated and injected into more Si-NCs in the second step. In this way, more Si-NCs have chances to be injected by hot electrons giving a larger memory window in comparison to the one achieved by using the conventional CHE program scheme. Furthermore, endurance and data retention characteristics using the proposed programming scheme and the FN erase method have also been investigated in this paper.

2. Experimental

Fig.1 gives the cross-section scanning electron microscope (SEM) images of the cell structure and Si-NCs. The tunneling oxide is 4nm thermally grown SiO_2. Si-NCs are formed with a two-step low pressure chemical vapor deposition (LPCVD) process [5] with a density of $1 \times 10^{11}/cm^2$ and an average diameter of 20 nm. The blocking oxide is 12 nm high temperature oxide (HTO). Then 200 nm polysilicon was deposited. The equivalent oxide thickness (~15 nm) of the gate stack is extracted from capacitance-voltage measurement. Memory characteristics are measured using Keithley 4200 semiconductor characterization system. The channel width and length of the measured device are 0.26 and 0.5 µm, respectively.

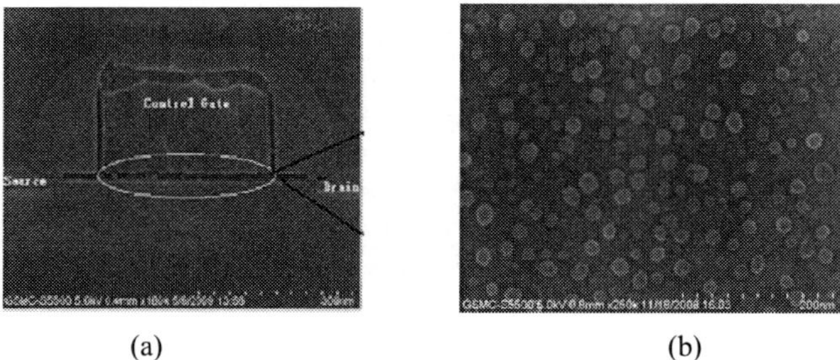

(a) (b)

Figure 1. (a) SEM image of the cross section of the cell structure and (b) Si-NCs

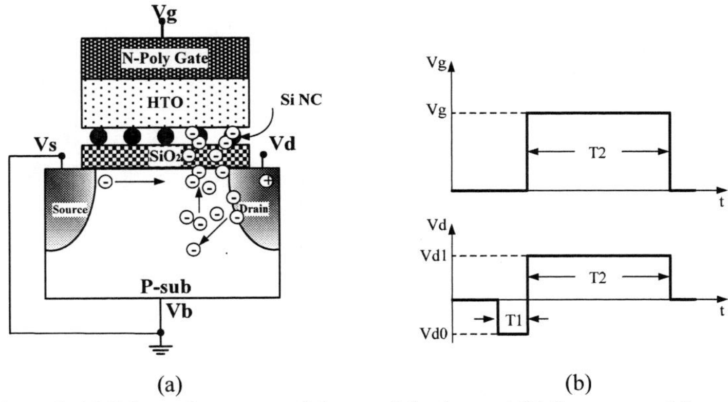

(a)   (b)

Figure 2. (a) Schematic process of electron injection and (b) the program bias setup

Fig. 2 shows the schematic process of electron injection and the programming bias setup. With source and p-well ground, the gate and drain voltage pulses are shown in Fig. 2(b) for the new programming scheme. In the first programming step, a negative drain voltage pulse V_{d0} is applied to the drain. Then in the second programming step, the gate and drain voltage are switched to a conventional CHE programming setup with positive bias V_g and V_{d1}, respectively. The width of the drain voltage pulse in the first step and that of gate and drain voltage pulses in the second step are denoted as T_1 and T_2, respectively.

3. Results and Discussions

In order to demonstrate that the proposed programming scheme can achieve a larger memory window, two programming conditions are compared: the conventional CHE programming scheme with $V_g = 9$ V, $V_{d1} = 8$ V (labeled as 'CHE') and the new programming scheme with $V_{d0} = -6$ V, $T_1 = 1\mu s$, $V_g = 9$ V, $V_{d1} = 8$ V (labeled as 'NEW'). It should be pointed out that the optimal pulse condition in new programming scheme is closely related to pn junctions and the device structure. With increasing forward drain-substrate pn junction voltage, more non-equilibrium electrons are injected into the junction space charge region in the first programming step. These non-equilibrium electrons are then driven to gate jointly by the positive gate and drain voltage in the second programming step. Our initial tests show the memory window increases with increasing the amplitude of V_{d0} and saturates at $V_{d0} = -6V$. To clearly compare two programming scheme, the $V_{d0} = -6V$ is chosen here. Certainly, a smaller V_{d0} value can be chosen in a practical implementation. Fig.3 shows the Id-Vg curves after programming under above two conditions with $T_2 = 500\mu s$. It is clearly seen that the new programming scheme can achieve a larger memory window over the conventional CHE programming setup. The memory window has been enlarged nearly 1V.

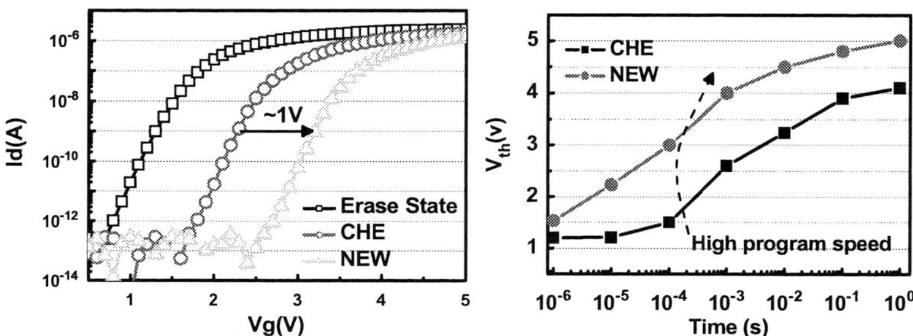

Figure 3. The Id-Vg curves at different programming conditions
Figure 4. Programming speed characteristics under the two programming conditions

In the following, using the above two programming conditions for V_{d0}, T_1, V_g, and V_{d1}, the effect of varying T_2 on V_{th} is examined. Fig.4 presents the programming speed characteristics under the two programming conditions. It can be observed that the new programming scheme increases the program speed 1000 times. To achieve 3V threshold voltage shift, about 1s programming time is needed for conventional CHE condition and only ~1ms programming time is needed for the novel programming scheme. The device reliability using the proposed programming method has also been examined by measuring endurance and date retention characteristics. Fig.5 shows the endurance test results up to 10^4 P/E cycles. The devices were cycled using a FN erase condition of V_g = -13V/1ms and two programming conditions: CHE and NEW. It can be seen that besides a larger memory window, using the new programming scheme also gives a threshold voltage shift similar to that using conventional CHE programming scheme. Therefore, the new programming scheme provides a large memory window without degrading device endurance characteristics.

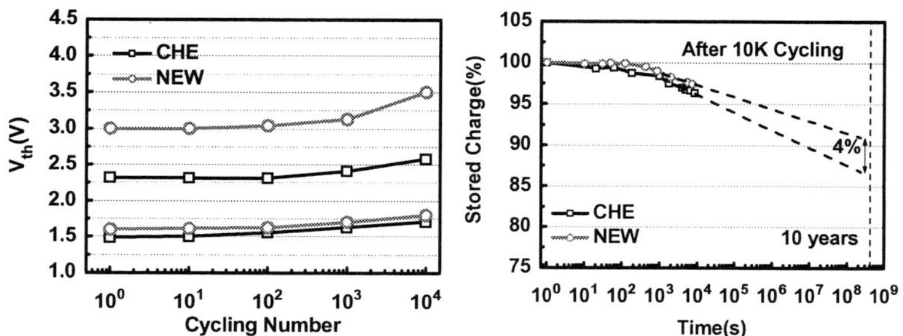

Figure 5. Endurance characteristics at different programming conditions
Figure 6. Date retention characteristics after 10K P/E cycling of the two operation conditions

Fig.6 gives the data retention characteristics at room temperature after 10K (10^4) P/E cycling under a same FN erase condition of V_g=-13V/1ms and two programming conditions (CHE and NEW). Obviously the data retention properties have also been improved greatly by using the new program scheme. Only 10% stored charge is lost in a device cycled under a FN erase condition and the new programming scheme if the retention time is extrapolated to 10 years. This is in contrast with the 14% charge loss in another device cycled under the same FN erase condition and the conventional CHE programming.

4. Conclusion

In summary, a new pn junction assisted programming scheme has been proposed for Si-NC memory. By using this new programming scheme, the device program speed has been improved 1000 times and the memory window has been enlarged nearly 1V. Furthermore, good endurance and the data retention properties using this new programming method have also been demonstrated.

Acknowledgments

This work is supported by National Basic Research Program of China (973 Program) under Grant No. 2010CB934204, National Natural Science Foundation of China under Grant No.60825403, Director's Fund of IMECAS, and Hi-Tech Research and Development Program of China (863 Program) under Grants 2008AA031403.

References

1. S. Tiwari, et al, *Appl. Phys. Lett.* **1377** 68 (1996).
2. R. Muralidhar, et al, *IEDM Tech Digest*, p. 601 (2003).
3. S. Saha, et al, *IEEE Trans. Electron . Devices*, **3049** 54 (2007).
4. Jane Yater, et al, *IEEE International Memory Workshop (IMW)*, p. 82 (2009).
5. Y.M. Wan, et al, *Materials and Processes for Nonvolatile Memories*, **257** 830 (2005).

ECS Transactions, 34 (1) 155-160 (2011)
10.1149/1.3567575 ©The Electrochemical Society

Dual Floating Gate Flash Cell Using Single Poly Processes

Xi Lin[a], Lei Liu[b], Xin-Yan Liu[a], Song-Gan Zang[a], Cheng-Wei Cao[a], Peng-Fei Wang[a],
and David Wei Zhang[a]

[a] State Key Laboratory of ASIC and System, Dept. of Microelectronics, Fudan University,
Shanghai, China,
[b] Oriental Semiconductor Co., Ltd, Suzhou, China

A novel memory device with dual floating-gate is investigated in this paper. The fabrication process of this device is compatible with the standard logic CMOS process flow using single gate poly. It can store 2 bits in a single cell without increasing the cell size. This study provides the fabrication process flow of the dual floating-gate devices. The transfer characteristics, programming, reading, and erasing performances are investigated. The crosstalk between these two floating-gates is also studied. Simulation results show a prosperous prospect of this device. It is promising for high density embedded FLASH applications.

Introduction

Due to capacitance coupling ratio and high operating voltage, it becomes a huge challenge for normal structure FLASH memory device to scale down to 30nm. To raise the FLASH storage density, scaling down may not be a competent choice. We may turn to promote novel memory cells based on different mechanisms. For normal FLASH structure, every cell can only store one bit using two-level storage. The nitride-ROM (NROM) device has a nitride layer for charge trapping. This NROM cell can store two bit, while it has the severe reliability problem due to the hot-hole injection during the erase operation (1), (2), (3). In this paper we design a new structure with two floating gates which can store two bits every cell and it does not increase the cell size. Because of the different structure, working mode of the new designed device is slightly different from that of normal FLASH. Peripheral circuits should be specially designed to support this new structure device correspondently. The fabrication process of this new structure device is compatible with the standard logic CMOS process flow, and this makes it promising to supplant normal FLASH. Device performance is investigated by Silvaco TCAD software.

<u>Device Structure and Working Principle</u>

The device structure is as shown in Figure 1. It has two separated floating gates named fgate and ngate. Each floating gate stores one bit. Length of the two floating gates is 40 nm, and length of the control gate is 100 nm. Oxide thickness under the floating gate and the inter-poly-dioxide (IPD) thickness is 6 nm. The doping concentrations for n^+ control gate, n^+ floating gate, n^+ source and n^+ drain are 1e20 cm^{-3}. P-type substrate doping is 4e17 cm^{-3}. There are also retrograde p^+ doping and p^+ HALO doping. This process is compatible to the standard logic processes with single poly. When the line of the control gate is formed, a wet process is performed to remove part of the oxide under the control gate. Then, the undercut part is filled up by a 6 nm oxide and doped poly-Si.

Device working principle is mostly the same to it of normal FLASH. But the operation principle is different. When programming a floating gate, the region near it should be treated as drain. When reading information from a floating gate, the region near the floating should be treated as source. That is because energy band structure in the channel near source region plays a vital role in I-V characteristic. It can be derived from the energy band state of a normal MOSFET working in saturation mode. If the floating gate near drain is programmed, the threshold voltage of the channel where is covered by the floating gate is raised, and threshold voltage of the remained channel is as initial value. When applying a gate voltage which is slightly larger than initial threshold voltage, the channel under the programmed floating gate remains depletion or weak inversion state. If a sufficient drain voltage is applied, the energy band is just similar with the energy band state of a normal MOSFET working in saturation mode. So we can still get considerable saturation current. If the floating gate near source is programmed, the channel is in "off" state when applying a gate voltage which is slightly larger than initial threshold voltage. So we can hardly get adequate current. This is the key principle to distinguish the stored two bits. Simulated energy bands shown in figure 2 also support this inference. Compare figure 2 (d) with figure 2 (a), we can find that there is a relatively large energy barrier between drain and channel in figure 2 (d), and channel controlled by ngate is still "off" when source voltage is 3 V. So the current is very small which is shown in figure 4. As shown in figure 2 (b) and (c), channel controlled by ngate turns from "off" state to "on" state when drain voltage increases from 1 V to 3 V. This characteristic can be found in figure 4.

<u>Simulation and Discussion</u>

When only one floating gate needs to be programmed, the gate voltage is set to 5 V and the drain voltage is set to 4 V. The charge density during this transient programming simulation is plotted in figure 3. We apply a proper voltage pulse length around 1.1e-7 s on drain to get a charge density around 1.1e-15 $C\cdot cm^{-2}$. After the programming operation, different voltage is applied on drain during read operation. As indicated in figure 4, it is found that the programmed floating gate makes no significant influence on I-V curve when Vd is sufficient large. When we apply different voltage on source, an obvious Vt shift is observed as shown in figure 4. Positive current means the current flow from drain to source. In figure 4, we still use the initial transfer curve considered from drain instead of the initial transfer curve considered from source. Because the only difference is the current value sign which means the current direction. To distinguish the two bits, 4 V drain voltage and 2.5 V gate voltage are recommended.

If the dual floating gates need to be programmed, we should program the dual floating gates one after another. We use 5 V gate voltage and 4 V source voltage. As indicated in figure 5, ngate has already been programmed at the beginning. Fgate is programmed during the programming transient simulation. Ngate also captures some charges during the programming transient simulation. We apply a proper voltage pulse length around 1.1e-7 s, and charge density of fgate and ngate is about 1.1e-15 $C\cdot cm^{-2}$ and 1.2e-15 $C\cdot cm^{-2}$. Vt shift is observed when applying voltages on drain side or the source side at reading operation (see Figure 6). It means that the two bits are stored.

To erase the dual floating gates with one action, we apply 0 V on gate while drain and source voltage is both 10 V. Erasing transient simulation results are shown in Figure 7.

Due to the FN tunneling, we should control the time of the erasing pulse to avoid over erasing. An easing time shorter than 1e-4 s is recommended for this device.

<u>Summary</u>

In this paper, we have designed a FLASH cell with dual floating gates. The operating conditions are studied. The most important advantage of this structure is the increased storage density without increased cell size. However, compared with normal FLASH, different structure results in different operating modes. New operating circuits should be designed to support the new device. The device performance has been investigated by simulation. Since the manufacturing processes are compatible to the standard logic process flow, this cell is promising for high density embedded FLASH applications.

Acknowledgments

This project is supported by The Program for Professor of Special Appointment (Eastern Scholar) at Shanghai Institutions of Higher Learning.

References

1. B. Eitan, P. Pavan, I. Bloom, E. Aloni, A. Frommer and D. Finzi, *Electron Device Letters, IEEE,* **21**, 11 (2000).
2. L. Larcher, P. Pavan and B. Eitan, *Electron Devices, IEEE Transactions on,* **51**, 10 (2004).
3. A. Shappir, E. Lusky, G. Cohen, I. Bloom, M. Janai and B. Eitan, *Electron Device Letters, IEEE,* **24**, 4 (2003).

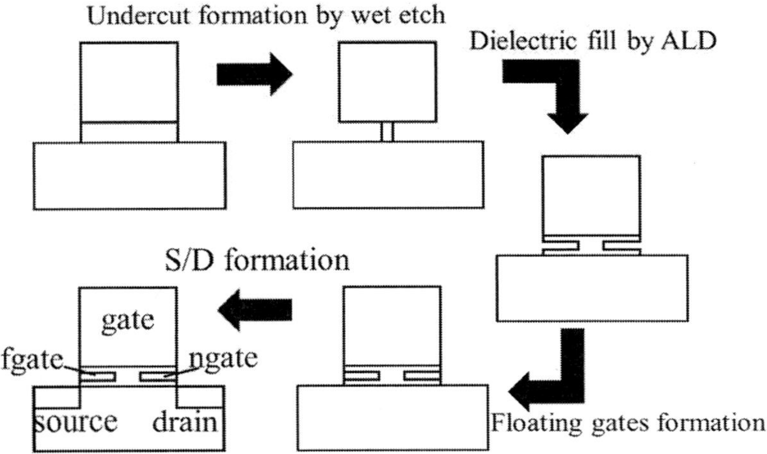

Figure 1. Device Structure and process flow.

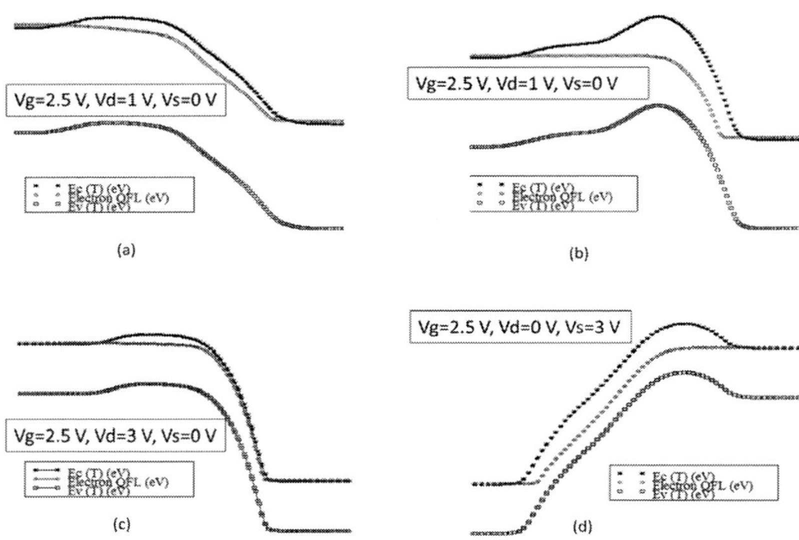

Figure 2. Simulated energy band diagrams of different states. Source, channel and drain are placed from left to right. Device in (a) is not programmed. Device in (b), (c) and (d) are ngate-programmed.

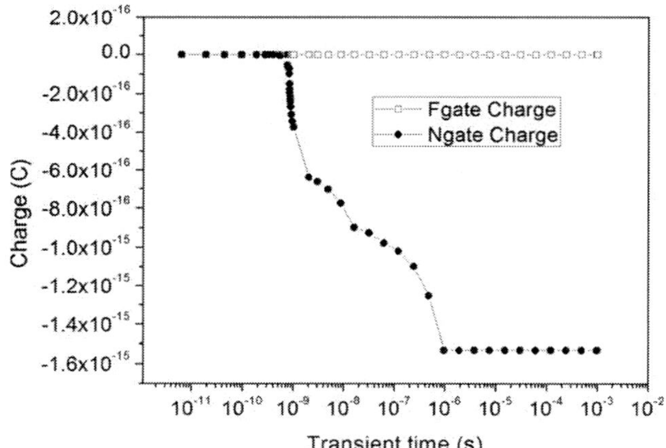

Figure 3. Ngate programming transient simulation. Time axis is in log scale. Fgate captures almost no charges during the ngate programming transient simulation. Gate voltage is set to 5 V, drain voltage is set to 4 V and source voltage is set to 0 V.

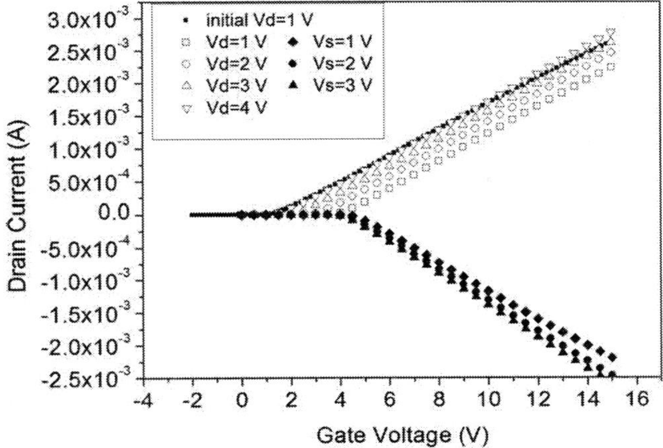

Figure 4.   Transfer characteristics curves with different bias conditions after ngate programming transient which is shown in figure (3). The only difference is the current direction which is indicated by the value sign. Minus sign means the current flow from source to drain. Positive sign means the current flow from drain to source. When Vd = 4 V, Vt shift caused by ngate is negligible. When applying voltage on source in the reading operation, there is an obvious Vt shift.

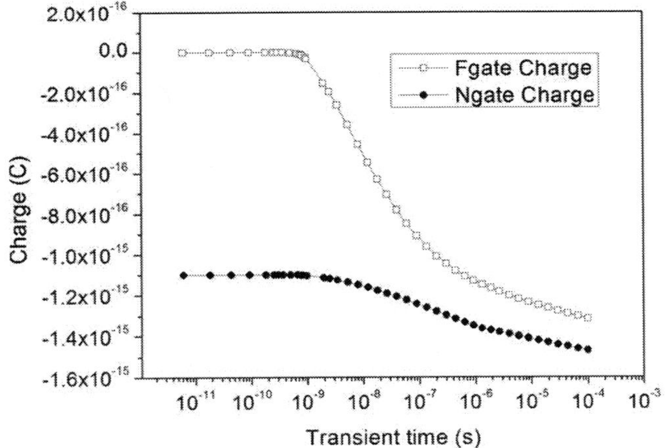

Figure 5. Dual floating gates programming transient simulation. Gate voltage is set to 5 V, source voltage is set to 4 V and drain voltage is set to 0 V.

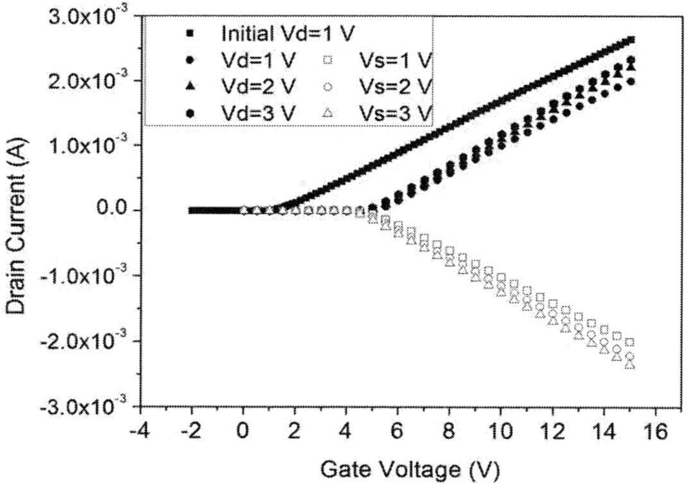

Figure 6. Transfer characteristics of the programmed dual floating gates after dual gates programming transient which is shown in figure (5). An obvious Vt shift is observed when reading both either from source or drain which means two bits are stored.

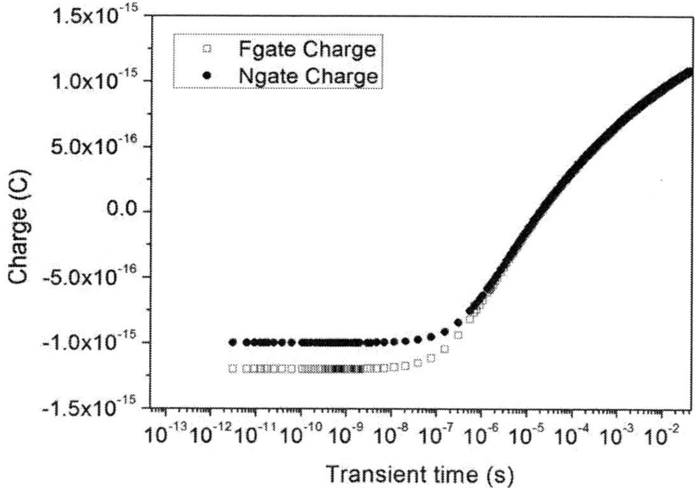

Figure 7. Erasing transient simulation. Erasing time should be controlled to avoid over-erasing.

ECS Transactions, 34 (1) 161-166 (2011)
10.1149/1.3567576 ©The Electrochemical Society

Anomalous Behaviors of Cubic GaInN Ternary Alloys

Nacir Tit

Department of Physics, UAE University, Al-Ain, United Arab Emirates
ntit@uaeu.ac.ae

The electronic and optical properties of cubic $Ga_{1-x}In_xN$ ternary alloys are theoretically investigated using the sp^3s^* tight-binding model, with the inclusion of spin-orbit interaction, versus composition and valence-band offset (VBO), stimulating the lattice-relaxation effects. Usually, the bowing of the band-gap energy in the common-anion ternary alloys should be vanishingly small as far as the virtual-crystal approximation remains valid. In contrary to this, the present alloys are shown to possess three unusual characteristics: (i) They possess rather strong bowing character with $b > 2.0$ eV; (ii) the bowing parameter is composition dependent $b = b(x)$; and (iii) the Stokes shift between emission and absorption is so large that can even reach 200 meV when $x \approx 0.5$. Two reasons are claimed to cause such behaviors; namely: the strong electro-negativity of nitrogen atoms and the large lattice relaxation as being induced by the large lattice-mismatch between the two constituents (GaN and InN). The favorable comparison with the available experimental data corroborates the above claims.

1. Introduction

The recent advent in growth techniques has paved the way to the nitride alloys to extend their range of applications beyond the telecommunication field to comprise the domain of photonics (1). There are already many new device structures which are commercially available and many more at the advanced research stage (1).

Understanding the chemical and physical properties of semiconductor alloys is essential and consists rather the only channel to make any further development in the side of technological applications. One among the approaches is to give priority to focus on understanding the properties of the common-cation and common-anion ternary alloys. From there, one can proceed toward studying the properties of more complex alloys such as quaternary alloys and beyond (2). However, despite decades of extensive studies, there are no commonly accepted explanations for the diverse observed behaviors of the bandgap energy (E_g) versus composition (*e.g.*, *bowing* (2-4), linearity (2-4), anti-crossing (2) and anomalous (5) behaviors).

Nitride alloys (In, Al, Ga)N exhibit many anomalous features particularly strongly pronounced in InGaN and InAlN ternary alloys. While the variation of band-gap energy versus composition is expected to be linear as these latter belonging to the category of common-cation ternary alloys, the behavior is rather anomalous. These alloys exhibit strong *bowing* whose parameter is composition dependent. Particularly, our present work focuses on investigating the origins of the bowing behavior in the $Ga_{1-x}In_xN$ ternary alloys using the sp^3s^* tight-binding method with inclusion of spin-orbit coupling effects.

2. Computational Method

The tight-binding (TB) Hamiltonian uses a minimal atomic basis set, which yields a great power to the method to deal with very large systems containing thousands of atoms. The parameters are usually obtained empirically by fitting the valence bands and low-lying conduction bands to those obtained via first-principle methods, while the bandgap energies to fit to the experimental values. In the present investigation, we use the sp3s* TB-models with the inclusion of the spin-orbit coupling effects, developed by Hernandez-Cocoletzi et al. (6). Furthermore, the validity of two main points is assumed: (i) the virtual-crystal approximation (VCA) in evaluating the supercell atomic structure. The energy error-bar in such calculations due to the neglect of atomic relaxation will also be estimated. (ii) the problem of energy reference between the alloy constituents is sorted out by taking the valence-band offset (VBO) into account. Furthermore, this latter quantity is considered to stimulate the atomic relaxation effects.

Using the eigen-values and eigen-vectors of the TB Hamiltonian, the total density of states (TDOS) and its components as the partial densities of states (PDOS) are calculated. TDOS is normalized to 10 electrons. The obtained PDOSs are, in turn, used to calculate the electric charge q_α and the ionicity I_α of the atomic species α (such as: Ga, In or N atoms).

3. Results and Discussions

<u>Ternary-Alloy DOS:</u>

Figure 1 displays the calculated TDOSs and PDOSs for the ternary $Ga_{1-x}In_xN$ alloys with increasing indium content: (a) x=0.25, (b) x=0.50, and (c) x=0.75. In each case, the VB-edge is taken as an energy reference and both the VB-edge and the CB-edge are indicated by the shown-vertical dotted lines. Figure 1 used VBO= 0.26 eV, which was obtained by Wei and Zunger (7). Furthermore, Figure 1 focuses on the near-gap energy region within 6.0 eV around the VB-edge as to study the variation of valency states versus indium content. We start by looking at the TDOS variation versus indium content x. The increase of the CB with increasing x reveals that the In-atoms are more electropositive than the Ga-atoms and, thus, more ionization of nitrogen atoms should be expected. By comparing the PDOSs of Ga-atoms and In-atoms within VBs (E $\leq$ 0 eV), especially at a concentration of x=0.50, one can clearly notice the arrangement taking place between the two cations in filling the VB-states and, consequently, in the compromise in losing their charges. Lastly, in the PDOS of N-atoms, one may notice the growth of the overall magnitude of VB with the increasing indium content. This reveals that N-atoms are getting more negatively ionized with the increasing In content. Therefore, it is interesting to analyze the variation of ionicity of each atomic species in the alloy versus In content.

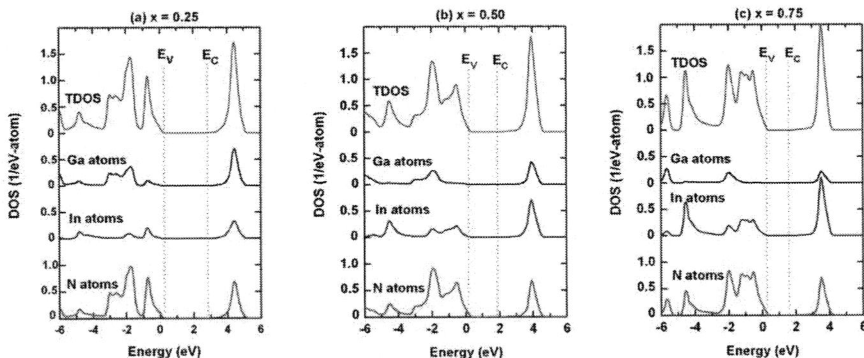

FIGURE 1. TDOS and PDOS components for $Ga_{1-x}In_xN$ with (a) x=0.25, (b) x=0.50 and (c) x=0.75. E_v is taken as an energy reference.

<u>Ternary-Alloy Charge Ionicity</u>:

The total charge on the upper shell of each constituent atom in the alloy is calculated by integrating its corresponding PDOS up to the Fermi level. From the calculated charge, the ionicity of each atom is calculated and presented in Figure 2. Three different VBO values are considered and presented in different panels: (a) VBO= 0 eV, which is usually predicted by the well-known common-anion rule of hetero-structures; (b) VBO = 0.26 eV, which was obtained by Wei and Zunger (7) using the first-principles all-electron band-structure method; and (c) VBO= 1.11 eV, which was recently obtained by Li and coworkers (8) using the same preceding method but by taking into account the deformation potential of core states. The charge ionicities of atomic species in the $Ga_{1-x}In_xN$ alloy are indicated by the following symbols: open circles for N atom; full triangles for Ga atom; and full squares for In atom. The charge neutrality is well fulfilled in every case, namely, as:

$$I_N + (1-x)I_{Ga} + xI_{In} = 0$$

Where I_N, I_{Ga} and I_{In} are the atomic-charge ionicities of N, Ga and In atoms, respectively.

Furthermore, in all three panels, the charge variation of nitrogen atom in the $Ga_{1-x}In_xN$ alloy seems to be linear and independent of VBO. In contrast, the ionicities of cation atoms (Ga and In) are found to be sensitive to VBO. Meanwhile, there exists a critical value of VBO (hereafter denoted by V_c) at which the slopes of ionicities of the latter two atoms change from being positive (Fig.2a) to negative (Fig.2c). It is noticeable that the ionicities of Ga and In follow the same variation as their corresponding curves are parallel (i.e., having the same slope). The rule of variation of the charge ionicities may be written as:

$$I_N = -0.29(1-x) - 0.624x$$

$$I_{Ga} = sx + 0.29$$

$$I_{In} = s(1-x) + 0.624$$

where s is the slope: s= 0.313 (V$_c$-VBO), with V$_c$ = 0.384 eV. At this latter critical VBO value (i.e., VBO=V$_c$), the ionicities of Ga and In atoms in the alloy are supposed to remain constant as in their corresponding bulk structures. This latter is not really a favorable case as the Ga and In atoms possess different electro-negativity characters and should rather couple to N in different ways. A competition in losing their charges especially in presence of very strong electronegative anion atoms, such as N-atoms, is very much expected. So, the slope "s" should be either positive or negative. For VBO>V$_c$ (as in Fig.2c), to expect the In atom to lose charge more than its state in the bulk InN, while the Ga atom to loose less charge than its state in the bulk GaN, is not really a probable case. On the other hand, for VBO < V$_c$ (as in Figs.2a and 2b), a compromise between the cation atoms (Ga and In) in losing their charges to the nitrogen atoms would be more physically favorable. So, our TB-calculation indicates VBO will be small in the present case of Ga$_{1-x}$In$_x$N alloys (i.e., VBO <0.38 eV). This is not only based upon the present discussion of the charge ionicity variation but also upon the modeling of the experimental data, which will be shown in next sub-section. Besides the fact that small or vanishing VBO values are consistent with the predictions of the common-anion rule.

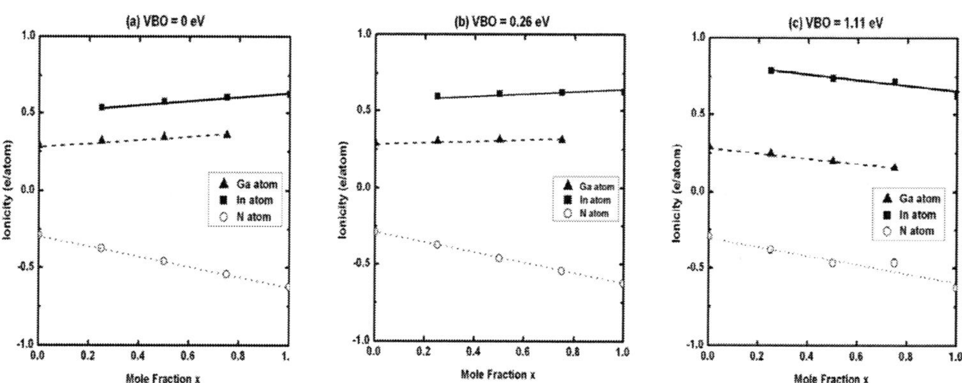

FIGURE 2. Charge ionization versus mole fraction x with different VBO: (a) VBO = 0, (b) VBO = 0.26 eV and (c) VBO = 1.11 eV.

<u>Modeling of Experimental Data:</u>

Using molecular-beam epitaxy (MBE), high-quality In-rich Ga$_{1-x}$In$_x$N films (0.5 ≤ x ≤ x 1.0) were grown on sapphire substrates (9), employing either an GaN or InN buffer layer and producing WZ alloys. The optical properties of the produced samples were characterized by optical absorption and photoluminescence spectroscopy at about 80 K. Based on the Abs data, with the exclusion of their proper PL data, besides using some experimental data of Ga-rich alloys (namely, due to the band-gap energy measurements by photo-modulated transmission and optical absorption (9), a *bowing* parameter B=1.43 eV was reported). However, one should emphasize here that if the PL data of Wu and coworkers (9) corresponding to the In-rich alloys were also included in the fitting, the *bowing* parameter would have been much greater than the preceding reported value.

More recently, Franssen and coworkers (10) reported their experimental results of pressure-dependence of PL of $Ga_{1-x}In_xN$ films in the full composition range $0 \leq x \leq 1.0$ at 80 K. Their PL data are shown in Fig.3 in open circles. They confirmed the clear deviation from the linear to the *bowing* behaviors with a parameter B even larger than the one reported by Wu and coworkers (9). In Fig.3a, we assumed a single *bowing* parameter to fit each theoretical or experimental data. For the displayed experimental data, we have performed a non-linear fitting using the functional form:

$$E_g = xE_g^{InN} + (1-x)E_g^{GaN} - Bx(1-x)$$

where we took $E_g^{GaN} = 3.30$ eV and $E_g^{InN} = 0.72$ eV. In Fig.3a, the result of the fitting of the optical absorption data, shown in dotted line, yields a bowing parameter B=1.703 eV. Whereas the fitting of the PL data (10), shown in small-dashed curve, yields B=2.297. The TB results are shown in Fig.3a by crosses. Using the above functional form, the best fit to our TB results is shown in solid line and yielding B=1.644 eV, in excellent agreement with the Abs data, and do coincide with the dotted line fitting the experimental absorption data. The PL data seem to predict an even higher *bowing* parameter and smaller VBO than 0.26 eV (see below). Consequently, the data shown in Fig.3a provide experimental evidence for the clear deviation from linearity to clear *bowing* character. Meanwhile, one may notice that TB result (cross) at x=0.25 lies below the solid line of TB fitting; whereas the cross at x=0.75 lies above the same solid line. This may reveal that the TB results corroborate the idea of *bowing* enhancement in the region of low In content. As a matter of fact the PL data, in the region of low In content, lye much below all fitting curves and reveal high *bowing* parameter. Here, it is worth trying to apply a composition-dependent *bowing* parameter, which decreases with the increasing *In* content.

It is worth noting the clear discrepancy between the PL and Abs experimental data. This discrepancy is known by the Stokes shift between emission and absorption spectra, respectively. Stokes shift might be a negative indicator of the quality of grown samples. From a theoretical point of view, it is not completely clear the origin of the Stokes shift, even if many models have been proposed (11). In the experimental data shown in Fig.3, the Stokes shift can reach 200 meV in consistency with the *ab-initio* calculations of Ferhat and coworkers (11). The formation of *InN* clusters (dot-like) should enhance the formation and recombination of the electron-hole pairs and the Stokes shift might be an indicator to that structural heterogeneity.

In our modeling, to assess the relaxation effects, we present in Fig.3b two extreme cases of VBO. The smallest value VBO=0, which is predicted by the common-anion rule. From there, we maximize the value of the *bowing* parameter by taking the one corresponding to lower *In* concentration (x=0.25). On the other hand, the largest value VBO=0.62 eV, which was recently reported by Moses and Van de Walle (12). From which we minimize the value of the *bowing* parameter by taking the one corresponding to the high *In* concentration (x=0.75). The curves corresponding to these latter two *bowing* parameters are shown in Fig.3b with a shaded area in between them. It is clear that the *bowing* parameter cannot be less than about 1.0 eV. However it should even exceed 2.362 eV particularly for low *In* content. The shaded area might represent the TB-theoretical error bar accounting for the lattice relaxation and the deviation from the VCA validity.

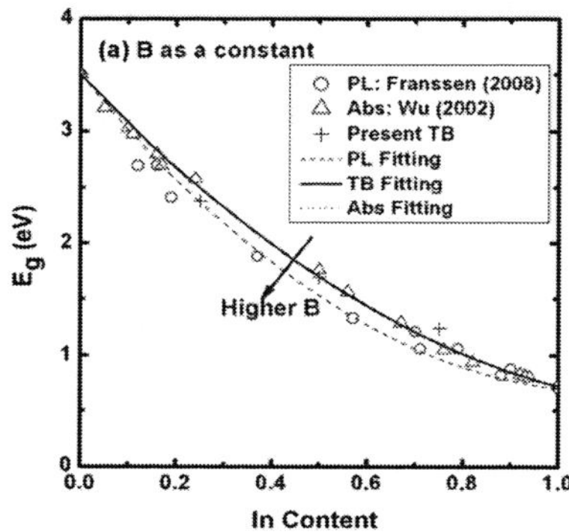

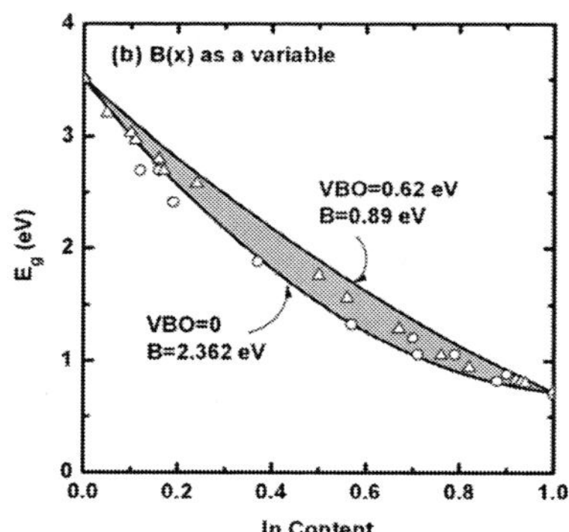

FIGURE 3. Comparison of the TB-results of $E_g(x)$ of $Ga_{1-x}In_xN$ with the experimental optical absorption and PL data. (a) B is constant and (b) B=B(x) as variable.

4. Conclusions

The *bowing* character in the $Ga_{1-x}In_xN$ ternary alloys has theoretically been investigated using the sp^3s^* tight-binding method with the inclusion of the spin-orbit interaction. Three unusual behaviors were under focus:

(i)   The *bowing* is rather strong and exists as a result of the pronounced electro-negativity of the nitrogen atoms.

(ii)   The *bowing* parameter is found to be composition dependent as a result of the large lattice-relaxation induced by the huge lattice mismatch between the alloy constituents, which is about 10% (between GaN and InN).

(iii)   The large Stokes shift between the absorption and emission properties can reach 200 meV when x=0.5 as been caused by the huge lattice relaxation.

The favorable comparison between the theoretical results and the experimental data corroborates the above claims.

5. References

1. M. Henini, in *Dilute Nitride Semiconductors,* Elsevier Science (2005).
2. I. Vurgaftman, J.R. Meyerand and L. Ram-Mohan, *J. Appl. Phys.* **89,** 5815 (2001)
3. H.F. El-Haj, S.J. Hashemifar and H. Akbarzadeh, *Phys. Rev. B 73,* 195202 (2006).
4. M.J. Seong et al. 1999 *Solid State Commun.* **112,** 329 (1999).
5. S.H. Wei and A. Zunger, *Phys. Rev. B 39,* 6279 (1989).
6. H. Hernandez-Cocoletzi et al., *Physica E* **41,** 1466 (2009).
7. S.H. Wei and A. Zunger, *Appl. Phys. Lett.* **72,** 2011 (1998).
8. Y.H. Li et al., *Appl. Phys. Lett.* **94,** 212109 (2009).
9. J. Wu et al., *Appl. Phys. Lett.* **80,** 4741 (2002).
10. F. Franssen et al., *J. Appl. Phys.* **103,** 033514 (2008).
11. M. Ferhat, J. Furthmuller, F. Bechstedt, *Appl. Phys. Lett.* **80,** 1394 (2002).
12. P.G. Moses, C.G. Van de Walle, *Appl. Phys. Lett.* **96,** 021908 (2010).

ECS Transactions, 34 (1) 167-171 (2011)
10.1149/1.3567577 ©The Electrochemical Society

0.18um Scalable 7~45V pLDMOS for Smart Power Application

Zhengchao Liu, Shushu Tang, James Shen, Chris Shao

Technology Development, Grace Semiconductor Manufacturing Corporation, Shanghai 201203, China

This work describes a 0.18um CMOS process based wide range scalable p-channel LDMOS transistor which targeted for power management application. The scalable pLDMOS is to use one process to cover different operation voltage devices, with the drain voltage scaling the device's specific on resistance (Rdson) must be always kept at a reasonable level and it is comparable with standalone technologies. The rated drain voltage can be continuously operated from 7V up to 45V, mean while the corresponding blocking voltage and Rdson are at 14V/6.5 mohm.mm^2 and 60V/102 mohm.mm^2 respectively.

1.0 Introduction

There're many new BCD technologies developed at 0.18um ~ 0.35um node by different companies (1,2,3,4) recently. These technologies can offer different operation voltage devices in a certain range but they are only some specific dots. In this work, a 0.18um logic process based scalable LDMOS is developed to use one process to cover a certain continuously operation voltage range from 7V up to 45V. Mean while the transistors specific on resistance (Rdson) are always kept at a reasonable level within this range. So designers can use any voltages within this range and they also have a chance to use different operation voltage devices in one chip at the same time without any additional cost, and all these devices are optimized at their rated operation voltage. From the foundry side, it means less supporting documents are necessary and it's easy to maintenance such kind of technology. This is a totally different way from the traditional method to develop a new high voltage technology. It can not only help the foundries to reduce the development cycle and cost but also can help the designers to shrink the design cycle and avoid risks.

Thanks to 0.18um core logic process, it's possible for designers to integrate high density logic circuit into DMOS technology when compared to other old BCD technologies. Such as high density NVM, SRAM, PWM controller circuit, it's also known as SoC design or Smart Power application. This is one of an important direction of BCD technology – high density BCD(5). Typically the voltage capability requested to power devices in high density BCD are from 5V to 50V, this work will cover most of this range.

Since it's a bulk wafer based low cost BCD process and the target market is smart power application. So analog devices are very important to this platform, such as BJT, MIM, High Resistor poly. For BJT, there're 7 different BJTs will be offered in this platform depends on the process combination. Even though they're parasitic BJTs but they have good performance, they can replace the EPI based BCD in most of the case.

Besides these, a modular architecture can provide a flexible platform for designers. They can easily select the components which they want to use in their design thus to save the mask layers. It's very critical for cost sensitive market now.

2.0 Device structure and power device characterization

2.1 Device structure

The scalable pLDMOS is fabricated on a bulk wafer based standard 0.18um logic process with two additional implantation layers. These two implantation layers are done before baseline implantation, so it's fully compatible with 1.8V/5V CMOS baseline process. As a dual gate process, the pLDMOS will share the same gate oxide as 5V device. The gate voltage is 5V for all the pLDMOS. Fig 1 shows the cross section of the scalable pLDMOS. The P-drift is sitting in a deep NWELL, with varying of drift width Lg, a voltage scalable pLDMOS can be implemented. It means the breakdown voltage is changing as Lg is varying, mean while the pLDMOS Rdson will be kept at a reasonable level within this range when compare to other standalone technologies. The tuning of this scalable pLDMOS is much more complicated than a single device. The balance among device Rdson, breakdown voltage and voltage scaling have to be considered.

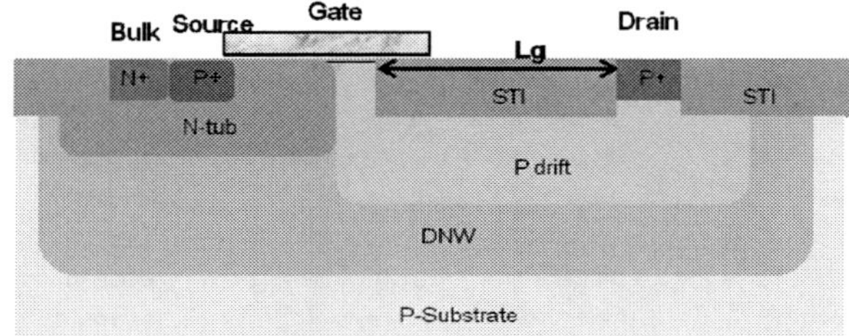

Fig 1 pLDMOS cross-section

2.2 Power device characterization

Four key important voltage nodes 45V, 24V, 12V and 7V are selected to stand for this scalable pLDMOS platform. Fig 2a~2d show the IDVD curves of the pLDMOS at rated operation voltage of 45V, 24V, 12V and 7V, all the devices have good performance at the full voltage range from Vg=0V to Vg=5V. Devices at other operation voltages have the same performance as these samples. It'll be very flexible for designers to use this technology because they can use any voltages in the range of 7V to 45V. From the foundry side, it can also benefit a lot from this technology. It can use one technology to serve very different application customers and save developing cost. Because there's only one time process development and it can also fast the time to market cycle.

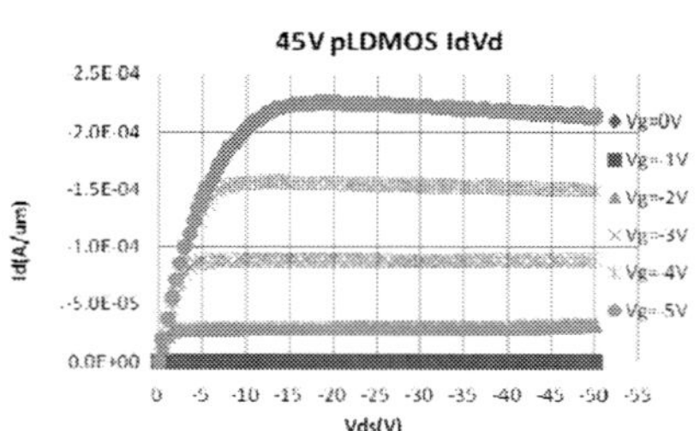

Fig 2a 45V pLDMOS output curve

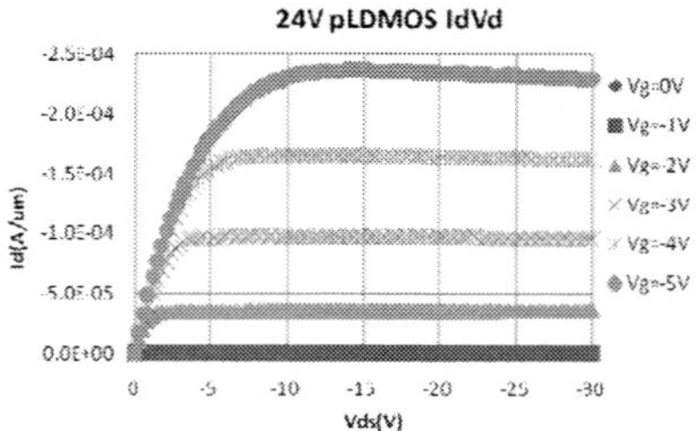

Fig 2b 24V pLDMOS output curve

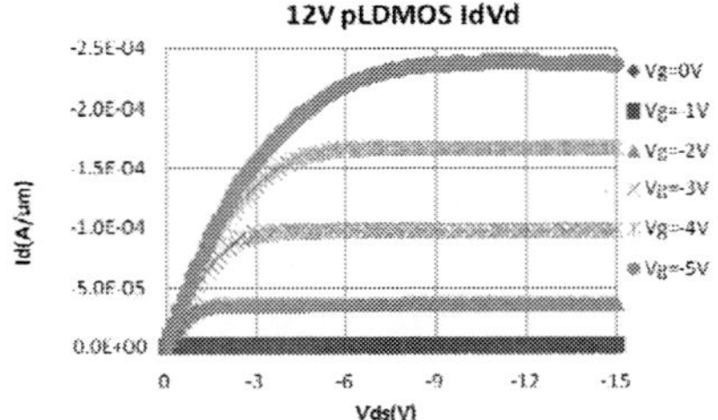

Fig 2c 12V pLDMOS output curve

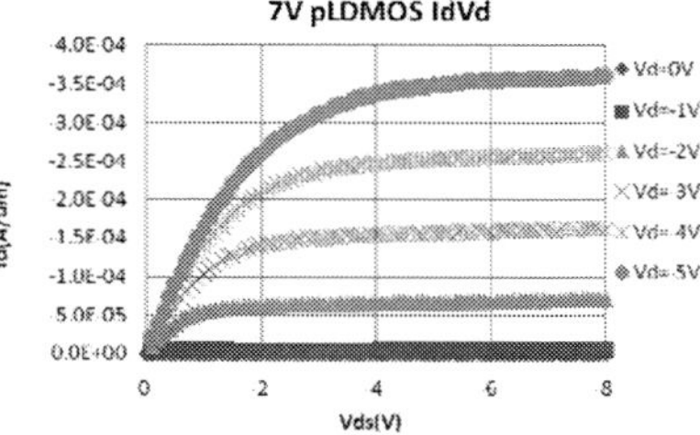

Fig 2d 7V pLDMOS output curve

Besides the output characteristics, blocking voltage and specific on resistance (Rdson) are two other important parameters for power devices. Fig 3 shows the device breakdown voltage under different operation voltage, the Bvds is increasing from 14V to 59V as the rated operation voltage is increasing from 7V to 45V. There's very good correlation between Bvds and rated operation voltage. Fig 4 illustrate the benchmark of Bvds vs Rdson, this work has very good scalable trend between Rdson and breakdown voltage. The Rdson are at 6.5mohm.mm^2 and 102 mohm.mm^2 when the corresponding blocking voltage is increasing from 14V to 59V. This result is much better than that which other foundries can offer. And it even can compete with some IDM companies.

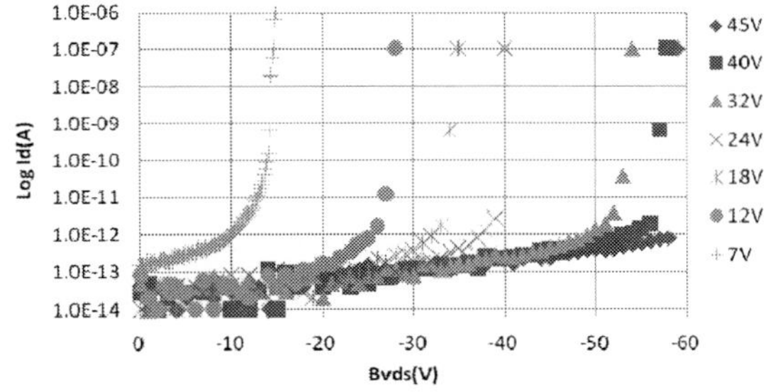

Fig 3 Bvds under different operation voltage

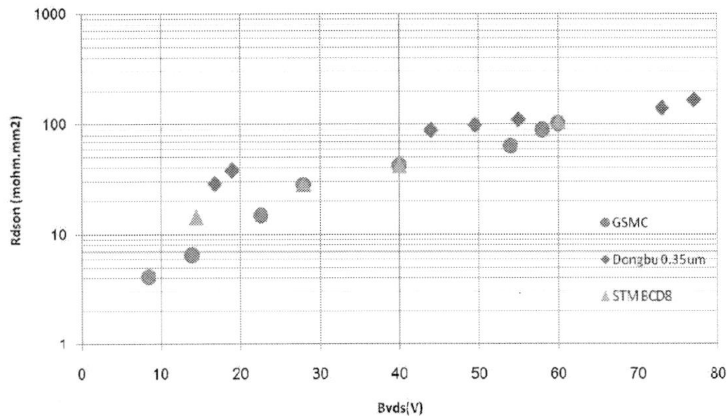

Fig 4 Benchmark Rdson trend vs Bvds

Nowadays, reliability is more and more important in power management application, sometimes it is even more important than performance. Electrical SOA is an important parameter to judge the robustness of a power device. Fig 5 is the measured Electrical SOA curve of a 45V pLDMOS under TLP pulse 100ns. The maximum drain voltage can as high as 90V before the parasitic PNP turns on and the current can sustain up to 7A/cm.

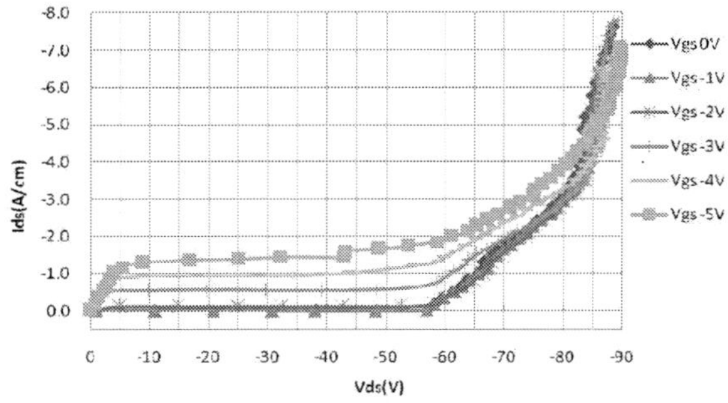

Fig 5 45V pLDMOS electrical SOA

2.3 BJT characterization

BJTs are key devices for analog application, in this platform high performance BJTs will be offered. Depending on the implant layer combination, 7 different BJTs can be formed. Here two BJTs are characterized. Fig 6a, 6b show the beta and BVceo of one vertical NPN BJT. This BJT's beta is as high as 60 and BVceo is 16V. Fig 7a, 7b show the beta and BVceo of one vertical PNP BJT, the beta can as high as 24 and BVceo can up to 55V. Even though these parasitic BJTs are formed on bulk wafer they still have good performance, they can replace the EPI based BJTs in most of the cases.

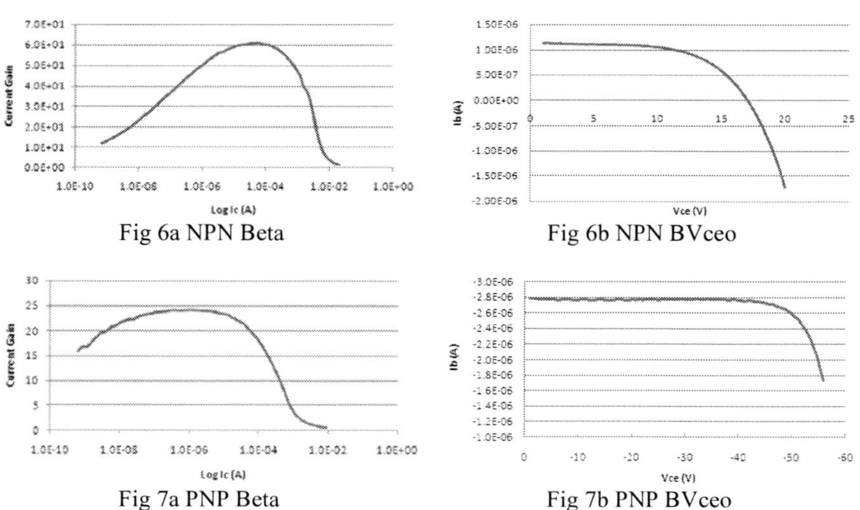

Fig 6a NPN Beta

Fig 6b NPN BVceo

Fig 7a PNP Beta

Fig 7b PNP BVceo

3.0 Conclusion

A low cost BCD used scalable pLDMOS is implemented on a bulk wafer based 0.18um logic platform. The operation voltage is wide enough to cover most of the smart power applications. The operating voltage can be from 7V up to 45V, the corresponding blocking voltage and Rdson are at 14V/6.5 mohm.mm2 and 60V/102 mohm.mm2 respectively. Besides superior device performance, this scalable pLDMOS has good electrical SOA performance too. For analog devices, high performance BJTs are offered in this platform. Even though they're parasitic devices the performance is comparable with epi based BJTs.

References

1. S. Pendharkar, R. Pan et al, *ISPSD*, p.419-422, 2004.
2. D. Riccardi, A. Causio et al, *ISPSD*, p.73-76, 2007.
3. T. Letavic, S. Sharma et al, *ISPSD*, p.108-111, 2009.
4. C.J. Ko, C.H. Cho et al, *ISPSD*, p.177-180, 2010.
5. C. Contiero, A. Andreini and P. Galbiati, *ESSDERC*, p.275-282, 2002.

ECS Transactions, 34 (1) 173-181 (2011)
10.1149/1.3567578 ©The Electrochemical Society

0.18 micron BiCMOS process with novel structure SiGeC HBT

Donghua Liu[a], Wensheng Qian[a], Xiong Bin Chen[a], Fan Chen[a], Jun Hu[a],
Sheng'An Xiao[a], Yungchung Wang[a], Tzu-Yin Chiu[a]

[a] Department of Technology Development , HHNEC, Shanghai 201206, P.R. China

This paper reports a manufacturable process of 0.18 micron SiGe
BiCMOS integrating novel structure SiGe HBTs and 0.18 micron
foundry-compatible CMOS devices. N-type SiGe HBT is newly
designed by removing deep trench isolation, N+ sink collector
pick-up and collector epitaxial growth. Instead, junction isolation,
collector pick-up of deep contact through field oxide and
implanted collector are used in SiGe HBT. The peak fT and fmax
of N-type high speed SiGe HBT are 120GHz and 110GHz,
respectively and the size of SiGe HBT is only half as conventional
one. CMOS front-end process steps except P+ source/drain
implants and RTA are prior to that of SiGe HBTs to minimize the
impact of CMOS thermal budgets on SiGe HBT.

Introduction

Silicon germanium (SiGe) BiCMOS process is now becoming the hot technology for
radio-frequency (RF) applications (1). This process offers high performance SiGe HBT,
standard CMOS and good integration of SiGe HBT, CMOS and passive devices to realize
RF, analog and digital functions on a single chip. The major application area of SiGe
HBTs is for amplifiers including power amplifiers (PA) and low noise amplifiers (LNA).
High voltage SiGe HBTs and high speed SiGe HBTs are convenient for fabrication in
one process flow and easy to form PA and LNA circuits on one chip, and the low noise
figure of SiGe HBT meets the requirements of LNA (2) (3). SiGe HBTs have some key
advantages over RFCMOS: 1. SiGe requires less aggressive technology nodes. 2. SiGe
HBTs have higher breakdown voltages. 3. SiGe HBTs have much higher gm. 4. SiGe
HBT's 1/f noise figure is two orders lower than that of RFCMOS. 5. SiGe HBTs have
better output conductance with same technology node. 6. PAs with SiGe HBTs have
higher power added efficiency (PAE). SiGe HBTs' advantage over GaAs HBTs is better
compatible with CMOS. It is convinced that SiGe BiCMOS technology will occupy more
and more wireless regions (4).

This paper presents a 0.18 micron manufacturable SiGe BiCMOS process which
integrates new structure SiGe HBTs and 0.18 micron foundry compatible CMOS. High
speed and high voltage SiGe HBTs exist simultaneously in the process to enable PA and
LNA integration. The device structure and process of SiGe HBT have significant
difference from conventional one. For high speed SiGe HBT, the cut-off frequency (fT)
and the maximum oscillation frequency (fmax) are 120GHz and 110GHz, respectively.
For high voltage SiGe HBTs, fT and fmax are 28GHz and 70GHz.

SiGe HBT Structure and Process

The basic structure of the new device concepts are depicted in Fig. 1. After a standard STI etch is complete, a lithography step is used to define a region surrounding the bipolar active area. A heavy implant is then introduced to serve as pseudo buried layer. The heavy implant is blocked from the active region by the STI dielectric stack and sidewall (Fig. 1a). After the STI oxide refilling and CMP (Fig. 1b), the bipolar active base area is opened for SiGeC epi layer growth (Fig. 1c). A SiGeC layer of 60nm with a base sheet resistance of about 2000 ohm/sq is grown. The base poly area is then lithographically defined and etched. An emitter region is defined and etched the oxide /nitride bilayer to expose the epitaxial SiGeC layer (Fig. 1d) and an undoped polysilicon layer is deposited followed by a high dose emitter implant. The emitter poly is then defined and etched (Fig. 1e). A high dose self aligned extrinsic base implant is then introduced. The dopant are activated by drive in and RTA. A self-aligned salicidation of emitter/base poly follows. After IMD was deposited, a deep collector contact over the pseudo buried layer was defined and etched (Fig. 1f). A second contact was opened to expose emitter and extrinsic base region. Standard barrier layer and W plug was deposited into the collector, base and emitter contacts. After W CMP, Al metallization follows (Fig. 1g). Fig.1(2) shows the layout comparison of new scheme HBT and conventional HBT. In new scheme one, the distance (x1) of deep contact of collector to base poly can be the minimum as process allowed. In conventional one, the deep trench isolation and NBL enlarge the device size as well as the AA used to pick up collector.

Fig. 2, 3 show the IV characteristics and gummel plot of a device with emitter size of 0.2um × 13.9um. Fig. 4 plots the current gain versus IC showing a constant current gain over 5 decades. The results show that the dc performance is reasonably ideal compare to that of a standard device with buried layer. We are able to generate a series of high BVCEO device by modifying the device layout. Fig.5 shows the fT/fmax versus Ic curves of devices with 0.2umx1.1um emitter window. The high speed and high voltage (BVCEO=7V) devices have measured fT/fmax of 120GHz/110GHz and 28GHz /40GHz respectively. The collector resistance is well within the normal range compare to devices with buried layer as table 1 showed. Other device parameters are shown in Table 1. The CCS parasitic of this device is highly competitive compare to trench isolated HBT bipolar. A CCS of 2fF is obtained for our minimum device with Ae = 0.18um x 1.1um. A potential issue with new device structure is the substrate pnp when the HBT goes into heavy saturation. Fig.6(1) and Fig.6(2) show the substrate pnp behavior of new scheme HBT device and conventional HBT device respectively.

SiGe BiCMOS Process Flow

Fig.7 indicates the overall BiCMOS flow and how the HBT device is integrated into the CMOS. For minimizing the impact of CMOS thermal budgets on SiGe HBTs, SiGe BiCMOS process flow follows the rule of "SiGe HBT after CMOS". However the process steps of N-type and P-type pseudo buried layers are done before CMOS fabrication. The pseudo buried layers is over-drive-in due to the additional thermal budget of CMOS. This issue can be solved by changing the dopants to heavier impurities, like As & In for N-type & P-type respectively. Other HBT relevant processes, from LC to DCT, can be integrated into the CMOS flow smoothly. The performance of N/PMOS has slight shift after HBT device merged in, but its impact can be minimized by optimizing the HBT process. Device is tuned back easily by LDD and halo implantation. Fig.8(1) and Fig.8(2) present the 1.8V N/PMOS output characters. Both are matched well with BL performance and their difference is less than 5%.

Conclusion

We have demonstrated an innovative new HBT process that eliminated conventional buried layer and collector epi growth with a high dose pseudo buried layer surrounding the active bipolar area. Trench isolation is unnecessary in the absence of collector epi and its associated lateral auto-doping. Very low CCS value can be achieved with inexpensive junction isolation. Diffused collector plug is replaced by deep via with W plug for a low RC and smaller device area. Good DC parameters and competitive parasitics are obtained. fT and fmax performance of 120GHz and 110GHz respectively have been measured. Process cost is significantly lowered along with sizable area reduction for the HBT devices. This device has also been integrated into CMOS process successfully with neglectable impact on the performance of CMOS.

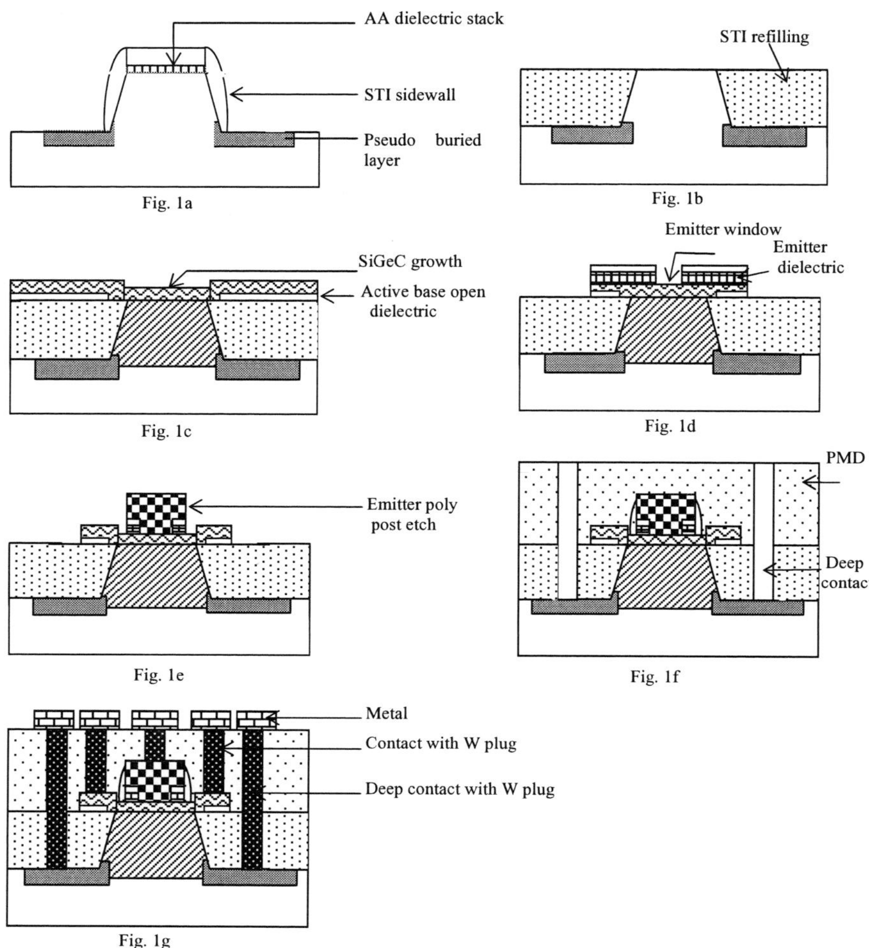

Fig. 1(1) Outline of the new HBT Fabrication with pseudo buried layer and W collector plug

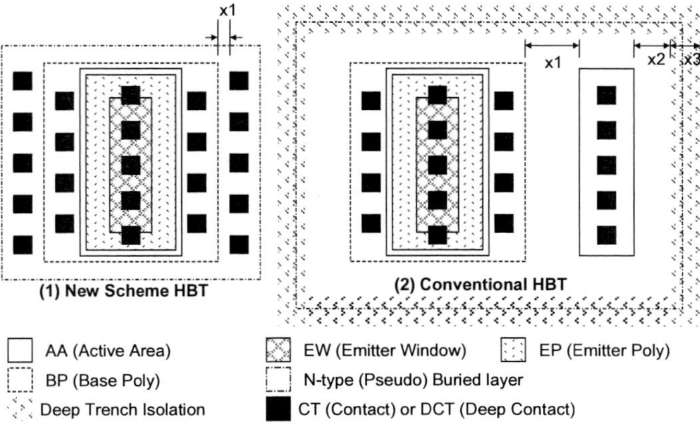

Fig. 1(2) Layout comparison of new scheme and conventional HBT

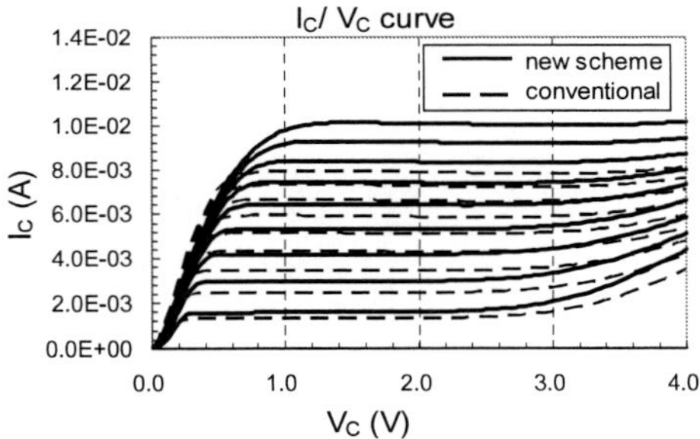

Fig. 2 IV Characteristics comparison with Ib=0uA to 90uA (10uA per step)

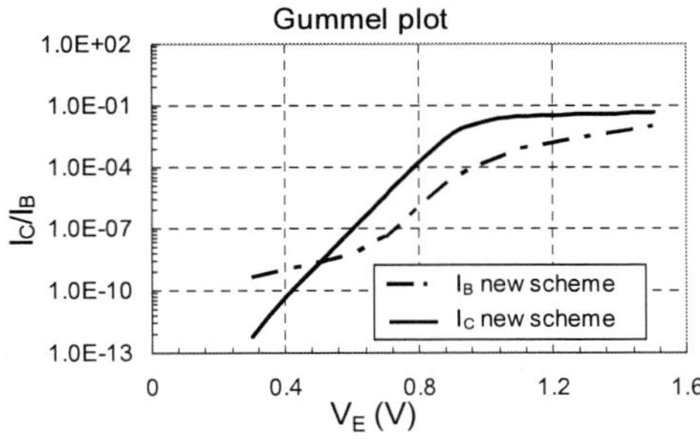

Fig. 3 New scheme gummel plot (Vb=0,Vc=0, sweep Ve)

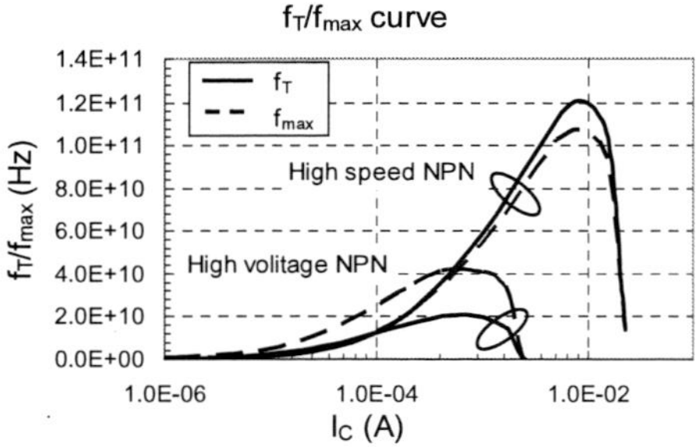

Fig. 4 Beta curve comparison(Vb=0,Vc=0, sweep Ve,Ic/Ib)

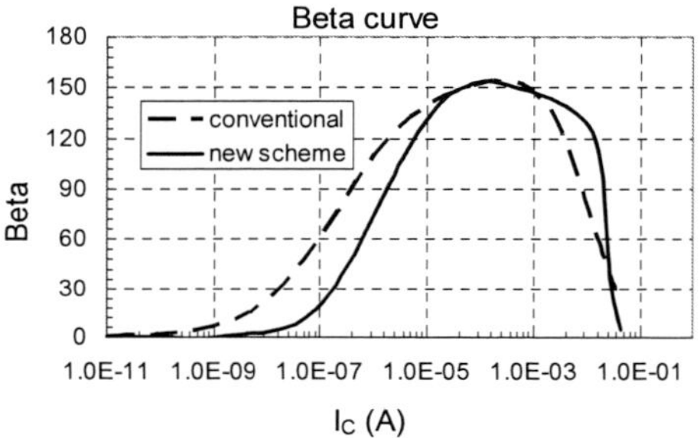

Fig. 5 fT/fmax curve (emitter window area= 0.22um2)

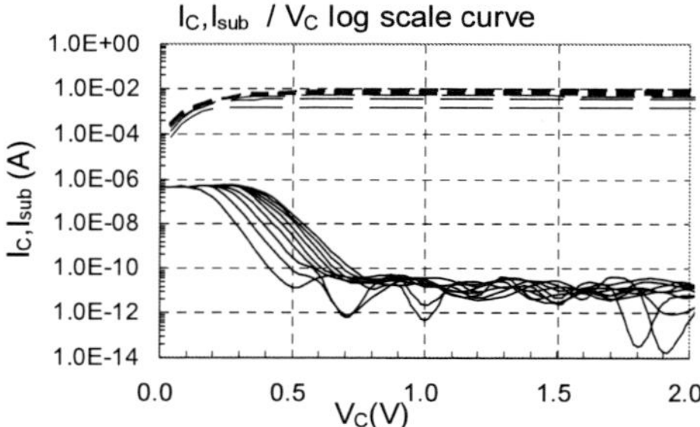

Fig. 6(1) New scheme HBT Is/Vc Curve Ib=0uA to 90uA (10uA step)

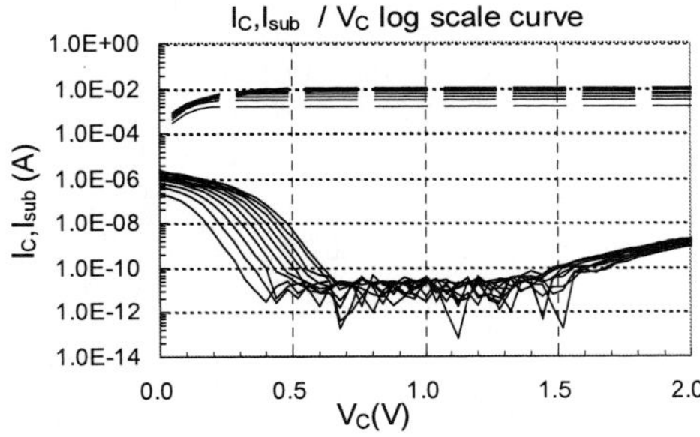

Fig. 6(2) Conventional HBT Is/Vc Curve   Ib=0uA to 90uA (10uA step)

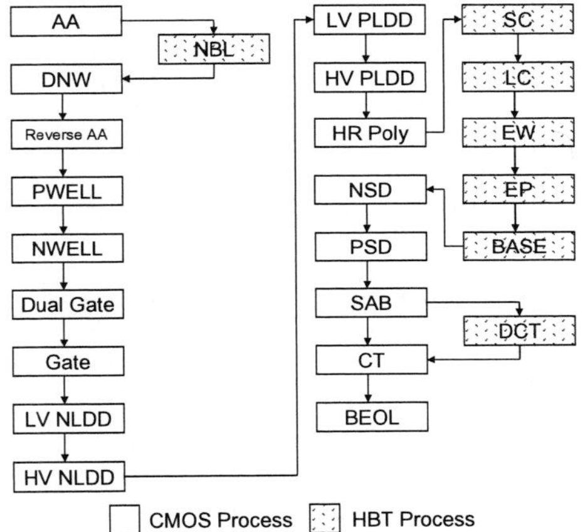

Fig. 7 The process flow of CMOS and HBT are integrated

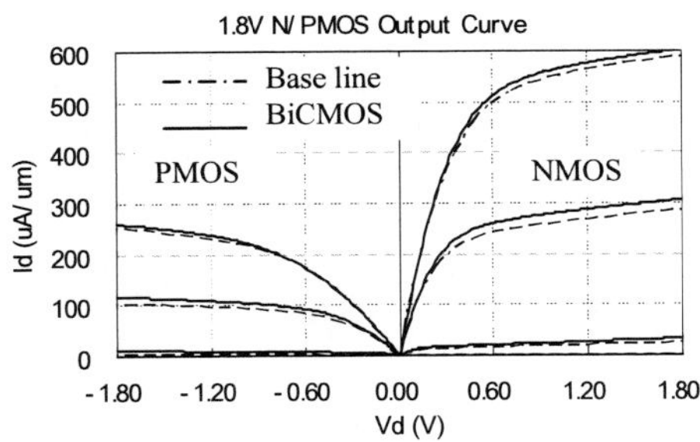

Fig.8(1) 1.8V CMOS input performance

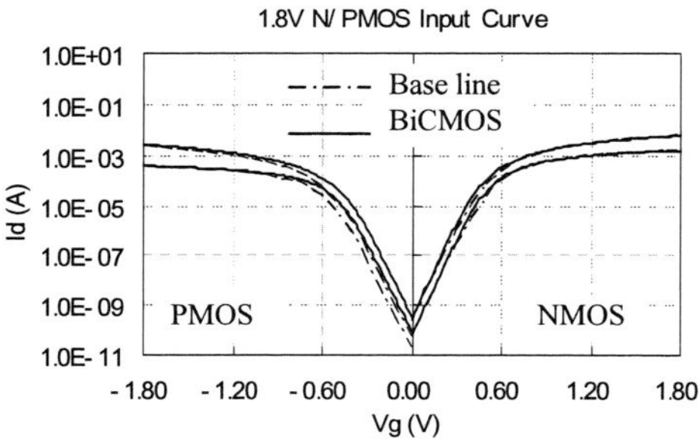

Fig.8(2) 1.8V CMOS output performance

Table.1 HBT device characteristics (60nm 20%Ge SiGeC base +1005°C final RTA thermal)

Parameters	Unit	Pseudo buried layer w/o trench (Ae=0.2um x 9.9um)		N-type buried layer with trench isolation (Ae=0.3um x 9.9um)		Measurement Condition
		High voltage	High speed	High voltage	High speed	
Beta	--	170	170	160	160	I_E=100uA
BV_{CEO}	V	7	2	6	2.3	$|V_{BE}|$=0.7V, Vc sweep @Ib=0V
f_T	GHz	28	120	30	71	V_{BC}=1v
f_{max}	GHz	40	110	65	75	V_{BC}=1v
C_{CS}	fF	9.0	9.3	12.2	12.2	Bias 0v
R_C(EBC 122)	ohm	42	36	37.6	33	$\triangle V_C/\triangle I_C$
V_A (Ib=10uA)	V	100	500	60	20	I_B=10uA , -X intercept

Acknowledgments

The authors would like to thank RF group and advanced module group of HHNEC for their great support on this SiGe HBT project.

References

1. Marco Racanelli and Paul Kempf, IEEE TRANSACTIONS ON ELECTRON DEVICES, **1259**, VOL. 52, NO. 7, JULY 2005
2. Zhenqiang Ma and Ningyue Jiang, IEEE TRANSACTIONS ON ELECTRON DEVICES, **875**, VOL. 53, NO. 4, APRIL 2006
3. J. F. W. Schiz and Andrew C. Lamb, IEEE TRANSACTIONS ON ELECTRON DEVICES, **2492**, VOL. 48, NO. 11, NOVEMBER 2001

4. Vijay S. Patri and M. Jagadesh Kumar, IEEE TRANSACTIONS ON ELECTRON DEVICES, **1725**, VOL. 45, NO. 8, AUGUST 1998

ECS Transactions, 34 (1) 183-188 (2011)
10.1149/1.3567579 ©The Electrochemical Society

Temperature Insensitive Clock Buffer and Its Application on Clock Tree

Meng Tie[a], Xia Li[a]

[a] China Chip Design Center, IBM Systems &Technology Group, Beijing 100193,
P.R.China

High power density and uneven power density distribution caused large thermal space gradient on VLSI. Since CMOS gate delay is highly relevant to temperature, the thermal gradient could impact the performance of IC by increasing clock tree skew on chip. To maintain circuit performance, we utilize thermal behavior discrepancy between P and N type CMOS transistors to reduce clock skew under this circumstance. First, this paper proposes a kind of temperature insensitive clock buffer with cross-coupled structure. Second, a clock tree is then built based on proposed buffer to reduce clock skew induced by thermal gradient. Unlike traditional methods, the proposed circuit can work under normal supply voltage and no special voltage converter is needed.

Background and related work

As semiconductor technology develops, the power density and total power consumption of integrated circuits are arising. According to ITRS(1), the power density of high performance processor has reached $120W/cm^2$. High power density can cause high temperature on die, which impacts performance. At the same time, temperature gradient is obvious in high performance chips, which is caused by some reasons. First, popular use of cache make the memory area ratio increased, the power density differences between logic gates and memory caused temperature gradient(2). Second, some low power techniques make the power density more uneven, such as multi-voltage, power gating and clock gating. Besides, low K dielectric material make the gradient more severe(3). In some processors, temperature gradient is reported as high as 77°C(4).

Since CMOS transistors and metal wire resistors are sensitive to temperature, thermal variation can affect circuit performance. Among all parts of circuits, clock signal distribution circuits are very critical. Temperature gradient can cause different clock signal arrival time to each flipflop, i.e. clock skew. This will not only happen during normal functioning but also among testing process(5). Unintended clock skew could cause setup time or hold time violations for synchronous circuits. During normal functioning, these timing violations can appear as performance degradation or functional failure, while during at-speed testing process some good paths could be incorrectly diagnosed as faults or vice versa.

Therefore temperature aware clock signal distribution is important. Several literatures have proposed different methods to reduce clock skew induced by temperature gradient. Temperature aware clock tree synthesis algorithm(6) can reduce 50% to 70% interconnect wire induced clock skew. It considers both even and uneven temperature distribution on chip to building clock tree, so that interconnect RC delay will cause minimum skew under both conditions. However, there are always a log of clock buffer gates on clock tree and gate delay holds as much as 75% of total clock latency(7). Thus this temperature aware clock tree cannot eliminate most of clock skew caused by temperature. Minz et. al proposed a clock tree synthesis algorithm which considers buffer delays under different temperature distributions. But their method only adapts two temperature circumstances so it cannot reduce clock skew sufficiently. Clock mesh is insensitive to environmental factors, such as temperature(8). But it is not appropriate to the designs with useful clock skew and it needs more routing resource(8).

Bota and Tawfik et. al proposed dual-VDD clock distribution scheme(5). They lower the supply of clock tree to a special voltage, which is called Temperature Insensitive Cross Point Voltage. Then the delay of a clock buffer almost no longer changes with temperature and the clock skew is minimized under any temperature distribution. With this technique, thermal induced clock skew can be reduced by 74%. However, lowered supply voltage makes clock signal more vulnerable to noise. Moreover, it needs extra dedicated power network which takes more design effort and routing resource.

In this paper, a cross-coupled structure based clock buffer, TICB (Temperature Insensitive Clock Buffer) is proposed, which can reduce the temperature dependence of clock buffer delay under normal supply voltage. An H type clock tree based on TICB is then simulated and compared with dual-VDD clock tree.

Temperature Insensitive Clock Buffer

A temperature insensitive structure dedicated for temperature sensor was proposed in (9). In this work, we simplify that structure, save the cell area and satisfy the balance requirement between rise and fall transition times to make it a clock buffer gate.

First, review the delay formula of a normal CMOS inverter which is sensitive to temperature. It can be expressed as(10):

$$T = \frac{2C_L V_T}{K\ (V_{DD} - V_T)^2} + \frac{C_L}{K\ (V_{DD} - V_T)} \ln\left(\frac{1.5V_{DD} - 2V_T}{0.5V_{DD}}\right) \qquad [1]$$

V_T is threshold voltage, C_L is the effective load of inverter, K stands for the transconductance, $K = \mu C_{ox}(W/L)$ where μ is mobility, Cox is gate capacitance per unit area, W stands for channel width and L stands for channel length.

During the parameters, mobility and threshold voltage are temperature dependent(11):

$$\mu(T) = \mu_0 \left(\frac{T}{T_0}\right)^{km} \qquad [2]$$

$$V_T(T) = V_T(T_0) + \alpha_T(T - T_0) \qquad [3]$$

Here, T_0 is reference temperature, μ_0 and $V_T(T_0)$ are mobility and threshold voltage under temperature T_0 respectively. When temperature rises, the change of mobility increases delay while the change of threshold voltage does oppositely. In the 65nm technologies, under normal supply voltage, thermal behavior of gate delay is mainly decided by mobility. Thus, gate delay increases as temperature rises.

According to the SPICE simulation of IBM 65nm technology, NMOS transistors are more sensitive to temperature than PMOS transistors. So similar with the work in (9), NMOS transistor can be cross-coupled in a gate to reduced thermal sensitivity.

In (9), the proposed structure consists six transistors including a pair of cross-coupled PMOS and NMOS transistors. We modified it as shown in Figure 1 left. In the modified circuit, there are two levels of inverters and one cross-coupled NMOS transistor, N2. On the right of Figure 1, it shows the relationship between current and temperature. Assuming a falling edge input at the gates of P1 and N2, P1 charges for node C while N2 tries to discharge node C at the same time. So the total charge current is $I=I_{P1}-I_{N2}$. It can be seen on the right graph of Figure 1 that, since I_{N2} is more sensitive to temperature than PMOS, the total current exhibits reversed temperature dependence opposite to the current of a single transistor. Therefore, the delay of the first level inverter has a negative temperature coefficient. The transistors shown in doted lines act as a load capacitance for point C. For a rising edge input, a coupled PMOS(not shown in graph) will work against N1. However, a rising edge input does not induce a total current with reversed temperature coefficient because PMOS is less sensitive to NMOS. Since a coupled PMOS cannot reverse thermal behavior, we cut off the coupled PMOS and only retain N2 transistor to save area and power.

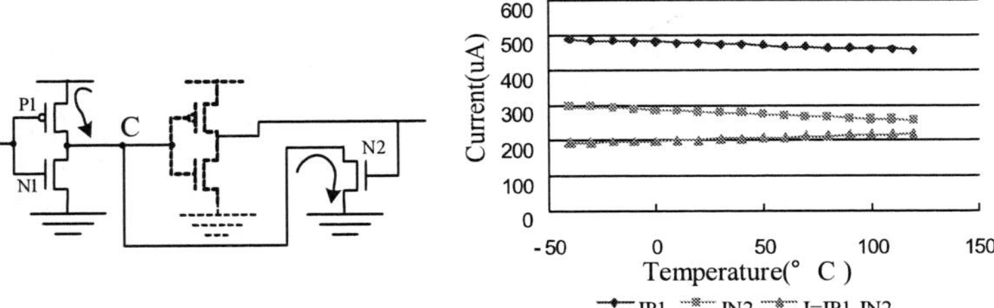

Figure 1. Clock buffer with reversed NMOS (left) and relationship between current and temperature (right)

On a clock tree, clock buffers are cascaded with each other and they always have the same size. Thus a special buffer which makes the clock tree insensitive to temperature can be designed. According to (9), with minimum channel length, when the channel width of P1 and N2 satisfies the following condition, the output low-to-high propagation delay T_{PLH} of first level inverter can have a negative temperature coefficient.

$$\frac{\mu_{0N}}{\mu_{0P}}\left(\frac{T}{T_0}\right)^{kmN-kmP} \cdot \frac{4V_{DD}-5V_{TN}}{4V_{DD}+5V_{TP}} < \frac{W_{P1}}{W_{N2}} < \frac{\mu_{0N}KI_N}{\mu_{0P}KI_P} \qquad [4]$$

In equation [4], α and KI are process dependent. Since output high-to-low propagation delay has a positive temperature coefficient, the total delay i.e. the sum of T_{PLH} and T_{PHL}, can be tuned to be temperature independent.

NMOS N2 in Figure 1 can only work for one single transition condition. To make clock buffer thermally insensitive to both rising and falling input, two NMOS transistors need to be coupled on the buffer, as shown in Figure 2. This could help to keep the duty cycle of clock signal. Since there is short-circuit current when P1 and N3 are both open, TICB consumes more energy than a normal buffer which is a shortcoming.

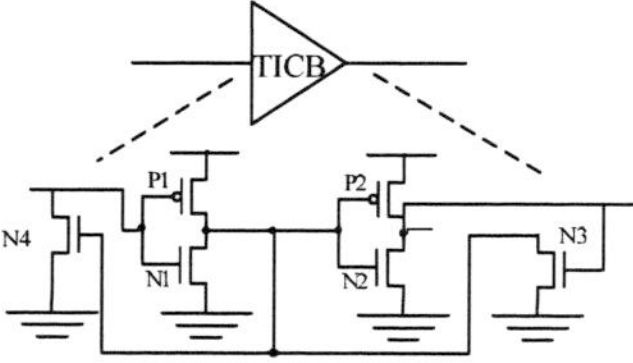

Figure 2. Schematic of TICB

Transistors with lower threshold voltage are more sensitive to temperature. Utilizing lower threshold voltage NMOS for N2 can reduce buffer area. Multiple threshold voltage option in a single chip is very well supported in popular foundries, such as IBM, TSMC and SMIC.

Clock Tree and Simulation

Based on the proposed TICB buffer, an H-type temperature insensitive clock tree is built and simulated under IBM 65nm technology. Assuming the clock tree is built in a typical circuit block with 10 thousands flipflops, there will be about 13 levels of clock buffers. Fanout number of each clock buffer is 2. Several conditions should be guaranteed: P2 transistor, as shown in Figure 2, can overwhelm two N4 transistors at the same time, P1 transistor can overwhelm N3, i.e. satisfying equation [4]. To ensure short transition on each flipflop, the last level of clock tree is implemented with normal clock buffer without coupled NMOS. In this way, half of buffers are normal buffers so that area and power overhead can be reduced.

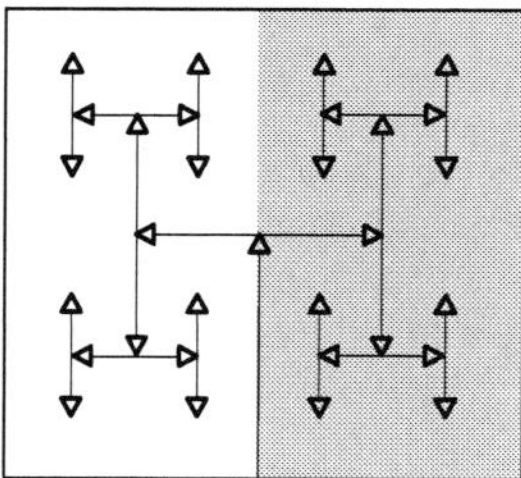

Figure 3. H-type clock tree; grey area stands for hot spot

The temperature distribution is assumed like this, as depicted in Figure 3, the right half grey area stands for hot spot where temperature is higher than normal temperature on the left side. Uniform temperature distributions in the two areas are assumed for simplicity. The normal temperature is defined as lowest working condition -40°C and room temperature 25°C. SPICE simulations are done at temperature gradient scan from

0°C to 80°C. The different thermal conditions on two parts produce clock skew. A normal H-type clock tree and temperature insensitive clock tree are both simulated.

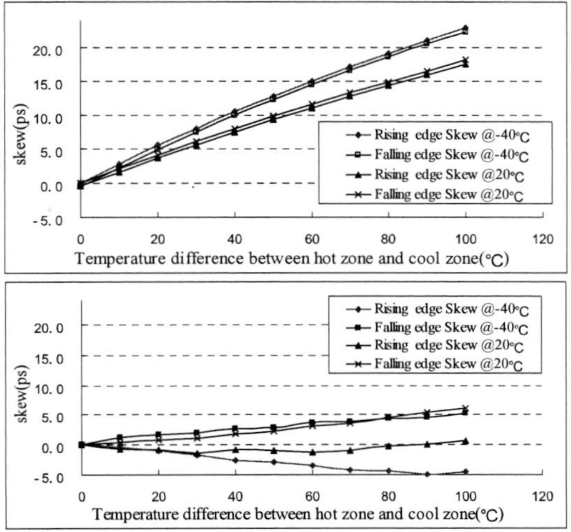

Figure 4 Skew of traditional clock tree(above) and temperature insensitive clock tree(below)

The simulated clock skew results under temperature gradient are shown in Figure 4. The maximum skew of normal clock tree caused by temperature gradient is 23.0ps while the maximum skew of our temperature insensitive clock tree is 6.1ps, reduced by 73.4%.

TABLE I. Data comparison of three clock trees.

	Normal clock tree		Temperature insensitive clock tree		Dual-VDD clock tree	
	Rising	**Falling**	**Rising**	**Falling**	**Rising**	**Falling**
Latency at -40°C(ps)	362.6	366.4	852.4	845.9	693.2	701.2
Latency at 20°C(ps)	377.6	381.4	848.9	849.4	699.3	707.4
Latency at 120°C(ps)	395.5	399.5	849.4	855.6	696.1	703.7
Transition time at -40°C(ps)	38.4	38.4	41.6	41.6	67.9	67.9
Transition time at 20°C(ps)	42.0	42.0	46.0	46.0	72.6	72.6
Transition time at 120°C(ps)	47.5	47.5	53.4	53.4	79.4	79.4
Area (um^2)	32764		34107		32764	
Power at 20°C(mW)	210.4		322.9		137.1	

As a reference, dual-VDD scheme(5) is also simulated with reduced supply voltage for clock tree. The comparison data are given in Table I, including clock latency, end point transition time, power and area for three kinds of clock trees. Both dual-VDD and this work can reduce clock skew under temperature gradient a lot. But normal clock tree has the shortest transition time and best balance for rising and falling latency.

There are still several shortcomings of this work. The proposed TICB clock tree consumes more area and energy, while Dual-VDD reduces power and takes same area with normal clock tree. And the proposed temperature insensitive clock tree has longer latency, which could potentially cause more latency variation when on-chip process variation is considered. These issues can be partially relieved by replacing more levels of clock buffers back to normal ones at the cost of more skew caused by temperature. Trade off is necessary. After all, future work is still needed for optimizing the proposed circuit.

Conclusion

In this work, a temperature insensitive clock buffer is proposed and a temperature insensitive clock tree is constructed. Simulation results show that, the thermal induced clock skew is reduced by 73.4% without changing supply voltage.

Acknowledgments

Thanks to the colleagues in China Design Center who helped with this work.

References

1. *International Technology Roadmap for Semiconductors.* http: //www.itrs.net.
2. T. Sato, et al. *On-chip thermal gradient analysis and temperature flattening for SoC design.* in *Proceedings of the 2005 Asia and South Pacific Design Automation Conference*: ACM.(2005)
3. J. Long, et al. *A self-adjusting clock tree architecture to cope with temperature variations.* in *Proceedings of the 2007 IEEE/ACM International Conference on Computer-Aided Design*: IEEE Press.(2007)
4. F. Mesa-Martinez, E. Ardestani, and J. Renau. *Characterizing processor thermal behavior.* in *Proceedings of the fifteenth edition of ASPLOS on Architectural support for programming languages and operating systems*: ACM.(2010)
5. S. Tawfik and V. Kursun. *Dual-VDD Clock Distribution for Low Power and Minimum Temperature Fluctuations Induced Skew.* in *8th International Symposium on Quality Electronic Design.*(2007)
6. H. Yu, et al. *Minimal skew clock embedding considering time variant temperature gradient.* in *Proceedings of the 2007 International Symposium on Physical Design*: ACM.(2007)
7. J. Yang, et al. *Sensitivity based link insertion for variation tolerant clock network synthesis.* in *Proceedings of the 8th International Symposium on Quality Electronic Design*: IEEE Computer Society.(2007)
8. A. Rajaram and D. Pan. *MeshWorks: an efficient framework for planning, synthesis and optimization of clock mesh networks.* in *Proceedings of the 2008 Conference on Asia and South Pacific Design Automation*: IEEE Computer Society Press Los Alamitos, CA, USA.(2008)
9. M. Tie and X. Cheng, *A cross-coupled-structure-based temperature sensor with reduced process variation sensitivity.* Chinese Journal of Semiconductors, **30**(4): p. 045002. (2009)
10. T. A. DeMassa and Z. Ciccone, *Digital integrated circuits*: Wiley.(1996)
11. I. Filanovsky and A. Allam, *Simulation/Modeling and Analysis-Mutual Compensation of Mobility and Threshold Voltage Temperature Effects with Applications in CMOS Circuits.* IEEE Transactions on Circuits and Systems-Part I-Fundamental Theory and Applications, **48**(7): p. 876-884. (2001)

ECS Transactions, 34 (1) 189-194 (2011)
10.1149/1.3567580 ©The Electrochemical Society

Low-power Design of Double Edge-triggered Static SOI D Flip-flop

Wan Xing[a], Jia Song[b], Du Gang[c]

[a,b,c] Department of Electronics Engineering and Computer Science, Peking University, Beijing 100871, China

In this paper, a double edge-triggered (DET), static SOI D flip-flop design suitable for low power and low area application is proposed. Silicon on insulator (SOI) is particularly good for low-power digital systems. Based on SOI technology instead of bulk silicon we realized a novel single edge-triggered (SET) static SOI D flip-flop using only ten transistors thus resulting in low-power consumption. Based on the SETDFF, a double edge-triggered D flip-flop is further constructed with an upper path and a lower path connected between the data input and an output node. Compared to single edge-triggered D flip-flop which processes data only at either the rising or falling transition of the clock, the DET D flip-flop doubles the rate of data thereby, increasing the data throughput. Simulation results indicated that the circuit is capable of significant power saving.

I. Introduction

Reducing power dissipation is one of the most important issues in very large scale integration design today. Flip-flops are some of the most frequently used elements in digital VLSI systems. In synchronous systems, flip-flops consume a large amount of power (1). About 30%-70% of the total power in the system is dissipated due to clocking network, and the flip-flops (2). A large portion of the clock power is used to drive these sequential elements. Reducing the clock power dissipation of flip-flops is thus a prime concern for the total chip power reduction.

The flip-flops can also be categorized based on whether they are dynamic or static, in nature of their operation. Compared to dynamic flip-flops whose charge stored at node capacitances leaking away in the transistor's 'OFF' state (clock stopped) and thus can produce faulty logic levels, static flip-flops maintain their state even when the clock is stopped and power is maintained. Though several significant contributions had been made to the art of realization of dynamic flip-flops, there was a still a need for static designs that further improve the relative power consumption (1).

D flip-flop is the most common configuration among various flip-flops. According to processing data at only one transition (either rising or falling transition) of the clock or at both of the two transitions of the clock, there are two flip-flop types consisting of double edge-triggered DFF (DETDFF) and single edge-triggered DFF (SETDFF). Compared to SET DFF, the DET D flip-flop doubles the rate of data processing or, alternatively halves the clock rate thereby, either increasing the data throughput or reducing power consumption in the clock circuit respectively (3).

Silicon on insulator (SOI) explained by the term" active silicon, separated from a silicon substrate by a silicon oxide insulator" is a very attractive technology for large volume integrated circuit production (3) and has been developed into a mainstream technology, especially for digital applications. Devices built on SOI have lower capacitances to substrate than comparable devices on bulk silicon and especially have a variety of advantages over bulk silicon for low power applications. The performance advantages associated with SOI have led to consideration of the technology for high performance applications especially for low-power applications (4). The advantages of SOI technology over bulk silicon for low-power applications are mostly related to the reduced capacitance and higher current capabilities of SOI, at lower off-state currents (5).

Because low-power large scale integration is an increasingly important and growing area of electronics and SOI circuits can save power dissipation over their bulk counterparts as mentioned above, it is worth mentioning that SOI circuits are considered to have significant improvement over conventional circuits in terms of power dissipation. Besides, SOI has also proved to be effective in various niche and growing market. Thus, it is significant to realize designs using SOI technology for low power applications and based on SOI we realized the proposed circuits and other circuits mentioned in this paper using SOI technology instead of conventional bulk silicon technology.

In this paper, a static SOI DET flip-flop has been proposed based on a SET DFF using ten transistors less than other existing SETDFF designs and consuming less power than the conventional designs.

The rest of this paper is organized as follows. The second section describes the existing static SET D flip-flop. In the next section, we explain the proposed DET D flip-flop. In the following section, we show the simulation results. Conclusions are drawn in the last section.

II. Existing Static Single Edge-Triggered D Flip-flop Designs

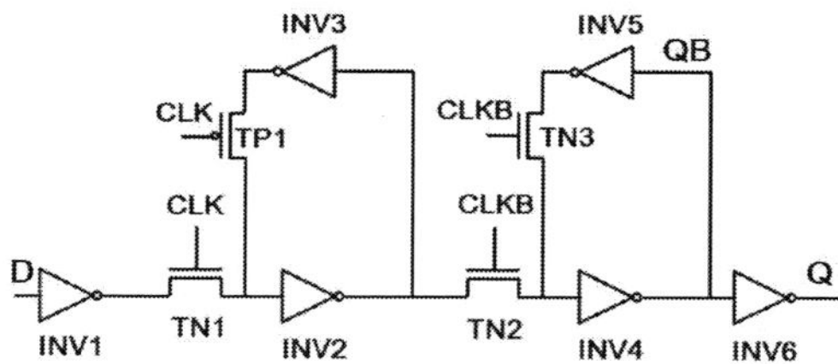

Figure 1. Circuit diagram of conventional static SET D flip-flop (SET1) (1)

A conventional single edge-triggered (SET) flip-flop typically latches data either on the rising edge or the falling edge of the clock cycle. As shown in Figure 1, the conventional static SET D flip-flop with 16 transistors (1) is a Master-Slave configuration. When the clock is stopped (grounded), the circuit is able to maintain the logic levels at Q and QB, which show a static behavior.

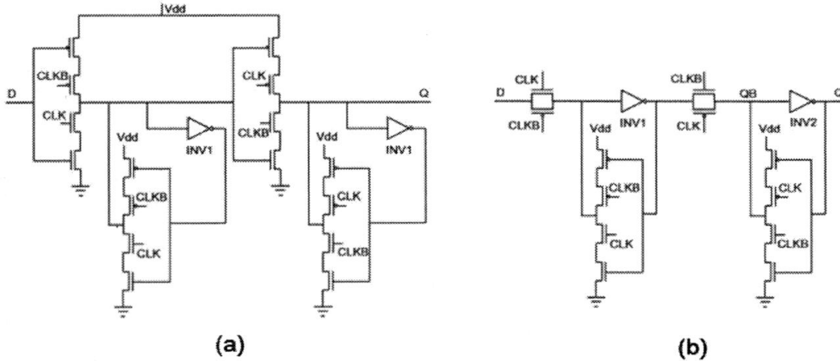

(a) **(b)**

Figure 2.   Circuit diagram of SET DFFs (a) with C^2MOS (SET2) (b) with TG (SET3)

The circuit design shown in Figure 2(a) (6) is a pseudo static C^2MOS flip-flop. Since the output can be maintain when the clock is stopped (permanently grounded), this circuit shows a static SET D flip-flop behavior. But the transistor count, which is 20, will bring extra power dissipation than previous design. But using Transmission Gates (TG1 and TG2) taking place of C^2MOS latches can overcome the limitation of previous design as shown in Figure 2(b) (1), which can reduce power dissipation and improve the design.

This design has a significant improvement in the power dissipation than the two previous designs but the transistor count is 16 and there is no considerable improvement in the area. Thus, to improve the power dissipation and area at the same time, a new static SET D flip-flop is illustrated in Figure 3 (1).

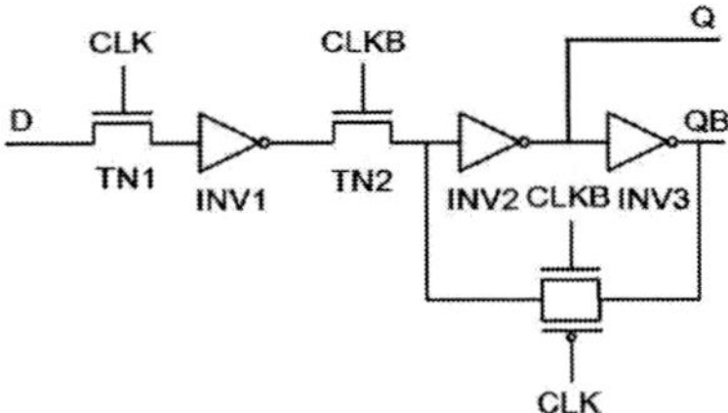

Figure 3. Circuit diagram of static SET flip-flop with 10 transistors (SET4) (1)

There are only 10 transistors in the circuit which is its main advantage and at the same time it can give power dissipation reduction. The Master section of the design consists of TN1 and Inv1 and the Slave section consists of TN2 and a weak feedback loop. Even if the clock is stopped (permanently grounded) the circuit is able to maintain the logic levels at Q and QB, which proves the fact that the proposed SET is static in nature.

After comparing several typically existing SET DFF designs, the circuit as shown above in Figure 3 is suitable for low-power dissipation and high chip density. Thus, we realized our proposed static DET DFF based on this SET DFF configuration.

III.   Proposed Design of the Static SOI DET D Flip-flop

The proposed design of static SOI DET D flip-flop is illustrated in Figure 4(4). The circuit, which can be triggered both at the rising edge and the falling edge of the clock, consists of an upper path and a lower path between the input node D and the output node Q. In nature, the upper path is a static SET DFF being triggered at the falling edge of the clock and the lower path is a static SET DFF being triggered at the rising edge of the clock. The two paths share two inverters INV2 and INV3 in the feedback loop to reduce transistor count to 16.

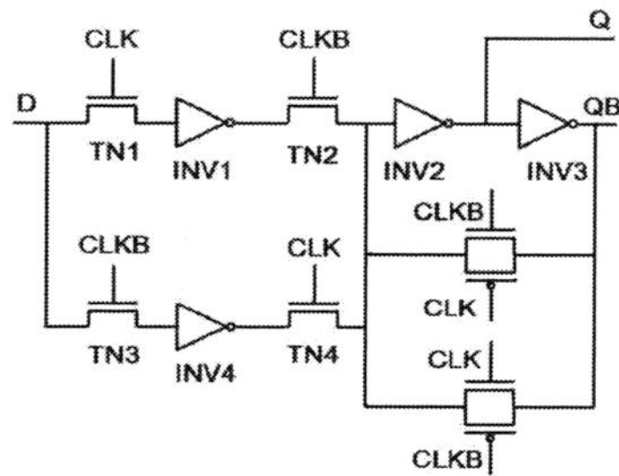

Figure 4. Circuit diagram of proposed static DET flip-flop

The proposed design consumes less power and saves area since the transistor count is only 16. The DET DFF is capable of capturing data at both of the edges of the clock. Thus, the same data throughput can be achieved with half of the clock frequency, comparing to its SET counterpart, to avoid a series of disadvantages when using of high clock frequency to improve performance, including increasing power consumption, clock uncertainties, and degradation of the clock waveform due to the non-ideal clock distribution, power supply noise and cross-talk (6).Or, if the clock frequency is constant, the DET D flip-flop doubles the rate of data processing. Moreover, when the clock is stopped (permanently grounded) the circuit shows a static behavior maintain the logic levels at output node Q and node QB.

IV.   Simulation Results

The designs are implemented using HSPICE tool and the circuits are simulated using 0.6um FD-SOI technology. To verify our results, we also simulate the circuits using 0.18um bulk silicon technology to compare simulation results of SOI circuits and their bulk silicon counterparts. Firstly, we measure the average power of the circuits of Figure 1, Figure 2(a), Figure 2(b) and Figure 3 to compare the power dissipation of the mentioned four static SET DFF designs. Secondly, we realize the DET DFFs based on the designs of Figure 1, Figure 2(a) and Figure 2(b) and measure their average power. And then their simulation results will compare with the proposed circuit.

As the average power consumption is related to the intersection point and the data sequence (3), so we adopted two different data sequences to compare the average power consumption of these flip-flops. Transient analysis waveforms of the proposed circuit with two different data sequences are shown in Figure 5. The result of measurement is shown in Table I and Table II.

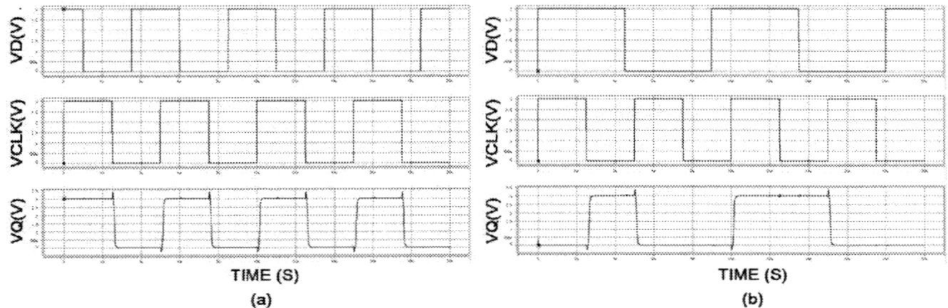

Figure 5. Waveforms for the proposed DET DFF with two different data sequences (a) DET DFF with data sequence 10101010 (b) DET DFF with data sequence 01001100

TABLE I. Power Consumption of Four SET DFFs With Two Different Data Sequences Using SOI

Design Of SET DFFs	Data Sequence	Average Power (uW)	SET4's Power Saving (%)
SET1(Figure1)	0000	472.20	69.52
SET2(Figure2(a))	0000	373.02	61.42
SET3(Figure2(b))	0000	225.21	36.10
SET4(Figure3)	0000	143.91	
SET1(Figure1)	1010	362.60	37.42
SET2(Figure2(a))	1010	508.30	55.36
SET3(Figure2(b))	1010	231.35	1.93
SET4(Figure3)	1010	226.88	

TABLE II. Power Consumption Between The Three DET DFFs Based On SET1, SET2, SET3 And The Proposed Design With Two Different Data Sequences

Design Of DET DFFs	Data Sequence	Average Power (uW)	Proposed Design's Power Saving (%)	Average Power Of Bulk Silicon (uW)	Power Saving Related To Bulk Si(%)
DET1(SET1)	10101010	1237.1	55.33	4677.2	73.55
DET2(SET2)	10101010	1771.1	68.80	5160.4	65.68
DET3(SET3)	10101010	635.31	13.01	3106.9	78.19
Proposed(SET4)	10101010	552.65		2578.7	78.57
DET1(SET1)	01001100	705.36	58.21	2518.1	72.00
DET2(SET2)	01001100	1037.3	71.58	3752.1	72.35
DET3(SET3)	01001100	321.67	8.36	1705.2	81.14
Proposed(SET4)	01001100	294.78		1572.3	81.25

The Table I shows comparison of power consumption of the four SET DFFs introduced previously when there are two different input data sequences. The result shows that the SET DFF with 10 transistors shown in Figure 3 consumes less power than the previous three and the power saving related to the previous ones can reach to 69.52% because of the fewer transistor count. The reduction of transistor count results in the total capacitance reduction. Thus the average power consumption reduces because of the total capacitance reduction. As shown in Table II the proposed design has a power reduction from 8.36% to 71.68% which is similar with the results in Table I. Moreover, related to the bulk technology, the SOI D flip-flops' power reduction is from 65.68% and 73.55% shown in Table II. It is significant that designs based on SOI technology are capable of low-power applications.

V. Conclusion

The proposed design is a static SOI double edge-triggered D flip-flop. To compare with the bulk silicon technology, SOI technology consumes less power which can reach 73.55%. Moreover, the proposed circuit's structure reduces power consumption than other designs and from simulation results the proposed design has a power reduction from 8.36% to 71.68% related to the other designs. Also, the DET D flip-flop doubles the data process rate under the same clock frequency with the SET configuration.

Acknowledgments

We are sponsored by National Science and Technology Major Project No. 2009ZX02305-003. We are grateful to Prof. Wang Yuan of Peking University, Dr. Wu Fengfeng of Peking University, Liu Limin of Peking University for constructive comments and help.

References

1. M. Sharma, A. Noor, S. C. TiwariandK. Singh, *An Area and Power Efficient Design of Single Edge Triggered D Flip Flop,* in Advances in Recent Technologies in Communication and Computing, 2009, p.478.
2. V. Stojanovic and V. G.Oklobdzija, *IEEE J. Solic-State Circuits,* **34**, 536 (2005).
3. Yu Chien-Cheng, *Design of Low-Power Double Edge-triggered Flip-Flop Circuit,*in Industrial Electronics and Applications, 2007, p. 2054-2057.
4. A. Marshall and S. Natarajan, *SOI Design: Analog, Memory and Digital Techniques,* p. 1, Kluwer Academic Publishers, Norwell (2002).
5. A. Marshall and S. Natarajan, *SOI Design: Analog, Memory and Digital Techniques,* p. 359, Kluwer Academic Publishers, Norwell (2002).
6. N. Nedovic, M. Aleksic, V. G. Oklobdzija, *Comparative analysis of double-edge versus single-edge triggered clocked storage elements,* in Circuits and Systems, 2002 (ISCAS 2002), p. v-105-v-108 vol.5.

ECS Transactions, 34 (1) 195-200 (2011)
10.1149/1.3567581 ©The Electrochemical Society

Effects of oxygen flow ratios and annealing on TiO$_x$ deposited by reactive magnetron sputtering

Lanlan Wang, Kailiang Zhang *, Qi Wang, Fang Wang, Xiaoying Wei

School of Electronics Information Engineering, Tianjin Key Laboratory of Film Electronic & Communication Devices, Tianjin University of Technology, Tianjin, China, 300384
*corresponding author, kailiang_zhang@163.com

Titanium oxide (TiO$_x$) thin films were deposited on Cu (111)/Ti/SiO$_2$/Si substrates by direct-current reactive magnetron sputtering at 3–15% oxygen flow ratios, then annealed by vacuum at 650 °C for 40min. The morphology, phase, and resistive switching behaviors of the as-deposited and annealed TiO$_x$ thin films were analyzed by atomic force microscope (AFM), X-ray diffraction (XRD), and semiconductor parameter analyzer (SPA), respectively. The results of X-ray diffraction (XRD) showed that annealing TiO$_x$ caused crystal structure changes compared to the as-deposited TiO$_x$ film. From XRD, the as-deposited TiO$_x$ films were practically amorphous. In contrast, the distinct crystalline peaks of anatase and rutile phases was detected at 10 F$_{O2}$% by 650 °C annealing and exhibited stable and superior resistance switching behaviors due to deep-level emissions of oxygen vacancies in the rutile and anatase phases.

Introduction

Since the first observation of bistable resistance states in the 1960s, many materials have been investigated in RRAM area. The binary oxide-based resistive random access memory (RRAM) has attracted growing interest as one of the most promising candidates for easing challenges of flash scaling down, owing to its good scalability and ease of fabrication. But the understanding of the switching mechanism and improving devices electrical characteristics has to be further researched for practical memory applications. The wide-band-gap titanium oxide material has intensively been studied because of its excellent resistive properties, such as high on/off resistance ratio, low on/off voltage, and long retention time. The practical application of the titanium oxide thin films covers in the dye-sensitized solar cell (DSSC) (1), photocatalysts (2), gas sensor (3), and antireflective coatings (4) and so on. Various techniques have been used for synthesis of the titanium oxide films, such as sol–gel deposition (5), spray pyrolysis (6), pulsed laser deposition (7), e-beam evaporation (8), chemical vapour deposition (9) and reactive magnetron sputtering (10). The reactive magnetron sputtering is a promising technique for depositing large scale and uniform titanium oxide thin films. The sputtering conditions, such as working pressure, gas partial pressure, and substrate temperature can be controlled for formation of films with different microstructures.

Thus, in order to obtain a high-quality TiO$_2$ film, more attention should be paid to the process of reactive sputtering. In this report, TiO$_x$ thin films were fabricated on Cu (111)/Ti/SiO$_2$/Si substrates by reactive sputtering. Moreover, the effects of oxygen flow ratios and annealing on the structure properties of TiO_x films were systematically investigated, and stable resistive switching behavior (RSB) was observed at 10 F_{O2}% after 650°C annealing with a narrow dispersion of the resistance states and switching voltages.

Experimental

TiO_x thin films were deposited with reactive sputtering system on Cu (111)/Ti/SiO$_2$/Si substrate. The oxygen flow ratios (F_{O2}% = F_{O2} / (F_{O2}+F_{Ar}) × 100%) were adjusted from 3% to 15% by mass flow controller for studying the effect of oxygen flow ratios on the formation of titanium oxide thin films. The base vacuum was evacuated to 3.0×10^{-4} Pa, and growth pressure was 1Pa. During the deposition of TiO_x thin films, DC power was maintained at 100W and substrate temperature was room temperature. Then certain films are annealed by vacuum at 650 °C for 40min. 500nm Cu thin film as the bottom electrode was deposited by DC magnetron sputtering.

The thickness of TiO_x thin films were measured by step profiler (Vecco, Dektak 150).The surface morphology of thin films was characterized by atomic force microscope (AFM). The crystalline phase of TiO_x thin films were identified by X-ray diffraction (XRD). The current-voltage characteristics were measured at room temperature with a semiconductor parameter analyzer (Agilent, B1500A) in voltage sweep mode. A current compliance (11) (CC) was used to protect samples from damages due to high currents.

Results and discussion

<u>Analysis for microstructure of TiO_x thin films</u>

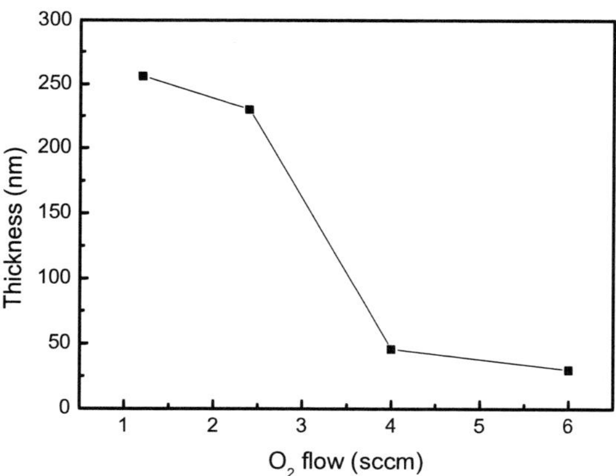

Figure1 Thickness curves of TiO_x thin films growth at different oxygen flow ratios

Figure 1 shows the changes of thickness for TiO_x thin films deposited at different oxygen flow ratios. As shown in Figure 1, the films thickness (such as 256nm, 230nm, 46nm, 30nm) are decreasing with the increasing of oxygen flow ratios(such as O_2=1.2sccm(3%),O_2=2.4sccm(6%),O_2=4sccm(10%), O_2=6sccm(15%), 1sccm=1.73×10^{-3}Pa·m^3·s^{-1}). It illustrates that the mode at the target surface is turned from transformation to reaction. In the transformation mode, target surface has higher sputtering capacity, which can results in higher deposition rate and more thick films. When transformation mode converts to reaction mode, the target is highly oxidized which leads to decrease of sputtering capacity and reduces deposition rate, so corresponding films turn to be thinner. It includes that the thickness of films depend on oxygen flow variations.

(a) 3%   (b) 6%

(c) 10%   (d) 15%

Figure2. AFM images of the as-deposited TiO$_x$ thin films

Figure 2 shows the AFM images of TiO$_x$ thin films deposited at different oxygen flow ratios. Topography of the surface (1μm×1μm) is scanned. It can be seen that the film deposited at 10 F$_{O2}$% is mostly smooth, uniform and dense. The roughness is measured to 1.03nm (rms) and the average grain size of the TiO$_x$ film presented in Fig. 2(C) is about 25 nanometers. As shown in Fig. 2(a) and Fig. 2(b), films growth at 3 F$_{O2}$% and 6 F$_{O2}$% have rough surface and untidy particles. The quality for the film deposited at 15 F$_{O2}$% rank only second to the sample 10 F$_{O2}$%, as illustrated in Fig. 2(d).

Figure 3 shows XRD patterns for Ti$_3$O$_5$ peak changes before annealing and the sample at 10 F$_{O2}$% by 650°C annealing. It is noted that all the as-deposited thin films without annealing are nearly amorphous except some weak Ti$_3$O$_5$ (004) peak appears while the crystalline peaks are observed after 650°C annealing. As shown in figure 3(a), Ti$_3$O$_5$ (004) phase decreases with increasing oxygen flow ratio as well as film thickness. The crystalline (103) anatase peak denoted as A(103) at 36.9° and the (210) rutile peak denoted as R(210) at 44.1° can be easily observed from the TiO$_x$ thin film formed at 10 F$_{O2}$% especially after 650°C annealing, as shown in figure 3(b). The intensity of rutile peak at 10 F$_{O2}$% by 650°C annealing is stronger than anatase one. The obtained anatase and rutile polycrystalline films are contributed to deep-level emissions of oxygen vacancies.

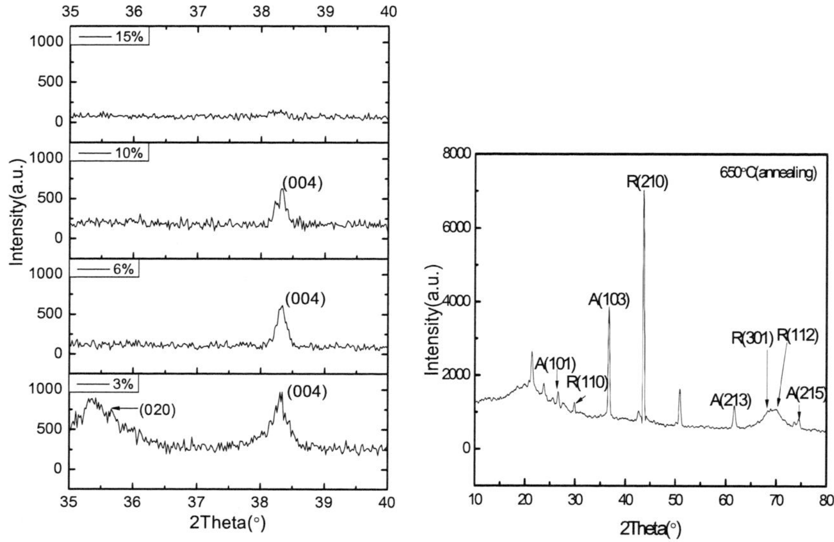

Figure 3 XRD patterns for (a) Ti_3O_5 peak changes before annealing (b) sample 10 F_{O2}% at 650°C annealing

<u>Analysis for resistive switching of annealed TiO_x thin films</u>

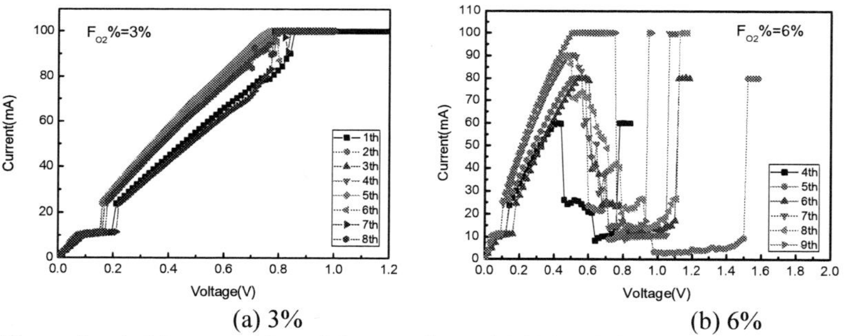

(a) 3%   (b) 6%

Figure 4 switching repetition of the samples at (a) 3% and (b) 6% oxygen flow ratios

For the film at 3 F_{O2}%, it shows very unstable resistive switching after forming process, as shown in figure 4(a).The called reset process for high resistance state (HRS) transition to low resistance state(LRS) hardly occurs, that is to say, the resistive switching is not reversible. When oxygen flow ratio is at 6 F_{O2}%, reversible unipolar resistance switching characteristics can be observed after initialization and reset current is about 10mA. Then go on scanning, the reset current reaches 60mA after three 10mA current occurring. After several sweeps, the film shows a relatively more stable reset process, but the value of reset current gradually increases, as shown in Figure 4(b).

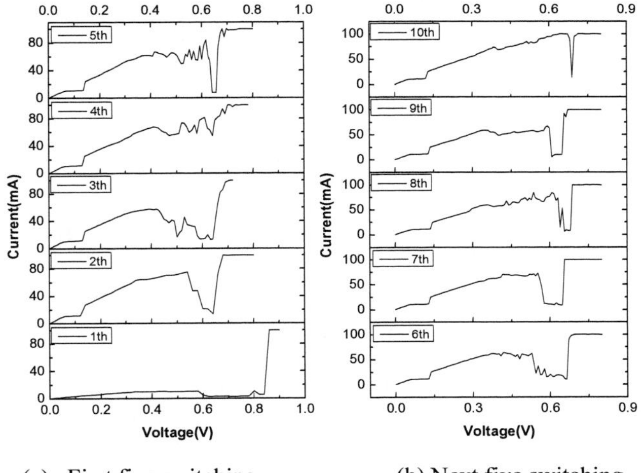

(a) First five switching (b) Next five switching

Figure 5 ten unipolar resistive switching (a) and (b) of the sample growth at 10 F_{O2}%

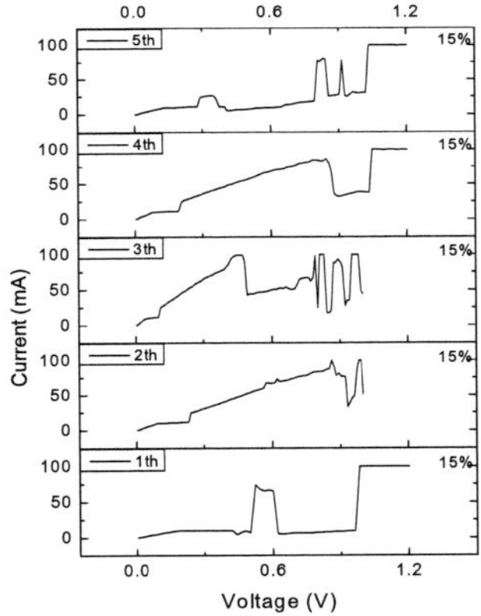

Figure 6 switching repetition of the sample growth at 15 F_{O2}%

For the film at 10 F_{O2}%, we observe good repetitive performance of unipolar resistive switching behaviors (RSB), as shown in Figure 5. The RSB for film at 10 F_{O2}% gradually tends to be stable with scanning number increases, the reset voltage has small fluctuation and is about 0.6V. It is clear that the film at 10 F_{O2}% shows better resistive switching characteristics compared to the samples at 3 F_{O2}% and 6 F_{O2}%. It is known that switching retention and repeatability are very important factors for the resistance random access memory (RRAM). Stable and repeatable properties are basis for achieving device application. For the sample at 10 F_{O2}% annealing by 650°C the resistance change ratio can reach about 100, which meets the requirement of developing memory device and provides the basis for the application in the future.

For the film at 15 F_{O2}%, the first scan shows a good unipolar RSB property reset voltage is about 0.6V, set voltage is about 1.1V, and forming voltage is 2.4V. However, during subsequent scanning processes, the film shows a very unstable resistance change characteristics, as shown in Figure 6.The reset processes appear extremely instable and reset voltage has large fluctuation.

Conclusions

In summary, we study effects of different oxygen flow ratios on switching properties of TiO_x. XRD analysis shows that a mixture of anatase and rutile phases is obtained at 10 F_{O2}% by 650 °C annealing. The results of step profiler show that the films thickness is decreasing with the increasing of oxygen flow ratio. The decrease of thickness depends on the reduced deposition rate. Electrical tests show that film sample at F_{O2}%=10% by 650 °C annealing shows a relatively stable resistance switching characteristic and highly repeatable characteristics of URS.

Acknowledgments

We acknowledge the financial support provided by the National Natural Science Foundation of China under Grant No 60806030, and Tianjin Natural Science Foundation under Grant No 08JCYBJC14600, No 10SYSYJC27700 and Tianjin Science and Technology Developmental Funds of Universities and Colleges under Grant No ZD200709.

References

1. O'Regan, and Grätzel, *Nature*, 353, 737 (1991).
2. Masaaki Kitano, Masaya Matsuoka, Michio Ueshima, Masakazu Anpo, *Appl. Catal. A-Gen.*, 325, 1 (2007).
3. Ibrahim A. Al-Homoudi, J.S. Thakur, R. Naik, G.W. Auner, G. Newaz, *Appl. Surf. Sci.*, 253, 8607 (2007) .
4. G. San Vicente, A. Morales, M.T. Gutie´rrez, *Thin Solid Films*, 403–404, 335 (2002).
5. K. Pomoni, A. Vomvas, Chr. Trapalis, *J. Non-Cryst. Solids.*, 354, 4448 (2008).
6. C. Natarajan, N. Fukunaga, G. Nogami, *Thin Solid Films*, 322, 6 (1998).
7. Masahiro Terashima, Narumi Inoue, Shigeru Kashiwabara, Ryozo Fujimoto, *Appl. Surf. Sci.*, 169–170, 535 (2001).
8. Lianchao Sun, Ping Hou, *Thin Solid Films*, 455 –456, 525 (2004).
9. Hongfu Sun, Chengyu Wang, Shihong Pang, Xiping Li, Ying Tao, Huajuan Tang, Ming Liu, *J. Non-Cryst. Solids.*, 354, 1440 (2008).
10. S. Boukrouh, R. Bensaha, S. Bourgeois, E. Finot, M.C. Marco de Lucas, *Thin Solid Films*, 516, 6353 (2008).
11. X.Cao, X.M.Li, X.D.Gao, Y.W.Zhang, X.J.Liu, Q.Wang, L.D.Chen, *Appl. Phys. A*, 97, 883 (2009).

CHAPTER 2

LITHOGRAPHY AND PATTERNING

ECS Transactions, 34 (1) 203-208 (2011)
10.1149/1.3567582 ©The Electrochemical Society

Robustness Enhancement in Optical Lithography: From Pixelated Mask Optimization to Pixelated Source-mask Optimization

Ningning Jia and Edmund Y. Lam

Imaging Systems Laboratory, Department of Electrical and Electronic Engineering,
The University of Hong Kong, Pokfulam Road, Hong Kong
elam@eee.hku.hk

Abstract

Optical lithography is facing a great challenge from the continuous shrinkage of industry node toward 22nm or below. The increasing sensitivity to process variations becomes a key problem hindering further progress. To address this problem, inverse lithography, which designs pixelated masks with appropriate algorithms, is favored for its capacity of exploring a larger solution space. This search capacity, however, is limited by the fixed source configuration. To this end, optimization of pixelated source and pixelated mask together has come up for further performance improvement. In this paper, a source-mask co-optimization (SMO) algorithm, which incorporates process variations into the optimization scheme, is introduced. To further improve the process robustness, we apply weighted total variation and aerial image intensity control as regularization. Simulation results show that we achieve greater pattern fidelity and enhanced process robustness by using SMO.

1 Introduction

In optical lithography, the 193nm water immersion lithography system is now on service. While further development of the shorter wavelength and higher NA (numerical aperture) is hindered for some practical reason, circuit feature size shrinkage is still continuing on the path to the ever smaller technology nodes beyond the Rayleigh resolution limit [1]. Meanwhile, printed features at small dimensions are easily affected by process variations (mainly dose and focus variations) for current low-k_1 optical lithography. Overcoming these physical limitations requires the use of computation: optimization and image processing techniques for solutions that are insensitive to process fluctuations [2]. One common approach to tackle this problem, known as optical proximity correction (OPC), makes pre-distortion on the mask topology so that the resultant circuit pattern matches the desired one [3]. Among these techniques, inverse lithography (IL) searches for a solution by solving an inverse imaging problem [4]. Compared to edge-based OPCs [5, 6], which calculate mask pre-distortion by adjusting the locations of feature edges, IL explores a wider solution space by manipulating pixels of the mask image.

Despite the theoretical ability to deliver superior wafer performance, practical implementation of inverse lithography remains a challenging task. Algorithms developed must conform to industry specifications [7], such as mask manufacturability and process robustness [8, 9]. Enhanced performance through multiple process conditions can be achieved by incorporating process fluctuations explicitly into optimization framework [9] or by the use of cost function defined by various specifications, *e.g.*, depth of focus, process window, process variation (PV)-band, etc. [10]. However, IL alone cannot completely solve the above imaging problems. The adjustment of illumination configuration is necessary for imaging systems at $k_1 < 0.75$ [3].

Many early works have been dedicated to the development of algorithms for illumination optimization. Burkhardt *et al.* explored the diffraction orders on the pupil plane, and found the optimal source configuration to enhance image contrast for phase-shifting contact masks [11]. Socha *et al.* developed a scheme based on Hopkins partially coherent imaging equations to determine the importance of illuminator areas for periodic patterns [12]. Below 45nm circuit feature sizes, imaging using standard techniques becomes difficult. Source optimization and the reticle pattern optimization have been integrated to extend the life of current immersion lithography toolset [13]. Further degrees of freedom for optimization are provided by customized diffractive optical element (DOE) realization of the pixelated source [14]. Together with pixelated mask used in IL, the so-called free-form source-mask co-optimization fits well into the inverse lithography framework. Therefore both the source shape and its pixel intensity can be changed for an optimal source computation.

In this work, a free-form SMO scheme based on the inverse source-mask synthesis technique is proposed. The source image and the mask are iteratively updated by minimizing a cost function, which includes the regularization on the aerial image to improve the robustness against dose variation. Experimental results of SMO and mask optimization with the traditional source are compared to illustrate the robustness enhancement made by SMO.

2 Inverse Lithography with Off-axis Illuminations

Under an off-axis illumination, the aerial image (denoted by I_a) formed by the partial coherent imaging system is modeled by the Hopkins equation [15]. The image intensity at (x, y) is given by

$$I_a(x,y) = \int_{-\infty}^{\infty} \cdots \int J(f,g)\hat{H}(f+f',g+g')\hat{H}^{\dagger}(f+f'',g+g'')\hat{M}(f',g')\hat{M}^{\dagger}(f'',g'')$$

$$\times e^{-i2\pi[(f'-f'')x+(g'-g'')y]} \, df \, dg \, df' \, dg' \, df'' \, dg'', \tag{1}$$

where $\hat{H}$ and $\hat{H}^{\dagger}$ denote the pupil function (optical transfer function, OTF) and its complex conjugate, respectively. $\hat{M}$ represents the mask spectrum. J is the source function, which is represented by a pixelated image, with each pixel denoting one discrete source point. A single source point gives rise to a projected image, and the collection of images from all source points contributes to $I_a(x, y)$.

As stated in [16], inverse lithography computes the mask pattern by solving an optimization problem. The solution can be obtained by minimizing a cost function, which has a general form of

$$M_{\text{opt}}(x,y) = \arg \min_{M(x,y)} \mathcal{D}\left\{\Gamma\{I_a(x,y)\}, \hat{I}(x,y)\right\}$$

$$\text{subject to} \quad M(x,y) \in \{0,1\}. \tag{2}$$

In the above equation, $M(x,y)$ represents the binary pattern on the mask and $I(x,y)$ is the target design, which is the desired pattern on the wafer. $\Gamma(\cdot)$ is the function approximating the photo-resist effect, which can be modeled as a soft thresholding by a logarithmic sigmoid function [8]. For IL optimization, the measure of closeness between the printed pattern and the desired circuit is not limited to a unique form. Here we denote it by $\mathcal{D}$ in general. A common method refers to the use of ℓ_1 or ℓ_2 norm in the form of $\|\Gamma\{I_a(x,y)\} - \hat{I}(x,y)\|$, which is regarded as the pattern fidelity term. Other than this, measure $\mathcal{D}$ may also include terms such as depth of focus (DOF), normalized image log slope (NILS), manufacturability, etc., which are called regularization terms.

3 Free-form Source and Mask Co-optimization

The source function J is fixed during optimization in IL. For SMO, both the source and the mask are variables to be optimized. The optimization problem described in Equation (2) then becomes

$$\{M_{\text{opt}}(x,y), J_{\text{opt}}(f,g)\} = \underset{\{M,J\}}{\arg\ \min}\, \mathcal{D}\left\{\Gamma\{I_a(x,y)\}, \hat{I}(x,y)\right\}$$

$$\text{subject to}\quad M(x,y) \in \{0,1\},\quad J(f,g) \geq 0. \tag{3}$$

Note that the source $J(f,g)$ takes nonnegative value, and is normalized by its total energy.

To enhance the process robustness, we explicitly add process variations into Equation (3). As in [9], focus and dose variations are first considered as random variables, and then a set of samples are taken to form the training set. Under the process conditions determined by the training set, J and M are iteratively updated to find the optimal solutions.

Meanwhile, to enhance the image contrast, we apply the total variation (TV) regularization, since the aerial image is piecewise smooth. To maintain the edge sharpness and avoid oversmoothing small features, a weight matrix based on the edge of the target image is added as a control of the strength of TV minimization. The weight matrix takes small values at the edge, where gradients are large, and it takes relatively large values elsewhere.

Other than giving the edge smaller smooth penalty, the image contrast can be further improved by adding another regularization term that manipulates the intensity of the aerial image. Here we choose to minimize $\|I_a(x,y) - t\hat{I}(x,y)\|_2^2$ at the edge, which means a weighted mean square error is minimized. The weight matrix is determined by the target image's edge, and $\frac{t}{2}$ is the threshold used in the photo-resist model. This penalty forces the intensities at the edge to reach a lower bound 0 when the phase of the target design is $0°$, and an upper bound t when the phase is $180°$. Thus intensities in the vicinity of the threshold are equally penalized. Since we minimize across multiple defocus conditions, it also helps to improve the depth of focus. The proposed form of $\mathcal{D}\left\{\Gamma\{I_a(x,y)\}, \hat{I}(x,y)\right\}$ is then given by

$$\mathcal{D}\left\{\Gamma\{I_a(x,y)\}, \hat{I}(x,y)\right\} = \sum_{x,y}\left(\Gamma\{I_a(x,y)\} - \hat{I}(x,y)\right)^2 + \lambda_1 \sum_{x,y}\left|W_1(x,y)\nabla I_a(x,y)\right|$$

$$+ \lambda_2 \sum_{x,y} W_2(x,y)\left(I_a(x,y) - t\hat{I}(x,y)\right)^2$$

$$= \mathcal{J}\left\{\Gamma\{I_a\}, \hat{I}\right\} + \lambda_1 \mathcal{R}_{\text{TV}}\{I_a\} + \lambda_2 \mathcal{R}_{\text{aerial}}\{I_a\}. \tag{4}$$

The measure $\mathcal{D}$ is composed of three terms, which relate to pattern fidelity ($\mathcal{J}$), TV regularization ($\mathcal{R}_{\text{TV}}$) and aerial image regularization ($\mathcal{R}_{\text{aerial}}$). The weight matrix of TV regularization and the aerial image penalty are denoted by W_1 and W_2, respectively.

In fact, $\|I_a(x,y) - \hat{I}(x,y)\|_2^2$ has been used as the fidelity term in [17], which also claimed that using $\|\Gamma\{I_a(x,y)\} - \hat{I}(x,y)\|_2^2$ would lead to further decrease of the pattern error. Note that the upper bound of the aerial image intensity over the whole image is difficult to

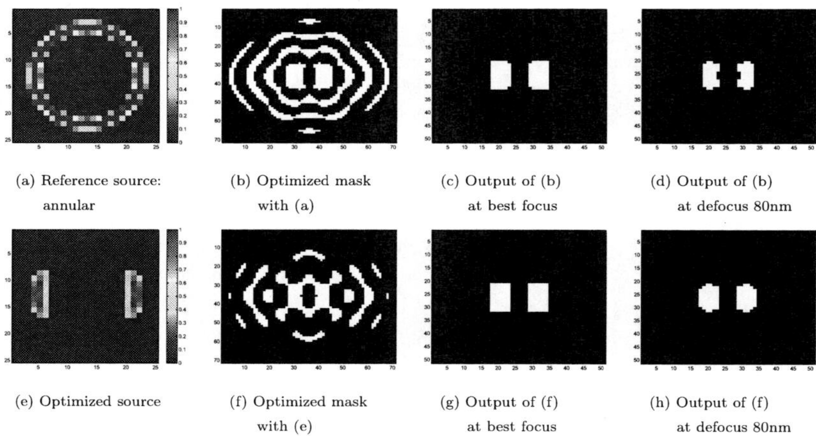

(a) Reference source: annular

(b) Optimized mask with (a)

(c) Output of (b) at best focus

(d) Output of (b) at defocus 80nm

(e) Optimized source

(f) Optimized mask with (e)

(g) Output of (f) at best focus

(h) Output of (f) at defocus 80nm

Figure 1: Results of reference source and SMO.

predict. Sometimes it might exceeds the target image intensity 1. Our penalty has no such concern, because it applies only on intensities near the edge that are mostly quite close to the threshold. With the fidelity term $\mathcal{J}$ minimizing the pattern error, $\mathcal{R}_{TV}$ and $\mathcal{R}_{aerial}$ help to enhance the process robustness further.

4   Results

We apply SMO on a target pattern with two rectangles, and compare the results with a reference annular source. The critical dimension of the test pattern is 50nm, which is the distance between two rectangles. The width of each rectangle is 60nm and length 110nm, with the mask pixel size $10 \times 10nm^2$. The image matrix size is 151×151. In the following we show truncated images for a clearer demonstration.

In Figure 1, the first row (figure (a), (b), (c) and (d)) displays the reference source, the optimized mask, the output at best focus and defocus, respectively. Following the same structure, the second row shows results of SMO. One can observe that, although the printed patterns in Figure 1 (c) and (g) show similar performance at best focus, the image quality of Figure 1 (d) deterioates more than that of Figure 1 (h) at a defocus 80nm.

To quantitatively exam the performance of SMO, we draw process windows [3] of the reference source and the optimized source. Four critical locations, which are numbered and marked with different colors in Figure 2 (a), are chosen for the process window calculation. To be specific, the width and length of the rectangle, as well as the space between two rectangles, are measured. Figure 2 (b) and (c) are calculated process windows for the reference source and SMO, respectively. In each figure there are four pairs of curves drawn by four different colors that are consistent with the colors used in Figure 2 (a). Each pair of curves with the same color represents the maximum doses and the minimum doses, between which the corresponding feature keeps its nominal size over the measured focus range. And the area bounded by the four pairs of curves defines the process window. The dose range specified by the maximum dose and the minimum dose is the exposure latitude [3]. One can observe that process window in Figure 2 (c) is larger than that in Figure 2 (b). For instance,

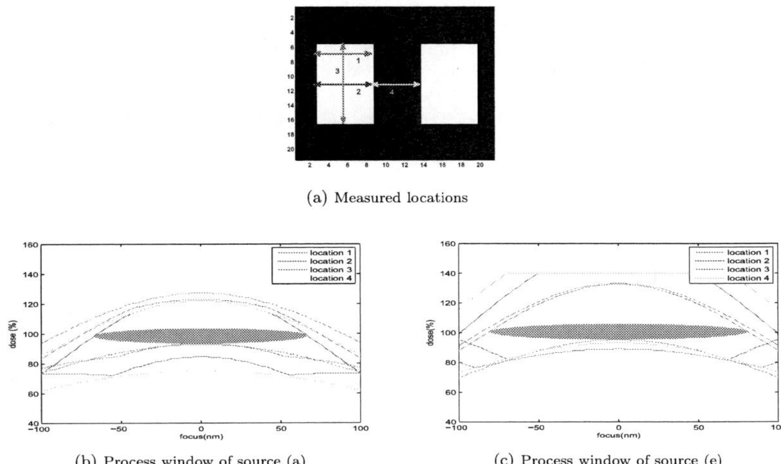

(a) Measured locations

(b) Process window of source (a)   (c) Process window of source (e)

Figure 2: Process window.

with a 10% exposure latitude, (c) exhibits a larger focus range than (b), as illustrated by the ellipses in (b) and (c). Thus the process robustness is enhanced by SMO. In addition, the optimized illumination configuration in this case is quite like a dipole source, which indicates that one can use the result of free-form SMO for parametric source optimization [18].

5   Conclusion

In this paper, we compare the performance of free-form source-mask co-optimization with that of pixelated mask design with a fixed traditional illumination. Experimental results show that SMO improves the printed pattern fidelity across multiple process conditions. Thus it enhances the process robustness. Further interest focuses on SMO for multiple circuit patterns and developing faster SMO algorithms.

Acknowledgments

This work was supported in part by the Research Grants Council of the Hong Kong Special Administrative Region, China, under Projects HKU 7134/08E, and by the UGC Areas of Excellence project *Theory, Modeling, and Simulation of Emerging Electronics*.

References

[1] M. Rothschild, "A roadmap for optical lithography," *Opt. Photonics News*, vol. 21, pp. 26–31, Jun. 2010.

[2] E. Y. Lam, "Regularization in inverse lithography: Enhancing manufacturability and robustness to process variations," in *China Semiconductor Technology International Conference (CSTIC)*, vol. 27 of *ECS Transactions*, pp. 427–432, 2010.

[3] A. K. Wong, *Resolution Enhancement Techniques in Optical Lithography*. Washington: SPIE, 2001.

[4] L. Pang, Y. Liu, and D. Abrams, "Inverse lithography technology (ILT): A natural solution for model-based SRAF at 45nm and 32nm," in *Photomask and Next-Generation Lithography Mask Technology XIV* (H. Watanabe, ed.), vol. 6607 of *Proc. SPIE*, p. 660739, 2007.

[5] J. Mitra, P. Yu, and D. Z. Pan, "RADAR: RET-aware detailed routing using fast lithography simulations," in *42nd Design Automation Conference*, pp. 369–372, 2005.

[6] A. Gu and A. Zakhor, "Optical proximity correction with linear regression," *IEEE Trans. Semicond. Manuf.*, vol. 21, pp. 263–271, May 2008.

[7] E. Y. Lam and A. K. Wong, "Computation lithography: Virtual reality and virtual virtuality," *Opt. Express*, vol. 17, pp. 12259–12268, Jul. 2009.

[8] A. Poonawala and P. Milanfar, "Mask design for optical microlithography — an inverse imaging problem," *IEEE Trans. Image Process.*, vol. 16, pp. 774–788, Mar. 2007.

[9] N. Jia and E. Y. Lam, "Machine learning for inverse lithography: Using stochastic gradient descent for robust photomask synthesis," *J. Opt.*, vol. 12, p. 045601, Apr. 2010.

[10] D. S. Abrams and L. Pang, "Fast inverse lithography technology," in *Optical Microlithography XIX* (D. G. Flagello, ed.), vol. 6154 of *Proc. SPIE*, p. 61541J, 2006.

[11] M. Burkhardt, A. Yen, C. Progler, and G. Wells, "Illuminator design for the printing of regular contact patterns," *Microelectron. Eng.*, vol. 41–42, pp. 91–96, Mar. 1998.

[12] R. Socha, M. Eurlings, F. Nowak, and J. Finders, "Illumination optimization of periodic patterns for maximum process window," *Microelectron. Eng.*, vol. 61–62, pp. 57–64, Jul. 2002.

[13] K. Tian *et al.*, "Benefits and trade-offs of global source optimization in optical lithography," in *Optical Microlithography XXII* (H. J. Levinson and M. V. Dusa, eds.), vol. 7274 of *Proc. SPIE*, p. 72740C, 2009.

[14] K. Lai *et al.*, "Experimental result and simulation analysis for the use of pixelated illumination from source mask optimization for 22nm logic lithography process," in *Optical Microlithography XXII* (H. J. Levinson and M. V. Dusa, eds.), vol. 7274 of *Proc. SPIE*, p. 72740A, 2009.

[15] A. K. Wong, *Optical Imaging in Projection Microlithography*. Washington: SPIE, 2005.

[16] L. Pang, Y. Liu, and D. Abrams, "Inverse lithography technology (ILT), what is the impact to photomask industry?" in *Photomask and Next-Generation Lithography Mask Technology XIII* (M. Hoga, ed.), vol. 6283 of *Proc. SPIE*, p. 62830X, 2006.

[17] A. Poonawala and P. Milanfar, "Double-exposure mask synthesis using inverse lithography," *J. Micro/Nanolith. MEMS MOEMS*, vol. 6, p. 043001, Oct. 2007.

[18] Y. Deng, Y. Zou, K. Yoshimoto, Y. Ma, C. E. Tabery, J. Kye, L. Capodieci, and H. J. Levinson, "Considerations in source-mask optimization for logic applications," in *Optical Microlithography XXIII* (M. V. Dusa and W. Conley, eds.), vol. 7640 of *Proc. SPIE*, p. 76401J, 2010.

ECS Transactions, 34 (1) 209-214 (2011)
10.1149/1.3567583 ©The Electrochemical Society

Mask synthesis for aerial image fidelity in optical lithography using a coarse-grid-approximation level-set approach

Yao Peng[a], Jinyu Zhang[a], Yan Wang[a], and Zhiping Yu[a]
[a] Institute of Microelectronics, Tsinghua University, Beijing 100084, China,

Inverse lithography technique, which treats mask synthesis as an inverse problem, has been considered as a strong contender to deal with the 45 nm technology node and beyond. But one problem exists in present pixel based mask synthesis algorithms: to describe mask patterns more accurately and to obtain better image quality, usually need finer grid representation, which results in extremely intensive computations. In this paper, a new method is proposed to mitigate this issue based on level-set method. This method utilizes the coarse-grid-approximation (CGA), which means the mask pattern defined on the fine grids is approximated using a new mask defined on coarse grids. The mask optimization is then performed on coarse girds and the final result is acquired by extracting mask patterns from coarse grids to fine grids. This method is demonstrated using a periodic array of contact holes pattern. Comparing with a gradient based algorithm and the traditional level-set algorithm, this method can provide almost the same image quality, but with over 100X running speed enhancement.

1. Introduction

Resolution enhancement techniques (RETs) are widely used to cope with the severe optical distortions in lithography, with the ever-decreasing technology nodes. Inverse lithography techniques (ILT) are proposed as an effective RET, which attempts to consider the mask synthesis as an inverse problem and compute an optimum mask given a desired image using rigorous mathematical approaches.

The pioneer work in ILT was by Saleh *et al.* [1]. After that, a number of methods [2-11] have been proposed to improve effectiveness and efficiency of ILT algorithm. Most ILT algorithms use pixel-based mask representation, or grid-based mask representation (One grid covers one pixel). However, with the technology node decreasing, finer grid mask representation is inevitable to describe thinner mask patterns, resulting in fast increase of computational cost. Suppose an $N{\times}N$ image with a grid size of 5nm, where N is the mask size. If using a 1nm grid size to get higher image resolution, the total image size should be $5N{\times}5N$. Thus, there will be over 25X growth of the running time.

In this paper, we propose an ILT algorithm to mitigate this issue. Comparing with present pixel based ILT algorithm, our method is capable to provide nearly the same effectiveness and much better efficiency. The framework of level-set approach is constructed in this algorithm. Different from traditional level-set shape optimization algorithm [12], we perform optimization on coarse grids using a coarse-grid approximation (CGA) to accelerate the mask synthesis. And the final result is acquired by extracting mask patterns from coarse grids to fine grids. We call this algorithm as CGALSA (coarse-grid-approximation level-set approach).

The objective of our mask synthesis is aerial image fidelity. Partially coherent illumination and binary mask are used. A periodic array of contact holes pattern is used in the numerical verification. To demonstrate effectiveness and efficiency of CGALSA, we also perform mask synthesis using gradient based algorithm (GBA)[7] and traditional level-set algorithm (TLSA)[11].

2. Methodology

The fundamental framework of CGALSA is inherited from the level-set approach, which was first developed by Osher and Sethian [12]. Due to the application of CGA, some revisions should be made on traditional level-set method.

2.1 Level-set based ILT framework

Aiming at the aerial image fidelity, the inverse lithography problem can be formulated as finding an optimum mask **m** that minimizes the cost function (CF), that is

$$\underline{\mathbf{m}} = \arg\min_{\mathbf{m}}\{F(\mathbf{m})\}$$

$$where \quad F(\mathbf{m}) = \sum_{x_1=1}^{M}\sum_{x_2=1}^{N}\left\| I(x_1,x_2) - I^*(x_1,x_2) \right\|^2 \tag{1}$$

where I^* denotes the ideal image intensity, and the size of mask matrix is defined by $M \times N$. (x_1, x_2) is the coordinates of the mask pixel. For simplicity, vector **x** is used to denote (x_1, x_2). The optimization problem defined in Eq.(1) can be expressed as

$$Find \quad D \quad to \quad minimize \quad F(\mathbf{m})$$

$$where \quad m(\mathbf{x}) = \begin{cases} m_{int} & for \quad \mathbf{x} \in D \\ m_{ext} & for \quad \mathbf{x} \notin D \end{cases} \tag{2}$$

where $m_{int} = 1$ and $m_{ext} = 0$ for binary mask. The level-set approach introduces an unknown function $\phi(\mathbf{x})$ to give the mask function a new description and the problem can be rewritten as

$$Find \quad \phi(\mathbf{x}) \quad to \quad minimize \quad F(\mathbf{m})$$

$$where \quad m(\mathbf{x}) = \begin{cases} m_{int} & for \quad \phi(\mathbf{x}) < 0 \\ m_{ext} & for \quad \phi(\mathbf{x}) > 0 \end{cases} \tag{3}$$

$\phi(\mathbf{x})$ is defined as the level-set function. Note that the boundary of D is just the solution of $\phi(\mathbf{x}) = 0$, which is also referred to as the zero level-set. Level-set based ILT doesn't directly optimize **m**, but optimize $\phi(\mathbf{x})$ using an evolution approach. Shen *et al.* [11] implemented mask optimization using traditional level-set method [12] for coherence image model. Luminescent Technologies performed mask and source optimization using level-set method [10] but the computational details are not published.

2.2 Application of CGA

The coarse grid approximation (CGA) is introduced to convert a mask defined on fine grids to a mask defined on coarse grids, namely, CGA mask. In this step, suppose that a large grid is the union of a number of small grids. The value of large grid is not restricted to 0 or 1, but equals to the mean of the all the small grid values. Fig. 1 shows some examples of the mask conversion from fine grids to coarse grids. Note that the CGA of the original mask is not binary, but in gray level. The traditional level-set optimization is then performed on the CGA mask. The intensity calculation is performed on the CGA mask and the kernel is adjusted to the same grid size accordingly. In each iteration, when the level-set function is updated, the CGA mask can be acquired using the interpolation method. The final optimized mask defined on fine grids can be acquired from the final optimized level-set function using interpolation method.

Due to the large decrease of the mask and kernel sizes during optimization, the time for the intensity calculation which is the most time consuming procedure can be greatly reduced. The improvement of the efficiency of the whole mask optimization is quite large.

3. Results and discussions

In this section, results of ILT using CGALSA will be shown and discussed. We will compare CGALSA to GBA and TLSA, on the optimized image quality and running time. All the computations are performed on MATLAB using an Intel-E7400-2.8G CPU and 2G memory PC.

The wavelength λ equals 193 nm. We use a simple annular light source (0.6/0.98), from which the convolution kernels and singular values are computed [13]. Ten largest singular values and corresponding kernels are used in the intensity computation. The grid size is 3nm×3nm and 15nm×15nm for fine and coarse grids, respectively. GBA and TLSA are performed on fine grids and CGALSA is performed on coarse grids.

Periodic array of contact holes, common in logic and SRAM circuits, is adopted as our numerical example. Fig.2 shows a contact holes pattern with critical dimension (CD) of 45nm and the pitch of 105nm. Black and white areas represent the opaque regions and clear openings, respectively. To observe the advantage of CGALSA on computation speed more clearly, we also perform ILT for an 8×8 contact holes array pattern.

Figs. 3-5 show results for 2×2 array of contact holes. Fig. 3(a) shows optimized mask patterns, using different algorithms. Fig. 3(b) provides the aerial intensity distribution for each mask pattern and final value of the cost function (CF). Results show that CGALSA can provide almost the same image quality as TLSA and GBA.

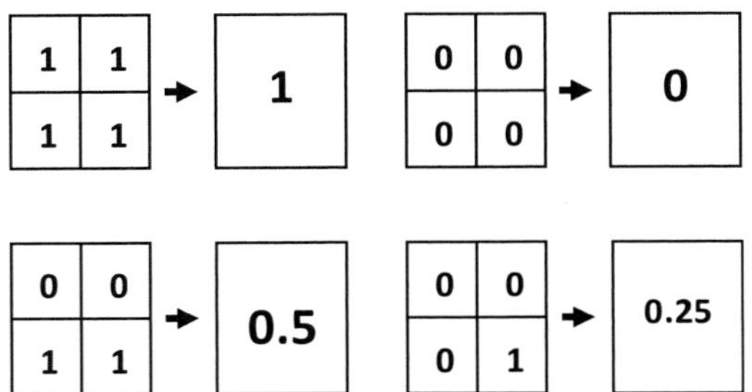

Figure 1: Examples of mask conversion from fine grids to coarse grids.

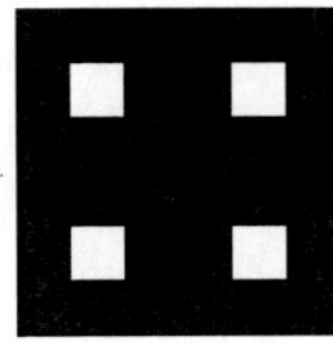

Figure 2: The 2×2 array of contact holes pattern.

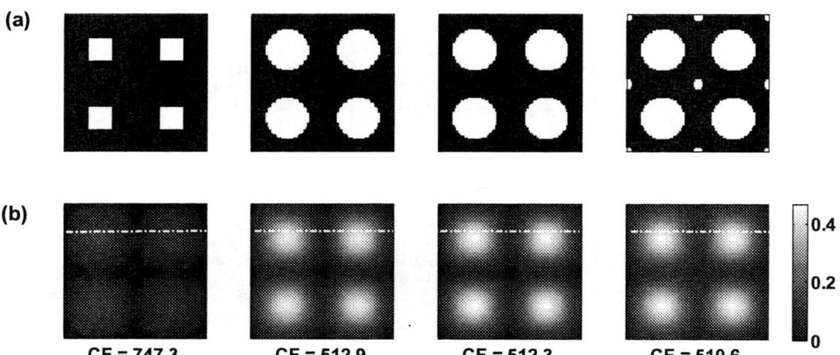

Figure 3: (a) Different mask patterns for 2×2 array of contact holes are shown. From left to right, the first one is the desired pattern without ILT, and the second to the fourth are the final optimized mask pattern using CGALSA, TLSA, and GBA, respectively. (b) Aerial intensity distributions and final CF are shown for the corresponding mask patterns.

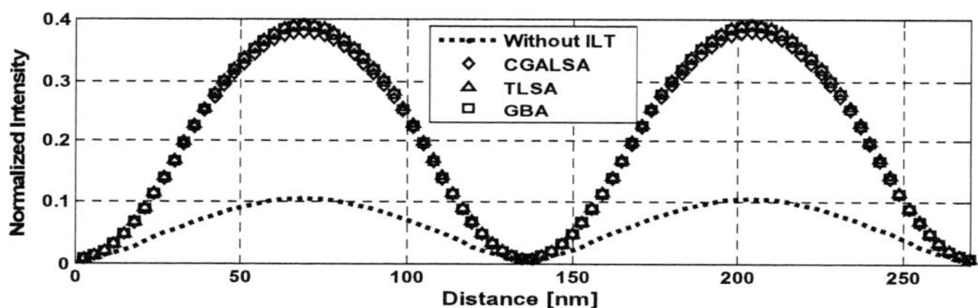

Figure 4: 1D intensity profiles along the cutline defined in Fig. 3(b). The dashed used the desired mask pattern, and others are computed using CGALSA (diamond), TLSA (triangle), and GBA (square), respectively.

Fig. 4 shows the 1-dimension (1D) intensity profile along the cutline in Fig. 3(b). It can be seen that nearly the same contrast enhancement is acquired using three different algorithms. Fig. 5 shows the convergence processes for all the three algorithms, together with the run time displayed. Great speed enhancement brought by CGALSA can be observed. Tab. 1 demonstrates the optimization performance using three algorithms, for 8×8 array of contact holes. All the three algorithms are terminated after 50 iterations. Comparing to CGALSA, TLSA and GBA bring very close improvement on image quality, but with more than 250X running time. CGALSA is comparatively effective and much more efficient than the other two algorithms.

Table 1: Optimized CF and running time for 8×8 array of contact holes

Algorithm	Optimized CF	Running time (seconds)
CGALSA	8201	3.2
TLSA	8198	855.6
GBA	8177	861.6

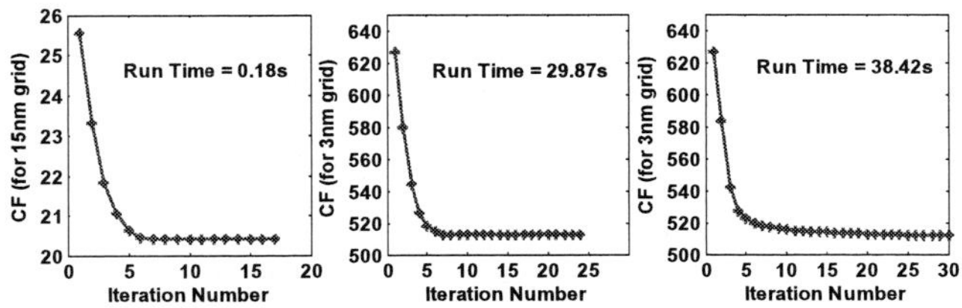

Figure 5: Convergence processes and running time for 2×2 array of contact holes using CGALSA, TLSA, and GBA (from left to right).

4. Summary

In this paper, we have proposed a highly efficient algorithm, CGALSA, to perform ILT optimization. A periodic array of contact holes patterns is adopted in the numerical verifications. Results show that CGALSA can provide sufficient improvements on image quality, and over 100X running speed enhancement.

Acknowledgements

The author would like to thank Leo Pang and Yong Liu from Luminecent Technologies and Dr. Min-Chun Tsai from Brion Technologies for their invaluable discussions. The support by NSFC, grant #60876073 and major national S&T program, grant #2009ZX02023-4-5 is greatly appreciated.

References

1. B. Saleh and S. Sayegh, "Reduction of errors of micro-photographic reproductions by optimal corrections of original masks", Opt. Eng. **20**, 781-787 (1981).

2. Y. Liu, A. Zakhor, "Binary and phase shifting mask design for optical lithography," IEEE Transaction on Semiconductor Manufacturing **5**, 138-151 (1992).

3. Y. C. Pati and T. Kailath, "Phase-shifting masks for microlithography: Automated design and mask requirements," J. Opt. Soc. Am. A **11**, 2438-2452 (1994).

4. A. Erdmann, R. Farkas, T. Fuhner, B. Tollkuhn, and G. Kokai, "Towards automatic mask and source optimization for optical lithography", Proc. SPIE **5377**, 646-657 (2004).

5. Y. Oh, J. C. Lee, and S. Lim, "Resolution enhancement through optical proximity correction and stepper parameter optimization for 0.12-μm mask pattern," Proc. SPIE **3679**, 607-613 (1999).

6. Y. Granik, "Solving inverse problems of optical microlithography," Proc. SPIE **5754**, 506-526 (2005).

7. A. Poonawala and P. Milanfar, "OPC and PSM design using inverse lithography: A non-linear optimization approach," Proc. SPIE **6154**, 61543H1–61543H14 (2006).

8. A. Poonawala and P. Milanfar, "Fast and low-complexity mask design in optical lithography: An inverse imaging problem," IEEE Trans. on Image Proc. **16**, 774-788 (2007).

9. J. Zhang, M. Tsai, W. Xiong, Y. Wang, and Z. Yu, "A highly efficient optimization algorithm for pixel manipulation in inverse lithography technique," in *Proc. of IEEE/ACM International*

Conference on Computer Aided Design (San Jose, CA, 2010), pp. 480-487.
10. L. Pang, Y. Liu, and D. Abrams, "Inverse lithography technology (ILT) for advanced semiconductor manufacturing," J. Exp. Mech. 22, 295(2007).
11. Y. Shen, N. Wong, and E. Y. Lam, "Level-set-based inverse lithography for photomask synthesis," Opt. Express **17**, 23690-23701 (2009).
12. S. Osher and R. P. Fedkiw, "Level-set method: an overview and some recent results," J. Comput. Phys. **169**(2), 463-502 (2001).
13. S. J. Chang and C. Chen, "Abbe-PCA-SMO: Microlithography source and mask optimization based on Abbe-PCA," Proc. SPIE **7520**, 75202G-1 (2009).

ECS Transactions, 34 (1) 215-221 (2011)
10.1149/1.3567584 ©The Electrochemical Society

A Fast OPC Algorithm for IC Layout Based on 1-D Cells after Optimization of Gap Distribution

B. Lin[a], C.-L. Xie[a], Z. Shi[a]

[a]Institute of VLSI Design, Zhejiang University, Hangzhou, Zhejiang, China

As VLSI technology scales down to 45nm process node, reliable printing becomes a persistent huge challenge to IC's manufacturability. The limitation of RETs now forces people to adopt a regular 1-D cell design methodology. A method of extending the line-ends and inserting dummies to optimize the gap distribution, with which the layout based on 1-D cells could get better printability, has been proposed in recent studies. In this paper, we present a fast OPC algorithm for IC layout based on 1-D cells after the previously proposed optimization of gap distribution. With this algorithm, during the OPC process, the extended parts and dummies of the layout are excluded from correction to save run time, and only the functional parts of the layout will be corrected. Experimental results on 45 nm process show excellent efficiency of this algorithm.

1. Introduction

As VLSI technology scales down to 45nm process node, reliable printing becomes a persistent huge challenge to the manufacturability when the application of EUV is still far from reality. According to the International Technology Roadmap for Semiconductors (ITRS) 2009, 193 nm immersion lithography technique will be used for years on 45 nm and sub-45 nm technology nodes, as Fig. 1 shows (1). By currently available Resolution Enhancement Techniques (RETs), there is great difficulty for printing complex 2-D layout in 45 nm and sub-45 nm. Therefore, it has become necessary to adopt a regular circuit patterns design methodology (2-5).

1-D gridded design refers to a layout style in which critical layers are drawn with 1-D lines on a coarse grid. The 1-D lines are straight lines looking like a grating pattern and can be in a vertical or horizontal direction. Fig. 2 illustrates a comparison between a conventional 2-D layout and the corresponding 1-D layout of a standard cell (2, 6). 1-D gridded designs are "litho friendly" and allow k_1 factor to be reduced significantly compared to 2-D random layouts. At the 45nm technology node, Intel adopted 1-D regular design rules for both SRAM poly layer and random logic poly layer to reduce gate length variation and improve yield (7).

The rest of this paper is organized as follows, Section 2 introduces preliminaries; Section 3 presents the fast OPC algorithm for IC layout based on 1-D cells after optimization of gap distribution; experiments and discussions are presented in Section 4; followed by conclusion in Section 5.

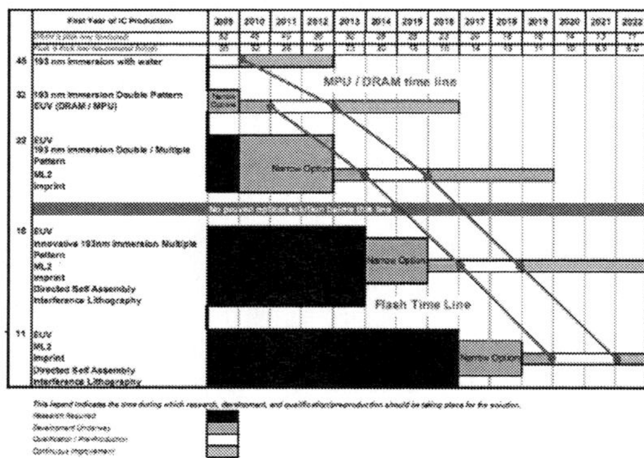

Fig. 1 Lithography exposure tool potential solutions

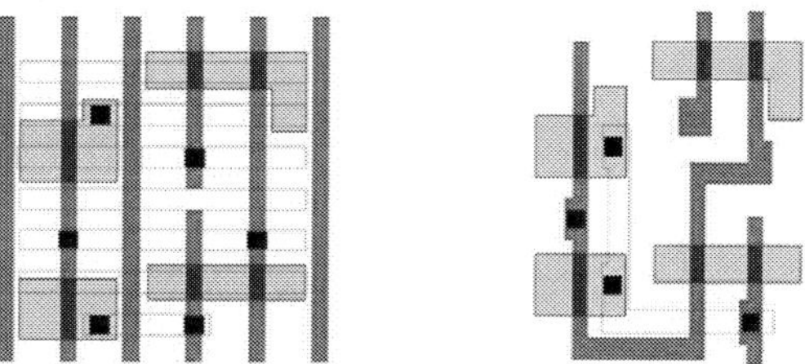

Fig. 2 1-D layout (left side) compared to 2-D layout (right side)

2. Preliminaries

The line width, spacing, and line-end gap are the three factors which determine the pattern in 1-D regular circuit. Line-end gaps affect printability greatly. Fig. 3 shows how line-end gaps affect line width roughness. Clear wavy shapes can be found when there is a gap nearby (7).

Zhang *et al.* (8, 9) have studied the relationship between the line-end gap distribution of 1-D cell and layout printability. Based on the gap distribution preferences, a method of extending the line-ends and inserting dummies to optimize the gap distribution, with which the layout gets better printability, has been proposed. Fig. 4 shows an example of an AOI21 gate, in which (b) is the 1-D layout of schematic (a). The optimization of gap distribution on metal 1 layer for better printability converts (b) to (c). Fig. 5 shows another example of gap distribution optimization. After the optimization, the simulation contour is much better than the result of original layouts.

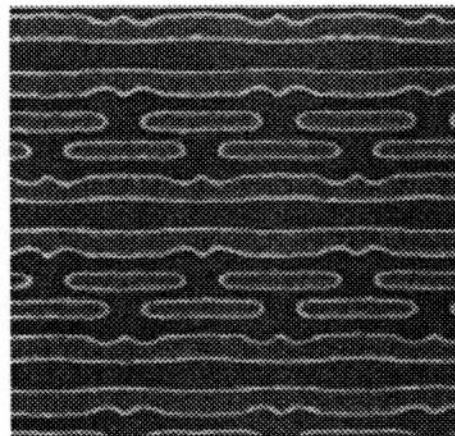

Fig. 3 An SEM image to show how line-end gaps affect line width roughness

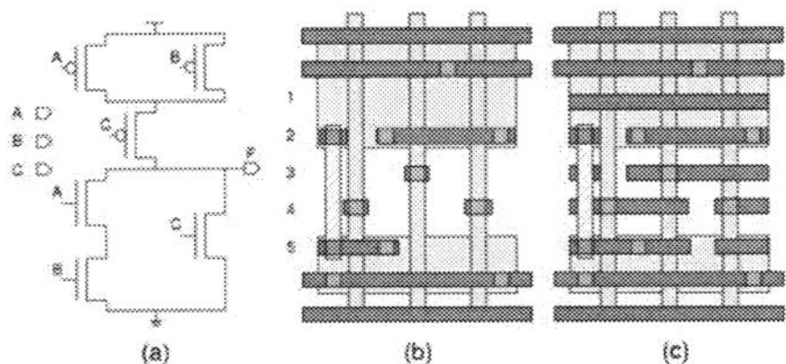

Fig. 4 An example of an AOI21 gate for the layout improvement targeting on printability of metal 1 layer

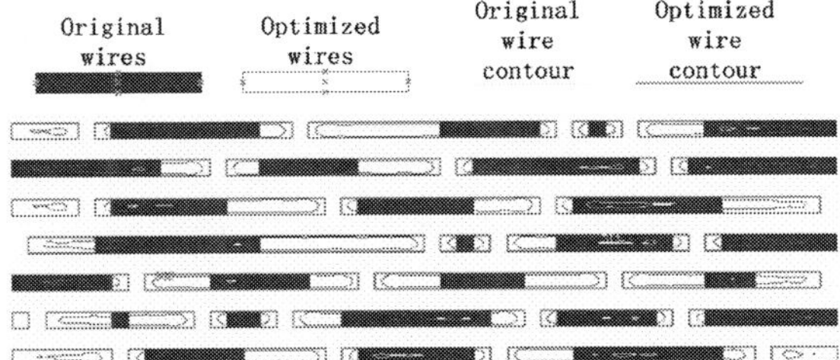

Fig. 5 A gap distribution optimization result and the corresponding simulation contours

3. Fast OPC Algorithm for IC Layout Based on 1-D Cells after Optimization of Gap Distribution

Although the optimization of gap distribution improves the layout printability, the EPE could be still unacceptable. Therefore, for IC layout based on regular 1-D cell with gap distribution optimization, OPC is always needed.

1-D cell gap distribution is optimized by extending the line-ends and inserting dummies. As Fig. 5 shows, after optimization, the optimized wires coverage ratio to layout is much larger than the original. However, in optimized layout, only the parts of original wires are the functional parts. EPEs on the functional parts are directly related to the IC functionality while EPEs on extended parts and dummies are not important in terms of the circuit functions. Therefore, we propose a fast OPC algorithm for IC layout based on 1-D cells after optimization of gap distribution here. With this algorithm, during the model-based OPC process, most of the extended parts and dummies of the layout are excluded from correction with empirical fixed offsets to save run time, and only the functional parts of the layout and the extended parts adjacent to them will be corrected, as Fig. 6 shows. The prerequisite for the usage of this algorithm is that both the layout of original wires before the optimization of gap distribution and the layout of optimized wires must be provided.

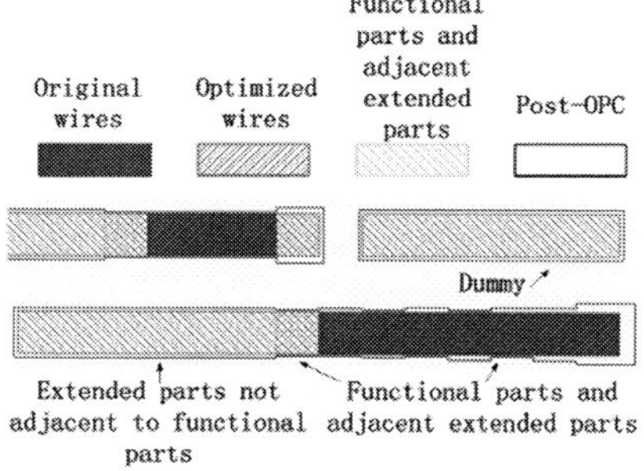

Fig. 6 Fast OPC algorithm for IC layout based on 1-D cells after optimization of gap distribution. Only the functional parts of the layout and the extended parts adjacent to them will be corrected.

4. Experiments and comparison

Two 45 nm layouts based on 1-D cells after optimization of gap distribution with area of 0.5 mm x 0.5 mm and 1 mm x 1 mm are used in the experiments and some results are shown here. The parameters of lithography model are λ= 193 nm, NA=1.27, sigma=0.95, sigma_in=0.89. All computations were performed on a Dell PowerEdge 2950 workstation (Xeon 2.5 GHz x 8 CPU and 16G Memory).
In these experiments, the edges of layout are dissected into small fragments with equal length of 45 nm which is the size of coarse grid in design for run time comparison.

Table1shows the run time comparison of conventional OPC and the proposed fast OPC respectively on the two layouts. With conventional OPC, all of the segments are involved in correction, while with fast OPC, only segments of functional parts of the layout will be corrected. We can observe that the fast OPC algorithm saves considerable run time compared to the conventional OPC algorithm.

Table II, Fig. 7 and Fig. 8 show the EPE comparison of conventional OPC algorithm and the fast OPC algorithm. From the figures, we can see that the latter gets slightly better EPE distribution result.

TABLEI. Run time comparison of conventional OPC and fast OPC.

CPU time spent on OPC (s)			
Layout 1 with area of 0.5 mm × 0.5 mm		Layout 2 with area of 1 mm × 1 mm	
Conventional OPC	Fast OPC	Conventional OPC	Fast OPC
142533	83308	511985	355546

TABLEII. EPE comparison of conventional OPC and fast OPC.

EPE (nm)	Layout 1 with area of 0.5 mm × 0.5 mm		Layout 2 with area of 1 mm × 1 mm	
	Conventional OPC	Fast OPC	Conventional OPC	Fast OPC
	Number of EPE distribution on original wires			
<-9	500576	80424	269655	90646
-9~-8	220381	26489	203751	31298
-8~-7	332037	30939	329151	43127
-7~-6	311387	32199	421673	54433
-6~-5	330505	39169	659687	60156
-5~-4	406483	53140	648264	99899
-4~-3	486606	98446	657242	161253
-3~-2	719602	235610	950077	320552
-2~-1	856206	953629	931671	1014302
-1~0	1664425	2084837	3300212	2889045
0~1	635164	2032707	1871715	3042912
1~2	466542	763138	649348	970415
2~3	759951	1926092	2276759	3081823
3~4	104758	475669	648704	838235
4~5	9822	33488	24895	15313
5~6	2846	2825	6738	117
6~7	2839	0	990	36
7~8	2813	0	2929	5
8~9	284	0	149	120
>9	38795	335	9132	205

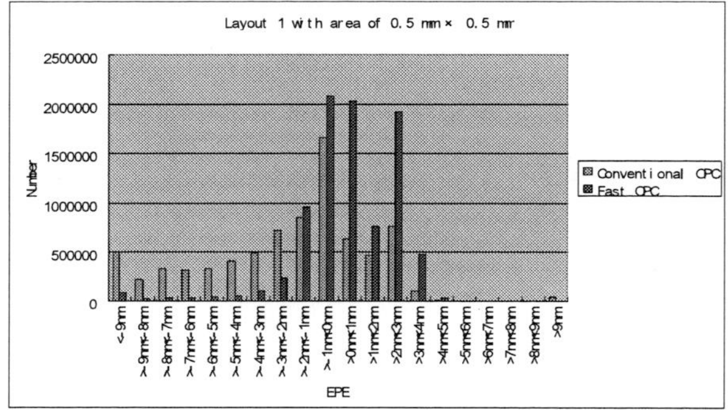

Fig. 7 EPE comparison on layout 1

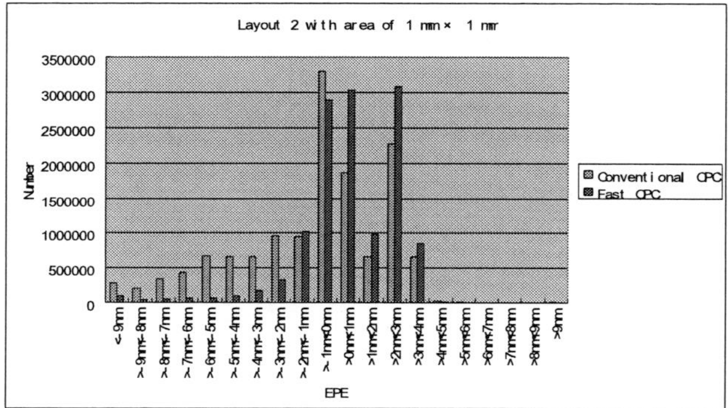

Fig. 8 EPE comparison on layout 2

5. Conclusion

This paper presents a fast OPC algorithm for IC layout based on 1-D cells after optimization of gap distribution. Experiments show that the proposed fast OPC algorithm has comparable accuracy performance to conventional OPC algorithm and significantly improves the OPC speed.

Acknowledgments

We would like to thank Hongbo Zhang at Department of Electrical and Computer Engineering University of Illinois at Urbana-Champaign for discussions and advices.

References

1. International Technology Roadmap for Semiconductors Lithography, p12, 2009 edition (2009).

2. R. T. Greenway, R. Hendel, K. Jeong, A. B. Kahng, J. S. Petersen, Z. Rao, and M. C. Smayling. Proc. SPIE 7271, 72712U (2009).

3. V. Kheterpal, V. Rovner, T. G. Hersan, D. Motiani, Y. Takegawa, A.J. Strojwas, and L. Pileggi. Proc. DAC '05, ACM, pp.353-358 (2005).

4. L. Liebmann, L. Pileggi, J. Hibbeler, V. Rovner, T. Jhaveri, and G. Northrop. Proc. SPIE 7275, 72750A (2009).

5. M. C. Smayling, C. Bencher, H. D. Chen, H. Dai, and M. P. Duane. Proc. SPIE 6925, 69251E (2008).

6. M.C. Smayling, H. Liu, and L. Cai. Proc. SPIE 6925, 69250B (2008).

7. C. Webb. Intel Technology Journal, Vol. 12(02), pp.121-130 (2008).

8. H. Zhang. ASP-DAC 2010, pp.838-842 (2010).

9. Hongbo Zhang, Martin D. F. Wong, Kai-Yuan Chao, Liang Deng, and Soo-Han Choi. Proc. SPIE 7275, 72751G (2009)

ECS Transactions, 34 (1) 223-230 (2011)
10.1149/1.3567585 ©The Electrochemical Society

Extension use of immersion lithography for the 22nm half-pitch and beyond

Reiji Kanaya

Nikon Corporation
201-9 Miizugahara, Kumagaya-city, Saitama, 360-8559, Japan

In double patterning process, exposure tools are required for better accuracy and productivity. NSR-S620D, which is Nikon immersion scanner, meets these requirements.
This paper discusses the current status of Nikon immersion scanner and possibility to extend use of immersion lithography for the 22nm half-pitch and beyond

Introduction

As device dimensions shrink 32nm half-pitch node, double patterning is required. For 22nm half-pitch node and beyond, EUVL may be needed. However it is facing some challenges such as a source power, mask defects and resist. EUVL will be delayed due to their remaining technical obstacles. In that case, double patterning will most likely take over 22nm half-pitch node and beyond. (See Fig.1)

In double pitch patterning, overlay error will affect the CD uniformity of spaces. For 32nm half-pitch node, overlay and CDU error 3sigma should be within 2.4nm. S620D can realize these specifications.

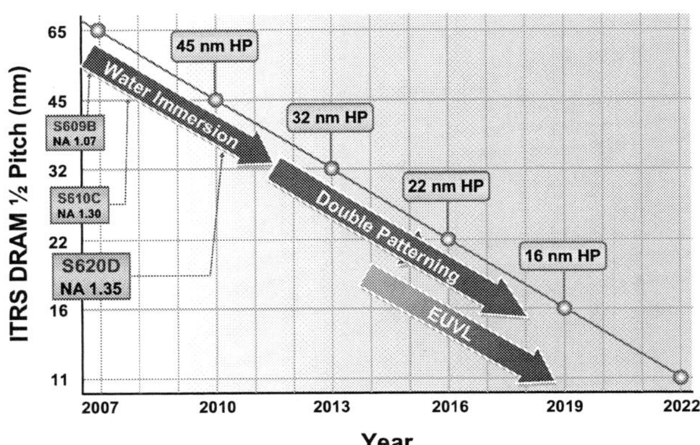

Fig.1 Nikon lithography Roadmap

1. NSR-S620D Tool Concept

There are three main technologies for S620D with "Streamlign" platform.
-Bird's eye control
-Stream alignment
-Moduler2 structure
With Bird's eye control, encoders and interferometers are both used to monitor the stage position. This technology reduces stage positioning errors and improves overlay accuracy. It helps to realize 2nm overlay which is one of S620D feature concepts.

With Stream alignment, five-eye FIA can measure EGA marks on the wafer and straight line AF can make the focus map during wafer loading. By reducing alignment and focus mapping time, high throughput accuracy is achieved. Our target for this is over 200wph.

Moduler2 structure makes tool setup, maintenance and upgrades easier. And it shortens the lead time. This technology helps to meet the target lead time of within 20 days for setup.

1.1 Bird's Eye Control

Bird's Eye Control is an encoder metrology system. Fig. 2 shows a schematic configuration of Bird's eye control. Encoders measure grating scales which are attached on the top of the wafer stage. The working distance between a grating plate and an encoder is only 2mm. Air fluctuation may be only slightly affected and Abbe error is negligible. It realizes stable the stage positioning.

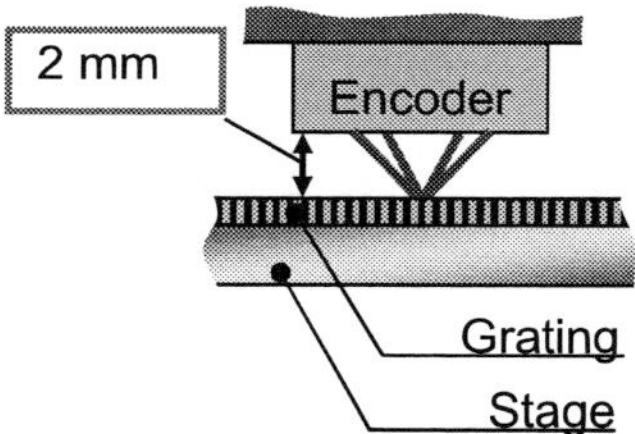

Fig.2 Bird's eye control (hybrid system)

However, high stage positioning accuracy wouldn't be achieved if encoder systems alone are used for monitoring the stage position. Therefore hybrid metrology that used both encoders and interferometers requires high accuracy. Table 1 compares with strong point of each system. With hybrid system, interferometer system compensate for the encoder system. For linearity, encoder system is affected in an adverse way because it is affected by the grating scale plate flatness. On the other hand interferometer system has perfect linearity. Thus interferometers calibrate the grating scales and compensate encoder linearity. In addition, the hybrid system provides good stability. The interferometers will take over immediately to maintain the accuracy for continuous exposure if the encoder servo is interrupted by a particle or water droplet on the scale. Then after the interrupting object moves away, servo control will speedily be switched back to the encoders again.

Table1 Comparison with strong point of each system

	IF	Enc	Hybrid
Linearity	Good	Bad	Good
Repeatability	OK	Good	Good
Longtime stability	Good	OK	Good

1.2 Stream Alignment

S620D has five alignment microscopes, which is called Five-Eye FIA, and Straight line AF that can measure the whole wafer height by only one scanning. Fig.3 illustrates schematic configuration of the Stream Alignment system. This optimized configuration can align a wafer and create an AF map while a wafer is being loaded on to the first exposure area. Accordingly, it enables dramatic reduction in the alignment time and achieves high throughput even with a large number of alignment sites.

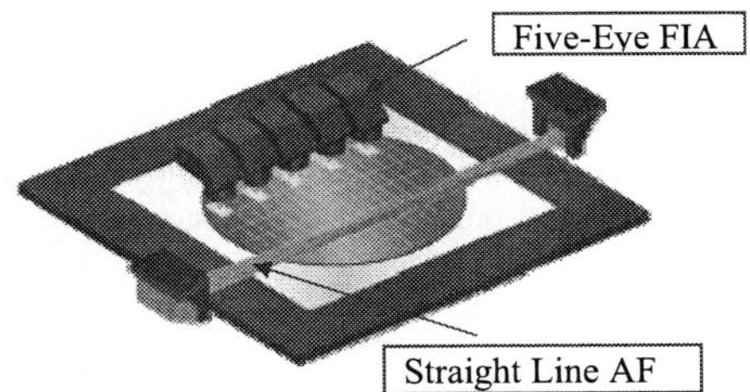

Fig.3 Stream Alignment system

1.3 Moduler2 structure

The main body consists of several independent modules, some of which can be further divided into sub modules. This structure provides us benefits of easy maintenance and short installation time. At the same time, this structure also enables to upgrade the parts easily for future. The upgrade of an individual module is possible, depending on various requirements of upcoming generations.

2. Latest Performance

2.1 Accuracy

Fig.4 shows scanning synchronization accuracy. We achieved good scanning synchronization results on S620D with 700mm/sec and zero settling time. X, Y and Z all meet the specifications. Even at the beginning of the exposure, the synchronization results show low MA and MSD. Reticle stage isolation, sky-hook isolation and motion control prevent vibration from transmittance. And Bird's eye control reduces stage positioning error.

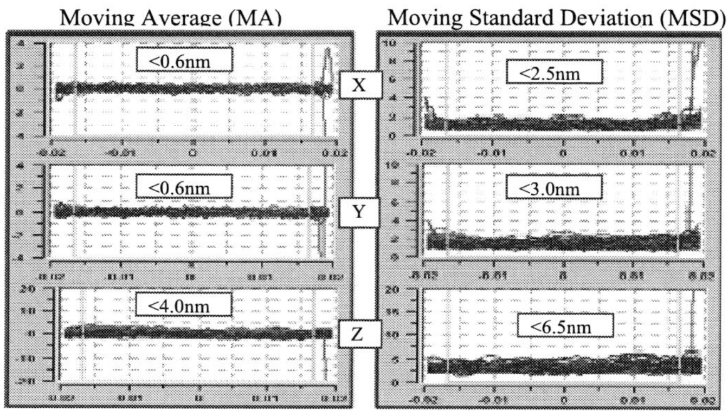

Fig.4 Stage Synchronization with 700mm/sec and zero settling time

Fig.5 describes overlay results through a lot and continuous exposure of 20 wafers at 320mm/sec. The overlay result for X and Y is less than 2nm. And the overlay result at 700mm/sec which is max speed is around 2nm (See Fig.6). We also have good overlay results with max speed. Stream Alignment measures multi-point alignment marks. It is not only fast but the measurement repeatability is also good.

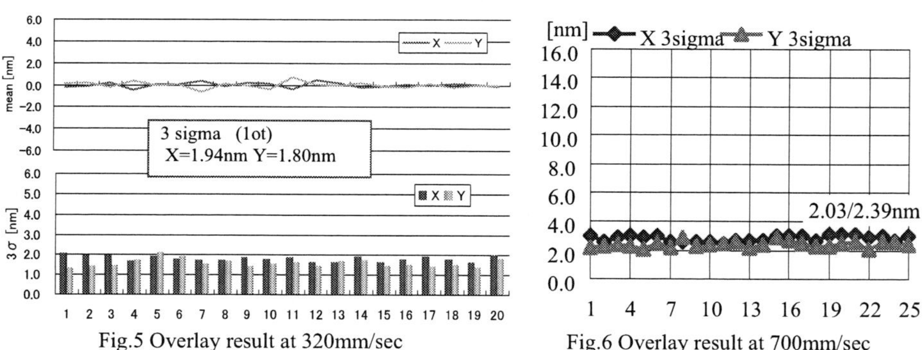

Fig.5 Overlay result at 320mm/sec   Fig.6 Overlay result at 700mm/sec

Fig.7 points the focus control repeatability result which is measured by Phase Shift Focus Monitor (PSFM) method. The result shows that 3 sigma is 14.3 nm on all shots include edge shots and 11.8 nm on all full field shots with 700mm/sec. It definitely meets the budget for the next semiconductor generation. Z sensors constantly bring out good focus repeatability less than 15nm.

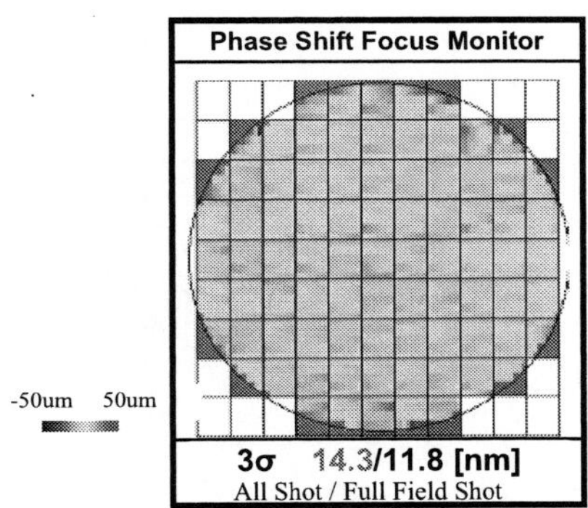

Fig.7 Auto Focus Repeatability Result by PSFM

2.2 Productivity

24 hours running test was carried out. In the result, 4,116 wafers completed running without error and interruption. The running test results on S620D compare advantageously with S610C. (See Fig.8)

These results prove the capability for 4,000wdp product on S620D.

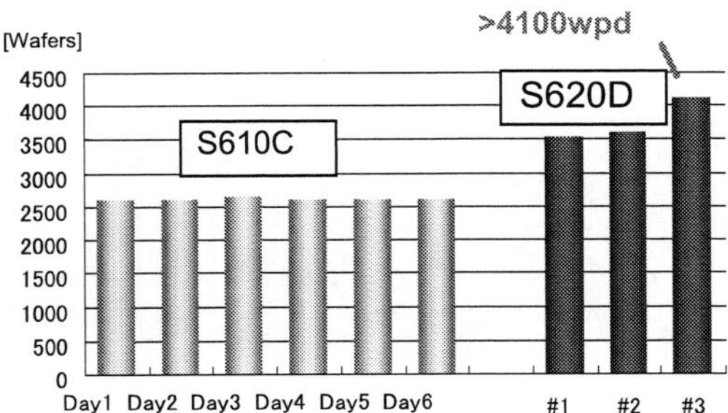

Fig.8 Comparison of 24 hours running test between S610C and S620D

2.3 Imaging

The numerical aperture of production lens for S610D is 1.35, and it has sufficiently low wavefront aberration. Fig.9 shows the total RMS of wavefront aberration on S610C and S620D tools. Total RMS on S620D is 0.7nm. The fine performance has been compared to S610C projection lenses.

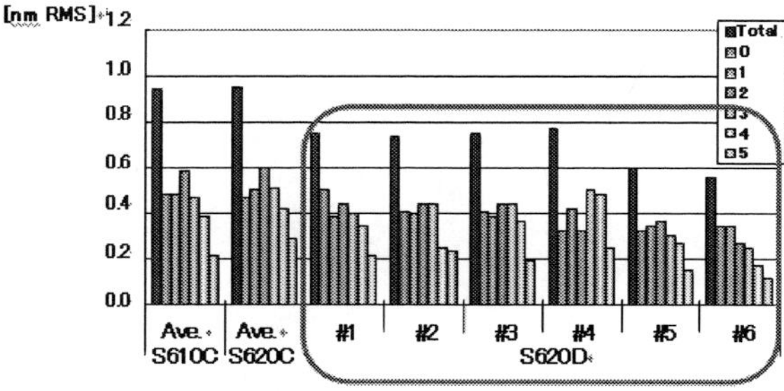

Fig.9 the Wavefront RMS

The lens of S620D has a function useful for the severe requirement of double patterning.

The schematic drawing of this function, which is the Adaptive 2theta Compensator, is shown in the left of   Fig. 10. A deformable mirror with piezo drive is introduced in the new catadioptric projection lens. By pushing the mirror mechanically, Zernike 2theta components Z5 and Z6 can be independently adjusted as shown in the right of Fig. 10. The deformable mirror controls thermal aberrations like astigmatism in dipole illumination. A feature of the Adaptive 2theta compensator is high speed response. The response time is within 1second. The compensator carries out static and dynamic adjustments.

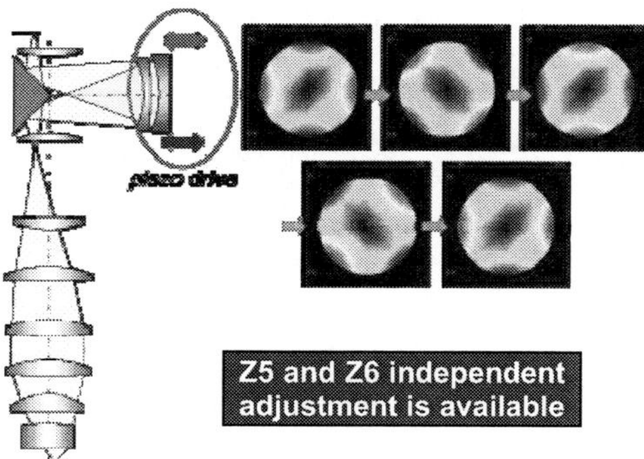

Fig.10 the Adaptive 2theta Compensator

3. Future prospects

We tested to expose 32nm L/S by S620D with pitch splitting double patterning. The conditions are follows: Dipole-Y, sigma 0.85/ Ratio 0.77, NA1.00 and LFL (Litho-Freeze-Litho).   SEM images of this are shown in Fig.11. The fine patterns are successfully resolved whole over the wafer. The CDU for pooled space across the wafer is 3.3nm and for pooled line is 2.5nm. These sufficiently meet the budget on the 32nm HP node. This achievement of difficult space CDU specification is evidence of high overlay accuracy.

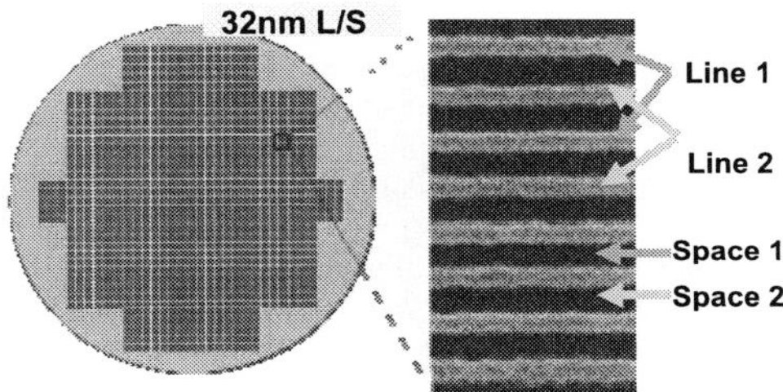

Fig.11 SEM images of double patterning

Resolution test for 22nm HP and 25nm HP was also performed. The SEM images are shown in Fig.12. 22nm L/S and 25nm L/S are exposed by S620D with pitch-splitting double patterning. Both the first and the second patterns are exposed successfully. The line pattern is formed clearly showing the potential of S620D for 22nm HP node.

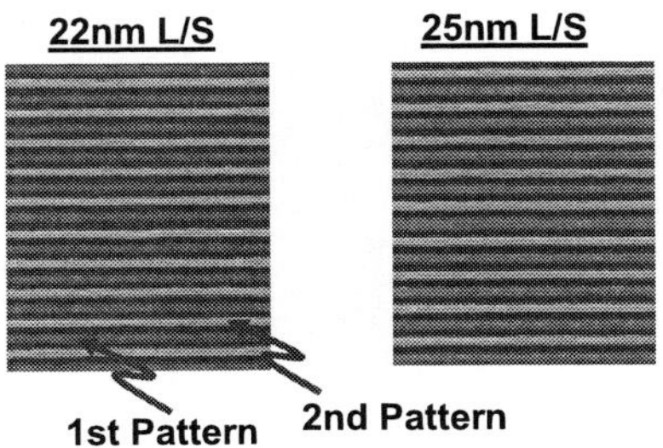

Fig.12 SEM images of double patterning

Conclusions

S620D with "Streamlign" Platform achieves the following results.
High Accuracy
-22nm L/S with double patterning
-Overlay accuracy <2nm
High Productivity
-High scanning speed up to 700mm/sec

-Throughput demonstration >4000wpd
Long Life Platform
 -Immersion extension is possible
NSR-S620D, Nikon Immersion scanner and possibility to extend use of immersion
lithography for the 22nm half-pitch and beyond

References

1. Motokatsu Imai, Hiroyuki Suzuki and Hirotaka Kohno: Environmentally Friendly
   Immersion Scanner, MNC2010
2. Hirotaka Kohno, Yuichi Shibazaki, Jun Ishikawa, Junichi:Latest Performance of
   Immersion Scanner S620D with Streamlign Platformfor Double Patterning
   Generation, 2010 SPIE, 7640-58
3. Takahisa Kikuchi, Yosuke Shirata, Masahiko Yasuda, Yasuhiro Iriuchijima,
   Kengo Takemasa, Ryo Tanaka, Andrew Hazelton, Yuuki Ishii: Double Patterning
   Lithography Study with High Overlay Accuracy, 2010 SPIE, 7640-16

ECS Transactions, 34 (1) 231-236 (2011)
10.1149/1.3567586 ©The Electrochemical Society

CYMER LPP EUV SOURCE SYSTEM DEVELOPMENT STATUS

Benjamin Szu-Min Lin[a], Bruno La Fontaine[b], David Brandt[b], and Nigel Farrar[b]

[a]Cymer Southeast Asia Ltd., 3F, No. 49, Lane 2, Kuang Fu-Rd., Sec. 2, Hsin-Chu, Taiwan, R.O.C.,
[b]Cymer Inc., 17075 Thornmint Ct., San Diego, CA 92127

In this paper, the technologies developed by Cymer to deliver clean EUV power will be reviewed. A high average power (>20kW) pulsed CO_2 laser is used to create a Sn plasma by irradiating 30μm Sn droplets at a repetition rate of approximately 50kHz. The EUV light emitted by the Sn plasma is collected using a 5sr multi-layer mirror (MLM) collector and refocused at the intermediate focus (IF). The integrity of the MLM collector is maintained over time using debris mitigation techniques based on H_2. A number of diagnostics are used to characterize the EUV light past IF. We will describe the performance of multiple laser-produced plasma (LPP) EUV sources, integrated with a 5sr MLM collector and debris mitigation.

Introduction

According to the International Technology Roadmap for Semiconductors (ITRS), EUV Lithography is widely considered for critical dimension imaging below 20nm half pitch generation after 193nm immersion lithography. The availability of reliable high power 13.5nm sources has been highlighted as one of the top issues among various EUV infrastructure barriers to enable the realization of EUV lithography during SEMATECH-held EUV symposia over several years. For ASML EUV pre-production exposure tools (NXE:3100), it is estimated that approximately 100W of clean EUV power is required at the intermediate focus (IF) to achieve a throughput of 60 wafers/hour (WPH), assuming a photoresist sensitivity of $10mJ/cm^2$. However, 100WPH is commonly accepted throughput for considering EUV as a cost-effect patterning solution compared to 193nm immersion Double Patterning techniques. As the NA of the EUV projection optical system increases to extend Moore's Law in the future, even more EUV power will be required because of the increase in the number of mirrors and the ensuing reduction of transmission through the system. So, a scalable EUV source architecture is needed to enable the evolution of EUV lithography during the life cycle of the technology. Laser-produced-plasma (LPP) sources are expected to deliver the necessary high power for critical dimension high-volume manufacturing (HVM) scanners for the production of integrated circuits in the post-193nm immersion era.[1]

The LPP development at Cymer includes three key modules, including a multiple-stage high-power high repetition rate CO_2 laser, a laser beam transport system (BTS), and an EUV source vacuum vessel. The LPP EUV system configuration is shown in Figure 1. The drive laser is a CO_2 laser with multiple stages of amplification[2] to reach the required power level of ~20 kW. It is operated in pulsed mode at 40-50 kHz with radiofrequency (RF) pumping from generators (not shown) operating at 13.56 MHz. The laser is typically installed in the sub-fab along with its RF generators and water-to-water heat exchangers. The source controller turns on and off bursts of pulses, which can be as long as several seconds, but will typically be 400ms for exposing a 26 x 33mm field size using 10mJ/cm^2 resist. The ratio of time when the burst is on to the burst period defines the duty cycle. The beam is expanded as it leaves the drive laser to maintain the energy density on the BTS mirrors within a certain operating range. Three turning mirrors are used to allow the beam to travel from the sub-fab to the fab through the waffle-slab floor with the needed flexibility for positioning the laser with respect to the source vessel (and scanner) on the floor above. The laser and BTS are completely enclosed and interlocked to meet laser class 1 requirements. The BTS delivers the beam to a focusing optic where the 10.6 micron wavelength light is focused to a minimum spot size defined by the numerical aperture of the focusing system. The focused beam propagates through a central aperture in the collector and strikes the droplet at the focus of the ellipsoidal collector mirror inside the vacuum space of the source vessel chamber. The droplet generator delivers liquid tin droplets of 30 micron diameter to the same position at 40-50 kHz repetition rate; both laser pulse and droplets are steered and timed to ensure proper targeting. The laser pulse vaporizes and heats the tin into a plasma. The EUV light emitted by the plasma is collected and reflected with the multi-layer coated ellipsoidal mirror to the intermediate focus (IF) where it passes through a small aperture into the scanner volume that houses the illumination optics. To ensure that no contamination can reach the scanner volume an IF protection module surrounds the aperture and suppresses flow or diffusion. Other modules on the source vessel include the droplet catcher which collects the unused droplets between the bursts, and metrology modules for measuring EUV energy and for imaging of droplets and plasma.

Four first-generation high-volume manufacturing (HVM I) sources have been shipped to our customers, including one to a chipmaker fab. Four other HVM I sources are presently under test in the cleanroom in our San Diego California facility, some of which will be shipped to our customers in the near future and some of which will be used for internal engineering studies.

LPP EUV Power Development Status

<u>EUV Power Development History</u>

Cymer started LPP EUV source development in 2004, and the main emphasis of past development activities has been on achieving substantial gains of the source output power by continuous improvement of laser intensity generated on the target droplet. In Q1, 2009[3,4], the production systems capability was reported to be an average power of approximately 20W (2% in-band at IF) over 18 hours continuous operation with 400ms burst duration and 80% duty cycle, which matches the required scanner stage scanning speed for a full field size. The dose accumulation during that run was more than 1MJ, which is enough to process ~250 wafers assuming a dose of 10mJ/cm^2. In Q3, 2009, the maximum EUV power has been increased to the level of 70W (2% in-band at IF) over one hour of continuous operation with conditions of 60% duty-cycle, 400ms burst duration, 30μm Sn droplet diameter, droplet position control on, and dose control off

(Figure 2). 70W at IF for 1 hour operation at 60% duty cycle produces a total energy of 150kJ per hour, and 24 hours equivalent is 3.6MJ or 900 wafers per day for 300mm wafers with photoresist sensitivity of 10mJ/cm^2. At SPIE Advanced Lithography Symposium 2010[5], Cymer showed raw EUV power above 90W obtained on an Engineering Test Stand LPP source[6] running at 80% duty cycle, as determined from measurements at plasma. IF-equivalent average powers above 90W were calculated using standard assumptions of 50% reflectivity for a 5 sr collector and for 90% optical transmission from plasma to IF. The power level was reached using a laser configuration with longer gain length compared to our standard production system and demonstrated the expected performance of the planned upgrade to HVM I sources. The open loop energy stability of a 10ms sliding window through the raw data is 5.4% (3σ). An estimated EUV power of 80W will be achieved after applying closed-loop dose control. These experiments were performed using tin droplets with 30 micron diameter at 50kHz repetition rate and 400ms burst duration, the standard production system parameters for our HVM I sources.

Recent EUV Power Development Status

At the EUVL Symposium in 2010, raw EUV power of 175W (~80W clean EUV power) was disclosed by Cymer using the Engineering Test Stand LPP source with longer drive laser gain length to increase laser power and an optimized pre-pulse technique to increase EUV conversion efficiency (CE). The EUV CE can be increased to ~3% typically using this pre-pulse technique. The total testing period lasted for 1.5 hours, and Figure 2 shows a 20 minute sample out of the total runtime. Power and dose stability qualification tests have been done under ~15W exposure power (assuming 65% SPF transmission), 40% duty cycle conditions, for a duration of 100 hours. The closed loop dose stability for a 10ms sliding window through the raw data is ~4% (3σ) over the entire 100 hour run, and the champion performance of < +/-0.1% is shown in Figure 3. These experiments were performed using tin droplets with 30 micron diameter at 50kHz repetition rate and 400ms burst duration, the standard production system parameters for our HVM I sources. The HVM I source has also demonstrated feasibility of controlling dose to within 0.3% for 98% of the dies on a wafer (125 fields per 12 inch wafer, Figure 4).

Droplet position stability in source is one of the important factors impacting dose control. Figure 5 shows the droplet position stability results measured over 7 days. Feedback control (active stabilization) in two dimensions was implemented during these 7 days. After examining these test results (<3μm 1σ), this illustrates the capability of meeting droplet position stability of <10μm 3σ.

Future Development Roadmap

The pilot product sources, HVM I, have been shipped for integration into preproduction EUV scanners in 2010, and a total of 4 HVM I sources have been shipped, including one to a chipmaker fab. In line with the expected requirements for increased EUV power,

HVM II and HVM III generations of LPP EUV sources are expected to be brought to market using higher power CO_2 laser technology with moderate improvements in CE and collection efficiency in 2011/Q3 and 2013/Q2, respectively.

Conclusion

Laser-produced plasma has been shown to be the leading EUV source technology with scalability to meet requirements from scanner manufacturers and provide a path toward higher power as the lithography tools evolve. Four HVM I sources have been shipped to our customers, including one to a chipmaker fab. The raw EUV power of 175W, equivalent to ~80W clean EUV power, was disclosed using longer driver laser gain length to increase drive laser power and an optimum pre-pulse technique to increase EUV CE close to 3%. The closed loop dose stability of a 10ms sliding window through the raw data is demonstrated to be ~4% (3σ) over an entire 100 hour run, with champion performance of < +/-0.1%. Droplet position stability meets the requirements of <10μm 3σ, which is necessary for good dose stability of the LPP sources. LPP source technology with power levels exceeding 350W is expected to meet the IF power requirement projected in the future, and to provide the much needed margin for photoresist sensitivity, spectral purity filters, optics degradation, process latitude, and overall equipment throughput.

Acknowledgement

The authors gratefully acknowledge the valuable contributions from Martin J. Neumann and David N. Ruzic of University of Illinois, Urbana Champaign, Marco Perske, Hagen Pauer, Mark Schürmann, Sergiy Yulin, Torsten Feigl and Norbert Kaiser of Fraunhofer Institut f. Angewandte Optik und Feinmechanik, Eric Gullikson and Farhad Salmassi of Lawrence Berkeley National Laboratory, Frank Scholze, Christian Laubis, Christian Buchholz and coworkers at PTB, and Mark Tillack and Yezheng Tao of the University of California at San Diego. We are also very thankful for the invaluable support and contributions, past and present, of many scientists, engineers and technicians involved in the EUV technology program at Cymer.

References

1. I. V. Fomenkov, D. C. Brandt, A. N. Bykanov, A. I. Ershov, W. N. Partlo, D. W. Myers, N. R. Böwering, G. O. Vaschenko, O. V. Khodykin, J. R. Hoffman, E. Vargas L., R. D. Simmons, J. A. Chavez, C. P. Chrobak, in: Proc. of SPIE Vol. 6517, Emerging Lithographic Technologies XI, M. J. Lercel, ED., 65173J, (2007).
2. I. V. Fomenkov, B. A.M. Hansson, N. R. Böwering, A. I. Ershov, W. N. Partlo, V. B. Fleurov, O. V. Khodykin, A. Bykanov, C. L.Rettig, J. R. Hoffman, E. Vargas L., J. A. Chavez, W. F. Marx, D. C. Brandt, in: Proc. of SPIE Vol. 6151, Emerging Lithographic Technologies X, M. J. Lercel, Ed., 61513X, (2006).
3. B. Lin, D. Brandt, N. Farrar, in: ECS Trans. Vol. 18, Issue 1, Photolithography, D. Huang, Ed., 391-396 (2009).
4. D. C. Brandt, I. V. Fomenkov, A. I. Ershov, W. N. Partlo, D. W. Myers, N. R. Böwering, N. R. Farrar, G. O. Vaschenko, O. V. Khodykin, A. N. Bykanov, J. R. Hoffmann, C. P. Chrobak, S. N. Srivastava, I. Ahmad, C. Rajyaguru, D. J. Golish, D. A. Vidusek, S. De Dea, R. R. Huo, in: these Proc. of SPIE Vol. 7271,

Alternative Lithographic Technologies, F. M. Schellenberg, Ed., 727103-1, (2009).
5. D. C. Brandt, I. V. Fomenkov, A. I. Ershov, W. N. Partlo, D. W. Myers, R. L. Sandstrom, N. R. Böwering, G. O. Vaschenko, O. V. Khodykin, A. N. Bykanov, S. N. Srivastava, I. Ahmad, C. Rajyaguru, D. J. Golich, S. De Dea, R. R. Hou, K. M. O'Brien, W. J. Dunstan, in: Proc. of SPIE Vol. 7636, Extreme Ultraviolet (EUV) Lithography, B. M. La Fontaine, ED., 76361I (2010).
6. D. C. Brandt, I. V. Fomenkov,, A. I. Ershov, W. N. Partlo, D. W. Myers, N. R. Böwering, A. N. Bykanov, G. O. Vaschenko, O. V. Khodykin, J. R. Hoffmann, E. Vargas L., R D. Simmons, J. A. Chavez, C. P. Chrobak, in: Proc. of SPIE Vol. 6517, Emerging Lithographic Technologies XI, M. J. Lercel, ED., 65170Q (2007).

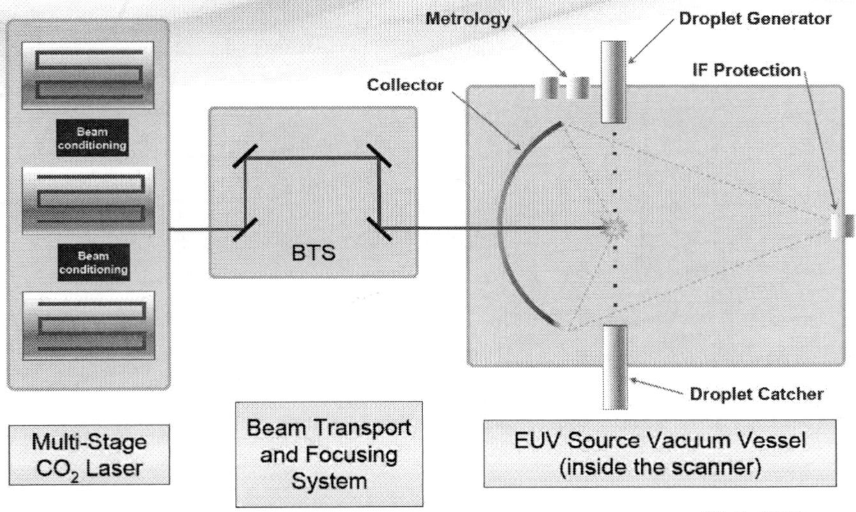

Figure 1. The simple LPP EUV system configuration. Three key subsystems include a multiple-stage high-power high repetition rate CO_2 laser, a laser beam transport system (BTS), and an EUV source vacuum vessel.

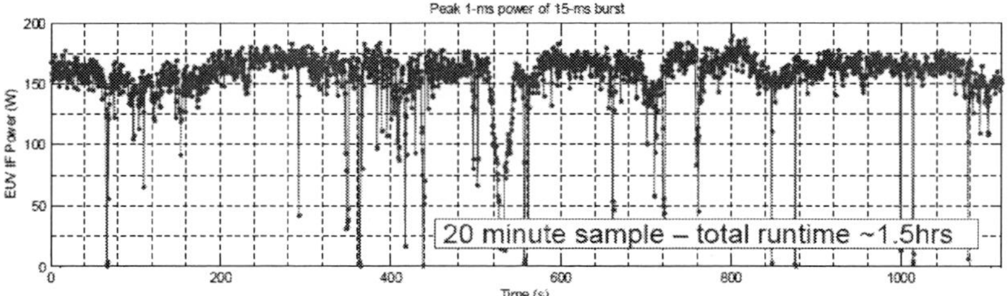

Figure 2. Raw EUV power of 175W (~80W clean EUV power) was disclosed with EUV CE of 3%. The total testing period lasted for 1.5 hours, and here showed a 20 minute sample out of the total runtime.

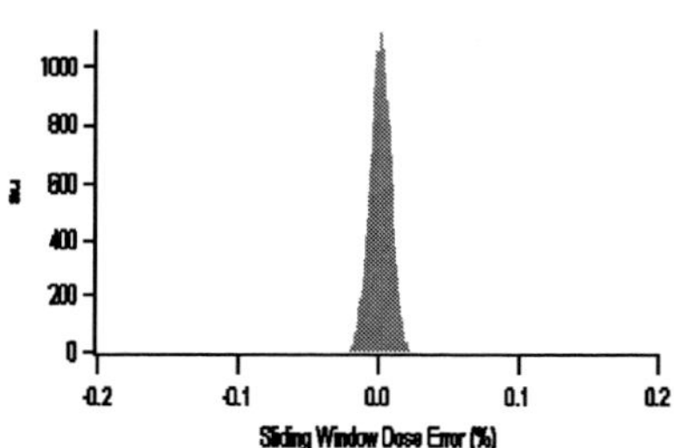

Figure 3. EUV dose stability under closed loop control (the champion performance of < +/-0.1%). Dose stability qualification tests have been done under ~15W exposure power.

Figure 4. The HVM I source has also demonstrated feasibility of controlling dose to within 0.3% for 98% of the dies on a wafer (125 fields per 12 inch wafer).

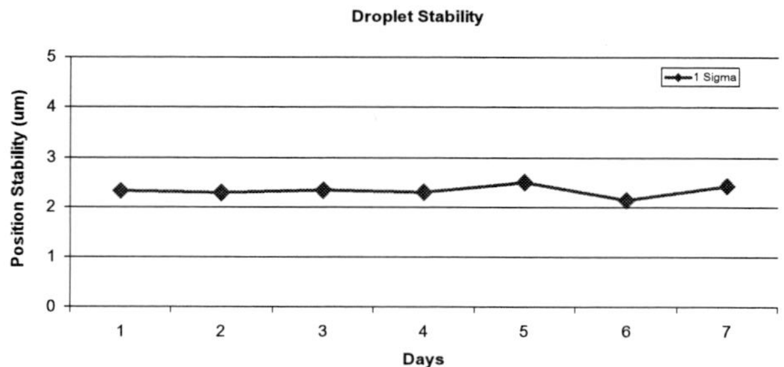

Figure 5. Droplet position stability results measured over 7 days. Feedback control (active stabilization) in two dimensions was implemented during these 7 days. After examining these test results (<3μm 1σ), this illustrates the capability of meeting droplet position stability of <10μm 3σ.

ECS Transactions, 34 (1) 237-242 (2011)
10.1149/1.3567587 ©The Electrochemical Society

Advanced Packaging Stepper for 300mm Wafer Process

Zhou Chang, Huang Ling
Shanghai Micro Electronics Equipment CO., LTD.

Abstract: With development of advanced package technology, requirements for lithographic process become tighter than before. When dealing with 200mm or 300mm wafer, stepper shows better uniformity of CD and overlay. SS B500 stepper can be used in RDL and Bump process. This paper introduces features of SS B500 stepper, such as wide band exposure lightsource, large field size, flexible alignment strategy, direct focus measurement, and proposes absolute overlay error concept and measurement mothod based on golden wafer. It also provides test data for CDU, UDoF, overlay for single machine and matching, thick photo resist performance, which shows specification of the stepper meeting well with advanced package process.

1 Advanced Packaging Technology and Lithography Tool

With the semiconductor industry grows rapidly, market is faced with several challenges in components packaging including performance, resolution, function and cost, which are the main driver of advanced packaging technology improvement and make the market booming. In current stage, big gap appears between the mainstream advanced packaging process, such as BGA, CSP, FCP, WLP and SIP, and the traditional process. A good example is widely used wafer-level bumping which can achieve smaller packaging/chip area ration. Bumping process contains gold bumping, solder bumping, pillar bumping and redistribution, and they require higher packaging technology, and is related to the manufacture of passive components (inductance, capacitance, resistor and so on). Lithography is one of the crucial process steps that ensure successful bumping and redistribution process for smaller IC dimension and more IO pins.

Comparing with the front-end lithography process, back-end lithography process focuses on some unique processes, such as thick resist (15-120um), ground or thicken wafer, without specialized alignment mark, and compliance with electroplating process. Presently, lithography tools for advanced package includes contact or proximity aligners and 1X stepper, both of them expose in broadband light and align based on machine-vision system. But stepper shows more advantage in CD uniformity, Overlay, Throughput and yield, and plays a more and more important role in advanced packaging process. When the requirement of packaging process (CD uniformity, Overlay accuracy) improves, or wafer increased to 8 or 12 inches, stepper is more attractive. SMEE SSB 500 series stepper is designed for advanced packaging process, and after optimization and improvement, more reliable process-compliance is achieved.

2 Advanced packaging stepper system

2.1 Compliance of thick resist process

Thick resist is widely used in back-end packaging application, and the thickness varies from 10um to 120um, so that high intensity and big usable depth of focus (UDoF) are primary requirement for advanced packaging lithography. SS B500 series' lithography use broad-band mercury lamp as light source, and the exposure wavelength is ghi line, which ensure the high efficiency of lamp house. The illumination system is optimized for higher coupling efficiency, and minimizes the optical components to achieve a high intensity. Meanwhile Kohler illumination structure is used to keep a good homogeneity. High intensity and good homogeneity guarantee the yield and CD Uniformity.

Projection Optics (PO) in SS B500 series' lithography is design to meet the special requirement of advanced packaging process. The exposure field is 44mm*44mm, which can cover 4 fields of front end reduction stepper, and it increases the throughput significantly; another advantage of this PO is big working distance, and it solve the problem of lens contamination coming from the resist evaporation in advanced packaging process; PO is double telecentric both in object side and image side, and the image quality is diffraction-limited level, ensure the big UDoF of 30um@3um L/S, which is compliance to thick resist process; Optimized UV coating is applied to increase the intensity in wafer, to achieve a $1500mw/cm^2$ in 44mm*44mm exposure area; Lens heating problem is also emphasized due to the high dose for thick resist, and PO can exposure in a high dose almost 20000mj, without any lens adjustment.

2.2 Compliance of back-end wafer

The variety of advanced packaging process leads to the diversity of back-end wafer, thickness, wraps, profile varies greatly. Further more, front-end wafer has different alignment marks, which should be handled in back-end stepper. Wafer stage in SS B500 series stepper is designed for big stroke of 2mm, and can handle all kinds of ground and thicken wafer. Wafer is supported by vertical structure in reasonable contact area before adsorption, in order to guarantee that either the wrapped up wafer or wrapped down wafer can be handled by the stepper. SS B500 series stepper is equipped with 4-points auto focus and leveling unit, the measurement area is in the center of exposure field, and bring the upper side of wafer exposure area into the best imaging focus, eliminate the defocus due to the wafer profile fluctuation by die to die focus and leveling. The stepper can achieved 500nm auto focusing repeatability. Alignment unit in SS B500 series stepper can align the reticle to wafer through the lens. Due to the difference between the measurement wavelength and exposure wavelength, the optical system in alignment unit is achromatic design to ensure wafer marks imaging clearly; alignment signal process is based on CCD sub-pixel image processing algorithms, it can capture the wafer alignment marks with different shape features, in other words, it can recognize the alignment mark by studying feature images, and special wafer alignment marks is not necessary. Test results shows that the stepper alignment repeatability is 150nm.

2.3 Overlay Control

Lithography process in advanced packaging usually applied at 2-4 layers (including redistribution layer), sometimes at 7 or more layers. The overlay is based on either the image in the front-end wafer or in the back-end wafer, so the advanced stepper should control the overlay for single machine and machine to machine. Because the overlay of the front-end machine is more accurate than the back-end machine, we use absolute overlay to evaluate the overlay of advanced packaging stepper. The absolute overlay use the golden wafer as test wafer, which is manufactured by front-end 90nm node equipments, including brown mark layer and corrosion resistance layer, the overlay mark's positioning error is smaller than 1/30 of machine overlay, so it can be neglected. By measuring the mark position with respect to the golden wafer as a measurement reference, the stage grid, alignment, lens magnification & distortion can be calculated and corrected, so that machine can get the best matching ability.

When manufacturing, SS B500 series stepper support global alignment with accurate mode, balanced mode and robust mode, in order to avoid the wafer mark alignment failure. It also support die to die alignment, the flexible alignment strategy and the setup of all the matching parameters (stage grid matching, alignment matching, lens matching) helps to achieve high accurate overlay relative to front-end wafer. Especially for the lens matching, magnification can be adjusted by movable lens, and distortion can be adjusted by choosing exposure system by movable blades to choose best matching exposure field inside 44*44mm. It should be noted that this kind of high accurate matching should be balanced with throughput.

3 Machine Performance Test

3.1 Process equipments

SS B500/10B stepper is the basic configuration of SS B500 series advanced packaging steppers, it can achieve 3um CD and 1um absolute overlay. The main process equipment used in performance test is listed below:

Table 1 Lithography process equipment

Equipment	Vendor	Type
Track	SOLIO TOOL	KS-S300
CD-SEM	Hitachi	S-8820
Overlay metrology system	KLA	Archer 10
Microscope	OLYMPUS	MX50L-R/T
Thin-Film Measurement Instruments	Filmetrics	F50-200
FE-SEM	Hitachi	S-8820
Resist cleaning machine	SNA	WMS

3.2 CD & Overlay test

The process parameter for SS B500/10B stepper CD Uniformity, UDoF and Overlay test with AZ Mir703 photo resist is as below:

Table 2 Lithography process parameter with AZ MiR 703

Photoresist	AZ MIR 703
Thickness	1.1um
Exposure Wavelength	ghi
Dose	$120mJ/cm^2$
Soft bake temp/time	90℃/60s
PEB temp/time	115℃/90s
Developer	2.38% TMAH
Development temp/time	22℃/ 120s
Post bake temp/time	115℃/60s

In CD uniformity test of SS B500/10B, 9 dies is exposed for each wafer, and for each die, CD of the horizontal and vertical lines in the central point and the four corner points is measured, CD Uniformity of 2um line and 3um line at best focus is calculated. Also, CD uniformity at 3um defocus position is calculated. Test data shows for CD Uniformity of 2um line and 3um line at best focus is less than 10% of CD, and for 3um line, the UDoF can achieve 30um, which meet the requirement of thick resist process.

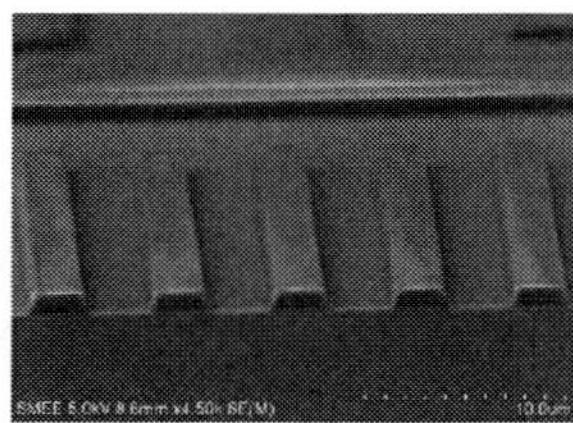

CD: 3um   *CD: 2um*

Figure 1 AZ MIR 703 test (Thickness 1.1um)

Table 3 CD uniformity data of 2um&3um lines

CDU TEST			mean	3sigma	CDU
2um		V line	2136.99	126.1674	5.90%
		H line	2171.42	143.0899	6.59%
		Both	2154.21	143.3429	6.65%
3um	TF-15	V line	3051.67	162.1617	5.31%
		H line	3034.25	133.8327	4.41%
		Both	3042.96	149.497	4.91%
	BF	V line	3108.54	148.9655	4.79%
		H line	3078.96	126.084	4.10%
		Both	3093.75	143.7517	4.65%
	TF+15	V line	3117.40	184.6306	5.92%
		H line	3066.68	168.7062	5.50%
		Both	3092.04	191.1638	6.18%

SS B500/10B stepper overlay test use golden wafer as test wafer, and choose 6 alignment marks for each wafer, use global alignment accurate mode, exposure area 44mm*44mm, exposure 1 layer, and the overlay machine measure the box in box position error as absolute overlay; expose 2 layers, the relative error is the single machine overlay. Test data shows that absolute overlay is less than 1um, and single machine overlay is less than 600nm.

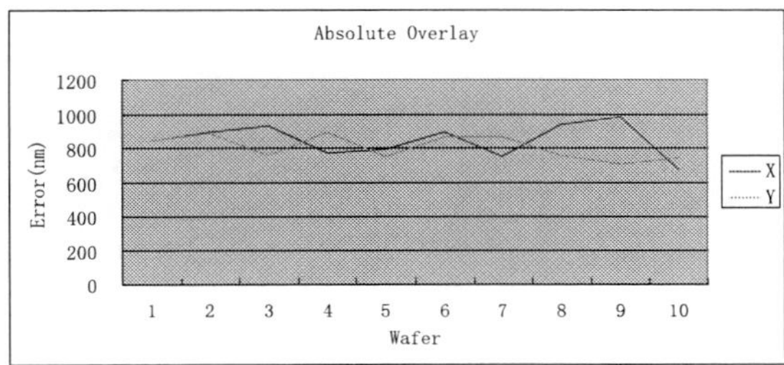

Figure 2 Absolute overlay test data (10 wafers)

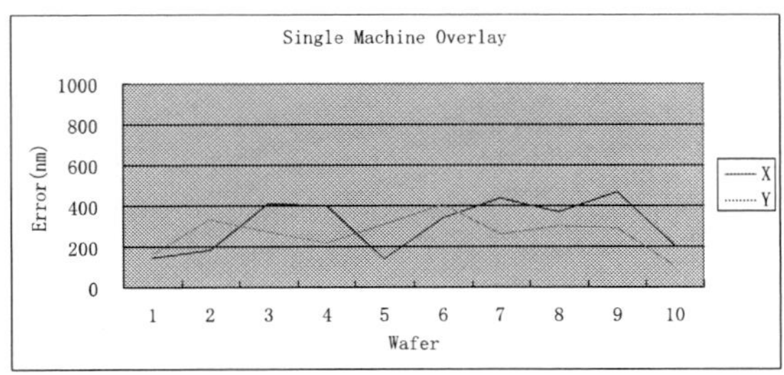

Figure 3 Single machine overlay test data (10 wafers)

3.3 Thick resist test

Thick resist test for SS B500/10B stepper use JSR 151N (negative photo resist), process parameters are listed below.

Table 4 Process parameter for thick resist

Resist	JSR 151N
Thickness	85um
Wavelength	ghi
Dose	1100 mJ/cm2
Soft bake temp/time	90℃/300s
Developer	2.38% TMAH

Development temp/time	22°C/180s
Post bake temp/time	115°C/90s

Test data shows that the stepper can achieve good performance for 85um thick resist.

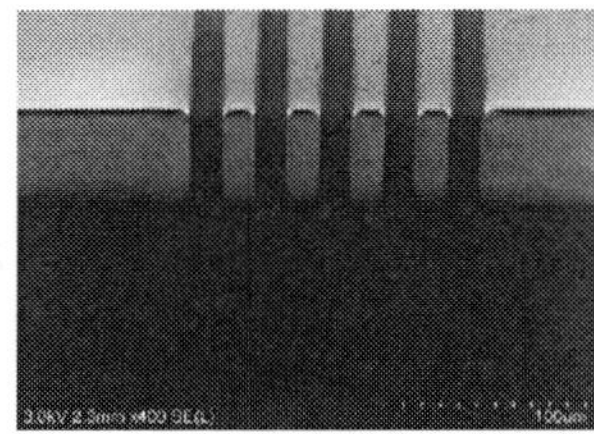

CD: 18um

CD: 15um

Figure 4 JSR151N test (Thickness 85um)

4 Conclusions

SS B500 Series stepper is a complex system and a precise equipment related to optics, mechanics, electronics and other sciences, after calibration, the equipment can exposure 2-3um lien with 10% CD Uniformity in big UDoF, and the overlay error is controlled less than 1um. The machine key performance specifications are within design budget, and meet the ITRS technical requirement for advanced packaging process, can be applied in production line for advanced packaging process.

Acknowledgments

Thanks Mr. He Rongming, CEO of SMEE for his great support for stepper development. Thanks all members of stepper project, especially Dr. Cai Liangbing, Dr. Mao Fangling, etc, for their hard work on stepper integration and testing. Also thanks national No. 02 grand project and client of SS B500 stepper for their great help.

Reference

1. ASSEMBLY AND PACKAGING, INTERNATIONAL TECHNOLOGY ROADMAP FOR SEMICONDUCTORS, www.itrs.net

ECS Transactions, 34 (1) 243-248 (2011)
10.1149/1.3567588 ©The Electrochemical Society

Foundry Efficiency Gains Through Common Photolithography Themes

James E. Lamb III, Chris Cox, Zhimin Zhu, David Drain, Daniel Sullivan

Brewer Science, Inc., Rolla, Missouri, USA

Here we examine photolithography methodologies to improve the efficiency of foundry and ASIC manufactures that have many part numbers with relatively low run volumes. In such cases, a fab must accommodate a large variety of device layers, topography dispersion, and layout parameters. These factors reduce the lithography process margin and limit design freedom and throughput in the lithography module. These factors also increase design complexity, optical proximity correction (OPC) design loops, mask cost, exposure dose, and number of material sets to meet layer requirements. Consequently, these factors impact the fab's overall efficiency. The primary method proposed to improve efficiency is to use a system that includes a planarizing etch transfer layer, an image transfer layer, and an image capture layer. By combining these elements, the lithography process becomes very predictable, simple, and repeatable, independent of the device.

1. Introduction

In this paper, we examine photolithography methodologies to improve the efficiency of foundry and ASIC manufactures that have many part numbers with relatively low run volumes. In such cases, a fab will need to accommodate a large variety of device layers, topography dispersion, and layout parameters. In turn, these factors reduce the lithography process margin, limit design freedom, and reduce throughput in the lithography module. At the same time, these factors increase design complexity, optical proximity correction (OPC) design loops, mask cost, exposure dose, and the number of material sets required to meet individual layer needs. Thus the above items impact the overall efficiency of the fab.

2. Theory and Simulation of Lithography Latitude

Lithography performance strongly depends on stack design. Often the substrate has a variety of topography, transparent dielectric layers, and multiple patterned device layers with a wide range of reflectivity. This situation leads to a wide range of topography and reflectivity dispersion that makes conventional lithography difficult, consuming much of the CD processing window.

Figure 1 represents a stack with different SiO_2 thicknesses on the same wafer. The bottom anti-reflective coating (BARC, ARC® 29 material) and the resist are assumed to be conformal coatings.

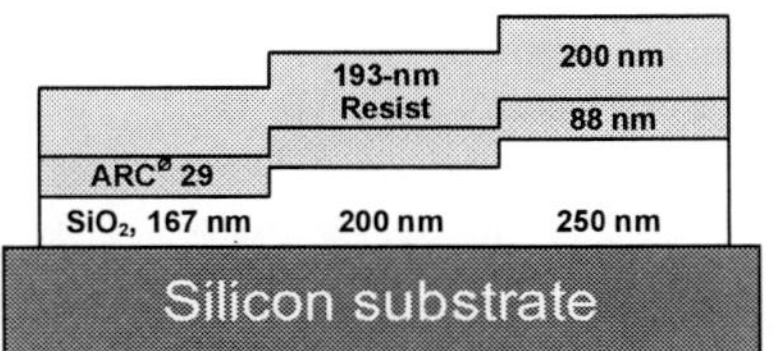

Figure 1. Lithography stacks on substrate with topography.

Figure 2 is the substrate reflectivity (%R) varying with oxide thickness simulated with PROLITH v11. The substrate reflectivity is 5.1% for an oxide thickness of 167 nm, 0.35% for a 200-nm thickness, and 2.1% for a 250-nm thickness. Thus, the lithography step is performed under very different optical conditions due to the underlying stack and topography. Even with a small change in oxide thickness from 200 nm to 235 nm, the reflectivity would change from 0.35% to 5.1%, leading to little or no process latitude.

Figure 3 shows the CD processing window simulated with PROLITH v11. These simulations are based on an ASML 1100 (ArF) scanner with settings of NA = 0.75, dipole 35Y, sigma = 0.89/0.65, and an 80-nm 1:1 mask. It is obvious that both windows shifted from each other in both depth of focus (DoF) and exposure dose, resulting in a smaller common window (dashed black lines).

Figure 4 is the exposure latitude with respect to DoF. If the exposure dose tolerance is set at ± 5%, then the individual stacks have a wide DoF, (~1000 nm). However, the DoF will shrink to zero, as shown with the black curve, if all three stacks receive the same exposure dose.

In this paper we recommend a multilayer patterning system, the OptiStack® system, as shown in Figure 5. In this figure, the etch transfer layer has four functions: topography planarization, optical isolation, high-aspect-ratio pattern transfer, and an easily strippable film following substrate etch transfer or implant. The second layer is a pattern transfer layer that has two key functions: transfer of a low-aspect-ratio organic image layer pattern into a high-

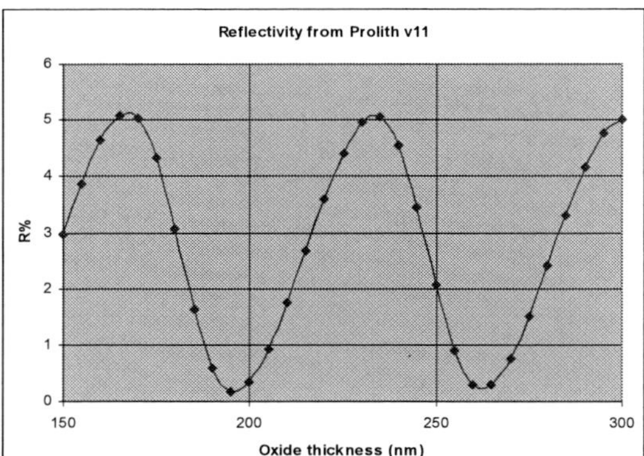

Figure 2. Substrate reflectivity varies with oxide thickness when using a single-layer BARC process.

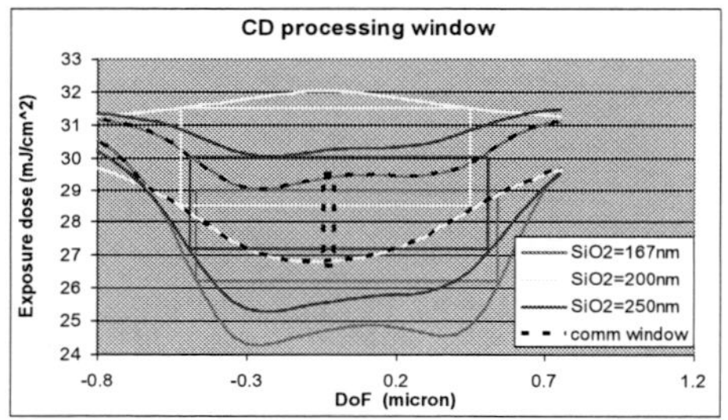

Figure 3. CD processing window: red for 167 nm, yellow for 200 nm, and blue for 250 nm of SiO$_2$ thickness; black dashed line is the resulting common window.

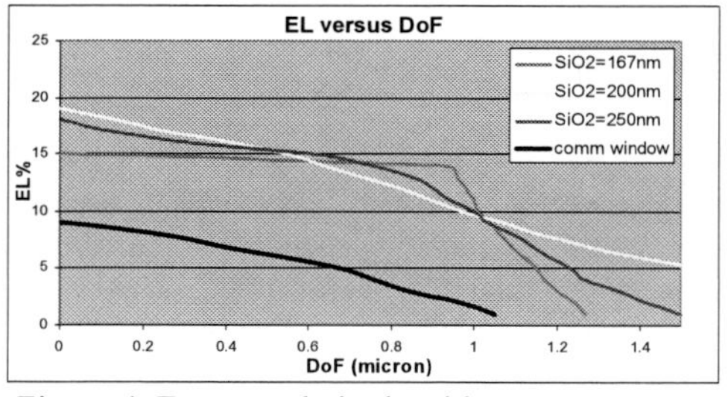

Figure 4. Exposure latitude with respect to DoF.

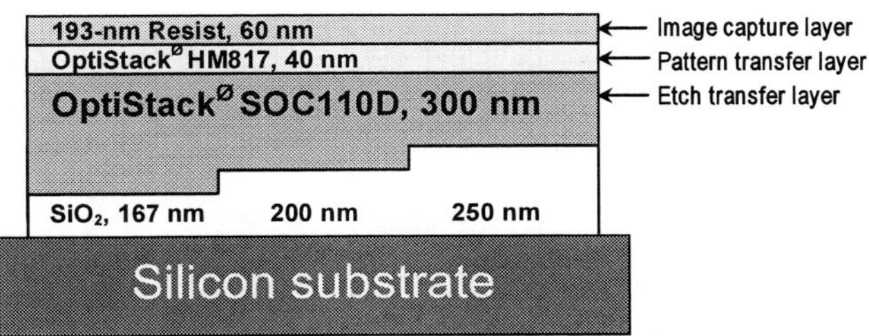

Figure 5. Recommended OptiStack® patterning stack.

aspect-ratio organic resist transfer layer and tuning of the ideal optical surface for lithography fidelity and process latitude. The top image capture layer is a thin version of a typical photoresist that has the primary purpose of capturing the lithography image and the ability to transfer that image by plasma etching into the thin pattern transfer layer. Thickness of the image capture layer is less influenced by the substrate topography, and, most importantly, the UV distribution is absolutely independent of the substrate reflectivity dispersion. Figure 6 is the reflectivity curve versus SiO_2 thickness. The oxide thickness does not change the optical property of the stack due to the isolation of the high-k OptiStack® SOC110D layer. In this case, the CD processing window will be identical over the entire topography of the wafer. Another big advantage of such a stack is that the image capture layer thickness is free of etch budget requirements and can be optimized for the maximum CD processing window.

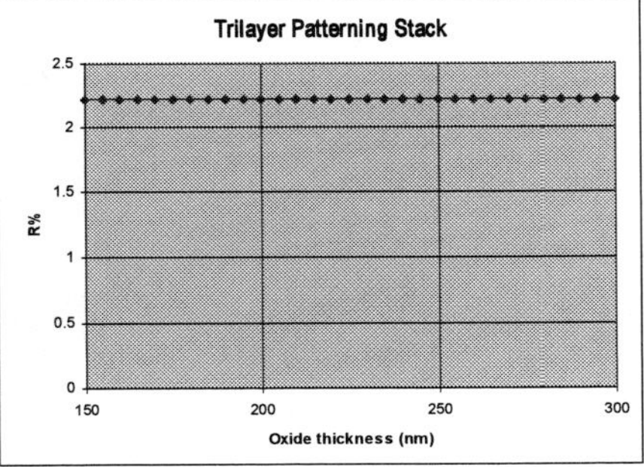

By eliminating reflectivity dispersion and minimizing topography dispersion, the user of the OptiStack® system can process with a common substrate independent of the device layer. The two key variables for any device layer are 1) the thickness of the etch transfer layer to meet the required aspect ratio for either etch or implant of the substrate layer and 2) the required etch transfer process to transfer the image into the pattern transfer layer. The implication is that the top two layers of the system are identical, independent of device layer or device type, thereby greatly simplifying the exposure and develop steps. This simplicity will have its greatest effect by allowing wide process margins, which in turn allow opening the design rules for a given exposure tool set operated by a fab. This simplicity also allows OPC to be calculated directly and incorporated into the design rules so no reiterative mask designs are needed for OPC feedback. All learning at any device layer can be applied to any and all other layers for the particular exposure tools used. Another advantage of implementing the parameters identified by the OptiStack® system would be a reduction in the overall number of materials that could be used across all devices and device layers, with only the etch transfer layer thickness varying as needed to achieve the planarization and aspect ratio required. Meeting these requirements would typically add one process layer per wafer and call for the additional etch capacity required for dry-etch pattern transfer. The impact on cost will be discussed in Section 4.

3. Experimental Validation

The lithography experiment for the two stacks shown in Figure 1 has been done at IMEC with ASML 1100 (ArF). Table I lists the experimental conditions.

Table I. Lithography conditions.

Parameter	Conditions
Resist:	AR1682J-15
Resist thickness (nm):	200
Resist coat (rpm/s):	1100/30
Target CD (nm/pitch):	80L/160P
PAB (□C/s):	110/90
Illumination mode:	Dipole35Y
NA:	0.75
Sigma (outer / inner):	0.89 / 0.65
Center dose/step (mJ/cm^2):	26 / 1
Focus offset/step (µm):	0 / 0.1
Reticle bar code:	TM99YCK%L1%M
PEB (□C/s):	110/90
Developer type/time (s):	OPD262/40

Figure 7 shows the results of top-down CD measurement, and Figure 8 shows the CD processing window. At ± 5% exposure dose tolerance, 200-nm SiO_2 has a 465-nm DoF, and 250-nm SiO_2 has a 585-nm DoF. The overlap of these two CD windows leaves 375 nm. Both simulation and experiment show the CD processing window shrinkage by substrate reflectivity variation.

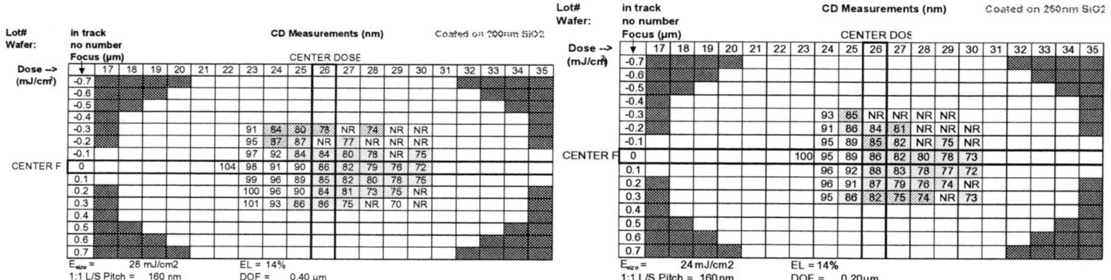

Lot# Wafer: in track, no number — CD Measurements (nm) — Coated on 200nm SiO2 — CENTER DOSE

Focus (µm) \ Dose → (mJ/cm^2)	17	18	19	20	21	22	23	24	25	26	27	28	29	30	31	32	33	34	35
-0.7																			
-0.6																			
-0.5																			
-0.4																			
-0.3							91	84	80	78	NR	74	NR	NR					
-0.2							95	87	87	NR	77	NR	NR	NR					
-0.1							97	92	84	84	80	78	NR	75					
0 (CENTER F)						104	98	91	90	86	82	79	76	72					
0.1							99	96	89	85	82	80	78	75					
0.2							100	96	90	84	81	73	75	NR					
0.3							101	93	86	86	75	NR	70	NR					
0.4																			
0.5																			
0.6																			
0.7																			

E_{size} = 28 mJ/cm2   EL = 14%
1:1 L/S Pitch = 160 nm   DOF = 0.40 µm

Lot# Wafer: in track, no number — CD Measurements (nm) — Coated on 250nm SiO2 — CENTER DOSE

Focus (µm) \ Dose → (mJ/cm^2)	17	18	19	20	21	22	23	24	25	26	27	28	29	30	31	32	33	34	35
-0.7																			
-0.6																			
-0.5																			
-0.4																			
-0.3								93	85	NR	NR	NR	NR						
-0.2								91	86	84	81	NR	NR	NR					
-0.1								95	89	85	82	NR	75	NR					
0 (CENTER F)							100	95	89	86	82	80	78	73					
0.1								96	92	88	83	78	77	72					
0.2								96	91	87	79	76	74	NR					
0.3								95	86	82	75	74	NR	73					
0.4																			
0.5																			
0.6																			
0.7																			

E_{size} = 24 mJ/cm2   EL = 14%
1:1 L/S Pitch = 160 nm   DOF = 0.20 µm

Figure 7. Focus and exposure matrix of top-down CD measurement results.

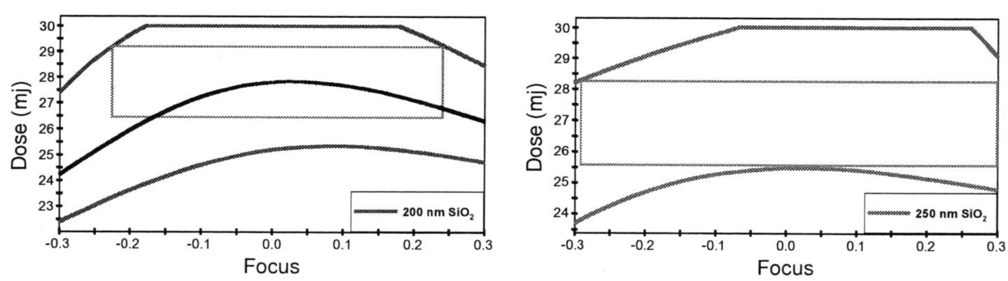

Figure 8. CD processing window. The graph at left is for the 200-nm SiO_2 thickness, and the graph at right is for the 250-nm thickness.

4. Cost of Ownership

Cost of ownership plays an important role in process materials and methods decisions. Process simplifications brought about by layer-to-layer synergy drive significant cost of ownership advantages for the OptiStack® system. Savings in mask engineering and manufacture are the greatest cost difference. OPC algorithms need only be determined once for all layers, rather than individually for each layer, which results in fewer mask corrections.

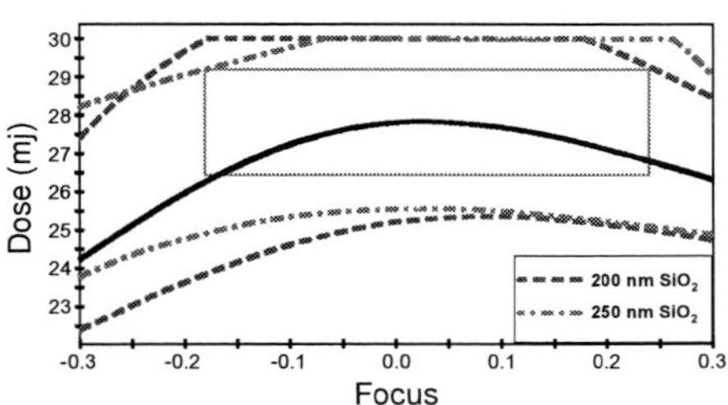

Figure 9. Overlap of CD processing windows.

Advanced devices with smaller critical dimensions benefit most from this system in that there are more layers at smaller critical dimensions, requiring greater mask design and production costs.

For the purpose of providing a cost of ownership estimate, we compared a process using the OptiStack® system on eight layers to a typical dyed resist process. Note that the OptiStack® system process would use the same OPC algorithm on all layers. More layers utilizing the same system will result in greater savings. We assumed that 2500 wafers [1] would be printed from each mask; this is said to be a typical usage - ASIC masks may be imaged on as few as 500 wafers. Mask design and mask production are the significant cost components in any analysis of lithography process cost of ownership [5]. For a high-performance mask set with 90-nm design rules, individual binary chrome-on-glass masks can cost $100,000 to produce, and phase shift masks can cost as much as $124,000 [6]. As a conservative estimate, the average mask cost was taken to be $70,000 because some layers can be printed using DUV tools and masks. Accounting for these and other costs, the OptiStack® system saves about $674 per wafer. The cost comparison is shown in Table II. "Other Optistack system costs" are caused by additional materials and processing. These are detailed for processing eight layers per wafer in Table III.

Table II. Cost of ownership comparison.

Mask Design		
Mask design cost	$	102,300
Layers per process converting to OptiStackØ system		8
Mask trials		2.25
Dyed resist mask design	$	3,682,800
OptiStackØ system mask design	$	2,659,800
Difference	$	(1,023,000)
Mask Production		
Mask production cost	$	70,000
Mask trials		2.25
Dyed resist mask production	$	2,520,000
OptiStackØ mask production	$	1,820,000
Difference	$	(700,000)
Wafer images per mask		2500
Design cost (savings) per wafer	$	(409)
Mask production cost (savings) per wafer	$	(280)
Other OptiStackØ system costs	$	86
OptiStackØ system net cost (savings)	$	(603)

Table III. Additional materials and processing cost versus single-layer resist.

Etch transfer material spin-on	$	25	New spin capacity and material
Pattern transfer material spin-on	$	15	New spin capacity and material
Spin on thin resist	$	(6)	Less material used
Resist pre-exposure bake	$	(3)	Less time due to thinner layer
Resist expose	$	(14)	Shorter exposure time
Resist post-exposure bake	$	(3)	Less time due to thinner layer
Etch transfer layer etch	$	36	Additional etch
Pattern transfer layer etch etch	$	36	Additional etch
Total cost per wafer	$	86	8 layers per wafer

Using this type of multilayer system is likely to bring even more benefits to the overall fab cost and efficiency. Other areas to be studied in future work are the impact of simplified inventory arising from the need for fewer custom materials by layer or device.

5. Conclusion

By combining these elements of the OptiStack® system, the lithography process becomes very predictable, simple, and repeatable, independent of the device. OPC can be well characterized and applied uniformly to all process layers in a single pass, reducing both the cost of design and mask-making loops. A single image capture layer and image transfer layer system can be used on all device layers for each exposure tool generation. Simply varying the thickness of the planarizing etch transfer layer achieves the aspect ratio needed for substrate etch transfer or implant protection. With the low-aspect-ratio image capture layer, process margins are greatly expanded, which potentially allows fabs with lithography tool limitations to expand their design rules to cover added device designs. Work will be continued to more fully identify the impact on overall fab efficiency and cost and to perform process testing to map the fab or ASIC manufactures conditions that can best take advantage of this technology.

Acknowledgments

Thank you to Heping Wang and Doug Guerrero of Brewer Science and IMEC for processing the wafers and for providing the support that made this paper possible.

References

1. Harry J. Levinson, *Principles of Lithography*, The Society of Photo-Optical Instrumentation Engineers, 2005.
2. Zhimin Zhu, Emil Piscani, Kevin Edwards, and Brian Smith, "Reflection control in hyper-NA immersion lithography," *Proceedings of SPIE*, vol. 6924, 2008, pp. 69244A-1 - 69244A-7.
3. Zhimin Zhu, Emil Piscani, Yubao Wang, Jan Macie, Charles J. Neef, and Brian Smith, "Thin hardmask patterning stacks for the 22-nm node," *Proceedings of SPIE*, vol. 7274, 2009, pp. 72742K-1 - 72742K-7.
4. Darron Jurajda, Enrico Tenaglia, Jonathan Jeauneau, Danilo De Simone, Zhimin Zhu, Paolo Piazza, Paolo Piacentini, and Paolo Canestrari, "Investigation of the foot-exposure impact in hyper-NA immersion lithography when using thin anti-reflective coating," *Proceedings of SPIE*, vol. 7273, 2009, pp. 72730Z-1 - 72730Z-10.
5. Vivek Bakshi, Ed., *EUV Lithography*, SPIE Press, Bellingham, WA, 2009.
6. B. Grenon and S. Hector, "Mask costs, a new look," *Proceedings of SPIE*, vol. 6281, 2006, 62810H.

ECS Transactions, 34 (1) 249-255 (2011)
10.1149/1.3567589 ©The Electrochemical Society

Use of DBARCs Beyond Implant

C. Washburn, J. A. Lowes, A. Guerrero

Brewer Science, Inc., Rolla, Missouri, 65401, USA

Almost a decade ago, the integrated circuit (IC) industry realized that traditional implant layers required new technology to achieve the specifications of the shrinking design rules and increased topography effects. In response, developer-soluble bottom anti-reflective coatings (DBARCs) were introduced as a solution. DBARCs offered the resolution and critical dimension (CD) control needed for the increasingly critical implant layers.

In this paper the benefits DBARCs can bring to other lithographic processes are reviewed. Back-end-of-line (BEOL) dual damascene processes are surveyed. Front-end-of-line (FEOL) processes are examined, including high-k metal gate (HKMG), double patterning, and spacer. All processes are defined, and the value that a DBARC brings to each process is examined, along with potential challenges.

Introduction

Advancements in technology beget more complex development. This inevitability is especially true in the semiconductor industry. As integrated circuit (IC) technology evolves, the materials and processes used to manufacture ICs become more sophisticated. This driving force exists for most materials used in the semiconductor industry, including bottom anti-reflective coatings (BARCs). In the early 1980s, BARCs advanced lithography patterning by reducing standing waves in photoresists by attenuating light (1). This success led to using BARCs to increase the substrate adhesion of photoresists. Subsequently, BARCs were adapted to fill vias and improve the dual damascene process (2).

Developer-soluble BARCs (DBARCs) are different than traditional dry-etch BARCs in that DBARCs are patterned with the resist during the develop step (1,3). The original application for DBARCs was to enhance the patterning of implant layers (3,4,5). These initial DBARCs developed isotropically and were adjustable by process modifications (6). DBARCs that developed anisotropically were brought to market next to answer the need for smaller features and less bias between the isolated and dense features (7,8,9).

Part of what drives technology growth is the ability of engineers to adapt knowledge from one focus area to another. An example is how dry-etch BARCs were adapted from use in line and space patterns on the front-end-of-line (FEOL) technology, to their use for filling vias and enhancing patterning in the back-end-of-line (BEOL) dual damascene process mentioned above. In this paper we will examine applications beyond implant in which DBARCs are being used. We will look at the use of a DBARC in the dual-damascene process. We will then look at the use of a DBARC as an adhesion promoter and will continue by reviewing a DBARC's role in advanced litho processes such as double patterning and spacer applications.

Dual Damascene Processes

The via-first dual damascene process is common in BEOL interconnect processes. The basic flow is to pattern vias in the interlayer dielectric (ILD), then to pattern trenches aligned with the vias. After the ILD has been etched, a single metal deposition is used to fill the vias and trenches to form the interconnects. Because the dual damascene process creates the wires that connect the transistors, improvements in uniformity, alignment, and dimension control have a direct positive impact on yield and manufacturing cost reduction.

Given DBARC's establishment in the FEOL use in implant steps, one of DBARC's initial implementations beyond the implant process was the BEOL dual damascene process. The first application was to use a non-photosensitive DBARC as a via-fill BARC (10). After the vias were formed, the DBARC was used to partially fill the vias in preparation to pattern trenches. This reduced the thickness bias seen with isolated versus dense via fields, so the resist was more planar when coated. Ultimately, the DBARC helped to reduce the iso-dense bias offset seen when patterning trenches (10).

The next adaptation of DBARC into dual damascene processes was to replace the dry BARC used to form vias with a DBARC (11). The typical challenge of developing a process that sufficiently controls undercut was not needed, as the undercut did not affect the transfer of the resist pattern into the ILD, as shown in Figure 1 (11).

Using a DBARC to pattern vias provides several advantages. When using a dry BARC and resist to form vias, reactive ion etching (RIE) is needed to transfer the pattern through the BARC; this step is commonly known as the BARC open etch. During the BARC open etch, the top of the resist can be damaged, and a loss of circularity in the via pattern can result. Because the resist pattern transfers through the DBARC during the develop step, the BARC open etch is not needed and circularity would be maintained. Thus, using a DBARC to pattern vias would increase the fidelity and uniformity of the via pattern as it is transferred into the ILD.

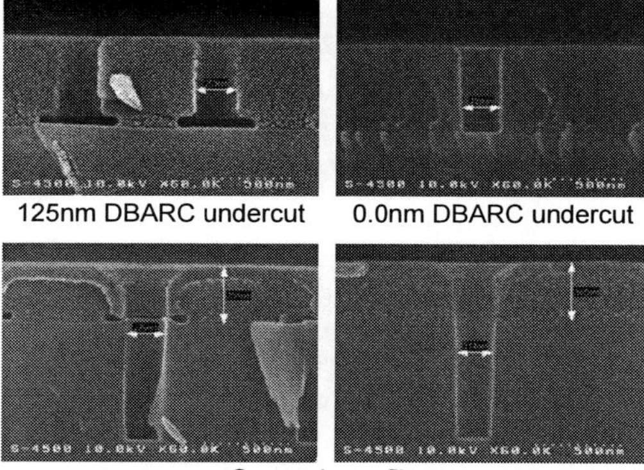

Figure 1: DBARC used in via formation (11).

The work on developing advanced photosensitive DBARCs (PS-DBARCs) for via patterning to realize these strengths has continued (12, 13). Features down to 80 nm have been shown with straight profiles. Because the PS-DBARC responds in many ways like a photoresist, one challenge to overcome is to establish what mask error enhancement factor (MEEF) will be required to accommodate the needed dose to clear both isolated and dense vias in the same exposure.

High-k Metal Gate (HKMG) Processes

The complementary metal oxide semiconductor (CMOS) process flow has adapted over the years to meet the challenges of each node. As the gate oxide thickness scaled with successive nodes, a new gate material was needed. The answer to this challenge was a metal gate, coupled with a high-k dielectric (HKMG) (15). The metal gate was needed because the common polysilicon gate was prone to defects at the low voltages used to switch the gate (16). The HKMG reduced gate leakage by increasing the transistor capacitance and allowed chips to function with reduced power needs.

Two process flows were developed to pattern the HKMG stack, a gate-first and a gate-last process. In the gate-first process, the thin capping metal in the HKMG needed to be patterned, and the process would be very sensitive. Photoresists did not adhere well to the metal layers, which was made worse by the wet chemistries needed to etch the capping metal (17). So to pattern the HKMG stack, a new process was needed.

DBARC was adapted as the enabler to pattern the metal capping layers in the gate-first HKMG process. DBARC protected the substrate by eliminating the use of RIE in the dry-BARC open etch step. Due to a DBARC's crosslinking ability, it increased substrate adhesion so the wet-etch process could transfer the pattern (9,17). An example of a process flow showing DBARC being used with HKMG technology is shown in Figure 2. The photoresist and DBARC are used in steps 2c and 2e, along with a wet etch to pattern the capping layers (17).

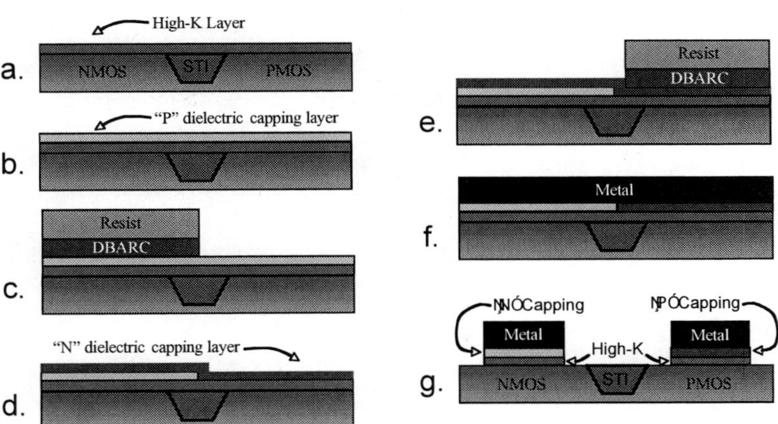

Figure 2: Gate-first HKMG process using DBARC.

As is often the case with shifts in technology, new issues emerged. One issue that DBARCs have faced many times is the thin layer of residue that is left behind after development. A significant amount of research has gone into quantifying and understanding the trends of residue produced by different DBARCs. The residue has been quantified using ellipsometry, X-ray photoelectron spectroscopy (XPS), and atomic force microscopy (AFM) (9, 18, 19, 20, 21, 22). Trends through processing conditions have also been evaluated. For non-photosensitive DBARC, BARC bake temperature emerged as the biggest process control for residue (20). BARC bake temperature was also found to affect certain PS-DBARCs (18). For PS-DBARC, the contributors to residue were more complex. The chemistry components, thickness, and exposure dose all affected the post-develop residue (9, 21, 22). In both PS and non-PS DBARCs, it was found that the substrate also influenced the residue (19, 20). With the diverse factors that are involved with residue, it is a challenge that will continue to garner focus, especially as HKMG processes move into production.

Advanced Patterning Processes

As the gate length of devices shrink, so does almost every other part of a device. The solutions that lithographers have employed to maintain pace with shrinking features have become more complex than the end devices being made. To produce smaller features and maintain dense pitches, a common approach is to pattern twice, with the second pattern set between the first, thus doubling the pitch of the final pattern. These processes are called double patterning. The most basic approach is to use RIE to etch the first pattern into the substrate, then to create a second pattern and use RIE to etch it into the substrate as well. This litho-etch-litho-etch (LELE) process can be expensive, as wafers need to travel between the litho and etch modules twice. A solution to this problem can be found by incorporating a DBARC into the process (23, 24, 25). An example of a double-pattering process using DBARC is shown in Figure 3.

A double patterning process such as this takes advantage of multiple DBARC features. The attenuation of light helps the photoresist form the litho patterns. The solubility of DBARC in developer allows the resist pattern to be retained in the DBARC without a RIE. Also, the resist can be selectively removed with a solvent, as the DBARC is crosslinked and not dissolvable in resist solvents. This process can use a non-photosensitive DBARC or a PS-DBARC.

One question that would need to be addressed is if a DBARC had sufficient RIE resistance to allow the pattern to transfer. To overcome this, a hardmask could be incorporated into the process (26). An example of how this might work is shown in Figure 4. Once the pattern is transferred into the DBARC, a RIE could be used to move the pattern into the hardmask. This process could be repeated for the second patterning step. In this case, the wafers would need to go through the etch module twice, so this process would only provide a limited advantage of removing the BARC open etch.

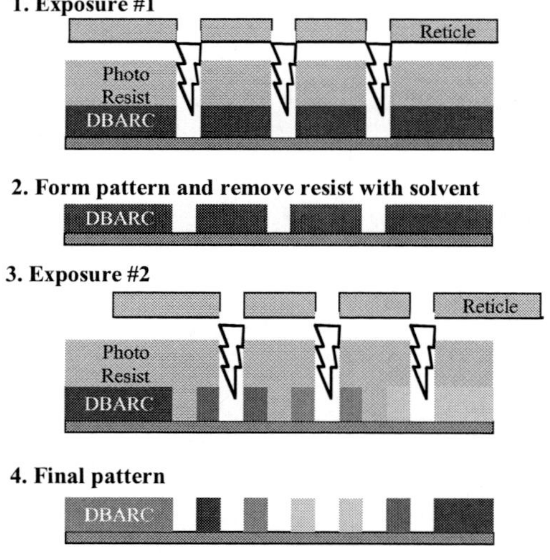

Figure 3: Double-patterning approach using a DBARC.

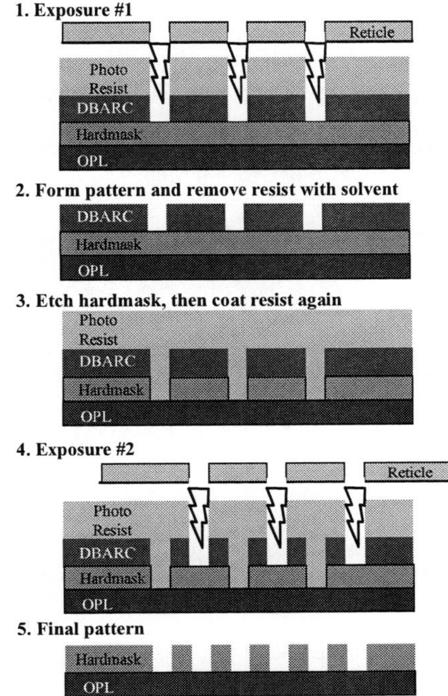

Figure 4: Double-patterning approach using a DBARC and hardmask.

Processes beyond double patterning are also using DBARC's unique capabilities. Spacer processes used in advanced patterning are a good example. Once lines are created using spacer technology, a loop is formed, as shown in Figure 5c. To trim the ends off the loop, an RIE is needed. An etch mask is needed that can easily be removed without damaging the delicate features. One approach is to use a DBARC and photoresist (27). This process forms spacer features as shown in Figure 5a, 5b, and 5c. In Figure 5d, the DBARC and resist are coated. Then in Figure 5e the RIE step removes the loop ends, while the DBARC and resist protect the rest of the features. Since the dissolution rate of DBARC is adjustable, a very high rate can easily be achieved. This high develop rate would allow DBARC to be removed from in-between the spaces as shown in Figure 5f.

DBARC Uses Beyond Semiconductors

This paper has examined several processes in which DBARCs are being used beyond implant. DBARCs have the potential to be used in processes to make devices beyond semiconductors. Lift-off layers are an example of such processes. Lift-off layers are commonly used when metal depositions are needed, and the surface of the substrate is sensitive. Figure 6 shows a basic lift-off process using a DBARC as the lift-off layer. For the process to work, some undercut is needed to allow the

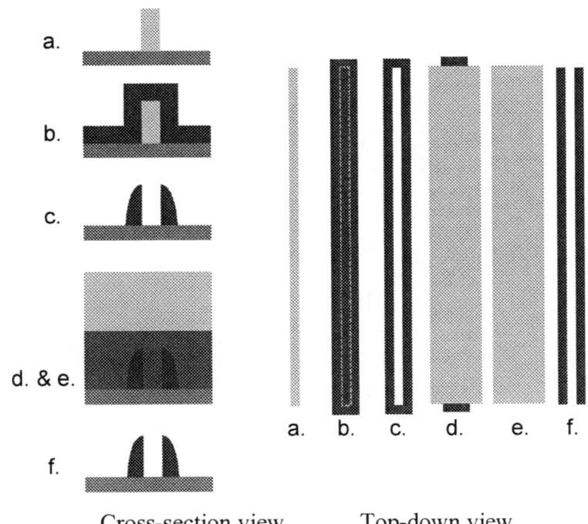

a. Photoresist line. b. CVD layer. c. RIE of CVD layer.
d. DBARC and resist pattern. e. RIE of exposed spacer.
f. Trimmed spacer after resist and DBARC removal.

Figure 5: Spacer process using a DBARC.

stripping solution to contact the DBARC, as shown in steps 6b and 6c. The stripping solution dissolves the DBARC, allowing the remainder of the litho stack to lift off, as shown in step 6d. A DBARC offers the advantage of providing smaller photoresist features, controlling dimensions over topography, and providing controlled undercut to allow lift-off. To prove the concept for large feature sizes, a thick DBARC and photoresist were used in figure 7. Different develop times were tested to create different profiles. An undercut profile was achieved with 40 seconds of develop time.

Conclusions

DBARC applications have matured over the past several years. The initial success of DBARCs in implant layers provided a knowledge base for it to broaden into other process areas. In BEOL dual damascene processes, a DBARC could be used as a partial fill BARC. DBARCs also had value to add to BEOL via processes, such as increased etch budget and improved pattern fidelity. This history paved the way for DBARCs to be used in FEOL applications other than implant. In the FEOL, DBARCs are being used in HKMG patterning to promote adhesion and reduce the use of RIE at the gate level. Advanced double-patterning schemes are also using DBARCs. The ability to retain a pattern in the DBARC because of its resistance to resist solvents has made DBARC a key component in current double-pattering patents and applications. Spacer technology has also benefited from using a DBARC's ability to easily adjust solubility. As use of DBARCs in the semiconductor industry has matured, it was proposed that other industries might benefit as well.

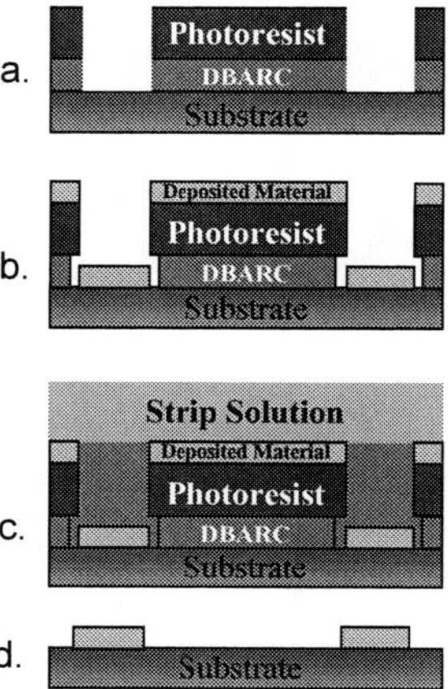

Figure 6: Lift-off process using a DBARC.

Figure 7: DBARC cross sections through develop time.

References

1. T. Brewer et al., *Journal of Applied Photographic Engineering*, vol. 7, no. 6, pp. 184-186 (1981).
2. B. Simmons et al., *INTERFACE '99: Proceedings of the ARCH Microlithography Seminar*, pp. 183-195 (1999).
3. X. Shao et al., *Proceedings of SEMICON China 2004 Technology Symposium*, March 17-19, (2004).
4. T. Katayama et al., *Proceedings of SPIE Optical Microlithgraphy XVII*, 5377, pp. 968-973 (2004).
5. D.C. Owe-Yang et al., *Proceedings of SPIE Advances in Resist Technology and Processing XXI*, 5376, pp. 452-460 (2004).

6. C. Washburn et al., *Proceedings of INTERFACE '06 Fujifilm Microlithography Symposium*, (2006).
7. D. Guerrero et al., *Journal of Photopolymer Science and Technology*, vol. 19, no. 3, pp. 343-347 (2006).
8. F. Houlihan et al., *Proceedings of SPIE Advances in Resist Materials and Processing Technology XXV*, 6923 (2008).
9. J. Lowes, et al., *ECS Transactions*, vol. 27, no. 1, pp. 503-508 (2010).
10. R. Subramanian et al., inventor; 2002 Sept 24. Developer Soluble Dyed BARC for Dual Damascene Process. United States patent US 6,455,416 B1.
11. C. Cox et al., *Proceedings of SPIE: Advances in Resist Technology and Processing XX*, vol. 5039, pp. 878-882, (2003).
12. T. Kudo et al., *Proceedings of SPIE: Advances in Resist Technology and Processing XXVII*, vol. 7639, (2010).
13. T. Kudo et al., *Journal of Photopolymer Science and Technology*, vol. 23, no. 5, pp. 731-740 (2010).
14. L. Vyklicky et al., *Journal of Photopolymer Science and Technology*, vol. 22, no. 1, pp. 17-24 (2009).
15. R. Chau et al., *Extended Abstracts of International Workshop on Gate Insulator (IWGI)*, pp. 124-126, (2003).
16. T. Hoffmann, "HKMG gate-first vs. gate-last options" International Electron Devices Meeting. Baltimore, Maryland (2009).
17. T. Schram et al., *Symposium on VLSI Technology Digest of Technical Papers*, (2008).
18. J. Cameron et al., *Proceedings of SPIE: Advances in Resist Technology and Processing XXVII*, vol. 7639, (2010).
19. D. Guerrero et al., *Journal of Photopolymer Science and Technology*, vol. 20, no. 3, pp. 339-343 (2007).
20. C. Washburn et al., *ECS Transactions*, vol. 18, no. 1, pp. 419-425 (2009).
21. J. Meador et al., *Proceedings of SPIE: Advances in Resist Technology and Processing XXVII*, vol. 7639, pp. 763926-1 - 763926-10 (2010).
22. J. Lowes et al., *Proceedings of SPIE: Advances in Resist Technology and Processing XXVII*, vol. 7639, pp. 76390K-1 - 76390K-11, (2010).
23. T. Paxton et al., "Reduced Pitch Multiple Exposure Process." U.S. Patent Application No. 11/387,047, Publication No. US 2006/0216653 A1, 2006 Sept 28.
24. T. Paxton et al., "Reduced Pitch Multiple Exposure Process." U.S. Patent Application No. 11/177,490, Publication No. US 2007/0003878 A1, 2007 Jan 4.
25. S. Dunn, "Method for Double Pattering a Developable Anti-Reflective Coating." U.S. Patent Application No. 11/543,365, Publication No. US 2008/0076074 A11, 2008 Mar 27.
26. V. Yu et al., "Double Pattering Method Using Metallic Compound Mask Layer." U.S. Patent Application No. 12/752,281, Publication No. US2010/0279234 A1, 2010 Nov 4.
27. B. Kim et al., "Method of Forming Semiconductor Device Patterns." U.S. Patent Application No. 12/480,807, Publication No. US 2010/0173492 A1, 2010 Jul 8.

ECS Transactions, 34 (1) 257-262 (2011)
10.1149/1.3567590 ©The Electrochemical Society

Development of Under Layer Material for EUV Lithography

Rikimaru Sakamoto[a], Bang-Ching Ho[a], Noriaki Fujitani[a], Takafumi Endo[a], Ryuji Ohnishi[a]

[a] Semiconductor Materials Research Department, Electronic Materials Research laboratories,
Nissan Chemical Industries, Ltd.635 Sasakura, Fuchu-machi, Toyama 939-2792, Japan

For the next generation lithography (NGL), several technologies have been proposed to achieve the 22nm-node devices and beyond. Extreme ultraviolet (EUV) lithography is one of the candidates for the next generation lithography. For lithography processes, the Line width roughness (LWR) and the pattern collapse of resist are the most critical issues for NGL, because of the small target critical dimension (CD) size and high aspect ratio. In this study, we design the new concept of EUV Under layer (UL) material to meet these requirements and study the impact of polymer design for pattern collapse behavior, pattern profile and LWR control by using EUV exposure tool.

1. INTRODUCTION

EUV Lithography is one of the most promising candidates for below hp22 device manufacturing. However, this technology has several issues that need to be overcome, especially development of high power EUV light source, manufacturing and inspection of multi layer mask and development of photo resist which has good balance for resolution-Line with roughness-Sensitivity (RLS) trade off are the most important issues. Regarding the development of Photo resist (PR) for EUV, improvement of LWR(10% of CD size), resolution limit below 20hp, high sensitivity for high through put (Eop<10mJ/cm2) and pattern collapse improvement are the major target.

Anti-Reflective coating material is the key technology for optical lithography (i-line, KrF and ArF), and prevents the reflection issue between PR and substrate interface. However, EUV (13.5nm) light can path through the materials without reflection at interface. According to this phenomenon anti-reflection property for EUV lithography is not necessary from the view point of optical behavior. However, UL material has been investigated to minimize the RLS trade off of PR. Recently, some reports discuss UL material can be applied as adhesion layer to prevent pattern collapse issue at small CD size and also work for improving resist profile and LWR[1,2,3]. So UL will be getting one of the key technology for EUV lithography. Based on this viewpoint, Nissan chemical have been developing UL material to improve the pattern collapse and control the resist profile to improve LWR. In the same time, for the EUV PR pattern, the thickness of PR is getting thinner because of suitable aspect ratio, the thickness of UL must be thinner, like 10nm thickness, to meet the requirement of etch process. How to control the coating property with ultra-thin UL is also one of the topic for the development of EUV UL.

In this paper, we will describe the design concept of UL material and discuss the material performance, including the pattern collapse margin, the resist profile, LWR and the investigation about ultra thin coating property.

2. EXPERIMENT

2.1 Preparation of materials

The UL material being used in this paper are listed in table 1. All of samples are the thermal crosslink material and include the base polymer, the crosslinker, the acidic catalyst and other additives. Propylene glycol mono methyl ether (PGME), Propylene glycol mono methyl ether acetate (PGMEA), Ethyl lactate (EL), etc. are used as solvent system.

The viscosity of materials were fixed for targeting 10nm or 5nm film thickness under 1500rpm/60sec. and 205 deg.C/60sec. bake condition. The formulated materials were filtrated with 0.01um UPE filter. Optical interference film thickness measurement tool (Nanospec AFT5100, Nanometrics) was used for thickness measurement. Ellipsometory tool (LAMDA ACE RE-3100, DNS) was used for measurement of below 10nm thickness.

Table 1. Under Layer Material

Under Layer	Polymer	Remarks
UL-1	a	High density unit-A
UL-2	a	High density unit-B
UL-3	b	Low density polymer
UL-4	a	High density, acid unit including
UL-5	c	Coating property improvement

2.2 Lithography

The performance of lithography was done by EUV alpha demotool, micro exposure tool and small field exposure tool. The process condition is listed below. EUV alpha demo tool (ADT, ASML) NA=0.25(σ 0.5), hp30nm, 28nm, 26nm, 25nm L/S=1/1, Micro exposure tool (MET) NA=0.3(σ 0.36/0.68 Quadrapole, hp30nm, 26nm L/S=1/1, , Small Field Exposure tool (SFET, Canon) NA=0.3 (σ 0.3/0.7 slit), hp45nm, 35nm and 32nm L/S=1/1 were used for EUV Lithography evaluation.

3. RESULTS AND DISCUSSION

3.1 Improvement of pattern collapse

Line pattern collapse issue is one of the most critical issue at small CD target for 22nm node and beyond. Because the contact area between the PR bottom and substrate is getting narrower, this issue is more severe when pattern size below 30nm hp. On the other hand, UL material can improve adhesion property between PR and UL and have capability to

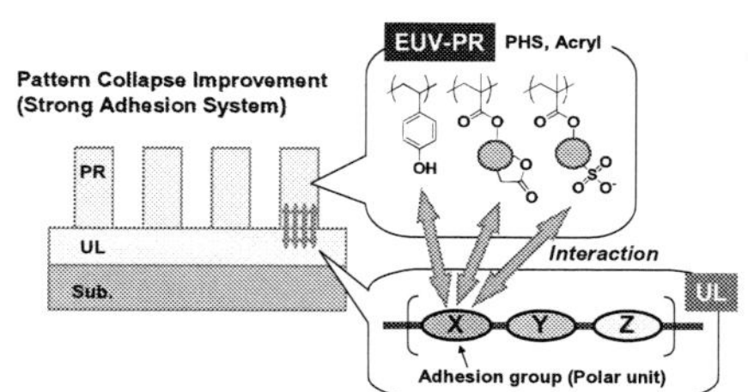

Fig.1. The concept of the improvement for the pattern collapse.

improve pattern collapse issue[1, 2, 3] on several reports. We investigated the pattern collapse improvement from the view point of chemical interaction between PR polymer and UL polymer. The new polymer structure was designed to have functional unit that have potential to interact the chemically and physically with the functional unit of PR, like hydroxyl styrene unit or Lactone or Bound PAG unit.

The data of pattern collapse margin with ADT was described in Fig. 2. UL-1 and UL-2 showed the better ultimate resolution compare to reference UL, UL-1 and UL-2 having new functional unit could make the PR pattern down to hp25nm. UL-1 and UL-2 showed wider pattern collapse margin at higher dose condition (smaller CD size). These result suggested to introduce the chemically interactive unit with PR could improve adhesion property between PR and Under layer interface and improve the pattern collapse margin.

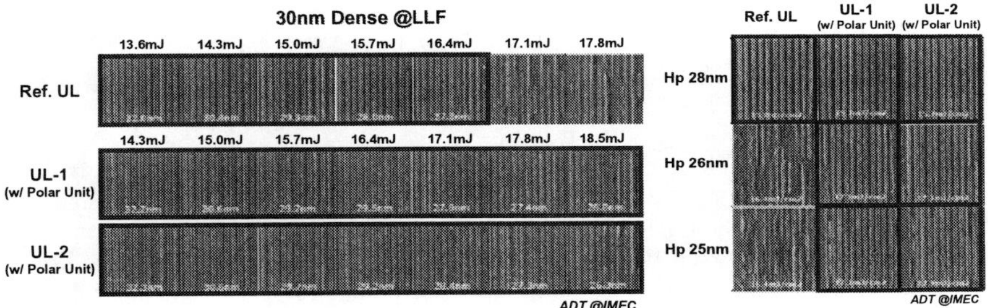

Fig.2. Comparison of Pattern collapse margin and Ultimate resolution in ADT evaluation.

3.2 The effect of UL for resist profile and LWR/LER.
3.2.1 The control of Resist profile.

The diffusion of acid being generated from PAG and quencher for chemical amplified PR can be the major reason to cause the footing or scumming of PR profile. For EUV Lithography, the diffusion of acid or quencher must be controlled more severely. UL design and performance was investigated in the point of diffusion control, especially increasing film density of base polymer was investigated for diffusion control from PR to Underlayer.

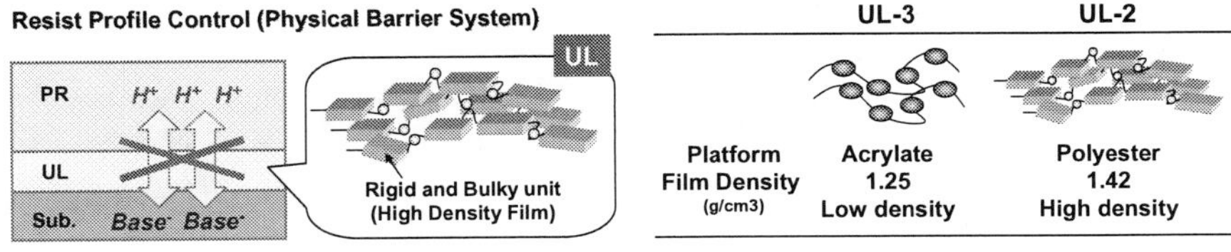

Fig.3. Concept of UL design for diffusion control.   **Fig.4.** Film density of Under layer samples

Fig. 5 shows the result of Lithography using SFET(Canon). The results are the comparison of resist profile with each hp45nm, 35nm and 32nm on UL-2 and UL-3, HMDS treated bare-Si as reference. HMDS treated substrate and UL-3 showed the footing profile at hp32nm, but UL-2 showed the vertical profile. UL-2 introduce the physically stackable unit into the base polymer and controlled the conformation of polymer chain. The film density of UL-2 showed 1.42g/cm^3 film density by XRR method On the other hand, the film density of UL-3 was 1.25g/cm^3 (Fig. 3 and Fig. 4). This result suggested that the high film density of UL can minimize the acid or quencher diffusion from PR into UL, and make the resist profile more vertical.

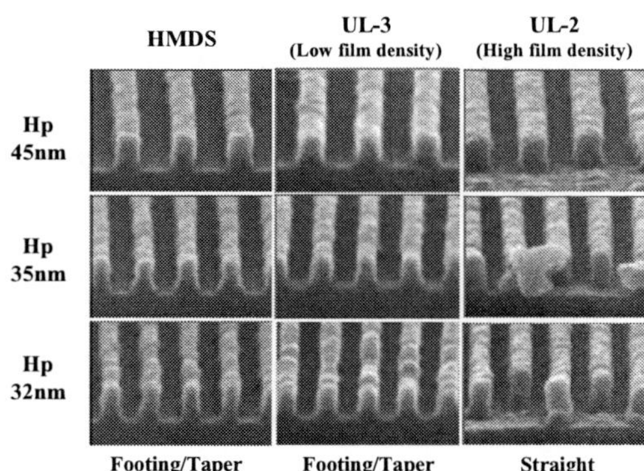

Fig.5. Comparison of resist profile with different UL in several hp.

3.2.2 The improvement of LWR/LER

The LWR/LER of resist impacts the device performance and is getting more critical, because the CD target got more critical. From the view point of resist profile, the shape of PR bottom will impact the LWR/LER. For example, the footing and scumming on the PR bottom will make the worse LWR/LER. Basically, PR bottom shape can be controlled by the surface acidity of UL and also some physical factor, for example, the high film density as 3.2.1. The surface acidity impact for LWR was investigated in this part, and we introduce the acid unit into the base polymer shown in Fig.6.

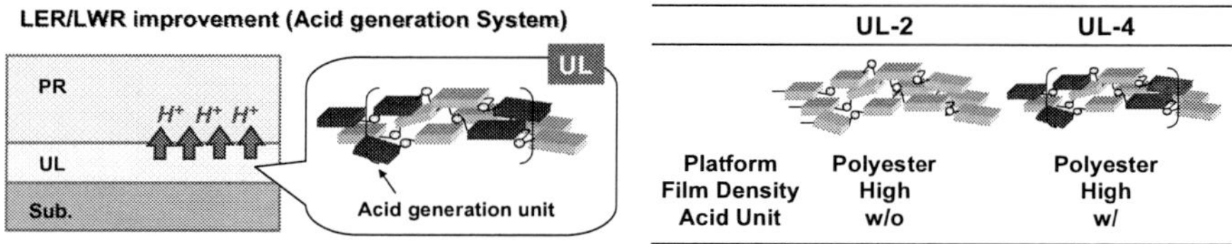

Fig. 6 The concept of material design for LWR control by introducing acidic unit.

Fig. 7 showed the EUV lithography result with ADT. For the hp30nm comparison, the LER value on UL-2 was 4.2nm, but 3.9nm was obtained on UL-4. UL-4 having the similar base polymer structure with UL-2 introduced the acidic unit into the base polymer but UL2 dose not have it. Based on X-SEM picture in Fig. 7, the resist profile on UL-4 is more vertical profile than on UL-2 at hp28nm. The results show that we introduce the acidic unit into the base polymer to improve the resist profile and more vertical, and also improve the LWR/LER.

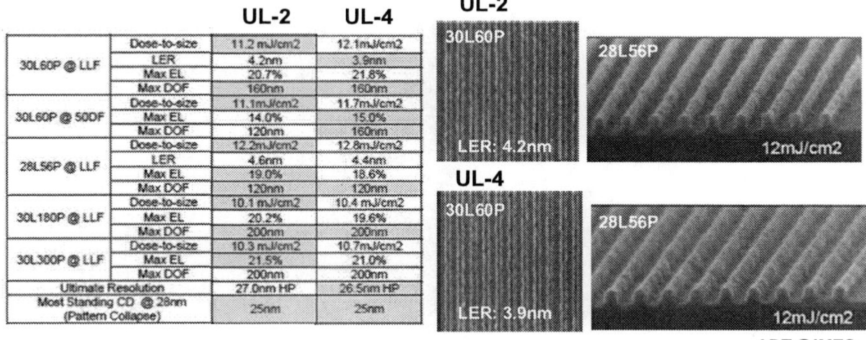

		UL-2	UL-4
30L60P @ LLF	Dose-to-size	11.2 mJ/cm2	12.1mJ/cm2
	LER	4.2nm	3.9nm
	Max EL	20.7%	21.8%
	Max DOF	160nm	160nm
30L60P @ 50DF	Dose-to-size	11.1mJ/cm2	11.7mJ/cm2
	Max EL	14.0%	15.0%
	Max DOF	120nm	160nm
28L56P @ LLF	Dose-to-size	12.2mJ/cm2	12.8mJ/cm2
	LER	4.6nm	4.4nm
	Max EL	19.0%	18.6%
	Max DOF	120nm	120nm
30L180P @ LLF	Dose-to-size	10.1 mJ/cm2	10.4 mJ/cm2
	Max EL	20.2%	19.6%
	Max DOF	200nm	200nm
30L300P @ LLF	Dose-to-size	10.3 mJ/cm2	10.7mJ/cm2
	Max EL	21.5%	21.0%
	Max DOF	200nm	200nm
Ultimate Resolution		27.0nm HP	26.5nm HP
Most Standing CD @ 28nm (Pattern Collapse)		25nm	25nm

Fig.7. The comparison of EUV Lithography performance with UL-2 and UL-4 in ADT.

3.3 The investigation of the coating property for ultra thin UL

The thickness of PR must be reduced because of suitable aspect ratio for the pattern collapse issue. In order to reduce the damage of PR during the UL open etch step, the thickness of UL should be reduced as well as PR. For hp26nm target, the thickness of PR will be around 40 to 50nm, and the thickness of UL will be 10nm or less than 10nm. We investigated the coating property of ultra thin UL, and discussed the thickness uniformity and pin hole on several kind of substrate.

3.4.1 The film thickness uniformity of UL on Bare-Si substrate.

Fig.8 showed the film thickness uniformity of the UL material UL-5 with 10nm and 5nm on Bare-Si substrate. The range (Maximum-Minimum) showed less than 2% for 10nm thickness and less than 1% for 5nm thickness was confirmed. This result shows that we don't have the problem of the thickness uniformity even 5nm UL thickness.

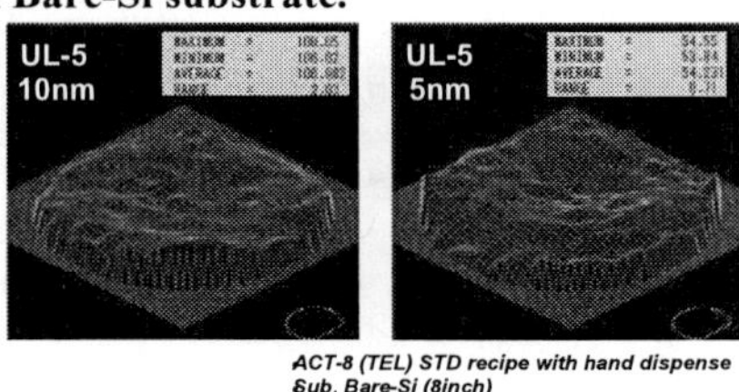

ACT-8 (TEL) STD recipe with hand dispense
Sub. Bare-Si (8inch)
FTK measurement with 300points

Fig.8. Film thickness uniformity of optimized UL.

3.4.2 The coating property on CVD Hardmask (HM) substrate.

Fig.9 shows the coating property of UL-5 with 10nm thickness on several kind of HM substrate. We didn't see any kind of defect by AFM analysis on UL-5, including pinhole like defect. Additionally, the good surface roughness value was observed on each kind of HM substrate.

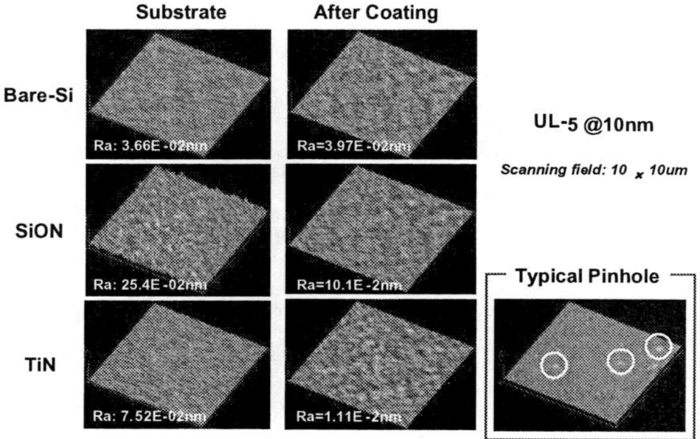

Fig.9. Surface image and roughness value by AFM analysis on HM substrate.

4. CONCLUSION

The UL material was investigated for improving the performance of EUV Lithography (Pattern collapse, resist profile control and LWR/LER control), and the coating property with ultra thin thickness was also investigated. Regarding the improvement of pattern collapse, we introduce the functional unit being interacted with PR polarity unit is very effective to improve the adhesion property between PR and UL interface and improvement of pattern collapse margin was confirmed. Using high barrier property film having the high film density showed improvement of the resist profile. Additionally, introduction of the acidic unit into base polymer showed more vertical resist profile and lower LER. To control the polarity and acidity of base polymer and film density of UL are the key parameter to achieve the good EUV lithography performance. In the same time, we could achieve the coating property of ultra thin UL with 10 and 5nm thickness on several kind of substrate.

From these result, the applying UL material for EUV Lithography has the benefit for the improving lithography performance without any disadvantage. In the future, we will develop the new UL material to meet the requirement of 22nm node and beyond device manufacturing.

5. ACKNOWLEDGMENT

We'd like to thanks for imec and Selete to help the EUV lithography evaluation.

6. REFERENCES

1) Hao Xui et al., Underlayer designs to enhance the performance of EUV resists, *Proc. SPIE* 7237, 72731J, 2009

2) K. Matsunaga et al., Development of resist material and process for hp-2x-nm devices using EUV lithography, *Proc. SPIE* 7636, 76360S, 2010

3) D. Kawamura et al., Pattern transfer process development for EUVL, *Proc. SPIE* 7273, 72731O, 2009

ECS Transactions, 34 (1) 263-267 (2011)
10.1149/1.3567591 ©The Electrochemical Society

Evaluation of 193 nm Photoresist Material at Advanced Immersion Nodes

Jing-an Hao, Yao Xu, Chang Liu, Qiang Wu, Xuelong Shi, Yiming Gu

Technology R&D, Semiconductor Manufacturing International Corp.
Pudong New Area, Shanghai, P. R. China 201203

Abstract

As 193 nm immersion lithography under single exposure moves toward the limit of 38 nm half pitch, more and more understanding has been achieved in the illumination. With the availability of same exposure tool and illumination condition, the key to the successful development of the photolithographic process will weigh toward good choice of and good process condition for the photoresist. Though different approaches exist for the evaluation of photoresists among major industrial players, the key understanding of the resist material has converged to some conservation quantity, which related to spatial resolution, line edge roughness, and resist sensitivity (RLS). Such quantity seems to be related to the formulation of modern chemically amplified photoresist. Although it is well-known that usual 193 nm photoresists needs a top coating for exposure tools that need water immersion since the water will interact with the acid and caused patterning failure, recently, topcoat-less photo resist process was developed for immersion lithography. The availability of such resist has simplified the resist process and saved cost, but may make the resist material more complicated. We have explored a number of normal immersion and topcoat-less immersion photoresists in the limit of 193 nm immersion lithography. We will present our data in the study of RLS and the resist profile.

1. Introduction

As CDs are pushed smaller toward the limit of 45 nm half pitches and lower and resists become thinner, the importance of understanding in the illumination system becomes greater. With the availability of standard exposure tool and illumination conditions, the key to the successful development of a suitable photolithographic process will weigh more toward the choice of photoresists and related process conditions. Chemically amplified photoresists used in advanced lithography nodes need to answer calls for more stringent requirements in advanced process characteristics. Although different approaches exist for the evaluation of photoresist from major industrial players, key understandings of the resist material have converged into some conservation quantity, which related to Resolution, Line Edge/Width Roughness (LER/LWR), and resist Sensitivity (RLS). Along with resolution, line width roughness and resist sensitivity are important parameters whose specifications have become very tight. It is now widely recognized that resolution, line width roughness, and resist sensitivity are fundamentally interdependent. Therefore, when evaluating or optimizing resist performance it is very important to take these three characteristics into consideration simultaneously. The use of a single figure-of-merit to judge resist performance with respect to line width roughness, resolution, and resist sensitivity has been proposed and evaluated [1-4].

Although the use of topcoat process has demonstrated good practical advantage for extending some conventional ArF dry resist platforms to immersion with good performance in low material leaching, high scanning speed, and immersion defect prevention [5-7], due to cost concern, recently topcoat-less resists have been developed for immersion lithography with a small amount of additives blended into the conventional resist material. These additives have been designed to separated within the resist and migrate to the resist surface to form a thin barrier, water-shedding film, which can effectively block the leaching of resist, achieve high scanning speed and comparable immersion defect level [8-9].

2. Experiment

In order to better understand the phenomenon of RLS tradeoff for topcoat-less resists, we chose two kind of topcoat-less immersion photoresists to evaluate the photolithographic process under single exposure in 193nm immersion lithography. In the experiment, the through-pitch process window and LWR have been tested. The ultimate goal is to determine the best photoresist candidate for the 32 nm lithographic process development.

For each pitch, a focus energy matrix was run to determine the process window (PW) with the selected resists. Exposures were carried out on immersion ASML1900i linked to a TEL track. Line width roughness was also measured for L/S pattern using an optimized CDSEM algorithm setting. All the tests were processed on the same substrate wafer with BARC material. The process tool settings are listed in Table 1.

Tool	NA	Illumination Mode
XT-1900i	1.3	Annular with Strong Sigma

Table 1: process tool setting

3. Results and Discussion

To best utilize the performance of selected photoresists, we did mask bias split to get maximum process window for each resist. For each pitch, the process window was determined using critical dimension (CD) target around 50 nm ± 10% nm with appropriate Energy Latitude (EL) limits. The process window versus pitch for each resist was shown in Fig. 1. The process window and LWR results were normalized for all data presented in this paper. The 100% in DOF is determined by production.

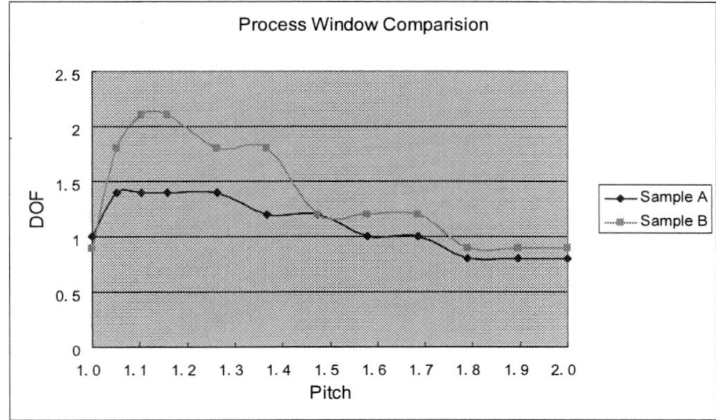

Fig. 1: Process window vs. pitch for 2 resist samples

For resist A, the process window is from 0.8 to 1.4; while the resist B is from 0.9 to 2.1. The process window range is 0.6 and 1.2 respectively. Both resists show the similar trend that the DOF goes up with pitch going up and drops down started at 1.5 times minimum pitch. The 1.5 x through 2.0 x minimum pitch range is also well-known as the "forbidden pitch". It is obvious that through pitch process window of sample B is significantly bigger than sample A, especially for the dense pattern area excluding the minimum pitch.

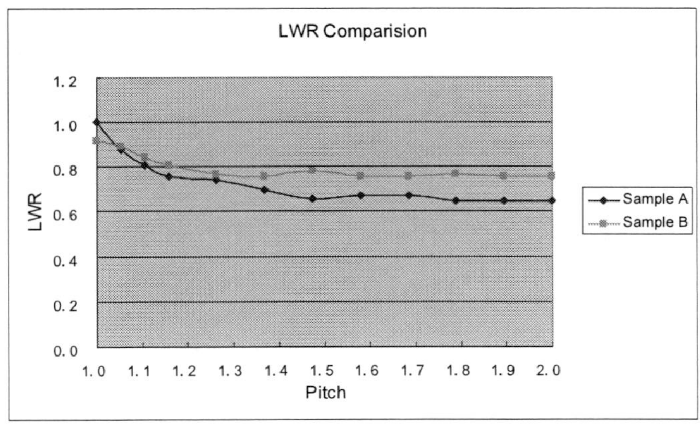

Fig. 2: LWR vs. pitch for 2 resist samples

The LWR versus pitch for each resist was displayed in Fig. 2. For resist A, the LWR window is from 1.0 to 0.65; while the resist B is from 0.91 to 0.76. The LWR range is 0.35 and 0.15 respectively. Both resists show the similar trend that the LWR goes down with pitch going up, then becomes nearly a constant started at 1.5 times minimum pitch. It is obvious that through pitch LWR of sample A is significantly smaller than sample B, especially around 1.5 times the minimum pitch range excluding the minimum pitch.

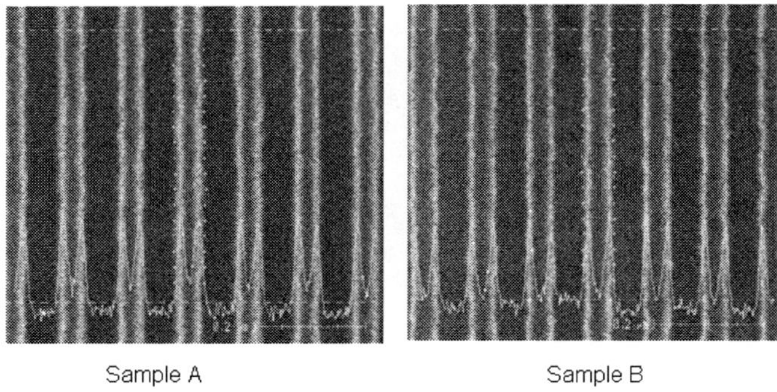

Fig. 3: Dense pattern's SEM image for each resist sample

The SEM image of the dense pitch under best energy and focus condition for each resist is displayed in Fig. 3.

Adding sensitivity, we have all information needed to construct the RLS triangle. In one representation, the RLS constant = $(Resolution)^3$ *$(LWR)^2$*Photo Sensitivity. As an example, we put the dense pitch's data in and calculate the constant for each resist sample. The results are summarized in Table 2.

PR Name	Resolution	LWR	Sensitivity	RLS
Sample A	1.00	1.00	1.00	1.00
Sample B	1.01	0.91	1.28	1.09

Table 2: Dense pattern's RLS data for 2 resist samples.

In this case, sample A shows a RLS constant, which is 9% smaller compared to that of sample B, which means we can get slightly bigger resolution and sensitivity while paying a small price in LWR if we choose sample A. Since this is measured at the minimum pitch, the LWR from sample B is not always better than that of sample A, especially at intermediate pitches.

4. Conclusion

In this paper, we have performed evaluation for two topcoat-less photoresists. We have discussed the through-pitch focus window and LWR. The RLS constant provides a clear analytic description of the tradeoff among Resolution, LWR, and Sensitivity. We have also shown the dense pattern results from two topcoat-less resists as an example.

Acknowledgements

The authors would like to thank the support of Fab 8 and TD colleagues.

References

[1] D. Van Steenwinckel, et., al., Proc. SPIE 6519, , (2007).
[2] R. L. Bristol, Proc. SPIE 6519 (2007).
[3] G. M. Gallatin, Proc SPIE 5753 (2005).
[4] G. M. Gallatin, Proc SPIE 6921 (2008).
[5] S. Kishimura et al., Proc SPIE 5753 (2005).
[6] F. Houlihan et al., Proc SPIE 5753 (2005).
[7] Y. Wei, Proc SPIE 6153 (2006).
[8] Steven Wu, Proc SPIE 6923 (2008).
[9] T. Naruoka, Proc SPIE 7273 (2009).

ECS Transactions, 34 (1) 269-275 (2011)
10.1149/1.3567592 ©The Electrochemical Society

Limit of Line End Shortening Correction under Single Exposure in 193 nm Immersion Lithography

Qiang Wu, Yao Xu, Jingan Hao, Chang Liu, Xuelong Shi, and Yiming Gu

Technology R&D, Semiconductor Manufacturing International Corp. Pudong New Area, Shanghai, P. R. China 201203

As the lithography starts to approach the limit of 193 nm immersion imaging capability under single exposure, successful development of a lithographic process rely on the good balance of the process parameters, such as the exposure latitude and depth of focus for the dense, semi-dense, isolated structures, MEF, LWR, proximity bias, etc. Such balance requires very good understanding of the inter-relationship among them and ways to give reasonable specification to the important materials such as photoresist and photomasks. Although, relatively speaking, 1-dimensional structures, such as, line and space or even square contacts are relatively easy to understand and model with accurate simulations, which is close to perfect, 2-dimensional structures are not easy to understand physically and the simulation for them are up to larger errors. Since previous study on this subject was either done at larger ground-rules or lack of good physical understanding in the simulation, this paper will use both experimental wafer measurement data and physical modeling to probe the limits for the OPC correction of line (or space) end shortening structures at around 50 nm dimensions. Several structures will be studied in theory and wafer data.

Introduction

Line end shortening is an important old topic in lithography [1-2]. It is caused by diffraction of light from the limited resolution in the optical system. The higher the numerical aperture, the better, or the less the line end shortening. Past studies have reviewed that the line ends tend to merge when they are at a small proximity when the mask error factor becomes very high. And the mask error factor is dependent on the optical resolution, partial coherence, photoresist acid diffusion length, linewidth, and neighboring proximity environment. We are discussing the limit at 100 nm pitch and 50 nm linewidth. What we trying to do here is to find out what to expect when we are at the minimum feature size in the 193 nm immersion era.

MEF of the Line End Shortening

The discussion starts with opposing line ends with more general proximity patterns, illustrated in Figs. 1(a) and 1(b).

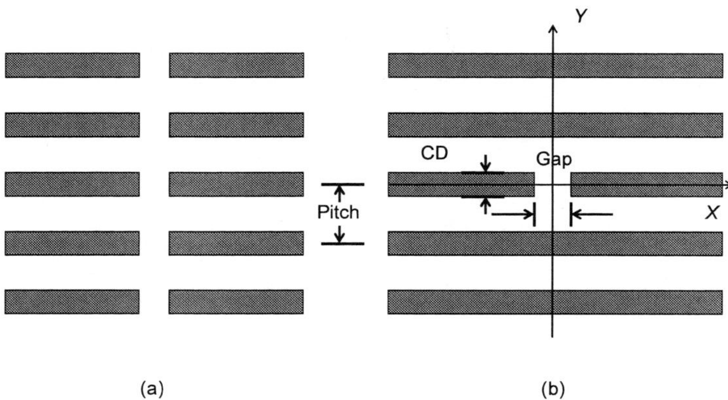

Fig. 1: Two different line end shortening patterns: (a) periodical array line ends; (b) isolated line ends placed within dense lines

Although it is quite easy to perform an aerial image simulation on the patterns, we choose not to do it since we would like to know the physics within the parameter space. Let us first assume coherent illumination (σ=0). And if we use the point spread function developed in an early publication [3], the aerial image amplitude can be written:

$$A(x, y) = \int M(x', y') PSF_D(x - x') PSF_D(y - y') dx'dy' \qquad [1]$$

where,

$$\boldsymbol{PSF_D}(x) \approx \sqrt{c}\, sinc\left[\frac{1.182}{a_e}(x)\right] \qquad [2]$$

and

$$a_e = \sqrt{a_I^2 + a^2} \qquad [3]$$

$$a_I = \frac{1.182\lambda}{2\pi NA} \qquad [4]$$

a is the photoacid effective diffusion length, which has been explored in detail in previous studies [4-5], and a_I is the blurring caused by diffraction., $M(x', y')$ is the mask function. Notice that the integration in the Y direction is independent of x both where the x is in the gap and non-gap areas. If our interest is in the X direction gap area, say $A(x,0)$, we can write eq. [1] as,

$$A(x,0) = \left[\int_{-\infty}^{-\frac{g}{2}} PSF_D(x-x')dx' + \int_{\frac{g}{2}}^{\infty} PSF_D(x-x')dx'\right]\left[\int_{-\infty}^{\infty} M\mid_{x\in(-\infty,-g/2)\,and\,(g/2,\infty)}(y')PSF_D(y')dy'\right]$$

$$+ \left[\int_{-g/2}^{g/2} PSF_D(x-x')dx'\right]\left[\int_{-\infty}^{\infty} M\mid_{x\in(-g/2,g/2)}(y')PSF_D(y')dy'\right]$$

$$[5]$$

Now if we only consider the mask error factor in the gap, we can take the differential of the eq. (5) in x and let $\delta A(x,0) = 0$, and notice that PSF_D is an even function, we thus get,

$$0 \equiv \delta A(x,0) = \left[PSF_D\left(x-(-\tfrac{g}{2})\right)(-\tfrac{1}{2})\delta g - PSF_D\left(x-\tfrac{g}{2}\right)(\tfrac{1}{2})\delta g + PSF_D(x-x')\mid_{-\frac{g}{2}}^{\frac{g}{2}}\delta x + PSF_D(x-x')\mid_{\frac{g}{2}}^{\infty}\delta x\right]\left[\int_{-\infty}^{\infty} M\mid_{x\in(-\infty,-g/2)\,and\,(g/2,\infty)}(y')PSF_D(y')dy'\right]$$

$$+ \left[PSF_D\left(x-\tfrac{g}{2}\right)\tfrac{1}{2}\delta g - PSF_D\left(x+\tfrac{g}{2}\right)(-\tfrac{1}{2})\delta g + PSF_D(x-x')\mid_{-\frac{g}{2}}^{\frac{g}{2}}\delta x\right]\left[\int_{-\infty}^{\infty} M\mid_{x\in(-g/2,g/2)}(y')PSF_D(y')dy'\right]$$

$$[6]$$

$$0 \equiv \delta A(x,0) = \left[-PSF_D\left(x+\tfrac{g}{2}\right)\tfrac{1}{2}\delta g - PSF_D\left(x-\tfrac{g}{2}\right)(\tfrac{1}{2})\delta g + PSF_D(x+\tfrac{g}{2})\delta x - PSF_D(x-\tfrac{g}{2})\delta x\right]\left[\int_{-\infty}^{\infty} M\mid_{x\in(-\infty,-g/2)\,and\,(g/2,\infty)}(y')PSF_D(y')dy'\right]$$

$$+ \left[PSF_D\left(x-\tfrac{g}{2}\right)\tfrac{1}{2}\delta g + PSF_D\left(x+\tfrac{g}{2}\right)\tfrac{1}{2}\delta g + PSF_D(x-\tfrac{g}{2})\delta x - PSF_D(x+\tfrac{g}{2})\delta x\right]\left[\int_{-\infty}^{\infty} M\mid_{x\in(-g/2,g/2)}(y')PSF_D(y')dy'\right]$$

$$= \left[-PSF_D\left(x+\tfrac{g}{2}\right)\tfrac{1}{2}\delta g - PSF_D\left(x-\tfrac{g}{2}\right)(\tfrac{1}{2})\delta g + PSF_D(x+\tfrac{g}{2})\delta x - PSF_D(x-\tfrac{g}{2})\delta x\right]\left\{\left[\int_{-\infty}^{\infty} M\mid_{x\in(-\infty,-g/2)\,and\,(g/2,\infty)}(y')PSF_D(y')dy'\right] - \left[\int_{-\infty}^{\infty} M\mid_{x\in(-g/2,g/2)}(y')PSF_D(y')dy'\right]\right\}$$

$$[7]$$

Since the integration in the Y direction is generally not zero, the terms in the first bracket must vanish. If we let $g_{wafer} = 2x$, then we get,

$$MEF = \frac{\delta\,g_{wafer}}{\delta\,g} = \frac{PSF_D\left(\dfrac{g_{wafer}-g}{2}\right) + PSF_D\left(\dfrac{g_{wafer}+g}{2}\right)}{PSF_D\left(\dfrac{g_{wafer}-g}{2}\right) - PSF_D\left(\dfrac{g_{wafer}+g}{2}\right)} \qquad [8]$$

This equation (Eq. [8]) indicates that the MEF for the gap is not dependent on the Y proximity under coherent illumination. Similar analysis holds for incoherent illumination, where the PSF_D must be replaced with PSF_D^2.

CD of the Line End Shortening

In line end shortening, we are only interested in the two dimensions: gap CD and line CD. Line CD is corrected by OPC and is very easy to understand. The gap CD, however, is not easy to understand and usually not simulated accurate enough. In Figs. 1(a) and 1(b), the line CD is basically the array line CD, its aerial image can be written,

$$A(x_0,y) = \int PSF_D(x_0-x')dx' \int M(x_0,y')PSF_D(y-y')dy'$$
$$= N\int M(x_0,y')PSF_D(y-y')dy' \qquad [9]$$

where N is the normalization factor of the point spread function. In the case illustrated by Fig. 1(a), Eq. [5] can be written,

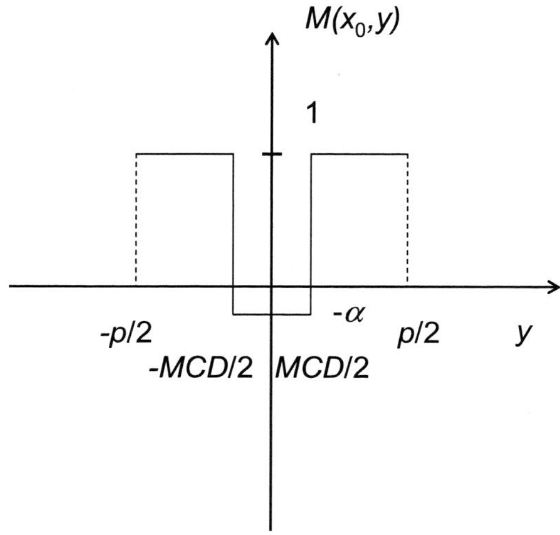

Fig. 2: Mask function for line ends. α is the attenuation factor for attenuated phase shifting mask, p is the pitch and MCD is the mask CD

$$A(x,0) = \left[\int_{-\infty}^{-\frac{g}{2}} PSF_D(x-x')dx' + \int_{\frac{g}{2}}^{\infty} PSF_D(x-x')dx'\right]\left[\int_{-\infty}^{\infty} M(x_0,y')PSF_D(y')dy'\right]$$
$$+ \left[\int_{-g/2}^{g/2} PSF_D(x-x')dx'\right]N' \qquad [10]$$

If we set the threshold equals to t, and let $A(x_t,0)= A(x_0,y_t)=t$, we get the new following equation,

$$N\int M(x_0,y')PSF_D(y_t - y')dy'$$
$$= \left[\int_{-\infty}^{-\frac{g}{2}} PSF_D(x_t - x')dx' + \int_{\frac{g}{2}}^{\infty} PSF_D(x_t - x')dx'\right]\left[\int_{-\infty}^{\infty} M(x_0,y')PSF_D(y')dy'\right] + \left[\int_{-g/2}^{g/2} PSF_D(x_t - x')dx'\right]N'$$
$$\qquad [11]$$

Consider taking the differential with respect to x_t and y_t of Eq. [11], and if we let wafer CD in Y $WCD=2y_t$, the left hand side of the Eq. [11] becomes,

$$N\int M(x_0,y')d(PSF_D(y_t - y'))\delta y_t = \left[NM(x_0,y')PSF_D(y_t - y')|_{-\infty}^{\infty} - \int PSF_D(y_t - y')dM(x_0,y')\right]\delta y_t$$
$$= 0 \pm \frac{1}{2}(1+\alpha)\sum_n [-PSF_D(WCD/2 - MCD/2 + np) + PSF_D(WCD/2 + MCD/2 + np)]\delta(WCD)$$
$$\qquad [12]$$

where, "+" sign represents line (dark) ends and "-"sign represents space (clear) ends. If we let $g_{wafer}=2x_t$, the right hand side of the Eq. [11] becomes,

$$\left[PSF_D(x_t - x')\big|_{-\infty}^{-\frac{g}{2}} + PSF_D(x_t - x')\big|_{\frac{g}{2}}^{\infty} \right] \left[\int_{-\infty}^{\infty} M(x_0, y')PSF_D(y')dy' \right]\delta x_t + \left[PSF_D(x_t - x')\big|_{-\frac{g}{2}}^{\frac{g}{2}} \right] N'\delta x_t$$

$$= \frac{1}{2}\left[PSF_D\left(\frac{g_{Wafer}}{2} - \frac{g}{2}\right) - PSF_D\left(\frac{g_{Wafer}}{2} + \frac{g}{2}\right) \right]\left\{ N' - \int_{-\infty}^{\infty} M(x_0, y')PSF_D(y')dy' \right\}\delta g_{Wafer}$$

$$[13]$$

Therefore, the differential relationship between δg_{wafer} and δWCD can be written,

$$\frac{\delta g_{Wafer}}{\delta(WCD)} = \frac{\pm(1+\alpha)\sum_n [PSF_D(WCD/2 + MCD/2 + np) - PSF_D(WCD/2 - MCD/2 + np)]}{\left[PSF_D\left(\frac{g_{Wafer}}{2} - \frac{g}{2}\right) - PSF_D\left(\frac{g_{Wafer}}{2} + \frac{g}{2}\right) \right]\left\{ N' - \int_{-\infty}^{\infty} M(x_0, y')PSF_D(y')dy' \right\}}$$

$$[14]$$

If $M(x, y)$ represents line (dark) end patterns, then, $N' - \int_{-\infty}^{\infty} M(x_0, y')PSF_D(y')dy' > 0$, and

we know $\sum_n [PSF_D(WCD/2 + MCD/2 + np) - PSF_D(WCD/2 - MCD/2 + np)] < 0$,

therefore, when the line CD increases, the gap CD will decrease. Same is true when $M(x, y)$ represents space ends. Similar derivation can be done for the line end patterns depicted in Fig. 1(b).

Similar analysis holds for incoherent illumination, where the PSF_D must be replaced with PSF_D^2 and use appropriate mask function. For partially coherent illumination, the equations will be between those of the coherent case and incoherent case.

Experimental Data

We have compared our calculation with experimental wafer data with patterns depicted in Fig. 1(a) at several pitches ranging from 95 nm to 140 nm. As we have expected, the experimental data fit quite well to the integrated Eq. [8],

$$g_{Wafer} = g_{Wafer}(0) + \int_{g_0}^{g} MEF(g_{Wafer}(0), g(0))dg \qquad [15]$$

which is not a function of pitch. However, there is some offset in gap CD on wafer between different pitches though the linewidth on wafer are all equal to 55 nm. In the fit, we have chosen to use a 60%:40% weighting in the coherent and incoherent MEF equations since we have used a quite aggressive annular illumination condition. The best-fit photoacid diffusion length is around 25 nm.

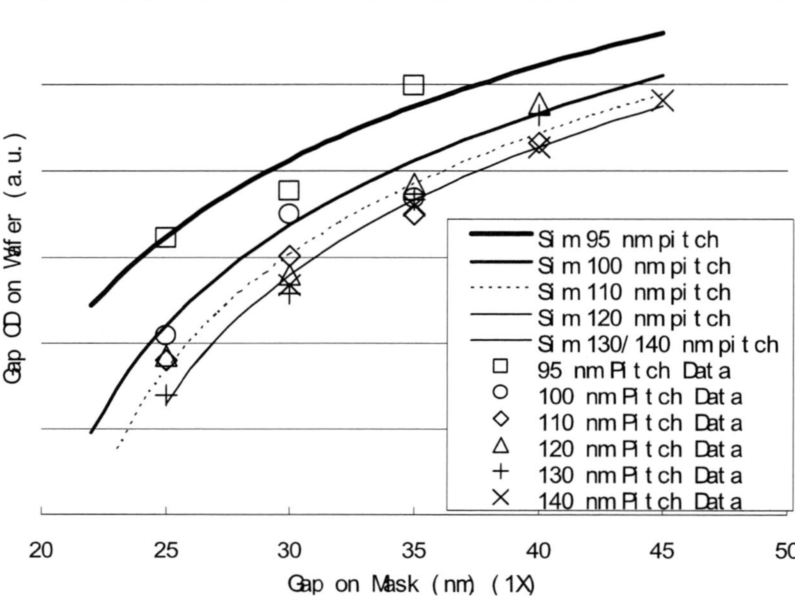

Fig. 3: Measured gap on wafer versus gap on mask with the condition: line CD printed on target of 55 nm. NA is 1.30 with annular illumination condition. The calculated MEF (Eqs. [8] and [15]) uses 60% coherent weighting and 40% incoherent weighting. The best-fit diffusion length is 25 nm.

This is surprising since, in the 1-dimension line/space simulation, we have been very certain that a diffusion length of 10 nm is routinely used for the aerial image calculation to very high precision in process window and MEF. In this simple calculation, we have not used vector algorithm, which may account for some loss of resolution, or loss of effectiveness of high NA. Nevertheless, the physics is here: line end shortening is a strong, if not the only, function of itself and the resolution limit.

In optical proximity correction, 1 important parameter is the minimum separation between edges such that they don't merge. In an early publication, this has been explored [6] with unexpectedly good accuracy. However, in the regime of 50 nm optical resolution, this is more complicated since we have observed merging where the MEF is not divergent, which means "pre-mature" bridging. Such "pre-mature" bridging may have been due to the resolution limit of the photoresist, which also shows up in the relatively speaking large, as compared with CD, line width roughness (LWR).

In this situation, we can still predict the line end merging boundary but not with a divergent MEF, but with experimental information. For example, for one photoresist, 35 nm may be the minimally resolvable wafer feature; for another photoresist, such limit may be extended to 30 nm.

Line End Shortening Correction Flow

From above analysis, we know that line end shortening or space end shortening is dependent on the line CD or space CD, but it is generally not dependent on the proximity around the line ends. The first step of correcting line end shortening is to do a good OPC such that all pitches can print at the same CD with same mask CD. Then we can probe the merging boundary for each representative linewidth, thus setting up a very good reference design rule. The knowledge we get from the learning will also help us check the OPC accuracy since the modeling in 2 dimensions may not be as accurate as the 1 dimension. As pointed out before [6], the prediction of the merging boundary before the mask is ordered can help in improving the turnaround time in the process development.

Conclusions

We have further explored line/space end shortening in simple analytical equations, we found that the line end shortening is a unique phenomenon, which is dependent on linewidth and gap clearance on mask and the gap CD on wafer. It is also dependent on the diffused aerial image resolution. We have compared our analysis with the experimental data at immersion technology nodes and a good agreement is found. As another output of the paper, we have proposed a line/space end shortening correction flow as a good supplement to the mainstream OPC flow.

Acknowledgment

The authors of this paper would like to thank the support from Semiconductor Manufacturing International Corporation (SMIC)'s production organization and the technology development center for their good support.

References

1. C.A. Mack, Proc. SPIE **4226**, 83 (2000).
2. X. Shi, R. Socha, J. Bendik, M. Dusa, and W. Conley, Proc. SPIE **3678**, 77 (1999).
3. Q. Wu, Proc. SPIE **6154-151** (2006).
4. T. Brunner, C. Fonseca, N. Seong, and M. Burkhardt, Proc. SPIE **5377**, 141 (2004).
5. Q. Wu, S. Halle and Z. Zhao, Proc. SPIE **5377**, 1510 (2004).
6. Q. Wu, J. Zhu, P. Wu, and Y. Jiang, Proc. Interface **2006** (2006).

ECS Transactions, 34 (1) 277-284 (2011)
10.1149/1.3567593 ©The Electrochemical Society

248nm Process Is Capable for sub 0.09 um Groundrules?

Lei Wang, Xiaobo Guo, Yufeng Tong, Honglin Meng, Bo Su, Shen'an Xiao

Shanghai Huahong NEC Electronics Company, Ltd.
No. 1188 Chuanqiao Road, Shanghai 201206 China

While the mainstream wafer production is at 0.065 and 0.045 um with 300 mm diameter wafers with ArF exposure tools systems, an idea to explore production feasibility under groundrules smaller than 0.09 um while maintain the cost advantages in KrF exposure tools systems becomes more and more popular and important to all companies including 300mm/200mm FAB.

But the k1 factor for sub 0.09um with current popular KrF exposure tools will be about 0.31, which has the same level of complexity in optical proximity correction compared to 0.045 um at 0.93 NA with 193 nm exposure tools. Several RET (Resolution Enhancement Technology) techniques have been proposed for low k1 case and some good results have been achieved. However, each RET techniques has their own merits and demerits. It is still attractive to find out a solution with current KrF exposure tools, popular illumination settings and cost effective masks.

In this paper, we will introduce our study for sub 0.09 um design rule with maximum 0.82NA KrF scanner tools. No special RET techniques are used and acceptable DOF, EL, MEEF, LWR and CD proximity are achieved. A novel photoresist optimization solution is both discussed.

I. INTRODUCTION

In past 30 years, the semiconductor development follow the Moor's law and the integration density increase too rapidly and cause the single chip cost decreases. The profit margin for each company especially for foundry becomes more and more poor. The ASP (Average Sales Price) for industry popular 0.09 um and sub 0.09 um technology product based on ArF exposure systems becomes more close to low end products and cost down is much important. As a result, the motivation to transfer 0.09um and sub 0.09um products to KrF exposure systems is increased [1-5].

In general KrF process can not meet the sub-0.09um technology node requirement due to the too low k1 [6-7]. Although some sub-0.11um results are also acceptable [1-3], the sub 0.09um design rule is still very difficult with current popular KrF tools. Because the 0.09um design rule had been close to single exposure limitation and each step for pitch shrink below it will become more and more difficult.

The most effective solution is increasing the NA to achieve larger k1 factor. Some good results have been demonstrated with new high NA KrF exposure systems [4-5]. But it is not to be a cost effective solution for current companies.

From the other side, as we all know there are several RET techniques [8-14] for low k1 case without increase NA. However each RET techniques has their own merits and demerits.

Double patterning is a very effective method to beyond the k1 physical limitation. But its cost is not competitive. It can be applied for high profit margin technology node but it is not to be a good solution in our case. [8~10]

Customer illumination settings such dipole and quasar can show perfect performance for optimized target features. But the non-optimized features will become worse. It needs the designer to avoid those non-optimized features or combining other techniques such as polarization exposures to improve non-optimized features performance. So it is a good solution for IDM companies but it is not good enough for fables IC designers and foundry companies. [11]

Alternating PSM and CPL technology can improve all features performance obviously. But the mask cost is not acceptable for most of IC designers and foundry companies except DRAM companies. [12-13]

Scattering bar technology can help to improve semi-ISO and ISO features performance. But it still can not resolve semi-dense such as forbidden pitches issues if the scattering bar can not be insert due to mask space limitation. On the other side, it will need complex OPC and tight control in mask making. [14]

All above current low k1 RET techniques focus on how to improve the aerial image intensity and image contrast. It is correct and very easy to understand. But unfortunately the poor aerial image in low k1 case is the physical limitation and it is not so easy to be beyond. That's why each solution which wants to beyond the limitation will have different problems.

Fortunately, lithography process is not formulated with optics principles only. When we can not get enough improvement from optical side, we still can get help from other regions such as photoresists characterization and selection.

In this paper, we will introduce our solutions to extend to sub-0.09um with popular illumination settings, cost effective mask and friendly OPC. No special RET techniques are used and the process cost is very competitive. The wafer result for critical layers – Gate poly, Metal1 – at sub 0.09 um technology node has been demonstrated. DOF, EL, MEEF, LWR and CD proximity are checked. A novel photoresists solution is both discussed.

II.   PHOTORESISTS SELECTION FOR LOW IMAGE CONTRAST

A classic idea in low k1 case will follow those steps: 1, enhance the aerial image intensity and image contrast; 2, select photoresist with relative slow photo acid response and short diffusion length. It can be easily understand if we can get good enough aerial imaging intensity and image contrast.

But for random IC design the aerial image intensity and image contrast is limited. If we need control the cost and don't want to use special RET techniques, we have to accept the much poor aerial image intensity and image contrast. In this case, we need to change our photoresist understandings.

On the other side, OPC (Optical Proximity Correction) is also very important but very difficult in low k1 case due to high MEEF. The basic OPC idea for CD proximity can be treated as the line/space ratio change from dark field to bright field. Dense patterns are more close to dark field and isolated patterns are more close to bright field.

For dense features the aerial image intensity is poor due to low k1 and the first problem is how to achieve enough photo acid. So the photo acid response needs to be very fast in this case. It is very different to normal understandings. Because the too fast photo acid response will cause MEEF and CD control capability issue if the aerial image intensity is strong. But in our case, it will be the better choice because our aerial image is limited.

Then for isolated features it will totally convert. The isolated features are more close to bright field. Too fast photo acid response will be a problem as normal understandings. How to resolve this issue? As we know, the isolated features DOF is very poor. It is because that the aerial image intensity decreases too fast when defocus occurs although it is very strong at best condition. So fast photo acid response can improve the DOF by enhance photo acid at defocus case. On the other side, the too low aerial image intensity in defocus case will need long diffusion when photo acid is not enough.

Now we have our solutions for photoresist. But it seemed unpractical because we need the diffusion length varies for different patterns and normally it is treated as the constant value for different IC designs. Fortunately some study had shown the photoresist diffusion length is not a constant value [15-16].

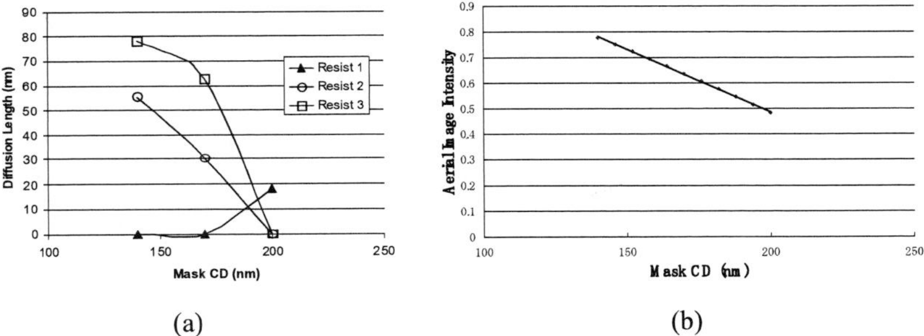

(a)   (b)

Fig. 1: (a) Diffusion length (b) The aerial image intensity at a pitch of 340 nm for three DUV photoresists.

Fig1 is an experiment for three different DUV photoresist is shown in reference paper [15]. Resist1,2,3 are typical line, space and contact hole optimized photoresist. From this figure we can find out that the diffusion length at dark field with poor aerial image intensity will increase for typical line optimized photoresist. On the other side, the diffusion length at bright field with strong aerial image intensity will increase too much for hole optimized photoresist. The resist2 will be a better choice with balance performance and meet our proposal.

III.   EXPERIMENTS

We studied typical line/space pattern for gate poly and metal-1 in photo process. The target features' design rule will follow sub 0.09um process - 0.09/0.13um gate poly and 0.11/0.11um metal1. A KrF excimer laser scanner is used as an exposure tool with maximum NA 0.82 and maximum out sigma 0.88 and maximum annular ratio 0.75. For minimum 0.20um pitch and maximum 0.82 NA, masks are fabricated using PSM (Phase

Shift Mask). A KrF chemically amplified positive resist (Resist2) is used. Organic BARC is used for gate poly layer and SiON DARC is used for metal layer.

Lithographic performance in terms of MEEF (Mask Error Enhanced Factor), DOF (Depth of Focus), EL (Exposure Latitude) and CD proximity effect is evaluated. MEEF is defined as the linear fitting curve for ADI CD vs. Mask CD from target CD -0.01 um to target CD + 0.01 um (1x) . DOF is defined as the focal range with +/- 10% CD variation with acceptable resist thickness loss. The optical proximity effect is calculated as the CD difference for lines with pitch change from 0.2um to 1.2um under best exposure condition.

For poly layer, in order to improve the process control capability, LWR (Line Width Roughness) is evaluated as additional check item. LWR is defined as the stand deviation of line CD variation through the line.

IV.   RESULTS AND DISCUSSION

3.1. Poly layer

Gate poly layer is used to define channel length of device. It is very critical to CD variation both on dense and isolated features. In low k1 case, how to improve dense features MEEF and isolated features DOF are most critical.

3.1.1 Process window and LWR

Displayed in Fig.2 is process window of dense and isolated features and Fig.3 is the top down SEM image. The dense DOF is 0.41um@8%EL, and non-scattering bar isolated line DOF is 0.24um DOF @ 7% EL. They are both acceptable.

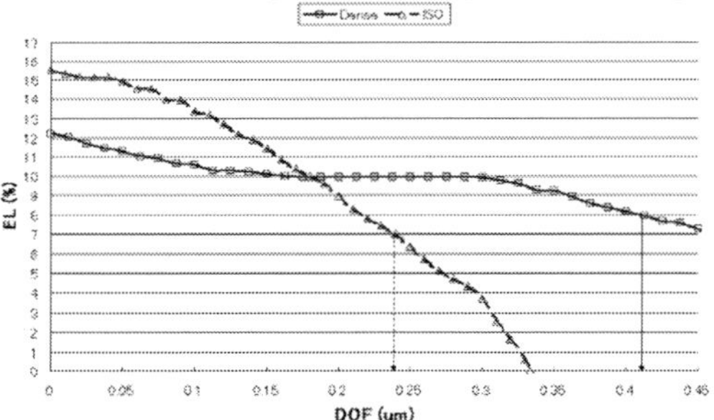

Figure. 2: DOF and EL

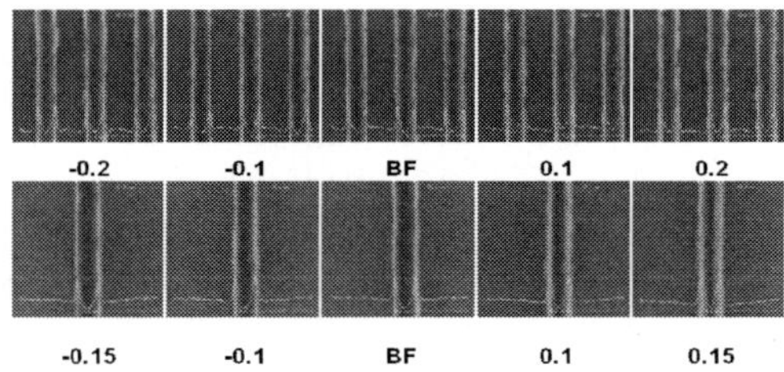

Figure. 3: Top view image depending on focus variation at poly layer

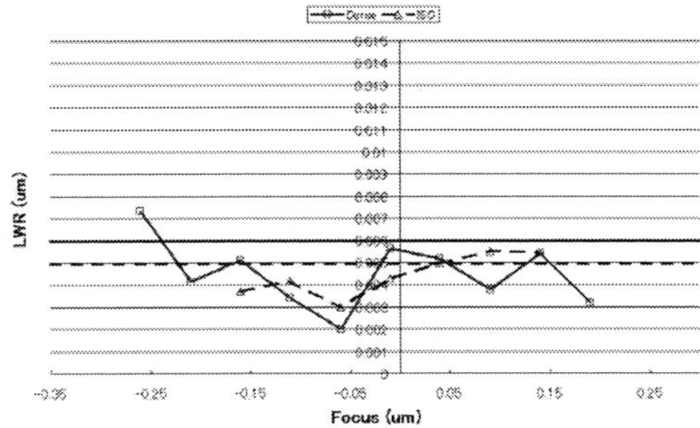

Figure. 4: LWR depending on focus variation at poly layer

LWR is also very important to poly layer for better CD control capability and device performance. We still checked the dense features and isolated features LWR performance for focus variation at best dose condition. From Fig.4 most of points are within 0.005um and all points within 0.006um when focus variation. It is comparable to ArF result and acceptable. The isolated line LWR increase at defocus case is also not found.

3.1.2 MEEF

Normally it is not so difficult to get acceptable dense DOF. But the MEEF is very hard to be improved and it is also the most difficult issues in low k1 case. The acceptable MEEF is the barrier from paper study to real production. We can demonstrate acceptable DOF for 0.20um or 0.18um pitch but the MEEF will be terrible.

The Fig. 5 shows the experimental ADI CD results for main features with line width from 0.09um to 0.12um. From the figure the MEEF is 2.3 and it is acceptable.

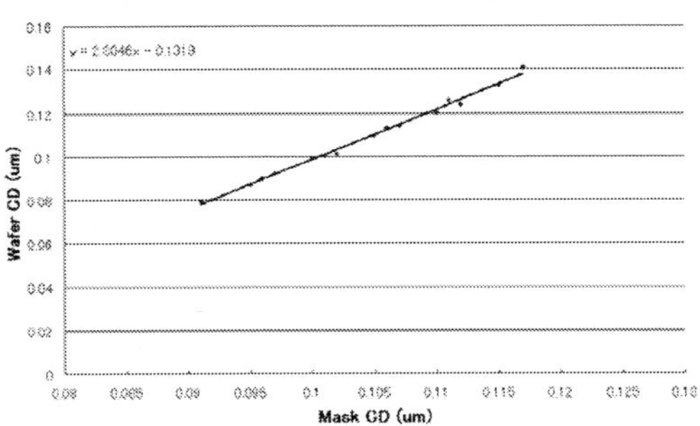

Figure. 5: MEEF comparison between high NA strong OAI and low NA middle OAI

3.1.3 CD Proximity

The optical proximity effect is compared with CD variation with pitch from Fig. 6. The post OPC range is within +/-0.005um and the maximum mask offset value is within 0.025um, which is friendly to OPC.

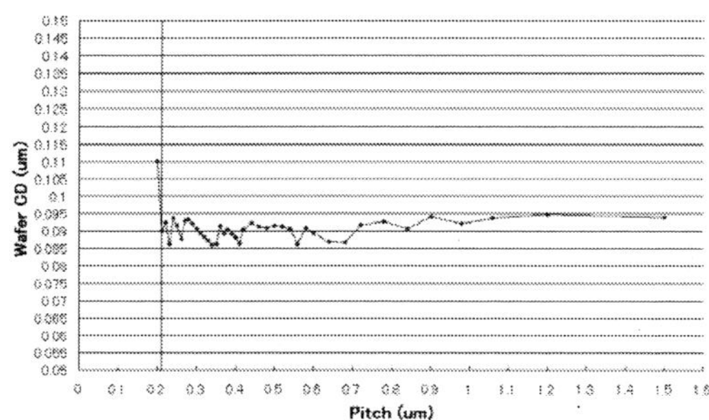

Figure. 6: Post OPC CD proximity

We had got the acceptable overlap process window, MEEF, CD proximity and LWR. The isolated line process window is around 0.3um, which is acceptable for popular scanner tools but maybe cause uptime loss due to tight tool control spec. If the scattering bar is acceptable from OPC side, it can be improved and tool control spec can be relaxed for productivity.

3.2 Metal layer

Different to poly layer, metal layer is defined to inter connect between different device. It needs high integrated density without opening and shorting. As a result, metal layer will focus on more tight pitch for equal line space and low requirement for isolated features. It needs photo process to improve resolution and isolated feature can be balanced by enlarging CD. On the other side, normally sub 0.09um BEOL interconnect will be copper process. So it is a dark field process and the short diffusion length and sensitive photo acid response to image intensity is needed. It is totally consistent to our assumption. In this case high NA and strong OAI is the best choice.

The process window of dense feature is shown in Fig. 7. The process window is larger enough. The DOF will be 0.45um @ 8% EL. The top view image depending on focus variation is shown in Fig. 8.

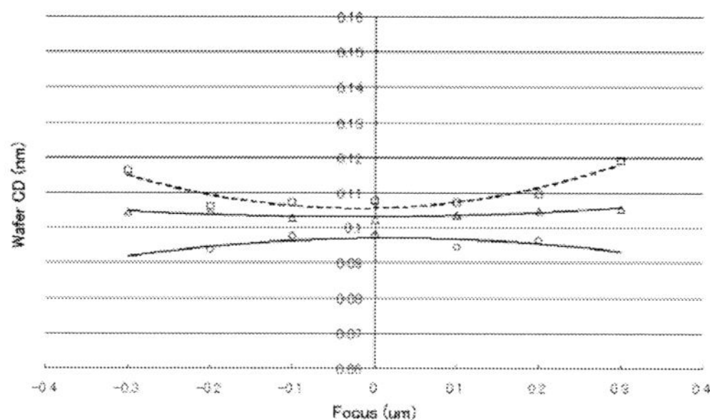

Figure. 7: Process Window

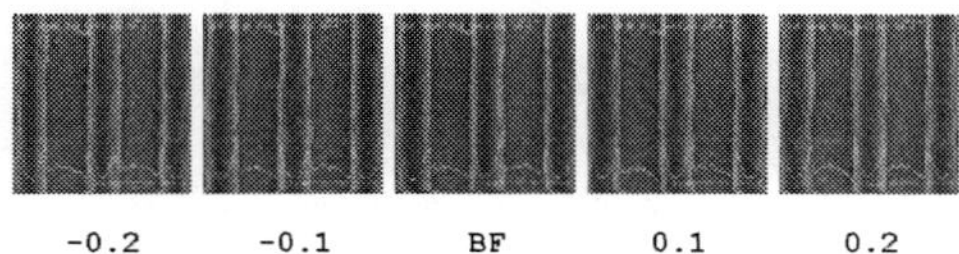

Figure. 8: Top view image depending on focus variation at metal layer

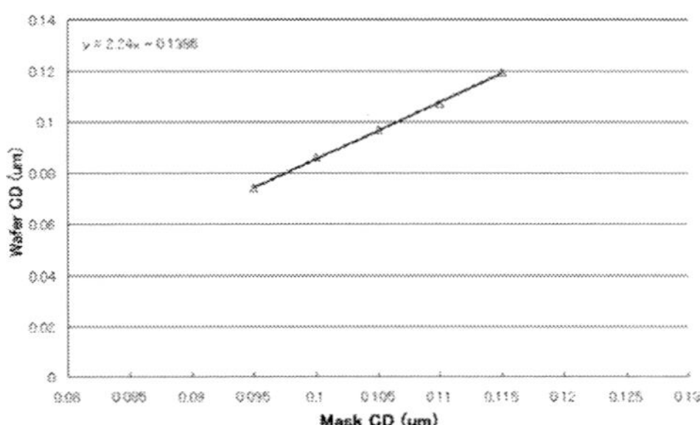

Figure. 9: MEEF

MEEF check result is shown in Fig. 9. The MEEF is 2.24 and which is acceptable.

From these results, KrF process is acceptable for metal layer with pitch extend to 0.22um. It means we can extend KrF process to sub-0.09 um BEOL layers and inter metal layers for 0.065um.

V.   CONCLUSION

We have performed a study of KrF process capability in critical layers of sub 0.09 um technology node. The result is not better than other RET techniques but it is still acceptable with low cost. A novel photoresist solution is both discussed. Fast photo acid response and variable diffusion length will be necessary. Consequently, we can expect KrF in sub-0.09um technology node by new higher NA KrF scanners, optimizing the process and improving the process margin.

ACKNOWLEDGEMENT

The author of this paper would like to thank the production division, the technology division, the third engineering division, the design service division, and the management teams of Shanghai Huahong NEC Electronics Company, limited for their kind support of the engineering work for this paper.

REFERENCES

1. Youngdoo Jeon, et. al., "90nm node Contact Hole Patterning through applying Model Based OPC in KrF Lithography", *Proc. SPIE*, **6924**, 6924-103 (2008) and references therein.
2. Lei Wang, "Process Window Analysis for KrF Photoresists Under Sub-0.11um Groundrules" , *ECS Trans.* **18** (1), 341 (2009)
3. Lei Wang, "Line Edge Roughness Analysis for KrF Photoresists under sub 0.11 um Groundrules" *ECS Trans.* **18** (1), 321 (2009)
4. Wim de Boeji, "Extending KrF lithography beyond 80nm with the TWINSCAN XT:1000H 0.93 NA scanner" *Proc. SPIE. Vol* **7140** , 71401B (2008)
5. Frank Bornebroek, "Cost effective shrink of semi-critical layers using the TWINSCAN XT:1000H 0.93 NA scanner" *Proc. SPIE. Vol* **7274** , 72743I (2010)
6. Shangho Lin, et. al., "How to Print 100nm Contact Hole with Low NA 193nm Lithography", *Proc. SPIE*, vol. **5376**, pp1091 (2004).
7. Jo Finders, et. al., "Can DUV Take Us Below 100nm", *Proc. SPIE*, vol. **4346**, pp153 (2001).
8. Stephen Hsu, et. al., "65nm Full-chip Implementation Using Double Dipole Lithography", *Proc. SPIE*, vol. **5040**, pp215 (2003).
9. Stephen Hsu, et. al., "Feasibility study of double exposure lithography for 65nm and 45nm node", *Proc. SPIE*, vol. **5853**, pp252 (2005).
10. Jungchul Park, et. al., "Application challenges with double patterning technology (DPT) beyond 45 nm", *Proc. SPIE*, vol. **6349**, 634922-1 (2006).
11. Mark Eurlings, et. al., "0.11um Imaging in KrF Lithography Using Dipole Illumination", *Proc. SPIE*, vol. **4404**, pp266 (2001).
12. Keeho Kim, et. al., "Optimization of process condition to balance MEF and OPC for alternating PSM: control of forbidden pitches", *Proc. SPIE*, vol. **4691**, pp240 (2002).
13. Toh, et. al., "Chromeless Phase-shift Masks: A New Approach to Phase-shift Masks", *Proc. SPIE*, vol. **1496**, pp27 (1990).
14. H.T. Lin, et. al., "Sub0.18um Line/Space Lithography Using 248nm Scanners and Assist Feature OPC Masks", *Proc. SPIE*, vol. **3873**, pp307 (1999).
15. Lei Wang, et. al., "The Characterization of Photoresist for Accurate Simulation Beyond Gaussian Diffusion", *Proc. SPIE*, **6519**, 6519-35 (2007) and references therein.
16. Lei Wang, et. al., "The Variation of the Effective Photoresists Diffusion Length for Different Line/Space Ratios", *Proc. ISTC* **2006**, pp679 (2007) and references therein.

ECS Transactions, 34 (1) 285-302 (2011)
10.1149/1.3567594 ©The Electrochemical Society

Study to Transfer 0.11μm DRAM
ArF Process to KrF Process in Litho

Joe Liu, Eric Yao, Erics Fan, Kent Chang, Ted Lv, Jeff Zhang,
Lee Liang, Jerry Hong, Mars Li
SMIC, 18,Zhangjiang Rd, Pudong, Shanghai 201203,PRC

ABSTRACT

In the mainstream microelectronics manufacturing process, lithography is the most complex, expensive and critical processes. In modern IC manufacturing process, the lithography has accounted for the total production cost 1/3, moreover along, with the technical unceasing development, the percentage which occupies also in the rise. [1] This is mainly because the smaller resolution capacity need the smaller wavelength of PHOTO exposure light source, which will lead the higher the cost machines. For example: the 193nm wave length ArF scanner is much more expensive than 248nm wave length KrF scanner.

To save costs, we need to seek some auxiliary technologies to improve the resolution of current machine and reduce the use of high-end exposure machine. Below article studied about how to transfer 0.11μm DRAM bite line (M0) process from ArF scanner to KrF scanner to achieve cost savings, maximize productivity purposes. The main content of this article are the following:

1. Introduction for the basic theory and application of IC manufacture's

lithography process , and description for the method to improve litho resolution.

2. Experiment with condition that light source wavelength increasing from 193nm to 248nm, and the numerical aperture increasing from 0.7 to 0.8, finding the most suitable KrF photo resist and related conditions, and collecting OPC mask data to re-production for KrF production mask. Finally established all the process conditions for KrF

3. Testing the KrF process on production to confirm if ArF process can replaced by KrF process, and method to overcome the shortcomings of the KrF process to enhance yield during mass production processes period.

After these series of tests, we successfully transferred 0.11μm DRAM M0 (Bit Line) layer lithography process from the ArF to KrF. And we optimized the KrF process condition to enhance yield to reach goal during mass production period. This change has the following several advantages:

1. Considerable cost saving in DRAM production by avoiding buying a new ArF Exposure and changing M0 lithography photo resists from ArF type to KrF type(resist costs reduced 76%).

2. Increased 33% capacity for 0.11μm DRAM.

3. We obtained the experience for the application of max NA= 0.8 KrF scanner. It can be extended to other high-end process and help us to better develop other 0.11μm product on KrF scanner.

This change already applied for mass production, and we have made 0.11μm DRAM final average yield above 85% with KrF process, which reached the international advanced level.

1. INTRODUCTION

In the mainstream microelectronics manufacturing process, lithography is the most complex, expensive and critical processes. In modern IC manufacturing process, the lithography has accounted for the total production cost 1/3, moreover along, with the technical unceasing development, the percentage which occupies also in the rise. [2] This is mainly because the smaller resolution capacity needs the smaller wavelength of PHOTO exposure light source, which will lead the higher the cost machines.

During sub-micron phase, the main wave length of lithography exposure light source are G-Line(436nm) and I-line(365nm) which are ejected by mercury-arc lamp with high voltage, and its' process limit is 0.25μm for I-Line. During deep sub-micron even nano-meter phase, the main lamp-house of lithography are high frequency laser KrF(248nm) and ArF(193 nm) which are characteristic spectrum of compound haloid . The process capacity of KrF tool type is 0.25μm, 0.18μm, 0.13μm. [3] The technology of below 13μm node (0.11μm for example) is base on ArF exposure flat. Also, with the development of technology, the KrF exposure tool of big numerical value aperture and relative resist were invented, the research of 0.11μm back to KrF exposure flat.

After 0.11μm DRAM (Dynamic Random Access Memory) introduced, the FAB realized to full loaded capacity in the second year. But a problem appeared: 0.11μm have 4 ArF layers(DT(Deep trench),AA(Active Area),P1(Volt line),M0(Bit Line)), and the lithography process were base on ArF(193nm) at the beginning of research, so ArF type tool is needed, but there were only 2 ArF tools, the capacity is 1200 pieces of wafers per day for each tool, so the throughput limit of DRAM in the factory is 18000 wafers every month. But a big amount of budget is needed if purchase another ArF tool, so process modification is the most important. At the same time, the FAB introduce another KrF tool (ASML850) with the biggest NA is 0.8, whose critical dimension limit is 0.11μm theoretically, (the biggest NA of quondam ASML750 KrF tool is 0.7, and the critical dimension limit of pitch is 0.13μm), so it is possible to change the ArF layer to KrF type tools. If change one ArF layer to KrF tool comes true, the throughput of DRAM could improve to 24000 pieces of wafers with no new ArF tool introduced. And the cost must be saved for KrF process is cheaper than ArF process.

2. EXPERIMENTS for KrF process condition

M0 layer was selected to change to KrF process from ArF.
The process condition of M0 layer in ArF process as follow:
Tool type: ASML 1100 (ArF)
Develop condition: LD nozzle, 70mm/s scan speed, 30s puddle
Reticle: ArF PSM
Exposure condition (illumination): NA = 0.7, Sigma = 0.85 / 0.55
Resist: GAR8105, Thickness: 2850A
Exposure energy: 25mj/cm2; Focus: -0.1μm

If change to KrF process, almost all process conditions has large difference expect develop condition. The most KrF condition was established by the following experiment.

2.1 Exposure condition

The resolution can be increased by Off-axis exposure , which had been used in ArF process. Off-axis exposure imaged by only +1 or -1 rank lamp-house super add 0 rank lamp-house in optical diffractive. Generally speaking, most of 0 rank lamp-house are caught by lens, but most of +1 and -1 rank lamp-house can not be caught by lens when exposure image is small enough. So high quality of imaging will depend on the quantity of +1 and -1 rank lamp-house pass lens.

A type of software named pupil which is provided by ASML was used in the experiment, it can calculate how much +1 lamp-house was caught by lens and used for exposure. The input interface of pupil as follow (Fig1):

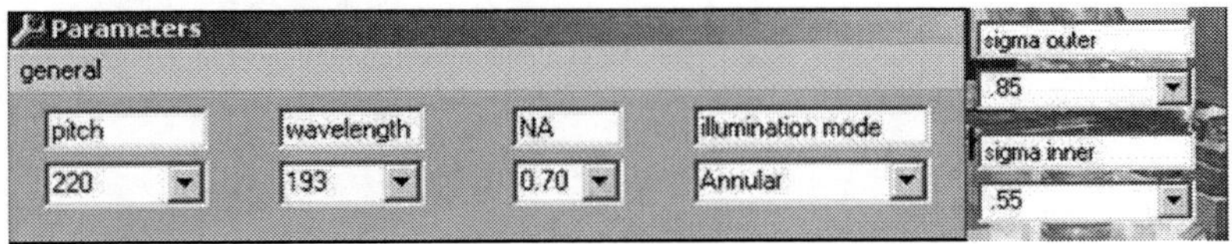

Fig 1.Input interface of Pupil (software of facula simulation)

Key the exposure condition of ArF tool in pupil:

NA=0.7, Sigma out/in=0.85/0.55, pitch (Line + Space) of M0=0.22μm. The result showed that there is 58% of 1 rank lamp-house is used for imaging, showed in Fig 2.

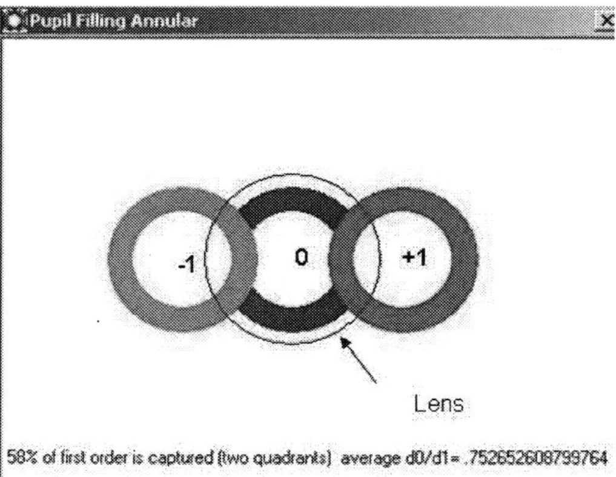

Fig 2. The facula of Off-axis exposure forM0 layer during ArF exposure condition

After modify the condition of Fig 1, change wavelength from 193nm to 248nm, new result showed in Fig 3, only 22% of 1 rank lamp-house is used for imaging, and it is not enough obviously. So exposure condition must be modified when M0 layer change to KrF process from ArF process.

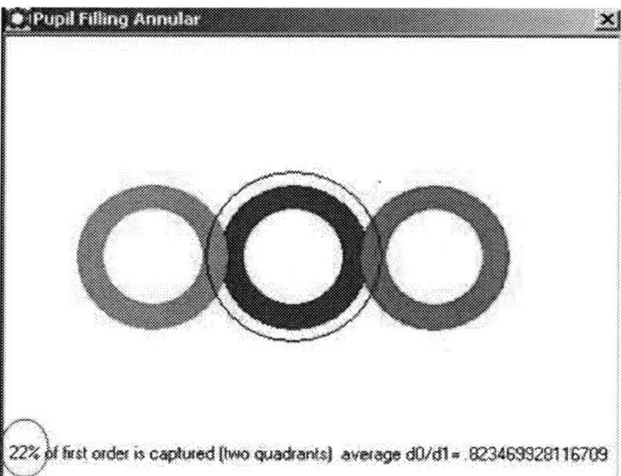

Fig 3. The facula of Off-axis exposure for M0 layer during KrF: NA=0.7

After modified NA to 0.8 from 0.7, calculation result showed that 46% 1 rank lamp-house is used for imaging, showed in Fig 4. The result is acceptable, though it is lower than ArF. The image should be high quality if the other condition of resist and mask are optimized.

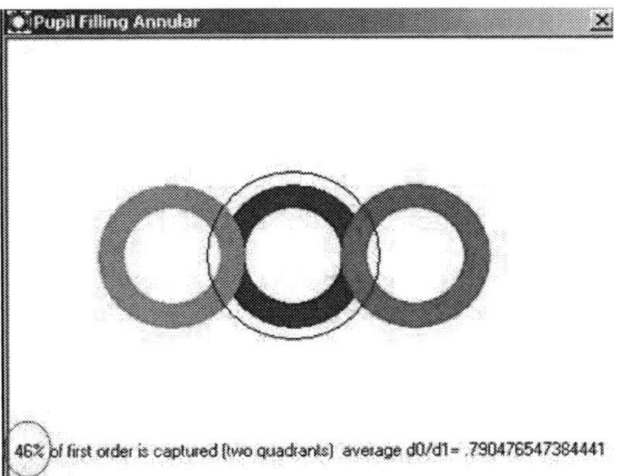

Fig 4. The facula of Off-axis exposure forM0 layer during KrF: NA=0.8

From the experiment result, the final exposure condition as follow: NA=0.8, other settings are same as before, Sigma out/in=0.85/0.55

2.2 Phase Shift Mask (PSM)

It is confirmed that ArF layers of 0.11μm DRAM must use PSM reticle, then if PSM still needed after change to KrF process? Base on the basic principle of PSM, 180 degree phase shift was determined by the thickness of Cr and wave length of exposure lamp-house. The exposure wave length changed to 248nm from 193nm after change to KrF process. So the thickness of Cr must be different when 180 degree phase shift occurred with 248nm lamp-house. Therefore, new KrF reticle is prerequisite.

There is another reason of producing new reticle that is OPC (Optical Proximity Correction) of reticle. The diffraction has big difference between 193nm and 248nm exposure wave length, then the primary OPC of ArF reticle must unsuitable to KrF. The OPC of KrF need be collected and then produce new KrF reticle.

Above all, a new KrF reticle for test was produced and then resist could be tested.

2.3 Resist

As we known before, resist was designed for appointed exposure lamp-house. GAR-8105 which is the resist for M0 layer is provided by FUJIFILM, it is for ArF tool type, namely, it must be photo under 193nm wave of length. A type of resist named M230Y which is produced by JSR was selected base on the information from resist vendor and the process of 0.11μm DRAM.

Incidence light would reflect when pass resist which is same as the theory of lens, showed in Fig 5.

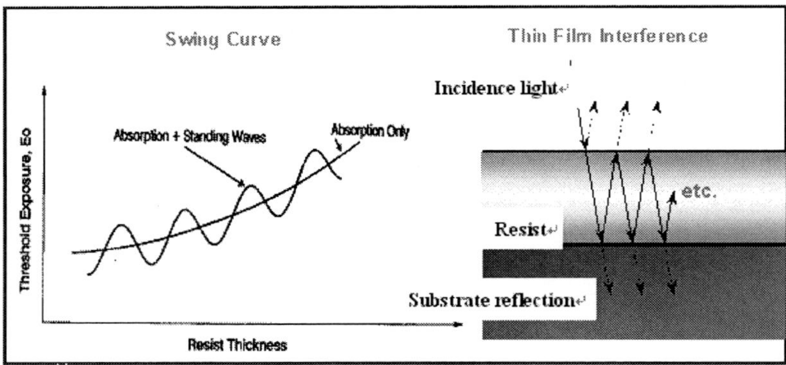

Fig 5. The absorbability and reflecting ability of beam of light pass resist

Because the light wave is sinusoid, if the thickness of resist is $\lambda/4n$ (λ is the wave length of incidence light), the reflected light is the least, then the absorbed light is the most; if the thickness of resist is $\lambda/2n$, the absorbed light in resist reach to the least. So the best thickness of resist should be $\lambda/2n$ or $\lambda/4n$, and the difference of absorbed light is the least when the thickness has a little shift. Then the uniformity of critical dimension could be well controlled. Showed in Fig 6, the thickness is same when thickness change from a to b and change from b to c, but the intensity of light has big different.

During actual practice, PR is a mixture of several chemicals, which is coated on wafers during spinning process, then is baked. Wave length of the incident lights can not predicted accurately before exposure, and it can only be estimated generally. So the thickness will be measured by another ways: Firstly measure the line width of several thickness continuously, then work out the sinusoid with scatter diagram of thickness and line width, finally defined the thickness according to actual needs.

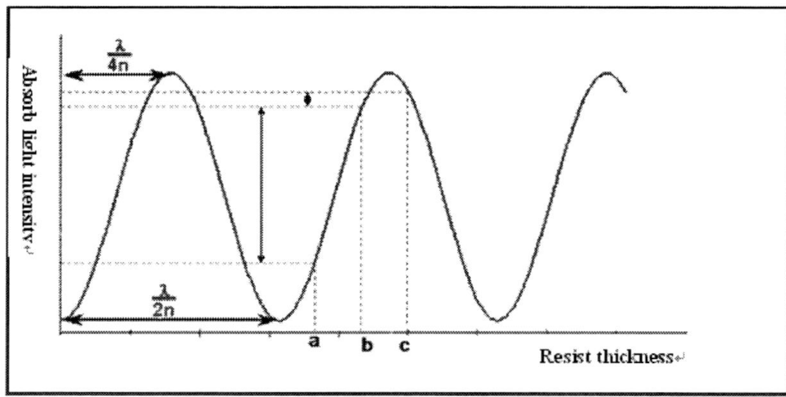

Fig 6. the influence of thickness of resist impacted to intensity of light.

For the original ArF process GAR8105 used 2850A (1A=0.1nm), KrF   process also need the same PR thickness 2850A or so. The thickness has been defined with 14 steps, each step is 50 A, and the range is form 2500A to 3200A. Then the spin speed is confirmed by formula. PR M230Y was coated on 14 dummy wafers, then the actual PR thickness was measured, as shown in Figure 4.1. Then the 14 wafers will be process with the Exposure condition, reticle and developer process which has described in the two previous chapters. Then CD was measured by SEM. There is two kinds of area to measure CD : Array Space), and ISO Line, shown in Table1: Column 5 and 6.

Table 1. PR Thickness vs CD (Critical Dimensions)

	Aim THK (A)	Coater RPM	Actural THK (A)	CD	
				Array space	ISO line
1	2550	5015	2564	0.1092	0.2361
2	2600	4825	2618	0.1063	0.2355
3	2650	1644	2659	0.0958	0.2402
4	2700	4474	2712	0.0879	0.2434
5	2750	4313	2754	0.0958	0.2351
6	2800	4150	2812	0.1155	0.2267
7	2850	4015	2851	0.1163	0.2258
8	2900	3878	2906	0.1194	0.2281
9	2950	3748	2951	0.1198	0.2306
10	3000	3624	3005	0.1215	0.2321
11	3050	3505	3049	0.1206	0.2332
12	3100	3393	3105	0.1188	0.2367
13	3150	3286	3149	0.1162	0.2394
14	3200	3185	3203	0.1108	0.2417

Figure 7 showed Scatter diagram of PR thickness and CD and draw out the sinusoid. With the thickness related to smaller CD of array space area, the thickness corresponding to **λ/2n** and **λ/4n** is 2700A and 3000A separately.

For the ArF Process (GAR8105) used the thickness 2850A , 3000A is selected finally to prevent over etching.

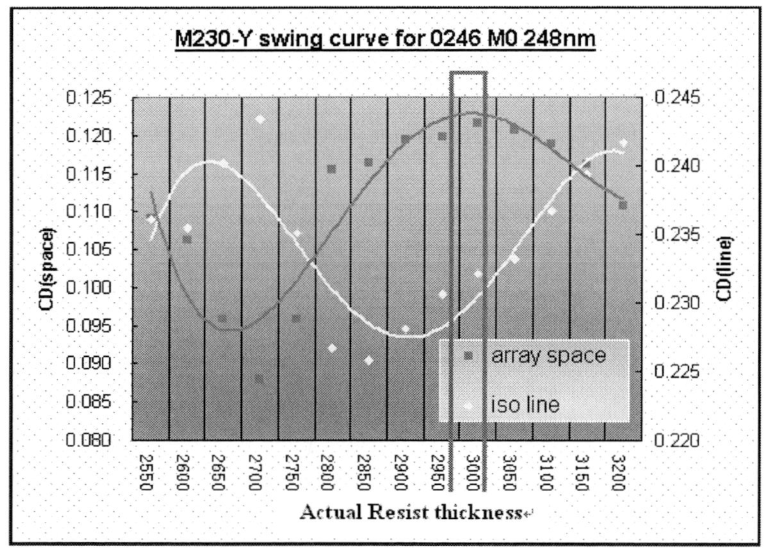

Figure 7. PR Thickness of M0 vs CD

So far, it is confirmed that PR is M230Y, thickness is 3000A.

2.4 Reticle OPC

After PR related condition are confirmed , it is the time to collect OPC data.

Figure 8 showed a SEM picture of a small area on these wafers with M0 patterns after Litho process. There are many patterns with different sizes and the pattern density also is deferent. In a rea A is line, the pattern density high and the distribution of Line and Space is 1:1. But in area B, the data is on the contrary, and in area C the data is among A&B.

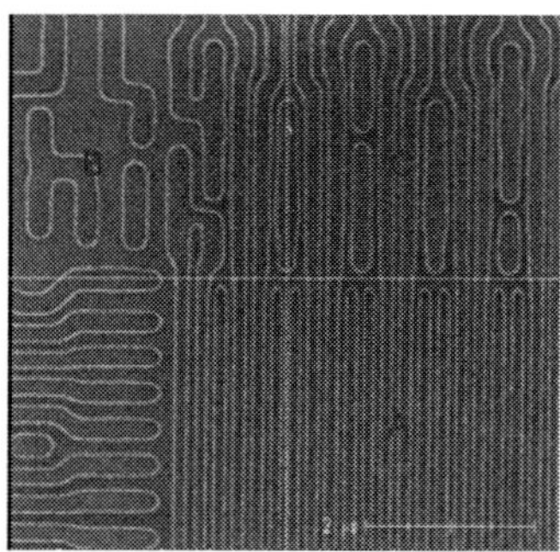

Figure 8. SEM Picture of a area M0 Layer

The complexity of these patterns, which is not unified in size, and distributed randomly, will induce the path of the exposure lights. The result is CD on wafers will be different from design. So the OPC is needed. Firstly, reticle and exposure condition and Photo Resist need to be tested. After Exposure and develop, process,

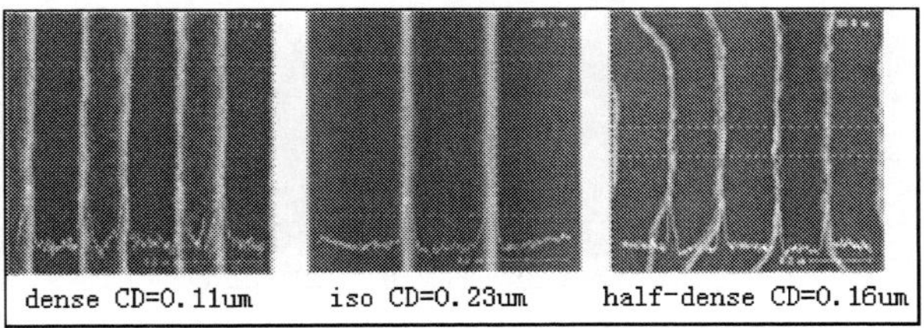

Fig 9. CD of Examples Using OPC

34 points were selected which were high repetitive, the measured CD in these areas with SEM, shown in Figure 9. Figure 10 showed the measurement results of OPC, in which we can find the margins between actual values and designs.

Then feedback these margin to the testing reticle, then the reticle correction will be gained, then formal reticle of production can be made.

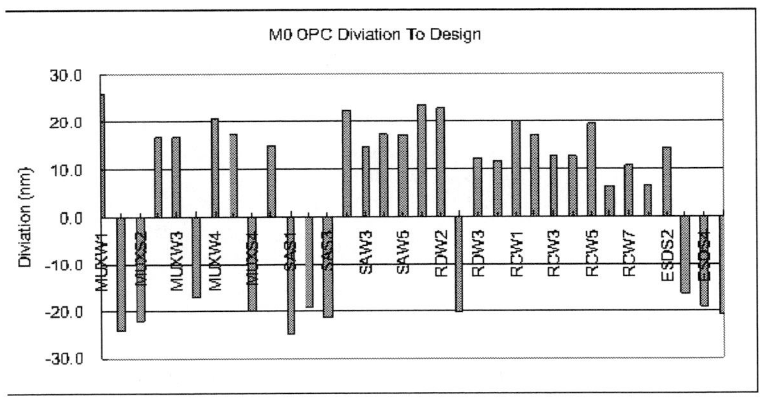

Figure 10. Measurement of OPC : Actual values' deviation to Designs

2.5 KrF Exposure and Focus

With KrF production reticle , best energy and focus would be confirmed. Usually

Matrix (FEM)is used , that is exposing the wafer by dies with different times, energy, and focus. Then measured the CD values of each dies with SEM. Table 2 is the testing results.

Table 2. KrF Energy and Focus Matrix

Focus \ Energy	32.5	33.5	34.5	35.5	36.5	37.5	38.5	39.5
-0.2			0.0971	0.1045	0.1054	0.1164		
-0.15		0.0968	0.0956	0.0987	0.1057	0.1101	0.1192	
-0.1	0.0876	0.0914	0.0964	0.1016	0.1062	0.1092	0.1159	0.1196
-0.05	0.0841	0.093	0.0956	0.1044	0.1073	0.1086	0.1145	0.1176
0	0.0876	0.0891	0.0948	0.1007	0.1049	0.1081	0.1169	0.1219
0.05		0.0903	0.0934	0.0996	0.1056	0.1115	0.1167	
0.1			0.0988	0.1031	0.1109	0.1167		

M0 layer CD is 0.107+/-0.06μm,   that is target is 0.107μm , upper limit is 0.113μm, lower limit is 0.101μm. The Data with yellow in Table 2 meet the requirements. The collusion is the best exposure energy is 36.5mj/cm2.

All these data are shown in Figure 11. By conic combination , it is drawn out that the best focus is -0.05μm。

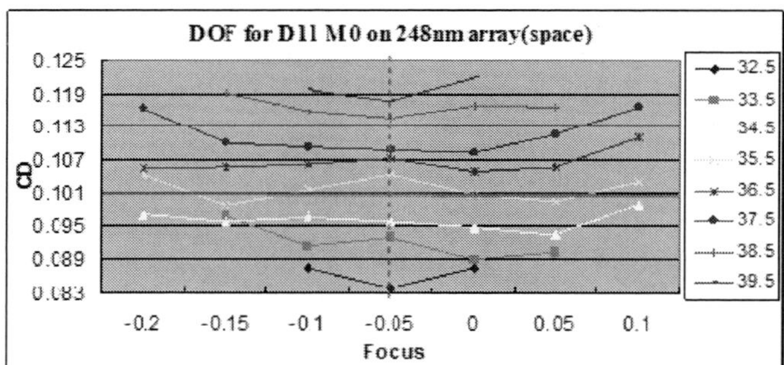

Figure 11. M0 Energy and Focus Matrix

Certainly the slicing up is done, the cross-section images of different specimen after developer under different focus condition. Figure 12 is the SEM pictures of cross-section. It is obvious that when Focus is -0.29, the image is a trapezium which top side is lower than the bottom (It is called as footing in Lithography) . But when focus is 0.19, the section is on the contrary, which top side is larger than bottom (It is called as T-top in Lithography) .

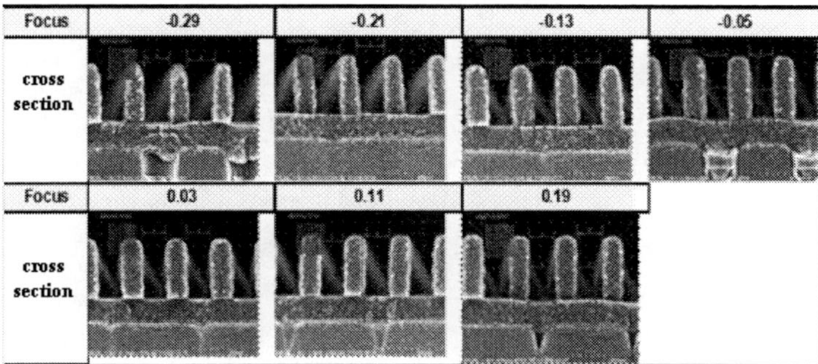

Figure 12, SEM Picture of PR after Developer

When the focus is -0.05, the cross-section image is good, it is very sharp. That strongly proved that -0.05is the best focus that KrF should select.

2.6 Summary of KrF Process's Condition

Summarized the experiments results of the previous sections ,   M0 Layer related KrF Process Condition as below:

Machine : ASML 850 (KrF)

Developer: LD nozzle, 70mm/s scan speed, 30s puddle
Reticle: KrF PSM

Illumination : NA = 0.8, Sigma = 0.85 / 0.55

PHOT Resist : M230Y, Thickness: 3000A

Exposure Energy : 36.5mj/cm2   Focus : -0.05μm

2.7 Summary

Through serials experiments and data collection , it approved that the possibility

of 0.11μm DRAM M0 layer expansion on KrF platform , and also found the related

best condition of KrF process, which provided the necessary basement and preparation of testing for products by KrF process..

3. KrF Process Application and Yield Improvement

3.1 KrF process qualification

With the confirmation of KrF Process condition, the trial-products can be started. According to the rules of foundry, any process change must follow the strict procedure, mass production can not start until trial-products related testing passed, in order to prevent risks. The production testing includes 2 steps , first is trail-products in miniature(STR, special test require), then is the cosmic production in a bit scale (MSTR:Mass Special Test Require).

25 pieces of wafers are placed in a sealed box (SMIF Pod) in FAB, which is called as Lot。 In miniature production, 3 lots were used, half of which is processed with KrF and the other were processed with ArF. At M0 layer stage, after comparing the CD post Litho and etch separately, checking overlay, and defects inspection, these 3 lots completed flow process. Finally the WAT and yield were tested. The results showed that yield and WAT parameter form KrF compared with ArF. Figure 13 showed the WAT comparison between these 3 Lots and STD (STD presents ArF

Process) .

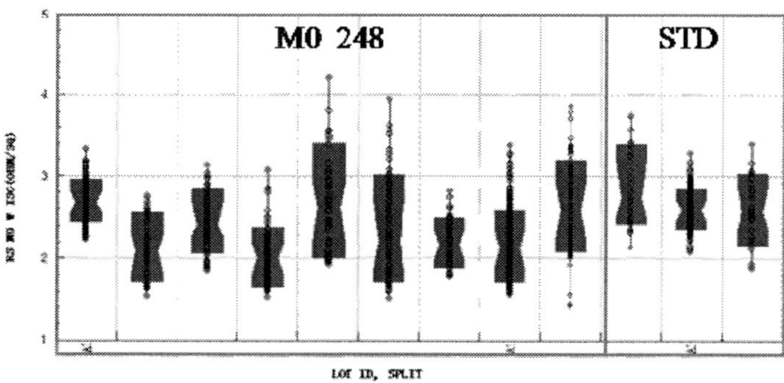

Figure 13. Trial-Products in miniature WAT Comparison

After WAT and yield testing, 3 lots had been sent to do reliability test under high temperatures within one month. The results showed that KrF and ArF are comparable.

Then experiment enters the 2nd phases, the wafers were produced cosmically in a bit scale, 16 lots were totally started. Figure 14 is the yield comparison of 16 lots with ArF.

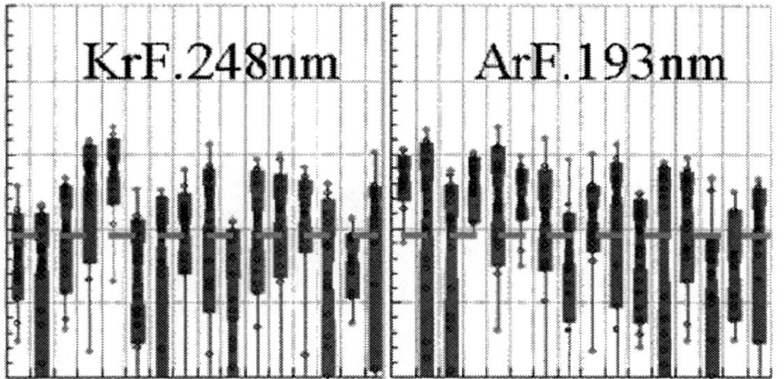

Figure 14. Yield of mass testing production

It is concluded that the yield of KrF process is comparable to ArF process.

3.2 Yield improvement on mass production

There is many problems during KrF mass production, which impacted the yield. But these issues were solved in different ways. Some examples will shown as below.

3.2.1 Line broken

Figure 15 and Figure 16 showed that some M0 lines were easy to broken due to the small size, which will decreased the yield.

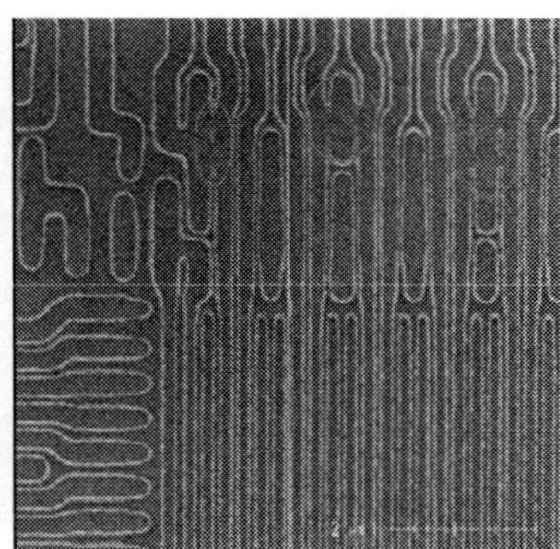

Figure 15. M0 layer's line marginally broken

Figure 16. M0 layer's broken line

According to this issue, we reevaluated the Energy and Focus matrix of M0 KrF Process. Scanned the related wafers and scanned with optical defects inspection tools, the related defects maps as shown below in Figure 17.

It was found that the best Focus is still -0.05, but when the energy 36.5mj or highethe defects density will increase sharply. Lower the energy is to lower the defects quantity. The final yield map of this wafer shown in Figure 18. At the same time, when energy is 36.5 mj or above, the yield is lower. When it adjusted to 35.5mj, the yield window is also larger.

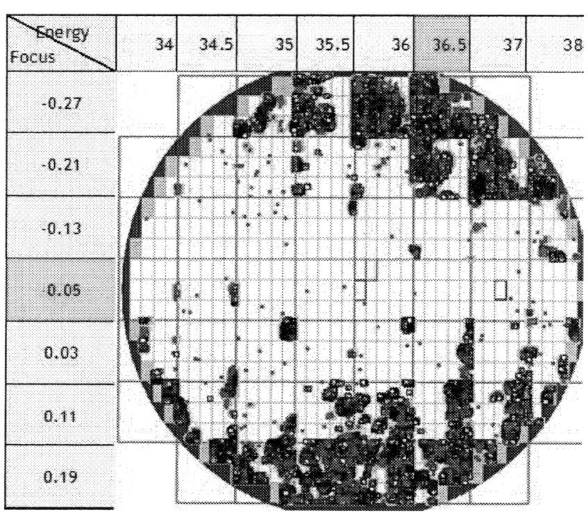

Figure 17. M0 KrF Matrix (FEM) Defects map

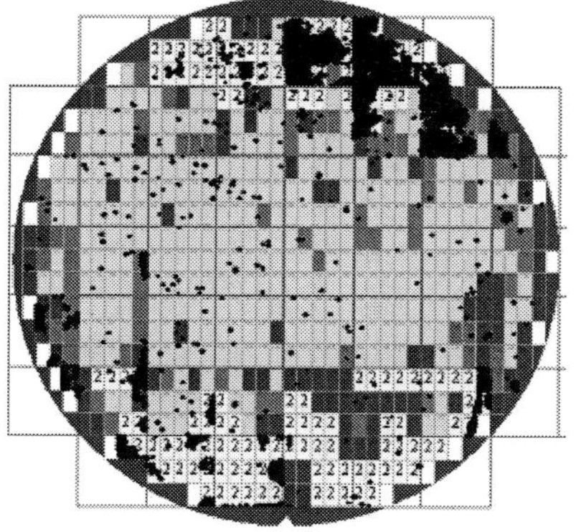

Figure 18. M0   KrF Matrix FEM) vs wafer yield

Decreasing the exposure energy is the vital to gian more safe windows and to prevent line broken. As we know, energy and CD corresponds to each other. To decrease the energy, it means to change CD target of M0 KrF. As for current CD target is 0.107μm, two lower CD target is selected: 0.102μm 和 0.098μm. Then experiments were performed the related yield showed in figure 19.

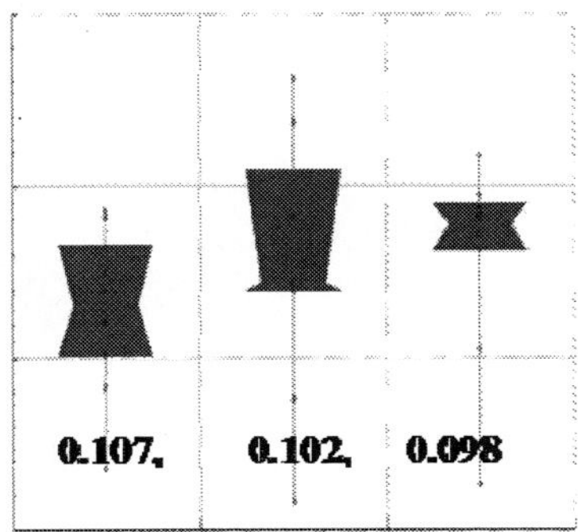

Figure 19. M0 KrF CD target vs yield

Figure 19 shoes that when target is 0.102μm, the yield is the highest. So the CD target of M0 KrF process is 102μm , and the problem is solved.

3.2.2 Defocus on wafer edges

The defects maps shown in Figure 20 were often found at defects scanning stages in production lines, and the defects density is extremely high on the left and right edges of wafers. Under SEM observation , all these defects were induced by defocus.

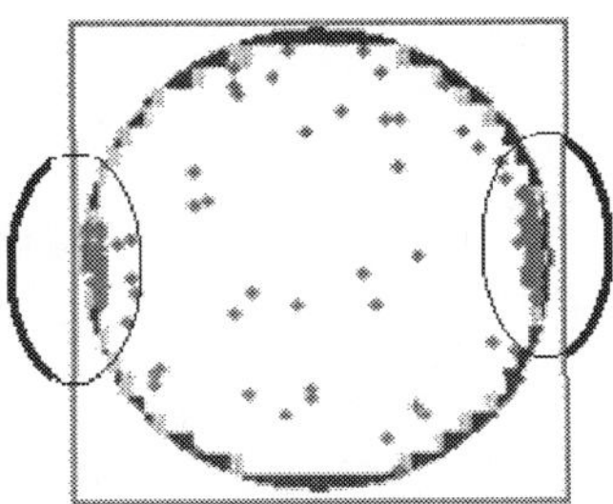

Figure20.DefectsMap

Figure21. The fatness of whole wafer

The flatness of wafers has been measured before exposure, and the flatness map in three dimensions is shown in Figure 21. The map shows that the thickness of wafer edge is much thicker than inner area, which is mainly induced by front-end processes such as : Film deposition , Etching , CMP and so on.

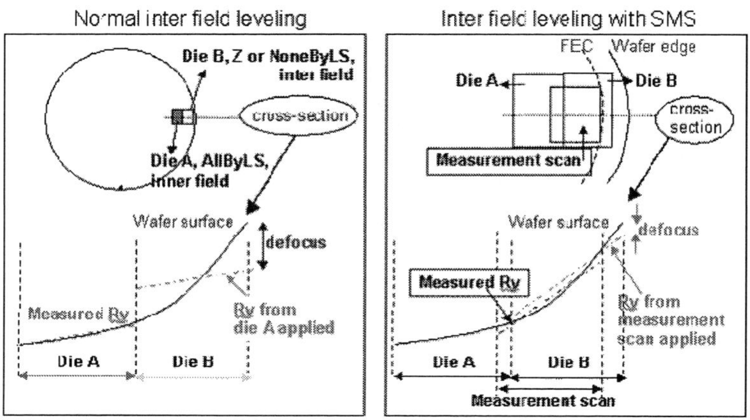

Figure 22. Normal Focus Sketch Map

Such worse flatness make great challenge to Scanner to find the best focal plan. Figure 22 is the Normal Focus Sketch Map. Die B is closest to wafer edge , the uniformity of flatness is worst. Die A is beside Die B, which is in the inner area. Under normal conditions, right part of Die B can not find the best focus plan due to too closer to wafer edges. So the plan only can be determined by the data from left parts. In the end, the focal plan will be tilt. (Ry is margin of focal plane height between left and right sides, being divvied by the width of Die) . If use the slope of Die A, which is not seriously fit to Die B , then the right parts of Die B will defocus.

To solve it, the SMS (Shift Measurement scan) function from ASML has been involved.   Although the right part of Die B is too close to the edge, there is no ways to measure the focal plan. The measured focal plan was shifted to left direction with some distance in order to avoid the issue area in Die B. With the information for Die A's right area and the majority of Die B's left areas, the slope of the new area can be confirmed, which is fit to Die B obviously. As the results, the defocus in Die B's right area was strongly decreased.

As shown in Figure 23, after comparison, the yield of dies on wafers' left or right edges has been strongly improved through turning on SMS function.(The thicker the color is, the higher the yield is).

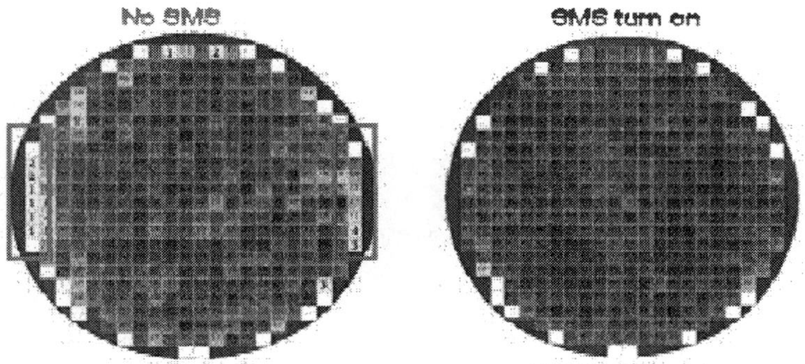

Figure 23. Yield Enhancement vs SMS (Shift Measurement Scan)

3.3 Summary

After solved above problems, the yield of 0.11µm DRAM, which process on KrF machines, has been improved obviously. As shown in Figure 24, the yield has increased 5% after defects density decreased, which has passed the yield from ArF..

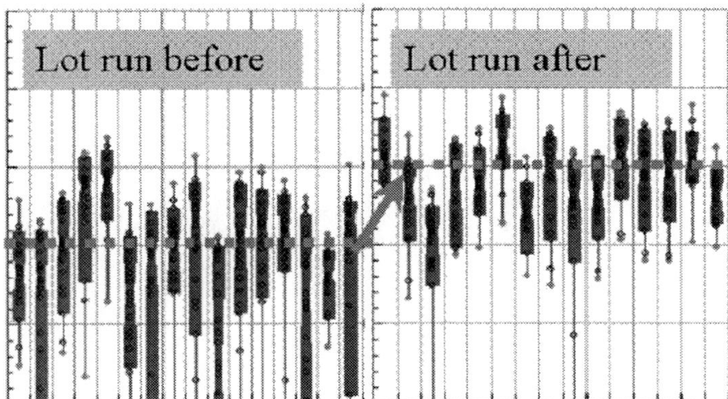

Figure 24. Yield vs Defects Density Reduction

Finally, KrF process replaced ArF process completely. It provides a fundamental base for 0.11µm Process expansion on KrF machines. That is to say, 0.11µm process can be initialed on KrF platform with high priority.

4. CONCLUSIONS

As mentioned above, we successfully transfer 0.11µm M0 layer process from ArF to KrF machines. The production capacity of foundry has been improved, the cost has been improved obviously, and the related yield also has been improved. Furthermore, this study has provided the platform for other 0.11µm processes' research.

ACKNOWLEDGMENTS

Authors would like to thank our supervisors Spencer Cai, Division head of S1-Litho, Eddy Lau, Department head of BJ-Litho for their encouragement and guidance. In additional, authors would give special thanks to Grace Gu from ASML China, and Richyard Xie from JSR SH for their technical support.

REFERENCES

[1] Stephen A.Campbell The Science and Engineering of Microelectronic Fabrication[M], Publishing House of Electronics Industry;   2003 Chapter 7,Chapter 8
[2] Michael Quirk, Julian Serda .Semiconductor Manufacturing Technology[M] Publishing House of Electronics Industry;   2004   Chapter13,Chapter14, Chapter15
[3] Takaharu Miura.Electron projection lithography tool development status[M].J.Vac.Sci.

Technol. 2002, B 20(6): 2622 - 2633

ECS Transactions, 34 (1) 303-307 (2011)
10.1149/1.3567595 ©The Electrochemical Society

Studying photoresist type for sub-32nm node dense SRAM 2nd GT layer

Yao Xu, Jingan Hao, Chang Liu, Xuelong Shi, Qiang Wu, and Yiming Gu

Technology R&D, Semiconductor Manufacturing International Corp.
Pudong New Area, Shanghai, P. R. China 201203

Abstract

With continuous shrinking of device dimensions, high performance, dense SRAM cells design has become more challenging than before, patterning quality is also more sensitive to fluctuations in lithography process. This inevitably calls for more aggressive lithography process conditions in terms of the required resolution. The tradeoff between process window optimization for random logic gates and dense SRAM is not always straightforward, and it sometimes necessitates design rule and layout modifications. In particular, patterning the small tip-to-tip gap for dense SRAM cells at gate level becomes extremely challenging. By delinking patterning of SRAM tip-to-tip from other geometries, one can optimize the patterning processes independently at the expense of cost. This can be achieved through a special double patterning technique that employs a combination of double exposure and double etch (DE2).

In this paper, we will show how a DE2 patterning process can be employed to pattern dense SRAM cells in the 32nm node. Besides illumination optimization, the selection of an appropriate photoresist and the development of good resist process conditions are found to play a key role to further improve the photolithographic process. The performance data, including process window, pattern profile accuracy, etc., from two different type photoresist samples are studied for SRAM 2nd GT layer lithography process. For comparison purpose, we also present a single exposure single etch result for such dense SRAM cells. In the 45nm node, the dense SRAM cell can also be printed to adequate tolerances and process window under single exposure (SE) with OPC. We present our data on DE2, which indicate that it can be used as an alternative solution to pattern dense SRAM with good process extendibility.

1. INTRODUCTION

While stepping into age of sub 32nm tech with the current mature immersion tool after decades of development in lithography, single exposure can barely meet the challenges in optical lithography of the space between two polysilicon line-ends in the special dense pattern area of SRAM. In order to ensure there is adequate coverage over the active area by poly lines to minimize electrical leakage, tip-to-tip gap patterns always determine the criterion of minimum area because of significantly line-end shortening of polysilicon lines. In order to give more weighting in the process window to the dense SRAM area, tradeoff has to be made for the more isolated features, such as random logic patterns.

This tradeoff is well-known for almost every technology node and it is more difficult to handle in 32nm node and beyond, which is due to restriction of SRAF (Sub Resolution Assist Features) size and limited optical resolution, the total process window budget is significantly smaller for logic devices. This leads to a greater drive on gate layer SRAM patterning technology which brings the double patterning to reality in advance.

2. RESULTS AND DISCUSSIONS

2.1. Comparison between single patterning and double patterning

Continuous logic device dimension shrinkage and dramatical line-end shortening of polysilicon lines after etch have eroded away conventional single exposure advantage (being fast with good overlay). It is clearly shown in Fig.1 that resolution limitation has caused line-end corner rounding at litho stage even after employing immersion exposure tech and obvious line-end shortening after etch.

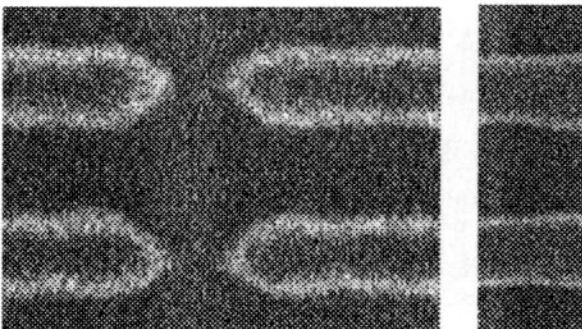

Fig.1: Traditional single exposure SRAM pattern after litho and etch.

High quality ultra-dense SRAM devices could be printed by a specific double patterning technique which separates SRAM area from other logic area. It combines double exposure and double etch (DEDE, DE2) while traditional SRAM patterns are divided to dense vertical lines and contact like trim slot dense patterns (Fig.2). Two major advantages have been brought up: 1) tighter SRAM pitch in AA with extremely small gate tip-to-tip variation, 2) corner rounding reduction of gate line ends. And the better pattern control brought by 1) and 2) will provide excellent coverage of gate over active area.

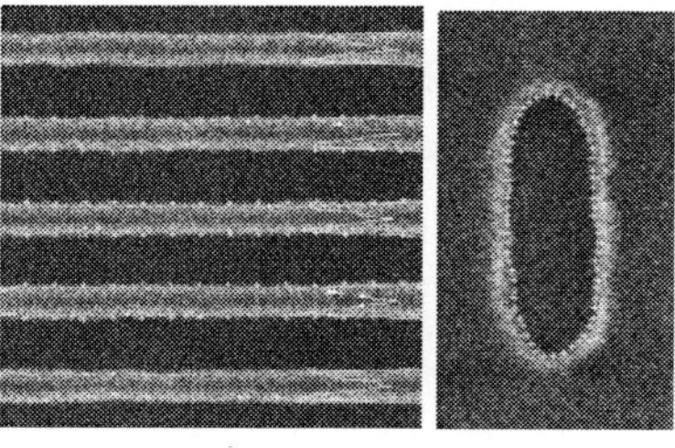

a   b

Fig.2: Traditional SRAM pattern is divided to two parts: a) 100 nm pitch dense lines, b) contact like trim slot pattern

After DE2 process (Fig.3), though patterning profile is still not perfect it is obvious that this technique could significantly improve line end profile in preventing corner rounding compared to the single exposure SRAM patterning of gate layer. The minimum poly tip to tip space can be achieved 70% of single exposure method.

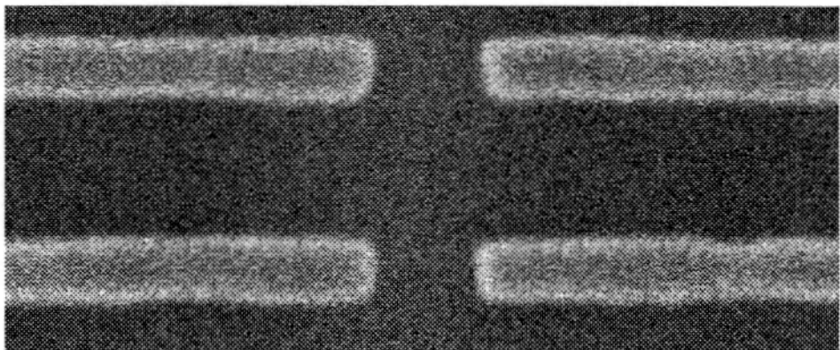

Fig.3: SRAM dense pattern line end after DE2

In a more general case, DE2 could also enable doubling of pitch frequency and it allows insertion of printed assistant feature to improve other logic pattern process window especially when the insertion SRAF is not possible due to space restrictions and possible assist printing. Although double patterning has the biggest drawback of cost addition which bring an additional litho and etch process, it is always a tradeoff between single exposure single etch (SE) solution and Double Exposure Double Etch (DE2) in terms of cost and device performance.

2.2. Comparison of different photoresist type

While the exposure conditions including illumination mode and exposure tool are limited, the quality of an image highly depends on photoresist. In general, there exist two types of photoresit divided by patterning feature—L/S and CH. One is to define line or space patterns and keep the pattern LWR or LER as small as possible. The other one is to pattern contact hole and make hole's bottom as clear as possible which commonly means higher sensitivity. This predicts that these two types photoresist active mechanism should be different.

While focusing on this trim slot layer, it should meets at least two requirements: 1) The pattern should be as strictly vertical as possible for two long edges and small end rounding for short edges; 2) Enough PW to keep proper X/Y ratio which dominate the final pattern quality after etch.

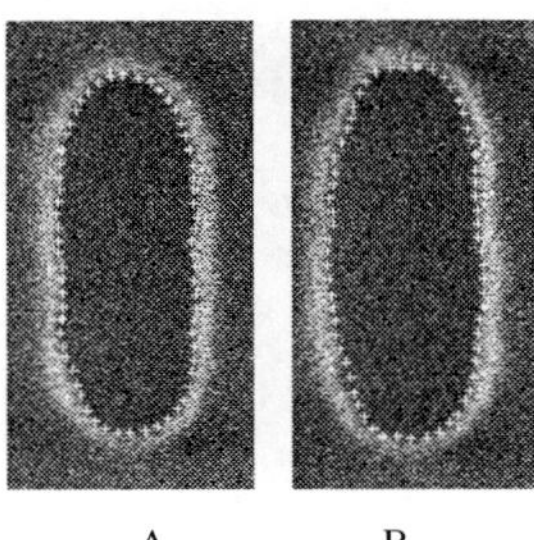

A   B

Fig.4: Trim slot pattern profile with different photoresist type

From the topview check, L/S photoresist A sample could achieve better profile while CH photoresist B sample shows relatively worse defining capability for such rectangular contact like pattern.

2.3. Litho process window

To ensure best image quality for this trim slot layer, high NA/SIGMA annular illumination mode with XY polarization was applied on ASML 1900i immersion tool after NA/SIGMA optimization with different illumination mode such as quasar, annular. But sample A L/S resist still cannot achieve satisfactory process window, while sample B gives a compromised performance with better PW and acceptable pattern profile.

Because A PW is marginal for manufacturing, B has been chosen as our baseline which needs future pattern profile improvement.

Sample	Description	DoF (in units of machine limit)
A	L/S	< 80%
B	CH	> 100%

Table I: process window in DOF for resists A and B. (100% is required by our production.)

3. SUMMARY AND CONCLUSIONS

The results of double patterning technique for logic device SRAM pattern on test wafers are presented for sub 32nm node gate layer. Such a technique can significantly decrease the space between two polysilicon line-end compared to single expose method in dense SRAM cells. DE2 ensures adequate coverage of polysilicon line ends over active areas due to absence of line-end shortening and corner rounding. We have discussed the difference and performance between L/S and CH type photoresists for this kind of trim slot layer.

4. ACKNOWLEDGEMENTS

The authors would like to thank SMIC SH FAB for metrology tool time support and all SMIC R&D litho and etch group colleagues.

REFERENCES

1. S. R. J. Brueck, S. H. Zaidi, Method and apparatus for extending spatial frequencies in photolithography images, U.S patent 6042998
2. I. Pollentier, P. Jaenen, C. Baerts, K. Ronse, "Sub 50nm gate patterning using line-trimming with 248 or 193 nm lithography," Future Fab Int. 12, 161, 2002.
3. C. Sarma et al., "Finding the right way: DFM vs. area efficiency for 65nm gate layer lithography", Proc. SPIE 6154, 61541L-1, 2006.
4. H. Zhuang, H. Wang, C. Yap, A. Gutmann, J. Lian, C. Sarma, L. Tsou, A. Gabor, U. Schroeder, S. Halle, K. Herold, H. Haffner, H. Lee, N. Rovedo, C. Chang, H. Ng, D. Shum, R. Wise, M. Hierlemann, and M. Proc. of SPIE Vol. 6924 692429-5 Ieong: "Patterning strategies for gate level tip-tip distance reduction in SRAM cell for 45nm and beyond", Proc. ISTC 2007, 154-159, 2007.
5. J. Meiring, H. Haffner, C. Fonseca, S. Halle, and S. Mansfield: "ACLV Driven Double-Patterning Decomposition With Extensively Added Printing Assist Features (PrAFs)",

Proc. SPIE 6520, 65201U, 2007.

6. H. Haffner, J. Meiring, Z. Baum, S. Halle: "Paving the way to a full chip gate level double patterning application", Proc. SPIE 6730, 67302C, 2007.

CHAPTER 3

DRY AND WET ETCH AND CLEANING

ECS Transactions, 34 (1) 311-318 (2011)
10.1149/1.3567596 ©The Electrochemical Society

Selective Removal of High-k Dielectrics

D. Shamiryan, and V. Paraschiv

IMEC, Kapeldreef 75, Leuven, 3001, Belgium

Continuous downscaling of integrated circuits brought an end to the era of SiO_2. In gate dielectrics, it is being replaced by materials with high dielectric constant, so-called high-k dielectrics. One of the challenges in the integration of the high-k material is removal of those materials selectively over the substrate. This work is one of the first attempts to review current state of the art of the high-k removal. Two main approaches are discussed: dry (plasma) removal and wet removal. The best results could be achieved by combination of both.

Introduction

As dimensions of semiconductor devices continue to shrink, conventional materials reach their limits in scalability. A good example is gate dielectric, where SiO_2 cannot be thinned down below 1 nm due to issues with high leakage current and dopant diffusion from the gate to the channel. In order to continue downscaling, new materials, so called "high-k" dielectrics were introduced in the gate stack. The idea behind the high-k dielectrics is the following: having higher permittivity they can provide the same high capacitance as thin SiO_2, but at larger thickness, thus preventing leakage and diffusion.

Introduction of new materials, however, brought technological challenges, e.g. during gate patterning where high-k dielectrics must be removed after the gate patterning. The main requirements for high-k removal are complete removal, high selectivity over the substrate and the gate stack and acceptable etch rate. Recent advances on high-k removal are summarized in a review.[1]

Approaches to removal of high-k dielectrics

To remove high-k dielectrics, one can use two main approaches: wet removal and dry (plasma) removal. Every approach has advantages and disadvantages and the best results could be achieved by combination of both.

Dry removal is anisotropic (in other words, can be done without lateral damage to the high-k dielectric); cost efficient, as it can be done just as an additional step in the gate etch recipe; it is also not sensitive to a crystalline state of the high-k dielectric (high-k is often crystallized upon anneal). The main drawback of the dry removal is requirement of volatility of etch products. Many high-k materials do not form volatile etch products and it is almost impossible to remove them dry without using elevated temperatures. Another disadvantage of the dry removal approach is that it is might be aggressive to the substrate.

Wet removal could be made very selective to the substrate, moreover, the range of acceptable etch products is wider since they don't need to be volatile, just soluble. On the other hand, the wet removal is isotropic by nature, so the lateral etching of high-k under the gate is unavoidable. Another disadvantage of wet removal is dependence of etch rate on the crystalline of the high-k. Very often crystallized (after anneal) high-k dielectric is virtually impossible to remove by wet chemistry.

Combination of both dry and wet removal seems to be the most promising method of high-k removal. First, dry removal etches most of the high-k, providing anisotropic profile and/or physical damage to the remaining high-k. If the dry etch stops before reaching the substrate, the selectivity is not a concern. Second wet step removes the remaining high-k with high selectivity over the substrate and since dry etch usually amorphize a crystalline high-k, the complete removal is possible. As a result, combination of dry and wet processes allows removal of almost all high-k dielectrics with high selectivity over the substrate and minimal lateral effects.

Dry removal

The first requirements for dry removal to be possible is volatility of the etch products. The volatility of an etch products is defined based on its saturated pressure at the temperature of the wafer during etching. The volatility limit is usually put at 0.1 mTorr, [2] but for high-k lower etch rates could be tolerated and we can set the volatility limit at 0.001 mTorr. Saturated pressure could be obtained directly from the JANAF tables.[3] If a compound is not listed in the JANAF tables, the saturated pressure could be calculated using Clausius-Clapeyron equation:

$$\ln\left(\frac{P_{atm}}{P_{etch}}\right) = \frac{\Delta H_{vap}}{R}\left(\frac{1}{T_{etch}} - \frac{1}{T_{atm}}\right)$$

[1]

where P is pressure, T is boiling points at different pressures (atmospheric and etch), R is universal gas constant and ΔH_{vap} is enthalpy of vaporization. A similar equation could be written for sublimating products with enthalpy of sublimation replacing melting point replacing boiling point. When enthalpy of vaporization or sublimation is known, [4] saturated pressures could be calculated directly, when they are not known, an estimation could be made based on the Trouton's rule[5] that enthalpy of vaporization is proportional to the boiling point. Similar observation could be made for enthalpy of sublimation and boiling temperature. Empirical dependencies could be expressed by the following relations:[1]

$$\Delta H_{vap} = 20.90 + 0.109 \cdot T_{atm} \tag{2}$$

$$\Delta H_{sub} = 32.28 + 0.194 \cdot T_{atm} \tag{3}$$

Estimation of volatility for etching at 60°C based on boiling and melting points is shown in Figure 1. This diagram could be used for quick estimation of possibility of plasma etching. E.g. $HfCl_4$ with a melting point of 432°C at atmospheric pressure and sublimation point of 317°C most likely will be volatile, while $LaCl_3$ with melting point of 859°C is not volatile at all. One can conclude that HfO_2 could be etched with Cl-based plasma, while La_2O_3 cannot be etched dry unless high-temperature cathode is used. The boiling and melting points of some etch products could be found in the Handbook of Chemistry and Physics

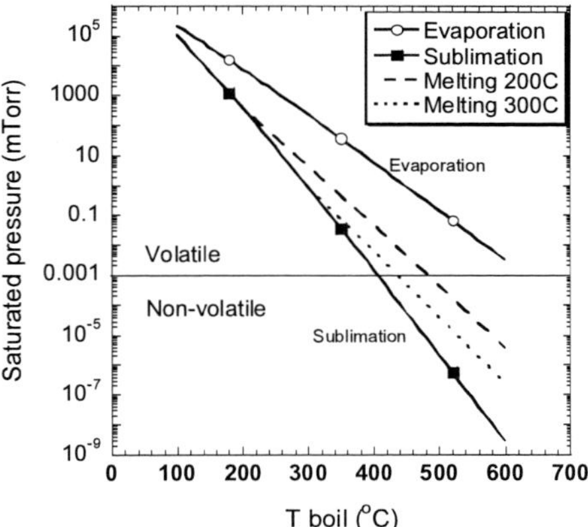

Figure 1. Saturated vapour pressure at 60°C as function of boiling temperature at atmospheric pressure.

Analyzing boiling points of etch products (typically halides), one can see that usually chlorides are the most volatile etch products for high-k dielectrics. Historically, first attempts to etch high-k were made by F-based plasma,[6,7] but it was quickly found that besides some high-k that form volatile fluorides, like TiO_2,[8] it is almost impossible to etch most of the high-k selectively – main etch mechanism is ion sputtering which does not provide selectivity over the substrate. Moreover, after etch a lot of non-volatile etch products (fluorides) were observed on the surface.[9,10] One can conclude that fluorine can break a metal-oxygen bond and a form a fluoride, but these fluorides are usually not volatile enough to be removed from the surface. Another concern for the fluorine-based chemistry is selectivity over the substrate. Si is known to be easily etched by fluorine-based chemistry and, therefore, selectivity is problematic.

Even though chlorides are the most volatile high-k etch products, it was found that Cl_2-based plasma is ineffective in etching high-k dielectrics. The etch rate was found to be quite low and comparable with the etch rate in fluorine-based plasma.[11,12] It should be noted that no chlorine was found on the high-k surface after exposure toCl_2-based plasma,[9,13] as well as no changes in binding energies of the metal elements of high-k dielectrics (e.g. Hf and Zr).[14,15] It can be concluded that, apparently, chlorine cannot break the metal-oxygen bond so the volatile chlorides are not formed. As in the case of fluorine-based plasma chemistry, selectivity over the substrate is an issue, even though Si etch rate in chlorine plasma is lower than that in fluorine based plasma, very low etch rate of high-k makes selective etching impossible.

Fortunately, there is a chlorine-based compound that allows selective plasma etching of high-k materials: boron tricholride (BCl_3). BCl_3-based plasma has several advantages for high-k etching. First of all, BCl_3 is able to react with metal oxides with formation of metal chlorides and $(BOCl)_3$., e.g. by the following reaction:

$$3HfO_2 + 6BCl_3 \rightarrow 3HfCl_4 + 2(BOCl)_3 \ (-62.7 \text{ kJ/mol}) \qquad [4]$$

The following mechanisms are believed to be responsible for etching: boron can reduce metal oxides by forming BO_x and/or $(BOCl)_3$ compounds, while reduced metal reacts with chlorine forming volatile halides. B-O bond (808.8 kJ/mol) is stronger than metal-oxygen bonds (typically 600-800 kJ/mol), $(BOCl)_3$ is a stable molecule with large negative heat of formation (-1630.2 kJ/mol). $B_xO_yCl_z$ compounds such as $(BOCl)_3$, B_2OCl_3, $B_3O_3Cl_2$, and B_2OCl_4 were observed by mass-spectrometry during etching of high-k with BCl_3/Cl_2 plasma,[16] optical emission lines of BO and BO_2 were observed in BCl_3/O_2 plasma mixture.[17]

Second advantage of the BCl_3-based plasma chemistry is its selectivity over Si substrate. Exposure of Si surface to BCl_3 plasma results in formation of a B-containing film [18] which passivates the substrate. Film deposition depends on plasma composition (e.g. presence of N_2 enhances the deposition leading to formation of a BN-like film, [19] while presence of oxygen may suppress film formation or lead to deposition of boron oxide at higher O_2 concentrations).[20,21] By tuning plasma parameters (chemistry, pressure, power) one can achieve rather selective removal of high-k, when the dielectric is etched but the Si substrate is passivated.

Wet removal

The main wet etch chemistry for high-k removal is HF-based. It has high selectivity over Si substrate, but selectivity over SiO_2 is an issue. Fortunately, it was found that different species in HF solution are responsible for etching of SiO_2 and high-k (in particular, HfO_2), and, therefore, selective removal of HfO_2 over SiO_2 is possible by tuning pH of the solution, its temperature, and fluoride concentration. [22]

SiO_2 is known to be etched by HF^{2-} and H_2F_2,[23] while etch rate of HfO_2 versus pH was found to be correlated with concentrations of undissociated HF and H_2F_2. The HfO_2 etch rate is the highest at pH between 0 and 2, where the concentrations of HF and H_2F_2 are increasing.

Studying the etch rate as a function of HF concentration and comparing the experimental results with the theoretical calculation of HF dissociation as a function of HF concentration, it was found that HF, and not H_2F_2, is the most probable species responsible for the HfO_2 etch. The observation that the HF concentration follows the same trend as HfO_2 etch rate strongly suggests that this is the main species responsible for etch.

All these results point to the etch mechanism similar to that of SiO_2 with only one change: HF is the active species. In Figure 2(a) we propose a mechanism in which the HF is coordinated by O of – OH group. Even in its protonated form, the oxygen still has a pair of free electrons that are available to coordinate the hydrogen. In this way the F is brought closer to Hf. Elimination of – OH group is closely coupled with addition of F (Figure 2[a]). After the – OH is replaced by F the next nucleophilic substitution can go easier because this time the nucleophilic reactants can reach Hf easily from the side opposite to the leaving group (Figure 2[b]). In this way the surface is again OH functionalized whereas the first HfO_2 molecule has now two F atoms already in place.

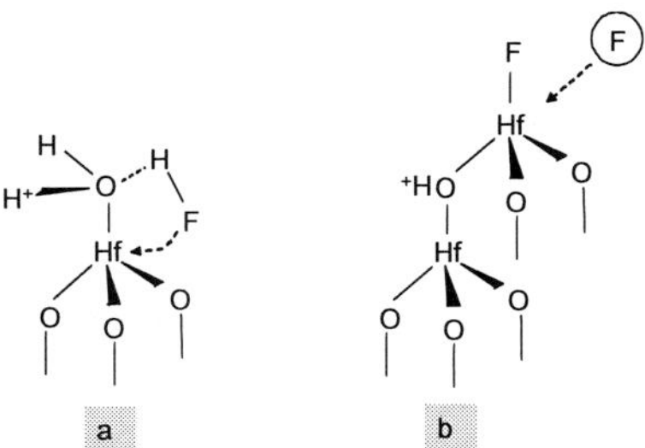

Figure 2. Etch mechanism proposed for HfO_2 etch in HF based solutions.

From the above consideration one can conclude that in order to achieve high selectivity over SiO_2, the HF dissociation should be reduced. HF dissociation can be changed by adding alcohol. Isopropanol, ethanol or even ethylene glycol, characterized by low dielectric constants, can reduce the HF dissociation. By replacing the H_2O with pure ethanol, the HF dissociation constant decreases by more than two orders of magnitude.[24] Indeed, as ethanol is mixed with water, the selectivity of HfO_2 over SiO_2 is greatly increases, as illustrated by Figure 3.

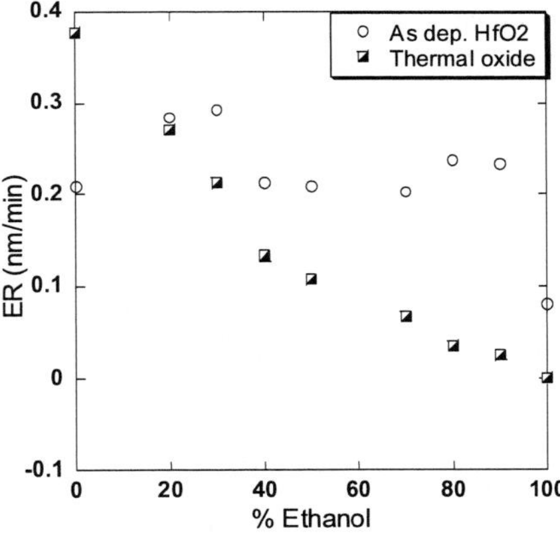

Figure 3. As deposited-HfO_2 and thermally grown SiO_2 ER variation as a function of the percentage of ethanol in the HF/H_2O/ethanol mixture ([HF]=0.05 mole/l; T= 25°C)

Although Hf is soluble in HF, the ERs of HfO_2 and $Hf_xSi_{1-x}O_2$ in HF solutions decrease drastically after a thermal treatment.[25,26,27,28,29] It is reported that the quality of the layer is highly improved after an annealing process. It is reported that the quality of the layer is highly improved after an annealing process. Crystallization and densification due to thermal treatment, are the main responsible factors for lower dissolution of Hf based layer in different aqueous solutions.[25] The effect directly depends on Hf concentration in case of $HfSiO_x$ layers. Saenger et al.[30] were able to remove the as deposited ALCVD HfO_2 with an ER of 8 nm/min in 10:1 HF. After annealing at 700°C, it became almost unetchable. Same behaviour was reported by M. Balasubramanian et al.[31] for the MOCVD HfO_2: after annealing at 1000°C, 1s in N_2 ambient became unetchable. In the case of ALD HfO_2 when the precursor was $Hf[N(CH_3)(C_2H_5)]_3[OC(CH_3)_3]$ the film was still etchable even after annealing for 1 min in N_2 at 800°C.[32] After analysis the microstructure of the annealed film the authors discovered that the film was still amorphous.

In device applications, the high-k layers are mostly thermally treated. This makes them very difficult to be removed wet selective towards an oxide layer. Improved selectivity might be achieved with an appropriate pre-treatment of the annealed HfO_2 layer prior to wet etching. This can be reached either by plasma damage or by ion implantation. It is known that ion implantation damage can increase the wet etch rate of SiO_2.[33] An increase of the etch rate after ion implantation has been reported by M. Balasubramanian et al.[34] after implanting MOCVD HfO_2 layer with As, BF_2, B or P. The best results were obtained for BF_2 and As implants when the implant energy was 5 keV. Still, only the first 4 nm were removed in the diluted HF solution (1:50). After this critical depth the etch rate became negligible.

Combination of wet and dry removal

One can see that neither dry nor wet removal alone yield desired results in terms of complete selective removal of high-k dielectrics. The combination of these methods, however, is quite promising. The main idea of the combined removal is the following. In the first dry removal step most of the high-k is removed by plasma, while the remaining high-k is damaged by ion bombardment. It should be noted that since ion bombardment is highly anisotropic, the high-k under the gate is not affected. Since the high-k is not removed completely, substrate damage is not an issue. In the second wet step the remaining damaged high-k is removed without any substrate damage. The whole sequence is illustrated in Figure 4

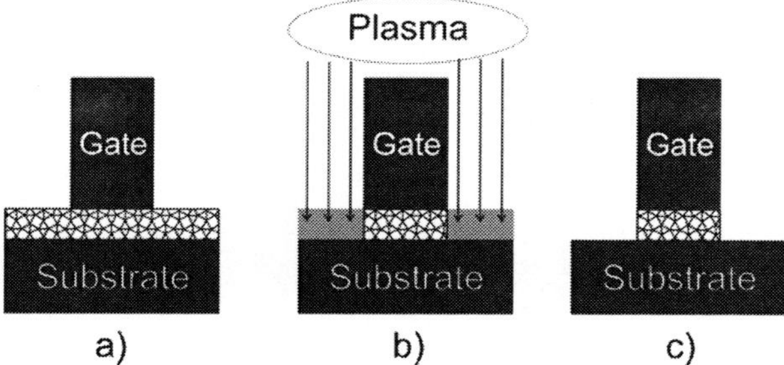

Figure 4. Schematic of combined dry/wet removal. A) – initial situation (crystalline high-k), b) – plasma removal damages the high-k, c) – wet step completely removes the damaged high-k.

It was shown that high-k dielectrics that were not possible to remove by wet chemistries, were removed wet after plasma damage by non-reactive Ar/O_2,[35] O_2,[36] Ar,[27] $HBr/Cl_2/O_2$[37]. It is more interesting, however, to use BCl_3-based chemistry that can partially remove the high-k and damage the remaining layer.[38]

Conclusions

Selective high-k removal is not straight forward, but possible. For the most popular high-k, HfO_2, both dry and wet removal could be used. The best results for dry removal are obtained using BCl_3-based plasma, which is able to etch high-k by reducing metal oxide producing volatile boron-oxygen-chlorine compounds and metal chlorides. Another advantage of this chemistry is formation of a passivating film on Si substrate that provides selectivity. The best wet chemistry is aqueous HF solution with pH between 0 and 2 and low dissociation of HF that could be achieved by adding alcohol to the solution. Low pH and low dissociation helps to make the solution selective over SiO_2 since the etching species are different (HF^{2-} for SiO_2 and HF for HfO_2). The main disadvantage of the wet etch is its inability to remove thermally treated (annealed) high-k.

The best results could be obtained by combing dry and wet removal. First, plasma partially anisotropically (without undercut) removes high-k layer and damages the remaining of the layer. Second, wet chemistry removes the damaged high-k with high selectivity over the substrate and other stack materials. Since the high-k dielectric is not damaged under the gate, lateral etch (undercut) is not an issue.

References

[1] D. Shamiryan, M. Baklanov, M. Claes, W. Boullart, and V. Paraschiv, *Chem. Eng. Comm.*, **196,** 1475 (2009).

[2] Handbook of Advanced Plasma Processing Techniques, Ed. By R. J. Shul and S. J. Pearton, Springer, (2000) p. 2

[3] JANAF Thermochemical Tables, 1985 Supplement 1, J. Phys. Chem. Ref. Data, **14** (1985)

[4] CRC Handbook of Chemistry and Physics, 85th edition, ed. D.R. Lide, (2004-2005)

[5] F. Trouton, Phil. Mag., **5,** 54 (1884)

[6] K. K. Shih, T. C. Chieu and D. B. Dove. J. Vac. Sci. Technol. B, **11,** 2130 (1993).

[7] J. A. Britten, H. T. Nguyen, S. F. Falabella, B. W. Shore and M. D. Perry. J. Vac. Sci. Technol. A, **14,** 2973 (1996)

[8] S. Norasetthekul, P. Y. Park, K. H. Baik, K. P. Lee, J. H. Shin, B. S. Jeong, V. Shishoda, E. S. Lambers, D. P. Norton and S. J. Pearton. Appl. Surf. Sci., **185,** 27 (2001).

[9] J. Chen, W. J. Yoo, Z. YL Tan, Y. Wang and D. S. H. Chan. J. Vac. Sci. Technol. A, **22,** 1552 (2004).

[10] K. Takahashi, K. Ono, and Y. Setsuhara. J. Vac. Sci. Technol. A, **23,** 1691 (2005).

[11] S. Norasetthekul, P. Y. Park, K. H. Baik, K. P. Lee, J. H. Shin, B. S. Jeong, V. Shishoda, D. P. Norton and S. J. Pearton. Appl. Surf. Sci., **187,** 75 (2002).

[12] T. Maeda, H. Ito, R. Mitsuhashi, A. Horiuchi, T. Kawahara, A. Muto, T. Sasaki, K. Torii, and H. Kitajama. Jpn. J. Appl. Phys., **43,** 1864 (2004).

[13] P. Y. Park, S. Norasetthekul, K. P. Lee, K. H. Baik, B. P. Gila, J. H. Shin, C. R. Abernathy, F. Ren, E. S. Lambers and S. J. Pearton. Appl. Surf. Sci., **185,** 52 (2001)

[14] L. Sha, R. Puthenkovilakam, Y.-S. Lin, and J. P. Chang. J. Vac. Sci. Technol. B, **21,** 2420 (2003)

[15] L. Sha, B.-O. Cho, and J. P. Chang. J. Vac. Sci. Technol. A, **20,** 1525 (2002).

[16] L. Sha and J. P. Chang. J. Vac. Sci. Technol. A, **21,** 1915 (2003).

[17] D. Shamiryan, A. Danila, M. R. Baklanov and W. Boullart, J. Vac. Sci. Technol. B, **28,** 302 (2010)

[18] K. Pelhos, V. M. Donnelly, A. Kornblit, M. L. Green, R. B. Van Dover, L. Manchanda, Y. Hu, M. Morris, and E. Bower. J. Vac. Sci. Technol. A, **19,** 1361 (2001).

[19] D. Shamiryan, V. Paraschiv, S. Eslava-Fernandez, M. Demand, M. Baklanov, S. Beckx, and W. Boullart. J. Vac. Sci. Technol. B, **25,** 739, (2007).

[20] K. Nakamura, T. Kitagawa, K. Osari, K. Takahashi, and K. Ono. Vacuum, **80**, 761 (2006).

[21] T. Kitagawa, K. Nakamura, K. Osari, K. Takahashi, K. Ono, M. Oosawa, S. Hasaka, and M. Inoue. Jpn. J. Appl. Phys., **45**, L297 (2006).

[22] Claes, M.; Paraschiv, V.; Boutkabout, H.; De Gendt, S.; Richard, O.; Lindsay, R.; Boullart, W. and Heyns M., 204th Meeting of the Electrochemical Society: 2nd Int. Symp. on High Dielectric Constant Materials, 13-17 October 2003; Orlando, FL, USA.

[23] D. M. Knotter, "Etching Mechanisms of Vitreous Silicon Dioxide in HF-Based Solutions", J. Amer. Chem. Soc., **122**, 4345 (2000).

[24] B. Garrido, J. Montserrat and J. R. Morante, J. Electrochem. Soc., **143**, 4059 (1996).

[25] M. Balog, M. Schieber, M. Michman and S. Patai, Thin Solid Films, **41**, 247 (1977).

[26] M. A. Q-Lopez, M. El-Bouanani, R. M. Wallace and B. E. Gnade, J. Vac. Sci. Tecnol. A, **20** (2002) 1891.

[27] J. Chen, W. J. Yoo, D. S. H. Chan and D. –L. Kwong, Electrochem. Solid-State, Lett., **7**, F18 (2004).

[28] B. Onsia, D. Hellin, M. Claes, A. Maes, S. De Gendt, M. Heyns, in Proceedings of the 6th International Symposium on Ultra Clean Processing of Silicon Surfaces, Oostende, Belgium, Sept 16- 18 (2002).

[29] J. J. Chambers, A. Rotondaro,M. Bevan, M. Visokay and L. Colombo, Abstract 1434, Meeting Abstracts, The Electrochemical Society and the International Society of Electrochemistry, Vol. 2001-2, San Francisco, CA, Sept 2-7, 2001.

[30] K. L. Saenger, H. F. Okorn-Schmidt and C. P. D'Emic, A selective etching process for chemically inert high-k metal oxides, in Novel Materials and Process for Advanced CMOS, edited by M. I. Gardner, S. De Gendt, J-P Marin and S. Stememr, (Mater. Res. Soc. Symp. Proc. 745, Warrendate, PA, 2003), p. 29.

[31] M. Balasubramanian, L. K. Bera, S. Mathew, N. Balasubramanian, V. Lim, M. S. Joo, B. J. Cho, Thin Solid Films, **101-102**, 462 (2004).

[32] M. Seo, S. K. Kim, K-M. Kim, T. J. Park, J. H. Kim, C. S. Hwang and H. J. Cho, ECS Transactions, **1**, 211 (2006).

[33] L. Liu, K. L. Pey, P. Foo, HF wet etching of oxide after ion implantation, Proceedings IEEE Hong Kong Electron Devices Meeting 29 june 1996.

[34] M. Balasubramanian, L. K. Bera, S. Mathew, N. Balasubramanian, V. Lim, M. S. Joo, B. J. Cho, Thin Solid Films, **101-102**, 462 (2004).

[35] W.-T. Chang, T.-E. Hseih, and C.-J. Lee. J. Vac. Sci. Technol. B, **25**, 1265 (2007).

[36] K. Saenger, H. Okorn-Schmidt and C. D'emic, MRS Symposium Proceedings, Novel Materials and Processes for Advanced CMOS, vol. 745m Pensylvania, p79, 2003.

[37] R. Mitsuhashi, M. Kubota, S. Hayashi, US 2003/0104706 A1, 2003.

[38] D. Shamiryan, V. Paraschiv, M. Claes and W. Boullart. Proc. of Defect in Advanced High-k Dielectrics NATO Research workshop, Saint Petersburg, Russia, July 11-14, 2005, ed. by E. Gusev, NATO Sci. Series, **220**, 331 (2006).

ECS Transactions, 34 (1) 319-323 (2011)
10.1149/1.3567597 ©The Electrochemical Society

Active Area Width and Topography Effects on Sub 45nm Poly Gate CD

Man-Hua Shen, Xiao-Ying Meng, Yi Huang, Hai-Yang Zhang, Shih-Mou
Chang, Kwok-Fung Lee, Yi-Ming Gu

Semiconductor Manufacturing International Corporation, No.18 Zhang Jiang Rd.,
Pudong New Area, Shanghai, 201203, P.R.China

ABSTRACT

Gate critical dimension (CD) is an important parameter in determining
the CMOS device performance. During the process of aggressive feature
size shrinking in 45nm node and beyond, aside from the recognized
topographical impact on the wafer-to-wafer CD variation, the active area
(AA) width has to be taken into consideration for the thorough analysis and
the rigid control of within-wafer gate CD variation.

Roughly 5%~8% within-wafer CD difference was observed on six gate
structures (the same defined gate with different AA widths) right after the
gate etch. Both topography wafer and blank wafer (deposited merely with
gate oxide plus poly-Si) were utilized to identify the potential key
contributors to such within-wafer CD variation. Results demonstrate that the
shallow trench isolation (STI) step height, one of notorious topography
factors, heavily relies on the local AA width. This leads to the instability of
poly-Si film thickness, therefore resulting in the inconsistency of gate
bottom profile. Despite, the optical proximity correction (OPC) based on the
post-gate etch CD could be utilized to assuage the AA width related CD
difference to some extent, the dependence of step height on AA width from
the synergetic action of CMP and WET process should be solved for the
fundamental solution to such within-wafer gate CD variation.

I. INTRODUCTION

As the semiconductor industry is moving towards 45nm node and beyond, the control
of gate CD becomes much more challenging. Not only the wafer-to-wafer CD variation
but also the within-wafer CD uniformity arouses more attention for the delivery of
desired device performance. As it is well-known, the topography effect from the
incoming wafer could induce the wafer-to-wafer CD variation in both lithography and
etch process. The nanometer-level accuracy in lithography can leverage the reduction of
the "topography effect" [1]. Besides, the inverse narrow-width effect (INWE) in the scaled
STI processes could not be ignored for its remarkable side impact on the function of
small device [2]. This triggers the requirement for not only the optimization of STI-
induced stress but also more tightening control of within-wafer gate CD variation. In this
work, we examined the role of AA-width in allaying the within-wafer CD variation after
gate etch on 40nm test vehicle.

II. EXPERIMENTS

We employed 40nm low-leakage vehicle to evaluate the impact of AA width on the within-wafer gate CD variation. Blank wafers were used as the reference for topography wafers to detect the potential key contributors to such CD variations. As shown in Fig. 1 (a), the topography wafers were prepared using the typical process flow including STI formation, poly-Si deposition and gate lithography process. Tri-layer patterning was applied for its proven superior pattern transfer fidelity on ArF photo resist. From the bottom to the top, it includes amorphous carbon, DARC (Dielectric Anti-Reflect Coating), BARC (Bottom Anti-Reflect Coating) and immersion photo resist. The gate etch was performed in Lam Kiyo series tool. Fig. 1 (b) demonstrates the post gate etch status for positive STI step height case. The typical AA width test keys (TK) used in this paper is illustrated in Fig. 1 (c). The AA width, W, is defined as the AA CD parallel to the final gate line. The investigated AA width ranges from 0.062um, 0.108um, 0.15um, 0.27um, 0.54um to 0.9um. The smallest one is close to gate CD. This AA width portfolio should be broad enough to monitor the actual AA width effect on poly gate CD in SRAM (Static Random Access Memory) area. The same tri-layer patterning scheme is used on blank wafer. Its corresponding film stack and post-gate etch state are illustrated in Fig. 2. Both ADI (After Development Inspection) CD and AEI (After Etch Inspection) CD were measured using Hitachi 9380II. Poly-Si height and STI step height were examined with TEM (Transmission Electron Microscope) for all six AA width splits.

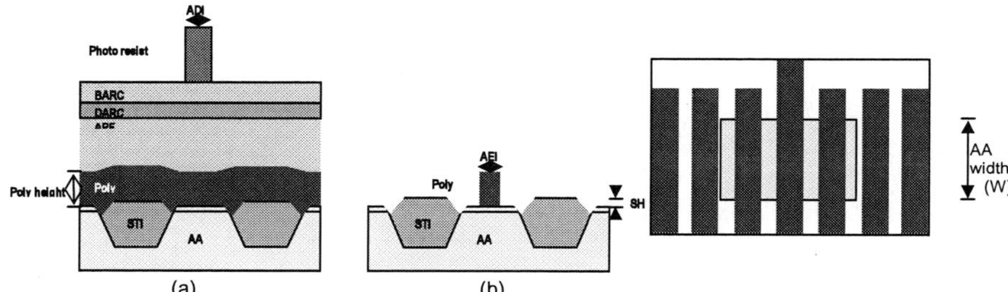

Fig. 1. Topography wafer: (a) Post gate lithography; (b) Post gate etch; (c) Gate pattern with AA-splits

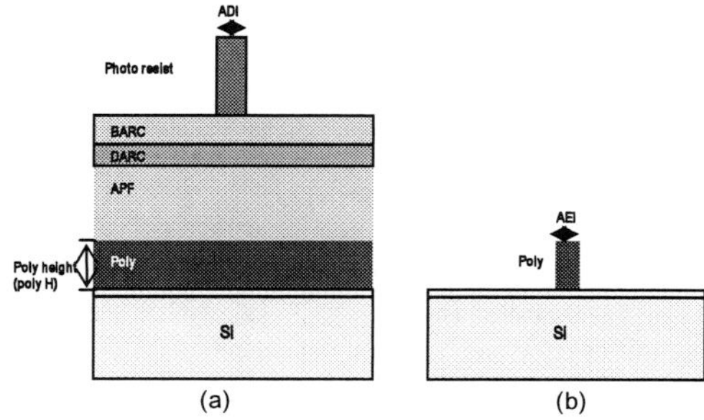

Fig. 2. Blank wafer: (a) Post gate lithography; (b) Post gate etch

III. RESULTS AND DISCUSSIONS

Table 1 (a) lists the impact of AA-width on ADI CD between blank and topography wafer. It demonstrates the ADI CD difference decays as the AA width increases if the bias at 0.9um AA width is excluded. The difference could be neglected when the AA width is larger than 270nm as it is close to the capability limitation of CD SEM tool. The slight difference among larger AA width splits might benefit from the interaction deduction between two AA/STI boundaries on lithography patterning. As shown in Table 1 (b), the etch performance of six splits on blank wafer shows ~2nm random variation but there is no clear etch bias trend. This indicates that it might come from the inherent measurement noise. The same splits end up with ~ 5nm (~5%~8% of gate CD) CD bias variation on topography wafers. Larger AA width is associated with the higher etch CD bias. If the effect of AA-width on ADI is removed, a linear relationship could be identified between AA width and the scaled AEI CD. As all the test keys feature the same pitch, the above phenomena could not be explained by RIE lag. To seek an effective solution to such local CD variation, we utilized TEM to disclose the possible root cause.

Fig. 3 plots the TEM results (along poly line direction) of STI step height, poly thickness at the middle of AA and at the boundary of AA/STI for all six AA width splits on topography wafers. Clearly, no remarkable poly height difference is observed at the middle of AA. This verifies again the local AEI CD variation does not stem from RIE lag. The third-order polynomial could be obtained between AA width and step height. The relationship between step height and poly height at the boundary of AA/STI is also non-linear. Nevertheless, the etch CD bias linearly depends on the corresponding STI step height. This indicates the current local CD variation inherits from the difference of STI step height, which is strongly associated with the AA width.

ADI CD(normalized)				Etch CD Bias		
AA Width (nm)	Blanket Wafers	Topography Wafers	Difference	AA Width (nm)	Blanket Wafers	Topography Wafers
AA Width 0.9	1	1-1.51%	1.51%	AA Width 0.9	5.6	4.8
AA Width 0.54	1	1-0.74%	0.74%	AA Width 0.54	6.6	4.7
AA Width 0.27	1	1-1.68%	1.68%	AA Width 0.27	5.8	3.3
AA Width 0.15	1	1-1.81%	1.81%	AA Width 0.15	4.3	2.6
AA Width 0.108	1	1+3.12%	3.12%	AA Width 0.108	5.3	-0.2
AA Width 0.062	1	1+3.87%	3.87%	AA Width 0.062	4.3	-0.3

Table 1. Comparison of (a) ADI CD and (b) Etch CD bias
between blanket wafer and on topography wafer with six different AA width splits

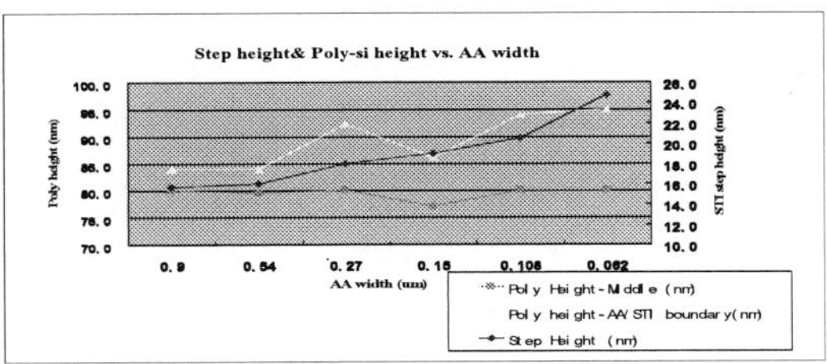

Fig. 3.   Step height and poly thickness among six different AA width splits

As compared in Fig. 4, the poly-Si film is almost pinching off on the AA area when AA width is 0.062um while the pinching-off phenomena dies out when AA width is 0.54um. Besides, we could also notice the severe pinching-off should come from the higher step height at narrow AA width. In order to disclose the underneath relationship, let's zoom in the poly gate etch process we used. Aside from the conventional film open steps on Barc, Darc and amorphous carbon, the combination of poly-Si main-etching (ME), a soft-landing-etching (SL), and an over-etching (OE) could result in poly gate CD variation if the poly-Si thickness varies.

ME step usually uses the polymer-lean process to achieve the vertical profile and leaves 20% of poly-Si film intact. This is to avoid the possible punch-through owing to the low selectivity of poly-Si over oxide in ME. Fig. 5 is one example of post-ME partial check for one-negative step height case. The remaining poly height on AA is much lower than that on STI area. Even the low-selectivity ME step could not deliver the "vertical" gate on STI, obviously, the gate CD is quite larger than those on AA area. This indicates the tapered gate already occurs in ME step when the variation of poly film thickness is above 15%. Moreover, more polymer gases are introduced in SL step for higher selectivity. This tends to result in the tapered profile for more bottom protection. The endpoint mode in SL step also could not alleviate the variation from poly-Si film thickness. This signals that the so-called local CD variation actually comes from the sidewall angle difference.

Fig. 6 compares the sidewall angle difference after one full gate etch. The gate line near STI is more tapered than the middle one (86 degree vs. 89 degree). Such local CD variation could not be compensated by either dosemap or poly trim step. The former is proven effective to improve the die-to-die variation while the latter is usually used to reduce the wafer-to-wafer variation in feed-forward APC mode. Even worse, the OPC could not completely solve this issue either for it is capable of reducing within-wafer CD variation but could not improve the side-wall difference, which will ultimately introduces the variation of poly gate top CD, thus impacting the subsequent spacer etch and silicide related steps. As we can see, it's imperative to reduce the variation of local step height from the synergic actions of CMP and WET processes.

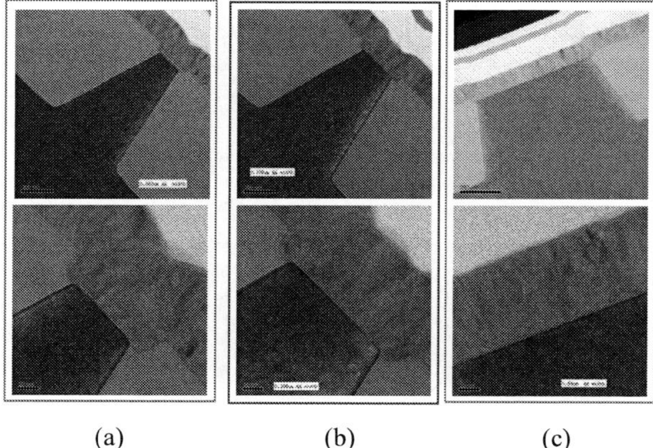

(a)   (b)   (c)

Fig. 4 Poly-Si deposition profile with different AA widths
(a) AA width: 0.062um; (b) AA width: 0.108um; (c) AA width: 0.54um

Fig. 5 Poly-Si remaining right after main etch step

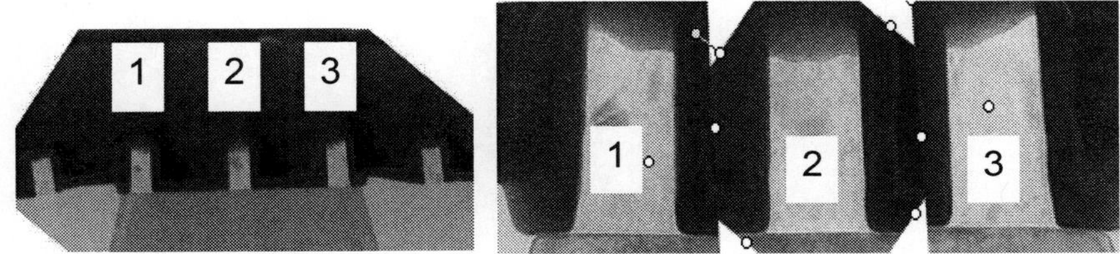

Fig. 6 Gate profile after full-etch (AA width: 0.062um)

IV. CONCLUSIONS

The role of narrow width AA on local CD variation is investigated. Narrow AA width is apt to form the higher STI step height, thus results in much thicker poly-Si film on AA surface. The severe pinching-off phenomena is observed in 0.062um AA width split on 40nm test vehicle. Even OPC could not thoroughly solve this kind of issue for such local CD variation actually originates from the sidewall angle difference. It's indispensable to improve the variation of local STI step height from the point view of the synergic actions in CMP and WET processes.

ACKNOWLEDGMENTS

Authors would like to thank Dr. Huang Song, Mr. Wang Jiagao of Lam Research China for the technical discussion and information, Mr. Feng Ji from the integration team of SMIC Technology R&D center for wafer preparation.

REFERENCES

1. I. Masaru, et al, "Effect of Topographical and Layout Factors on Gate CD Modeling for MOS Transistor Area", IEEE Transaction on semiconductor manufacturing, Vol. 22, No. 2, May, 2009.
2. K.Youngmin, et al, "Trench Isolation Step-Induced (TRISI) Narrow Width Effect on MOSFET", IEEE Electron device letters, Vol. 23, No. 10, October, 2002.

ECS Transactions, 34 (1) 325-328 (2011)
10.1149/1.3567598 ©The Electrochemical Society

Reverse Phase Solution for Mesa chamber Uniformity Improvement

Qing Ge [a], Ying Huang [a], Xiaoduo Tang [a]

[a] Semiconductor Technology Group Applied Materials China Globe Account

Applied Materials® newly launched advanced 300mm plasma etcher, Mesa® chamber implemented many key hardware modifications to provide much more robust process knobs to improve the depth uniformity and CD uniformity for critical applications of <32nm technology. The new features including DMC,MRAD and others features focus on the innovation of Top ICP Source to optimize ion flux distribution uniformity to improve the Etch Rate uniformity and CD uniformity. As a result, the whole wafer uniformity can achieve 1~1.5nm (3σ). This article will focus on two among the new features: DMC and MRAD which can solve the special distribution map to further improve uniformity significantly. The discussion includes hardware fundamental introduction and plasma simulation results as well as supporting wafer results.

Introduction

With semiconductor technology nodes shrinking to 32nm and beyond, the plasma etch process faces an enormous challenge to control the whole wafer CD uniformity, depth uniformity and profile uniformity. Various techniques have been adopted in advanced 300mm etch chamber design including tunable gas distribution, tunable source power and dynamic multiple zone ESC temperature control. With all those features implementation, the whole wafer CD uniformity <3nm (3σ) becomes achievable in mass production. But for some critical applications, further improvements are expected. Further improvement of the uniformity requests to solve some special distribution map. M-shape is one typical map often seen for a wafer processed in 300mm ICP top source chamber which has dual coil design. Randomly skewed map also create difficulty to further improve the uniformity because it is very hard to eliminate the skewed distribution map by tuning center to edge process knobs. Mesa™ Chamber is the latest version of DPSII AdvantEdge™ serial plasma etch chamber. Mesa™ chamber focuses on ICP source assembly innovation to optimize ion flux distribution uniformity.

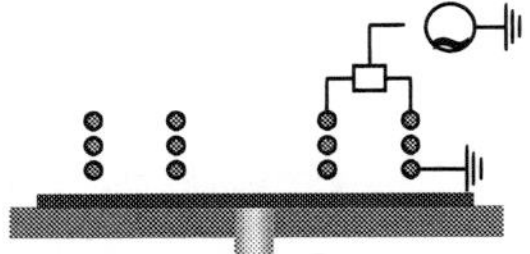

Fig.1. Schematic of ICP Source

DMC is one of the new features, which stands for Dual Mode Current source. It adopts reverse phase current to eliminate M-Shape to further improve the whole wafer uniformity. Fig.1. shows a typical standard DPSII AdvantEdge™ Series Plasma Etch chamber top source design. The source coil is divided into inner and outer parts with different diameter and the source current ratio between the inner and outer parts is tunable with various DC (Divide Capacitor) setting to provide dynamic control of ion flux uniformity from wafer center to edge. Mesa™ chamber provides two current mode settings: one is Standard Current Mode another is Reverse Current Mode. In the Standard Current Mode, inner and outer coils have in-phase current (Fig.2), but Reverse Current Mode applied 180 degree phase shift between inner and outer coils which is called out-phase current (Fig.3). The current phase differences will have different interference behavior between inner coil and outer coil and generate a different electromagnetic field distribution map at effective chamber area.

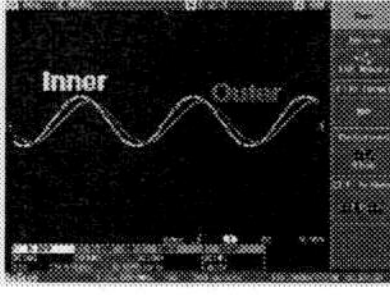

Fig.2 Standard Current Mode

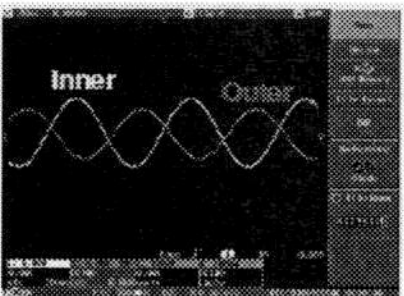

Fig.3 Reverse Current Mode

MRAD is another new feature to solve the random skewed distribution map typically seen in a conventional 300mm etch chamber (possibly by asymmetric chamber design or other unknown reasons from previous integration steps). MRAD stands for Motorize RF Assembly Dial. Because the skew effect occurs in different direction randomly,, MRAD enables the function to dynamically control the tilt height and azimuth of RF Source assembly to change the plasma center and achieve symmetric etch rate maps. Motorized control will enable the function controlled by recipe and step. Fig.4. shows the illustration of MRAD.

Fig.4 illustration of MRAD®

Experimental

To prove the benefit of Reverse Current Mode, Plasma simulation and blanket etch rate tests are done in a Mesa chamber. Fig.5. shows the plasma simulation results that indicate different ion distribution maps.

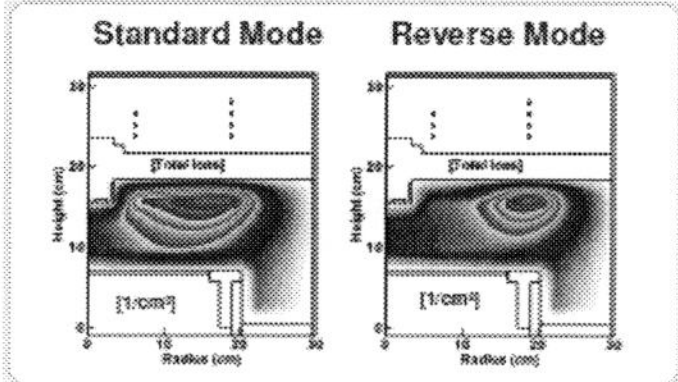

Fig.5 Ion Density Distribution Map

When the Standard Current Mode was used, the peak of ion density appeared somewhere between wafer edge and wafer center .But under the same process regime, when the Reverse Current Mode was used, a different ion distribution map was observed. The peak of ion density moved towards the wafer edge. As a result, the wafer edge etch rate is brought up and wafer middle etch rate dropped accordingly. Thus, the M-shape distribution map should have been eased or eliminated. We used the typical STI main etch recipe to check blanket poly etch rate with the different current modes. As illustrated in Fig.6, using the Standard Current Mode results in an M-Shape distribution map, whereas using the Reverse Current mode eliminates the M-Shape distribution map.

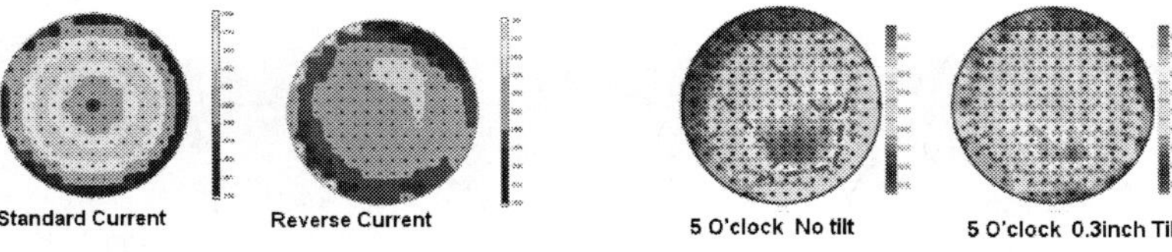

Fig.6 Blanket Etch Rate Comparison   Fig.7 Pattern Wafer CD Skew effect

To prove the benefit of MRAD, we collected the pattern wafer CD full map without source tilt and with 0.3 inch tilt at 5 O'clock orientation (see Fig.7). Without any source tilt, the CD map showed an obvious skew effect point at 5 O'clock orientation. With 0.3inch source tilt at 5 O'clock applied at mask open step, the Final CD distribution map becomes concentric. This test shows that MRAD provides a robust process knob to solve asymmetric distribution map to further improve the uniformity.

Result and Discussion

Through the top source hardware innovation Mesa™ Chamber can significantly improve the etch rate uniformity and CD uniformity for the critical application of <32nm technology. Fig.8. shows the etch rate performance with M-shape elimination and other optimizations. 49pts data measurements range within 100A/min for a nominal etch rate of 3800A/min. Fig.9. shows the 32nm TiN metal hard-mask open pattern wafer performance. Critical Dimension Uniformity (CDU) is within 1.2nm (3σ) based on a mean CD below 50nm on the full wafer.

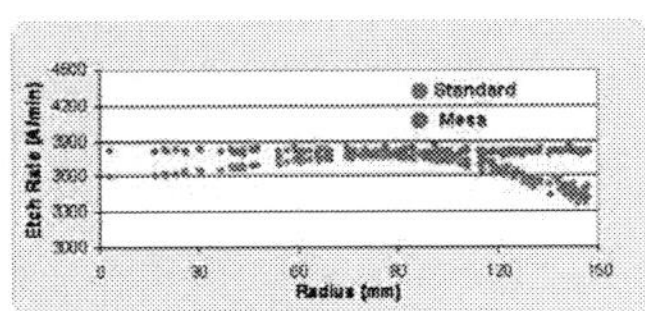

Source	Standard (G5)	Mesa
CD Mean	4x nm	4x nm
CDU 3σ	2.1 nm	1.1 nm
CD Range	3.0 nm	1.6 nm

Fig.8 Etch rate performance   Fig.9 MHM CD Performance

To further understand the difference of standard current mode and reverse current mode, we need to further study and discuss the electromagnetic field generated by source coil and the interference effect between inner and outer coils, since we know that the plasma is excited by the Electromagnetic Field.

Fig.10. is the simplified cross section view of source coil assembly. The illustration indicated the source coil and current direction with "⊗" and " ⊙". "✗" and "•" indicate the E-field direction. The circular broken lines indicate the B-field generated by alternating current flow in the source coil. In standard current mode (Fig.10.left hand side), inner coil and outer coil flow in-phase current so the E-field originating from outer and inner coils have the same direction, which results in an add-up effect between inner and outer coil area. The add-up E-field will generate more ions at corresponding area inside process chamber, because the plasma excitation is mainly dictated by E-field distribution. That's why the plasma simulations show the peak of ion density exists in between inner and outer coils. Oppositely, in reverse current mode, out-phase current will cancel-out E-field effect between Inner and outer coils (Fig.10 right hand side). As a result, the peak of ion density moves towards the chamber wall and brings up the edge ion density. So we saw the M-shape distribution map is eliminated.

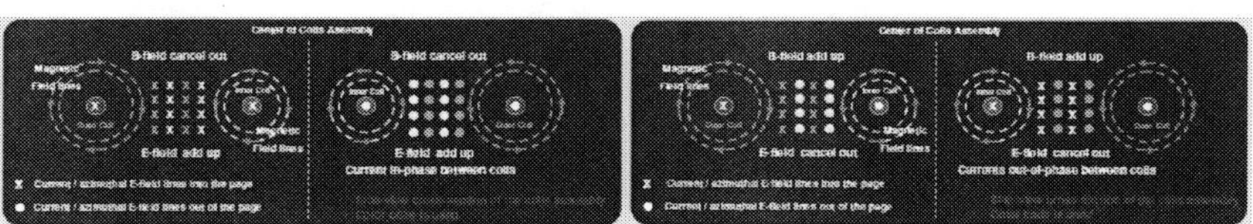

Standard Current ModeReverse Current Mode

Fig.10 Cross-Section view of Coil Assembly

Conclusion

Mesa™ chamber adopts reverse current solution to eliminate M-Shape distribution map successfully, thus enlarging the process window for depth and CD uniformity tuning. At the same time, Mesa™ chamber integrated many hardware innovations including but not limited to DMC and MRAD. All those innovations make CDU <1.5nm achievable for mass production of 32nm technology and beyond. Based on Applied Materials® announcement, Mesa chamber can be integrated within the newly released Centris™ platform. The whole system can deliver very competitive productivity performance and lower cost solution to volume production customers.

Acknowledgments

Special Thanks to Applied Materials Etch engineering and technology group for preparing the introduction material. Thanks to Applied Materials conductor etch process development group for test support and data preparation.

ECS Transactions, 34 (1) 329-334 (2011)
10.1149/1.3567599 ©The Electrochemical Society

Plasma Etch Challenges for Porous Low k Materials for 32nm and Beyond

C. Labelle[a], R. Srivastava[b], J. Arnold[c], Y. Yin[c], M. Ishikawa[d], Y. Mignot[e], H. Yusuff[f], J. Linville[a], D. Horak[c], N. Fuller[c], R. Patz[g], A. Darlak[g], K. Zhou[g], Y. Zhou[g], J. Pender[g]

[a] GLOBALFOUNDRIES, 2070 Route 52, Hopewell Junction, NY 12533
[b] GLOBALFOUNDRIES Singapore, 2070 Route 52, Hopewell Junction, NY 12533
[c] IBM Research, 257 Fuller Rd., Albany, NY 12203
[d] Toshiba America Electronic Components, Inc., 257 Fuller Rd., Albany, NY 12203
[e] STMicroelectronics, 257 Fuller Rd., Albany, NY 12203
[f] IBM Microelectronics, 2070 Route 52, Hopewell Junction, NY 12533
[g] Applied Materials, 255 Fuller Rd., Albany, NY 12203

The challenges facing back-end-of-line (BEOL) etch are becoming increasingly difficult as shrinking dimensions are compounded by new materials integration. As critical dimensions decrease, key dimension-related etch challenges include CD control, trench litho stack aspect ratios, RIE lag, and LER. The move to low k and ultra low k dielectrics requires the etch to consider the sensitivity of the films to compositional modification, polymer interactions with the pores, and diffusion effects possible with porous materials. As the minimum pitch reaches sub-100nm, new interactions of the materials with the critical dimensions need to be considered. The ability of low k materials to be patterned at the required aspect ratios without losing structural integrity will be the key challenge for future technologies' success. This paper will review the key etch challenges as a function of dimensions vs. materials and highlight where their interactions which will drive future work.

Introduction

Back-end-of-line (BEOL) etch has been challenged in recent years to achieve many things: critical dimensions (CDs) on the order of $\leq$ 65nm, increasing via CD shrink to compensate for lithography limits, and most importantly, to date, inclusion of low k or porous low k dielectrics (1). The issues associated with integrating low k dielectrics have been numerous for etch, most important being the sensitivity of the film to compositional modification during the etch and/or ash (2). Etch development on the current generation of materials is generally maturing, and while more subtle issues are now being discovered, the next wave of etch challenges is coming in the form of interactions of dimensional issues with newer, lower k, materials issues.

The patterning approach for BEOL dual damascene etching is evolving with technology node. Table I gives a snapshot of one possible evolution of patterning schemes. Technology shrinkage pushes the patterning approach in new directions, with via-first-trench-last (VFTL) moving to trench-first-hard-mask (TFHM) to enable better top CD control and provide self-aligned via (SAV) capability, while lithography limitations require one to move from standard litho-etch (LE) patterning to LLE or full LELE patterning.

TABLE I. Patterning Approach by Technology (L=litho, E=etch)

Technology node	45nm	32nm	28nm	22/20nm	14nm
Minimum pitch (nm)	130	100	90	< 90nm	< 80nm
Trench patterning	LE	LE	LE	LLE/LELE	LELE
Patterning approach	VFTL	VFTL	TFHM	TFHM	TFHM
Self-aligned via (SAV)	Non-SAV	Non-SAV	Non-SAV	SAV	SAV

Figure 1 shows a typical VFTL patterning scheme, with key etch issues highlighted at each step. A similar figure can be envisioned for TFHM patterning. Figure 2 shows one possible trench double patterning scheme, with key etch issues also highlighted. These challenges, as well as some not highlighted in these figures, are categorized as a dimensional or a materials challenge in Table II.

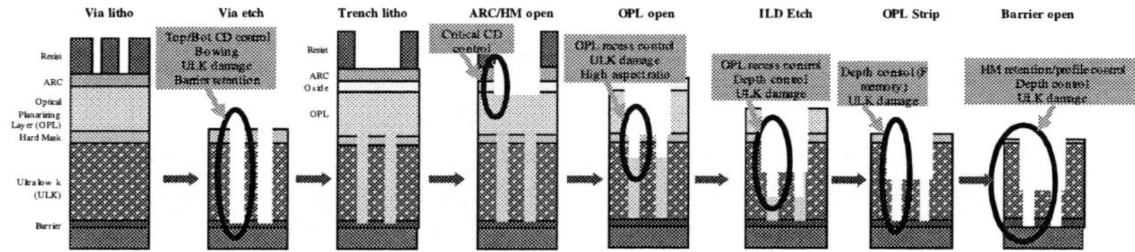

Figure 1. VFTL patterning scheme flow, with etch challenges highlighted at each step.

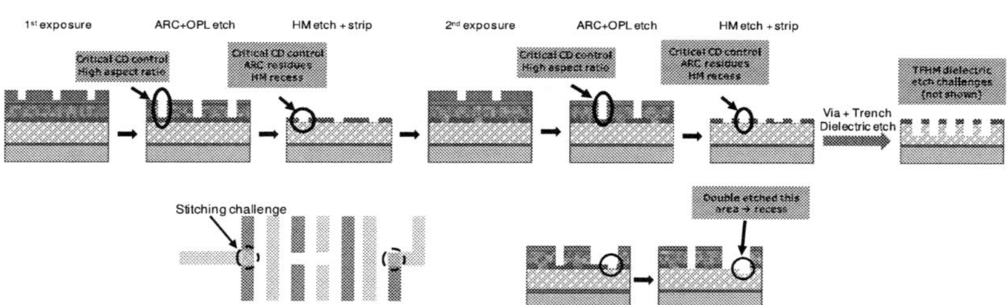

Figure 2. LELE TFHM patterning scheme flow, with etch challenges highlighted at each step.

This paper will review some of these challenges and highlight the key new challenges ahead that result from an interaction of progressively smaller dimensions and progressively lower k materials.

TABLE II. BEOL etch challenges categorized by dimension or material.

Dimensions		Materials	
CD control	Aspect ratios	Rough etch front	Top profile control
CD shrink	Flop over	Polymer management	CD shrink
Tighter tolerances	Wiggling	RIE lag	Double patterning
RIE lag	Top profile control	Flop over	HM selectivity
Through-pitch CD control	Double patterning	Porous materials interactions	Through-pitch CD control
LER	Via self-alignment	Wiggling	Via self-alignment
		ULK damage (etch/ash) → CD control	

Experiment

All etch experiments were completed on state-of-the-art 300mm dielectric etch tools at either IBM East Fishkill or Albany Nanotech, both in NY. Dielectric materials were deposited at either IBM East Fishkill or Albany Nanotech. PECVD SiOC-based films with dielectric constants ranging from 3.0 to 2.2 were used (3). Film stacks generally consisted of a barrier material, the SiOC-based dielectric, and a hard mask. A trilayer lithography scheme was used to pattern all wafers, with different technology CD targets.

Discussion

<u>Dimensions</u>

Table III shows the critical CD parameters for several technology nodes. Both CD targets, as well as the uniformity requirements, decrease with each technology node, and it will be a major challenge to meet 22/20nm and beyond requirements. In addition, via CD measurement is a major challenge for TFHM schemes where vias have to be measured inside trenches.

TABLE III. CD targets and 3σ uniformity requirements per technology node.

Level	Parameter	Technology				
		45nm	32nm	28nm	22/20nm	14nm
M1	CD	65nm	50nm	45nm	<45nm	<35nm
	3s Uniformity	3nm	2.5nm	2nm	1.5nm	1nm
Vx	CD	65nm	50nm	45nm	<45nm	<35nm
	3s Uniformity	3nm	2.5nm	2nm	1.5nm	1nm
	Shrink	~20nm	~25nm	~30nm	~35nm	~40nm
Mx	CD	65nm	50nm	45nm	<45nm	<35nm
	3s Uniformity	3nm	2.5nm	2nm	1.5nm	1nm

The aspect ratio (AR) of the lithography film stack also becomes an issue as CDs decrease. For the VFTL scheme in Fig. 1, the OPL AR can potentially increase from ~4.5 to 7.5 from the 45nm to 22nm node, and flopover of the OPL is a concern at that high an AR. An example of this OPL flopover issue is shown in Fig. 3, where the bending of the OPL shadow-masked the underlying dielectric. Etch chemistry and process regime optimization was able to mitigate this issue.

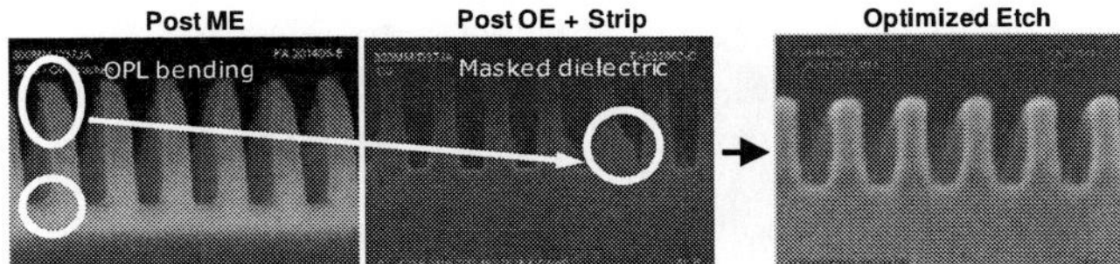

Figure 3. Example of OPL bending during 32nm M1 etch.

RIE lag and LER are also key challenges when going from one technology node to the next. As technology CDs decrease, the RIE lag between min. CD and wide CDs will increase unless the etch is re-optimized. Innovative approaches to the etch are needed to succeed with this challenge at each generation, and will be even more important for double patterning schemes, where misalignment of the two litho exposures can result in variable CDs. Similar optimization is needed at each technology node to minimize LER.

CD shrink is a focus for via etch, with each technology node requiring additional shrink from etch to compensate for lithography limitations. Figure 4 shows the difference in shrink performance for a nominal via vs. a nominal x 2 size via on the same mask. While the nominal via is close to CD target, the 2x via is ~35nm smaller than target, illustrating a very steep through-CD shrink response. Via shrink bias will be a key issue for future technology nodes, and may require rethinking of the via designs.

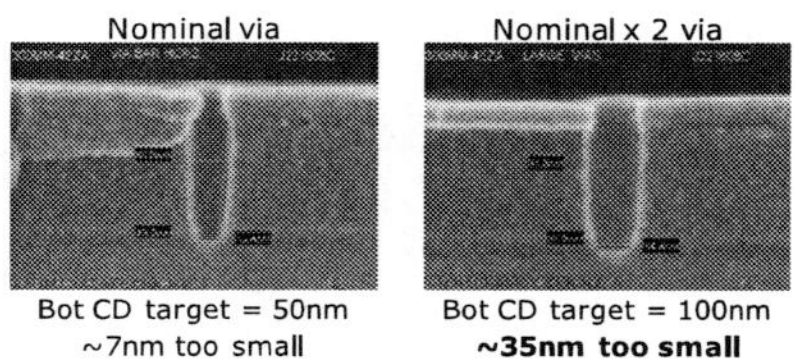

Figure 4. Via CD shrink at two different target CDs on the same mask.

New etch challenges appear as technologies move to sub-100nm pitches. The OPL aspect ratio is a major concern and has shown more severe pattern collapse behavior than observed at previous nodes. Figure 5 demonstrates the challenges for sub-100nm pitch trench structures. OPL collapse or wiggling is predominant and can be triggered even after a clean OPL transfer etch (post-ME image). Optimization of the LER is critical, as LER alone can induce the OPL pattern collapse. Successful pattern definition has been achieved at sub-100nm pitch, but the CD window for no pattern collapse is very small (~8nm in example).

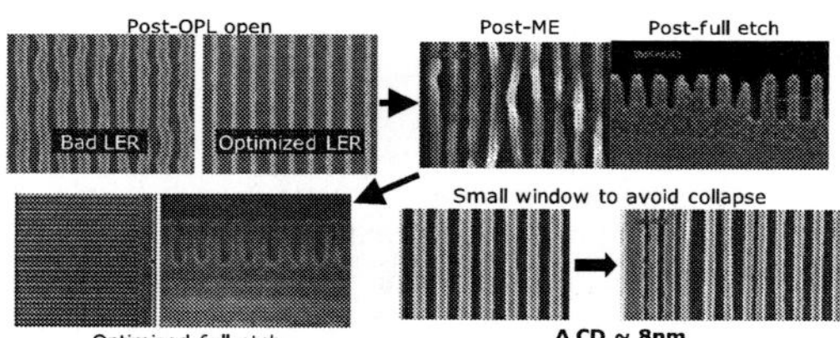

Figure 5. Sub-100nm pitch pattern collapse challenges. Process optimization can yield viable sub-100nm pitch structures, but the CD window for no pattern collapse is small.

<u>Materials</u>

As dielectric materials move from low k to porous low k (k=3 → k=2.4 → k=2.2), the materials-related challenges for etch are clear: dielectric damage (i.e., compositional modification, most often demonstrated as C depletion), rough etch front, and porosity effects. Each of these issues could be a paper in of itself. Therefore, only a few highlights will be given here.

Figure 6 shows the progression of ash optimization to go from 30nm/side of hard mask (HM) undercut after DHF removal of the damage layer to ~3.5nm/side. Figure 7 shows the effect of dielectric constant on the dielectric damage behavior. Modifying the dielectric film composition (k=2.2 #2 vs. #1) can influence the degree of compositional modification induced by the plasma etch conditions.

Dielectric etch front optimization is needed for porous low k dielectrics, as the porosity provides a micro-masking mechanism if the plasma etch polymers are not managed carefully. In addition, as the dielectric constant is decreased, pitting of the dielectric surface is observed. Modification of the plasma parameters can successfully reduce the pitting and generate a smooth etch front.

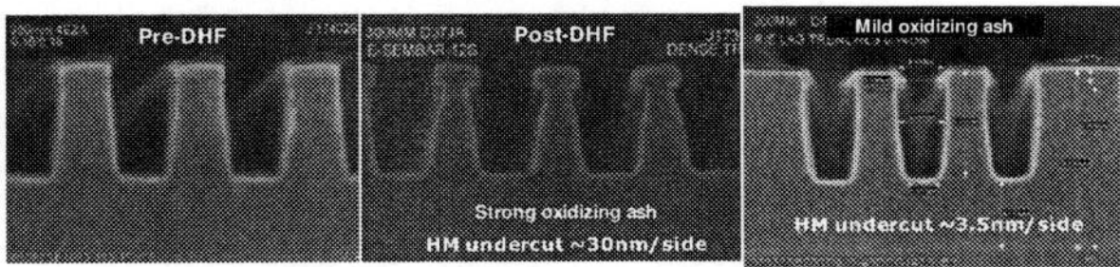

Figure 6. Exposure of a porous dielectric etch to DHF reveals significant dielectric damage. Ash optimization was successful in reducing HM undercut to ~3.5nm/side.

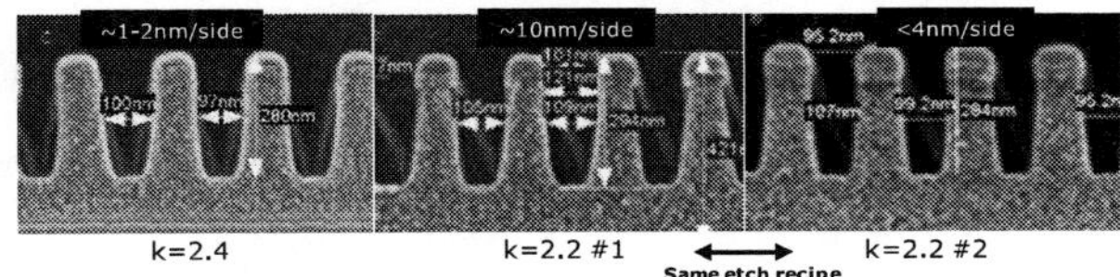

Figure 7. Hard mask undercut as a function of dielectric material.

<u>Dimensions + Materials</u>

The interaction of dimensional and material-related challenges is the driving force behind many of the current BEOL etch challenges. If the same dielectric damage performance was targeted each node, then the % of the CD taken up by damage would be unacceptably high. Rather, the % CD must be held constant, indicating that sub-4nm/side HM undercut performance is necessary for $\leq$ 140nm CDs. In addition, for $\leq$90nm pitch technologies, it is increasingly seen that metallization is much more sensitive to even a small degree of HM undercut and therefore the requirement becomes zero HM undercut.

Via-to-line (V/L) spacing has become increasing critical through the technology nodes. The V/L spacing must be maximized to pass reliability and dielectric damage heavily influences the final space CD (Fig. 8). This issue has driven the transition to TFHM schemes, which in its most basic form will allow better via/trench top CD control and improve V/L spacing. In addition, TFHM patterning, in conjunction with the right via etch chemistry, enables the use of a self-aligned via (SAV), which is the optimal solution to maximizing V/L spacing.

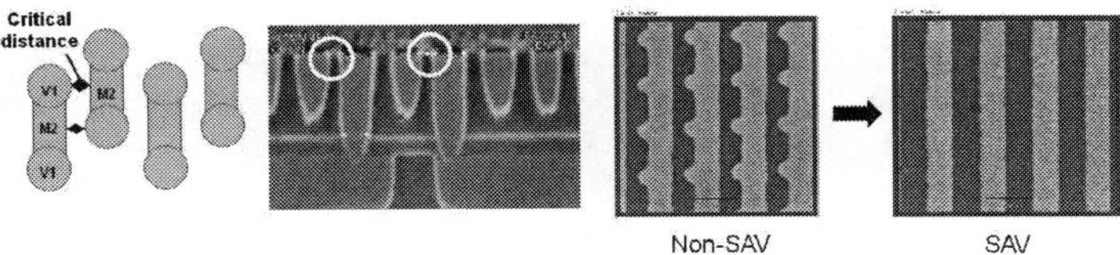

Figure 8. Schematic of via-to-line spacing. XSEM shows critical CD affected by dielectric damage. SAV improvement also illustrated.

A new phenomenon observed at sub-100nm pitch is dielectric flopover, often induced by a wet chemical exposure. Dielectric flopover can also occur without WETs exposure if the aspect ratio is too high. If the CD is pushed too high, then OPL flopover happens and the dielectric is masked. Figure 9 shows the different regimes for flopover for a sub-100nm pitch structure (VFTL example). In addition, the stress of the HM in a TFHM scheme can induce wiggling of the dielectric (Fig. 10). Finally, for double patterning, CD variation caused by litho misalignment can make the lines more susceptible to WETs-induced flopover. Therefore, structural integrity of sub-100nm pitch structures is the number one concern for these technologies and is driving etch, films, and integration work to resolve.

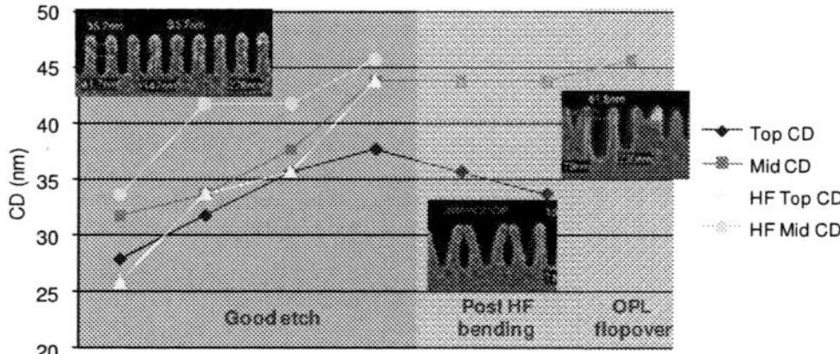

Figure 9. Different structural integrity regimes for sub-100nm pitch structures.

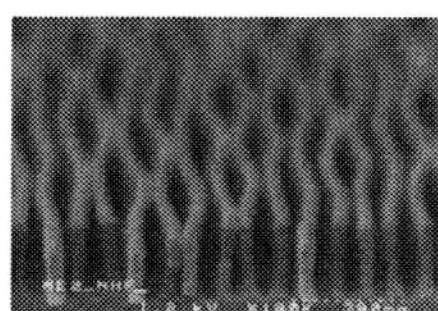

Figure 10. TFHM HM stress-induced dielectric wiggling.

Conclusions

BEOL etch is increasingly challenging as technologies progress. Both the required dimensional and materials targets define the etch challenges. The interaction of dimensions with the materials is creating the biggest challenges (i.e., 0nm/side HM undercut, via-to-line spacing control). At sub-100nm pitch, dielectric flopover due to both the tight pitch and the porous low k dielectric is a structural integrity problem that will determine the success or failure of future technology nodes.

Acknowledgments

This work has been supported by the independent Bulk CMOS and SOI technology development projects at the IBM Microelectronics Div. Semiconductor Research & Development Center, Hopewell Junction, NY 12533. This work was performed by the Research Alliance Teams at various IBM Research and Development Facilities.

References

1. S. Sankaran, *et al.*, *International Electronic Devices Meeting*, (2006).
2. T.J. Dalton, *et al.*, *IEEE International Interconnect Technology Conference*, **154** (2004).
3. A. Grill *et al.*, *IEEE International Interconnect Technology Conference*, (2008).

ECS Transactions, 34 (1) 335-341 (2011)
10.1149/1.3567600 ©The Electrochemical Society

Dry Etch Process Effects on Cu/low-k Dielectric Reliability for Advanced CMOS Technologies

Jun-Qing Zhou[a], Wu Sun[b], Hai-Yang Zhang[a], Min-Da Hu[a]. Fan Li[a], Xing-Hua Song[a], Shih-Mou Chang[a], Kwok-Fung Lee[a]

[a] Semiconductor Manufacturing International Corporation, No.18 Zhang Jiang Rd., Pudong New Area, Shanghai, 201203, P.R.China
[b] Semiconductor Manufacturing International Corporation, No.18 Wen Chang Rd., BDA, Beijing, 100176, P.R.China

Cu and low-k dielectric based back-end-of-the-line (BEOL) interconnects is indispensable in advanced CMOS technologies for its significant improvement of chip resistance-capacitance (RC) delay. In this paper, we investigate the effects of dry etch process on the reliability of copper interconnects such as electromigration (EM), time dependent dielectric breakdown (TDDB) and stress migration (SM). Both the via height of its vertical part and the bevel profile of via isolation in dual damascene (DD) structure are detected to remarkably impact the final EM performance. Main contributors for TDDB include the low-k sidewall damage from ash, the interface necking right after the diluted HF (DHF) wet owing to insufficient sidewall passivation and the bowling profile from non-optimized integration process. EM enhancement related profiles are usually associated with smaller line top critical dimension (CD). This might degrade TDDB performance. We proposed the optimized etch method to address this trade-off issue. SM performance is normally defined by the mechanical and thermal properties of copper interface in DD structure. It can also be improved in etch process by post etch scheme such as N_2/H_2 treatment from the point view of polymer residue removal and copper interface repair.

INTRODUCTION

The continued scaling of advanced CMOS technology is driving the need for low-k dielectric and copper metallization in BEOL to reduce the RC delay. All these pose the challenges to the long-term built-in reliability of interconnects from the point view of the increased current carry requirement and the weak intrinsic break-down strength of low-k materials. Three dominant wafer-level reliability concerns consist of EM, SM and TDDB

The common approaches to EM improvement rely on interface engineering [1], which usually includes either the optimization of barrier/seed processes, or the introduction of more robust interface material such as self-aligned CoWP [2] on Cu wire. The downstream EM could leverage the former method while the latter might compound the already problematic low-k dielectric TDDB reliability for the formation of metallic particle in low-k film during CoWP process. Nevertheless, due to the well-known structure dependent EM property, the former has to be coupled with dry etching on solving upstream EM issue. Moreover, it is tough in trench etching to achieve the desired chamfer at via-to-trench transition area without compromising TDDB performance. TDDB is an index to characterize the gradual degradation of interconnects. Post chemical mechanical polish (CMP) clean and the pre-treatment in liner deposition are often used to improve the TDDB [4]. In addition, the strict spacing control between via and trench has to be realized at the minimum pitch area for low-k–based interconnects. This is also one of reasons for the metal hard-mask patterning scheme to be popular in 45nm and beyond. Both EM and TDDB are current accelerator factors while SM is temperature accelerator factor to evaluate the chip reliability. SM is normally related to the mechanical and thermal properties of copper interface in DD structure. Post-etch residue is one of key concerns in SM enhancement but the residue removal in dry etching could result in trench CD shift, thus impacting TDDB. In brief, precise etched profile and residue-free etching are the fundamental items to achieve the on-target EM, TDDB and SM.

EXPERIMENTS AND DISCUSSIONS

In this work, the via-first all-in-one dry etching scheme is utilized together with LTO (low temperature oxide)-based tri-layer patterning to demonstrate the critical role of dry etching on wafer-level BEOL reliability enhancement. All dry etchings are performed on sub65nm test vehicle in Lam Flex series tool. TTF (time to failure) was checked in EM test at 300 °C/0.17mA on upstream test structure and in TDDB test at 125°C/25V on via-to-trench test key of comb type. In SM test, resistance shift is monitored after a baking at the temperature of 225 °C for 168 hrs. The proposed solutions can be directly extended to metal hard mask open scheme.

Fig 1 shows the metal1-via1-metal2 structure for upstream EM test. Generally, barrier thickness and coverage at the chamfer of via to trench transition area have shown the remarkable impact to upstream EM [3]. Here, we studied the effect of post-etch via isolation on upstream EM from the point view of its height and shape. Fig. 2 illustrates all the via isolation profiles used to see the corresponding upstream EM difference. The highest via isolation height (y) could be detected in Fig 2 (e) and (f). It's almost 600A higher than Fig 2 (a) (the lowest one). Fig 2 (b), (c) and (d) show the via isolation height falling in between Fig 2 (a) and Fig 2 (f). More specifically, compared with Fig 2 (f), Fig 2 (e) gives the slightly lower via height but much rounding chamfer at via-to-trench transition area. Fig 2 (b) is the same as Fig 2 (c) except for the difference at chamfer of via isolation. The via isolation in Fig 2 (d) is slightly higher than Fig 2 (c). All these profiles come from the trench etch optimization in terms of plug height, selectivity between plug and low-k materials in main etch (ME) step, and the low-k etch rate in liner remove step. Plug refers to the organic material filling in via hole, aiming to protect the stop layer from being attacked in the early of trench etch. In brief, the chamfer profile is determined by the joint actions of the above three factors while the via isolation height is mainly related to ME process time. To well control the plug height, we introduced one extra N_2/H_2-based step right before ME step to adjust the plug height. The insertion of this step before hard-mask open will result in severe tapered profile at dense area for its much consumption at the top of bottom photo-resist, ending up with tapered hard mask.

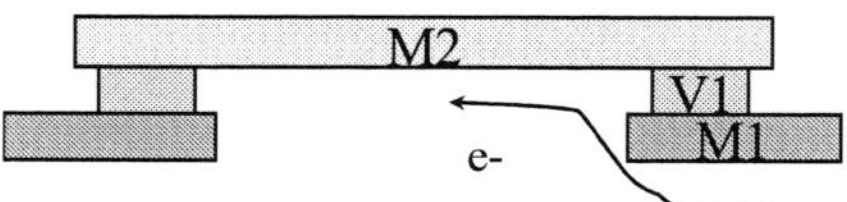

Fig.1. Upsteam EM test structure

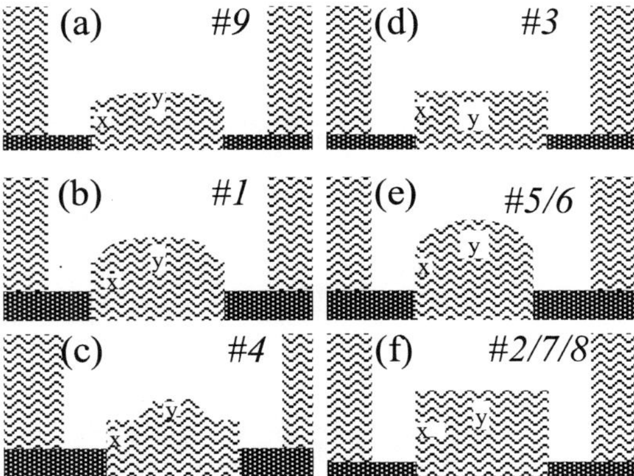

Fig. 2. Different via isolation profiles of DD structure.
#No. refers to the wafer number shown in Fig3. X is via height with 900A
range, Y stands for via isolation height with 600A range.

The cumulative distribution function (CDF) of TTF is shown in Fig.3 for the above six profile splits. The same soft barrier/seed process is applied to reduce its bombardment at the chamfer of via-to-trench transition area and the via bottom. The severe bombardment tends to form either one spike or two spikes at via bottom, depending on the via CD and the bombardment strength. Clearly, split (a) outperforms all other five splits for its lower via isolation and rounding chamfer. Its being slight better than split (b) stems from its lower via height (x). Nevertheless, the lower via height is also associated with the deeper trench depth, thus impacting the sheet resistance (Rs). The difference among split (b), (c) and (d) indicates the importance of rounding chamfer at via-to-trench transition area. Similarly, split (e) is superior to split (f) for its rounding chamfer. Briefly speaking, not only the ratio of via isolation height over trench depth but also the chamfer of via-to-trench transition area need optimize for EM enhancement. This phenomenon can be explained from the point view of barrier coverage and the alleviation of local maximum current density. The smaller aspect ratio of via, the better barrier coverage [3]. Via isolation barrier coverage will be also improved if the rounding chamfer is achieved. Besides, the weakest point at upstream structures is located at the chamfer. The larger quantity of materials will be removed at this area if its shape is not desired. Therefore, EM enhancement will benefit from the rounding chamber to some extent.

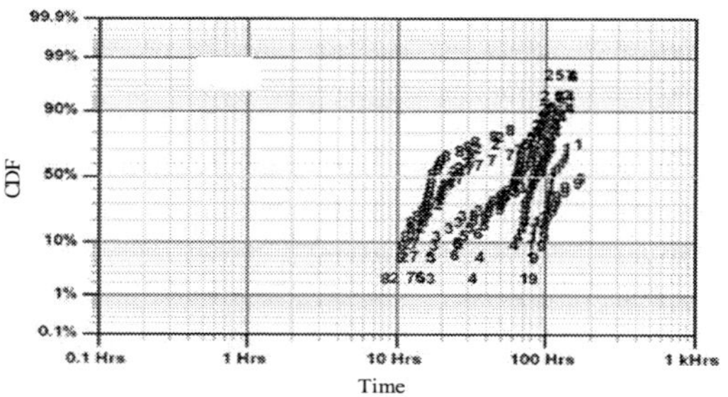

Fig 3. TTF of different via isolation profiles.

As show in Fig 4, TDDB test key is with the metal-via-metal structure, in which s refers to trench CD and l stands for the distance between via and trench. *l* represents the weakest point at such TDDB test key. Hence, TDDB failure can be partly ascribed to the marginal spacing distance (l) between via and trench. This is especially true for the case when via CD is larger than trench CD as shown in Fig 4(b). This issue has been solved by metal hard-mask open scheme to some extent. As CD could not be too small for the sake of gap-filling concern, the strict profile control from both via etch and trench etch is imperative from the point view of dry etching.

Fig 5 indicates the influence of via etch and ash process to via CD and its profile. With more C_4F_6/N_2 gas flow ratio in ME step and low-pressure *in-situ* strip, the profile in Fig 5

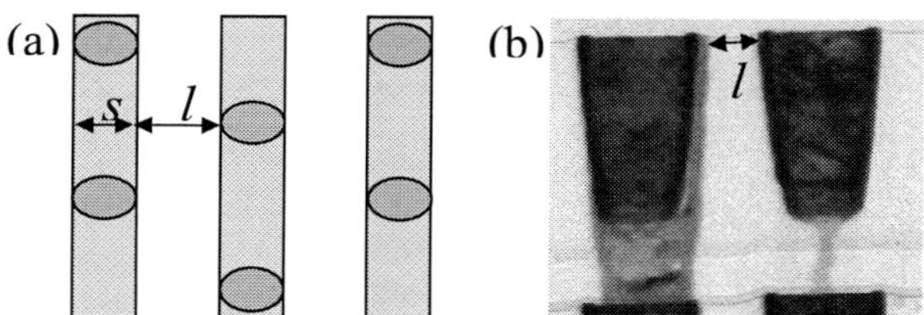

Fig 4. TDDB test structure (a) Top view and (b) TEM cross-section

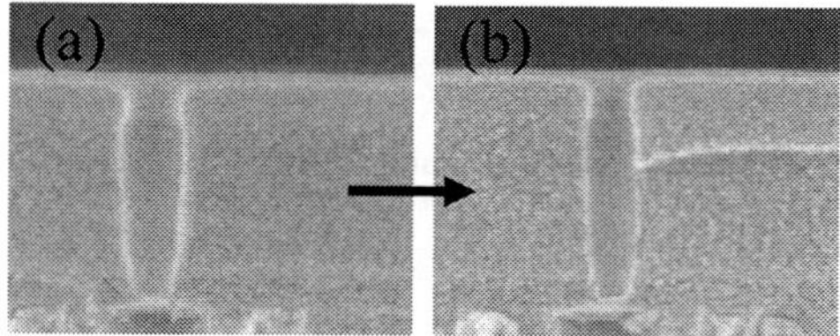

Fig 5. Via profile (a) non-optimized etch/ash process   (b) Optimized etch/ash process.

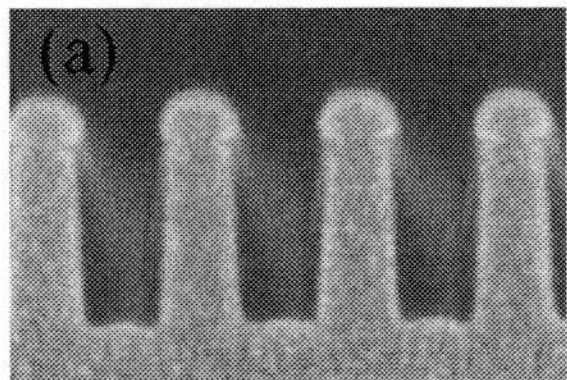
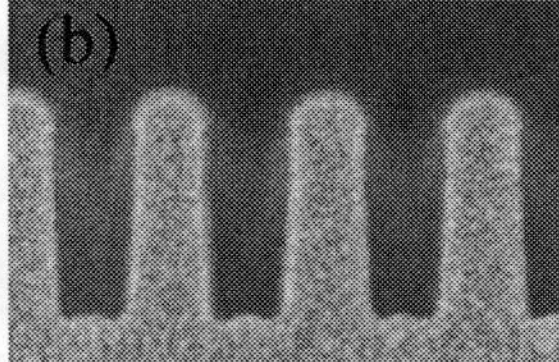

Fig 6. Trench profile (a) Non-optimized etch/ash process (b) Optimized etch/ash process.

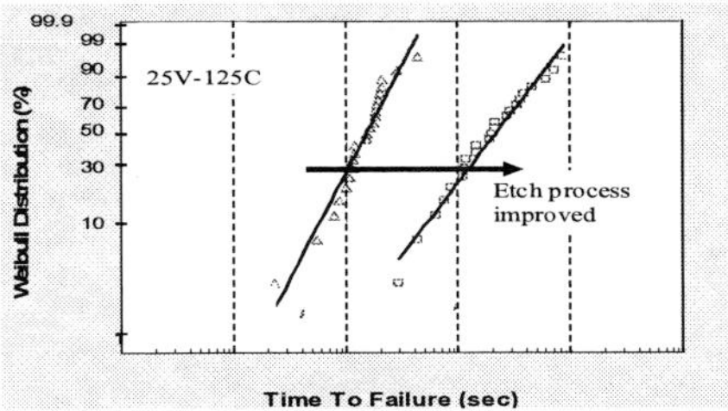

Fig 7. Etch process effect on TDDB

(b) becomes more vertical than Fig 5 (a). This can be attributed to less polymer generation and low-k damage in via etch. The bowing profile improvement will definitely increase the TDDB window. Nevertheless, via CD could not be arbitrarily reduced for it will drastically trigger the via open issue. Fig 6 demonstrates the trench profile improvement with the help of the optimized trench etch process. By increasing the sidewall polymer protection in low-k etch step and the CO_2 flow rate in ashing step, the sidewall pull back has been greatly alleviated after DHF clean. Finally, the distance between via and trench defined in Fig 4 (a) is wide enough to achieve the improved TDDB performance for low-k interconnects as shown in Fig 7. However, as mentioned in EM section, the tuning in ME step might impact the EM performance. Hence, the etching optimization for TDDB improvement could not be fully decoupled from that for EM enhancement. On etching side, there is a trade-off relationship between EM and TDDB.

Fig.8 is the schematic of SM test structure. It consists of two-layer metals, connected by a single via. Thousands of this unit form the entire via chain SM test structure. The stress temperature 225 °C is the worst SM performance temperature determined on the given test vehicle. SM failure could result in either electric circuit open or high electric resistance. Aside from the poor mechanics of low-k dielectrics and Cu diffusivity, polymer residue at via bottom also easily leads to the poor adhesion due to locally deteriorated coverage of barrier metal. It is one of main SM failure causes that are related to etch. We simplified the SM improvement by only optimizing the post etch treatment.

As shown in Fig 9 (a), the heavy polymer residue is detected after trench etch even with CO_2-based post etch treatment (PET). We identified it as polymer residue for it will be fully removed with high-flow rate O_2 strip as seen in Fig 9 (b). It should stem from the polymer residue accumulated in via during trench etch. Post wet clean finally carries them out of the via and part of them remains at the trench bottom. The heavy polymer at

Fig 8. Schematic of SM test structure

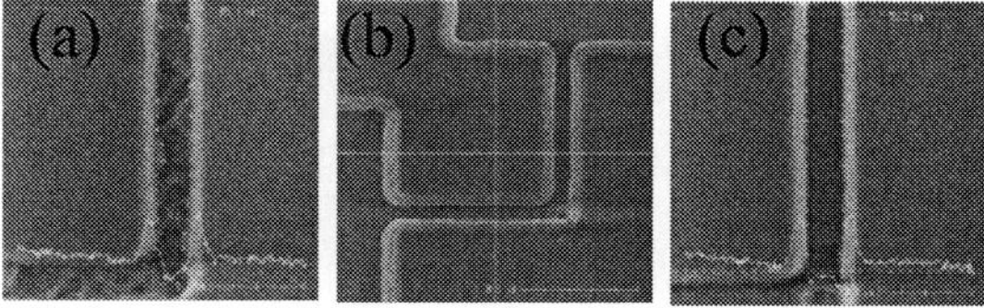

Fig 9. Post trench etch (a) Baseline CO_2 PET, (b) High flow O_2 strip, (c) PET I

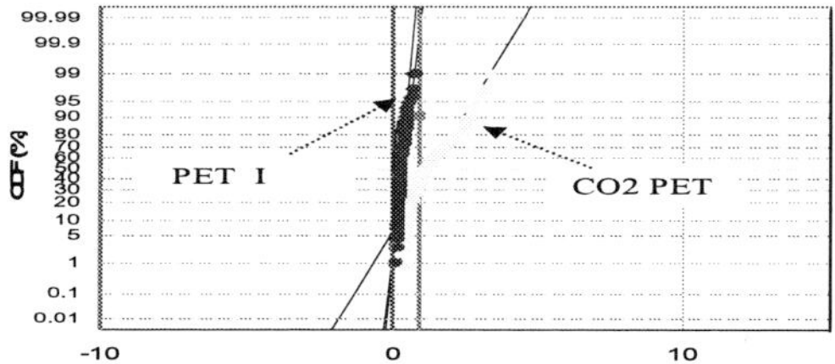

Fig 10. The effect of PET on R shift on via chain structure

via bottom is no doubt the root cause, partly from the poor clean capability of CO_2-based PET. Polymer lean PET I w/o CO_2 is evaluated and delivers the residue-free performance as shown in Fig 9 (c). Fig 10 compares the R shift between PET I and CO_2-based PET. The remarkable SM improvement can be observed for PET I. This verifies the appropriate PET selection is an effective and feasible knob for SM improvement. Once again, PET tuning could not be completely isolated from EM and TDDB for PET will impact the trench CD to some extent. The linear correlation between inline post-etch CD and treatment time of PET I is illustrated in Fig 11. Despite SM could be greatly improved merely with PET tuning, however, it could not be fully decoupled from EM and TDDB improvement for the ultimate polymer condition strongly depends on the joint actions from EM and TDDB optimization.

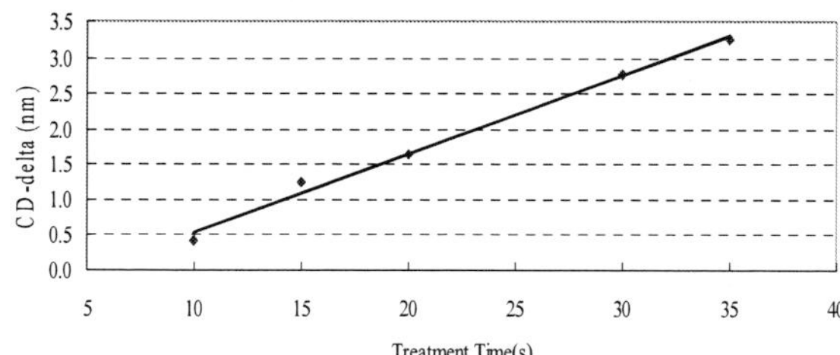

Fig 11. The correlation between trench CD and treatment time

CONCLUSIONS

Dry etch effects on Cu/low-k dielectric reliability such as EM, TDDB and SM have been evaluated with the via first patterning scheme on a sub65nm test vehicle. Both the lower via isolation height and the rounding chamfer could improve EM reliability. More vertical via profile, less low-k damage in trench etch are keys for TDDB enhancement. SM reliability can be recovered by introducing polymer-lean PET other than CO_2 PET. There exists a trade-off relationship for the dry etching optimization actions among EM, TDDB and SM. Roughly, the trench etch tuning for better EM and SM behavior might degrade TDDB.

ACKNOWLEDGMENTS

Authors would like to thank Mr. Jiangang Liu, Mr. Jihong Zhang and Mr. Junwen Huang of Lam Research China for the technical discussion.

REFERENCES

1. A. Fischer, et al, "Process optimization the key to obtain highly reliable Cu interconnects", International Interconnect Technology Conference, 2003.
2. O. Aubel, et al, "Process options for improving electromigration performance in 32nm technology and beyond", 47th Annual International Reliability Physics Symposium, Montreal, 2009.
3. K. Lee, et al, "Via processing effects on electromigration in 65nm technology", 44th Annual International Reliability Physics Symposium, San Jose, 2006.
4. F. Chen, et al, "Addressing Cu/Low-k dielectric TDDB-reliability challenges for advanced CMOS technologies", IEEE Transactions on electron devices, Vol.56, NO.1 Jan 2009.

ECS Transactions, 34 (1) 343-348 (2011)
10.1149/1.3567601 ©The Electrochemical Society

New Al Post-Etch Residue Remover with Al Surface Passivation Function

Joyce C.Y. Wei , Micky Huang

EKC Technology, DuPont Electronics and Communications, Hsinchu, 30078, Taiwan

A new Al post-etch residue remover with Al surface passivation function was developed. It is capable of removing polymer and inorganic residues with good compatibility to thin Aluminum, TiN, TaN, NiSi, and various silicon oxide films. A monolayer of passivation on the Al surface can be deposited *in situ* to prevent further oxidation and corrosion of the thin Al film. The passivation layer can be easily removed by a short plasma clean or baking to prepare the pristine Al surface for the successive processes. This solution is recommended for cleaning in FEOL high k/metal gate processes and contacts.

Introduction

In the fabrication of integrated-circuits (IC), liquid chemicals are used for wafer cleaning in various processes to remove particles or residues so that the substrate surface can be best prepared for the subsequent process steps. During wet cleaning, the unwanted substances must be removed from the underlying substrate materials without altering their properties. Ineffective cleaning can lead to wafer yield loss or reliability issue. The requirements for wafer surface cleaning are different depending on the preceding process and the materials involved. In general, in front-end-of-line (FEOL) cleaning processes, the substrate materials are mostly silicon, silicon oxides and silicon nitrides, which are robust enough to withstand harsh thermal and chemical operations, thus the conventional wet cleaning methods often employ mixtures of inorganic acids or ammonia with hydrogen peroxide (1). In the recent years, there has been a lot of development work on finding new materials for transistors such as replacing poly-silicon gate by metals with the implementation of high-k dielectrics (2, 3, 4, 5). This poses new challenges for wet cleaning because these newly-introduced FEOL substrates may not be able to endure such aggressive cleaning operations. In addition, the tolerance for substrate material loss is getting smaller as the device feature size keeps shrinking.

In this work, a post-etch residue remover was developed, which is compatible with the materials being used in the FEOL processes, including Al, TiN , TaN, NiSi and silicon oxide. In addition, a new functionality is designed into the remover for metal surface conditioning. It adopts a similar approach as in the control of metal corrosion by using corrosion inhibitors that bind to the surface to block corrosion reactions. However, surface conditioning is more than just corrosion retardation. An ideal surface conditioner is a chelater which forms an ordered thin monolayer on the pristine metal surface after the residue is removed from the metal. It should remain on the surface to protect the metal from corrosion and oxidation, and yet be readily detached upon plasma treatment or baking so that a fresh metal surface is prepared for the following processes.

The metal of interest in this work is Aluminum, and several types of ligands, including ionic surfactants (4), carboxylic acids (5) and poly-phenols (6) can adsorb onto the Al surface through different binding mechanisms. Of these chelators, it has been shown that long chain *n*-alkanoic acids (7) can form organized self-assembled monolayers on the Al surface, and thus are used in our current design.

Experimental Section

<u>Design of the Al post-etch residue remover</u>

The remover chemical design concept is depicted in Figure 1, showing the main components and their functions inside the dotted square. Additional Al surface passivation functionality can be achieved through the formation of a self-assembled monolayer of Al-chelating compound.

Pattered Si wafers with FEOL high-k/metal gate structures including Al, TiN, NiSi and TEOS were used to optimize the ratios of the components in the remover chemical to achieve complete residue removal with minimal etching on the substrate materials.

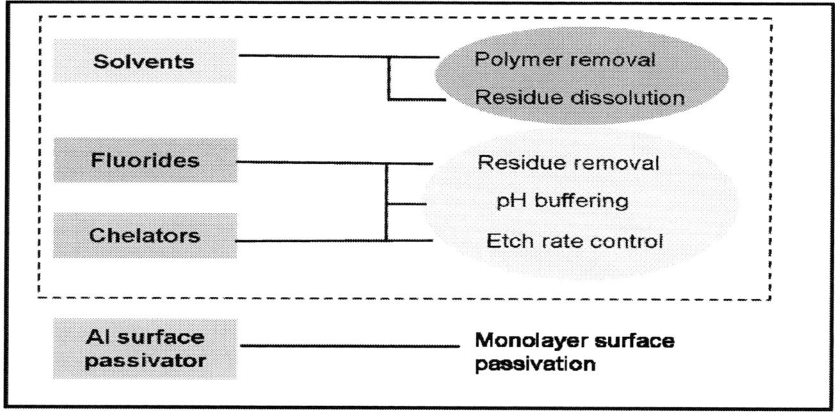

Figure 1. Al post-etch residue remover design concept. Inside the dotted-line box are the basic functionalities, where as Al surface passivation is the added new feature to fulfill the corresponding surface preparation requirement.

<u>Selection of Al surface passivator</u>

A long-chain *n*-alknaoic acids was chosen and doped into the remover solution and diluted to different concentrations. Blanket Al thin-film wafers were immersed in these Al passivator-doped chemical solutions at room temperature for several minutes, rinsed with DI water and blown dry with nitrogen gas.

X-ray Photoelectron Spectroscopy (XPS) was employed to check for surface conditions of thin-film Al after chemical treatment.

Contact angles of water on the chemically treated Al wafers were also measured to verify the formation of the surface passivation layer.

Results and Discussion

Mechanism of Al Surface Corrosion

One of the mechanisms proposed for Al corrosion after dry etch and wet clean is illustrated in Figure 2. The residual etch halide gas may form defects on Al surface after aging in ambient environment. These defects may be removed along with Al oxide by conventional APM (ammonium peroxide mixture) cleans or post-etch residue removers. However, during wafer queue time until the next process step, the corrosive halide gas may redeposit onto the Al surface, and metal-halide defects may grow back in the ambient environment.

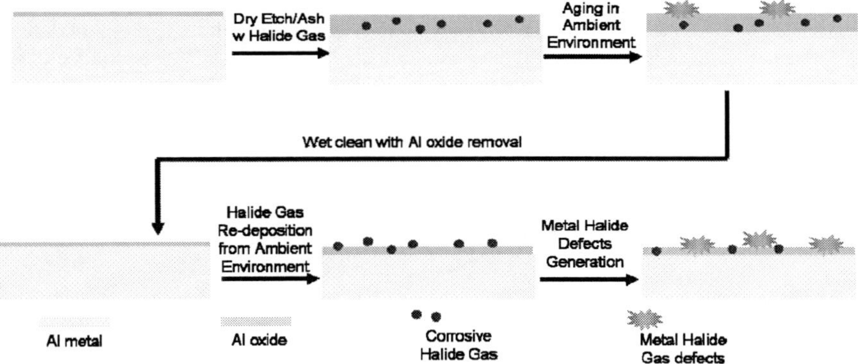

Figure 2. Al surface corrosion mechanism without surface passivation.

This mechanism hypothesis is supported by XPS analysis and contact angle measurement data shown in Table I.

TABLE I. XPS analysis and contact angle of Al surface after chemical treatment

Wafer surface treatment	XPS results of elemental concentration (%)								Contact angle
	C	O	N	F	Al	Cl	Br	Cu	
Post-etch, before wet clean	17.1	37.2	0.8	12.3	30.4	1.9	0.4	ND	86.99
EKC clean (no passivator)	12.7	40.1	ND	1.8	45.1	ND	ND	0.2	29.55
APM clean	19.4	42.7	0.6	0.6	36.4	0.3	ND	0.1	30.87

After the dry etch and ash processes, the Al surface is oxidized and the residual etch gas molecules may diffuse into the oxide layer. XPS data confirm the existence of F, Cl and Br on the Al surface. The high contact angle value also indicates a hydrophobic Al oxide surface. Immediately after EKC or APM clean, the relative concentrations of halides remaining on the Al surface decreased significantly, and as a result, relative concentrations of other elements such as O and Al increased. It is worth pointing out that the concentration of Carbon on post-EKC clean surface is lower than that of post-APM clean surface, while as Al signal on post-EKC clean surface is higher than that on the post-APM clean surface. This suggests that EKC remover has a better residue removal capability than APM. The contact angles are similar for these two chemically-treated surfaces, and the low values also indicate a more hydrophilic Al metal surface.

Self-assembled Monolayer of Passivator on Al Surface

In the new Al post-etch remover, a long-chain alkanoic acid capable of forming a self-assembled monolayer on the Al surface in the remover chemical was selected as the Al passivator. Contact angles of a series of dilutions were measured to determine the best doping concentration range for forming the monolayer. As shown in Figure 3, upon increasing of the passivator concentration, the molecules start to attach onto the Al surface until they form an ordered monolayer on the surface, and the contact angle reaches a maximum value that is above 90 degrees. When more passivating molecules are doped into the solution, they start to attach onto the first monolayer to form randomly oriented secondary layers, and as a result, the contact angle will start to decrease. Precipitation of a very thin film of the alkanoic acid can also be visually detected at higher concentrations. Therefore, by using contact angle measurement, the best chelating concentration range for monolayer formation can be determined.

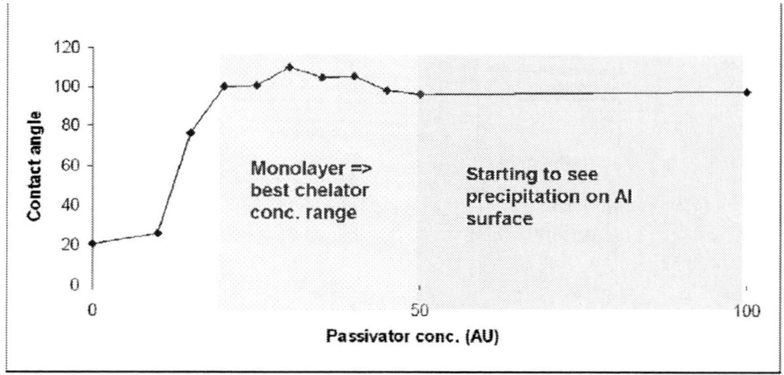

Figure 3. Contact angle as a function of Al surface passivator concentration.

Mechanism of Al Surface Conditioning

Al surface protection after wet clean is desirable so that defect growth or corrosion can be suppressed to permit a longer wafer queue time, which would create a wider production process window.

Based on the previously discussed mechanism for Al corrosion (Figure 2), a designed mechanism for Al surface conditioning is constructed with supporting evidence from XPS and contact angle data.

As depicted in Figure 4, after dry etch and ash, Al surface is cleaned by the novel EKC remover with Al passivating feature. It not only removes the residues and the halide contaminants but also deposits a passivation layer on the metal surface. This is supported by the large contact angle (Table II) and the reduced XPS signals of F, Cl and Br. The formation of an organic monolayer on the metal surface also explains the increase of Carbon concentration.

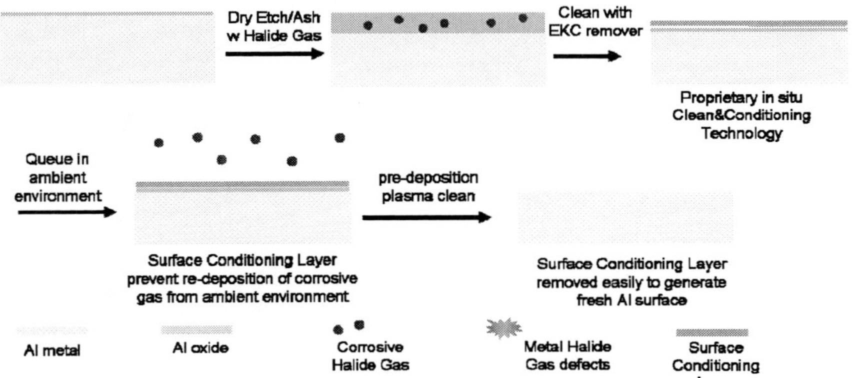

Figure 4. Mechanism of Al surface passivation and conditioning.

Now that a surface passivation layer is in place, it would prevent the deposition of the corrosive halide molecules and also retard the oxidation of Al surface. This would prolong the process queue time control. However, this organic thin film should be easily removed to produce a fresh Al metal surface that is ideal for the successive processes such as deposition. To verify this design, 10 seconds of Ar ion sputtering was applied to the passivated Al wafer surface, and XPS data show a significant drop of Carbon concentration, providing evidence that the monolayer has been removed and that fresh Al metal surface is generated, which is also supported by the increased signals of Al. Therefore, this new technology is better described as Al surface conditioning rather than just surface passivation or corrosion inhibition because it also generates a fresh, pristine Al metal surface ready for the next process.

TABLE II. XPS analysis and contact angle of Al surface

Wafer surface treatment	XPS results of elemental concentration (%)								Contact angle
	C	O	N	F	Al	Cl	Br	Cu	
Post-etch, before wet clean	17.1	37.2	0.8	12.3	30.4	1.9	0.4	ND	86.99
EKC clean (with passivator)	22.8	33.6	ND	3.0	40.5	ND	ND	0.1	100.84
EKC clean+ ion sputtering	1.3	36.7	ND	3.1	58.8	ND	ND	0.2	-

Conclusion

We have demonstrated in this work a novel design for Al post-etch residue remover which not only can remove residues with compatibility to various substrate materials, but can also deposit *in situ* a self-assembled monolayer on the Al surface to protect the surface from defect growth and corrosion during aging in an ambient environment. In addition, this passivation layer can be readily desorbed by a short sputtering step to reveal a fresh metal surface ideal for the subsequent processes. This new surface conditioning technology is recommended for FEOL cleaning processes such as high k/metal gate or contacts where thin metal substrates like Al are prone to corrosion.

Acknowledgments

The authors would like to thank Dr. Robert M. Hsu for his contributions in sharing his experiences in wet clean chemistry and metrology development.

References

1. S. Wolf, *Microchip Manufacturing,* Lattice Press, California (2004).
2. S. Beckx *et al*, *Microelectronics Reliability*, **45**, 1007 (2007).
3. R. Chau, S. Datta, M. Doczy, B. Doyle, J. Kavalieros and M. Metz, *IEEE Device Letters,* **25**(6), 408 (2004).
4. M. Batouti, *J. Mater. Sci. Technol.,* **15**(1), 39 (1999).
5. A. Martell, R.J.Motekaitis and R.M.Smith, *Polyhedron,* **9**(2/3), 171 (1990).
6. P. Karlsson, A.E.C. Palmqvist and K. Holmberg, *Adv. In Colloid and Interface Sci.,* **128-130**, 121 (2006).
7. Y.-T. Tao, *J. Am. Chem. Soc.,* **115**, 4350 (1993)

ECS Transactions, 34 (1) 349-353 (2011)
10.1149/1.3567602 ©The Electrochemical Society

WAT and VBD Distribution Improvement on Low-K Trench All-in-one process

Jeff Song[a], David Duan[a], Andrew Liu[a]
Cheng Lien Huang[b], Hao Zhijie[b], Jemmy Hendrianto[b], Huang Junwen[c]

[a]Semiconductor Manufacturing International Corporation, No. 18 Wenchang Rd., BDA, Beijing 100176, P.R. China
[b]Lam Research Corporation, Beijing branch, No. 15 Ronghuazhong Rd., BDA, Beijing 100176, P.R. China

Abstract

WAT (Wafer Acceptance Test) and VBD (Voltage Breakdown) distribution are not merely a function of critical dimension and oxide thickness as were seen on conventional BEOL process on larger technology node. The complexity of electrical and reliability parameter becomes more on advanced technology node, i.e. 65nm and sub 65nm, as its k value and Cu surface plays some significant role on the parametric distribution result. This study is addressed to improve the WAT and VBD distribution by minimizing the Cu surface and low-k damage, such that the impact to the electrical parameter is small enough that can be negligible, hence the WAT and VBD will be more predictable and controllable. Optimization on this study would be focused on the chemical selection of the PR stripping and PET (Post Etch Treatment) steps, which are the critical steps for low-k and Cu surface damages.

Introduction

In scaling interconnect toward the 65nm node, low-k/Cu interconnects with a narrow pitch less than 200nm have been developed. With the narrower pitch thus smaller trench CD, the optimization of the process polymerization level and preservation of the k-value became extremely crucial for the WAT and VBD distribution. Sufficient polymerization level in barrier removal process is essential for obtaining a desired vertical profile. However, excessive polymer, i.e. CFx radicals, degraded the underlying Cu. These happened because the plasma-surface interactions were not controlled. Hence, a quantitative control over the amount of CFx radical species by O2 flow optimization in barrier removal process has been conducted[1].

We also investigated the excessive O2 in PR stripping process deteriorates k value of the dielectric materials, indicated by VBD distribution becomes more divergent. The substitution of O2 to CO2 gas as PR stripping chemical reactants is proven to considerably improve the VBD distribution.

As for the post barrier removal treatment, H2 gas is shown to be an effective Cu surface cleaning chemical[3]. The post cleaning is critical due to Cu seed barrier requires high purity, smooth interface, good adhesion to the barrier metal and low thickness concurrent with coherence for ensuring void-free fill[7]. The idea of using H2 plasma cleaning is to remove light impurities from metallic materials, as well as volatile hydrides formed. Thus, in this study, we investigated how effective the H2 treatment to remove oxides and etch residues that contaminate the metal surface at the bottom of via.

Experiments

Wafer Acceptance Test improvement:

1. After optimizing the barrier removal process by adjusting the power, pressure, and gas ratio to meet the required profile, then we did some split tests on the post etch treatment to see how the Rc_chain behaves with different Cu surface processing. We processed the same batch of wafers in the same process chamber, by using two condition splits, i.e. one is without H2 treatment, and the other one is with H2 treatment. The table below is the Rc_chain Standard Deviation data (after normalized), and we can clearly see the standard deviation is improved by around 15% after introducing H2 plasma treatment at post barrier removal process. This finding is very interesting since other post etch treatment other than H2 gas that we had tested has never given a significant improvement as being seen on H2 plasma treatment.

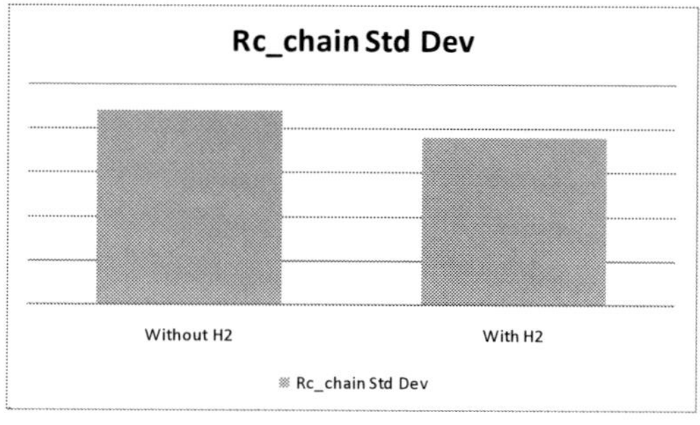

2. A study of barrier removal process with different O2 gas flow also was conducted in this experiment. Before we did the electrical split, we first make sure that the profile of barrier film is vertical, so that we can eliminate the discrepancy of WAT value induced by profile change. What we are interested in this study is how the Cu surface behaves differently with increasing O2 flow, detected by electrical test. As can be seen in the chart below, the higher the O2 flow during the barrier removal process, the higher the standard deviation of the via resistance (Rc) would be. The finding indicates that O2 gas amount is crucial for controlling the extent of oxidation of the Cu surface, which makes the barrier removal process so critical and process optimization is absolutely necessary when dealing with this process.

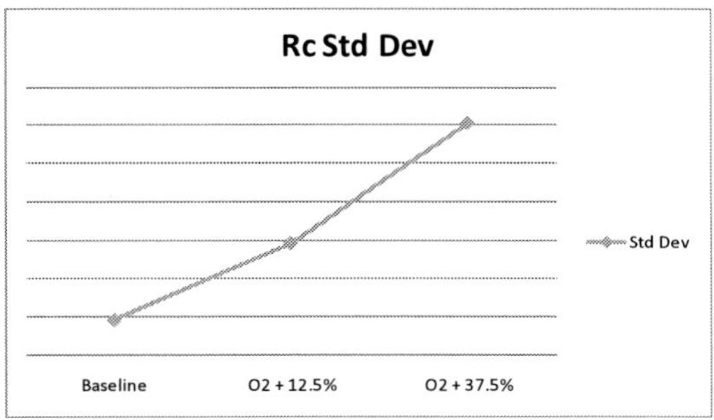

Voltage Break Down Improvement:

Our study shows that the preservation of k-value on low-K process is very crucial to improve the VBD distribution. High amount of O2 plasma is widely used for stripping the photoresist in etch technologies. However, when dealing with low-k process on advanced technology node, excessive O2 plasma could impact the k-value, which is eventually could influent the VBD distribution performance. We did a study on different photoresist stripping process, i.e. O2 and CO2 plasma. The finding is interesting since the Vbd distribution is significantly improved by implementing CO2 plasma instead of O2 plasma for photoresist strip process. The CO2 might be already widely studied on sub 65nm technology, but how it impacts to VBD was not widely studied before.

Results and Discussions

a. The impact of Cu surface to WAT (via resistance):

The Cu surface condition can be detected by electrical test, which is indicated by different Rc standard deviation observed with different Cu surface treatment. As being mentioned by other publications, H2 plasma treatment can effectively clean the impurities on Cu surface, as well as it can reduce the level of surface oxidation. Clean Cu surface is possibly done by O2 plasma treatment, but O2 plasma worsened the Cu oxidation. That is why H2 plasma treatment provides the benefit, which is able to take care both the surface cleanliness and surface purity. Still within the scope of oxidation, excessive amount of O2 gas during the barrier removal process can be harmful to the via resistance distribution. It is sometimes difficult for a process engineer to make a decision to reduce the O2 gas, since O2 gas is good for meeting the vertical profile requirement in barrier removal process, meanwhile it also oxidized the Cu surface badly. Hence, eventually a process engineer has to reduce the O2 amount without doubt, and take care the barrier verticality profile by adjusting other parameters, such as power, pressure, etc.

b. The impact of photoresist stripping chemistry to VBD:

O2 plasma used during photoresist stripping is able to react with low-K film, which etches away the carbon material in low-K film, and eventually the low-K film became oxygen rich. This oxygen rich low-K film became similar to silicon oxide, which has higher k value. The higher k value could influence the breakdown voltage (VBD) value. If the O2 plasma itself has certain etch rate map that is not uniform enough, the extent of low-k damage also would become non-uniform, that's why it could induce larger distribution of breakdown voltage.

Conclusions

To improve the via resistance distribution, we cannot only focus on CD and profile, but the materials surface reaction has to be taken into consideration. Hence, adding an H2 post etch treatment is crucial to clean the Cu surface as well as to reduce the oxidation on Cu. Excessive Cu oxidation after barrier removal etch could induce excessive Cu loss during post-etch wet clean, thus creating uncontrollable factor which impacting Rc uniformity. Surface reaction between low-k and O2 plasma is also significant in influencing the breakdown voltage distribution. CO2 plasma is used to substitute the O2 plasma in photoresist stripping process to avoid worsened Vbd distribution due to the uncontrollable low-k damage.

Acknowledgements

The authors thank Su Xingcai (Lam Research), as well as Zhang Xin (SMIC) for their technical consultancy and assistances during this work. This work is a collaboration product between SMIC Beijing and Lam Research China on the continuous improvement project.

References

1. K. Yatsuda, T. Tatsumi, K. Kawahara, Y. Enomoto, T. Hasegawa, K. Hanada, T. Saito, Y. Morita, K. Shinohara, and T. Yamane, *Manuscript (to be submitted).*
2. K. Yatsuda et al., *Advanced Metallization Conference* (2003) Asian Session 34, *and Materials Research Society Book.*
3. Juhnwan O, Seongwook Lee, Jae-Bum Kim, and Chongmu Lee, *Journal of the Korean Physical Society*, Vol. 39, December 2001, pp S472-477.
4. R. Knizikevcius, V. Kopustinskas, *Microelectronic Engineering* Vol. 83 (2006) p. 193-196.
5. T.K. Kang and W.Y. Chou, *J. Elecrochemical Society*, 151, G391 (2004).
6. M.T. Wang, Y.C. Lin, and M.C. Chen, *J. Electrochemical Society*, 145, 2538 (1998).
7. R.L. Jackson, E. Broadbent, T. Cacouris, A. Harrus and T. Walsh, *Technical papers.*

ECS Transactions, 34 (1) 355-359 (2011)
10.1149/1.3567603 ©The Electrochemical Society

Clean mode Al etch process development for defect reduction

Fan Qiang[a], Cheng Lien Huang[a], Jemmy Hendrianto[a],
Jeff Song[b], Mil Lv[b], Ken Wang[b], Cris Shi[b],

[a]Lam Research Corporation, Beijing branch, No. 15 Ronghuazhong Rd., BDA, Beijing 100176, P.R. China
[b]Semiconductor Manufacturing International Corporation, No. 18 Wenchang Rd., BDA, Beijing 100176, P.R. China

Abstract

Clean mode chamber condition controlling provides a good method to improve Al etch defect performance. In the same time, it brings significant difference of physical performance such as CD, side wall, corrosion etc. In this study, in order to maintain clean chamber condition, Wafer-less Auto Clean(WAC) is introduced to clean out by-products deposition and optimized to achieve stable particle performance during full MTBC(mean time between clean). To compensate process shift because of chamber condition change, Al etch process and PR strip process are optimized. Cl2/BCl3 ratio, bias power and Cl2 gas flow in DARC step were found having significant impact on CD controlling. PR strip process adjusting has been proved to have great help to enlarge corrosion process window. One question that different Cl2/BCl3 ratio affect plasma sustaining when etching through TiN/Al interface was found during this study and need further efforts to find out the detail mechanism.

Introduction

Chamber condition controlling is the most important factor of Al etching process defect maintenance, which is main challenge for productivity enhancement of Al backend technology. Good chamber condition controlling is required not only to be able to keep the stable process performance but also good defect level during whole cycle of chamber wet clean. To achieve longer MTBC and keep good defect level as well, could help to enhance tool uptime, reduce yield loss.

The VersysTM Metal 2300 design[1] is compatible with use of O2 based waferless auto clean (WAC) and waferless recovery steps. Clean mode provides a good method to improve Al etch defect performance. In the same time, it brings significant difference of physical performance such as CD, side wall, corrosion etc.

In this study, in order to maintain clean chamber condition, WAC is introduced after every wafer to clean out by-products deposition. WAC time is adjusted to achieve stable particle performance during full MTBC. Al etch process and PR strip process were optimized to compensate process shift because of chamber condition change. Cl_2 gas flow[2] and Bias power in DARC step has been found having significant impact on CD controlling. Because clean mode chamber condition made less sidewall by-product deposition to lead poor corrosion performance, Al-etch process gas ratio and bias power optimization are used to bring more side wall passivation. PR strip process was adjusted by using O2 heat up step and adding Vapors[3] in strip step to enlarge corrosion process window.

Experiment

2.1 Process optimizing

When WAC was introduced, by-products deposition on the chamber wall was removed by WAC which leads CD getting smaller, sidewall protection getting worse, and PR remaining less. In order to control CD, different Cl2 flow and Bias power have been tried in DARC step.
To improve PR remaining, OE2 step TCP power and pressure were adjust to improve PR selectivity.
Based on OES(Optical Emission Spectroscopy)curve, according different thickness of Al-Cu layer WAC time was optimized to make sure chamber was cleaned completely.
Sidewall protection getting worse when chamber condition changed can also lead corrosion window narrow. Optimizing PR strip recipe was chosen to improve corrosion performance. O2 heat up step was used to replace N2 heat up step. Vapor was added to PR strip step

2.2 Methodology

Oxide and PR ER tests were use 4K thermal oxide wafer and 30K I-line PR wafer. Film thickness measurements were conducted on KLA. Selectivity calculation is (Oxide Etch Rate)/(PR Etch Rate). CD measurement was conducted on Hitachi CD SEM. Cross section profile was checked on Hitachi 5200 SEM.
Corrosion test was conducted as dry box 72hrs test (after wet clean). Defect was scanned by KLA F5X, reviewed by SEMVison.

2.3 System configuration

Process development was carried out on VersysTM Metal 2300 system. Standard process chamber combined with Microwave stripper.

Result and Discussion

3.1 DARC step adjusting effects on CD controlling
DARC step Bias power and Cl2 flow were adjusted to check CD bias response. Result is showed in chart1. Bias power in DARC step has positive correlation with CD. Increasing bias power brings higher Vpk, which means there will be high Iron bombardment on the wafer. As a result, PR erode passivation will increase to made DARC profile taper and CD get larger.

One the other hand, chlorine flow in DARC step recipe can significantly affect Bias power influence on CD controlling. When 60sccm Cl2 was used in DARC step, 30W bias power change only leads 2nm CD bias change. In contrast, same bias power change can lead 16nm difference. Normally, Cl2 provide more chemical-like etch, which is isotropic. Compare 40sccm Cl2, 60sccm Cl2 recipe has less isotropic etching. And also, higher total flow Decreases mean free path and reduce ion bombardment. Lower ion bombardment makes less carbon polymer comes from PR erosion. Therefore, CD did not change obviously.

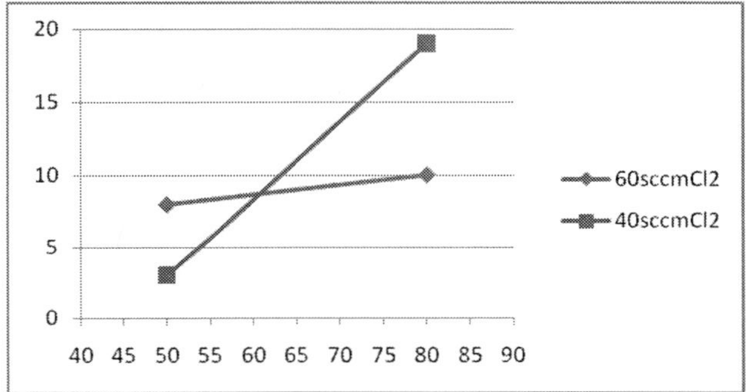

Chart 1: DARC step Cl2(sccm) and bias power (W) influence on CD bias(nm)

3.2 PR selectivity improvement
PR erosion is the most important carbon polymer sources for Al etch. Bias voltage is the most effective index for PR ER. Selectivity test of OE2 step results are shown in Chart2.

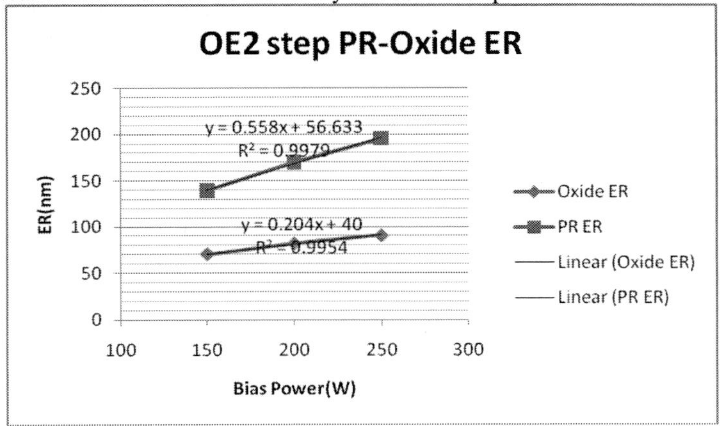

Chart 2: OE2 step PR-Oxide ER test

From Chart2, when Bias power was reduced, PR Etching rate dropped faster than oxide which means PR selectivity get improved.
When bias power was keep at 120W, increasing TCP power from 700W to 1000W can made PR ER from 134nm/min to 123nm/min. this test proved that PR ER is strongly correlated with Bias voltage. Higher TCP power will lead higher plasma density which will made less ion bombardment.

Lower bombardment could help to gain more PR remaining; it can also improve PR profile, which to get more protection on DARC.

3.3 Strip recipe optimization

Using O2 pin up heat up step; Using VoDM only passivation; adding Vodm into strip step have been done as strip recipe optimization to resolve the corrosion issue, which came out after Main etch recipe changed. The pre-heat step plays a critical role in corrosion control. The goal[4] of baking out a significant amount of the chlorine is accomplished by gradual heating of the wafer under certain gas flowing conditions, at relatively high pressure and no power. The pins-up condition provides the slow gradual heating that the wafer needs.

Adding H_2O in passivation and strip step can remove most of the chlorine that is trapped on the sidewall polymers and on the remaining photo-resist surface.

Conclusion

Process will shift when Al-metal etch chamber condition was converted from deposition mode to clean mode, because of different polymer condition on chamber wall. To compensate the process shift, recipe optimization is needed.

DARC step bias power has significant impact on CD controlling, especially at low Cl2 flow recipe. OE2 step PR etching rate has linearity correlation with bias voltage. Increase TCP power and decrease bias power is effective for improve PR to oxide selectivity. Heat up step use pi-up mode, adding H2O in passivation and strip recipe, can enlarge corrosion window. After converted to clean mode, Chamber MTBC has been improved significantly from 30hrs to current 70hrs(chamber will not be open until defect fail)

Questions

During recipe development there are two questions we met:

1. Different Cl2/BCl3 ratio affect plasma sustaining when etching through TiN/Al interface was found and need further efforts to find out the detail mechanism

2. Another stable plasma status was found with one recipe, which has lower OES intensity and higher Bias voltage, but similar etching rate, compare with normal process. It is suspected to be a kind of double layer[5] need further study.

References

1. J Brett Richardson CEE-27706 Revision: A,Date: 12-Jan-2001
2. Greg Goldspring, CEE-21000 Revision: A,Date: 6-Dec-98
3. Cl Level Effects on Corrosion for Various Metallization Systems. N. Parekh and J. Price, J. Electrochem. Soc. Vol. 137, N° 7. July 1990).

4. Corrosion on Plasma Etched Metal Lithography from Processing and Environmental Factors. K.E.Mautts, Electrochemical Society Proceedings, Vol. 97-31, page 240, 1997.N.

5. Plihon,a! C. S. Corr, and P. Chabert, APPLIED PHYSICS LETTERS 86, 091501 s2005d

ECS Transactions, 34 (1) 361-364 (2011)
10.1149/1.3567604 ©The Electrochemical Society

Dummy Poly Silicon Gate Removal by Wet Chemical Etching

T. Yang, H. X. Yin, Q. X. Xu, C. Zhao, J. F. Li, D. P. Chen

Key Laboratory of Microelectronics Devices & Integrated Technology, Institute of
Microelectronics of Chinese Academy of Sciences, Beijing, 10029, China

For the gate last integration scheme, dummy poly silicon gate
removal is one of the indispensable processes either for a high-k
first or a high-k last route. In this paper, experimental results of
dummy poly silicon gate removal using TetraMethyl Ammonium
Hydroxide (TMAH) chemical etching are presented. The
preliminary results show that the poly silicon removal rate was
highly sensitive to the wet etch conditions. By optimizing the wet
etch conditions, high selectivity of poly silicon with respect to
SiO_2, Si_3N_4 and hafnium silicon oxynitride (HfSiON) was obtained.
A gate trench of critical dimension (CD) of about 70 nm without
poly silicon residue was fabricated using optimized conditions.
The excellent etch capability of the optimized wet etch process was
also demonstrated by controlling trench profiles.

I. Introduction

As CMOS logic devices are scaled in their dimensions from 45 to the 32 nm node, a lot
of advanced process technologies may have to be introduced to overcome barriers
introduced by pitch scaling, such as the extension of immersion lithography, mobility
enhancement technology, ultra-shallow junction technology, etc (1). In these technologies,
the metal/high-k (MHK) gate stack is of great importance in many aspects such as
significantly decreasing the gate tunneling current and gate resistance, eliminating the
depletion effect of the poly silicon, the boron penetration problem for PMOSFET; and
alleviating Fermi-level pinning effects (2). Two integration schemes for MHK
engineering can be used. The gate first scheme, which is similar to the conventional
CMOS process, and the other is gate last scheme. By now, the gate last scheme have
prevailed over gate first schemes because the thermal stability of both the high-k and
metal gate stack can be maintained by avoiding the high temperature source/drain
activation anneal of the MHK stack (3). For the gate last scheme, we try two different
integration routes, high-k first and metal gate last route (high-k first) and pure gate last
route (high-k last). For the high-k first/metal gate last route, high-k is deposited before
removing the poly silicon dummy gate, and the deposition of the gate metal is done after
the removal. For the high-k last route, the high-k was deposited after removing the poly
silicon dummy gate, covering the bottom and the sidewalls of the gate trench. For the
gate last integration scheme, the dummy poly silicon gate removal is an indispensable
process. Both complete poly silicon removal and perfect trench profile control are keys
for the dummy poly silicon gate removal process, which will in turn affect the subsequent
deposition of the high-k for high-k last scheme or metal gate material for high-k first
scheme. This paper presents a series of experimental results of dummy poly silicon gate
removal by using TetraMethyl Ammonium Hydroxide (TMAH) chemical etching in
order to find an appropriate process to remove dummy poly silicon gate in the gate last
integration scheme.

II. Experimental Procedure

The samples were prepared on 4-inch P-type (100) oriented silicon wafers with a resistivity of 8-12 Ω-cm. Hafnium silicon oxynitride (HfSiON) was selected as the high-k material, which was prepared by co-sputtering Hf (99.999%) and Si (99.999%) targets for deposition on silicon substrate with a thin SiO_2 layer (~1nm) on top in Ar/N_2 ambience, followed by a rapid thermal anneal at 900 °C for 30 sec (4). Other dielectric layers were deposited using conventional chemical vapor deposition (CVD) and furnaces. Blanket wafers of the respective films were prepared to monitor the wet chemical etch properties, which are summarized in Table I. Poly silicon gates of critical dimension (CD) of about 70 nm were fabricated by the joint effort of a JEOL6300 E-beam tool and a LAM Rainbow 4520 tool. The isolation spacers were composed of Si_3N_4 and a low temperature silicon oxide (LTO). LTO was also used as the inter layer dielectric (ILD) in this paper. The electronic grade TMAH aqueous solution was used as the wet chemical for the poly silicon dummy gate. This is usually used as silicon substrate wet etch in the fabrication of micro-electromechanical systems (5). TMAH concentrations were varied from 10 to 25 vol%, and the processing temperatures were from 50 to 80 °C. All wet etch tests were processed in a quartz tank.

The thickness of the HfSiON layer was measured by a GES5E ellipsometer, and the thickness of the other dielectric layer by a NanoSPEC elcometer. The poly silicon gate removal and trench profile after TMAH wet etching were evaluated by carrying out cross-sectional scanning electron microscopy (SEM) characterization on a Hitachi 4800.

Table I. Five Types of Films for Wet Chemical Etching

Films	Preparing Method	Layer Thickness (Å)
Poly silicon	LPCVD	4000
SiO_2	Furnace	100
LTO	LPCVD	6000
Si_3N_4	PECVD	1000
HfSiON	Co-Sputtering PVD	30

III. Results and Discussion

The etch rate (ER) of poly silicon on blanket wafers in 10 vol% TMAH at different process temperature is shown in Fig .1 (a). The ERs of poly silicon on blanket wafers in TMAH solution with different concentrations etched at 80 °C is summarized in Fig. 1 (b). One can see that the processing temperature had a remarkable impact on the ER, while the TMAH concentration had little effect on the ER. The ER increased by about 3 times when the temperature changed from 50 to 80 °C in 10 vol% TMAH. The ER remained unaltered as the TMAH concentration increased from 10 to 25 vol% at 80 °C process temperature.

The final wet etch condition was chosen as 10 vol% TMAH and 60 °C process temperature. Under this condition, the ER for thermal silicon oxide, silicon nitride, LTO, and HfSiON were also checked. For all these dielectrics, the ER selectivity were higher than 300. For HfSiON and silicon nitride, the selectivity was over 1000, as shown in Table II.

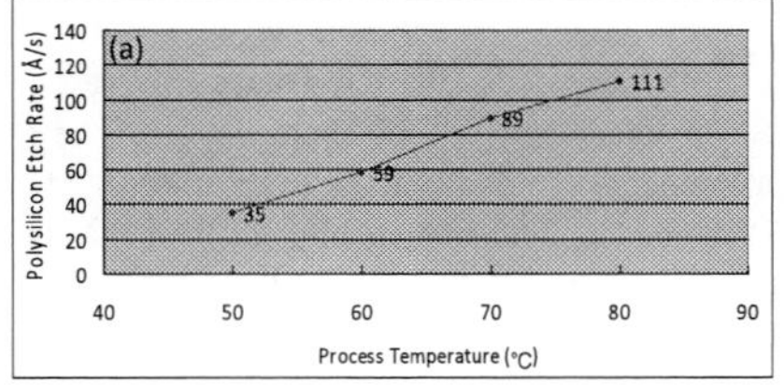
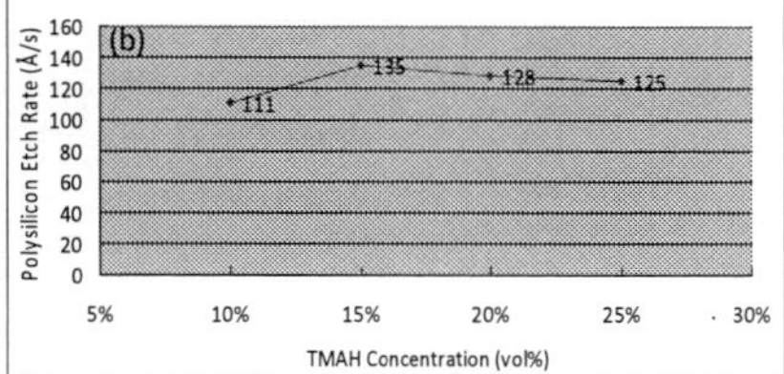

Figure 1. The etch rate of poly silicon (a) in 10 vol% TMAH at different process temperature, (b) in TMAH solution with different concentrations at 80 °C.

Table II. ER Selectivity of Poly Silicon with respect to Other Dielectric Layers in 10 vol% TMAH solution at 60 °C

Poly vs HfSiON	Poly vs Nitride	Poly vs Thermal Oxide	Poly vs LTO
2671	1469	392	367

It was found that the pre-removal of silicon oxide in dilute HF solution was very important to trigger the wet etching of the poly silicon dummy gate in TMAH, since a very thin layer of silicon oxide will completely prevent the attacking of poly silicon from TMAH. Two cross-sectional SEM micrographs for samples prepared by high-k first and high-k last procedures are shown in Fig. 2(a) and Fig. 2(b). Obviously, for both high-k first and high-k last procedures, poly silicon gate trenches with CD of about 70 nm were produced. The poly silicon dummy gate was completely removed and no poly silicon residue can be found, even at the corners of trench. In addition, no over etching was observed and good trench profiles were obtained due to the high ER selectivity of poly silicon with respect to other materials mentioned above. The ideal transistor electric performance were obtained for both high-k first and high-k last procedures by using TMAH remove dummy poly silicon gate, which will be published by our research group.

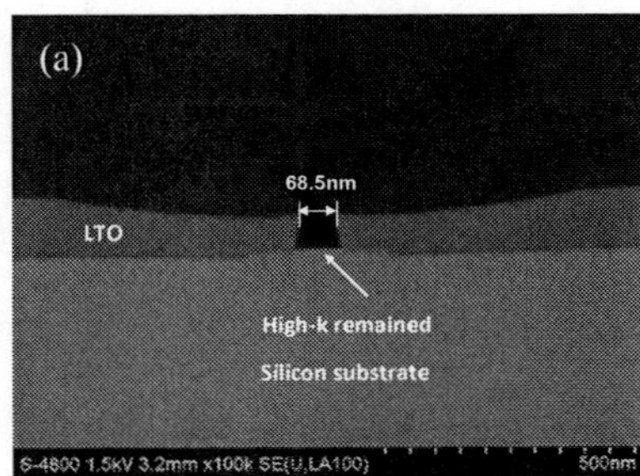

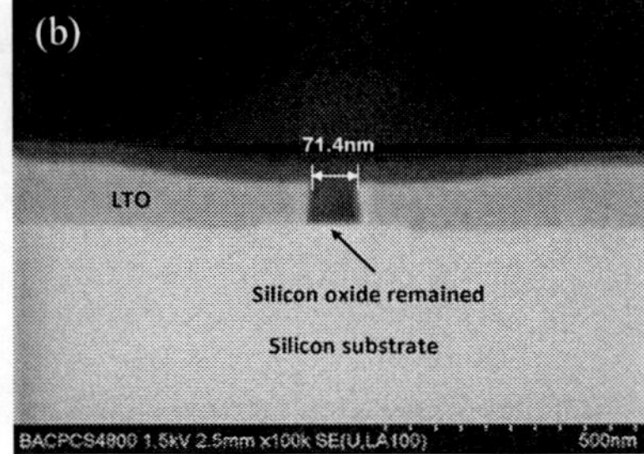

Figure 2. Cross-sectional SEM images of gate trenches after dummy poly silicon gate removal (a) in high-k first procedure, and (b) in high-k last procedure.

IV. Conclusions

A method of using TMAH as a chemical for removing the dummy poly silicon gate to form a gate trench was investigated for both high-k first/metal last and high-k last/metal last routes. It is found that poly silicon can be etched effectively by dipping the wafer in TMAH aqueous solution in a range of concentrations between 10 to 25 vol% after a dilute HF dipping for removing the native SiO_2 on the top of poly silicon. The etching rate is much more sensitive to the process temperature than to the TMAH concentration. High selectivity of poly silicon with respect to thermal SiO_2, Si_3N_4, LTO and HfSiON was achieved in 10 vol% TMAH solution at 60 °C. Gate trenches with CD of about 70 nm were successfully fabricated using the selected conditions. The dummy poly silicon gate was removed completely and a good trench profile was observed. The above results demonstrate that the TMAH wet chemical etching is a promising method for the dummy poly silicon gate removal in the gate last integration scheme.

Acknowledgments

The authors acknowledge B. L. Yang, Y. Y. Zhao, G. B. Xu, J. Luo, T. T. Li and G. L. Wang for their technical and experimental contributions. This work is supported by the State Key Development Program for Basic Research of China (No. 2006CB302704).

References

1. H. M. Wu, G. H. Wang, R. Huang and Y. Y. Wang, *J. Semiconductor*, **29**, 1637 (2008).
2. E. P. Gusev, V. Narayanan, M. M. Frank, *IBM J. RES. & DEV.*, **50**, 387 (2006).
3. K. Mistry, C. Allen, C. Auth, B. Beattie, D. Bergstrom, M. Bost, M. Brazier, M. Buehler, A. Cappellani, R. Chau, C.-H. Choi, G. Ding, K. Fischer, T. Ghani, R. Grover, W. Han, D. Hanken, M. Hattendorf, J. He, J. Hicks, R. Huessner, D. Ingerly, P. Jain, R. James, L. Jong, S. Joshi, C. Kenyon, K. Kuhn, K. Lee, H. Liu, J. Maiz, B. McIntyre, P. Moon, J. Neirynck, S. Pae, C. Parker, D. Parsons, C. Prasad, L. Pipes, M. Prince, P. Ranade, T. Reynolds, J. Sandford, L. Shifren, J. Sebastian, J. Seiple, D. Simon, S. Sivakumar, P. Smith, C. Thomas, T. Troeger, P. Vandervoorn, S. Williams, K. Zawadzki, *Electron Device meeting, IEDM,* 247 (2007)
4. G. B. Xu, Q. X. Xu, *Chinese Physics B,* **2**, 768 (2009)
5. B. S. Kalpathy, V. Arum, S. Ganesh, *Microelectronic Engineering,* **77**, 230 (2005)

ECS Transactions, 34 (1) 365-370 (2011)
10.1149/1.3567605 ©The Electrochemical Society

Theoretical And Experimental Development Of Advanced Dopant-Sensitive Systems

P. L. Zhang[a], L. C. Zhang[b], Y. D. Ye[c], Y. Yang[b]

[a]Qingdao Feiyang Vocational & Technical College, Qingdao, Shandong 266111, CHINA
[b]No. 55 Research Institute, China Electronics Technology Group Corporation, Nanjing, Jiangsu 210016, CHINA
[c]Analytical Lab, Nanjing University, Nanjing, Jiangsu 210093, CHINA

Abstract

X-ray diffraction of antimony doped single crystal (100) surfaces with concentrations of 1.4 to $4.7 \times 10^{18}/cm^3$ are reported. The lattice constant is the same as for Si. The anisotropic etching changed the fourfold rotational symmetrical (100) surfaces into the six-fold rotational symmetrical (111) surfaces. The etch rate of the Sb doped (100) substrate is 100 times faster than the pure silicon (100) when etching in 20wt% KOH at 80-100°C. The advanced wet etch facets that correspond to the crystallographic planes and the advanced dopant-sensitive etching system were measured.

Introduction

The anisotropic etching of the silicon substrate using the wet chemistry is enabled by direct chemical reaction of the etchant with the silicon atoms (1, 2).

Alkaline etchant compositions can be used primarily for etching along crystallographic orientations. Since the bond density and surface energy varies with crystallographic orientation and the alkaline etchants interact with the silicon crystal planes preferentially, specific etchants can be designed for etching along different planes.

In anisotropic etching, the etchant, usually slow-acting, removes atoms that are more weakly bound in preference to those that are more strongly bound. As a result, the advancing surface develops facets that correspond to crystallographic planes (3).

These etches are called anisotropic because the etching rate in the <100> direction is high and in the <111> direction is low. Etch ratios for these two directions can be as high as 100-600 to 1 (1, 2, 4-12). Various factors have been put forth to contribute to this directionally dependent etch rate, but none explain the phenomena fully. Tables displaying the materials in the experiment to show that the (111) crystal plane is the most dense, with three of the four covalent bonds below the plane, and thus less accessible to the etchant; and also the surfaces with the highest bond density etch fastest (1, 2).

The <100> surface orientation changes to the <111> surface orientation after etch in our experiment. Microscopically, the fourfold rotational symmetry with respect to the very top (100) surfaces changes to a 6-fold rotational symmetry with respect to the very top (111) surfaces after the anisotropic etch. To the etchant, the etch stop becomes the (111) surface.

Theoretically and experimentally, six-fold rotational symmetry of the silicon (111) surfaces can be found from the famous silicon surface 7x7 reconstruction in <111> direction (7, 8) under vacuum. The XRD results of Si (111) by some scientists support this claim as well. Another school of thought is that alkaline etching at the (100) surface peels off the surface layer by layer based on the broken step structure (7).

The silicon substrate with antimony, the n-type substrate, is made by Czochralski (CZ) method. Most likely, one antimony atom is covalently bonded to the silicon atom, where the antimony atom is near the ionization of its own two electrons for bulk conductivity.

In our experiment the continuous transition from fourfold rotational symmetry (100) surface to threefold (in fact it is six-fold) rotational symmetry (111) surface occurs with the continuous chemical etching reaction. The transition is gradual.

The chemical bond of silicon is directional that once the stronger and more resistant (111) surface is exposed to the etchant, the etching stop shows up. The chemical bond energy between silicon and antimony is not tested yet in our experiment, but we performed XRD experiment to verify the crystal structure of the heavily doped silicon substrate. Excellent XRD results indicated that no crystal lattice structure change occurred. The lattice constant of the doped silicon stays the same as the undoped one (13, 14). Our XRD experiment also confirmed that the antimony atoms are located at the substitutional sites of the diamond structure. Anisotropic wet etch occurs on this doped substrate which preserves the same lattice constant but changes the elasticity with respect to the pure silicon. This may cause the etch rate difference. Therefore we developed the advanced dopant-sensitive systems to monitor the impurity level of the ion implantation and the diffusion process.

Further more the etchant can be carefully designed for the slow and controllable chemical reaction with the silicon substrate. Mostly, the etchant can be designed to be sensitive to the amount of the impurity in the substrate, such as boron, phosphorous or antimony, with slow directional chemical reaction.

We designed the photomask specifically for the further study on the advanced dopant-sensitive systems.

Experimental

<u>Substrate:</u> The maximum antimony impurity in the single crystal silicon substrate is $1.4 - 4.7 \times 10^{18}/cm^3$ through the CZ method. The electrical resistivity of our substrate is $0.01\text{-}0.02\,\Omega \cdot cm$ and the off-orientation is $< 1^\circ$ (15).

<u>XRD:</u> Rigaku D/MAX-RA (Cu, K_α, K_β) X-ray diffractometry is used for the study and analysis of antimony doped silicon. The experimental result is compared with XRD of Si from the references (13, 14).

<u>Photolithography:</u> AZ1500 and AZ5214 from Clariant and the negative resist from an un-specified manufacturer are used for the resist film formation. The resist goes through the spin-coating and other photolithography process procedure. The primary flat and the secondary flat of the wafer are used for the rough alignment for the wafer (100) surface with respect to the photomask.

<u>Wafer Cleaning Solution and Alkaline Etchant:</u> Potassium hydroxide (KOH) and 25% Ammonia solution (NH_4OH) from Tianjin No. 1 Chemical Reagent Company and Tianjin Zhengcheng Chemical Product Manufacturing Limited Corporation were used respectively. We introduce KOH solely for the off-line usage and for reference purpose in circuitry manufacturing because of the metallic ion contamination (MIC) issue. Ammonia stability at room temperature was monitored during the process. Ammonium hydroxide (NH_4OH) is used in this experiment for the wafer cleaning.

Results and Discussion

<u>Silicon Substrate:</u> For the dopant antimony, its solid solubility limits in silicon at $1100^\circ C$ is $5 \times 10^{19}/cm^3$ (1, 16) and $2 \times 10^{19}/cm^3$ at room temperature (1, 2). In our experiment the maximum antimony impurity in the single crystal silicon substrate is $1.4 - 4.7 \times 10^{18}/cm^3$ at room temperature. The resistivity in silicon is determined by the

concentration of the active charge carriers, which in turn is determined by the concentration of the doped antimony atoms located at the substitutional sites. Table 1 provides some data about our silicon substrate, germanium and antimony.

TABLE 1   Resistivity and Molar Volume of Si (15), Ge (1,2) and Sb (12)

Element	Electrical Resistivity	Molar Volume
Silicon doped with Sb	0.01-$0.02\ \Omega \cdot cm$	$12.06\ cm^3$
Antimony	$40\ \mu\ \Omega \cdot cm$	$18.19\ cm^3$
Germanium	$0.05\ \Omega \cdot cm$	$13.63\ cm^3$

XRD: The correlation between the doping impurity concentration (dopant concentration), the electrical resistivity and the carrier density has been investigated and studied. We performed XRD experiments which confirm that the lattice constant of the doped silicon stays the same as the undoped one for this heavily doped substrate (13, 14). Excellent XRD results indicate that no crystal lattice structure change occurred. Our XRD experiment also confirmed that the antimony atoms are located at the substitutional sites in the diamond structure -almost all the impurity antimony atoms are located at the crystal lattice positions. That means the maximum activation of the conductivity. From our dicing experiment the brittle property of antimony is transferred to the silicon substrate after introducing it to silicon. Table 2 provides some XRD data. Figure 1 provides the diffractogram.

TABLE 2   XRD of Sb Doped Silicon Substrate at <100> direction:

2θ (degree)	D (Å)	Counts	FWHM
32.940	2.717	517	0.080
69.120	1.358	98859	0.120
69.320	1.354	55739	0.120

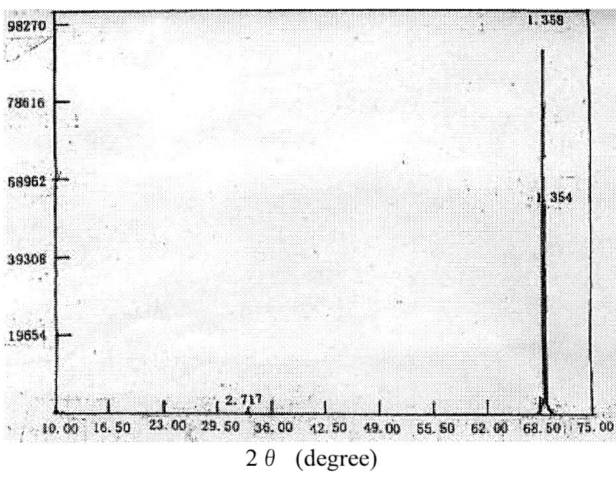

2θ (degree)

Figure 1.   X-ray diffractogram of (100) single crystal silicon with the dopant concentration of $1.4 - 4.7 \times 10^{18}/cm^3$.

Most likely, one simple antimony atom is covalently bounded to the silicon atom, where antimony atom is near the ionization of its own two electrons to contribute to the conductivity of the bulk. (Future spectroscopic study can be applied to verify whether antimony is a cluster of four atoms or two atoms, or just one simple atom).

Usually it is hard to quantify, especially to separate the three parameters, the doping impurity concentration (dopant concentration), the electrical resistivity and the carrier density, not to mention that it is hard to differentiate the impurity at the substitutional sites from the interstitial sites.

<u>Mechanism of Si Etching Reaction in Aqueous KOH</u>: The reaction between KOH and Si is normally understood to result in the dissolution of Si by formation of K_2SiO_3 and H_2. The equation is as following (17):

$$Si + 2KOH + H_2O \rightarrow K_2SiO_3 + 2H_2 \uparrow \qquad [1]$$

The etch rate is affected by the nature of the silicon, the doping levels, the thermal treatment and other parameters, such as the concentration (1, 2, 4-7, 9-12, 17). Table 3 also provides the etch rate of KOH with respect to its concentration.

TABLE 3.   Etch Rate of Si (100) in aqueous KOH at 80-100°C:

Concentration	Etch Rate	Concentration	Etch Rate	Concentration	Etch Rate
<10%-30%	Increase	~ 30%-40%	Maximum	> 40%-45%	Decrease

Optical microscopy is used to observe the morphology change on the wafer. Our laser reflectometry using the diode laser was not sensitive enough to distinguish the subtle difference between the doped silicon wafer and the undoped wafer, nor the subtle difference between the fourfold rotational symmetry at (100) surface and the six-fold rotational symmetry at the (111) surface.

The <100> surface orientation changes to the <111> surface orientation after etch in our experiment. Microscopically, the fourfold rotational symmetry with respect to the very top (100) surfaces changes to 6-fold rotational symmetry with respect to the very top (111) surfaces after the anisotropic etch. To the etchant, the etch stop becomes the (111) surface. In our experiment the continuous transition from fourfold rotational symmetry (100) surface to threefold (in fact it is six-fold) rotational symmetry (111) surface occurs with the continuous chemical etching reaction. The transition is gradual.

<u>Ammonium Hydroxide (NH₄OH) and Alkaline Etchant KOH</u>: The wafer cleaning chemical can be a weak etchant, while the preferential etching, also called defect etching, which targets the crystal defects of the single crystal silicon, is a stronger etchant. It can yield small pits on the silicon wafer after the blanket etching. The slope etching or the ultraslope etching, can be of importance towards the MEMS manufacturing. If etching proceeds exclusively in one direction (e. g., only vertically), the etching process is said to be completely anisotropic, such as the plasma etching.

<u>Doping Level</u>: The measurement of Sb impurity is also performed spectroscopically. The precision of the (111) surface as the etch stop can be of the metrology usage, such as the verniers.

The exact initial starting point of etch and the exact time to transfer the wafer into the rinsing water quickly is relatively hard to control and repeat with our wet etch experiment. Sometimes the step height measurement works fine after etch for the big open space. The big fiducial marks for the average etch rate monitoring are not suitable for the production due to the loading effect, the micro-loading effect and the limited space at the scribe lines on the wafers.

Dopant-sensitive systems are even more important for the ion implantation process. The amorphization of the substrate and then the annealing of the substrate to re-stabilize the crystal structure can re-establish the lattice structure in the process. The higher the amount of the implant impurity, the higher the etch rate during the etch process.

The dopant concentration of our substrates is incorporated into the etch rate monitoring system using the specially designed mask for etch rate monitoring. The purpose is to develop the measurement baseline using one wafer. The etch rate of the Sb doped (100) substrate is 100 times faster than the pure silicon (100) substrate. The anisotropic etching of this silicon uses 20wt% KOH at 80-100°C.

Anisotropic Etch: Alkaline solution can be utilized for etching a silicon (100) surface, a high degree of anisotropy and the extremely precise sidewall angle of 54.7° is consistently obtained, leaving behind the intersecting (111) planes or the reversed pyramid (or reversed truncated pyramid). In another word KOH stops at silicon (111) surfaces. In the structure of the silicon crystal lattice, the (111) planes are oriented 54.74° relative to the (100) planes (1, 2).

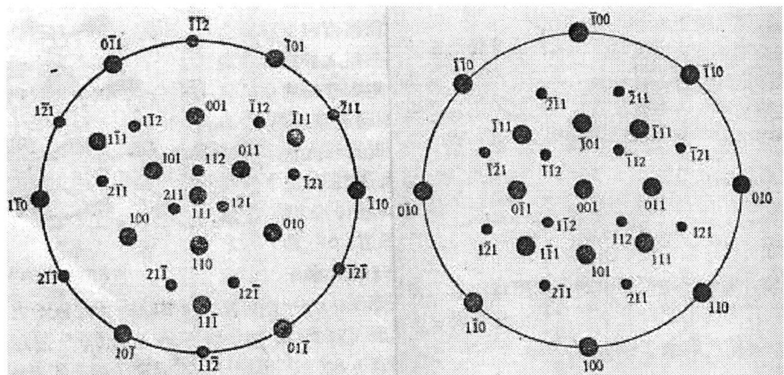

Figure 2.   Rotational symmetry projections of single crystal silicon. On the left it is the projection at <111> direction. It contains three-fold (in fact six-fold) rotational symmetry. One dot represents two spots. On the right, it is the projection at (100) surface. It contains fourfold rotational symmetry. One dot represents one spot.

TABLE 4 Rotational symmetry, bond & atomic density & d distance between the crystal planes, Å

Crystal Plane	(100)	(111)
Covalent Bond Density	$4/a^2$	$2.3/a^2$
Rotational Symmetry	Fourfold	(Threefold) Six fold
Atomic Surface Density	$2/a^2$	$2.3/a^2$
Distance between the crystal planes	$(1/4)a=1.36$	$(\sqrt{2}/4)a=2.35$ (l) $(\sqrt{3}/12)a=0.78$ (s)

The operation (of course not the simple rotational operation) at Si (100) surface re-generated the whole diamond structure (Figure 2). But (100) planes are weaker than (111) planes facing the foreign etchant. Etching stops at the (111) surface. The (111) planes possess unusually high atomic densities. The covalent bond that connects two atoms is normal to those (111) planes. Thus the pattern of the (111) planes is two closely spaced (0.78Å) and tightly bound planes followed by a relatively large space with a low bond density. The paired (111) planes can be treated as a single plane. The silicon diamond structure can be described as a pair of interpenetrating fcc lattices with the displacement of 1/4 distance at the body diagonal place. Although the areal density of atoms located at the paired (111) planes appreciably exceeds that of any other plane, and the paired planes do indeed appear to function as one in some situation. Table 4 presents the results of our calculations.

Conclusion

XRD confirmed that antimony doped silicon can maintain the same lattice constant. The anisotropic etch of the <100> direction with fourfold rotational symmetry is changed into a six-fold rotational symmetrical <111> direction. The etch rate is 100 times faster for the Sb doped (100) substrate compared to the undoped silicon (100) substrate, using 20wt% KOH at 80-100°C. The advanced etch develops facets that correspond to crystallographic planes.

Acknowledgments

The authors appreciate the help provided by Dr. Hu of Institute of Optoelectronics, Chinese Academy of Science and by Dr. Chen, Dr. Sze and their students of Department of Electronics, Nanjing University.

References

1. S. A. Campbell, *The Science and Engineering of Microelectronic Fabrication, 2nd edition*, Chapter 5, 19, (references therein), Oxford University Press, (2001).
2. J. Z. Zhang, *Chemical Mechanical Polishing Technology of Single Crystal Silicon Wafers,* p.120, p.164, p.180, Chemical Industry Press, (2005), (in Chinese).
3. R. W. Warner, B. L. Grung, *Semiconductor Device Electronics,* p. 49, Publishing House of Electronics Industry, (2005), (in Chinese).
4. Semiconductor Institute, Academy of Science, *Semiconductor Diagnostic Technique*, Chapter 1, Science Press, (1984), (in Chinese).
5. M. Quirk, J. Serda, *Semiconductor Manufacturing Technology*, p. 75, Prentice Hall (2001).
6. S. Franssila, *Introduction to Microfabrication*, p.221, Publishing House of Electronics Industry (2005), (in Chinese).
7. M. Elwenspoek, H. Jansen, Silicon Micromachining, Chapter 1, 2, (references therein), Chemical Industry Press, (2007), (in Chinese).
8. R. Wiesendanger, G. Tarrach, D. Burgler, H. -J. Guntherodt, *Europhys. Lett.*, **12**, 57 (1990).
9. H. Seidel, L. Csepregi, A. Heuberger, H. Baumartel, *J. Electro. chem. Soc.*, **137**, 3612 (1990).
10. M. Elwenspoek, *J. Electrochem. Soc.*, **140**, 2075 (1993).
11. O. J. Glembocki, E. D. Palik, G. R. de Guel, and D. L. Kendall, *J. Electro. chem. Soc.*, **138**, 1055 (1991).
12. H. Xiao, in *Introduction to Semiconductor Manufacturing Technology*, p. 270, Prentice Hall (2001).
13. JCPDS 27-1402.
14. American Society for Testing and Materials, "Powder Diffraction File (inorganic phases)", Penny-Ivania, Swarthmore Press, 1984.
15. Guosheng Company Nanjing Office sales manager, personal communication, 2010.
16. F. A. Trumbore, *Bell Systems Tech. J.,* **39**, 210 (1960).
17. K. E. Bean, *IEEE Trans Electron Devices, ED-25*, 1185 (1978).

ECS Transactions, 34 (1) 371-376 (2011)
10.1149/1.3567606 ©The Electrochemical Society

Ultrapure Water-Related Problems and Waterless Cleaning Challenges

Takeshi Hattori
(hattori@stanfordalumni.com)

ECS Fellow

Ultrapure water can cause several serious problems in modern semiconductor manufacturing. With device geometry shrinking and becoming more complex, conventional aqueous cleaning and subsequent drying tends to collapse high aspect-ratio nano-structures due to the high surface tension of water. The high resistivity of water can cause flow electrification during wafer rinsing and the high reactivity of water with oxygen and silicon forms watermarks on silicon surfaces. Furthermore, in the FEOL, some high-k gate capping materials dissolve in water while, in the BEOL, Cu is corroded in water during rinsing. If high-permittivity water remains in porous low-k interlayers after aqueous treatments, the k-value of the film increases. The solutions to all of these problems will be discussed. Finally, as the ultimate solutions, waterless cleaning techniques are described and discussed, which include HF vapor cleaning, cryogenic nitrogen aerosol cleaning, supercritical carbon-dioxide cleaning, and pinpoint cleaning such as nano-tweezer pickup.

Introduction

Semiconductor fabrication plants (fabs) are generally located at the sites where abundant clean water is available. Ultrapure water (UPW) plays an essential and important role in semiconductor manufacturing for cleaning and rinsing the surfaces of silicon wafers and photomasks as well as diluting aqueous chemicals used for polishing etching, and cleaning the surfaces (1,2). Recently, in the most advanced photolithography area, UPW has been inserted between the final lens and the silicon substrate to enhance resolution. Wafer edge and bevel cleaning as well as wafer backside cleaning becomes indispensable before immersion lithography as well as after film deposition. As new processes and materials have been introduced in the most advanced semiconductor manufacturing, the versatility of UPW has extended ironically against the general trend toward dry processing.

The yield in semiconductor manufacturing is affected by the quality of UPW. Table I compares ITRS requirements for UPW contaminant quality levels for the years 2011 and 2024 (3). UPW has already been purified to the utmost level and it is generally the cleanest fluid available in the manufacturing process, so that the UPW roadmap is very stable over time except the critical particle size to be controlled.

As semiconductor device geometry shrinks to the nanometer range, it becomes clear that UPW causes several serious problems in the most advanced semiconductor manufacturing. In this paper, these UPW-related problems will be described and practical solutions to these problems will be given. This paper will then describe and discuss several waterless cleaning techniques as the ultimate solutions to the problems, which includes elevated-temperature low-pressure HF vapor cleaning, cryogenic aerosol cleaning, supercritical carbon-dioxide cleaning, and pinpoint cleaning such as nano-tweezer pickup.

Table I. ITRS requirements for UPW contaminant quality level.

Requirement Item	Year 2011	Year 2024
Resistivity at R.T. (MOhm-cm)	18.2	18.2
Total oxidizable carbon (ppb)	<1	<1
Bacteria (CFU/liter)	<1	<1
Total silica (ppb) as SiO_2	<0.3	<0.3
Number of particles > critical size (#/L)	4000 (>20nm)	4000 (>4.5nm)
Dissolved oxygen (ppb)	<10	<10
Dissolved nitrogen (ppm)	8~18	8~18
Critical metals (ppt, each)	<1.0	<1.0
Other critical ions (ppt each)	<50	<50

UPW-Related Problems and their Solutions

<u>High Resistivity of Water</u>

During water rinsing of silicon wafers, water jet impacted against the surface of a silicon wafer can generate streaming electrification on the silicon surface with highly resistive water, resulting in the degradation of gate oxide integrity as well as accelerated electrostatic adhesion of particles onto the wafer surface. Figure 1 shows the surface voltage distributions on a silicon wafer for three different rotation speeds of spinning wafer stage when UPW is poured onto the wafer. It can be seen from the figure that the surface voltage increases over the whole wafer with an increase of rotation speed. The use of carbon-dioxide dissolved water can prevent the electrostatic-charge generation even at the highest speed. Connecting the wafer to ground is another solution but it will be sometimes difficult to modify the tool-component materials. Air-ionizers do not work well on fast-spinning wafers.

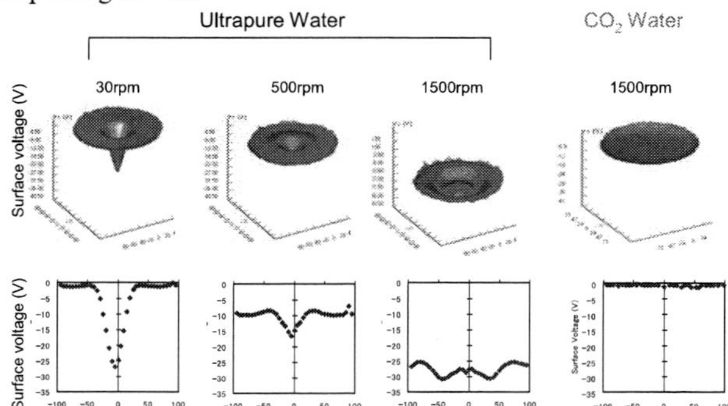

Figure 1. Surface voltage distributions on silicon wafers after UPW or CO_2 water pouring for different rotation speeds in single-wafer spin cleaning.

High Reactivity of Water with Silicon

In the aqueous processing of silicon wafers, drying after water rinsing is a critical step in terms of the prevention of both particle adhesion and watermark (or drying spot) formation. The easy reaction of water with oxygen in air and substrate silicon at room temperature forms silicic acid (SiO_2 xH_2O), resulting in, after wafer drying, the formation of its residues or so-called watermarks (or drying spots) on the silicon surface, The watermarks cause prevention of both etching and epitaxial growth.

It has become increasingly more difficult to reduce the watermarks to the extent desirable as the device geometry has shrunk. Conventional drying methods such as centrifugal spin drying, IPA vapor drying, and even Marangoni (or Rotagoni) drying can not meet the stringent requirement for watermark prevention in the most advanced wafer processing lines for nano-device fabrication. In order to prevent watermark formation, various novel methods have been employed, which include drying in an inert-gas ambient with a seal plate (4) and more preferably in a sealed chamber in the single-wafer spin cleaning system and blowing ultra-dry air with a dew point of -96°C or below along the wafer surface in the batch immersion cleaning equipment. Very recently, the replacement of water with liquid IPA during rinsing and subsequent wafer drying with IPA evaporation is employed in some of the most advanced lines, but the excess consumption of IPA is unfavorable from the environmental viewpoint. Introduction of non aqueous cleaning will be the ultimate solution.

High Permittivity of Water

In the BEOL, the k value of interlayer insulating films has been being decreased from 4 (SiO_2) through 3 (SiOC) and 2 (ULK) toward 1(airgap) in order to reduce RC time delay of LSI circuits and increase the system performance. If water remains in porous low-k interlayer films after wet cleaning and drying, the effective k-value of the film increases due to the high k value of water (k=78), causing delayed signal propagation in the circuits. The use of thermal curing or non-aqueous processing will be able to solve the problem.

Dissolution/Corrosion of Metals in Water

In the FEOL, some high-k gate capping materials such as La_2O_3 dissolve in water while, in the BEOL, Cu is corroded in water during wafer rinsing. These metal losses result in the degradation of electrical characteristics of semiconductor devices. The solution to the former issue is to employ organic-solvent cleaning and that for the latter case is to lower the oxidation-reduction potential of UPW by eliminating dissolved oxygen in UPW or more preferably by using electrolyzed reducing water or Hydrogen water. It should be noted that UV treatments of water to eliminate total organic carbons in on-site UPW plants generate unexpected hydrogen peroxide in the ppb range in UPW, which can cause metal corrosion as well as chemical oxidation of the silicon surface.

High Surface Tension of Water

The last but not least problem of aqueous cleaning is the high surface tension of rinsing water (5), which was able to be completely ignored when the device geometry was big enough, so that the surface tension of water had virtually no influence on the mechanical strength of the circuit structures. However, with semiconductor-device geometry shrinking and becoming more complex, conventional aqueous cleaning/drying tends to collapse high-aspect-ratio fragile photoresist and ULSI device patterns due to the comparatively high surface tension of water during wafer drying, as shown in Figs. 2. Pattern collapse occurs not only in thin gate-stack and STI structures in the FEOL but also in porous ultra-low-k interlayer structures in the BEOL for most advanced devices. In many cases, conventional aqueous cleaning/drying will be inadequate and impractical for free-standing MEMS structures and very high aspect-ratio LSI structures. Aqueous cleaning/drying should be further modified or avoided in the future.

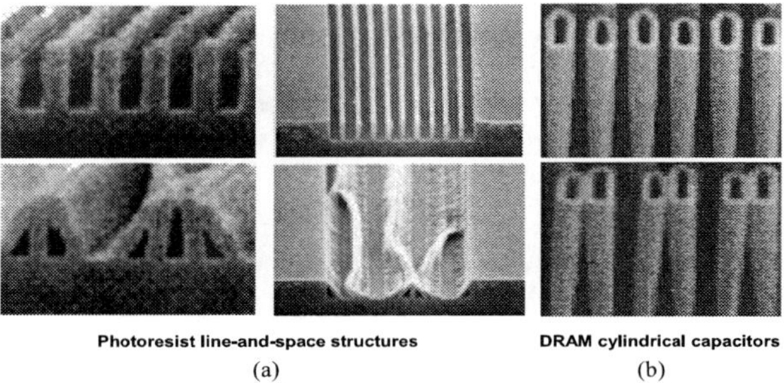

Figure 2. SEM images of typical collapse/stiction of nano-structures: (a) a normal 100-nm line-and-space photoresist pattern (upper) and a collapsed 70-nm line-and-space photoresist pattern (lower), and (b) a normal DRAM cylindrical capacitor array (upper) and a collapsed DRAM cylindrical capacitor array (lower).

Wafer Cleaning without Using Water

Finally, as the ultimate solutions, several techniques of wafer cleaning without using water will be demonstrated and discussed (5).

HF Vapor Cleaning

Vapor-phase etching/cleaning to use a controlled mixture of anhydrous HF/MeOH has been employed for etching thin silicon dioxide films without causing pattern collapse or stiction. The chemical reactions produce not only residues but also water as a product of etching which can sometimes cause stiction of fragile MEMS structures such as RF switches having air-filled cavity structures. In such a case, an elevated-temperature, low-pressure HF vapor treatment is preferable in order to prevent water condensation on the silicon surface (5). Temperature and pressure should be carefully selected to prevent the condensation of water on the etched surface as well as to prevent non-volatile residues left behind. We have developed the dry-clean version of SCROD cleaning (single-wafer spin cleaning with repetitive use of ozonated water and diluted HF)(4,6) to employ ozone gas, HF vapor, and a certain physical force instead of conventional ozonated water and liquid HF, reducing the consumption of chemicals and water drastically (7).

Cryogenic Aerosol Cleaning

Cryogenic aerosol-based cleaning is one of the most promising dry cleaning techniques that can remove particles and process residues. Conventionally used carbon-dioxide and argon aerosols, however, can cause the destruction of fragile fine structures. High aspect-ratio structures can be cleaned without causing damage by employing nitrogen as a source of aerosols, whose force of impact against the structures is smaller than argon and CO_2 due to its smaller molecular weight (5,8). Our simulation based on molecular dynamics showed that the pressure and temperature of impacting nitrogen aerosols against the surface reach the critical point of nitrogen. Hence resulted supercritical nitrogen fluid will be considered to generate relatively large shear force to remove particles out of silicon wafers even when the force of the impacting nitrogen is smaller than argon and CO_2 due to the smaller size of the aerosols (8).

Supercritical Fluid Cleaning

Supercritical carbon dioxide (SCCO$_2$)–based technology has been proposed for various steps in device fabrication, such as wafer drying, cleaning (5,9), and metal deposition . SCCO$_2$ has zero surface tension like gas, and thus, can penetrate easily into deep trenches and vias, and also enables cleaning and drying without pattern collapse or stiction, as shown in Fig. 3. We have successfully stripped high-dose ion-implanted photoresists by using SCCO$_2$ without damage and silicon recess (9,10). SCCO$_2$ can also be applied for stripping the post- etch photoresists on ultra low-k films by adding chemical formulations without causing damage and k-value change (9,11). In NEMS fabrication, etching of nanometer-thick sacrificial SiO_2 layers and *in-situ* drying in SCCO$_2$ can form nano-gap beam structures without stction (12). The potential of particle removal using SCCO$_2$ has also been shown (13). The unique properties and capabilities of SCCO$_2$ will continue to draw interest for various applications in the future semiconductor and nano-electronics industries (5,9).

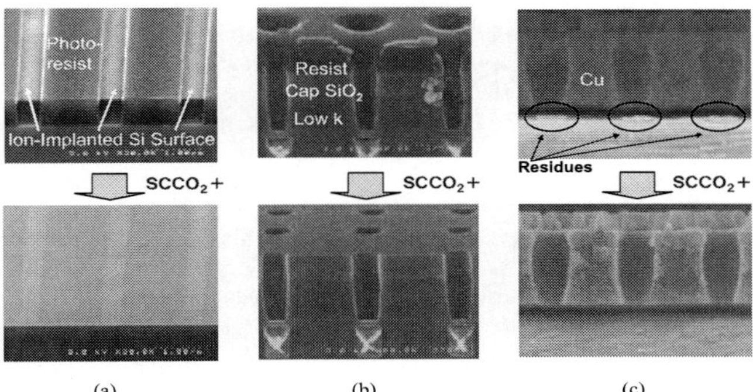

Figure 3.   SEM images of (a) the high-dose ion-implanted silicon surface, (b) the post-etch via structure, and (c) the post ESL(etching- stop-layer)- etch porous ultra-low-k film on a blanket copper film, each before and after processing in co-solvent/chemical additive-containing supercritical carbon dioxide.

<u>Pinpoint Cleaning</u>

While all the methods described above clean the whole surface of a wafer in a single operation, novel local-area or pinpoint cleaning, in which contaminants are targeted one by one without affecting the surrounding non-contaminated areas of the wafer surface, will become necessary as the required level of cleanliness gets even higher. Several pinpoint cleaning methods have been proposed, which include femt-second laser illumination, laser shockwave irradiation, AFM nano-probe sweeping, and nano-tweezer pick-up (5). Figure 4 shows nano-tweezers, made of silicon by using NEMS technology, successfully picking up a 50nm particle from the wafer surface. without causing damage. The tweezer pick-up tools have already been applied to particle removal from the surfaces of both extreme-ultraviolet lithography (EUVL) masks and 35-mm-film sized CMOS image-sensing chips (5).

As we work to develop non-aqueous/dry cleaning methods, we are encouraged by knowing that these methods are environmentally-friendly. There will be more research challenges and opportunities in the damage-free waterless cleaning technologies for fragile fine nano-structures in the future.

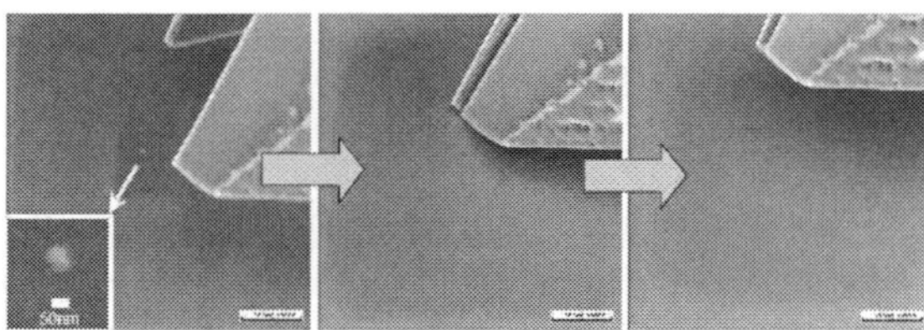

Figure 4. Picking up a particle on a silicon wafer by using silicon nano-tweezers.

References

1. T.Hattori, "Trends in Wafer Cleaning Technology," in *Ultra Clean Surface Processing of Silicon Surfaces-Secrets in VLSI Manufacturing*, pp. 437-461, Springer, Heidelberg and New York (1998).
2. T. Hattori, *ECS Proceedings,* **PV97-35**, p.3, The Electrochemical Society (1977).
3. *International Technology Roadmap for Semiconductors 2010Update*, ITRS (2010).
4. T. Osaka *et al.*, *ECS Proceedings,* **PV 2001-26**, p.3, The Electrochemical Society (2001).
5. T. Hattori, *ECS Transactions*, **25**, no.5, p.3, The Electrochemical Society (2009).
6. T. Hattori, *J. Electrochem. Soc.*, **145**, p.3278 (1998).
7. T. Osaka *et al.*, Extended Abstracts, Spring 2003 Meeting, Japan Society of Applied Physics, **2**, p.840 (#27a-P2-5) <in Japanese>.
8. H. Saito *et al.*, *ECS Proceedings,* **PV2003-26**, p.289, The Electrochemical Society (2004).
9. K. Saga and T. Hattori, *Solid State Phenomena*, **134**, p.97 (2008).
10. K. Saga *et al.*, ECS Transactions, 1, no.3, p.277, The Electrochemical Society (2005).
11. M. Korzenski *et al.*, *Solid State Phenomena*, **103-104**, p.193 (2005).
12. K. Saga *et al.*, *Solid State Phenomena*, **103-104**, p.115 (2005).
13. M. Korzenski *et.al.*, *Solid State Phenomena,* **103-104**, p.193(2005).

ECS Transactions, 34 (1) 377-382 (2011)
10.1149/1.3567607 ©The Electrochemical Society

Dry-Etch Fin Patterning of a sub-22nm node SRAM Cell: EUV Lithography new Dry Etch Challenges

E. Altamirano-Sánchez[a], Y. Yamaguchi[b], J. Lindain[b] N. Horiguchi[a], M. Ercken[a], M. Demand[a] and W. Boullart[a]

[a]IMEC, Kapeldreef 75, 3001 Leuven, Belgium.
[b]Lam Research, Fremont California, US.

This work describes the dry etching process for patterning crystalline Silicon (c-Si) on a 6T-SRAM cell using Extreme UV (EUV) lithography. The target fins consist of straight structures (30nm height and 15nm of critical dimension) patterned on a sub-22nm node with 80nm fin pitch. Scaling down the fin pitch had a direct influence on the fin critical dimension and profile. We found out that the fin etching process developed for a 22nm node process with 90nm fin pitch was no longer functional for patterning fins on a sub-22nm node with 80nm fin pitch. In order to pattern straight fins with 15nm of critical dimension, we had to improve the selectivity towards EUV photo resist, redesign the patterning stack thickness and eliminate the typical softlanding step used in previous nodes.

Introduction

FinFET devices are considered one of the most promising candidates for allowing continued SRAM scaling thanks to their improved Short Channel Effects behavior (1). Recently, we presented the dry etching challenges for patterning a 32nm node $0.186\mu m^2$ (124nm fin pitch) and a 22nm node $0.099\mu m^2$ (90nm fin pitch) SRAM cells (2), which were exposed using 193nm immersion lithography (193i). We have observed that at these dimensions the dry etching process window was dramatically narrowed. The progress of full field extreme ultraviolet (EUV) scanners makes the exposure of a 16nm node possible. As preparation for a 16nm node SRAM cell, imec's EUV alpha demotool (ADT) has exposed a 80nm pitch SRAM cell (22nm node; $0.072\mu m^2$) allowing dry etching to investigate and develop a new patterning strategy for the future 64nm pitch cell (16nm node; $0.055\mu m^2$). The EUV Photo Resist (PR) has different etch resistivity and thickness compared to a 193i PR, bringing new challenges for dry etching. In this work we will present the dry etching process development for patterning fins on SOI (silicon-on-insulator) with 15nm target critical dimension (CD).

Experimental

The dry etching was carried out in a Lam Research Kiyo 2300 reactor. This is a transformer-coupled plasma (TCP^{TM}) reactor that allows a separate control of plasma power and substrate bias.

The test material consisted of 300mm Silicon on Insulator (SOI) wafers with 145nm buried silicon dioxide (BOX) and 30nm thick (100) crystalline Silicon (c-Si). The fin structures were patterned in the (110) direction using the stack described below.

For patterning the SOI we use a dual hard mask (HM) composed of two layers. The stack from bottom to top is:

-70 or 40nm of Amorphous Carbon Layer (ACL) deposited using chemical vapor deposition (CVD). This layer was used as hard mask two (HM2).

-25 or 15nm of silicon oxycarbide (SiOC) deposited using CVD. This layer was used as a HM one (HM1)

Lithography was performed using EUV (13.4nm) on the ADT from ASML, with NA (Numerical Aperture) = 0.25, sigma = 0.5, conventional illumination. The litho stack consisted of 24nm of under layer (UL) and 65nm of photo resist (PR).

After the dry etching was completed, the organic residues on the wafer were stripped in an Axiom™ reactor and cleaned on Oasis Clean™ chamber both from Applied Materials.

Top down inspections were done using Scanning Electron Microscopy (SEM). The critical dimension (CD) was measured using a Hitachi SEM S-9380 tool. The cross section samples were inspected using a Philips Nova 200 SEM. Some samples were also inspected using Transmission Electron Microscopy (TEM).

Discussion

<u>Dry-Etching development</u>

Before discussing the sub-22nm fin etch process, we present a brief description of our dry etching process for patterning fins on a 22nm node SRAM cell. This will make the discussion easier in this short paper and let us stress on the main differences of the etch process between a pattern printed with 193i lithography and one printed with EUV lithography (EUVL).

<u>Dry etching fin patterning process for a 22nm node 6T-SRAM cell (193i lithograpy).</u>
A schematic representation of the dry etch fin patterning process for a 22nm cell is detailed in figure 1 and figure 2. Lithography prints 45nm CD lines with a final PR height of ~70nm. The dry etching process starts with BARC opening using a $Cl_2/O_2/He$ plasma chemistry, followed by a PR trim step based on $Cl_2/O_2/He$ plasma chemistry with no bias (not shown). After this, a DARC opening step is applied using a CF_4/CH_2F_2 plasma chemistry and a PR hardening step is also applied in order to make the PR more etch resistant and prevent any PR wiggling (Fig. 1). The HM or ACL was etched using HBr/O_2 plasma chemistry. The c-Si etching was done in three steps: 1) a ME based on a CH_2F_2/SF_6 plasma chemistry, which etches ~80% of the silicon, 2) a softlanding using a HBr/O_2 plasma chemistry (at low pressure), which etches the remaining Si and expose the BOX. Since the foot of these fin structures cannot be corrected by the softlanding due to a lack of ion deflection, we stopped the softlanding on endpoint (3). Therefore, we used the OE (HBr/O_2) step to tune the profile; the advantage of this step is that it is carried out at high pressure resulting in a very high selectivity towards the BOX and high bias (three times larger than the bias in the standard softlanding) providing high energetic ions that do not deflect prematurely towards the sidewalls.

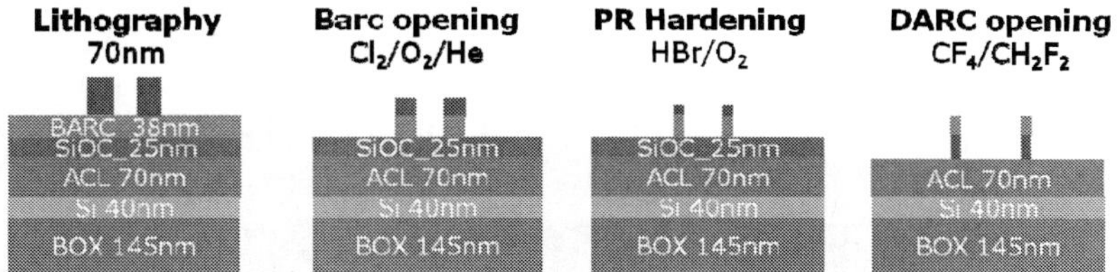

Figure 1. Schematic representation of the patterning stack after lithography, BARC opening, PR hardening and DARC opening (90nm fin pitch).

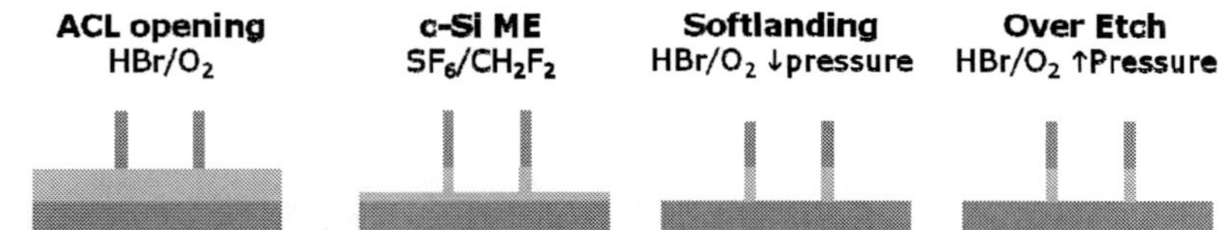

Figure 2. Schematic representation of the ACL and c-Si (ME, Softlanding and Over-etch) (90nm fin pitch).

<u>Dry etching fin patterning for a sub-22nm node 6T-SRAM cell (EUV lithography).</u> One drawback of EUV lithography is the PR budget which is reduced after exposure down to ~50nm. Figure 3 shows the evolution of the lithography stack (PR+BARC or UL)[1], it can be observed that as the fin pitch is scaled down, the PR budget is dramatically reduced.

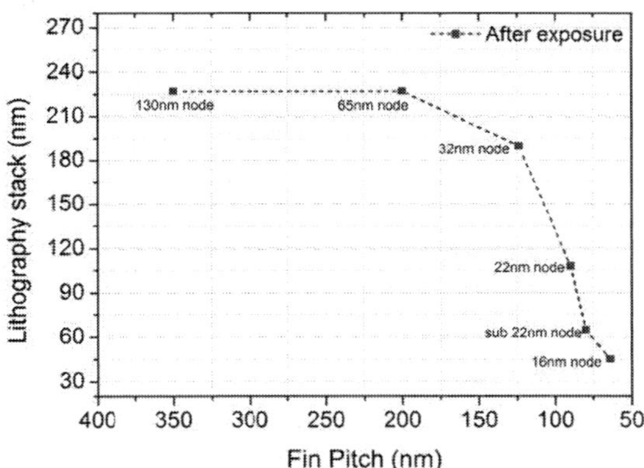

Figure 3. Evolution plot of the lithography stack after exposure. For 193i, the stack consist of BARC and PR and for the EUVL, the stack consist of UL and PR.

A low PR budget has implications on the etching process i.e. the opening of the 24nm of UL using Cl$_2$/O$_2$ or HBr/O$_2$ chemistries, as we have done it for the previous nodes,

---

[1]   In EUV, the lithography community does not talk anymore about BARC, because its antireflective properties are not needed anymore. Therefore, we name it now under-layer (mainly there for sticking purpose and/or avoiding of resist poisoning by the stack).

seems not to be an option since the PR height is reduced creating a narrow process window for further steps (Fig. 4b). On the other hand, for the 193i PR nodes, before the DARC opening, we used to apply plasma pretreatment (PPT) using Ar or HBr, which makes the PR more etch resistant. We have observed that this PPT decreases the PR height due to cross-linking and graphitization effect (4).

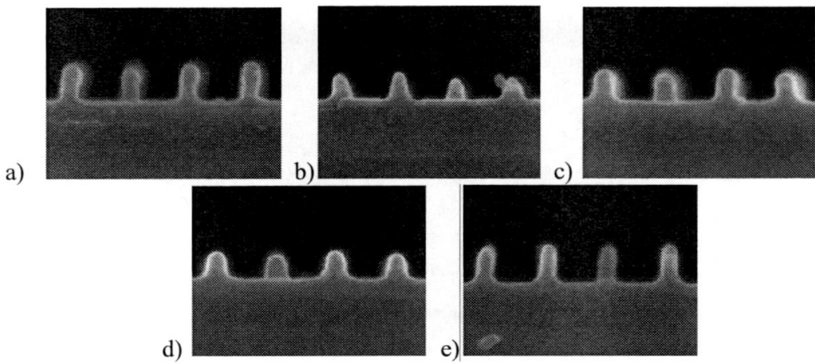

Figure 4. PR XSEM pictures of PR height: a) after exposure (~52nm), b) after Cl_2/O_2 UL opening (~30nm), c) after 30s of Ar PPT (~40nm), d) after 30s HBr PPT (~30nm) and e) after CH_2F_2/O_2 UL opening (~53nm). (Scale 100nm________). All these processes were applied to the EUV PR separately.

After the issues mentioned above, we did not consider to apply any of the PR PPT, because even at short times (<30s) of PPT the EUV PR height was decreased by ~35% (Fig. 4c-d). Hence, by avoiding a typical PPT and using a $CF_4/CH_2F_2/O_2$ chemistry for opening the UL, the EUV PR height budget is preserved as much as possible for transferring the pattern into the HM1 (Fig. 4e). However, even using a high selective chemistry towards the PR for opening the UL, the PR budget is still marginal for patterning 25nm of SiOC (this thickness was used in the previous 22nm node, Fig. 1). One of the main advantages of EUV over 193i lithography is that the patterning stack can be tailored for dry etching because reflection control on EUV lithography (EUVL) is not an issue anymore. Since the thickness of the layers involved on the patterning stack can now be determined by the needs for dry etching, the narrow process window faced in the 22nm node is considerably opened. Therefore the thickness of the SiOC or HM1 was reduced to 15nm.

The HM1 opening was carried out using CF_4 chemistry and the proper amount of CH_2F_2 in order to preserve the PR. For the ACL opening we used a lean Ar/O_2 chemistry avoiding any kind of sidewall passivation during this step in order to reach the 15nm critical dimension target.

In figure 2, it can be observed that the c-Si patterning for the 22nm node was done using a ME, softlanding and overetch steps. The ME etches ~80% of the c-Si and the profile after this step has a bottom tapered shape. The softlanding plasma conditions (8mtorr/350W/-65V/HBr/O_2) were designed for correcting this bottom tapered profile (also called foot). The ion deflection in the softlanding step is the key parameter for correcting this "foot". It is based on the principle of the notching effect, where the charging of the newly exposed dielectric surface deflects ions and subsequently the ion trajectories diverge towards the bottom of the fins (5, 6). We have observed that the c-Si patterning is strongly affected by the reduction of the open space between fins. Figure 5 shows the evolution of the open space as the fin pitch is scaling down from a 130nm node to 16nm node. We have also observed that the aspect ratio (AR) of the features after the ME (second drawing in fig. 2) may also have an impact on the c-Si patterning, especially during the SL and OE steps (3).

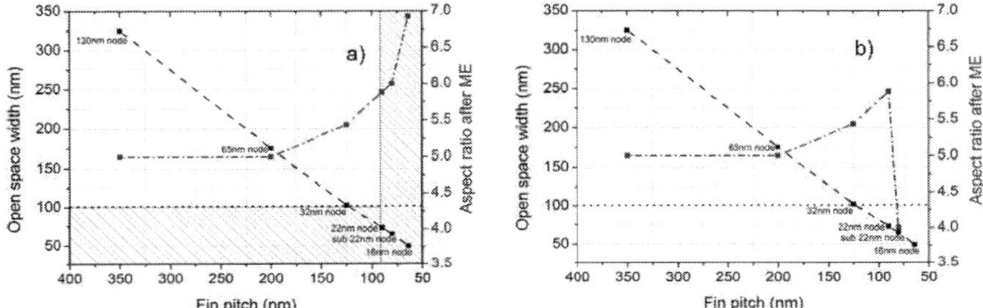

Figure 5. Evolution plot of open space between fins as the fin pitch is scaled down. a) Aspect Ratio after ME step (ACL + 80% of the c-Si thickness) maintaining the 70nm ACL thickness; b) by decreasing the ACL thickness from 70 to 40nm, the AR dropped from 6 to 4 for the sub-22nm node process.

One way to open our process window for patterning the c-Si was to decrease aspect ratio of the features after ME by reducing ACL thickness from 70nm to 40nm (Fig. 5b). We also found that, for such thinner layer of c-Si (30nm), the SL was no longer needed since the overetch step could correct the fin profile with a high selectivity towards the BOX. In figure 5, we present the TEM cross sections of the fins after dry etching, strip and wet clean.

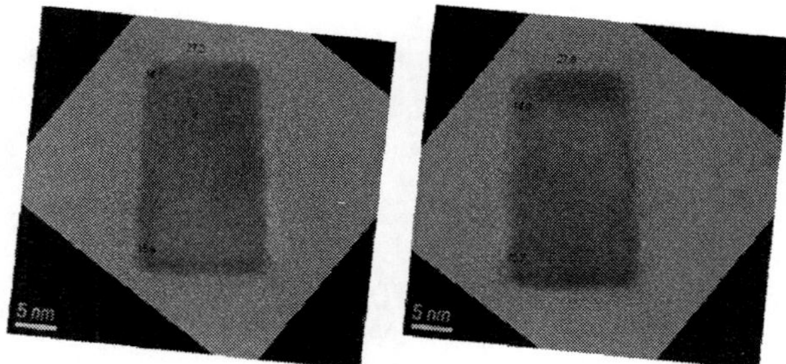

Figure 7. TEM profiles of the c-Si fins after dry etching strip and wet clean.

Conclusions

We developed dry etching process for c-Si fins on a $0.072\mu m^2$ 6T-SRAM cell with 80nm fin pitch. For patterning straight fins with a target CD of 15nm and with smooth sidewalls, it was necessary to redesign the original stack (SiOC and ACL thickness) and redevelop the UL, HM1 and HM2. In order to preserve the CD after HM1 the ACL opening process was done using an Ar/O$_2$ plasma chemistry. Straight profiles were achieved by eliminating the softlanding and replacing it by a longer overetch step. This dry etching process reached the CD target of 15nm with a 3sigma CD uniformity of 3.5nm.

References

1. Y.K. Choi, IEEE Electron Device Lett. **23(1)**, 25 (2002).
2. E. Altamirano-Sánchez, *DPS* Proceedings *p239, Busan, Korea (2009)*
3. E. Paragon, J. Appl. Phys. **105**, 094902 (2009)
4. E. Paragon, J. Vac. Sci. Technol. B **22(4)**, 1869 (2004).
5. Kinoshita, J. Vac. Sci. Technol. B **14(1)**, 560 (1996)
6. A. Steven, J. Vac, Sci. Technol. A **19(5)** 2197 (2001).
7. E. Altamirano-Sánchez, submitted to Microelectronic Engineering (2011)

ECS Transactions, 34 (1) 383-388 (2011)
10.1149/1.3567608 ©The Electrochemical Society

Effect of O_2/Ar ratio on Etching of Diamond Films by MPCVD

Shasha Wang, Kailiang Zhang *, Taofeng Zhang, and Jun Ren

School of Electronics Information Engineering, Tianjin Key Laboratory of Film Electronic & Communication Devices, Tianjin University of Technology, Tianjin, China, 300384, *corresponding author, kailiang_zhang@163.com

Because of its good properties, diamond can be used as protective film of optical components, optical window materials and radiator materials of high power optoelectronic components. But its rough surface has been an obstacle for widespread application. In this paper, etching of diamond films by ICP (Inductively Coupled Plasma) oxygen plasma which was used to improve the surface of the diamond films. After evaluating etch rates under different O_2/Ar conditions, the etch mechanism is investigated. The results show that the surface roughness is reduced after etching. Etching rate increases with higher O_2/Ar ratio. At the same time, when etching with a high O_2/Ar ratio, the surface roughness is low. In conclusion, O_2 and Ar plasmas, especially O_2 plasmas, have a significant effect on etching of diamond films.

I. INTRODUCTION

Diamond has superior properties such as hardness, high thermal conductivity, chemical inertness, high stiffness, high carrier mobility etc. The rough surface of the deposited diamond has been an obstacle for a widespread application (1).

Etching can play an important role in many aspects of processing and in the preparation of the diamond film, not only patterning of the film, but also non-diamond film surface etching. Research in the field of at home and broad mainly in the following, such as, studies by Zeng-sun Jin on microwave oxygen plasma etching of diamond films indicate that with the increase of oxygen concentration, with the enhanced role of etching and the different diamond crystal to produce a selective etching, in order to tend to form highly oriented films and the formation of larger convex surface relief (2). The Wei-dong Man research group used DC glow oxygen discharges to etch diamond films, and their study showed the diamond film main graphic and etching mechanism, without mention of the surface modification effect (3). D S Hwang investigated ICP (Inductively Coupled Plasma) plasma etching of diamond films with O_2 and Ar, and form different graphical layout based on Al etching mask (4). In order to achieve high etch rate in a conventional oxygen plasma it is necessary to increase the bias power of the plasma, which can cause damage to the surface of the film. The ICP etch technology is widely used in integrated circuit technology.

ICP plasma systems can provide high-density plasmas. The same etching rate can be used under low bias voltage. It is easy to control the density and energy of plasma, so it is conducive to large area etching, well uniformity, less damage and pollution. Using this strategy is able to overcome etching defects produced by the common method of oxygen plasma.

II. EXPERIMENT

2.1 Preparation of Diamond Films

The samples of diamond film were prepared on the mirror silicon substrates by Microwave Plasma Chemical Vapor Deposition (Japan SEKI Company, AX6550 type). The gas source was included Hydrogen, Methane and Oxygen. The specific parameters of deposition process are shown in Table 1.

Table 1 The deposition parameters of diamond films

Experimental parameters	Specific parameters
Microwave power	4000 W
Gas	$H_2:CH_4:O_2=562:30:9$ (sccm)
Gas pressure	70 torr
Substrate size	Diameter=50mm, Thickness=1.5mm
Deposition time	6 hours

2. 2 Etching of Diamond Films

The etching of the diamond films in oxygen/argon plasmas was conducted in a high-density plasma reactor (Institute of Microelectronics, ICP-98C type). The detailed process conditions were as follows: Bias and ICP source power at 100W and 500W respectively, He gas flow of 50sccm, O_2 gas flow of 30sccm and an Ar gas flow of 10sccm to 40sccm. The etch time was fixed to 10min and the working pressure was varied between 0.7Pa and 1.0Pa.

2.3 Testing and Characterization

The samples were rinsed in an ultrasonic acetone bath for about 10min, cleaned with deionizer water and dried. The surface morphology of the diamond films was analyzed by atomic force microscope (Agilent Technology, 5600LS). The composition of the diamond films was studied by micro-Raman spectroscopy (Tianjin Port East Technology Co, Ltd., LRS-5 type). The surface roughness of diamond films was tested by a step profiler (Veeco Instruments Inc, DEKTAK 150 SURFACE PROFILER)

III. RESULTS AND DISCUSSION

3.1 Surface morphology and Raman Spectra of Diamond film as-deposited

Figure1 shows the sample before etching using AFM topography. The deposition of the polycrystalline diamond results in grains of varying shapes. Edges and corners are clearly visible in Fig. 1, with an average particle size of more than a dozen µm, the surface roughness was measured to be 0.634µm. Figure 2 shows the characteristic Raman peak of diamond ($1332.5cm^{-1}$) very clearly.

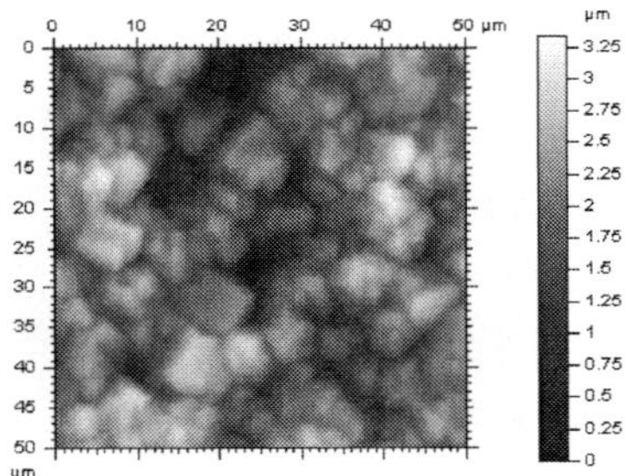

Figure 1 The AFM topography of diamond film before etching

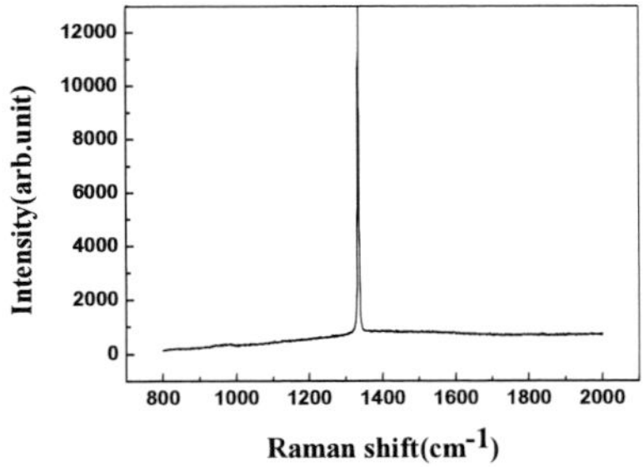

Figure 2 The Raman spectra of diamond film before etching

3.2 Effect of O_2/Ar ratio on etching rate and surface of diamond films

Experiments were conducted where the effect of the oxygen/argon ratio, which was varied from 30/10, 30/20, 30/30 and 30/40, affected the surface morphology as shown in Figure3.

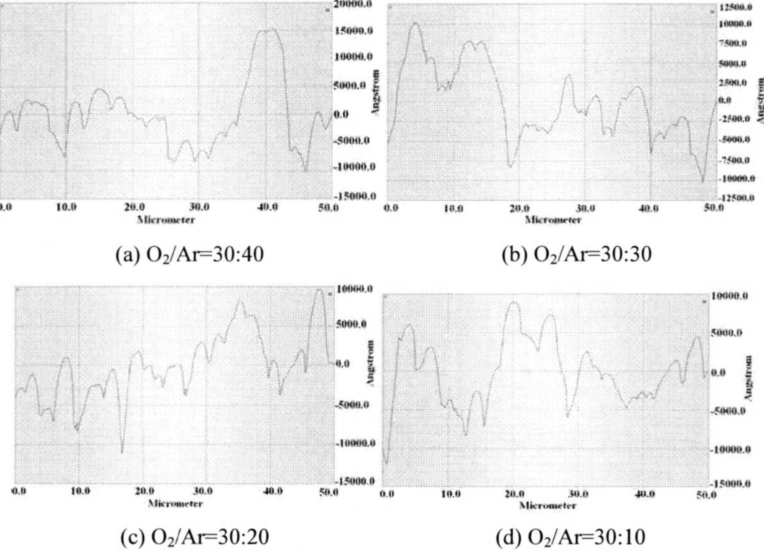

(a) O_2/Ar=30:40

(b) O_2/Ar=30:30

(c) O_2/Ar=30:20

(d) O_2/Ar=30:10

Figure 3 The surface profile measured by the surface profiler

The surface roughness of the experiments was measured repeatedly and the data shown in Figure 4 reflects the average values. When the oxygen/argon ratio was 30/20, the surface roughness was the lowest with 213.53nm. Meanwhile, Figure 5 shows the etch rates of diamond films under different oxygen/argon ratios. The etch rate is fast at the same process condition where we obtained surface smoothing. It appears that the surface roughness and the etch rate are indirectly proportional to each other.

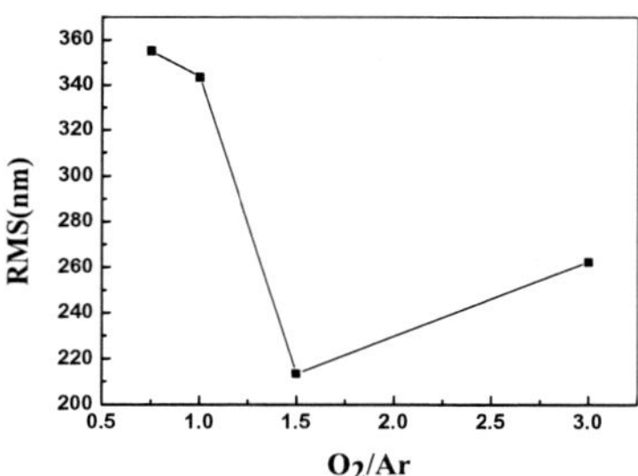

Figure 4 Surface roughness of diamond films after etching

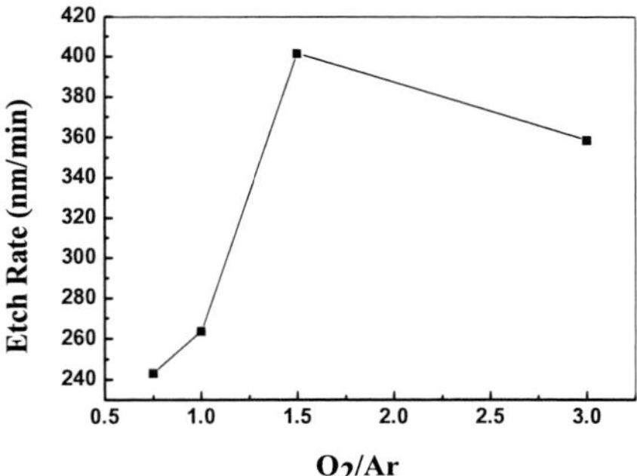

Figure 5 Etching rate of diamond films under different O_2/Ar conditions

3.3 Evaluation of etching mechanism by O_2/Ar

During the experiment, there are two effects contributing to the etch mechanism, physical etching and chemical etching (5).

Argon plasma is played a major role in the physical etching of the experiment. Physical etching is the use of glow discharge of gases argon dissociate into charged ions, and then accelerate the ions by the bias voltage, high-energy ion bombardment and the substrate surface is etched by etching atoms in the material(6).Oxygen plasma is played a major role in the chemical etching of the experiment. Chemical etching is the use of glow discharge oxygen gas to dissociate into ions, electrons and highly reactive neutral particles. In general, with plasma etching in the same chamber, physical reaction and chemical reaction are working simultaneously.

From Figure 5, we can see that the etching rate increase fast with the O_2/Ar ratio increases, indicating that the oxygen plasma etching (chemical etching) plays a more important role than Ar plasma etching (physical etching). This may be because of the structural stability of diamond, which cannot easily be removed with the argon plasma etching. Experimental results in the aspects of surface roughness, which is shown in Figure 4. It shows that when oxygen content relatively is high, the ultimate roughness also reduced obviously. It can be demonstrated more than that chemical etching caused by the oxygen plasma of etching diamond films is in the role of the predominant position.

IV. CONCLUSION

Using high-density ICP plasma etching of diamond films with different oxygen/ argon ratios, we examined the surface morphology of thin films, which the sharp structure of the surface is etched to remove the defects and surface roughness is reduced after etching. When etching with an O_2/Ar ratio of 3/2, and the surface roughness is the lowest. Etching rate increases with higher O_2/Ar ratio. In conclusion,

O2 and Ar plasmas, especially O2 plasmas, have a significant effect on etching of diamond films.

ACKNOWLEDGMENTS

We acknowledge the financial support provided by the National Natural Science Foundation of China under Grant No 60806030, and Tianjin Natural Science Foundation under Grant No 08JCYBJC14600, No 10SYSYJC27700 and Tianjin Science and Technology Developmental Funds of Universities and Colleges under Grant No ZD200709.

REFERENCES

[1] B. H. YANG et al., *J. Optoelectronics ·Laser*, **19**(5), 625-627(2008).

[2] X. Y. LV et al., *New Carbon Materials*, **9**, 191-194(2003).
[3] X. F. ZHENG et al., *Diamond &Abrasives Engineering*, **1**, 35-38(2007).
[4] D. S. Hwang et al., *Diamond Relat. Mater.*, **13**, 2207-2210(2004).
[5] T. Yamada et al., *Diamond Relat. Mater.*, **16**(4-7),996-999(2007).
[6] R. E. Rowles et al., *Diamond Relat. Mater.*, **6**, 791-795(1997).

ECS Transactions, 34 (1) 389-394 (2011)
10.1149/1.3567609 ©The Electrochemical Society

Porous SiOCH integration:
Etch challenges with a trench first metal hard mask approach

N. Possémé [a], T. David [a], T. Chevolleau [b], M. Darnon [b], Ph. Brun [a], M. Guillermet [a], J.P. Oddou [c], S. Barnola [a], F. Bailly [c], R. Bouyssou [c], J. Ducote [b], R. Hurand [a], C. Vérove [c] and O. Joubert [b]

[a]CEA-LETI, *MINATEC campus*, 38054 Grenoble, France
[b]LTM-CNRS/UJF, 38054 Grenoble, France
[c]STMicroelectronics, 38926 Crolles cedex, France
Contact : nicolas.posseme@cea.fr

The use of a metallic hard mask approach for porous dielectric film integration implies for patterning processes, different difficulties like dimensional control, bottom line roughness and residue formation or chamber conditioning. In this paper we propose to present these issues and associated solutions for p-SiOCH integration in dual damascene structure using a trench first metallic hard mask approach.

I.   Introduction

The choice of copper/ultra low-k interconnects architecture is one of the key building blocks for integrated circuit performance, process manufacturability and scalability. Many low-k materials have been proposed as silicon containing (silica-based and silsesquioxane) or non silicon containing materials [1]. Finally, from the 90 nm technological node, the semiconductor manufacturers opted to go with methyl groups (Si-CH_3) containing organosilicate materials close to the well known silicon dioxide (SiO_2) [2]. The SiOCH trench patterning can be achieved using different hard mask strategies (metallic or organic hard masks), all presenting different advantages and drawbacks in terms of selectivity, faceting, etc. [3]. But from the 45nm interconnect technology node, with porosity introduction into SiOCH materials, the metallic hard mask strategy appears to be a promising integration scheme compared to a full organic hard mask approach. Indeed with porosity introduction, the dielectric film presents a high sensitivity upon fluorocarbon (FC) etching [4] and ashing plasma exposures [5] which can lead to dielectric constant increase and reliability degradations. In this context, the metallic hard mask integration provides a greater compatibility with porous dielectric materials than the organic hard mask approach since no ashing step is required. However, in all cases, etching processes involved in dual damascene structures fabrication generate serious issues.

In this paper we propose to present and discuss the porous SiOCH (p-SiOCH) integration difficulties using a trench first metallic hard mask integration scheme (TFMHM) with the following stack: starting from the silicon substrate: a silicon-carbonitride layer deposited by plasma-enhanced-chemical vapor deposition (PECVD SiCN) / porous carbon-doped silicon-oxide (PECVD p-SiOCH, 25% porosity, k = 2.5) / PECVD silicon dioxide (SiO_2) / titanium nitride deposited by physical vapor deposition (PVD TiN).

We will focus on problems encountered during metal hard mask (MHM), via and line etching steps with associated solutions.

I. Metal hard mask opening

A thin metallic hard mask layer is deposited on the top of a dense dielectric layer (SiO_2) which encapsulates the underlying ultra low-k material. The trench photolithography is performed using a positive tone photoresist coupled with a bottom antireflective layer (BARC) (Fig. 1(a)). The BARC and titanium nitride layers are etched using a chlorine-based chemistry in a dedicated transformer-coupled plasma (TCP) (Fig. 1(b)).

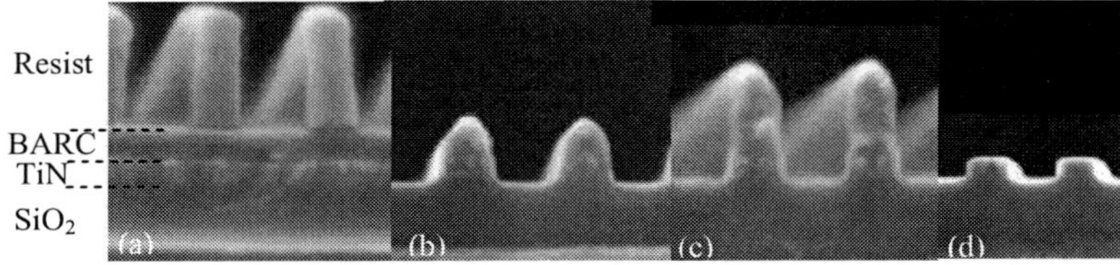

Fig. 1: Sequence of metallic hard mask opening from line lithography (a), BARC and hard mask opening with either chlorine only (b) or with methane addition (c) chemistries followed by the resist removal with an O_2 based plasma (d).

The control of line dimensions is fine tuned during this step by playing with the different plasma parameters (like RF bias voltage) or TiN over-etch time. But the low etching resistance to resist during the TiN etching combined with the shrink in resist thickness at each new technological node makes this step very critical. The solution to overcome this issue is either to develop new metallic hard mask etch chemistries, like methane-based chemistries allowing a two-times selectivity improvement to the resist (Fig. 1(c)), or to transfer the line into the metallic hard mask using a tri layer approach (not shown here). In this later case, the line dimension is fine tuned by playing on the CF_4/O_2 ratio during the SiARC opening. Then the organic planarized layer (OPL) and TiN layers are opened using NH_3- and chlorine- based chemistries, respectively. After TiN opening, the remaining resist (for the conventional approach) or OPL (for the tri-layer approach) are removed using an O_2-based (Fig. 1(f)) or $He/H_2/O_2$ downstream plasmas, respectively.

II. Partial via opening

The shallow hard mask topography, is then planarized with an organic BARC and the via lithography is performed (Fig. 2(a)).

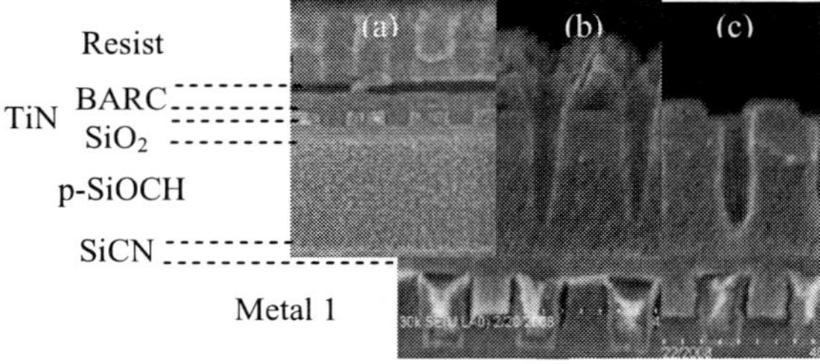

Fig. 2: Sequence of partial via etching from via lithography (a), BARC/SiO_2/p-SiOCH etching in FC based plasma (b) followed by resist removal in O_2 plasma (c)

The via is partially etched into the p-SiOCH material using a fluorocarbon (FC) chemistry in a capacitive coupled plasma (CCP) etcher (Fig. 2(b)). Similarly to MHM etching, the problem during this step is the low selectivity to the resist which becomes a major problem to control the via dimensions. Furthermore, a highly physical etching step used to sputter the TiN in the misaligned vias can lead to arcing in the plasma chamber. The solution is the use of a tri layer approach previously described. In this case the via dimension is fine tuned by the CF_4/O_2 ratio during the SiARC opening then the OPL and dielectrics (SiO_2+p-SiOCH) are open using NH_3- and FC-based chemistries, respectively. The remaining resist or OPL are removed by an O_2 (Fig. 2(c)) or NH_3 plasma, respectively.

III. Via-line etching

The last step in the p-SiOCH via-line patterning using a TFMHM approach is the via-line etching processed with the same fluorocarbon (FC) based chemistry followed by the etch-stop layer (SiCN) opening with a $CH_3F/CF_4/N_2/Ar$ chemistry. But this step must be optimized in order to overcome the different issues that we will present.

III.1 Bottom line roughness and hard mask faceting

The two first problems encountered during the line-via etching, using low polymerizing FC chemistry at substrate temperatures above 50°C, are the hard mask faceting in via-line misaligned structures (Fig. 3(a)) and bottom line roughness formation (Fig. 3(b)). In the first case, it is difficult to control the line-via dimensions which can lead to short defects while the second issue degrades the time dependent dielectric breakdown.

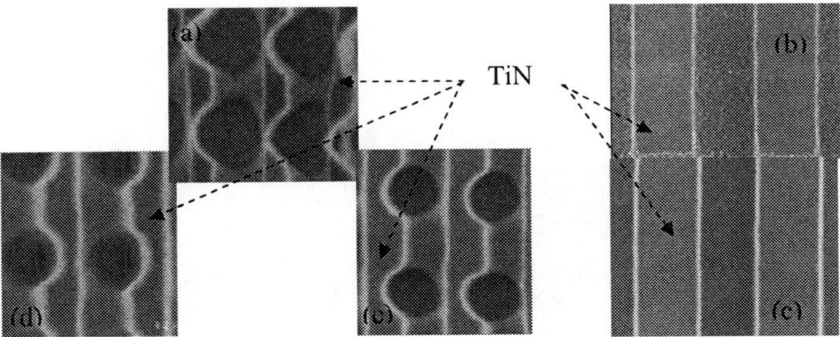

Fig. 3: Top view of problems encountered during via/line etching using a low polymerizing chemistry at substrate temperature above 50°C: hard mask faceting (a), bottom line roughness (b). High polymerizing chemistry at substrate temperature above than 50°C allows to solve both issues (c) while developing low polymerizing chemistry at substrate temperature lower than 50°C improves TiN faceting (d).

III.1.1 Mechanism of TiN etching and bottom line roughness formation

Previous studies on the p-SiOCH etching in fluorocarbon-based plasmas using TiN as a hard mask reported that the TiN etching is mainly driven by an ion-assisted chemical etching mechanism, and that the etch rate, increasing with temperature, is controlled by the fluorine concentration in the plasma gas phase, and the ion bombardment density and energy [3]. The p-SiOCH roughness formation is explained by different mechanisms like intrinsic roughness or micromasking (coming from reactor wall conditioning or FC species), increased by the low volatility compounds (TiF_x) coming from the sputtered hard mask.

III.1.2 Solutions

The solution to limit the metallic hard mask faceting during line-via etching is to develop process at temperature lower than 50°C to reduce TiN etching reaction (Fig. 3(e)). However, this can lead to profile distortions [3] and bottom line roughness increase due to (in both cases) the condensation of non-volatile TiF_x compounds coming from the TiN mask sputtering. Finally, using a metallic hard mask approach, the best way is to develop etching processes at substrate temperature above 50°C (to limit sputtered metal species condensation leading to micromasking) with a highly polymerising chemistry to protect the MHM (Fig. 3(d)) and prevent the p-SiOCH surface roughening (Fig. 3(c)) [6].

III.2 Residues formation

The metallic hard mask approach suffers also from specific issues such as the growth of metallic residues on top of the mask after dielectric materials etching in fluorocarbon-based plasmas (Fig. 4(b)). These residues grow with air exposure time and are not removed after conventional wet cleaning steps (basic solution + 0.1%HF) (Fig. 4(b)). They prevent a good metal barrier deposition conformity, which generates via or line opens (Figs. 4(d) and 4(e), respectively) after chemical mechanical polishing (CMP) and strongly impacts the electrical yield (Fig. 4(c)).

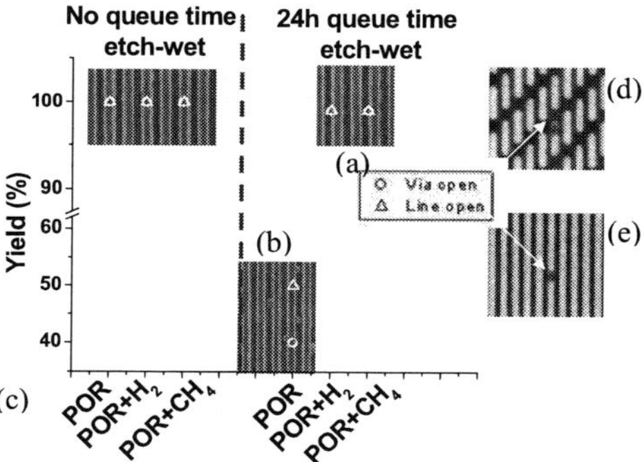

Fig. 4: Study of residue formation with (a) and without (b) in-situ post etch treatment addition after etching and one day air exposure. Effect on yield as a function of the waiting time between etch and wet steps (c) with the formation of defects like via (d) and line (e) opens.

III.2.1 Mechanism

The mechanism of formation was attributed to fluorine species reacting with the titanium nitride and catalyzed by air moisture. More precisely, fluorine species present on the TiN surface after plasma exposure react with air moisture from the atmosphere leading to HF formation. HF reacts with the partially oxidized TiN surface and forms metallic salts through the following sequence [7]:

$$4F + 2H_2O \rightarrow 4HF + O_2$$
$$HF + TiO \rightarrow TiOFH$$

Therefore, the solution to avoid the metallic residue formation is to act on the acid formation by either removing the fluorine or to avoid the reaction with the air moisture.

III.2.2 Solutions

Different *in situ* post-etch plasma treatments such as hydrogen- and methane-based plasmas are efficient to avoid residue formation after one day air exposure (Fig. 4(a)), strongly improving the yield (Fig. 4(c)). This efficiency was explained by the fact that hydrogen-based plasmas act on the kinetic of residue formation by removing the fluorine concentration on top of the titanium nitride layer. With methane-based chemistries, the formation of a carbon and nitrogen based passivation layer on top of the titanium nitride layer prevents the reaction with air moisture [8].

The compatibility and benefit (on reliability) of methane PET addition with porous SiOCH integration using TFMHM has been demonstrated for 45nm interconnect technology node [9].

III.3 Dielectric line undulation

With the constant scaling down of dimensions, etching process steps are facing new issues. Indeed, with patterning of sub-50 nm porous SiOCH trenches using a metallic hard mask approach, a line undulation phenomenon, also called line wiggling, is observed (Fig. 5(b)). Mechanical simulations performed using the ANSYS™ code [10] show this line undulation is due the compressive residual stress of the metallic hard mask (Fig. 5(a)) combined with the p-SiOCH film thickness (not shown). The Young modulii of the metallic hard mask and the p-SiOCH also play a role on the wiggling formation but in second order compared to the previous parameters (not shown).

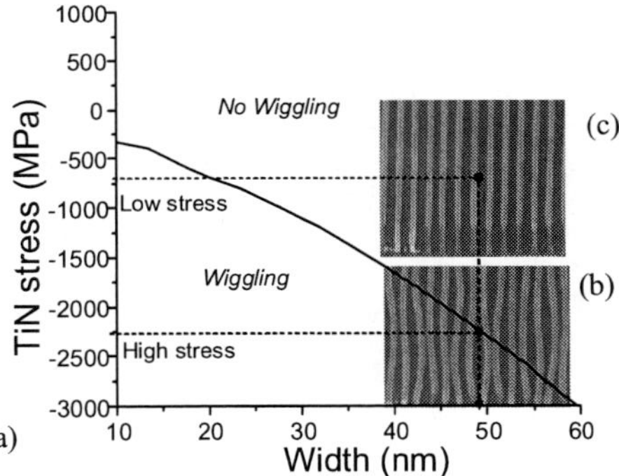

Fig. 5: Comparison of the line wiggling formation between simulation prediction (a) and experimental results as a function of the line dimension for the stack 20nm SiCN/ 290 nm p-SiOCH/ 40nm SiO_2 with high (b) and low (c) TiN residual stress.

Therefore, one solution to continue the porous material integration with a metallic hard mask for advanced interconnect technology nodes is to develop TiN materials with lower residual stresses and thicknesses (Fig. 5(c)). In addition, the effect of the p-SiOCH film thickness on line undulation will be minimized for advanced technology node, since the p-SiOCH thickness continuously decreases (aspect ratio almost constant).

IV. Chamber conditioning

The use of a TFMHM has also a strong effect on process stability and reproducibility. Indeed, with this metal hard mask approach, a drift of the line etching processes (etch rates, uniformity) and lower mean time between cleans (compared to the use of an organic hard mask) are observed when using non heated top electrode with a standard clean procedure (O_2 plasma). The main cause is the chamber defectivity increase after several RF hours of etching process (Fig. 6(a)) due to the formation of micromasking on silicon top electrode induced by Ti etched by product redeposition.

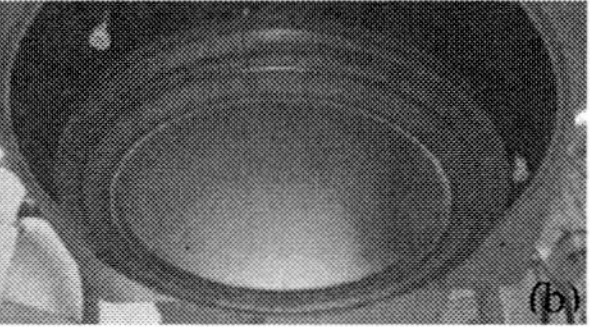

Fig. 6: Observation of (a) non heated and (b) heated (>100°C) top electrodes after more than 500 Rf hours processes.

A solution to delay this chamber defectivity is to heat the top electrode (above 100°C) to limit Ti species redeposition and thus silicon micro-masking (Fig. 6(b)). A second possibility is to clean up the reactor walls after each wafer being etched by developing appropriate plasma clean processes. For instance, Chevolleau et al. propose to use a SF_6 plasma to remove the fluorocarbon layer and Ti based species followed by an O_2 flash plasma to clean up the chamber walls from the remaining carbon [11].

V. Conclusion

Major difficulties of the p-SiOCH material integration using a TFMHM integration have been presented from the etching point of view. The first problem encountered during the hard mask and partial via etching is the low selectivity to resist leading to dimensional control issue. The use of a tri layer approach can solve this problem. While during via-line etching we are dealing with hard mask faceting, bottom line roughness and residue formation or line undulation. In this case the use of TiN hard mask with a low residual stress and thickness but also the development of highly polymerizing chemistries followed by *in situ* post etch treatment addition at high substrate temperature (above 50°C) are required. Finally, with the use of a metallic hard mask, reactor cleaning strategies must be revisited to allow better process stability and reproducibility.

References

[1]   D. Shamiryan et al. materials today, pp 34-39 (2004)
[2]   A. Grill et al. proceedings IITC 2004, pp. 54–56.
[3]   M. Darnon et al Microelectronic Engineering 85, 2226–2235, (2008).
[4]   N. Posseme et al, J. Vac. Sci. Technol. B Vol. 22, No. 6, Nov. 2004
[5]   N. Posseme et al, J. Vac. Sci. Technol. B, Vol. 25, No. 6, Nov/Dec 2007
[6]   Bailly et al, J. Appl. Phys. Vol. 108, No.1, 014906, Jul 2010
[7]   N. Posseme et al J. Vac. Sci. Technol. Vol 28, No. 4, 809-816, Jul 2010
[8]   N. Posseme et al J. Vac. Sci. Technol. B 29(1) Jan/Feb 2011
[9]   M.Darnon et al, Appl. Phys. Lett. 91, 194103 (2007)
[10]   N.Posseme et al, IITC proceedings Pages: 240-242, Jun 2009
[11]   Chevolleau et al, J. Vac. Sci. Technol. B Vol. 25, No. 3, May-Jun 2007

ECS Transactions, 34 (1) 395-398 (2011)
10.1149/1.3567610 ©The Electrochemical Society

Plasma Etching Parameters Impact To Low-k Damage

Jihong Zhang, Huiyuan Pei, LH Cheng
Lam research (Shanghai) Co., Ltd, Shanghai 201203, China

Mechanical strength of low-k material greatly decreases; material composition change will also give dry etching a lot of problems which have never met before. This paper explores the approaches to solve new problems when plasma etching low-k material, such as avoiding low-k damage and other defects, keeping anisotropic and needed selectivity; finding out the process parameters with the biggest influence, such as gas species, pressure, RF frequency, power, etc. Applies the process trend to plasma etching low-k material (BD) of 40nm dual damascene trench etch and achieves the critical target.

Process Parameter Influence to Low-k Damage

The traditional O2-based plasma etching process will cause severe damage to porous low-k materials. Oxygen will replace Carbon in low-k materials, forming volatile organosilanol and absorb moisture, causing K value increases. Sidewall surface composition of low-k materials is changed by chemistry of plasma etching, and can be removed by HF. The thickness of sidewall pull-back can indicate the degree of low-k damage in plasma etching.

<u>Gas Species</u>

<u>O2</u>
High-flow O2 causes severe damage to low-k material.
0" HF dip: CD 87nm
30" HF dip: CD 105nm, HF loss 9nm/side
120" HF dip: CD 117nm, HF loss 15nm/side

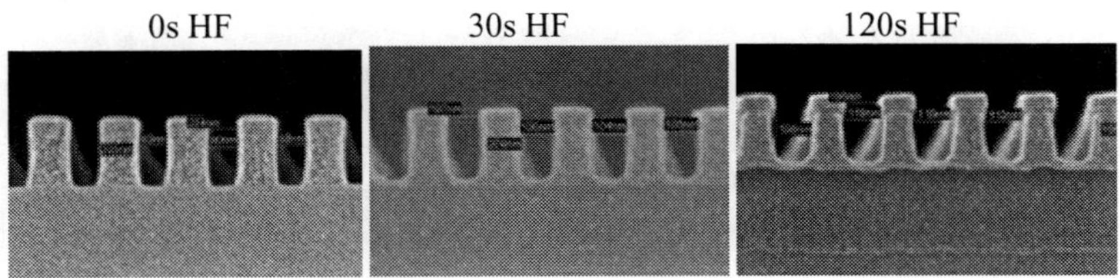

Figure 1. O2 impact to low-k damage

<u>CO2</u>
CO2 has little impact to low-k damage.

0" HF dip: CD 87nm
30" HF dip: CD 89nm, HF loss 1nm/side
120" HF dip: CD 92nm, HF loss 2.5nm/side

0s HF   30s HF   120s HF

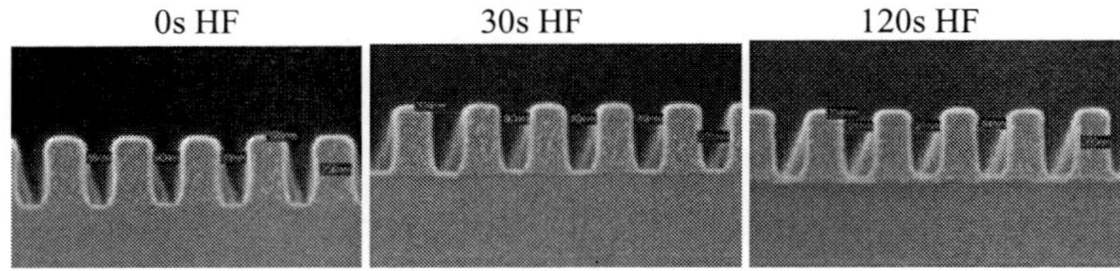

Figure 2. CO2 impact to low-k damage

NH3
NH3 has little impact to low-k damage.
0" HF dip: CD 88nm
30" HF dip: CD 90nm, HF loss 2nm/side
120" HF dip: CD 90nm, HF loss 2nm/side

0s HF   30s HF   120s HF

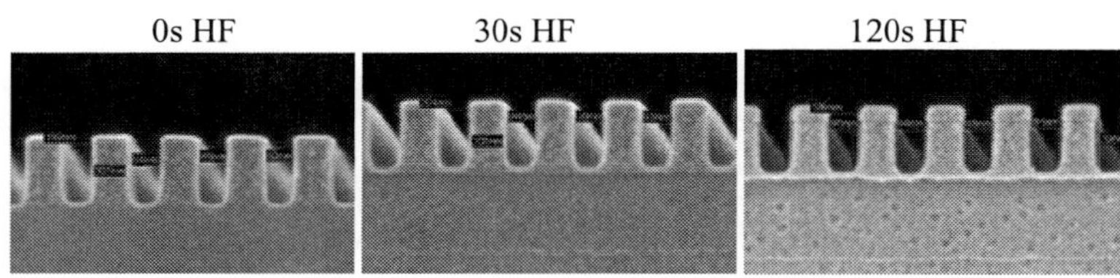

Figure 3. NH3 impact to low-k damage

CO2+CH4
The combination of CO2 and CH4 is better than O2 but worse than pure CO2.
0" HF dip: CD 87nm
30" HF dip: CD 99nm, HF loss 6nm/side
120" HF dip: CD 98nm, HF loss 5.5nm/side

0s HF   30s HF   120s HF

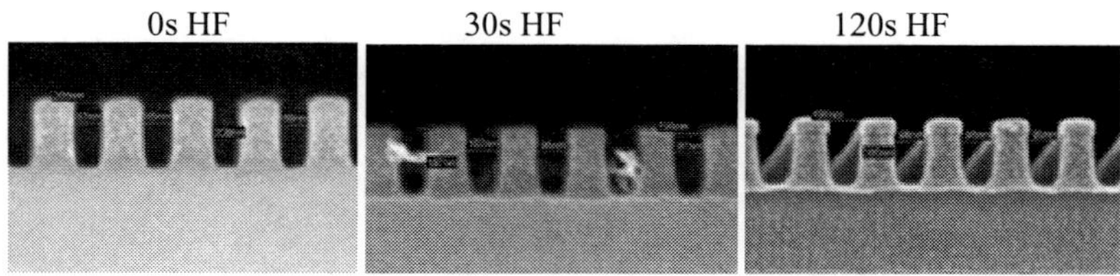

Figure 4. CO2+CH4 impact to low-k damage

Pressure and RF Frequency

Lower pressure provides lower radical/ion ratio (with higher frequency). Moving to lower pressure regime lowers sidewall damage however there are other influential factors.

At higher pressure, lower sidewall damage with the lower frequency. Trend is to lower sidewall damage at lower pressure. Minimum sidewall damage is at 10mT, with 60MHz or 27MHz. Frequency has a strong influence especially at higher pressure: lower frequency, lower sidewall damage.

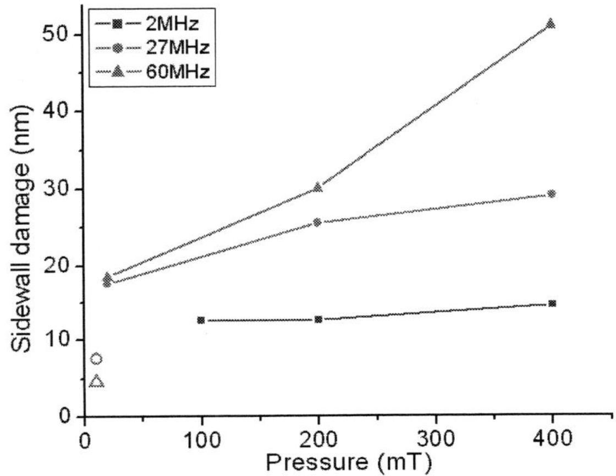

Figure 5. Pressure and Frequency influence to low-k damage

40nm DD Trench Etch with BD

<u>Process Requirement</u>
    <u>Dense area</u>: profile vertical, sidewall no pull-back, no bowing, no damage, to keep reliability acceptable.
    <u>Dual Via area</u>: corner rounding, no terrace, no fence, to make Cu filling easier.
    <u>Single Trench area</u>: profile vertical, no polymer deposition.

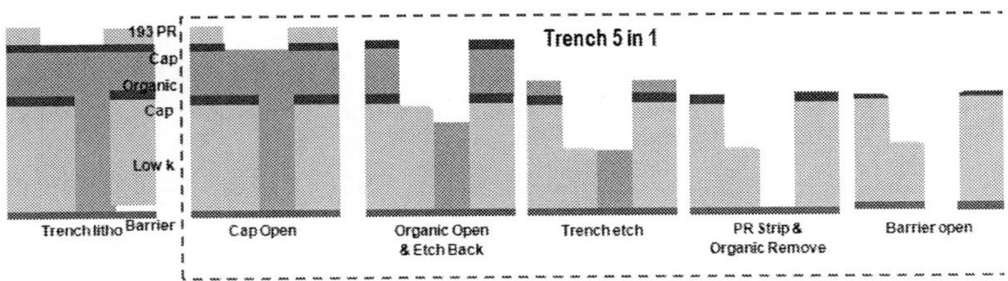

Figure 6. Dual Damascene Approach with Tri-layer Mask

<u>Result:</u>

After a series of experiments, finally got a satisfactory recipe for 40nm DD low-k etch. Several key points can be summarized as:

1) Low-k damage
2) Polymer by-product control
3) BPR top shape maintain

These key points can influence each other, can't modify any one without impact others, so the balance point must be found out.

When etching low-k material of 40nm DD trench, N2/H2 causes less low-k damage than CO2, but etch rate is also less than CO2; HM step generates more polymer and protect the low-k material sidewall to avoid damage; low pressure can significantly reduce low-k material sidewall damage while keep uniformity good.

Acknowledgments

Lower pressure causes lower low-k damage; lower RF frequency causes lower low-k damage in high pressure regime; O2 causes the most severe low-k damage; CO2 or NH3 can be instead and improve a lot; polymer deposition on sidewall can protect low-k material to be damaged.

References

1. Dr. Stanley Wolf, Introduction to dual-damascene interconnect processes [J], Silicon processing for the VLSI era Vol. 4, 2002, Page 674-679
2. D.L. Keil, B.A. Helmer, and S. Lassig, Review of trench and via plasma etch issues for copper dual damascene in undoped and fluorine-doped silicate glass oxide [J], Journal of Vacuum Science & Technology B: Microelectronics and Nanometer Structures, September 2003, Volume 21, Issue 5, Page 1969-1985
3. Andrew Li and Reza Sadjadi, Trench Etch for OSG Low-k Dual Damascene [R], Shanghai, Lam Research, 2003
4. W. W. Lee and P. S. Ho, Low-Dielectric-Constant Materials for ULSI Interlayer Dielectric Applications [J], MRS Bulletin, 1997, October, 19
5. Yun-Sang Kim, Peter Ta-Chin Wei, George R. Tynan etc. Selective Plasma Etching for High-Aspect-Ratio Oxide Contact Holes [J], Japanese Journal of Applied Physics, 1998, Vol. 37, Page 327-331
6. Yi Zheng, Wen Zhu, TzuFang Huang, etc. Plasma Enhanced CVD Low-k Black Diamond[TM] Film Formation Process For 65nm Technology Node, Dielectric systems & modules, 2004

ECS Transactions, 34 (1) 399-403 (2011)
10.1149/1.3567611 ©The Electrochemical Society

Prevention of AlCu Line Galvanic Corrosion after Fluoride Containing Stripper Cleaning: A Case Study

Victor Luo, Jason Chang, Kevin Shi
Semiconductor Manufacturing International (Shanghai) Corp., Shanghai, 201203, China

Bing Liu, Libbert Peng, Andrew Wang, Justan Sun
Anji Microelectronics (Shanghai) Co., Ltd. Shanghai, 201203, China

In this paper, the status of galvanic corrosion on AlCu lines during the AlCu post etch and ashing residue removal process is studied with a fluoride containing stripper in details, including DIW rinse condition and dry process. The experimental data show that galvanic corrosion on AlCu lines is not observed in the chemical cleaning process, but inclined to be caused by the rinse and dry processes in this case. Different scenarios in the rinse process, such as overflow or quick dump rinse (QDR), the times of quick dump rinse cycles, CO_2 bubbled DIW and metal protection solution rinse, are studied. At the same time, two kinds of dry processes are also evaluated. The results show the optimized rinse and dry processes can eliminate the galvanic corrosion on AlCu lines without the change of wafer manufacture prevenient process and cleaning solution.

Introduction

Cleaning processes become more and more critical and challenging along with the continuous shrinking of semiconductor device size. During this process the unwanted residues such as sidewall polymer must be removed completely while the substrate materials and patterns should be kept intact. Fluoride containing strippers are very attractive to end user due to its low operation temperate, low cost, ease of rinse in water and ease of disposal compared to that of the hydroxylamine based chemical(1,2,3,4). AlCu metal line is easily subject to corrosion if not properly passivated (5). More importantly, the control of galvanic corrosion stands as a major challenge in this cleaning process.

In this paper, the case of galvanic corrosion on AlCu lines during the AlCu post etch and ashing residue removal process is studied in details with a fluoride containing stripper. The experimental results show optimized rinse and dry process can address the galvanic corrosion concern on AlCu lines in this case without the change of the wafer manufacture prevenient process and cleaning solution.

Experimental

AlCu metal wafer film stacks consist of Ti/TiN/AlCu/Ti/TiN/Oxide. The metal wafer is processed at 40°C for 20min with IDEAL Clean 960plus as the fluoride containing stripper. Beaker test is carried out in a temperature well controlled water-bath with 50ml

IDEAL Clean 960plus solution in a 250ml beaker, the metal line coupon wafer is immersed in the beaker for the required time, picked up by a tweezer and then manually rinsed in 2 beakers with 500ml de-ionized water (DIW) by sequence, finally dried with room temperature N_2 gun for SEM analysis. Wet bench research is processed on Universal Plastics wet bench. Hitachi 4700 is applied for the SEM cross section evaluation.

Results and Discussions

Case Description

The coupon beaker test result of AlCu metal line wafer with IDEAL clean 960plus as stripper is shown in Figure 1(A). Figure1 (A) shows the residue is totally removed and the metal line is smooth and no obvious corrosion. Wet bench test results of the same wafer and same stripper with hot (160°C) N2 dry before DIW rinsing condition optimization is shown in Figure 1(B), which shows the residue is free but galvanic corrosion appears.

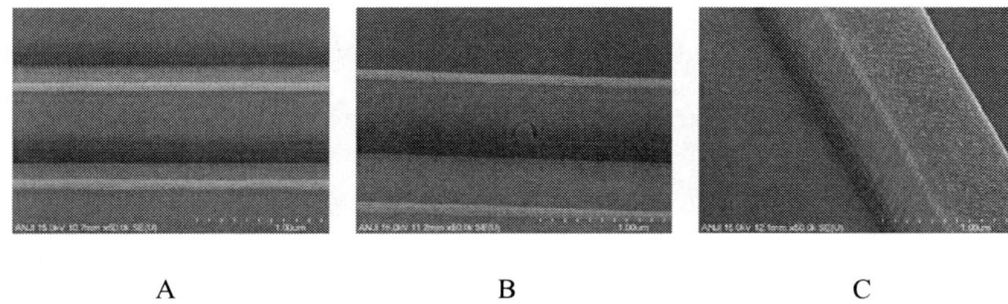

A   B   C

Figure1 SEM picture of the cleaning performance and controlled wafer
A. beaker test performance; B. wet bench test performance without process
optimization; C. controlled wafer

Case Study

In general, galvanic corrosion on AlCu lines can be resulted from low quality metal deposition, ineffective or inefficient etching and ashing process, long queue time after etching & ashing, chemical cleaning, rinsing process and dry process. From the result in Figure1, it is easily deduced the galvanic corrosion do not happen in the chemical cleaning process because the same kind of wafer and stripper are used and the wafer queue time of wet bench test is the same as that of the beaker test. The obvious difference is the rinse process and dry process, which are the root cause for galvanic corrosion.

<u>Rinse Process Study</u> In order to understand why the galvanic corrosion happens in the rinse and dry process, it is worthy to know whether the AlCu metal line has any change when the IDEAL clean 960plus stripper is diluted in DIW. The detailed data are shown in Figure 2, which shows there is an AlCu etch rate mountain shape peak when the IDEAL clean 960plus stripper is diluted in DIW. The AlCu etch rate is nearly no change when the weight ratio of DIW to IDEAL clean 960plus stripper is no more than 0.25. However,

if the weight ratio is between 0.25 and 0.95, the Al etch rate increases more and has a mountain shape peak. But when the weight ratio is more than 0.95, the AlCu etch rate is recovered and well controlled. So it is very important to study the effect of different rinse scenarios on the AlCu metal line corrosion including galvanic corrosion and general corrosion. The detailed results are listed in Table1.

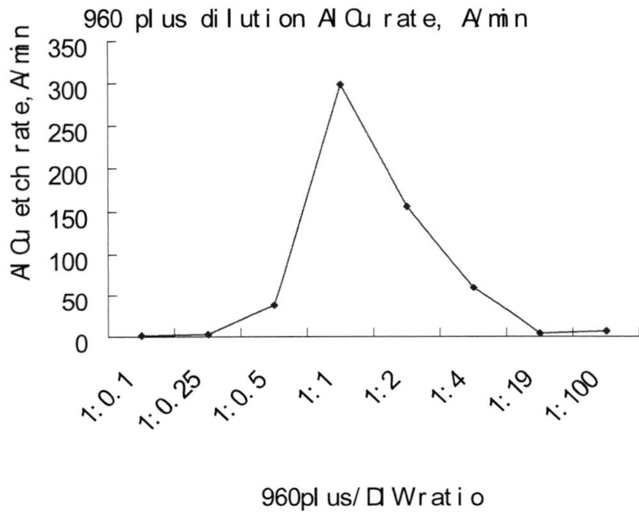

Figure 2 AlCu etch rate vs. the IDEAL clean 960plus/DIW ratio

Table I shows the single DIW overflow rinse process is not suitable for AlCu metal line post etch and ashing cleaning. Because both the general corrosion and galvanic corrosion happened in this kind of wafer in this case. It is also shown the QDR rinse process is better than the single overflow rinse process if just considering corrosion impact. The general corrosion is eliminated by QDR rinse process although the galvanic corrosion still exists, which reveals the galvanic corrosion shouldn't be addressed just by quickly passing the dilution peak in Figure2 but the general corrosion can. Table I also shows CO_2 bubbled DIW is helpful to suppress the galvanic corrosion. Even only use CO_2 bubbled DIW QDR 30s once, and then followed by un-bubbled DIW QDR twice, the AlCu metal line is kept intact. If extending followed un-bubbled DIW QDR to thrice, the galvanic corrosion occurs again. In another word, only CO_2 bubbled DIW QDR once can not support an enough window to extend the un-bubbled DIW QDR time from twice to thrice, and long un-bubbled DIW rinse has an obvious side-effect to galvanic corrosion. Based on those results, we recommend using CO2 bubbled DIW always instead of just bubbled DIW in 1st QDR. In addition, Table I also shows the water based metal protection solution further extends the galvanic corrosion prevention windows than CO_2 bubbled DIW. The AlCu metal line is not attacked even just QDR 30s by water based metal protection solution once and followed by un-bubbled DIW QDR thrice.

TABLE I. the effect of different rinse scenarios on the Al Cu metal line corrosion.

Experiment	Rinse Process	Corrosion information	
		General corrosion	Galvanic corrosion
1	DIW / overflow / 6mins	Y	Y
2	DIW / QDR / 6mins	N	Y
3	CO_2 bubbled DIW / QDR / 6mins	N	N
4	CO_2 bubbled DIW /QDR 30s once / DIW QDR twice	N	N
5	CO_2 bubbled DIW /QDR 30s once / DIW QDR thrice	N	Y
6	Water based metal protection solution / QDR 30s once / DIW QDR thrice	N	N

Other experimental conditions are same.
Y: corrosion occurred; N: free corrosion

<u>Dry Process Study</u> Comparing the beaker tests and wet bench tests in this case; the dry process is also different. The room temperature N_2 is used in the beaker test while the hot $N_2(160°C)$ is used in the wet bench test. The effects of N_2 temperature change on the AlCu metal line corrosion are listed in Table II.

TABLE II. The effect of N_2 temperature change on the Al Cu metal line corrosion.

Experiment	N_2 temperature,°C	Corrosion information	
		General corrosion	Galvanic corrosion
1	160	N	Y
2	25	N	N

The two experiments are processed on wet bench and other experimental conditions are same.
Y: corrosion occurred; N: free corrosion

Table II shows the room temperature N_2 dry yields better prevention corrosion performance than hot N_2 dry. In the wet bench test, there is a small amount DIW and air on the AlCu metal line surface when wafer is moved to dry. Under high temperature, DIW and air, the AlCu metal line corrosion is easy to occur. So if the AlCu metal line is not handled properly in dry process, galvanic corrosion may still appear.

Summary

In this case, the rinse process and dry process are optimized to suppress AlCu metal line galvanic corrosion. The results show the optimized rinse and dry process can yield enough process windows to prevent the galvanic corrosion occurring on AlCu lines without change of the wafer manufacture prevenient process and cleaning solution. The detailed mechanism study is in progress.

Acknowledgments

The authors would like to thank Luck Zhang for high quality X-SEM picture collection.

References

1. Robert Small, Simon Kirk and Mihaela Cernatl, *J. Vac. Sci. Technol. B,* **20(2),** 635 (2002).
2. Bing Liu, Libbert Peng, Julia Peng, Joey Yu and Shumin Wang, *Semiconductor Technology,* **635,** (2009).
3. Libbert Peng, Bing Liu, Shumin Wang, David Yin, Todd Hogan, *Surface Preparation and Cleaning Conference,* (2008)
4. Libbert Peng, Bing Liu, Yong Gong, Shumin Wang, *9th International conference on Solid-State and Integrated-Circuit Technology ,* E4.7 (2008)
5. Francesco Pipia, Annamaria Votta, Gloria Obetti, Enrico Bellandi, Mauro Alessandri and Thomas Nolan, *Solid State Phenomena,* **134,** 367 (2008)

ECS Transactions, 34 (1) 405-408 (2011)
10.1149/1.3567612 ©The Electrochemical Society

Highly Selective Etching Solutions for Advanced Logic Technologies

Xin-Peng Wang[a], Hai-Yang Zhang[b], Shih-Mou Chang[b], Kwok-Fung Lee[b]

[a]Semiconductor Manufacturing International Corporation, No.18 Wen Chang Rd., BDA, Beijing, 100176, P.R.China
[b]Semiconductor Manufacturing International Corporation, No.18 Zhang Jiang Rd., Pudong New Area, Shanghai, 201203, P.R.China

ABSTRACT

As logic semiconductor moves to 65nm node and beyond, several process-enhancement technologies have come into prominence with the help of dry etch processes. In high-aspect ratio contact etch (CT) process, the sufficient over-etch during dielectric oxide etch without too much loss of SiN-based stop layer is required to maintain the enough process window for less silicide loss loading and leakage. Stress memorization technology (SMT) is designed to enhance the P-MOS performance. It removes the stressed nitride and stops on one thin oxide layer. The enough oxide remaining is also imperative to avoid silicon damage. Besides, the loading effect of nitride thickness in SRAM area has to be compensated. Highly selective etching process is indispensable to fulfill these tightening and newly-born process requirements. In this paper, we address the effect of etch parameters including pressure, chemistry, gas ratio and power on the etching selectivity between nitride and oxide film in the above two processes. Scanning Electron Microscope (SEM) is utilized to compare the final etched profile. Results demonstrate besides the etch chemistry and power ratio, the etcher selection also plays the vital role on etching selectivity and etch rate.

I.INTRODUCTION

As the advanced logic technology is scaled to 65nm node and beyond, contact etch has started to act as a key role for the robust function of integrated circuit. To ensure the low contact resistance and less leakage, aside from the introduction of Ni-silicide, the loading of Ni-silicide loss after CT etch has to be minimized not only between the square and rectangular CT hole but also between landing on poly gate top and active area. This indicates the highly etching selectivity of oxide over nitride in CT main etch (ME) is imperative.[1,2] SMT has become one of strain-silicon techniques to enhance n-MOS performance since 65nm node. In SMT, the stressed SiN on P-MOS has to be removed for the hydrogen under oxide can not flush during the spike anneal in the presence of the SiN capping layer. This ends up with boron deactivation and exo-diffusion in source and drain of P-MOS. Another challenge in SMT etch stems from the thickness variation of SiN layer between dense and ISO area. With the down-scaling of CMOS, this becomes much worse for the thinner SiN film at small pitch, which is usually associated with punch-through risk and degrades P-MOS performance for the sake of SiGe loss. High etching selectivity between the stressed SiN and its oxide stop layer is also required.[3,4,5]

In this work, we investigated the etching selectivity behavior between nitride and oxide in the above mentioned two typical etch processes. Results indicate the desired selectivity could not be achieved without the appropriate etcher. The identified trend could be extended to other crucial processes such as spacer etch, stress proximity technology etch and dual stress liner formation.

II. EXPERIMENTS AND DISCUSSIONS

Etch of high aspect ratio CT hole is usually related to high-power process to avoid the etch stop phenomena. Highly energetic ion bombardment might cause device damage. Capacitively coupled plasma (CCP) reactor such as TEL SCCM series and Lam Flex series can be utilized for its medium plasma density. Nevertheless, due to the design difference, Lam Flex series are with two or three frequency powers at bottom electrode while TEL SCCM series utilizes two separate RF generators (one at top electrode, another at bottom electrode) and/or DC superposition at the top electrode. The identified selectivity dependence needs to be re-evaluated if different etcher is used. Regarding SMT etch or the similar thin film removal, inductively coupled plasma (ICP) reactor such as Lam Kiyo series and AMAT DPS series is a good choice for more accurate process control.

Condition	Pressure, mTor	Sourece power, W	Bias power, W	Etch chemistry gas, sccm					Etch rate, A/min		Etch selectivity, Oxide:Nitride
				C4F8	C4F6	CH2F2	Ar	O2	Oxide	Nitride	
I-1	a	b	c	f	0	0	d	e	2430	260	9.3
I-2	a	b	c	0	f	0	d	e	2250	110	20.5
I-3	a	b	c	0	0	f	d	e	1760	340	5.2
I-4	a	b	1.2c	0	f	0	d	e	2300	130	17.7
I-5	a	b	0.8c	0	f	0	d	e	2120	105	20.2
I-6	a	b	c	0	f	0	d	1.25e	2310	165	14.0
I-7	a	b	c	0	f	0	d	0.75e	2210	85	26.0

Table 1 Normalized recipes, corresponding etch rate and selectivity of oxide over nitride

Condition	Pressure, mTor	Sourece power, W	Bias power, W	Etch chemistry gas, sccm					Etch rate, A/min		Etch selectivity, Nitride:Oxide
				CHF3	CH2F2	CH3F	He	O2	Nitride	Oxide	
II-1	A	B	C	F	0	0	D	E	190	55	3.5
II-2	A	B	C	0	F	0	D	E	215	46	4.7
II-3	A	B	C	0	0	F	D	E	130	8	16.3
II-4	A	B	1.2C	0	0	F	D	E	160	26	6.2
II-5	A	B	0.8C	0	0	F	D	E	70	4	17.5
II-6	A	B	C	0	0	F	D	1.25E	146	8	18.3
II-7	A	B	C	0	0	F	D	0.75E	115	11	10.5

Table 2 Normalized recipes, corresponding etch rate and selectivity of nitride over oxide

Table 1 summarizes the normalized contact etch recipes and corresponding etch rates and selectivity of oxide over nitride from TEL SCCM series chamber. I-2 is the baseline. Table 2 shows the normalized SMT etch recipes and corresponding etch rate and selectivity of nitride over oxide from Lam Kiyo series chamber. II-3 is the baseline. All the etch rates come from the blanket nitride/oxide wafers with the splits based on the specific step in baseline recipe. Clearly, the quite different gases are employed in two typical etch processes. This comes from the fact the etch depth is drastically reduced as the contact hole shrinks. Any polymerizing gas with H should be precluded. CT ME step hinges on $C_4F_6/O_2/Ar$ and its selectivity of oxide over nitride ranges from 5 to 25 while SMT over-etch (OE) step relies on $CH_3F/O_2/He$ and its selectivity of nitride over oxide shows the similar range. In both cases, the lower C/F ratio gas gives lower selectivity, the higher H ratio gas tends to increase the selectivity of nitride over oxide, O_2 could suppress the oxide etch rate to some extent. Besides the diluting role, unlike He, Ar is also used to enhance the bombardment.

Fig.1 compares the effect of bias power (+-20%) on selectivity between oxide and nitride in both cases. In CT ME, the impact of bias power could be neglected while the selectivity of nitride over oxide increases as bias power decreases in SMT OE. Such reverse trend could not be explained by the bias power strength. The difference of chemical bond energy between Si-O (8.3eV) and Si-N (4.6eV) might indicate that SMT etch is more sensitive to bias power. Namely, it's ion bombardment dominant process. Besides the high selectivity, the corresponding impact on the etched profile has to be examined. Fig 2 demonstrates the role of bias power in forming the post-etch SiN boundary. The split II-5 with -20% bias power exhibits much worse SiN boundary on structure wafers, although it features the similar selectivity to split II-3. It might be due to the less ion direction for lower bias power. Fig 3 reveals the slight profile angle variation owing to different bias powers. In brief, high bias power delivers the more vertical profile.

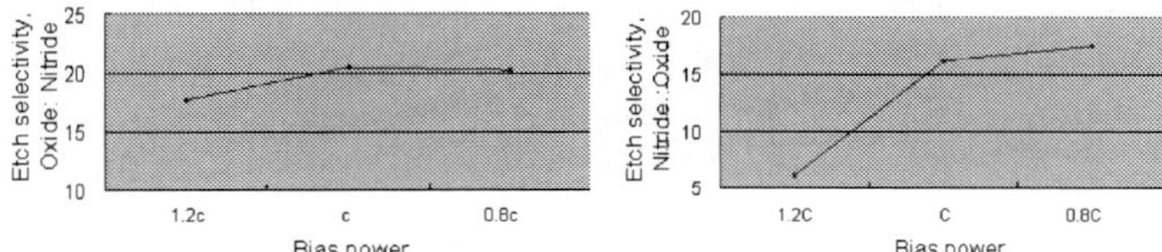

Fig. 1 The effect of bias power on selectivity: oxide/nitride (left); nitride/oxide (right)

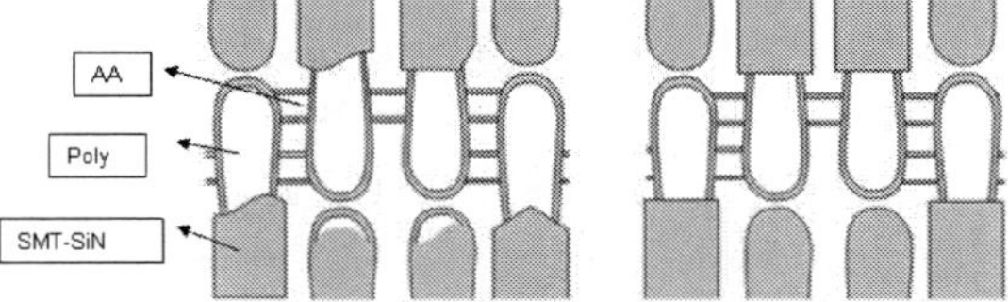

Fig. 2 The effect of bias power on SRAM line edge roughness: II-5 (left) vs II-3 (right)

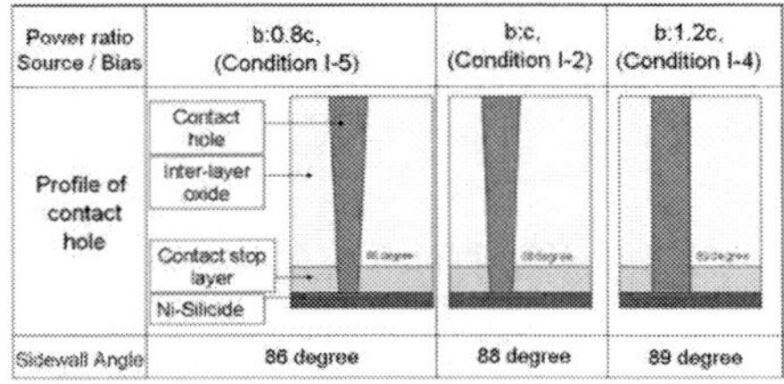

Power ratio Source / Bias	b:0.8c, (Condition I-5)	b:c, (Condition I-2)	b:1.2c, (Condition I-4)
Profile of contact hole	Contact hole / Inter-layer oxide / Contact stop layer / Ni-Silicide		
Sidewall Angle	86 degree	88 degree	89 degree

Table. 3 The effect of bias power on the sidewall profile of contact hole

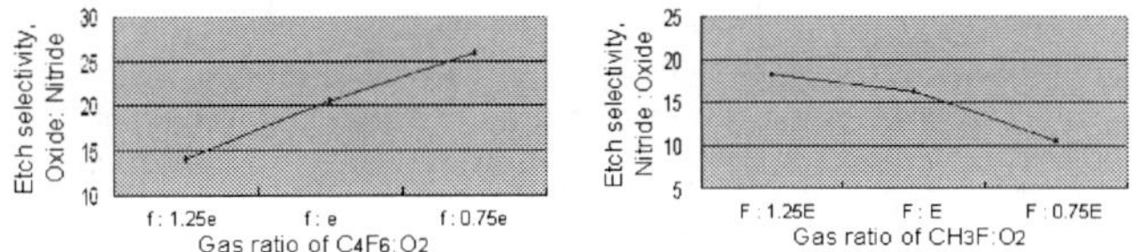

Fig. 3 The effect of O_2 gas ratio on selectivity: oxide/nitride (left); nitride/oxide (right)

Aside from the bias power, O_2 is another effective knob to adjust the selectivity. Unlike the etching gas-dependence bias power, the well-matched trends are observed in Fig 3 for both CT ME and SMT OE. This means more O_2 will increase the selectivity of oxide over nitride no matter what etching gases are employed. This can be explained by oxygen reacting with carbon to free more fluorine coupled with its suppression of oxide etch rate.

The reconciliatory clue can be seen for 5% increase of oxide etch rate and 25% increase of nitride etch rate with 50% higher O_2 flow rate. However, the less O_2 ratio in CT ME might suffer the open issue and/or the polymer residue issue. Fig.4 compares the effect of O_2 on the depth and profile of contact hole. The 25% higher selectivity from I-7 is at the expense of 20% less etch depth and bowing profile. Hence, the selectivity enhancement could not be fully isolated from the over-all performance evaluation.

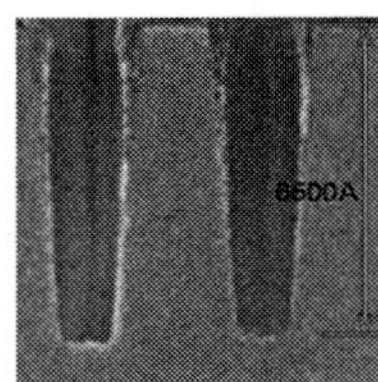

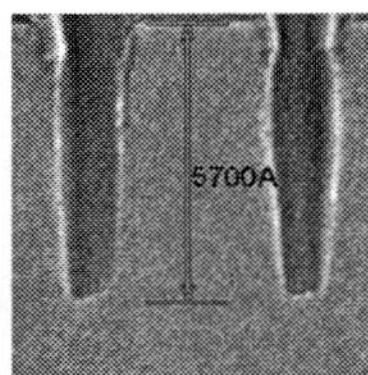

Fig.4 The effect of O_2 on etch depth and profile of contact hole: I-2 (left) vs I-7 (right)

III. CONCLUSIONS

In this work, we investigated the etch selectivity between silicon oxide and silicon nitride in CT etch and SMT etch. The effect of etch parameters such as etch chemistry, bias power and gas ratio on selectivity are evaluated. Results show the utilization of different etching gases rely on the specific process. The bias power is strongly etching gas dependent and shows the reverse trend in CT etch and SMT etch. The O_2 gas ratio exhibits the similar behavior, independent of main etching gas type. Normally, higher O_2 gas ratio is related to higher selectivity of nitride over oxide. Nevertheless, the higher selectivity from either bias power or O_2 might be at the expensive of other process performance.

ACKNOWLEDGMENTS

Authors would like to thank their colleagues at etch department of technology research and development center for technical assistance and useful discussions during the course of this work.

REFERENCES

1. M. Chen, et al, "Effect of Finger Pitch on the Driving Ability of a 40-nm MOSFET With Contact Etch Stop Layer Strain in Multi-finger Gated Structure", Electron Devices, Vol.57, 2010.
2. X. Wang, et al, "Impact of Etching Chemistry and Sidewall Profile on Contact CD and Open performance in Advanced Logic Contact Etch", CSTIC, 2010
3. C. Ortolland, et al, "Stress Memorization Technique—Fundamental Understanding and Low-Cost Integration for Advanced CMOS Technology Using a Nonselective Process", Electron Devices, Vol.56, 2009.
4. J. S. Kim, et al, "Infinitely high etch selectivity during CH2F2/H2 dual-frequency capacitively coupled plasma etching of silicon nitride to chemical vapor-deposited a-C", Journal of Vacuum Science & Technology A: Vacuum, Surfaces, and Films, Vol.28, 2010.
5. S. Lee, et al, "Ultrahigh selective etching of Si3N4 films over SiO2 films for silicon nitride gate spacer etching", Journal of Vacuum Science & Technology B: Microelectronics and Nanometer Structures, Vol.28, 2010.

ECS Transactions, 34 (1) 409-414 (2011)
10.1149/1.3567613 ©The Electrochemical Society

Discovering Practical Use of Sensor Wafers in CCP Reactors

A.P. Milenin[a], M. Demand[a], W. Boullart[a], P. Arleo[b]

[a]IMEC, Kapeldreef 75, 3001 Heverlee, Belgium
[b]KLA-Tencor, Seven Technology Drive, Milpitas, CA 95035, USA

This paper reviews two types of sensor wafers that fit into the niche of spatially resolved in situ plasma diagnostics. The effect of both the incident power and the backside He flow on wafer surface temperature map was studied with the temperature sensor wafer. Another sensor wafer is designed to measure the distribution of displacement current passing through the wafer. Calibrated at 13.56 MHz, it is capable of working in a wider range of source frequencies which has successfully been demonstrated in CCP reactors. It was found that the particular type of plasma reactor and the process chemistry can impact the level of correlation between the sensor and the blanket etch rate uniformity data.

Introduction

In plasma etching, there are many important parameters that have to be precisely controlled during the process. The number of important process parameters is usually balanced with the number of sensors installed on the tool, and both numbers typically increase from one tool generation to the next. In most cases, we have single point data acquisition systems providing a chamber-averaged data-set on each process and/or plasma parameter. Nowadays this approach may not be adequate, especially regarding the area of process uniformity.

For example, spatially resolved wafer-level data collection is the key to monitoring the effect of tool knobs such as center weighting, tuned gas flow, independent power biasing of the segmented TCP coil, etc. Yet such detailed levels of metrology systems are currently not present onboard any industrial plasma etch tool to date. Thus, the ex situ blanket wafer measurements must still be performed. In this context, a re-usable in situ wafer-level metrology proxy could greatly speed up the process characterization, development, and chamber re-qualification time. In this paper we aim to present such a wafer-level spatially-resolved sensor system as a metrology proxy targeted to improve the traditional approach.

Experimental

The 300-mm sensor wafers are autonomous units equipped with sensors, programmable data collection and storage systems, a battery with an inductive charging module, and a wireless data transmission system. The PlasmaVolt™ X2 (PVx2) sensor wafer provides spatial information on displacement RF current flowing through the wafer in 14 separate locations. The PlasmaTemp™ (PT) provides a wafer temperature map during the etching process in 42 localized points. More details on PVx2 can be found in

(1). The etch experiments presented here were conducted in industrial capacitively coupled plasma (CCP) reactors: 2300® Exelan® chambers (Lam Research).

PT Sensor Wafer

The sensor wafer can be used in four different ways: *static* – using saturated steady-state data, *dynamic* – using, for example, the decay front from initial moments after plasma-off, *direct* – using the temperature data only, and *indirect* – using the temperature profile to draw a conclusion, for example, on the local ion current density. In this paper we mainly consider the *"static - direct"* combination, which is the most popular way of the sensor use.

The importance of substrate temperature control is quite obvious; in uncontrolled plasma etching without any active cooling, the wafer temperature can abruptly go over a hundred degrees Celsius (2, 3). In modern tools the wafer temperature control is typically accomplished by means of the He backside flow that helps to keep the wafer's temperature determined by the temperature of the electrostatic chuck, which is cooled with use of a fluid media. Thus, the peak temperature is well-controlled and determined by chucking consistency and closed-loop helium flow/pressure parameters. But there is another important parameter - the temperature uniformity. This parameter can be used in a controlled way to tune the etch uniformity, as it can be seen in tools equipped with multiple zone chuck systems. The PT sensor wafer opens a way to monitor the temperature uniformity and to check its response on different recipe parameters that is very helpful to know in the process development.

Effect of Power and Backside He Cooling on Wafer Temperature

The effect of substrate temperature increase on power level used was studied in a BEOL CCP reactor equipped with both 2- and 27-MHz power generators. The C_4F_6/CO/CH_2F_2/O_2/Ar - based dielectric etch process was utilized at 10 $^{\circ}$C of chucking electrode set-point with the He backside pressure of 30 Torr. As shown in Fig 1, the temperature increase of about $\Delta T \approx 3.2$ $^{\circ}$C can be expected for each 500W power increase of the 2MHz generator. Taking into account that the chuck temperature was set at 10 $^{\circ}$C, it is clear that the wafer can easily be heated up by the plasma to more than 25 $^{\circ}$C higher than the set-point and even further in high power dielectric etch regimes.

The limited uniformity change on 2MHz power variation, which is expressed in 3σ (Fig. 1), speaks for the efficient enough heat dissipation on the whole area of electrostatic chuck at 30 Torr of He backside pressure under low and moderate powers. At higher power levels there is an indication of increasing challenges, as the 3σ value goes up. Such challenges of uniform heat dissipation can be better visualized with a dedicated etch experiment.

Fig. 2 shows the temperature map of the wafer during the dielectric etch process for three different backside He pressures. The chuck temperature was set at 20 $^{\circ}$C and the dielectric etch recipe parameters were kept constant as: 25mTorr/C_4F_8/O_2/Ar $1250W_2$/$750W_{27}$. As shown in Fig. 2, the He backside pressure determines not only the net wafer temperature but also the temperature uniformity map. The maximum recommended backside He pressure of 30 Torr, which corresponds to 14.5 - 15 sccm of He flow for the type of chuck used in the study, provides the best cooling efficiency.

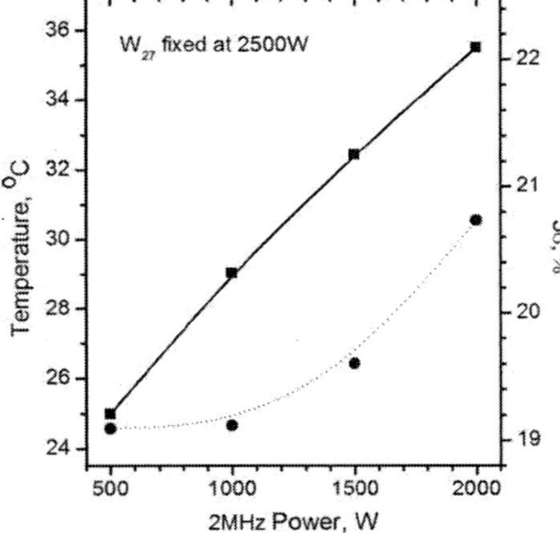

Figure 1. Mean temperature change (solid line) and temperature uniformity expressed via 3σ (dotted line) as a function of 2MHz power applied in the $C_4F_6/CO/CH_2F_2/O_2/Ar$ - based dielectric etch recipe at 45mTorr. The power of 27MHz generator was kept fixed at 2500W.

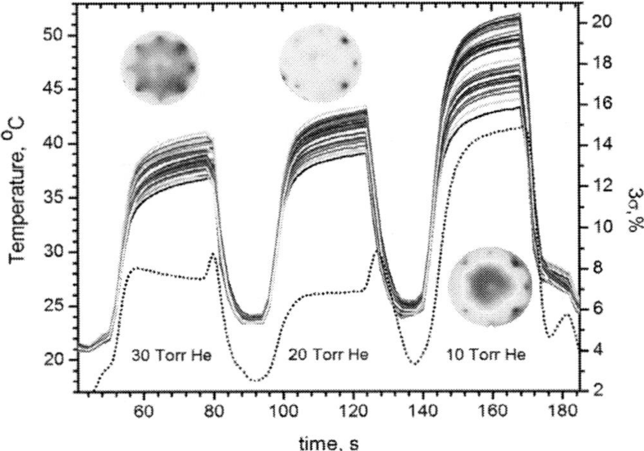

Figure 2. Temperature dependence on backside He pressure in a dielectric etch recipe. Each solid line represents the temperature variation measured by a single sensor unit in all 3 steps of the recipe. The dashed line is the temperature uniformity expressed via 3σ (in percentage). The inset images show a typical 2D temperature map for each recipe step.

Thus, 30 Torr of He provides the minimum wafer temperature, which is increased by only 17 °C during the process with respect to the nominal chuck temperature. It is interesting to notice that an improved 3σ can be achieved with somewhat lower backside He pressure of 20 Torr. But further decrease of He backside pressure results in doubling of 3σ at 10 Torr. By lowering He pressure for a specific recipe step, one can try to temporarily increase the substrate temperature, which may give some advantage in selectivity or etch rate for a particular material system etched in CCP reactor, but as it is shown for the case of 10 Torr backside He pressure the uniformity of the process can be strongly affected due to the limited heat dissipation ability in the center part of the wafer. In case of multiple zones chuck system the non-uniformity could be partially compensated by separate adjustment of He backside pressure, e.g., applying slightly higher backside pressure in the inner part of dual zone chuck. This effect can be easily monitored and optimized by the PT sensor wafer.

PVx2 Sensor Wafer

This sensor wafer has 14 capacitive detectors surface-mounted on the wafer. Each one is connected to an RF peak detection unit calibrated at 13.56 MHz. Covered with a 100-μm thick polyimide layer, the sensor is capable of measuring only the signal proportional to the displacement RF current flowing through the wafer, as it is discussed in (1,4). The direct use of the RF current distribution is not widely practiced in the plasma etching community, yet an attempt of its correlation to the other plasma parameters looks reasonable. As it has been shown in (1), the blanket wafer etch rate uniformity (ERU) trends may be successfully predicted in a TCP reactor using the PVx2 sensor wafer for mainly ion-assisted processes. In the current study we focused on establishing the correlation between PVx2 data and blanket wafer ERU measurements in a CCP reactor.

PVx2 and ERU in CCP

As we know, the ERU pattern is a result of the superposition of many components. Among them are: spatial ion energy distribution, plasma density distribution, heat dissipation uniformity, active radical distribution etc. Thus, if we know the 2D surface distribution of each single component and their "weighting functions" in a particular process, we should be able to describe the etching result precisely. PT and PVx2 are able to provide the spatially-resolved information on some of the mentioned components and if the particular etch process is mainly determined by the measured one (its weight function is close to one and others are close to zero), we should be able to see a good data correlation with blanket ERU.

This was checked for several recipes with a most typical combination of power generators used: 2+27 MHz. The ERU was studied with respect to pressure on blanket wafers of 600-nm PE-CVD SiO_2. The thickness measurement was performed with an ellipsometer using a standard 49-point algorithm. The recipes were standardized with respect to the other variables by using the simplified recipe editor and the recipe details are listed below:

$$Ar/1250\ W_2/750\ W_{27}, \qquad [1]$$
$$Ar/6.6\%O_2/1250\ W_2/750\ W_{27}, \qquad [2]$$
$$C_4F_8/O_2/Ar/1250\ W_2/750\ W_{27}, \qquad [3]$$
$$CHF_3/CF_4/O_2/Ar/100W_2/300W_{27}. \qquad [4]$$

Fig. 3a shows the 3σ dependence on pressure for dielectric etch recipes, while Fig. 3b represents the uniformity trends for the Ar-based recipes. If we rule out the local perturbations observed in the 3σ dependence of pure Ar and C_4F_8-based etch chemistries on pressure, one can find a satisfactory trend agreement between the PVx2 and the ERU data sets for the case of dielectric etches and pure Ar.

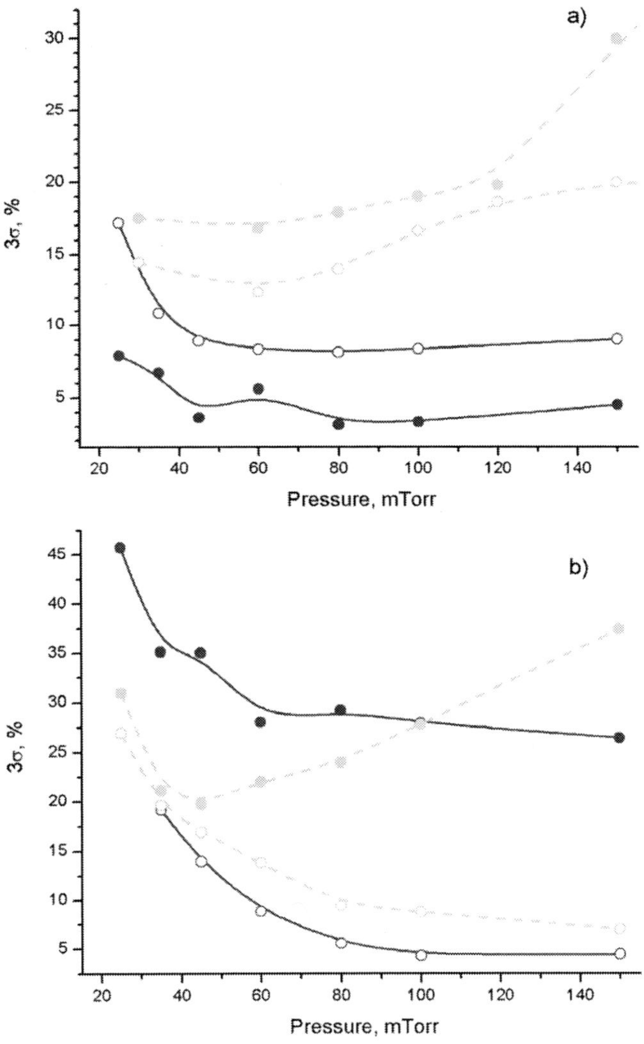

Figure 3. ERU (solid circles) and RF current uniformity (open circles) on pressure:
(a) – solid lines are for the dielectric etch recipe [3], while dashed lines are for [4];
(b) – solid lines are for pure Ar, [1], while dashed lines are for Ar diluted with O_2 [2].

Thus, both the PVx2 and ERU data indicate that the best uniformity in the Ar recipe can be achieved at highest pressures, while the C_4F_8-based etch chemistry shows global minimum at 80-90 mTorr and another chemistry, [4], shows the best uniformity at around 60 mTorr.

However, the agreement can be easily broken by adding 6.6% O_2 to the Ar chemistry, as it is shown in Fig. 3b. The same poor agreement between blanket wafer ERU and PVx2 was observed in the case of a small localized gas flow modification implemented by using the tuning gas lines of the CCP reactor.

Though the detailed analysis of local perturbations in 3σ dependence of the ERU (shown in Fig. 3a) and the discrepancies observed for $Ar+O_2$ plasma (shown in Fig. 3b) will require a dedicated study, it can be stated that unlike in the TCP reactor, the CCP reactor has several RF energy sources attached to the bottom electrode that provide an amplitude-modulated RF current flowing through the system. But these generators can control different parameters: the higher frequency controls the density and the low frequency is meant to provide ion energy control. The challenge of using a broad-band RF detector to measure the RF current in a CCP is that we could not separate the effect of each frequency in the measured pattern of RF current uniformity. Thus, only if the component measured by PVx2 is the dominant one in the amplitude-modulated RF current and if it determines the final ERU, the absolute correlation between PVx2 and ERU can be established.

Conclusions

Compared to the blanket wafer ERU tests, the sensor wafers are capable of providing valuable sets of uncomplicated information on the plasma components that define etch processes. Thus, the direct in situ information about water temperature and RF current distribution can be obtained using PVx2 and PT sensor wafers. Though the level of deconvolution is sometimes limited, especially with PVx2 in the CCP reactor for the processes utilizing two and more sources of power, it has been shown that the information coming from the sensors can readily be used in the recipe development either in direct or in complementary way.

In order to further strengthen the current sensors value, we suggest to enlarge the sampling frequency range of the PT sensor and to provide the frequency-separated analyses by filtering the components of each source frequency for the PVx2 sensor combining it with an increased resolution by doubling the number of sensors mounted on the wafer.

References

1. A.P. Milenin, J.F. de Marneffe, H. Struyf, W. Boullart, P. Arleo, *Advanced Semiconductor Manufacturing Conference (ASMC)*, pp. 34-38, IEEE/SEMI (2010).
2. C.T. Gabriel and E.K. Yeh, *Solid State Technology Magazine*, **42**(10), (1999).
3. R.J. Visser, *J. Vac. Sci. Tech.*, **A7** (2), (1989).
4. A.P. Milenin, J.F. de Marneffe, H. Struyf, and W. Boullart, *ECS Trans.*, **13** (8), 17 (2008).

ECS Transactions, 34 (1) 415-420 (2011)
10.1149/1.3567614 ©The Electrochemical Society

Study on Silicon Sieve Holes Array for Future Lithography Application

Weihua Si [a], Ming Yin [a], Jian Qin [b], and Zewen Liu [a*]

[a] Institute of Microelectronics, Tsinghua University, Beijing 100084, China
[b] School of Physics and Microelectronics Science, Hunan University, Changsha 410082, China
*corresponding author: liuzw@tsinghua.edu.cn

Abstract: Silicon sieve with nano-scale holes array used as atomic lithography aperture had been realized with a method combining of deep dry etching and silicon anisotropic wet etching. The influence of etchant concentration and temperature on the etching rate is carefully studied. Unusual bath temperature at 40°C was used in the wet etching process. With a sputtered layer of Cr (300nm) as the wet etching mask and a layer of Al (700nm) as the deep dry etching mask nano holes array with 108nm feature size are obtained on single crystalline silicon.

1. Introduction

Fabrication of nano scale aperture has been much interested due to its possible application in near field optics, biology and lithography [1-4]. Small aperture may play a crucial role in future lithography for particle or beam formation and limitation [2]. Many through-hole fabrication methods had been tried elsewhere such as the wet etching and e-beam combination method [1], deep Reactive Ion Etching (D-RIE) method [3] and focused ion beam (FIB) method [4]. The drawbacks of these methods are not only their expensive and complex equipment but also the difficulty in large scale manufacturing. In some case a further reducing of the size of the holes is necessary [4]

Alternative approach for fabrication of extremely small through holes array is possible based on the anisotropic single crystalline silicon wet etching. Massive processing of silicon wafers with various wet and dry etching processes are essential in many silicon based micro-mechanical structures [5] [6]. From the seventies of last century, various silicon cavities and silicon beams had been realized using anisotropic silicon wet at both micro and nano scale [7] [8]. Theoretically, holes with dimension in 10 nm can be realized with anisotropic wet etching process. But the thickness of the silicon substrate is normally in several hundred microns and factors affect silicon etching quality including silicon etching selectivity, rate of etching and surface tolerance [9][10], which will limit the holes density and the resolution. In this paper, a two step fabrication process is designed as a first trial. In the first step, deep dry etch process is used, and then wet etching process is carried out in the second step. In order to obtain accurate etching rate, the influence of etchant concentration and temperature on the etching rate are analyzed. Nano-scale silicon holes arrays with diameter on the order of 100 nm are obtained using optimized etchant concentration and unconventional temperature. The results reveal that the fabrication process proposed in this paper is very promising to realize extremely small large scale nano holes array. This holes array can be used as key element in atomic lithography or other particle lithography system as an aperture to define the beam size as illustrated in Fig.1. With a help of a precision mechanical and controlling system, patterned structures can be also realized if a relative movement between substrate and the holes mask is well performed in x and y direction.

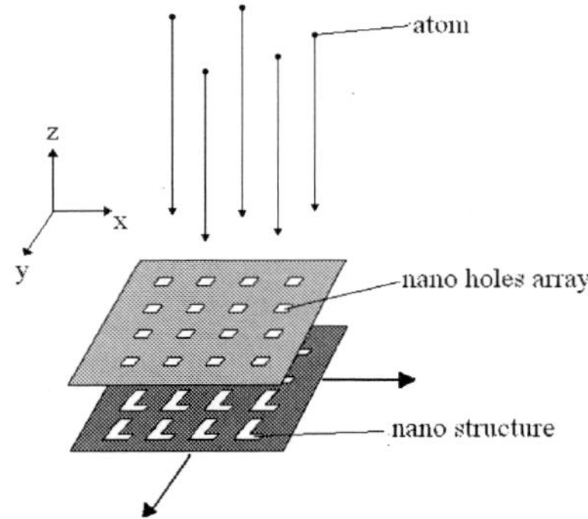

Figure 1. Schematic view of particle lithography using nano holes array

2. Process design

The intrinsic anisotropic property in silicon wafer wet etching is utilized in process design to obtain nano holes. Figure 2 shows the main fabrication steps of the nano silicon hole arrays fabrication. In the first step mask layers on both sides of the silicon substrate are deposited and then patterned using normal photolithographic process in order to format the etching windows for dry and wet etching. In the second step, ICP deep dry etch process is used for back size thinning. In the following step, silicon wet etching anisotropic using KOH solution is carried out to obtain the nano holes arrays structures. The mask layer is removed at last after the nano silicon hole arrays are realized.

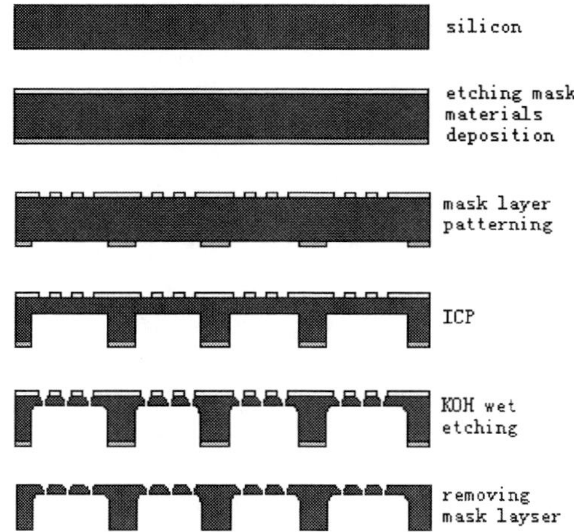

Figure 2. Process of nano silicon hole arrays fabrication

The process started from a 3 in P-type (100) single crystalline silicon wafers. Wafers are cleaned with standard RCA solvent. Then a layer of Cr (300nm) [11] is deposited as the wet etching mask on top side of the silicon and a layer of Al (700nm) as the deep dry etching mask on another side of the substrate. The feature size of mask for dry etching is in millimeters and is in tens microns for wet etching. So the mask patterns are fabricated easily by standard photolithography. Exposed Cr is removed in ceric ammonium nitrate solution and Al is removed in phosphoric acid in order to form open windows.

Cr is used as mask layer in wet etching, due to the good adhesion between Cr material and Si substrate and its stable chemistry property (nearly no reaction in KOH solution), which insure fastness in wet etching . Furthermore, it can easily be removed in ceric ammonium nitrate solution. Meanwhile Al is a standard materials used as mask layer in deep dry etching

3. Experiment

The etching solution is 33% KOH solution with the addition of IPA (KOH: water: IPA $=50$g: 100ml: 10ml). The wet etching is carried out in magnetic stirring apparatus with the temperature variation of $\pm 1°C$.

The deep dry etching is carried with ICP facility (Alcatel A120) and the etching rate to silicon is 7um/min.

The depth of the etched silicon is measured with surface profile measuring system and the structures are measured with scanning electron microscope (SEM).

4. Result and discussion

4.1 Etching rate

To fabricate silicon nano mechanical structure with good size and uniformity control, it is important to master the etching rate of silicon accurately. The main principle of the chemical reaction between silicon and the solvent in aqueous base solution is:

$$Si + 2OH^- + 2H_2O \rightarrow SiO_2(OH)_2^{2-} + 2H_2$$

Specifically the reaction is decomposed into several processes including reactive ion diffusion to the silicon surface, silicon surface adsorption reaction ion or non-reactive ion, surface reaction, precipitation reaction product and reaction product diffusion from surface to the solution[12]. With the increasing of the KOH solution concentration, hydroxide concentration increase, which contributes more on silicon surface adsorption and increases the reaction sites so that the etch rate increases. Nevertheless with the continuing increasing of the KOH solution concentration, the concentration of free water reduces and hydroxyl that participates in the real reaction decreases, so that the rate of reaction decreases [13]. Figure 3 shows a measured relationship between the etching rate and the concentrations at different temperatures of the KOH solution bath. With the increasing of solution concentrations, the etching rate firstly increases, then reaches a maximum at 30% (wt) and then it reduces gradually. From the different temperature curves we can know that etching rate increases rapidly with temperature. It is expected since the temperature dominates the reaction active energy. Temperature increasing also contributes to speed the diffusion rate of reactive ion to the silicon surface and remove the reaction product from silicon surface to solution. Importantly from Figure 3 we can know that at the higher temperature, the etching rate fluctuations change more with the concentration variation. This makes more difficult to control the process at high temperature solution. Low-temperature bath will be preferred if nano-scale structure is important. In fact in the micro system device fabrication, the etching temperature is often 60°C or even higher.

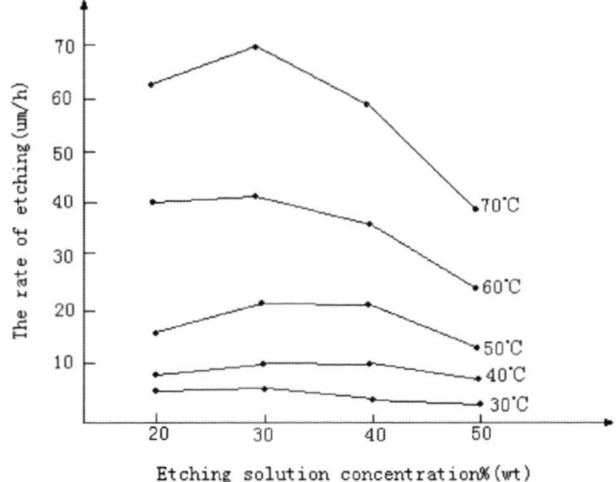

Figure 3 Relationship between the etching rate and the concentrations at different KOH solution temperature

4.2 Anisotropic wet etching

(100) silicon substrate is used in the experiment. A conical concave structure with square open window is formed when it is exposed to the KOH etchant solution due to the anisotropic etching between (100) and (111) crystallographic planes. The very big etching selectivity is because of the different number of dangling bonds and back bonds structure in single crystalline silicon. In the conical concave structure, the (111) crystallographic planes consist of four border surface.

Figure 4 SEM picture of the micro holes array

Figure 4 shows the SEM pictures of the holes array in micro-scale. Which is formed with wet etching at a temperature of 60°C. Figure 4-1 shows the overall structure of micron holes array, Figure 4-2 shows one conical structure of one array element with hole feature size of 1.6 microns and Figure 4-3 shows the detail of the hole. From the photographs we can found that very good conical holes array can be realized with small feature holes size. Further observation find that there is a small pit between two (111) planes (indicated with the upper red arrow in Figure 4-3), which is normally ignored in the micro scale structure processing. The pits will greatly affect the final result in the case that a nano-scale hole is expected. The formation mechanism of the pit is currently not clear and a further study on both experimental and theoretical analysis is needed. From figure 4-3 we can also see a certain thickness in the through hole (indicated with the lower blue arrows), which is different with the theoretical anticipation. Both phenomena indicate a big challenge to fabricate nanoscale holes array. In the following nano through holes fabrication experiment special bath temperature is used to perform the etching process.

4.3 Fabrication of the nano holes array

From the fabrication of micro silicon holes array, we found that it is difficult to obtain nano holes using the conventional temperature and process during the wet etching. Furthermore a reduction of substrate thickness is also necessary for the purpose of control of the holes size and enhancement of the through holes density. ICP deep etch is carried out on to the (100) single crystalline silicon to reduce the substrate thickness. Low-temperature etching is performed to control nanometer diameter holes in the course of wet etching in KOH solution. Figure 5 shows SEM pictures of the result. Figure 5-1 is the overall structure consisting in a 14x14 holes array and the pitch between two elements is 10 microns. Figure 5-2 shows the nano hole in the center of a conical concave cavity, which illustrates the holes array is successfully realized, and the feature size of the hole is 108 nm. Figure 5-3 shows the detail of the hole in the element. From the picture we can find a further reduce the hole size is possible.

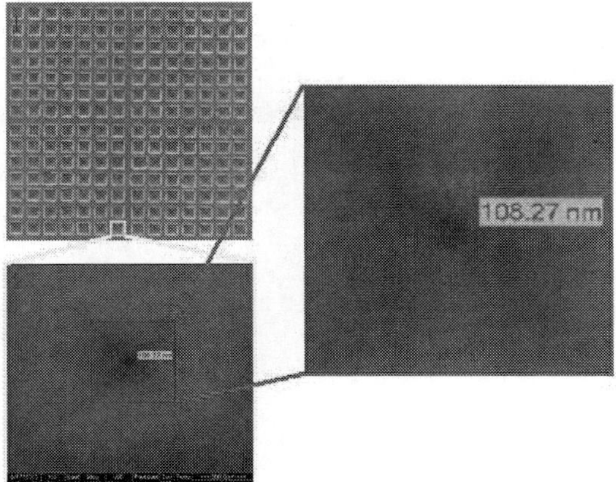

Figure.5 SEM picture of the nano holes array

5 . Conclusion

Anisotropic wet etching of silicon plays an important role in MEMS processing. The etching rate of single crystalline silicon strongly depends on the crystallographic planes. Based on a research on silicon anisotropic wet etching and the analysis of wet etching of the results, a fabrication process for realization the nanoscale through holes array is performed and a minimum feature size of 108nm nano holes array have been obtained with well controlled process parameters. The impact of small pits between the two (111) crystallographic planes is less important when an unusual low etching temperature is used. It can be seen that when the process is well controlled, realization of holes array with smaller feature size is possible. Further experiments on patterning structures with obtained holes array and on reducing the holes dimension should be carried out in future work.

Acknowledgments

This work was jointly supported by the National Basic Research Program of China (973 Program, Grant No. 2007CB310504) and National Natural Science Foundation of China (NSFC, Grant No. 10772180).

References

1. D.W. Kim, Y.C. Kim, V.C. Bui, et al., Microelectronic Engineering, 83, p. 2493–2498 (2006).
2. M. L. Taylor, A. Alves, P. Reichart, et al., Nuclear instruments & methods in physics research section B-beam interaction with materlals and atoms, 260(1), p. 426-430 (2007).
3. M. Kamakura, H. Shiroishi and T. Taura, et al, Proceedings of the 13th international display workshops, 1-3, p. 1545-1548 (2006).
4. Z. Jcak, Y. GuoLiang, Journal of Manufacturing Science and Engineering, 132, p. 011005-1-011005-8 (2010).
5. J. Y. Park and H. C. Lee, 33rd European Microwave Conference, 3, p. 907-910 (2003).
6. S. Calvez and J. M. Hopkins, Journal of Crystal Growth, 268, p. 457-465 (2004).
7. P. Pal, K. Sato and S. Chandra, Journal of micromechanics and micro engineering, 17(10), p. R111–R133 (2007).
8. FSS. Chien, C. L. Wu and Y. C. Chou, et al., Applied physics letters, 75(16), p. 2429-2431(1999).
9. H. Tanaka and S. Yamashita, Sensors and Actuators A, 114, p. 516-520 (2004).
10. C. R. Yang and P. Y. Chen, Sensors and Actuators A, 119, p. 271-281 (2005).
11. L. Nie, T. L. Shi and G. L. Liao, et al., Semiconductor Technology, 30(12), p. 26-18(2005).
12. C. Moldovan and R. Iosub., IEEE Semiconductor Conference, 2, p. 353-356 (1998).
13. K. Zhang, H. S. Gu and G. Hu, et al., Journal of Hubei University, 29 (3), p. 255-257(2007).

ECS Transactions, 34 (1) 421-426 (2011)
10.1149/1.3567615 ©The Electrochemical Society

Controlled Etching of III-V Materials with Optical Emission Interferometry (OEI)

C. Johnson, D. Johnson, R. Westerman, D. Geerpuram, L. Martinez, and J. Plumhoff

Plasma-Therm LLC, 10050 16th Street North, Saint Petersburg, Florida 33716, USA

For plasma etching processes, Optical Emission Interferometry (OEI) combines the advantages of Optical Emission Spectroscopy (OES) and laser interferometry. Radiation emitted by the plasma is reflected from the substrate surface and monitored using a multi-wavelength spectrometer. Analysis of the OEI signal can provide endpoint data, as well as etch rate and etch depth information for both single and multi-component materials. Examples of these applications are described.

Introduction

Optical techniques are used extensively to monitor plasma-based processes because they are both relatively easy to implement and are non-intrusive. OES is commonly used to monitor emission from species in the plasma which can then be used to determine etch end point. Interference techniques, which measure the intensity of a laser reflected from a substrate surface, give information about the reflecting surfaces and can be used for endpoint detection, as well as real time etch rate determination. OEI combines the two techniques by monitoring the plasma emission which is reflected from the substrate surface.

Here we describe applications where OEI may be used during the etching of III-V materials, which include endpoint detection and process monitoring as well as real time etch rate and etch depth measurement.

Equipment

A schematic of the instrumental set-up used to make all OEI measurements is shown in Figure 1. The samples were etched using a Plasma-Therm Versaline etch system fitted

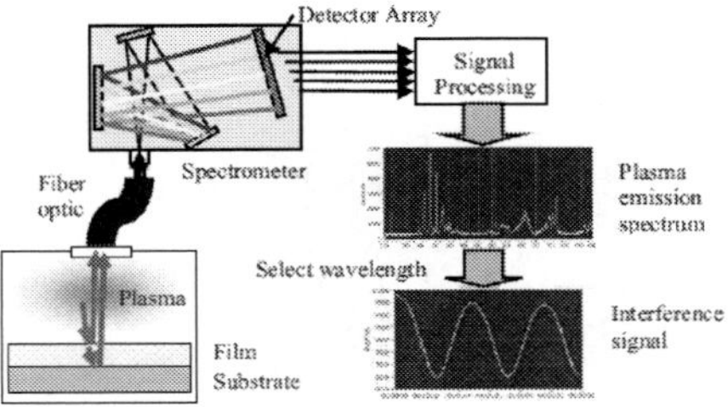

Figure 1. Optical Emission Interferometry schematic

with an ICP source. A sapphire window located in the top of the source was used to view the plasma radiation which was reflected normally from the substrate surface. A collimating lens focused this radiation into an optical fiber, giving an effective viewing area on the sample of approximately 2cm diameter. The fiber optic was coupled to a Ocean Optics USB 2000+ UV-VIS spectrometer, which was set to acquire a complete spectrum at one second intervals. The spectral data was further processed using Plasma-Therm's EndpointWorks software. The specific functions used will be described in more detail in the following sections.

<u>Results and Discussion</u>

<u>OEI being used as OES.</u> The radiation received by the detector is not only that which has been reflected from the substrate, but also emission from species in the plasma. The plasma emission is dependent on the material being etched and can be used to detect endpoint, similar to conventional OES. An example is shown in Figures 2 – 4, and described in the following.

An AlGaN/GaN epi structure with a ~500nm upper layer of AlGaN was etched using a Cl_2-based process. A portion of the spectrum is shown in Figure 2, which shows both Ga emission at 403.3nm and 417.2nm, and Al emission at 394.4nm and 396.2nm.

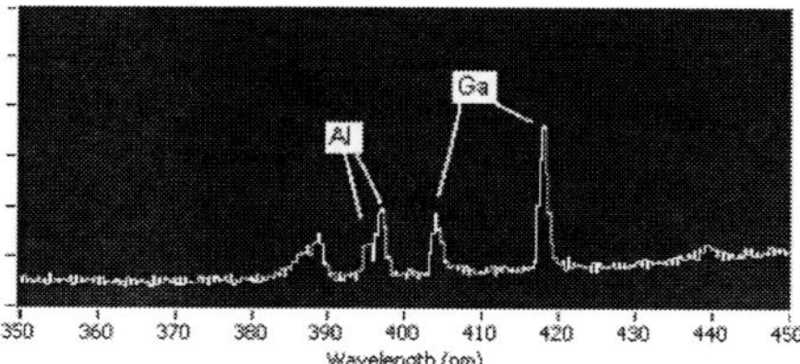

Figure 2. Partial OEI spectrum from AlGaN/GaN etch

When the Ga emission at 403nm is monitored versus time there is an initial increase as the etch starts but is then essentially constant because the Ga content of the structure varies only slightly (Figure 3). However, when monitoring the Al emission at 396nm, the signal increases and decreases as the Al-containing layers are etched (Figure 4). Using this technique it is easy to detect an increase or decrease in Al signal and trigger an endpoint. When etching multi-layer structures, it is then possible to stop the process at a specific layer boundary.

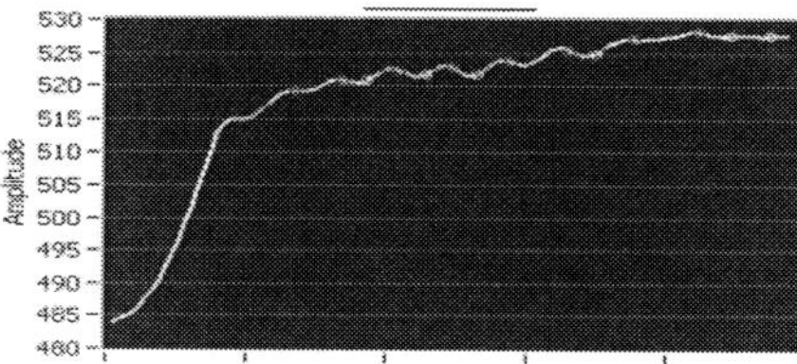

Figure 3. Ga emission at 403.3nm versus time

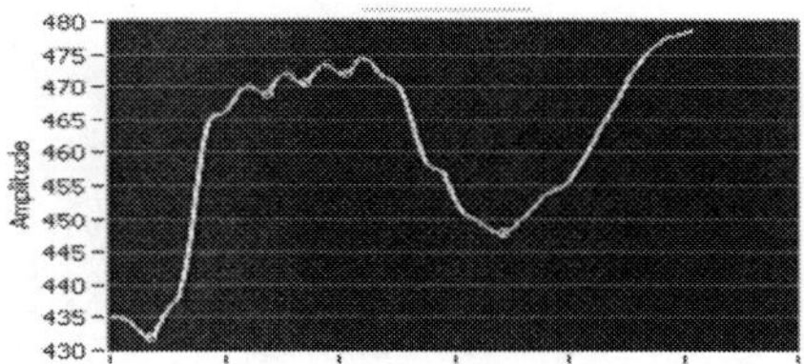

Figure 4. Al emission at 396.2nm versus time

<u>OEI for Etch Rate and Thickness Measurement.</u> In Figures 3 and 4 a cyclical variation is seen superimposed on the emission signal. This is due to interference caused by reflections from the boundaries within the multi-layered structure. This is more evident at wavelengths which do not otherwise change due to chemical changes within the plasma. An example from the same AlGaN/GaN etch is shown in Figure 5, where the plasma emission at 555nm is monitored versus time and the interference cycle is clearly seen.

The film thickness change (d) for one cycle calculated by:

$$d = \lambda/2 *n_f \tag{1}$$

where:
λ is the wavelength at which the interference is monitored
n_f is the refractive index of the layer at that wavelength.

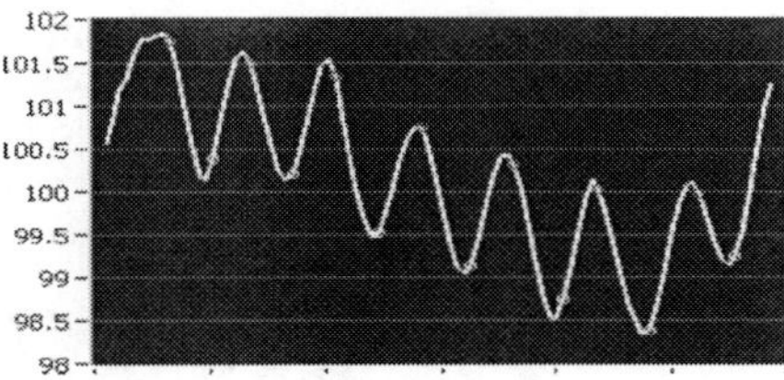

Figure 5. Interference signal from emission at 555nm

For the example given, the refractive index of the layer is 2.3 at 555nm, therefore the thickness change corresponding to one complete interference cycle is 120.7nm. By measuring the elapsed time between each cycle the etch rate of the AlGaN/GaN material can be calculated. A simple algorithm is used to detect and count peaks/valleys of the signal. Cycles can then be interpolated and, by using the calculated etch rate, a real time measure of etched depth is given.

This technique is extremely useful when it is required to remove a given thickness of material, but no etch stop layer present. Figure 6 illustrates the removal 7.5µm of a GaN epi layer at a rate of approximately 2µm/min. The upper plot shows the interference signal measured at 555nm and the lower plot shows the calculated etch depth versus time. The etch process is terminated once the depth exceeds the specified etch depth (7.5µm for this example).

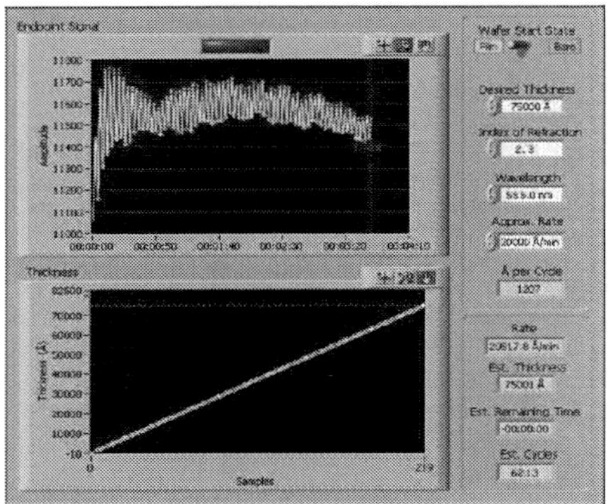

Figure 6. EndpointWorks display of removal of 7.5µm of GaN epi, monitored using OEI

In the example given it is assumed that the etch rate remains constant with time, and it is the value of the average calculated etch rate that is used to interpolate between cycles. However, when etching layered structures which etch at different etch rates it is necessary to use the instantaneous calculated etch rate rather than the average value. This value can be updated at the completion of every half cycle of the interference signal.

This also highlights another capability of OEI which is the ability to monitor etch rate versus time and adjust a process based on the detection of an etch rate change. An example of this is shown in Figures 7 and 8. A structure which consists of an approximately 300nm thick GaP layer on top of a AlInGaP layer was etched using a BCl_3/Cl_2 process. In Figure 7 the interference signal observed at 650nm is shown, and in Figure 8 the etch rate calculated at every half cycle is plotted versus time. It is clear from this plot that there is a sharp drop in etch rate as the Indium containing layer is reached.

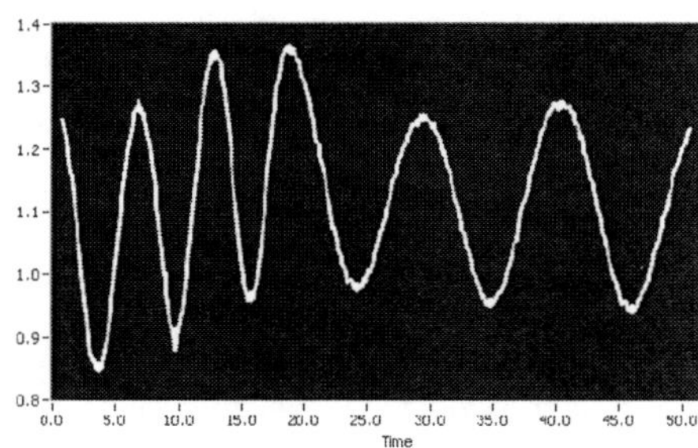

Figure 7. Interference signal from GaP/AlInGaP multilayer structure

This change can be used to trigger a process change, for example adjusting the process parameters to optimize the etch for the second material, or to simply terminate the process.

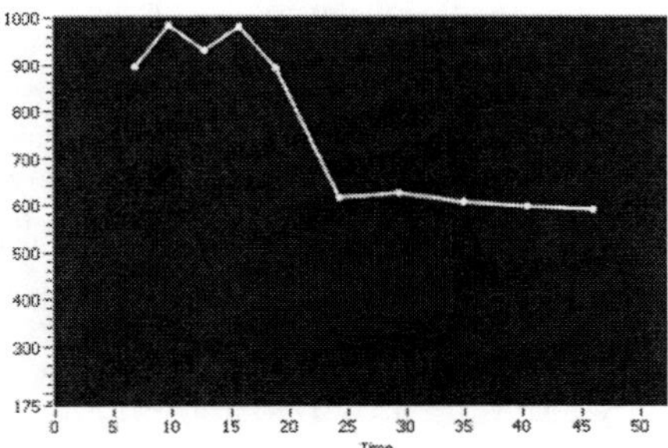

Figure 8. Etch rate of GaP/AlInGaP multilayer structure versus time

<u>OEI for multiple etch rate measurements.</u> The data shown in the above examples was obtained using samples of un-masked material. The interference signal obtained closely approximates the theoretical response expected, which is a simple sine wave. However, when more than one material is present within the reflectance acceptance angle of the collection optics, separate interference signals will be generated by the different materials. This results in a signal consisting of superimposed sine waves, each generated according to Equation 1 for the different materials. The time for an individual cycle is dependent on the etch rates of the different materials.

As an example, in order to demonstrate this capability, we prepared samples which consisted of a silicon dioxide layer patterned with a silicon nitride layer: the pattern was a simple grating structure with a nominal 50/50 line space ratio. The sample was etched using an SF_6-based process, with the process parameters adjusted to etch the silicon nitride approximately 3 times faster than the silicon dioxide. The interference signal obtained during the etching of this structure is shown in Figure 9 and illustrates the complex nature of the interference signal.

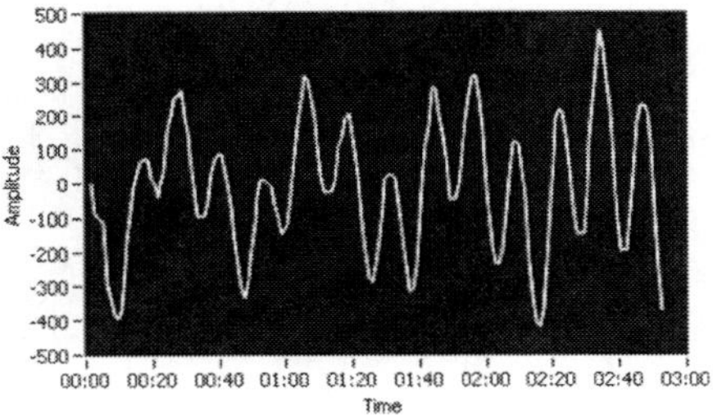

Figure 9. Complex interference signal from simultaneous SiN/SiO$_2$ etch

Rather than using a simple peak counting algorithm, this interference signal was analyzed using a Fast Fourier Transform (FFT) routine. In its basic form this gives an output which indicates the magnitude of signal at the different frequencies contained within the raw data. For example, it can be seen from Figure 9 that there is an obvious cyclical signal with a period of approximately 15 sec, or 0.067Hz.

However, it is more useful to convert the frequency data to etch rate information. From the frequency data, the cycle time is calculated, which in turn is used to calculate etch rate using equation 1. The output of the FFT can then be presented in the form of the etch rate of the components contained within the interference signal. Such an output is shown in Figure 10, which shows two distinct peaks with etch rates of approximately 500nm/min and 150nm/min. These correspond to the etch rates of the nitride and oxide films respectively. Note that these rates are obtained from completely in-situ measurements, and avoid the usual need for pre- and post-etch measurements.

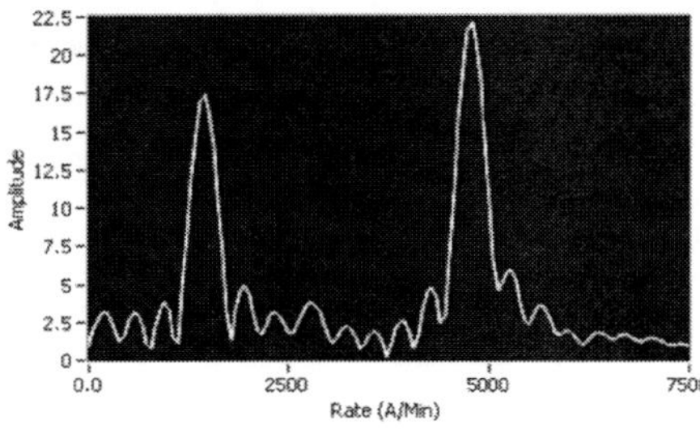

Figure 10. Fourier Transform of interference signal from figure 9

Conclusions

It has been shown that OEI is a useful technique which can be used to monitor III-V etch processes. In its simplest form it can be used to determine process endpoint much like OES. When the interference component of the signal is analyzed, both etch rate and etch depth information can be found. For simultaneously etched layers, a FFT analysis permits etch rates of the individual components to be calculated. This type of analysis should prove extremely useful during process development or for routine process monitoring, where maintaining etch rates within defined limits is important for process stability.

Future work will attempt to extend the FFT capability for more commonly encountered structures such as III-V materials masked with photo-resist or other mask materials, and will also investigate what effect the mask/open area ratio has on the accuracy of the etch rate determinations.

References

1. Gary S. Selwyn, *Optical Diagnostic Techniques for Plasma Processing*, AVS monograph series, pp.120-124 (1993).
2. Mucha et al., *Introduction to Microlithography,* ACS Professional Reference Book, pp.470-480 (1994).
3. A.N.Zaidel et al., *Tables of Spectral Lines,* IFI/Plenum (1970).

ECS Transactions, 34 (1) 427-432 (2011)
10.1149/1.3567616 ©The Electrochemical Society

Low Silicon and SiGe Loss in High Dose Implant Resist Strip

Xiao-Ying Meng, Man-Hua Shen, Yi Huang, Hai-Yang Zhang, Shih-Mou
Chang, Kwok-Fung Lee

Semiconductor Manufacturing International Corporation, No.18 Zhang Jiang Rd.,
Pudong New Area, Shanghai, 201203, P.R.China

ABSTRACT

The loss of both silicon and silicon germanium in resist strip process has attracted much attention for generic device development. Such loss mainly stems from the substrate re-oxidation during strip. In this paper, we focus on the high-dose implant resist strip to investigate the re-oxidation of substrate Si and SiGe without compromising the removal capabilities of photoresist (PR) and residue. Besides, the impact of ashing gas ratio on SiN offset spacer critical dimension (CD) is also addressed. Specifically, we studied the substrate re-oxidation performance under various ash process parameters such as pressure, gas ratio, pin position, process time in crust and over-ash (OA) steps. Aside from the re-oxidation effect, we also compared the defect removal capabilities with the above ashing conditions and the dependence of SiN offset spacer CD on ash gas ratio. Results indicate A) The high pressure in preheat step coupled with the low pressure in main-step can effectively preclude Si/SiGe loss. B) For SiGe substrate, the lower ratio of O2/forming gas (FG) exhibits the different impacts between N-MOS and P-MOS implant condition. C) Less OA time brings to better performance for residue removal. D) Continuous offset spacer CD shrink is noticed in each step of lightly doped drain (LDD) loop. The optimized strip scheme delivers not only a residue-free process with the minimum substrate loss but also the improved device performance on both N-MOS and P-MOS.

I. INTRODUCTION

In logic device manufacturing, the plasma ashing step is indispensable after each dopant implantation step in LDD loop to remove the remained photoresist. However, the presence of oxygen in ashing process could oxidize the silicon substrate, the oxidized part will be removed in the subsequent wet-clean step, thus forming the silicon recess (also termed as silicon loss) adjacent the gate. Due to the aggressively reduced overall device size, the LDD implant regions were not that much larger and deeper as before, the silicon loss in the LDD regions can have a remarkable negative impact on device performance such as negative bias temperature instability (NBTI). Similarly, the SiGe loss is another critical index to monitor the P-MOS performance for 65nm node and beyond.

The high dose implant (HDI) is normally utilized in either LDD or source/drain formation. Its high energy and high dose species can drive H and OH groups out of resist, resulting in a carbonized crust layer on resist. Crust layer usually includes the implanted species As, B or P. Even worse, dopant ions can also cause silicon substrate being sputtered and attached to resist sidewall. The combination of the above both phenomena poses the increasing challenge for the conventional oxygen plasma to strip the post-HDI photoresist. Therefore, the high dose implant stripping (HDIS) has to leverage more aggressive gases like FG or CF_4 to ensure the complete removal of resist and residue. This means more substrate re-oxidation and material/dopant loss incurred in front-end-of-line (FEOL) ashing and cleaning. The allowable total material loss has been tightened as device is being scaled down. It becomes more difficult to maintain residue removal capacity while meeting the stringent oxidation/loss specifications of the ultra shallow junctions.

This paper addresses the development of HDIS process development based on O_2 and FG. In addition to the investigation of Si/SiGe loss, resist and residue removal efficiency from HDIS, its impact on the offset spacer CD are also discussed.

II. EXPERIMENTS AND DISCUSSIONS

All the HDIS experiments were conducted in Mattson Superma tool. As shown in Fig 1, it is an inductively coupled plasma system with the special pin-up/down function to control the wafer temperature. AMAT SEMVision G3 (scanning electron microscope) is employed to check the surface cleanness. The substrate loss is *in-situ* determined by in-line thickness measurement tool Optiprobe 7341XP before and after plasma strip, and the offset spacer CD is measured by HITACH CD-SEM S-9380. Finally the offline TEM (transmission electron microscope) is used to double confirm the substrate damage and the offset spacer CD difference.

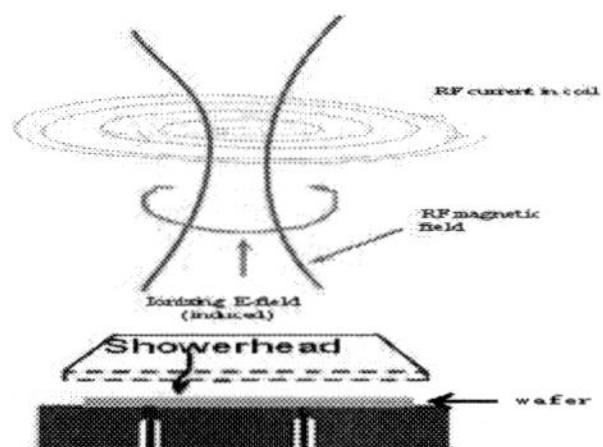

Fig.1 Schematic of Mattson Superma

1. The Effect of Ash Process Splits on Substrate Loss and Residue

Han et al [2] discovered the higher ratio of O_2 in O_2+FG helps both crust breakthrough and bulk resist removal: When O_2 ratio increases from 0% to 10%, the process time for crust and resist removal decreases drastically; and then it reaches its saturation when O_2 ratio continuously increases. Besides, it is well-known that O_2 plasma also enhances the oxidation of the substrate, thus tending to result in more substrate loss. The ratio of O_2/FG we evaluated ranges from 5% to 20%.

Parameter / Substrate	Pre-heat Step Pressure ↑	Crust Step O2 ratio ↓	Crust Step Pressure ↓	Crust Step Pin Position ↑	Over Ash Step Process Time ↓
Si Loss	↓	↓	↓	↓	↓
SiGe Loss	↓	NMOS ↑ / PMOS ↓	↓	↓	↓
Residue	tiny	tiny	tiny	tiny	clean

Table 1 Substrate impact of ash process parameters under O_2+FG plasma

Various ash process parameters such as pressure at pre-heat step, pressure, O_2 ratio, pin position at Crust step, and over-ash time are studied. Their effects on substrate Si (N-MOS) and SiGe (P-MOS) loss are summarized in Table1. As expected, the higher pressure in pre-heat step delivers less substrate consumption for its higher temperature.

The lower pressure at crust step shows the similar trend to pin-up position for less substrate loss. This can be ascribed to their moderate reaction temperature. Less OA time gives the reduced substrate loss. However, the OA time could not be decreased arbitrarily without considering the removal capabilities of photoresist (PR) and residue. Regarding SiGe substrate loss, the lower O_2 ratio split exhibits the different impacts between N-MOS and P-MOS. This perhaps comes from different species implanted into photoresist. Besides, SiGe loss might be different from the silicon loss for the better solubility of GeO_2 in aqueous solution.

We also studied the defect removal capabilities with the above ash conditions, and found most of splits with tuning actions in pre-heat and crust step still suffer the tiny residue issue. The EDX (Energy dispersive X-ray) only detected Si and O species, as shown in Fig. 2. Residue should be the oxidized silicon, thus being tough to be removed by wet strip. Surprisingly, when we reduced OA time from two or three hundreds percentage to less than one hundred percentage, such residue disappears. This indicates the higher temperature and the higher O_2/FG gas ratio in OA process play the critical role in forming the final not-removable oxidized silicon.

 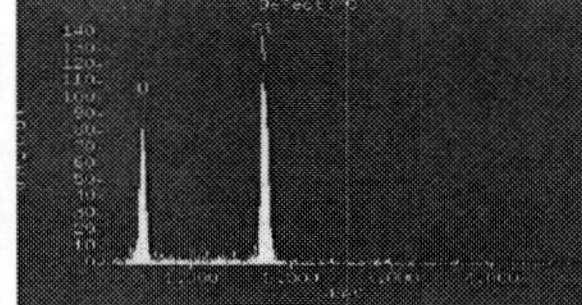

Fig. 2 a: Defect after ash +wet strip.　b: EDX O and Si species.

We usually evaluate the ash process splits from the point view of residue and substrate loss as the above discussed. Ideally, free residue and ITRS-required substrate loss are expected. Table 2 compared the two index difference in terms of both in-line and off-line data between baseline and the final optimized ash condition based on Table 1. Results show, after the LDD loops, the optimized ash process reduces ~40% SiGe loss, ~50% Si recess and achieves the residue free performance.

		Baseline (normalized)	Optimized
Inline	Oxide Growth on Si Pad (A	1	53%
Measurment	SiGe Loss (A)	1	73%
TEM	Si Loss (A)	1	48%
	SiGe Loss (A)	1	60%
SEMVision	Residue Count (ea)	1	0

Table 2 Inline and TEM substrate loss comparison

2 The Effect of O_2/FG Ratio on Offset Space CD

We noticed the CD shrink of offset spacer during LDD loop. This phenomenon might be related to SiN loss during numerous ash steps. To clarify this conjecture, we firstly checked the film loss on blanket SIN and OX film (with neither IMP nor photoresist). Table 3 compares the film loss after one-time ash, four-times ash and nine-times ash, respectively. Almost no SiN loss is detected and 2%~3% (3~4A) oxide growth is noticed even after nine-times ash. Nevertheless, on the full-loop wafers, we found offset spacer CD gradually shrinks after each ash step in LDD loop. Offset spacer CD mainly comes from the thickness of its outer layer SiN. Hence, there is no doubt offset spacer CD loss

	Pre-data	Asher*1	Asher*4	Asher*9
SIN wf (A)	1	1+0.13%	1+0.32%	1+0.51%
OX wf (A)	1	1+2.65%	1+2.77%	1+3.93%

Table 3 Normalized blanket SIN and OX film loss only with ash process

originates from the SiN film loss. As the above mentioned, ash process itself couldn't cause SiN film loss. Therefore, we concluded such loss should be formed by the joint action of IMP and ash. In addition, as shown in Fig.3, the shrink of offset spacer CD is different between P-MOS and N-MOS even for the single ash step case. BF_2 is usually utilized for P-MOS related implant. It could provide the fluorine ions. This will result in SiN loss when hydrogen group is driven out from the photoresist during HDI processes. If this is the case, the SiN loss will not be significant for As is designed for N-MOS related implant. Fig 3 shows the reconciliatory clue, namely, offset spacer CD loss does not depend on O_2/FG ratio on N-MOS while it is linearly inversely proportional to O_2/FG ratio on P-MOS. The latter is associated with the deduction of hydrogen when O_2/FG ratio is higher. The normal LDD loop should include more than five P-MOS LDD steps. It is equal to the capability of reducing more than 10 nm CD loss for offset spacer if O_2/FG ratio is around 0.1. This will result in the remarkable negative impact to P-MOS device performance, thus impacting the yield and reliability. Fig. 4 compares such severe SiN CD loss between $O_2/FG=2$ and $O_2/FG=0.1$ right after the LDD loop. As expected, all SiN film is completely removed for $O_2/FG=0.1$ case. As such, the optimization of ash process is not limited to residue free and substrate loss minimization, the SiN CD shrink should be also considered. This is the trade-off issue for lower O_2/FG ratio reduces the substrate loss but increases the SiN CD loss.

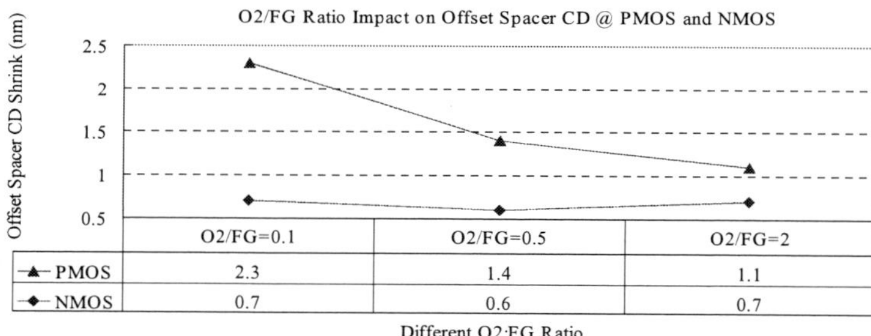

	O2/FG=0.1	O2/FG=0.5	O2/FG=2
PMOS	2.3	1.4	1.1
NMOS	0.7	0.6	0.7

Fig. 3 The dependence of offset spacer CD shrink on O_2/FG ratio at P-MOS and N-MOS

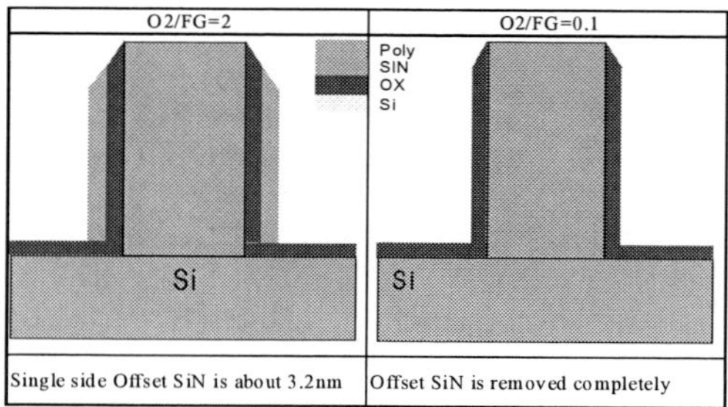

Fig. 4 SiN Thickness comparison between two O_2/FG ratios right after LDD loop

3. Improved electrical behavior

The N-MOS and P-MOS Idsat vs Ioff characteristics are shown in Fig 5 (a) and (c), respectively. The optimized ash split (O_2/FG=2) outperforms baseline in both plots. Moreover, P-MOS exhibits more remarkable improvement than N-MOS. This is consistent with the fact the substrate loss deduction could effectively reduce the source-drain extension dopants consumption. Better P-MOS gain benefits from its SiN CD recovery. This can be explained by the vertical fringing electric field effect from offset spacer. The similar ~5% achievement can be seen in Fig 5 (b) and (d) for Vt vs Idsat.

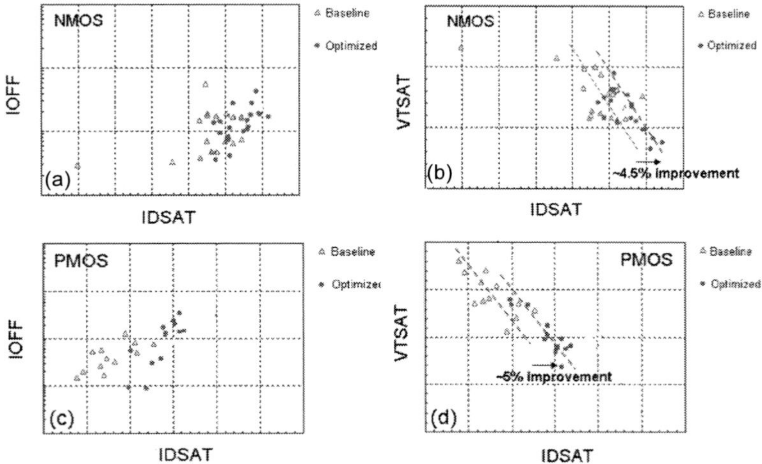

Fig. 5 (a) Ion-Ioff curve of N-MOS; (b) Ion-Vt curve of N-MOS;
(c) Ion-Ioff curve of P-MOS; (d) Ion-Vt curve of P-MOS;

CONCLUSIONS

We investigated the effect of ash process in LDD loop in terms of substrate loss, residue and its effect on offset spacer CD. Results reveal both high pressure in preheat step, low pressure and pin up in main-step can effectively preclude Si/SiGe loss. Less OA time could bring to better performance for residue removal and substrate loss to some extent but OA time could not be reduced arbitrarily. Offset spacer CD loss is linearly inversely proportional to O_2/FG ratio at P-MOS while it is insensitive at N-MOS. Such difference stems from different implant dopants. Electric data shows the optimized ash process outperforms baseline by ~5% gain in Idsat-Vt plot for both N-MOS and P-MOS.

ACKNOWLEDGMENTS

Authors would like to thank Ms. Linda Sui of Mattson for the technical discussion and information, Mr. Ferris Chen and Ms. Nikki Li from the integration team of SMIC Technology R&D center for wafer preparation.

REFERENCES

1. M.Ameen, et al, *Solid State Phenomena,* Vol. 134, 2008

2. K. Han, et al, Non-fluorine plasma strip of HDI resist for 45 nm node, PESM 2007

ECS Transactions, 34 (1) 433-438 (2011)
10.1149/1.3567617 ©The Electrochemical Society

Effluent Management for Non-Oxidizing Plasma Strip Processes

S. Luo, C. Waldfried, O. Escorcia, I. Berry,
P. Geissbühler, A. Srivastava, D. Roh

Axcelis Technologies Inc., Beverly, MA 01915, USA

This paper reports on the techniques employed to control the re-deposition of the partially dissociated organic photoresist (PR) by-products in advanced non-oxidizing strip processes developed to meet the PR removal requirements of future technology nodes. System features, such as the design of heated process chamber walls and an on-board, RF-based oxygen plasma effluent abatement system are described in detail. The performance of these features to prevent or eliminate hydrocarbon buildup and manage effluent with non-oxidizing strip processes is also presented and discussed.

Introduction

Continuously shrinking device nodes employ ultra shallow junction and high-K/metal gate stack technologies in order to fabricate high speed transistors. Optimization of the PR and residue removal process step after high-dose-implantation has become essential in order to preserve proper dopant activation and to avoid substrate loss and/or gate stack material modifications (1, 2). As a result the industry has started to utilize non-oxidizing strip chemistries for these critical PR and residue removal steps, in replacement of the traditional oxidizing chemistries (3, 4). However, the non-oxidizing chemistries, such as Forming gas (N_2:H_2), lack the oxidizing species to efficiently remove the long chain polymers of the PR. A non-oxidizing plasma only partially dissociates the resist into short-chain hydrocarbon by-products, as opposed to an oxidizing plasma that converts the PR into gaseous by-products such as CO_2, CO, H_2O or the like. These non-oxidizing strip by-products tend to condense on cold surfaces such as chamber walls, vacuum lines, valves, pumping lines, pumps, and exhaust lines. As more and more wafers are processed, this condensation leads to buildup of hydrocarbons. This buildup on the surface inside of the process chamber is a potential source of particulate contaminants to wafers in the process chamber. Having the buildup occur inside of valves, pumps and pumping lines, etc can cause pre-mature pump and valve failure, require frequent maintenance and might even pose a safety hazard. Conventional process chamber cleaning and system maintenance are done "off-line" and the strip tool is unavailable for production use, lowering the utilization of the tool. For this reason, there is a need to reduce the buildup and the frequency of maintenance, and the prevention and control of organic byproduct re-condensation has become a critical feature for the implementation of non-oxidizing strip chemistries.

In this paper two techniques are investigated and employed to reduce the re-condensation of the partially dissociated PR by-products of non-oxidizing strip processes. Elimination or reduction of hydrocarbon buildup in the process chamber is done by heating the chamber wall. An afterburner where a secondary RF-based oxygen plasma is formed downstream of the process chamber converts the short-chain hydrocarbons to gaseous by-products, which can be pumped out and eliminate the buildup on surfaces of any components downstream.

Experimental

All studies shown in this work were carried out on a 300mm, three-module, six-chamber Axcelis IntegraES dry-strip system, with a microwave-driven, remote plasma source and a load-locked platform design which incorporates active wafer cooling. The IntegraES is designed specifically for non-oxidizing PR plasma strip processes which enables "zero" metal and substrate oxidation and loss. Figure 1 shows a schematic diagram of the features designed to control non-oxidizing ashing effluent and by-product buildup.

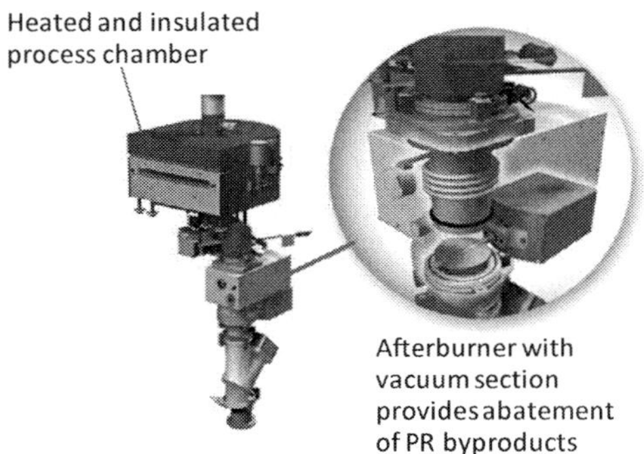

Figure 1. Schematic of heated chamber and afterburner.

The heating element increases chamber interior surface temperature to greater than 60°C during wafer processing. For safety reasons the chamber is thermally insulated to prevent burns when servicing. In addition, thermal insulation reduces the cost-of-ownership of the tool by reducing the power needed to maintain an elevated temperature within the chamber. To control the effluent downstream, an afterburner assembly is coupled to the exhaust conduit, wherein the exhaust conduit comprises a gas port intermediate to the process chamber and the afterburner assembly. A secondary RF plasma is formed inside afterburner, where oxygen is injected to abate any ashing effluent into gaseous by-products. An important element in operation of the afterburner is to keep the injected oxygen from back-streaming into the process chamber. This is achieved by keeping a sufficient pressure in the region above the afterburner (so that the gas flow is in the viscous or transition flow regime) and keeping process gas flow sufficiently high to prevent the back-streaming of the injected oxygen. The portion of the exhaust conduit between the chamber and afterburner is also heated to prevent hydrocarbon buildup at this location.

Three non-oxidizing strip chemistries, which include Forming gas-only (FG-only, $N_2:H_2$) and two Axcelis proprietary non-oxidizing chemistries (COD-01, COD-02), were tested with the features described above. Multiple resist wafers were stripped with these chemistries and the performance of process chambers (ash rate, particle adders) was examined to determine the efficacy of chamber wall heating to reduce/eliminate re-condensation and the proper chamber wall temperature needed. Resist hydrocarbon buildup downstream of afterburner was monitored at several locations along the exhaust line by measuring deposition on Si wafers coupons taped at these locations. Inner surfaces of process chamber and exhaust line parts were also visually examined for resist buildup after marathon tests, in which resist strip marathon was performed to simulate the IC manufacturing environment.

Experimental Results and Discussion

The objective of this work is to provide a method to eliminate the non-oxidizing strip PR by-product buildup inside the process chamber wall, which could cause particulate contamination. Figure 2 shows the chamber wall pre and post stripping ten 1.8μm blanket I-line resist wafers with FG-only process in a cold-wall chamber. A visible resist buildup occurred even after ashing only ten resist wafers with non-oxidizing FG-only process. Similar buildup inside the chamber was observed when COD-01 and COD-02 were used to strip resist wafers in a cold-wall chamber.

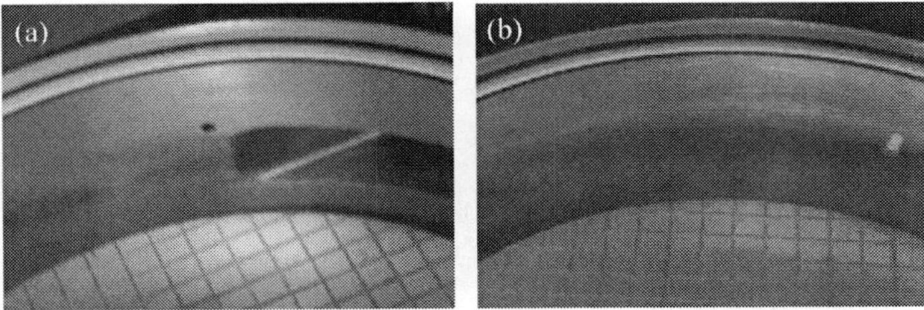

Figure 2. Photographs of chamber wall: (a) pre and (b) post FG-only strip in cold-wall process chamber.

Several techniques were tested to eliminate/reduce the re-condensation of PR by-product on a cold chamber wall. The results demonstrated that heating of the chamber wall is an effective way to achieve this goal. When the process chamber wall was heated to greater than 60°C, the PR hydrocarbon buildup was substantially reduced. However, in order to completely eliminate the buildup within the interior of the process chamber, the wall temperature should be maintained at 140°C or higher. Figure 3 shows small amount of buildup with 120°C chamber walls, but completely no buildup at 140°C chamber walls with the FG-only process. For COD-01 and COD-02 processes, there was no buildup at 140°C chamber wall as well.

Since resist buildup inside the chamber would lead to chamber performance shift and particle issues, the heated chamber configuration was also evaluated during extensive marathon testing by checking resist ash rate (AR) and particle adders on wafers, with some of the data being shown in Figure 4. In Figure 4 (a), particle adders were measured after increasing number of blanket coated 2.8μm I-line resist wafers (up to 1525) that were processed to endpoint with a FG-only plasma. All data points, except for one, fell in the tool particle specification. Figure 4 (b) shows AR and ash non-uniformity (NU), which were regularly checked during the marathon testing in which 500 blanket coated 2.8μm I-line resist wafers were stripped with the COD-02 process. Throughout the marathon test, both AR and NU were consistent and well within tool specification. In both tests, the process chamber wall was visually checked and showed no resist buildup.

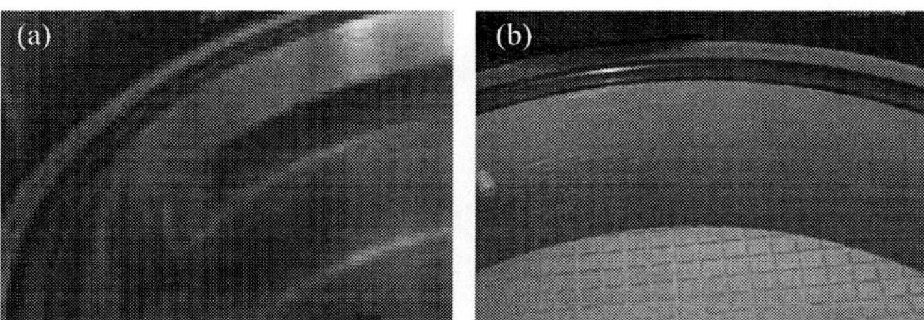

Figure 3. Photographs of chamber wall: (a) after striping 100 2.8μm resist wafers with 120°C chamber wall, (b) after striping 500 2.8μm resist wafers with 140°C chamber.

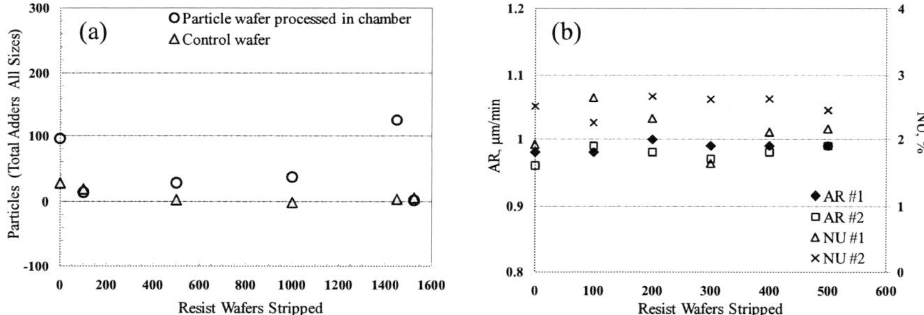

Figure 4. Performance tracking of 140°C chamber wall marathon testing: (a) particle adders of FG-only process, (b) AR and NU of COD-02 process.

Another important goal is to reduce the ashing byproduct buildup in the exhaust lines. While there are other techniques besides utilizing an afterburner assembly which can control the effluent, such as cold trap and heating of vacuum and exhaust lines, the afterburner has proven to be the best solution due to its efficiency and low cost (5). To verify the efficiency of the afterburner, several silicon chips were placed on a wire mesh which was then installed in the vacuum line downstream of the afterburner. After stripping ten blanket 2.8μm I-line resist wafers in a 140°C chamber with the FG-only process, the vacuum lines were opened to check for deposition on the silicon pieces. Figure 5 shows resist deposition on silicon pieces with the afterburner turned off, while no resist deposition was found on the chips with the afterburner on. Up to 2.1μm resist deposited on Si-chips was measured with the afterburner off.

Several marathon tests were conducted with the afterburner turned on and tuned for different non-oxidizing processes which were used to cycle blanket resist wafers in 140°C process chamber. The vacuum and exhaust lines were opened and checked at the end of each test. Figure 6 shows some pictures of the exhaust line and parts which clearly demonstrated that, with the effluent abatement in the afterburner, the downstream hydrocarbon byproduct buildup was completely eliminated. Tool particle and AR performances with afterburner on were also monitored. The data shown in Figure 4 above were all collected with the afterburner on, which indicated no interference of afterburner to the tool performance.

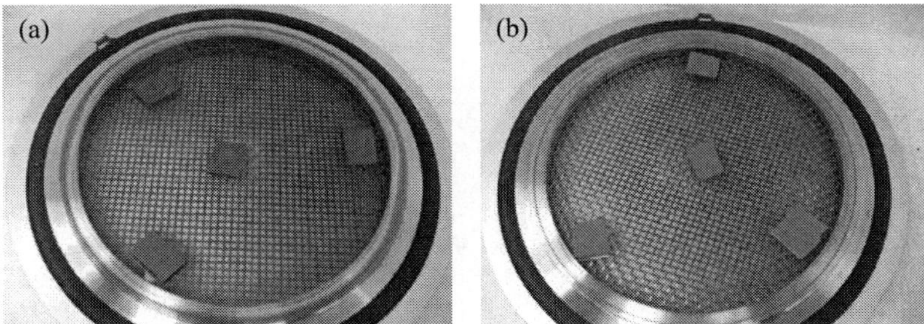

Figure 5. Resist strip byproduct buildup in vacuum line: (a) visible buildup with the afterburner off, (b) no buildup with the afterburner on.

Figure 6. Photographs of exhaust line downstream of afterburner: (a) after cycling 8000 resist wafers with FG-only process, (b) after cycling 1000 resist wafers with COD-01 process, (c) after cycling of 500 resist wafers with COD-02 process, and (d) heavy buildup with no afterburner.

Conclusions

Two features to control the buildup of resist strip by-products in non-oxidizing strip processes were designed and investigated. Heating the process chamber wall >60°C reduced the hydrocarbon buildup inside chamber, while temperature >120°C completely eliminated the buildup. The afterburner assembly showed good efficacy to break down the ashing effluent and significantly reduced the re-deposition on surfaces of exhaust line parts. These two techniques make the non-oxidizing strip processes feasible to meet the requirements of wafer cleaning of the 28nm technology node and beyond.

Acknowledgments

The authors would like to thank Armin Huseinovic, Patrick Mayfield and James DeLuca of Axcelis Technologies, Inc. for their valuable work in system designing and data collecting.

References

1. F. Arnaud, H. Bernard, A. Beverina, R. El-Farhane, B. Duriez, K. Barla, and D.Levy1, *Solid State Phenomena*, **103-104**, UCPSS 2004.
2. S. Garaud, R. Vos, D. Shamiryan, V. Paraschiv, P. Mertens, J. Fransaer, and S. De Gendt, *Solid State Phenomena*, **145-146**, UCPSS 2008.
3. I. Berry, C. Waldfried, K. Han, S. Luo, R. Sonnemans, and M. Ameen: *IEEE Junction Technology*, IWJT 2008.
4. S. Luo, O. Escorcia, C. Waldfried, D. Roh, and I. Berry III, *Study of Alternative Plasma Strip Techniques for Advanced Photoresist Removal*, ECS Trans.-CSTIC 2010.
5. A. Srivastava, P. Sakthivel, T. Buckley, and A. Becknell, *A New Plasma Source for Destruction of Organic Material in the Post Chamber Hardware, and Implications for Process Endpoint Detection for Specific Low-k Applications*, 51th AVS Symposium, November 2004.

ECS Transactions, 34 (1) 439-443 (2011)
10.1149/1.3567618 ©The Electrochemical Society

Wafer Backside Particle Reduction By Optimizing AC3 Coating for Poly Etch Chamber

Bao-Shan Ma[a], Wen-Jun Liu[a], Feng Niu[b], Jian-Jun Xia[b],
Lien-Huang Cheng[a] and Kun-Yueh Liang[b]

[a] Lam Research (Shanghai) Co., Ltd, Zhangjiang Hi-Tech Park, Pudong, Shanghai 201203, China
[b] Semiconductor Manufacturing International Corporation, No.18 Gaoxin Road 4th, East-lake High-Tech Development Zone, Wuhan, China

Abstract

As semiconductor devices are being scaled down to CD (Critical Dimension) of 65nm and below, the control of chamber condition in Poly Etch chamber is critical to achieve stable CD and defect performance from wafer to wafer, lot to lot. AC3 (Advanced Chamber Condition Control) coating is widely used after traditional WAC (Waferless Auto Clean): an $SiCl_4$ coating step is added after WAC to coat a thin layer on both chamber wall and ESC chuck. This layer maintains chamber in good condition for every wafer to achieve stable Etch Rate, CD and defect performance, while low quality coating makes wafer backside particle high. This paper introduces two key factors for $SiCl_4$ coating quality control which affects wafer backside particle performance. One is $SiCl_4$ gas overshoot, the other one is ESC temperature effect. The wafer backside particle high can finally cause wafer frontside high defect count.

Introduction

As the advanced technology is scaled to 65nm node, wafer backside particle control has become important for defect performance, especially for those steps which have subsequent weak clean process. The particle migrates from wafer backside to frontside during subsequent batch clean process, causing defect issue.

AC3 is widely used in LAM Poly Etch chambers: after traditional WAC, an $SiCl_4$ coating step is added afterwards to coat a thin layer on both chamber wall and ESC chuck. This layer not only serves as chamber protection, but also maintain chamber in good condition for every wafer to achieve stable Etch Rate, CD and defect performance, while low quality coating causes wafer backside particle high.

This paper introduces two key factors for $SiCl_4$ coating control. One is $SiCl_4$ gas overshoot. $SiCl_4$ is a high condensation gas, when it's not properly heated or MFC is bad, it tends to have clogging somewhere. When the gas is turned on during coating step, gas burst causes overshoot around 30-50% of nominal set point, excessive coating layer is deposited on ESC. Such coating will adhere to wafer backside when wafer sits on ESC and is chucked by high voltage. After Etch, wafer will normally go to next step: wet clean process. This is a batch process, backside particles will be loosen and migrate to wafer surface, causing surface particles.

The other one is Temperature effect: different thermal expansion between coating layer and ESC. Thermal expansion makes coating loosen on ESC surface and adhere to wafer backside by chucking voltage. In production line, different products run at different ESC temperature: 60C, 40C or 20C. When there is process switch, temperature up and down cause bad adhesion of $SiCl_4$ coating to ESC and coating may stick to wafer backside.

Experiment

Current issue: a spike is observed when turning on $SiCl_4$ gas, which can be 30% to 50% higher than nominal set point. In this case, wafer backside presents a high count of particles (>2000 counts). After wet clean process, flat shape defects are detected on the front side of the wafer. With $SiCl_4$ gas overshoot, wafer backside particle count can easily reach more than 2000, and up to 10000.

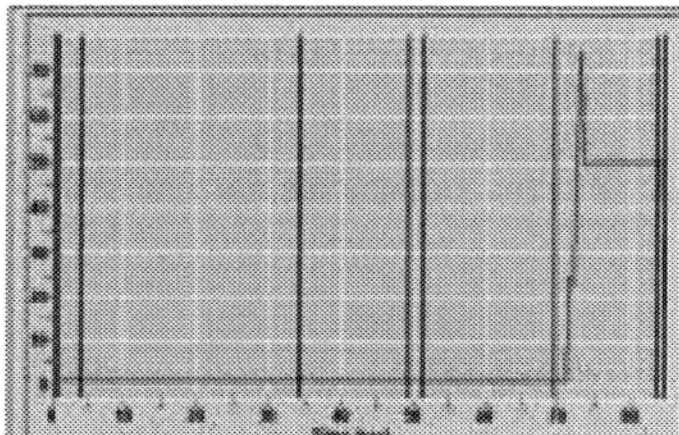

Figure 1: Evolution of $SiCl_4$ gas flow (Red Line) during $SiCl_4$ coating.

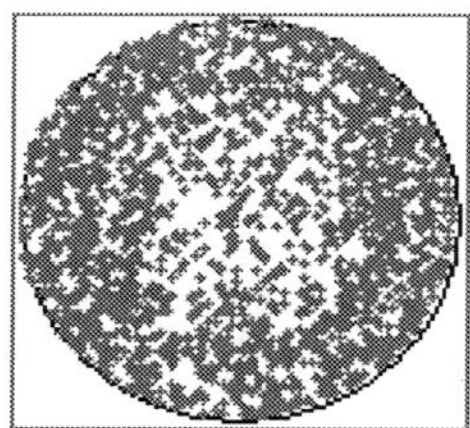

Figure 2: Map of backside particles. Typically, more than 2000 particles are measured.

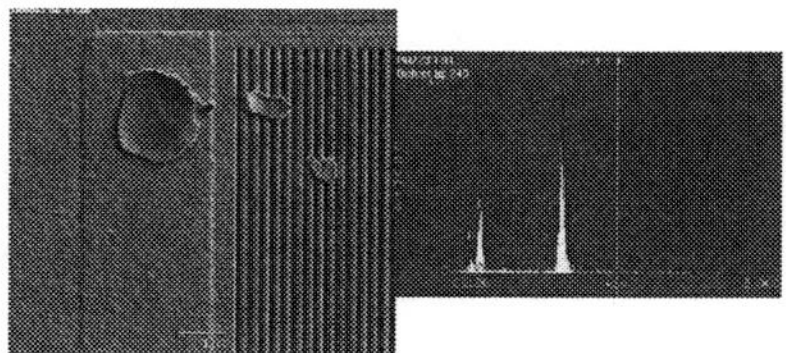

Figure3: Frontside defect observed by SEM. EDX shows Si and O which is same composition as backside particles.

Several experiments are carried out to minimize backside particle count. One is adding a stabilization step before plasma is turned on during the coating step. $SiCl_4$ gas still presents an overshoot and the number of backside particle is still high. Another approach is adding a gas flow ramp up such as a ramp up step with 20% of nominal gas flow for 3 seconds. In this later case, an overshoot is still observed, but it remains below the nominal flow of $SiCl_4$. As a consequence, the number of backside particle remains low. The other factor is temperature effect when there is process switch between different temperatures. The high number of backside particles is due to different thermal expansion between coating and ESC. One test is using several seasoning wafers when there is process switch. The seasoning wafer can remove loosen coating from ESC with chucking force, serving as a sacrificial dummy wafer. Then seasoning wafer is then sent for recycling to clean its backside.

Result and Discussion

3.1 $SiCl_4$ Gas Ramp Up Effect

Based on analysis, whenever there is $SiCl_4$ gas overshoot, the backside particle count will be high. After wet clean, frontside particle also becomes high. Backside particle is the same type as frontside defect. This helps us understand correlation of defect source to root cause: 100% of gas overshoot wafers have high backside particle count, causing high surface defect. Excessive coating layer on ESC is identified as particle source.
When gas flow is set at 50sccm, overshoot reaches up to 65 to 75sccm, so there could be up to 50% overshoot. A gas ramp up step with 20% of set point for 3 seconds is added. In this case, 10sccm of $SiCl_4$ is injected for 3 seconds. Typically overshoot is small for small gas flow. Even with a 100% overshoot, the gas flow does not exceed 20sccm. So overshoot is much lower than process with no ramp up. As a result, wafer backside particle is greatly reduced.

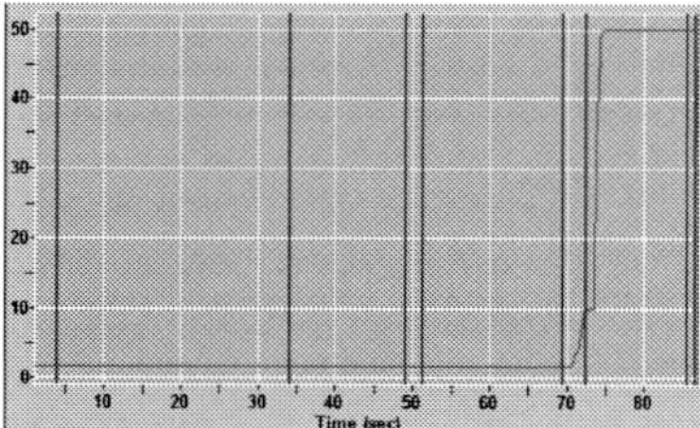

Figure 1: Temporal evolution of $SiCl_4$ gas flow (Red Line) when 20% gas ramp up for 3s is added during $SiCl_4$ coating

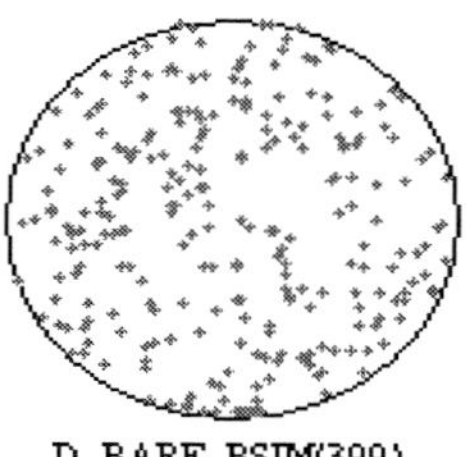

Figure 2: Map of backside particles on a wafer etched after AC3 with SiCl$_4$ ramp up.

3.2 ESC Temperature Effect

Same phenomenon is also observed when switching between different processes performed at different temperatures (60C, 40C or 20C). When temperature is switched, the number of backside particles is also high for the first few wafers. This is due to different thermal expansion between coating layer and ESC surface which makes coating loosen, then adhere to wafer backside by chucking voltage. Seasoning with several dummy wafers can minimize this effect.

Other experiments can also be carried out to eliminate particle source to minimize this effect. One is using long WAC time to consume coating layer before each lot starts. Then coat a fresh layer after WAC. This way, all the coating on ESC is removed. Also after each lot completes, a post WAC may be done to maintain ESC in a clean mode. These experiments can serve as long term solutions to enhance productivity, which could eliminate the need for seasoning wafers.

Conclusions

In this work, we solved the issue of backside particles by introducing a 20% SiCl$_4$ gas ramp up step during the SiCl$_4$ coating. In addition, seasoning wafers must be used to clean ESC when there is temperature switch. Long WAC time consuming excessive coating layer before ESC temperature switching can also minimize wafer backside particle. By using above methods, backside particle count can be greatly reduced to a controllable region, typically less than few hundreds, preventing defect issue highlight.

Acknowledgements

Authors would like to thank Xiao Jun at LAM and KC Wong at SMIC for their technical assistance during the course of this work.

References

1. F. Kroninger, et al., Appl. Surface Sci., Vol. 63, No. 1–4, pp. 93–98, January 1993.
2. R. Miura, et al., Proc. of the Eleventh Int'l Conf. on Ion Implantation Tech., pp. 174–177, June 1996.

3. F. Beaudoin, et al., J. Vac. Sci. Technol. A, 16, 3, pp. 1976–1979, May-June 1998.
4. W. Frutiger, et al., Proc. of the Eleventh Int'l Conf. on Ion Implantation Tech., pp. 346–349, June 1996.
5. W. Au, et al., Proc. of ASMC, pp. 294–297, 1998.

Comments:

This paper deals with the issue of back side particles. During the seasoning of the chamber in $SiCl_4$ gas, bursts of $SiCl_4$ gas may occur when the gas is injected, yielding a too thick deposition of $SiCl_4$. Furthermore, when the ESC temperature is changed, differences in dilatation coefficient between ESC and $SiCl_4$ polymer lead to polymer delamination and sticking to the backside of the wafer, which creates particles.

The authors propose two solutions to this issue: 1) gas ramp up during the SiCl4 prevents the gas flow to overshoot above 20scmm, which prevent the too high deposition of SiCl4.

2) seasoning wafers may be used after changing the ESC temperature to remove any polymer from the ESC.

3) Long WAC (waferless auto clean) time consuming excessive coating layer before ESC temperature switching can also minimize wafer backside particle.

From my understanding, $SiCl_4$ plasmas do not deposit polymer. On the contrary, they are very efficient to clean reactor walls since $SiCl_4$ plasmas can remove AlF3 from the chamber (see Ramos et al Plasma Sources Sci. Technol. 16 (2007) 711–715). On the contrary, as soon as O_2 is added in the plasma, SiOCl polymer can build up very rapidly. Does the author use O_2 as well in the plasma?

Yes, lower ratio of O2 is used with $SiCl_4$ in the plasma for coating. For company confidential, I have not listed the detail recipe.

ECS Transactions, 34 (1) 445-451 (2011)
10.1149/1.3567619 ©The Electrochemical Society

The Study of Dry Etching Process on Plasma Induced Damage in Cu Interconnects Technology

Jun-Qing Zhou[b], Hai-Yang Zhang[b], Wu Sun[a], Xing-Peng Wang[a], Min-Da Hu[b]. Fan Li[b], Li-Ya Fu[a], Shih-Mou Chang[b], Kwok-Fung Lee[b]

[a] Semiconductor Manufacturing International Corporation, No.18 Wen Chang Rd., BDA, Beijing, 100176, P.R.China
[b] Semiconductor Manufacturing International Corporation, No.18 Zhang Jiang Rd., Pudong New Area, Shanghai, 201203, P.R.China

ABSTRACT

As complementary metal-oxide-semiconductor (CMOS) is rapidly shrunk and the plasma processing steps are increased in number at 65nm node and beyond, plasma-induced damage (PID) has become a serious reliability problem in state-of-the-art semiconductor manufacturing.

In this paper, we only focus on the PID issues originating from the dry etching processes in back-end-of-the-line and utilize the in-line test structure to in-situ debug the possible PID impact from the specific layer. The effects of various etching process parameters such as power, power frequency, magnetic field and process time are examined. Results demonstrate metal 1 (M1) etch, passivation etch and Al-pad etch play the critical role in PID performance on the given test vehicles. Nevertheless, their sensitivities are quite different between N-MOS and P-MOS. Basically, N-MOS degradation strongly depends on Al-pad etch while P-MOS PID performance is closely related to M1 etch and passivation etch. More specifically, the over-etch percentage in M1 etch is one of effective knobs in alleviating PID while it is not in passivation etch. All the above discrepancies indicate the single PID failure mechanism could not be applied. We tentatively explain the existing phenomena from the point of view of electron shading effect (ESE), reverse electron shading effect (RESE), plasma non-uniformity, photo-conductive effect and ultra-thin charging collector during the interface switch in etching.

Keywords: Plasma induced damage, BEOL, antenna ratio, M1-etch, passivation etch, Al-pad etch

I. INTRODUCTION

PID [1,2] has been noticed since the introduction of plasma processing in silicon IC manufacturing. It is one of the important reliability concerns in both front-end-of-the-line (FEOL) and back-end-of-the-line (BEOL) process development. As CMOS is being scaled down to 65nm node and beyond, more interconnect layers are required to ensure the valid connection of the hierarchical wiring network. Hence the final cumulated PID from each plasma related process step in BEOL becomes more prominent. The charge collected from interconnect network could flow through the connection between the interconnect down to the gate electrode of a MOS transistor and might cause the "Fowler-Nordheim"(F-N) tunneling current through the gate oxide, generating electron traps to degrade across the entire layer or at some weak points in gate oxide.

Plasma processes in BEOL manufacturing include plasma enhanced-chemical vapor deposition (PE-CVD), sputtering and dry etching. The PID and corresponding solutions for the first two processes have been addressed by Matsunaga [3] and Pan [4]. In brief, the pre-clean step in CVD and rapid plasma clean (RPC) in sputtering have been identified as the two root causes in BEOL related PID cases. The optimized pre-clean step and less RPC are the proven directions toward PID free process. Regarding dry etching, many previous studies do not differentiate or are not aware of the PID dissimilarity between N-MOS and P-MOS. Either ESE or RESE model is not appropriate to explain the existing PID phenomena at 65nm node and beyond. Besides, the advanced function of etchers and the potential side-effect from improving other reliability indexes also need special attention to re-identify the effective and usable knobs for PID reduction and improvement. The effect of the scaled critical dimension (CD) and thicker Al-pad thickness on PID performance is also not negligible.

In this work, we disregarded the latent uncertainty from layout, investigated all BEOL related dry etching processes on two given test vehicles and spotted M1 etch, passivation etch and Al-pad etch as the possible PID resources. The corresponding qualitative solutions are provided for the reduction of these BEOL etching induced PID.

II. EXPERIMENTS

To investigate the potential PID from BEOL dry etching processes, as illustrated in Fig1, we employed two PID test keys (TK) with the antenna of comb type for all splits in this paper. Antenna refers to the interconnect area connected to the gate electrode of the MOS transistor. Antenna ratio (AR) is defined as the ratio of the interconnect area attached to gate electrode over the active device area. The maximum allowable AR is the upper-limit of the antenna size for device to be immune to antenna effect. AR is stated for each layer separately. Protection diode (PD) is usually designed against plasma-charging. Its valid insertion could eliminate the PID to some extent. However, PD might hide the actual root-causes of potential PID. Hence, it is only used to decouple the cumulative PID.

Fig 1 (a) is the schematic of cumulative PID test-key. The final accumulated PID from the full-loop wafer can be detected at each layer using the structure of Fig 1 (b). It can measure the PID from M_n layer by using antenna at M_{n-1} layer. As core transistors are more robust than input/output (I/O) transistors due to their thicker gate oxide, we only pay attention to the PID performance on I/O MOS devices. An Agilent 4072 parametric tester is used to collect and analyze the gate leakage current (Ig) through the gate oxide. We completed all related trench etch, passivation etch, Al-pad etch in Lam Flex series, AMAT E-max and Lam versys metal plasma etch reactors, respectively.

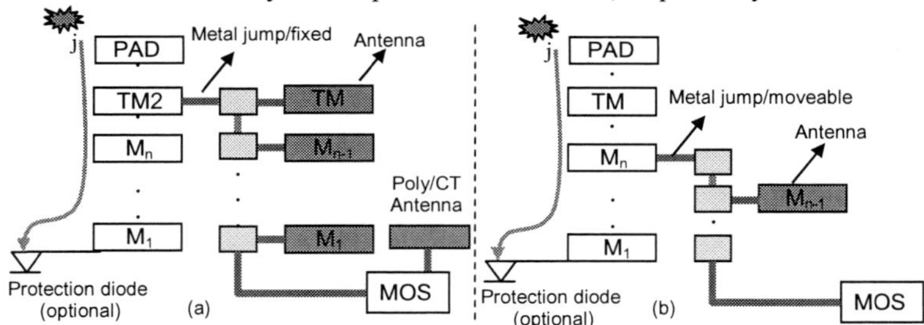

Fig 1. Schematic of PID test-key (TK) structure a) accumulated; b) layer-by-layer

III RESULTS AND DISCUSSIONS

Fig 2 shows the distribution maps of -log (Ig) on contact (CT) antenna for the in-line PID results of M1-main etch (ME) time splits. Clearly, the PID damage is located at the wafer center and strongly depends on the length of process time and AR. The worst PID is associated with the longest process time and the highest AR. To decouple the PID from CVD liner, we summarized the effect of Mx-liner on PID in Fig 3. The similar PID location is detected. Meanwhile, it manifests low power and thinner liner in CVD process could induce less PID. This is consistent with the reported "spreading antenna effect" [3]. Moreover, as CT antenna is more sensitive than Mx antenna, there is no guarantee for the PID-free Mx-liner used at M1 layer. We could notice this from their huge AR difference. Three power frequencies (60MHz/27MHz/2MHz) have been employed in BEOL etch to deal with the challenges from new materials and continuously downscaling devices. The cumulative plots of –log (Ig) in Fig 4 demonstrate again the lower ratio of 27 MHz over 2MHz tends to result in worse PID, and the worst PID performance is observed with extra 50% over-etch (OE). All the above signals 1) liner should be removed for better understanding PID from M1-etch; 2) the combination of multiple-frequency power in M1-etch should be re-assessed; 3) the process time is a key factor for PID deduction.

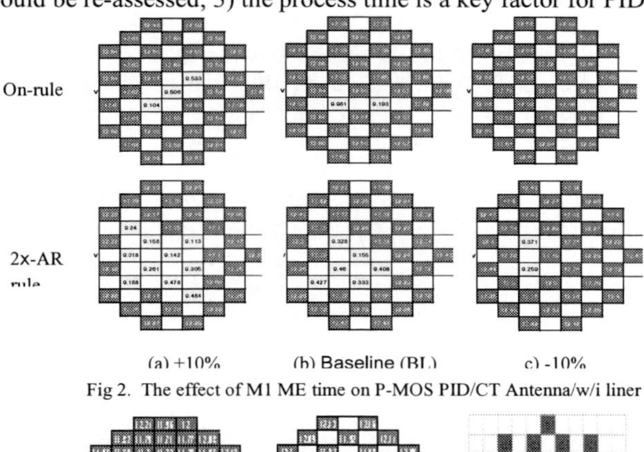

Fig 2. The effect of M1 ME time on P-MOS PID/CT Antenna/w/i liner

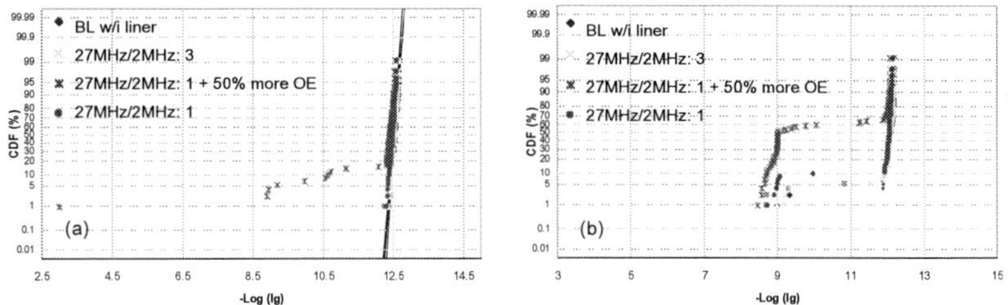

Fig 3. The impact of liner splits on P-MOS PID in Mx

Fig 4. The impact of M1 splits on P-MOS PID /Contact Antenna. (a) on-rule; (b) 2x AR

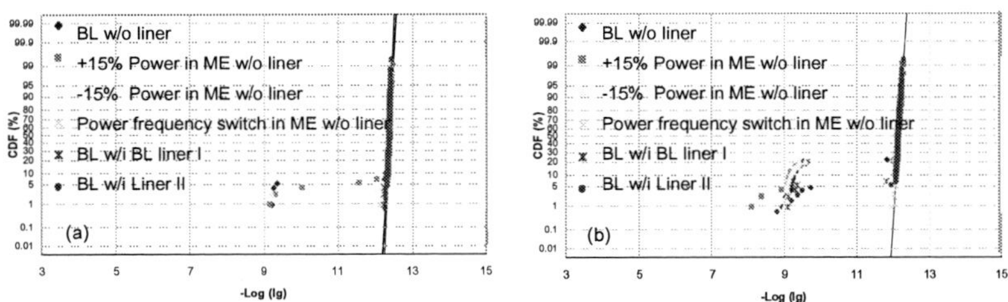

Fig 5. The impact of M1 splits on P-MOS PID /Contact Antenna. (a) on-rule; (b) 2x AR

As shown in Fig 5, liner II-based M1 outperforms three no-liner power splits (+-15%) on on-rule antenna. Nevertheless, it is similar to BL on 2x AR condition. This indicates the PID from dry etching is not negligible. Similar undesired PID performance among power splits shows the power itself in dry etching is not a critical index. The only PID-free split on both on-rule and 2x AR TK comes from the power frequency split, in which the power injected at 60MHz is replaced by 27MHz, the high ratio of 27MHz over 2MHz is applied. These results mean the PID-free performance could be achieved on the given test vehicle by optimizing the process time and the combination of power frequencies.

All mentioned PID at M1 are related to P-MOS transistors. Two possible PID mechanisms for BEOL etch are listed in Fig 6. As illustrated in Fig 6 (a), due to the scaled trench CD at M1, ESE could result in the F-N current. High frequency power could increase the plasma density while low frequency power could generate the ions with higher speed. The net effect is more ions accumulated at trench bottom, thus damaging the gate oxide. Besides, the over-sensitivity of CT antenna ends up with its susceptibility to the longer exposure time to plasma. The photo-conductive effect from the end of inter metal deposition (IMD) can be disregarded in this case.

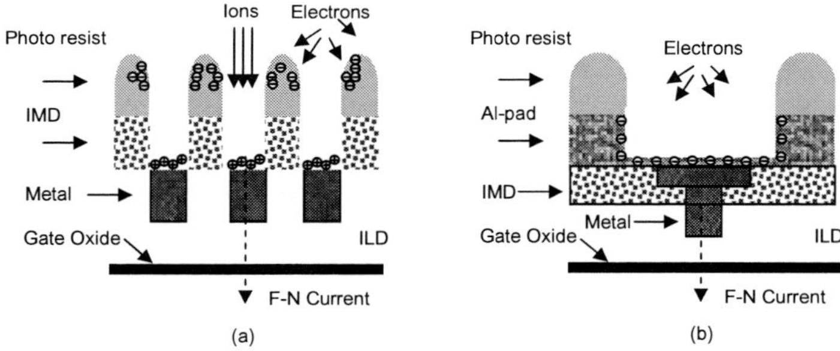

Fig 6. Electron shading effect (a) vs Reverse electron shading effect (b)

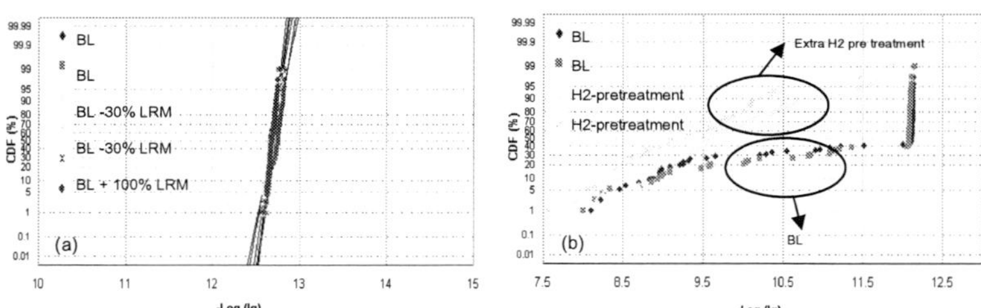

Fig 7. The impact of PAS1-LRM on PID: (a) LRM process time; b) additional H2-pretreatment

Fig 7 (a) compares the PID performance of different process times in liner remove step of passivation 1 (PAS1) etch. Unlike M1, almost no difference is detected even for 100% extension in process time. This could be partially attributed to the much lower sensitivity of its substrate (copper). Fig 7 (b) shows the final cumulative PID result after passivation 2 (PAS2) etch with the TK described in Fig 1 (a). The large tails are caused by the magnetic field in PAS2 etch. We will address more details on this point later. In addition, both wafers with additional H_2-pretreatment in PAS1-etch exhibit much more severe tails. The introduction of H_2-pretreatment after PAS1-LRM could greatly reduce the copper hill-lock issue [5] but worsens P-MOS PID for its providing too much positive ions.

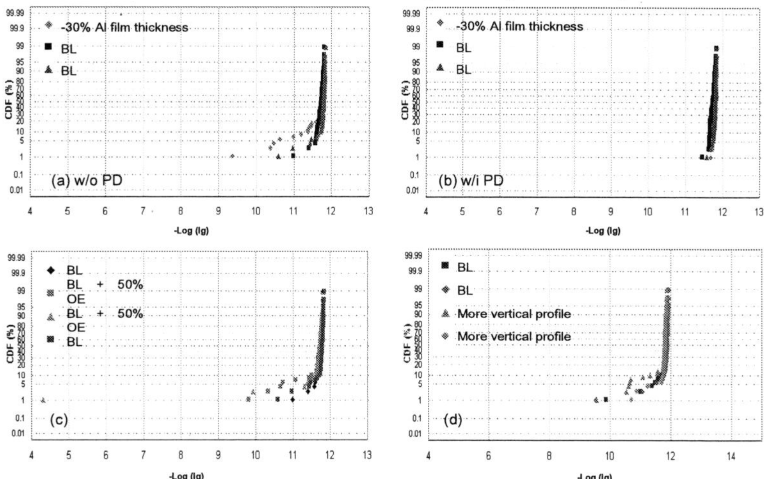

Fig 8. The role of Al film thickness (a/b), OE percentage (c) and Al-line sidewall (d)
in Al-pad etch on N-MOS PID

The role of Al film thickness, OE percentage and Al-line sidewall on PID is examined in Fig 8. As is well known, Al film thickness becomes thicker for better heat transfer as CMOS is being scalded down. Usually, the PID is thought to be correlated with the longer exposure time of aluminum during the etching. However, Fig 8 (a) shows the thinner Al-film results in worse PID. In "-30%" film thickness split, the ME time is greatly reduced while the OE time is kept intact. This demonstrates the PID is not related to the time of Al exposure but depends on the OE time. Namely when the surface is conducting in ME, no charge is built up. Fig 8 (c) shows the reconciliatory clue for its 50% more OE split. The PD inserted at TM2 delivers the PID free performance as shown in Fig 8 (b). It manifests that the PID comes from the PAD loop and the role of valid PD is not negligible. Unlike M1-etch and PAS1-etch, N-MOS show higher PID sensitivity to all Al-pad etch related splits. The possible mechanism is demonstrated in Fig 6 (b). RESE, coupled with ultra thin charging collector and photo-conductive effect could be used to explain the above phenomena. Firstly, the much larger spacing at Al-pad etch provides the platform for RESE, which results in poor PID on N-MOS. At the end of endpoint in Al-pad etch, the unmasked ultra-thin aluminum film becomes the charge collector. As the sidewall angle is not a sensitive factor as shown in Fig 8 (d), we can conclude the effect from the area of exposed vertical sidewall can be ignored. The slight worse PID from vertical split may be attributed to the larger area of ultra-thin charging collector. Hence, the sensitivity of OE should come from the photo-conductive effect on PAS1 film.

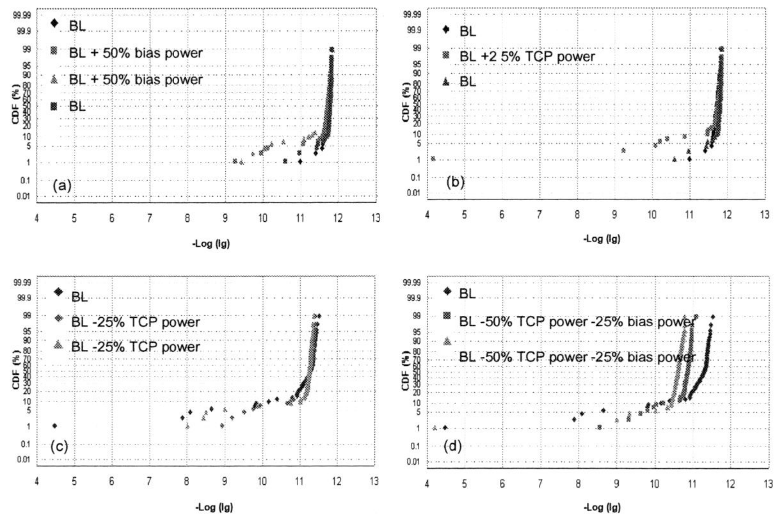

Fig 9. The role of TCP power and bias power in Al-pad Etch on N-MOS PID

We summarized the effect of TCP power and bias power on N-MOS PID in Fig 9. Clearly, both TCP power and bias power could deteriorate the PID when the higher one is used. However, the -25% lower power in Fig 9 (c) does not show any improvement. Even worse, after we simultaneously reduce the TCP power by 50% and the bias power by 25%, a significant Ig shift could be detected in Fig 9 (d). Therefore, we could conclude PID does not depend on the absolute value of either TCP power or bias power. The synergetic plasma uniformity is worthy of more attention for PID deduction.

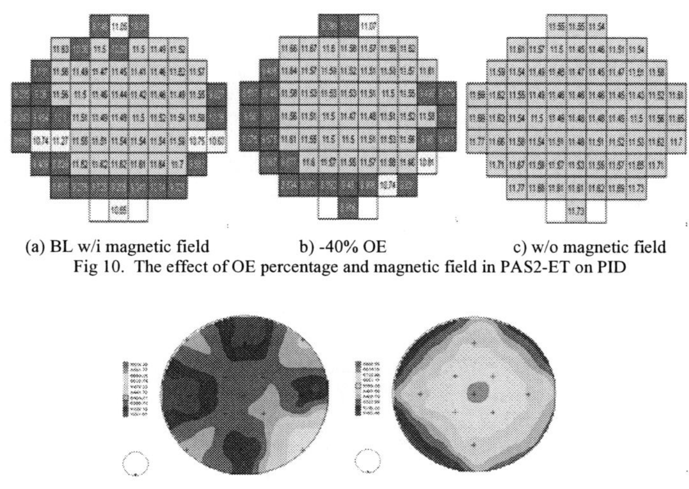

(a) BL w/i magnetic field   b) -40% OE   c) w/o magnetic field

Fig 10. The effect of OE percentage and magnetic field in PAS2-ET on PID

(a) Range: 300A; Uniformity: 2.3% (b) Range: 654A; Uniformity: 5.6%

Fig 11. ME etch rate (a) w/i magnetic field and (b) w/o magnetic field

As shown in Fig 10, the split with 40% OE reduction in PAS2 etch does not show any improvement while the removal of magnetic field (MF) achieves the PID-free process. This is consistent with Pan [4] who ever reported that PID is strongly related to the magnetic field in PAS2-etch. The insensitivity of OE time is similar to that in PAS1-etch, namely, the effect of copper exposure time can be neglected. Fig 11 demonstrates the etch rate map before and after the removal of the MF. Obviously, the introduction of MF improves the etch rate uniformity. However, it exacerbates the PID at wafer edge. This can be explained by that the MF increases the plasma density at wafer edge while the higher etch rate at no-magnetic field condition comes from the high ion energy at wafer center. This means the plasma uniformity other than etch rate uniformity requires special care in allaying the PID. Last but not the least, PAS2-etch related PID only occurs on P-MOS. This also demonstrates the RESE does not work unless it is coupled with either ultra thin "charging collector" or worse plasma uniformity.

CONCLUSIONS

We demonstrate that M1 etch, passivation etch and Al-pad etch play the critical role in PID performance on two given test vehicles. The underneath mechanisms are categorized into two groups from the point of view of the damaged MOS type. Al-pad etch could degrade N-MOS reliability for the joint actions of RESE, ultra-thin charging collector and photo-conductive effect. P-MOS degradation from M1 etch differentiates itself from PAS1-etch and PAS2-etch for its over-sensitive substrate. Either the advanced features such as different frequency, MF or the post-treatment with highly positive ions should be carefully evaluated before being introduced into process.

ACKNOWLEDGMENTS

Authors would like to thank Dr. Huiyuan Pei of Lam Research China, Dr. Zhenghao Gan of SMIC/T-R&D reliability group, Mr. Ying Huang of Applied Materials China, Mr. Xiaodong Zou of SMIC/T-R&D integration group for the technical discussion, Mr. Jing Pan, Mr. Xinghua Song of SMIC/T-R&D integration group for sharing PID data.

REFERENCES

1. M. Kamei, et al, "Bias frequency dependence of pn junction charging damage induced by plasma processing", Thin Solid Films 518, 2010.
2. A. Martin, "Review on the reliability characterization of plasma induced damage", J. Vac. Sci. Technol. B 27 (1), Jan/Feb, 2009.
3. N. Matsunaga, et al, "Spreading antenna effect of PID in damascene interconnect process", IEEE Transaction on electron device, Vol, 54. No. 6, June, 2007.
4. J. Pan, et al, "Plasma-process induced damage on 65nm logic VLSI manufacturing", ICSICT, 2008.
5. M. Ooka, et al, "Effect of H_2 addition during Cu thin film sputtering", International symposium on VLSI Technology, system, and applications, 2006.

CHAPTER 4

THIN FILM TECHNOLOGY

ECS Transactions, 34 (1) 455-465 (2011)
10.1149/1.3567620 ©The Electrochemical Society

Selective Epitaxial Growth: Trends in a Modern Transistor Device Fabrication

A. Hikavyy, W.Vanherle, J. Dekoster, L.Witters, T.Hoffman and R. Loo

Imec, Kapeldreef 75, B-3001 Leuven, Belgium

Although selective epitaxial growth (SEG) of Si and SiGe is a recognized technology used by the semiconductor industry, continuing device scaling and appearance of new architectures impose new, more stringent and challenging requirements to it. SiGe growth with high Ge concentrations, layers integration in 3D structures, growth on alternatively oriented Si wafers and decrease of thermal budget are only few trends which we would like to address in this article. Problematic points raising concerns from the epi point of view are highlighted and solutions used at Imec are described in detail together with once proposed in literature.

Introduction

Selective epitaxial growth (SEG) of Si and SiGe is a rather mature technology at present. Nevertheless, although SiGe SEG is extensively used in HBT manufacturing, its use in a modern high performance CMOS transistors production flow is usually limited to SiGe source/drain strain booster introduced at 90nm technology node [1] and Si raised source/drains in DRAM used to decrease parasitic resistance. To the best of our knowledge Ge concentrations between 15-40% are used for embedded source/drains, which give considerable positive effect on p-MOS device performance without a strong deviation from the conventional Si flow. It should also be mentioned that SiGe buried channels with relatively low Ge content of ~25% are currently considered in the gate first technology in order to correct unfavorable V_t shift caused by high-k dielectric.

Although SEG involvement in the current CMOS device production is limited, this situation changes as devices are scaled and different new architectures are proposed. Looking at those advances a few general trends can be highlighted:
- Integration of high Ge content SiGe in different device schemes like 32 nm replacement metal gate pMOS [2], SiGe buried channel pMOS [3], implant free quantum well pMOS [4] in order to either further improve device boost or to benefit from its higher, compared to Si, mobility.
- Integration of Si or SiGe SEG in FinFet [5], FD SOI, planar Tunnel FET structures in order to decrease access resistance and, if possible, to boost performance by the additional strain.
- Growth of SiGe on alternatively oriented (110) Si wafers in order to benefit from higher mobility observed for this plane (see [6] and references therein). It also becomes crucial since we see a strong move towards 3D structures in which growth on (110) planes is important.
- All presented above SEG applications, in one or another way, require decrease of epi thermal budget. This goes in two directions. In the first place it concerns pre-epi bake, usually done at 800°C [7]. This temperature is not compatible with

modern devices integration due both: Si reflow (in the case of 3D structures) and rather pronounced impurities diffusion. On the other hand deposition temperature of both Si and SiGe tends to become lower either due to higher Ge content [8] or due to more stringent requirements (relaxation of underlaying SiGe layers, dopants diffusion, etc).

Although a few very nice reviews on SiGe technology and its possibilities have been published recently (see for example [9]) the actual problems and challenges of the epitaxial growth are usually omitted. In this paper we would like to discuss the trends presented above together with typical challenges for epi and current advances made at Imec or elsewhere in the each mentioned direction.

Integration of high Ge content SiGe

Incorporation of SiGe source/drain (S/D) strain booster at 90 nm technology node was a turning point for embracing SiGe technology by the manufacturers dealing with the high performance logic devices. Following 65, 45 and 32 nodes repeatedly incorporate SiGe gradually increasing Ge concentration from 15 up to 40% [9]. Although there are some concerns about defects appearance in the channel area during high temperature anneals under large stresses [10], a careful processes optimization in the gate first and the use of gate-last approaches seem to be effective ways to avoid them. Taking into account that metastable SiGe can be grown way above its critical thickness even higher Ge concentrations can be probably incorporated as p-MOS is further scaled down.

On the other hand a number of alternative pMOS architectures are being at present intensively investigated. Both SiGe buried channel and implant-free quantum well devices, which also can be implemented in 3D configuration, show a rather strong potential [3, 11, 12]. These devices not only benefit from the high Ge content SiGe channel, which concentration can be increased up to 70%, but also potentially can be additionally boosted by embedded/raised SiGe source/drains.

High Ge content SiGe can be relatively easy grown in a conventional RPCVD or UHVCVD epi tools [13]. Conventional precursors as SiH_4, SiH_2Cl_2, Si_2H_6 in combination with GeH_4 can be used for this purpose. In order to avoid elastic relaxation (surface roughening) and assure high quality of the final layers, temperature has to be carefully chosen. Generally it has to be lowered down as concentration increases [8,14].

In Fig.1 a typical dependence of the final Ge concentration in the layers grown with DCS and GeH_4 at different temperatures is presented. Although generally similar dependences can be found for other sets of precursors as well (SiH_4/GeH_4, Si_2H_6/GeH_4, etc.), differences in the Ge concentration at which layers start to relax at given temperature are seen [8]. For example, strained SiGe layers with Ge concentration up to 60% can be grown at 500°C with DCS/GeH_4 gases whereas in the case of SiH_4/GeH_4 combination, layers with Ge concentration higher than 45% show a strong relaxation via islanding at this temperature.

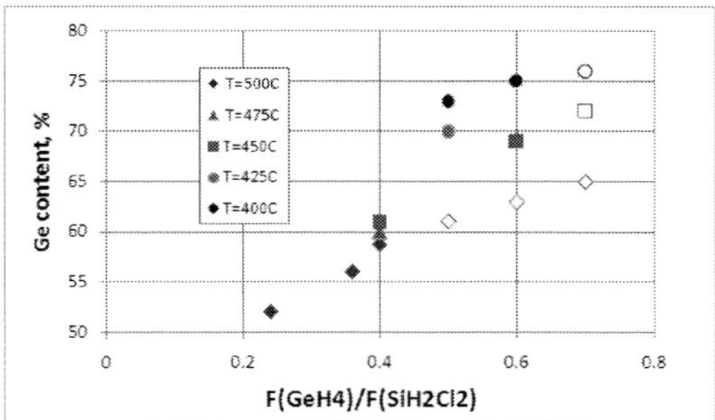

Fig.1. Ge concentration as function of precursor ratio for layers grown at different temperatures. Closed symbols show strained layers whereas open symbols represent layers which show surface roughening when the same temperature is used.

Use of low temperatures, typically below 500°C, leads to rather low growth rates which are in the range of 0.5-1 nm/min for silane and DCS based processes. This growth rates are still acceptable since the required layers thickness usually does not exceed 10 nm. It should be noted that growth rate for silane and DCS based SiGe processes is not constant in this thickness range. Although, incubation time and a change of the growth rate due to the surface change from Si to Ge reach one are expected within few monolayers, during the initial stages of SiGe growth, such behavior during the further growth is not well understood. High angle annular dark field scanning TEM and EDS showed that Ge content is uniform over the layers thickness, hence Ge concentration variation cannot account for such growth rate behavior. This leads to a conclusion that more subtle effects, for example surface temperature change, may play a significant role in this case.

Eventually a few words should be said here concerning layers quality monitoring. The most straightforward and direct way to get both layers thickness and Ge content in the case of fully strained layers on blanket wafers is HRXRD. When layers become too thin and thickness determination via HRXRD is not reliable anymore, XRR configuration becomes useful. Unfortunately both techniques cannot be applied in the case of the patterned wafers. Use of micro X-ray sources specially developed for analyzing patterned wafers is difficult due to limited layers thickness. This leaves spectroscopic ellipsometry the only valuable in-line metrology option for the moment. Careful calibration of ellipsometer models with XRD and TEM gives satisfactory results.

Integration of Si or SiGe SEG in advanced device architectures

It is expected that new device architectures like FinFets, FD SOI, Tunnel FET, etc. at some point might substitute standard planar devices taking over the most valuable practices developed for the planar once.

When it comes to epi, both raised Si (SiGe) and embedded SiGe S/D together with SiGe buried channel (SiGe implant free quantum well) are good candidates for further use. It has been shown previously that the use of SEG for raised (S/D) yield high

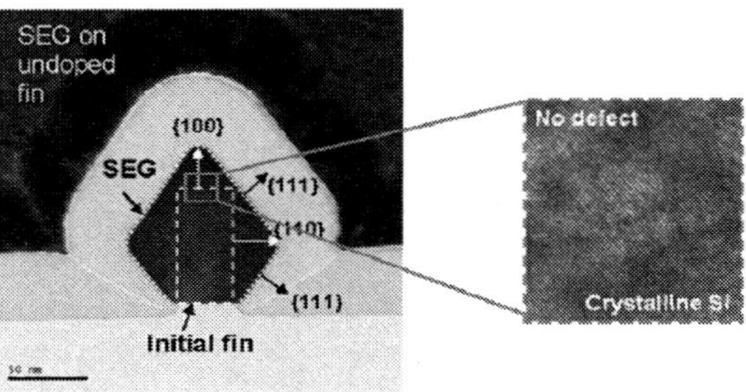

Fig. 2. TEM picture showing typical SEG of Si on a fin (SOI). Facets created during the growth are indicated.

performance FinFets by reducing the access resistance if Si is used [15] which further can be combined with additional boost due to strain if SiGe is applied [16].

After Si SEG, the fins initial rectangular shape is changed to a so-called diamond one which originates from the combination of different factors: fin orientation, the orientation dependent growth rate of the Si, fin corners and oxide presence which facilitate faceting (Fig. 2).

Usually raised S/D for FinFets are deposited after HDD implantation which adds an extra difficulty of varying Si growth rate (GR) on B and As (P) doped Si surface. The GR difference and eventually epi layer quality depend very much on the dopant implantation conditions. It was found that in the case of sub-20 nm wide FinFets, after an amorphizing implantation, surface proximity suppresses crystal regrowth and promotes formation of twin boundary defects and eventually can lead to regrowth of polycrystalline Si [17]. If implantation optimization is not done SEG in these regions will result in a rough polycrystalline layer.

Another difficulty appears from the fact that pre-epi bake treatment at temperatures ~800°C and higher is not desirable for FinFet structures. The reason for this is two-fold: on one hand typical bakes at reduced pressure and temperatures $\geq$800°C cause so-called Si reflow, which degrades 3D structures; and on the other hand dopant diffusion, especially B, becomes too high to accept for the short-channel devices.

In the case of Si, SEG without pre-epi bake is relatively straightforward. It is related to rather low Si sensitivity to the interface contamination and somewhat high growth temperature (typically ~ 750°C). Although it is not high enough to remove surface contamination completely, it still helps reducing it considerably (see Fig. 7).

Situation changes dramatically when SiGe is deposited. If the pre-epi bake is omitted, relaxed or very rough layers are obtained. It can be remedied by an increased HF dip duration (Fig. 3), incorporation of a buffer Si SEG layer, an addition of a mild 750°C bake or a combination of above mentioned options.

An incorporation of a long HF dip seems to be a rather simple way to fix the layers quality problems, though oxide loss should be taken in consideration. The positive effect can be rather simply explained by an additional dissolution of the remaining surface SiO_x suboxides after the main SiO_2 removal. Using this approach both SiGe raised S/D and SiGe buried channels were successfully deposited on the FinFet structures.

In the case of fully depleted SOI devices difficulties of SEG integration are somewhat similar to the FinFet case. The task usually is to grow raised Si (SiGe) S/D on a ultra thin Si (sub-10 nm) resting on buried box oxide. We found that for our standard Si SEG conditions (T=750°C, low pressure) high quality Si could be grown only when starting Si thickness is around 6 nm or higher. If thinner starting Si layer was used Si agglomeration occurs during the temperature ramp up. This obviously puts a strong restriction on the standard 800°C pre-epi bake application, which eventually was omitted by us without detrimental effect on Si epi layer quality confirmed by TEM and excellent device performance [18]. A rather important effect observed during Si growth is

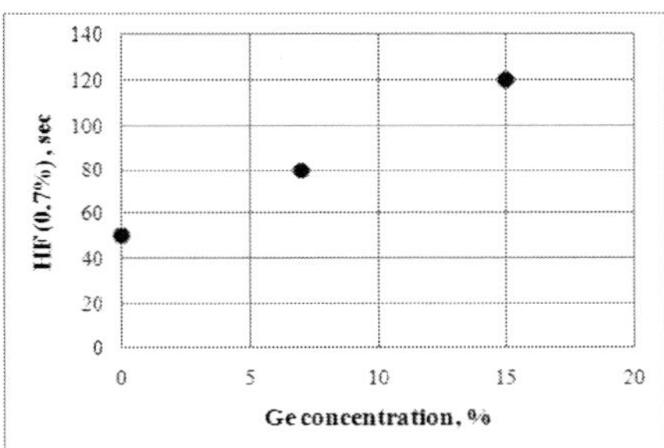

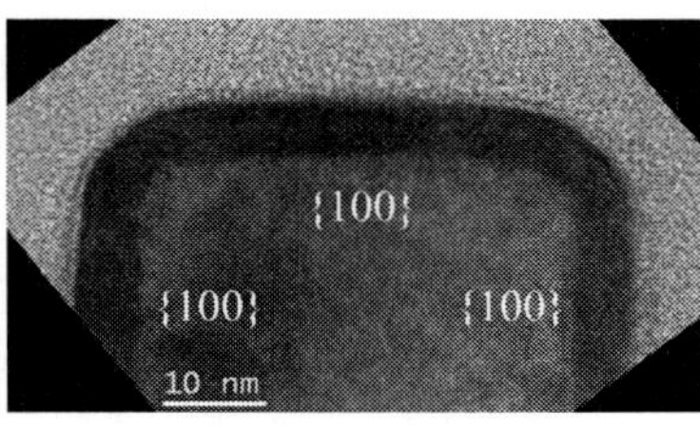

Fig. 3. The graph shows a dependence of the necessary HF dip in order to grow high quality SiGe layer. Layers were grown in S/D areas at 750°C without bake. The right picture shows a buried $Si_{55}Ge_{45}$ layer deposited on a fin with all sides having {001} plane. Deposition was done at 500°C without bake.

Si GR dependence on the starting Si thickness which is explained by the differences in starting Si surface temperature. It is expected that eventually GR will level up after deposition of a layer with certain thickness. Nevertheless, for typical target thickness of 30 nm the GR variations must be taken into account.

Growth of SiGe on alternatively oriented Si wafers

Although it has been known since the late 1960s that hole mobility's are higher in bulk Si (110) than in bulk Si (100) [19] the real interest of VLSI community to this effect was attracted only relatively recently by demonstration of substantial mobility increase in both Si and SiGe based devices [20, 21]. Additional interest to Si and SiGe growth on (110) Si comes from the fact that 3D structures as FinFets possess (110) side walls (Fig. 2) and basic knowledge of epi growth there would be beneficial.

It is rather surprising that in spite of the increased amount of articles dealing with devices built on (110) bulk Si involving growth of both Si and SiGe, studies on basic growth and morphology of thin and strained layers appeared recently are practically limited to few works [22-24]. On the other hand already those show that growth of both Si and especially SiGe on (110) surface is not straightforward. It is related to the slower

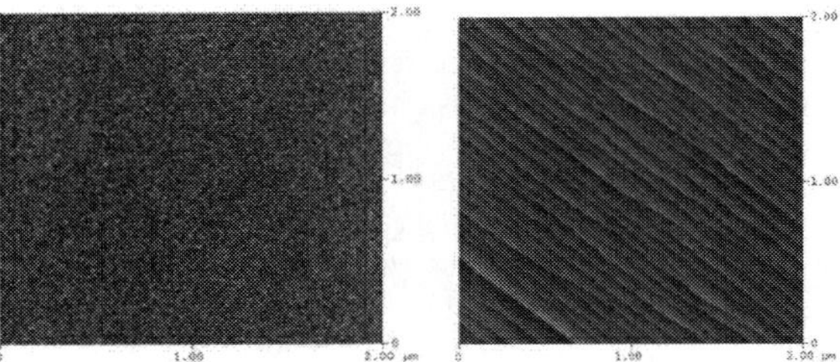

Fig. 4. AFM pictures of two (110) Si samples (same wafer); left – untreated, right – after 1050C, 30 sec, H_2 bake.

hydrogen desorption which causes the decrease of growth rate and substantially lower compared to (100) Si critical SiGe thickness. An additional difficulty comes from the fact that, both low (800°C) and high (1050°C) temperature pre-epi H_2 bakes lead to formation of step arrays which can be explained by a slight off-cut of the initial surface with consequent Si atoms diffusion to more energetically favorable atomic step sites at these temperatures [24] (Fig. 4). Mentioned step arrays will further show up after Si or SiGe epitaxial growth making judgment about the epi layers quality rather difficult.

In our work we used both SiH_4 and DCS as Si precursor for growing SiGe with high Ge content (above 40%). Our growth results do follow trends observed for low Ge contents in [22] for silane and in [25] for DCS. Namely, we see that in the case of silane based processes Ge concentration Si is ~ 20% higher, whereas for DCS based processes it is ~30% lower on (110) Si compared to (100) Si. Morphology of the layers with the same concentration and thickness deposited at low temperature (~500°C) for both types of process is roughly similar with RMS values in the range of 0.3-0.5 nm.

Additional comparison of silane and DCS based processes at 600°C showed that processes with silane are more flexible and react much stronger on the change of such main parameters as GeH_4 flow and temperature (Fig. 5). The difference in the processes kinetics behavior, as in the case of (100) Si, is most probably attributed to Cl presence in the DCS based processes. It is well known that Cl desorption plays a more important role than H desorption, when Cl–containing species are employed, because the desorption energies of HCl and $SiCl_2$ are greater than of H_2. The activation barrier for H and Cl transfer from Si to Ge on (100) Si, for example, are 1.0±0.1 eV/mol and 1.2±0.2 eV/mol respectively (see [26] and references therein) and on (110) Si they are expected to be even higher.

Integration of SiGe processes on (110) Si patterned wafers may have extra complications connected with HCl etch and consequent SiGe growth, which is a typical route for buried SiGe channel devices. Our preliminary experiments with HCl etch at low pressure of 40 Torr and 800°C showed that etch rate compared to (100) Si is 3 times higher and rather strong faceting at the edges is observed (Fig. 6a). It must be pointed out that faceting was very different from the one observed in [25], which probably is related to different process conditions. We have evidence that going to higher temperatures

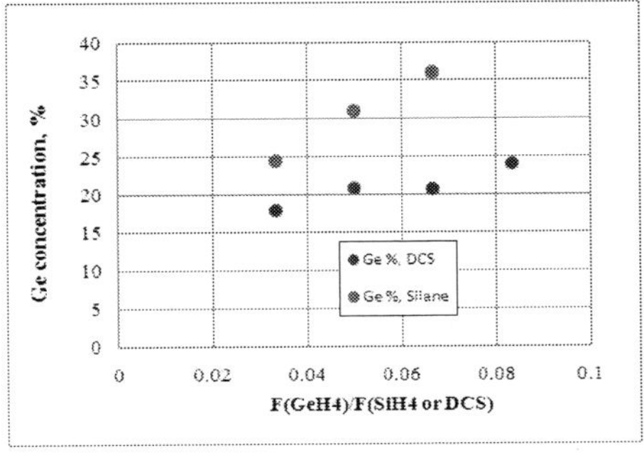
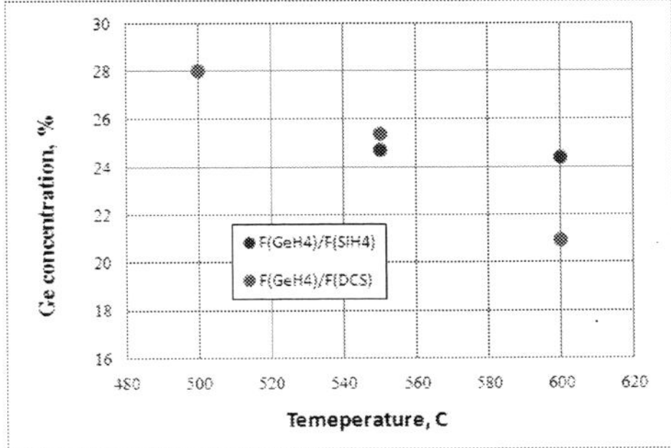

Fig. 5.   Ge concentration as function of precursor flow and temperature (at fixed precursor ratio).

(mass-flow etch regime) should solve this problem (Fig. 6b). The consequent growth of $Si_{50}Ge_{50}$ layer with either silane or DCS based process shows an additional unwanted effect, namely thickening at the edges with STI (Fig. 6c). Presence of HCl etch prior the SiGe growth does not influence the thickening suggesting that it is an intrinsic effect at the Si/STI edge. This thickening depends on the deposition time and can be 4 times thicker than the layer in the centre of the active window. A close examination at the STI/Si edge shows that this thickening starts at early stages of deposition and has a rather regular shape, which probably can be connected with elastic relaxation. Being energetically favorable place for incoming ad-atoms during further growth, SiGe growth rate in this place will increase. Influence of loading effects which also can lead to the SiGe layer thickening is for the moment not clear. Also, rather regular shape of the thickening shows appearance of {100} and {320} planes which also might lead to increased SiGe growth rate at this location.

Overall growth of SiGe on (110) Si is very different and more difficult than on (100) Si and further investigation of basic growth and optimization on patterned wafers is needed for successful application of such layers by industry.

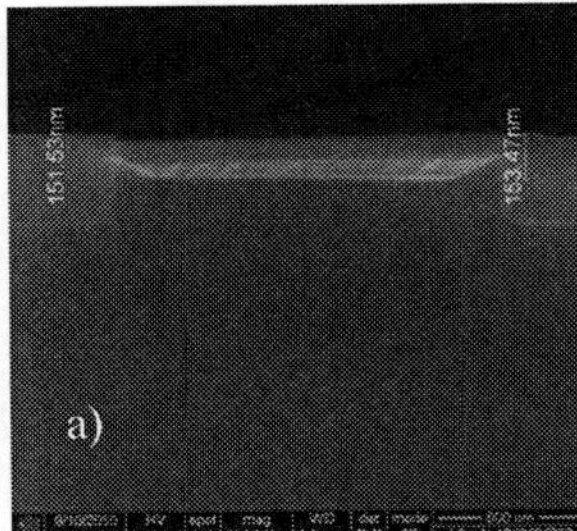
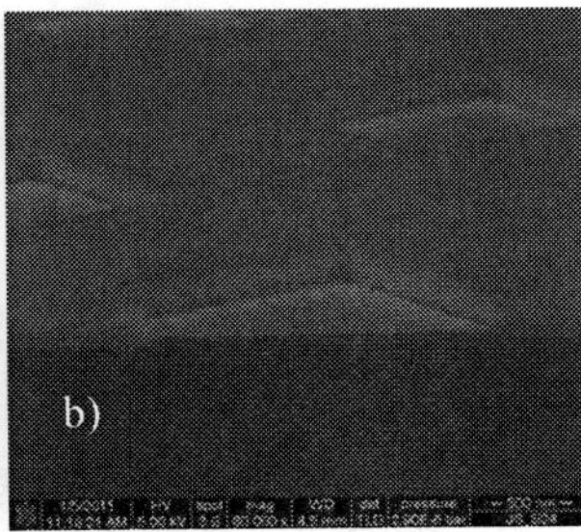
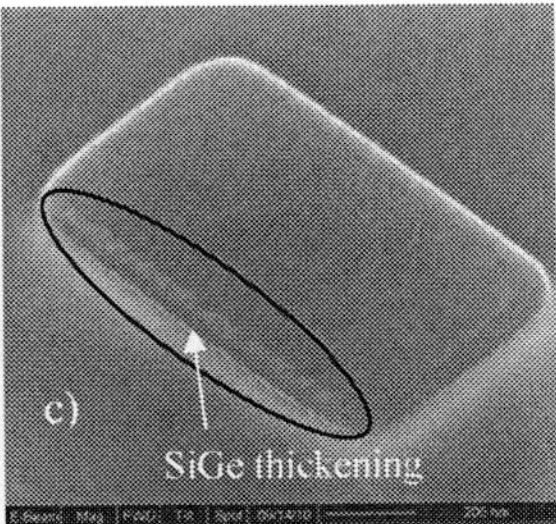

Fig. 6.   Typical processing done on (110) wafers. a) HCl etch at 800°C, b) HCl etch at 900°C, c) $Si_{50}Ge_{50}$ layer grown on a patterned wafer.

Decrease of epi thermal budget

As presented above, practically all modern SEG applications, in one or another way, require decrease of epi thermal budget which consists of both pre-epi bake and processing itself.

In order to grow high quality epi layers the starting Si surface must be free of O and C. In practice it means that native oxide formed on Si surface is removed by means of ex situ wet-chemical HF dip [27] optimized for minimal oxide loss, followed by an H_2 bake inside the epi-reactor at the temperature sufficiently high to facilitate volatilization of the remaining silicon oxide:

$$SiO_{2(s)}+Si_{(s)} \longleftrightarrow 2SiO_{(g)}$$

The lowest temperature fulfilling the above condition for modern AP/RP CVD tools is at present around 800°C (Fig. 7). This temperature is closely linked to the moisture level in the epi tool which is in the range of 10-100 ppb and to the bake pressure which is limited by 10 Torr in the state of the art epi reactors.

Use of UHV/CVD tools allows to bring O_2/H_2O partial pressures to the mid-10^{-10} Torr allowing contamination free (below SIMS sensitivity) Si surface after bakes ~ 700°C [13] showing that by using conventional H_2 bakes it is very difficult to go beyond this temperature.

As we mentioned before, at few instances H_2 bake was either skipped or used at 750°C without dramatic deterioration of layers quality. It is explained either by the layer material contamination sensitivity, for example Si is less sensitive than SiGe, or by the fact that long HF dips and mild bakes at 750°C allow to bring overall contamination

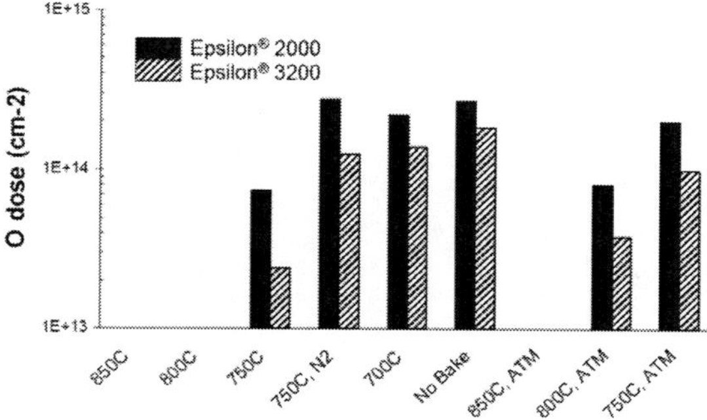

Fig. 7. O contamination at the substrate/SiGe epi interface for various pre-epi bake conditions performed in ASM 200 and 300 mm epi reactors. The bake was always done for 2 min. at either 40 Torr or atmospheric pressure. With one exception H_2 was used as carrier gas during the bake [28].

level much lower than no bake case (see Fig. 7). Such result although being acceptable in some cases is not ideal in general and alternative cleaning procedures would be welcome.

This situation is well recognized by both epi community as well as by the tool suppliers. An integration of an in-situ pre-epi clean which would substitute pre-epi bake or considerably decrease its temperature has been in consideration for more than a decade though no real commercial tool implementation was until now available. At present, two different approaches seem to be the most perspective. In the first place use of HF-vapor etch of SiO_2 was proposed [29]. Further combination of HF-vapor with acetic acid vapor used as catalyst made the process controllable and allowed to eliminate the H_2 bake resulting in a contamination free interface between Si wafer and epi-Si [30]. The second approach is based on removal of native oxide through a low-temperature two step dry chemical clean process (Siconi^{TM} clean) developed by Applied Materials. In the first step of this process, a remote plasma source generates the NH_4F etchant species from NF_3 and NH_3 which are condensed into a solid by-product on the cooled wafer surface through a reaction with native oxide. In the second step an in-situ anneal is performed to decompose the by-product using convention and radiation heating. Latest reports from Applied Materials show that this type of clean allows to bring bake temperature to as low as 130°C [31] allowing growth of relaxation-free, high Ge content (40%) epitaxial SiGe films beyond the critical thickness. [32].

If the first, HF-vapor based approach, has been pursued by the ASM and still has to be implemented and introduced to the market, the second proposed by Applied Materials is currently available. The drawback of both proposed cleaning methods is a cluster implementation, potentially leaving some of the existing epi RP/AP CVD machines not compatible with new additional cleaning chambers.

Although the pre-epi bake for the moment accounts for the largest part of epi thermal budget, the decrease of epi processing temperature can also be seen as an obvious trend. The reasons for this spun from the more stringent requirements due to complex device architecture (underlaying layers relaxation, dopants diffusion, etc.) to economical (more processed wafers per hour). If, as it was shown previously, high Ge concentration SEG processes tend to occupy temperature range around 500°C and lower, situation with Si epi is somewhat different. Conventional selective Si epi processes which use HCl as an etchant precursor typically apply temperatures in the range of 700°C and higher. Partially this is due to necessity to keep a decent growth rate, which depends on temperature, and partially to ineffectiveness of HCl etchant below 650°C. Selective low temperature Si epitaxial processes call for new precursors effective at low temperatures (~500°C) combined with new etchants. A good example of such processes can be a cyclic process proposed in [33] for deposition of SiCP applying trisilane and chlorine or a mix of dichlorodisilane (DCDS) and chlorine as the etchant [34] allowing selective growth of high quality Si layers with adequate growth rate of Si at temperatures in the range of 500-600°C. It should be pointed out that this process can be used for applications ranging from Si Cap, raised S/D for FinFets and FD-SOI devices to embedded SiCP to provide additional boost for nMOS transistors.

Conclusions

During the last few years interest to epi layers integration in advanced transistor structures has increased substantially. SEG becomes one of the most important manufacturing steps enabling a considerable device improvement. In this article we described few the most obvious trends observed over the past years. It should be mentioned that the list of the trends presented here is not exhausted. Incorporation of Ge and III-V based devices on Si platform, investigation of new materials like GeSn, synthesis of new precursors being just few examples.

References

1. T. Ghani, M. Armstrong, C. Auth, M. Bost, P. Charvat, G. Glass, T. Hoffmann, K. Johnson, C. Kenyon, J. Klaus, B. McIntyre, K. Mistry, A. Murthy, J. Sandford, M. Silberstein, S. Sivakumar, P. Smith, K. Zawadzki, S. Thompson and M. Bohr, *IEDM Tech. Dig.*, 978 (2003).
2. P. VanDerVoorn, M. Agostinelli, S. Choi, G. Curello, H. Deshpande, M.A. El-Tanani, W. Hafez, U. Jalan, U. Janbay, M. Kang, K. Koh, K. Komeyli, H. Lakdawala, J. Lin, N. Lindert, S. Mudanai, J. Park, K. Phoa, A. Rahman, J. Rizk, L. Rockford, G. Sacks, K. Soumyanath, H. Tashiro, S. Taylor, C. Tsai, H. Xu, J. Xu, L. Yang, I. Young, J. Yeh, J. Yip, P. Bai, C. Jan, , *VLSI*, p.137 (2010)
3. J. Mitard, L. Witters, M. Garcia Bardon, P. Christie, J. Franco, A. Mercha, P. Magnone, M.Alioto, F. Crupi, L.-Å. Ragnarsson, A. Hikavyy, B. Vincent, T. Chiarella, R. Loo, J. Tseng, S. Yamaguchi, S. Takeoka, W-E. WangP. Absil and T. Hoffman, *IEDM Tech. Dig.*, p. 10.6.1, (2010).
4. G. Hellings, G. Eneman, M. Meuris, *US 2009/0283756*
5. N.Collaert, et al., Abstract book, ICSI-5 Marseille, Abstract book, pp.129. (2007).
6. J. W. Pan, P. W. Liu, T. Y. Chang, W. T. Chiang, C. H. Tsai, Y. H. Lin, C. T. Tsai, G. H. Ma, S. C. Chien and S. W. Sun, *IEDM Tech. Dig.*, p.1 (2006)
7. R. Loo, A. Hikavyy, F. Leys, M. Wada, B. De Vos, A. Pacco, M. Bargallo Gonzalez, E. Simoen, P. Verheyen and M. Caymax, *Proc. of UCPSS, Diffusion and Defect Data Part B (Solid State Phenomena)*, 145-146, 177 (2009).
8. A. Hikavyy, L. Witters, S. Brus, T. Hoffmann, and R. Loo, *5th International SIGe Technology and Device Meeting*, Abstract book, 1916152, (2010).
9. K. J. Kuhn, A. Murthy, R. Kotlyar, and M. Kuhn, *ECS Transactions*, 33 (6), p.3, (2010)
10. V.Moroz and M.Choi, *ECS Transactions*, 33 (6), p.21, (2010)
11. P. Majhi, P. Kalra, R. Harris, K. J. Choi, D. Heh, J. Oh, D. Kelly, R. Choi, B. J. Cho, S. Banerjee, W. Tsai, H. Tseng, and R. Jammy, *IEEE ELECTRON DEVICE LETTERS*, 29, .1, p.99, (2008)
12. G. Hellings, L. Witters, R. Krom, J. Mitard, A. Hikavyy, R. Loo, A. Schulze, G. Eneman, C. Kerner, J. Franco, T. Chiarella, S. Takeoka, J. Tseng, W.E. Wang, W. Vandervorst, P. Absil, S. Biesemans, M. Heyns, K. De Meyer, M. Meuris, T. Hoffmann. *IEDM Tech. Dig.*, p. 10.4.1, (2010).
13. T. N. Adam, S. Bedell, A. Reznicek, D. K. Sadana, R. J. Murphy, A. Venkateshan, T. Tsunoda, T. Seino, J. Nakatsuru, S.R. Shinde, *ECS Transactions*, 33 (6), p.149, (2010)
14. A.Hikavyy, R.Loo, L.Witters S. Takeoka, J. Geypen, B.Brijs, C. Merckling, M.Caymax and J. Dekoster, *ECS Transactions*, 25 (7), p.201, (2009).

15. J. Kedzierski, M. Ieong, E. Nowak, T. S. Kanarsky, Y. Zhang, R. Roy, D. Boyd, D. Fried, H.-S. Wong, *IEEE Trans. Elctron Devices*, 50, 952, (2003).
16. P. Verheyen, N. Collaert, R. Rooyackers, R. Loo, D. Shamiryan, A. De Keersgieter, G. Eneman, F. Leys, A Dixit., M.Goodwin, Y.S.Yim, M. Caymax, K. De Meyer, P. Absil, M. Jurczak, S. Biesemans, *VLSI Technology Digest of Technical Papers*, p.194, (2005).
17. R. Duffy, M. J. H. Van Dal, and B. J. Pawlak, M. Kaiser and R. G. R. Weemaes B. Degroote, E. Kunnen, and E. Altamirano *APPLIED PHYSICS LETTERS* 90, p.241912, (2007).
18. Z. Lu, N. Collaert, M. Aoulaiche, B. De Wachter, A. De Keersgieter, J. G. Fossum, L. Altimime and M. Jurczak, *IEDM Tech. Dig.*, p. 407, (2010).
19. T. Sato, y. Takeishi, and H.Hara, Japan J. Appl. Phys. 8, p. 588, (1969).
20. M. Yang, E. P. Gusev, M. Ieong, O. Gluschenkov, D. C. Boyd, K. K. Chan, P. M. Kozlowski, C. P. D'Emic, R. M. Sicina, P. C. Jamison, and A. I. Chou *IEEE ELECTRON DEVICE LETTERS*, 24, 5, p.339, (2003).
21. J. W. Pan, P. W. Liu, T. Y. Chang, W. T. Chiang, C. H. Tsai, Y. H. Lin, C. T. Tsai, G. H. Ma, S. C. Chien and S. W. Sun, *IEDM Tech. Dig.*, p. 1, (2006).
22. M. L. Lee, D. A. Antoniadis and E. A. Fitzgerald, *Thin Solid Films*, 508, p.136, (2006).
23. J.M. Hartmann, M. Burdin, G. Rollandl, *Journal of Crystal Growth*, V 294, 2, pp. 288, 2006
24. V Destefanis, J M Hartmann, J L Huguenin, V Delaye, M P Samson, P Boulitreau, Y Morand, P Brianceau, C Arvet, P Gautier, S Monfral and T Skotnicki, *Semicond. Sci. Technol.*, 25, 045014, (2010).
25. V Destefanis, PhD thesis, (2009).
26. P. Tomasini , M.Bauer, N.Cody and C.Arena, *Journal of Applied Physics 99,074904 (2006)*.
27. R. Loo, C. Walczyk, P. Verheyen, R. Rooyackers, F.E. Leys, G. Eneman, D. Shamiryan, P.P. Absil, T. Delande, A. Moussa, H. Bender, C. Drijbooms, L. Geenen, M. Caymax, J.W. Weijtmans, R. Wise, V. Machkaoutsan, P. Tomasini, C. Arena, J. McCormack, S. Passefort, H. Sorada, A. Inoue, B.C. Lee, S. Hyun, S. Jakschik and S. Godny, *ECS Trans.*, 3 (7), 453, (2006).
28. F. Leys et al., *ISCI-5*, book of abstracts p. 427 (2007).
29. A. B. Storm, H. Sprey, J. W. Maes, E. Granneman, R. de Blank, E. Rohr, M. Caymax, and M. Heyns, *Solid State Phenom.* 65–66, p.225, (1998).
30. I. J. Raaijmakers, H. Sprey, A. Storm, T. Bergman, J. Italiano, and D. Meyer, *J. Vac. Sci. Technol.* B 17.5., p.2311, (1999).
31. Applied Materials Extends Epi Leadership with Innovative Critical Pre-Clean Technology http://www.appliedmaterials.com/news/articles/
32. J. Park, M. Ishii, R. Balasubramanian, Y. Kim, and S. Kuppurao, *ECS Transactions,* 33 (6) p.523, (2010).
33. M. Bauer, D. Weeks, Y. Zhang, V. Machkaoutsan., *ECS Transactions*, 3 (7), p.187, (2006).
34. S. G. Thomas, M. Bauer, M. Stephens, J. Kouvetakis, *Solid State Technology*, p.12, (2009).

ECS Transactions, 34 (1) 467-472 (2011)
10.1149/1.3567621 ©The Electrochemical Society

Electron States at Interfaces of Semiconductors and Metals with Insulating Films

V. V. Afanas'ev, M. Houssa, and A. Stesmans

Laboratory of Semiconductor Physics, Department of Physics and Astronomy,
University of Leuven, Celestijnenlaan 200D, B-3001, Leuven, Belgium

Evolution of electron energy barriers at interfaces of A_{IV} (Si, Ge), $A_{IV}B_{IV}$ (Ge_xSi_{1-x}, SiC), and $A_{III}B_V$ (GaAs, $In_xGa_{1-x}As$, InP, GaSb) semiconductors with wide-gap oxide insulators is overviewed. Over a wide spectrum of studied materials no significant interface dipole contribution to the barrier height is found. Instead, it appears that the energy position of the top of the O2p-derived oxide valence band remains nearly constant, which can be used to evaluate the interface band alignment between oxides and semiconductors as well as between dissimilar oxides materials. By contrast, barriers for electrons at metal/oxide interfaces appear to be strongly affected by the interface processing indicating that near-interface oxide charges in combination with a resulting polarized metal surface may provide significant dipole contribution to the interface barrier height.

1. Introduction

Interfaces of insulating oxides with metals and semiconductors are encountered in a wide spectrum of electron devices including logic and memory integrated circuits. Because the conventional SiO_2-based insulator fails to meet scaling requirements in terms of specific capacitance and gate leakage current, transition to high-permittivity metal oxide insulators becomes mandatory for the future generations of the devices. Usually, the dielectric permittivity and the bandgap width are considered to be the most important parameters of a candidate insulator (1-4). However, the electron and hole injection rates at the interfaces are determined by the corresponding energy barriers rather than by the gap width. Therefore, the interface band offsets represent the parameters that should be considered to evaluate the insulator performance potential. Yet, most of the barrier height (band offset) values are extracted from theoretical calculations (1,3) because reproducible fabrication of metal/oxide and semiconductor/oxide interfaces poses a formidable technological challenge. The goal of this overview is to present results of the internal photoemission (IPE) and photoconductivity (PC) experiments [5-7] aimed at experimental determination of energy barriers for electrons and holes at interfaces of semiconductors and metals with a broad variety of oxide insulators. This enables us to reveal several common features in evolution of the interface barrier heights. Among others, a marginal influence of interface dipoles on the semiconductor/oxide band alignment is observed, which contrasts with considerable dipole barrier components revealed at the metal/insulator interfaces.

2. Experimental

Reliable quantification of barrier heights at semiconductor/insulator or metal/insulator interfaces represents a significant experimental problem by itself because the carrier emission rate in thermal equilibrium becomes negligible for the barriers in excess of 2 eV under typical experimental conditions. The widespread photoelectron spectroscopy approach based on referencing the valence band (VB) edge or the metal Fermi level to the core level in the contacting materials is long known to suffer from the insulator charging effects [8,9], leading to systematic errors in the measured band offset values. These problems can be solved by using a combination of PC and IPE measurements on metal-oxide-semiconductor (MOS) or metal-insulator-metal (MIM) capacitors [5-7]. The basic idea of the technique is clarified in Fig. 1 schematically showing electron transitions excited under illumination of the interface with a flux of monochromatic photons of energy hv. When the latter exceeds the energy barrier for electrons (Φ_e) or holes (Φ_h), the corresponding charge carriers are injected into the insulator. With a bias of appropriate polarity applied to the opposite electrode, drift of these carriers generates a photocurrent in an external circuit that can be measured as a function of hv. The spectral thresholds of IPE can be found from the analysis of quantum yield (defined as photocurrent normalized to the incident photon flux) as a function of photon energy hv. In the case of a metal photoemitter the barrier height is measured directly; In the case of a semiconductor/oxide interface, the conduction band (CB) and VB offsets are calculated by subtracting the semiconductor bandgap $E_g(sc)$ from the measured barriers Φ_e and Φ_h, respectively. In the same spirit, the oxide bandgap $E_g(ox)$ can be determined from the spectral threshold of the intrinsic PC. More details on the IPE/PC measurements and examples of yield spectra can be found elsewhere [5-7].

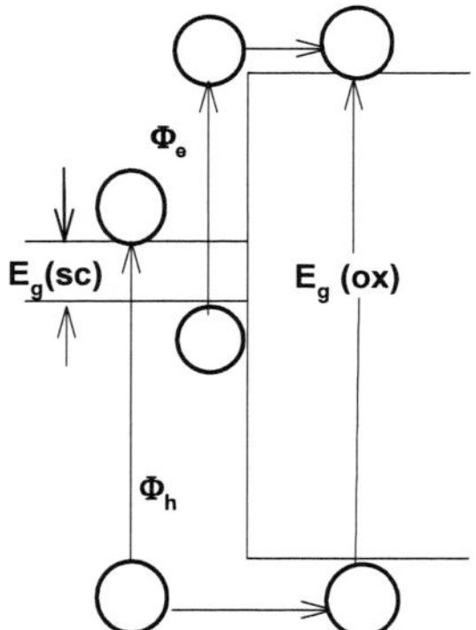

Figure 1. Schematic energy band diagram of the semiconductor/oxide interface indicating the electron transitions in the case of electron IPE (threshold Φ_e) and hole IPE (threshold Φ_h) from the semiconductor into the oxide, as well as the oxide PC [threshold $E_g(ox)$].

3. Properties of Interface Barriers

<u>Silicon/Oxide Interfaces</u>

Interfaces of silicon with insulating oxides are most widely studied and can be considered as a model case when addressing later interfaces of other semiconductors. Several features of the interface barrier behavior are reproducibly observed (6,7) and are briefly summarized below:

- Neither variation of oxide deposition method nor change of the Si surface plane orientation from the traditional (100) to (111) affect the interface barriers, suggesting a negligible (<0.1 eV) contribution of structure-sensitive interface dipoles. It thus appears that the band alignment is predominantly determined by the bulk density of states (DOS) of Si and that of the oxide;

- From the energy position of the oxide CB and VB edges relative to the bandgap edges in Si (Fig. 2) one may conclude that the measurable difference between barriers at the interfaces of amorphous (open symbols) and crystallized (filled symbols) oxides is encountered only for oxides of the cations with small r_i (Al, Mg, and Sc). Most of the crystallization-induced gap widening in these compact oxides occurs through the downshift of the top VB edge;

- In the oxides with $r_i > 0.07$ nm crystallization has no measurable effect on the Si/oxide band alignment. The results for Ti and Ta (cf. Fig. 2) also suggest that the oxide VB top remains at the same energy irrespective of the oxide gap width;

- Comparison of electron and hole IPE spectra indicates that the 'band tailing' observed in amorphous oxides is caused by the 'tails' in the oxide CB while the VB DOS is marginally sensitive to the crystallization (6).

These results can be understood if recalling that the top of the VB and the bottom of the CB in the insulating oxides are derived from the lone-pair 2p orbitals of bridging O

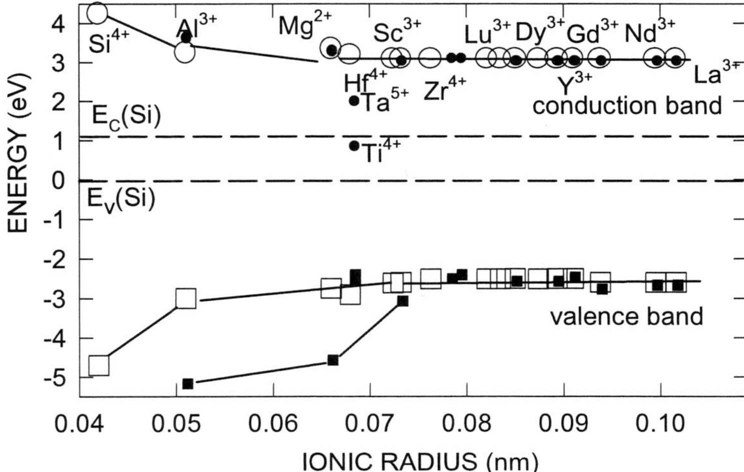

Figure 2. Energy position of the oxide bandgap edges as a function of cation ionic radius referenced to the bandgap edges of the Si crystal. Open and filled symbols correspond to amorphous and crystallized oxides layers, respectively.

atoms and s, p, or d states of the metal cations, respectively (10). It is likely that, with increasing cation radius, interaction between 2p-states of the neighboring oxygen anions come to a constant value resulting in the 'universal' oxide VB top energy at 2.5 eV below the Si VB top. Then it becomes possible to use this energy level to evaluate band alignment between Si and different oxides as well as between dissimilar oxides by using the known bandgap value of Si and the same energy of the oxide VB top. In particular, this result explains why the oxide gap width correlates well with the CB offset ant the interfaces with Si and which can thus be used as a guide for the insulator selection (2).

Germanium/Oxide Interfaces and Other Group IV Semiconductors

As compared to silicon, interfaces of germanium with high-permittivity oxides exhibit similar insensitivity of interface barriers to the type of sample processing and Ge surface orientation, suggesting negligible impact of interface dipoles (11). This allows one to use the interface barrier measurements to monitor the composition-dependent semiconductor DOS variation in Ge_xSi_{1-x} alloys (12) revealing that with increasing Ge concentration the semiconductor VB top is predominantly affected. This behavior may be used for fabrication of quantum well (QW) structures for hole channels. In a similar way, band alignment between different polytypes of SiC and high-k oxides can be evaluated from the SiC/SiO_2 interface energy barriers indicating a low VB offset as a common problem (13). The latter may possibly be solved by inserting an ultrathin SiO_2 interlayer (IL) between SiC and the metal oxide to block injection of holes.

Despite the seeming similarity to the case of silicon, interfaces of Ge also reveal an extra feature in the barrier physics: Oxidation of Ge during oxide deposition leads to formation of IL with a narrow bandgap ($E_g \approx 4.3$ eV) and low band offsets (14). This IL not only limits the thickness scalability of the insulation stack, but also enables easy charge injection into the insulating stack under high-field stress causing charge instability. As a possible way to minimize the impact of Ge oxidation, interface doping with La can be applied which results in a wider-gap La germanate IL (15). Another approach consists in passivation of Ge with a few monolayers of silicon. Remarkably, even the latter IL does not lead to any measurable interface dipole (16).

$A_{III}B_V$/Oxide Interfaces

Analysis of the band alignment between $A_{III}B_V$ semiconductors and oxide insulators reveals, again, absence of significant change in dipoles associated with changing semiconductor surface orientation, e.g., from (100) to (111)A or (111)B face in GaAs. Also, despite complex interface chemistry, variation in the semiconductor surface preparation does not lead to dipole formation in most of the cases. The bulk DOS of the $A_{III}B_V$ material appears to be one of the major factors affecting interface band offsets. The CB offset is found to be mostly affected by substitution of A_{III} group cation: In the $In_xGa_{1-x}As$/oxide structures the semiconductor VB top remains at the same energy in the range $0 \leq x \leq 0.75$ (17,18) indicating that the gap narrowing occurs entirely through the CB shift. By contrast, replacement of As anions by P leads to a significant (≈ 0.6 eV) downshift of the VB top in InP (19), while in GaSb the VB top appears to be at 0.4 eV higher than that in GaAs (20). These composition-dependent band edge variations revealed by IPE indicate great potential of $A_{III}B_V$ materials for both electron and hole QW channel engineering.

Similar to the Ge/oxide interfaces, oxidation of $A_{III}B_V$ surfaces may lead to formation of a narrow-gap IL resulting in low-energy thresholds for charge injection. At the interfaces of GaAs with Al_2O_3 and HfO_2 an IL with $\approx$4-eV wide gap is typically encountered (21) which can be associated with the formation of Ga_2O_3 (22). The tickness of this IL depends on the processing chemistry, suggesting a possibility for optimization. Indeed, in the case of molecular beam epitaxy of Al_2O_3 on (100)GaSb no measurable IL is found thanks to a low pressure of oxygen (20). Remarkably, at the (100)InP/Al_2O_3 interface no substantial amount of In in the IL is detected suggesting the formation of a P-rich oxide that delivers large band offsets and, as a consequence, good insulation (19).

<u>Metal/Oxide Interfaces</u>

In sharp contrast to the semiconducting electrodes, barrier height at interfaces between conductors and oxides is highly sensitive to the sample processing. For instance, in the case of heavily-doped polycrystalline Si electrodes on HfO_2, the band alignment is affected by the Si doping type, and can be traced back to barrier perturbation by charged defects located in the oxide near the poly-Si/oxide interface (23). Moreover, if considering the same metal, the effective work function found from the IPE energy threshold and the electron affinity of the insulator appears to be sensitive to the type of insulating oxide and subsequent processing (24-26). This result suggests an additional contribution to the interface barrier associated with charges in the oxide leading to the appearance of a dipole-like barrier component. Easy in-diffusion of metal ions into the near-interface oxide layers and polarization of the metal electrode that provides the charge of opposite sign are likely to be the major reasons for the drastic difference in barrier behavior between semiconductor and metal interfaces (26). The most remarkable example of barrier variation is presented in Fig. 3 which shows the IPE/PC spectra measured in Si/TiN_x/HfO_2(10 nm)/TiN_x and Si/TiN_x/Hf/HfO_2(10 nm)/TiN_x ($x\approx$1) MIM structures. The spectra reveal that the replacement of the TiN_x bottom electrode by TiN_x/Hf(10 nm) stack results in a lowering of the barrier height at the opposite HfO_2/TiN_x interface by almost 1 eV (cf. spectra taken under negative bias). This can be

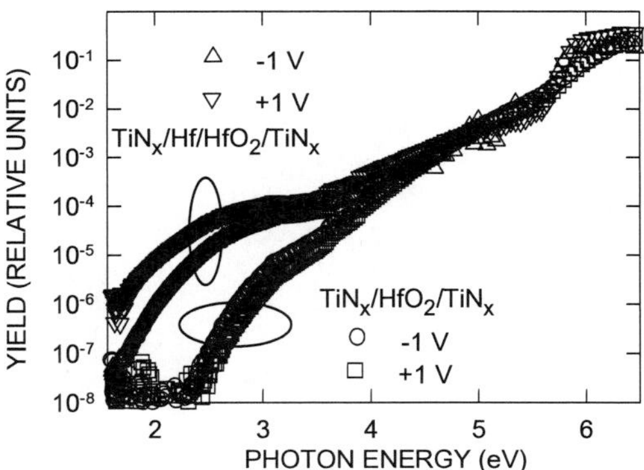

Figure 3. Quantum yield as a function of photon energy in TiN/HfO_2/TiN and TiN_x/Hf/HfO_2/TiN_x samples. The signal observed under positive (+1 V) and negative (-1V) bias corresponds to electron IPE from the bottom and top metal electrode, respectively, into the oxide.

explained by oxygen scavenging by the bottom Hf metal resulting in the formation of positively charged defects at the interfaces of MIM structure leading to the barrier lowering.

4. Conclusions

The overviewed results of IPE/PC spectroscopy regarding energy band alignment at interfaces of metals and semiconductors with insulating oxides reveal several fundamental mechanisms which affect the interface barrier height. These include the composition- and structure-dependent DOS of the materials in contact, the presence of additional chemical phases in the form of an IL, and generation of charged defects close to the interface. Using the revealed trends, the semiconductor/oxide band offsets can be predicted in many cases on the basis of the oxide gap width. However, this approach is not applicable to the metal/insulator contacts because of substantial, up to 1/3 of the total barrier height, defect-related dipole component which instead will require a case-to-case interface barrier analysis. It also becomes clear that IPE spectroscopy offers unique possibilities in analyzing interface barriers even in multilayered insulating stacks making IPE the method of choice when characterizing fundamental band alignment at interfaces of insulating materials with semiconductors and metals.

References

1. J. Robertson, *J. Vac. Sci. Technol. B*, **18**, 1785 (2000).
2. D. G. Schlom and J. H. Haeni, *MRS Bull.*, **27**, 198 (2002).
3. J. Robertson, *Rep. Prog. Phys.*, **69**, 327 (2006).
4. D.G. Schlom, S. Datta, and S. Guha, *MRS Bull.*, **33**, 1017 (2008).
5. V. K. Adamchuk and V. V. Afanas'ev, *Progr. Surf. Sci.*, **41**, 111 (1992).
6. V. V. Afanas'ev and A. Stesmans, *J. Appl. Phys.*, **102**, 081301 (2007).
7. V. V. Afanas'ev, *Internal Photoemission Spectroscopy*, Elsevier, Oxford (2008).
8. W. M. Lau, *J. Appl. Phys.*, **65**, 2047 (1989).
9. E. Bersch et al., *J. Appl. Phys.*, **107**, 043702 (2010).
10. See, e.g., G. Lucovsky, *J. Non-Cryst. Solids*, **303**, 40 (2002).
11. V. V. Afanas'ev et al., *Mater. Sci. Semicond. Proc.*, **11**, 230 (2008).
12. V. V. Afanas'ev et al., *Appl. Phys. Lett.*, **94**, 172106 (2009).
13. V. V. Afanas'ev et al., *J. Phys.Condens. Matter*, **16**, S1839 (2004).
14. V. V. Afanas'ev et al., *Appl. Phys. Lett.*, **92**, 022109 (2008).
15. V. V. Afanas'ev et al., *Appl. Phys. Lett.*, **93**, 102115 (2008).
16. V. V. Afanas'ev and A. Stesmans, *Mater. Sci. Semicond. Proc.*, **9**, 764 (2006).
17. V. V. Afanas'ev et al., *Appl. Phys. Lett.*, **94**, 202101 (2009).
18. N. V. Nguyen et al., *Appl. Phys. Lett.*, **96**, 052107 (2010).
19. H.-Y. Chou et al., *Appl. Phys. Lett.*, **97**, 132112 (2010).
20. V. V. Afanas'ev et al., *unpublished.*
21. V. V. Afanas'ev et al., *Appl. Phys. Lett.*, **93**, 212104 (2008).
22. V. V. Afanas'ev et al., *Appl. Phys. Lett.*, **89**, 092103 (2006).
23. V. V. Afanas'ev et al., *Appl. Phys. Lett.*, **86**, 072107 (2005).
24. V. V. Afanas'ev et al., *Appl. Phys. Lett.*, **86**, 232902 (2005).
25. L. Pantisano et al., *Appl. Phys. Lett.*, **89**, 113505 (2006).
26. S. Shamuilia et al., *J. Appl. Phys.*, **104**, 073722 (2008).

ECS Transactions, 34 (1) 473-478 (2011)
10.1149/1.3567622 ©The Electrochemical Society

Atomic-layer Deposition of Lutetium Aluminate Thin Films for Non-volatile Memory Applications

C. Adelmann[a], J. Swerts[a], T. Conard[a], B. Brijs[a], A. Franquet[a], A. Hardy[b], H. Tielens[a], K. Opsomer[c], A. Moussa[a], M.K. Van Bael[b], M. Jurczak[a], J.A. Kittl[a], and S. Van Elshocht[a]

[a] Imec, B-3001 Heverlee, Belgium
[b] Institute for Materials Research, Inorganic and Physical Chemistry, Hasselt University, B-3590 Diepenbeek, Belgium
[c] Department of Solid State Sciences, Universiteit Gent, B-9000 Gent, Belgium

Thin $Lu_xAl_{2-x}O_3$ films were deposited by atomic-layer deposition using $Lu(thd)_3$, and TMA in combination with O_3 as oxidizer. High-quality dielectric films were obtained with good process control. The full range of the Lu/(Lu+Al) composition was found to be accessible. The films showed bulk density and low roughness. As a result, this process enables the study of $Lu_xAl_{2-x}O_3$ as dielectric in advanced non-volatile memory devices.

Introduction

In recent years, high-κ material research has gained enormous interest since alternative dielectrics were required to replace SiON in complementary metal-oxide-semi-conductor (CMOS) transistor applications to enable their continuous scaling (1). This work has lately led to the integration of HfO_2-based dielectrics into commercial CMOS devices (2). More recently, high-κ dielectrics have also been evaluated for integration into non-volatile memory (NVM) devices such as NAND flash devices (3).

For standalone NVM applications, high-κ materials are of interest to address issues related to the scaling of the device architecture. In current devices, the control gate wraps around the floating gate in order to improve capacitive coupling. When cell dimensions are reduced, the spacing between adjacent devices becomes too small for a wrap-around control gate. Cell planarization can address this issue but entails the constraint for a higher dielectric permittivity for the separating interpoly dielectric (IPD) to compensate for the capacitance reduction. Al_2O_3 has been identified as a potential replacement IPD for conventional $SiO_2/Si_3N_4/SiO_2$ stacks, demonstrating good retention (4).

As potential materials beyond Al_2O_3, lanthanide aluminates have been identified by first principles calculations as promising IPD candidates because they combine large band gaps with high dielectric constants (5,6). The band gap of lanthanide aluminates generally increases with atomic number, whereas the dielectric permittivity decreases; hence, the system is ideally suited to examine different trade-offs between band gap and dielectric permittivity.

Atomic-layer deposition (ALD) is a deposition technique that has become the method of choice for the deposition of dielectrics for microelectronic applications because of its high degree of control over the deposition process and conformal deposition at low temperatures. An ALD process is therefore ideal for the deposition of dielectric thin films in semiconductor devices. In this paper, we demonstrate an ALD process for the deposition of $Lu_xAl_{2-x}O_3$ and demonstrate that this process can lead to high quality dielectric films with good process control.

Experimental Details

$Lu_xAl_{2-x}O_3$ was deposited by atomic-layer deposition (ALD) on 300 mm Si (100) wafers in an ASM Pulsar[®] 3000 reactor, connected to an ASM Polygon[™] 8300 platform. $Lu(thd)_3$ and trimethyl-aluminum (TMA) were used in combination with O_3 as oxidizer. The deposition temperature was 300°C.

Before ALD, a chemical SiO_2 layer (about 1.1 nm thick) was grown on the Si wafers in an O_3/H_2O solution ("IMEC clean,", Ref. 7). Dielectric stack thicknesses were determined by spectroscopic ellipsometry using a KLA-Tencor Aleris system. A single-layer model was used to model the dielectric stack, which contains both the high-κ layer and the interfacial SiO_2 layer. The within-wafer (WiW) nonuniformity was determined by 49-point maps of the wafer with an edge exclusion of 3 mm.

The Lu/(Lu+Al) concentrations were determined by elastic recoil detection (ERD), inductively-coupled plasma atomic-emission spectrometry (ICP-AES), and x-ray photoelectron spectrometry (XPS). ERD was performed with a time-of-flight telescope in combination with a Si detector. Timing gates based on the electrostatic mirror design were used with a flight path of 845 mm. The ERD set-up was installed at a 2 MV Pelletron accelerator. A 6 MeV Cl^{4+} ion beam was used; the scattering and recoil angles were 40°, whereas the exit angle (sample surface to detector) was 5°. For ICP-AES, the films were dissolved in a 20% HNO3/0.8% HF mixture. The resulting solution was then analyzed using a Perkin-Elmer Optima 3000 spectrometer. X-ray photoemission spectrometry (XPS) was carried out using a Theta 300 system from Thermo Electron with monochromatic Al Kα radiation.

X-ray diffraction (XRD) was performed in a Bede MetrixL system; additional temperature-dependent *in situ* XRD (IS-XRD) was performed in a in a Bruker D8 x-ray diffractometer during ramp-annealing. All measurements used Cu Kα radiation. The contamination in the layers was assessed by time-of-flight secondary-ion mass spectrometry (TOF-SIMS) using an ION–TOF IV instrument with a Ga^+ source operating at a beam energy of 15 keV (area of 80 μm×80 μm) for analysis and a Xe^+ beam with an energy of 350 eV (area of 400 μm×400μm) for depth profiling. Atomic-force microscopy was performed using a Veeco Nanoscope IV Dimension 3100 in tapping mode.

Results and Discussion

To deposit ternary $Lu_xAl_{2-x}O_3$, an *n:m* ALD supercycle was defined, which consisted of *n* $Lu(thd)_3/O_3$ cycles followed by *m* repetitions of TMA/O_3 cycles. The numbers *n* and *m* were varied to adjust the Lu/(Lu+Al) mole fraction *x* of $Lu_xAl_{2-x}O_3$. Identical O_3-pulse and post-O_3 N_2-purge lengths were used for both the $Lu(thd)_3/O_3$ and TMA/O_3 cycles. To study the saturation properties of the ALD process, the influence of $Lu(thd)_3$ on the growth per cycle was examined after deposition of Lu_2O_3. The data in Fig. 1a showed saturation behavior of the growth per cycle, indicating self-limiting $Lu(thd)_3$ adsorption and thus ALD behavior. To study the ALD behavior of $Lu_xAl_{2-x}O_3$, a 4:1 supercycle was used. The effects of the TMA and O_3 pulse lengths are depicted in Fig, 1b and 1c, respectively. In both cases, saturation behavior was observed, indicating an ideal ALD behavior of the $Lu_xAl_{2-x}O_3$ process.

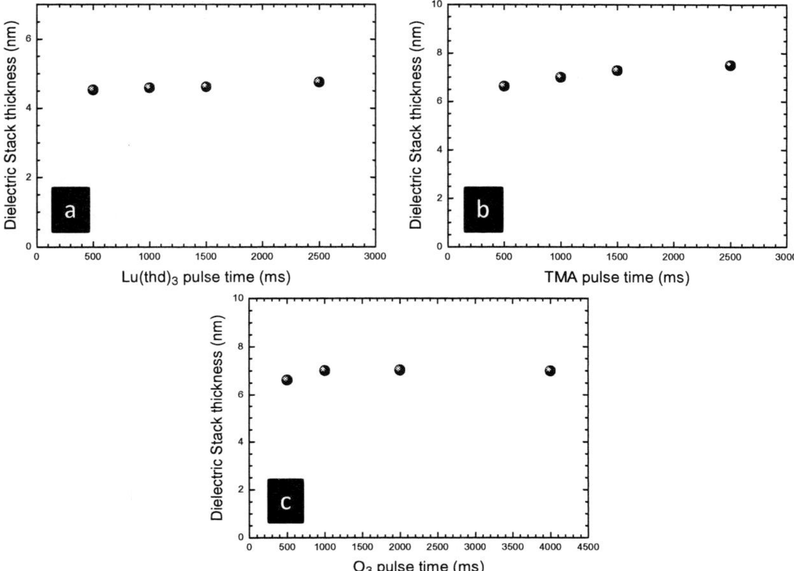

Figure 1: Saturation curves: (a) Dielectric stack thickness after 200 cycles of Lu(thd)$_3$/O$_3$ as a function of Lu(thd)$_3$ pulse time. (b) Dielectric stack thickness after 40 4:1 supercycles as a function of TMA pulse time and (c) as a function of O$_3$ pulse time.

Figure 2 shows growth curves for Lu$_x$Al$_{2-x}$O$_3$ deposited using both 3:1 and 7:1 supercycles. The resulting dielectric stack thicknesses increased linearly with the number of su-percycles. The WiW non-uniformity was found to be independent of the film thickness, which demonstrated the stability of the process (data not shown).

Figure 3a shows the Lu/(Lu+Al) mole fraction x of the Lu$_x$Al$_{2-x}$O$_3$ films as a function of the n:m Lu(thd)$_3$/O$_3$:TMA/O$_3$ cycle ratio. Generally, good agreement was obtained between the different techniques. The Lu/(Lu+Al) mole fraction increased monotonously with the cycle ratio; however, because of the lower growth per cycle for Lu(thd)$_3$/O$_3$ (see below), a strong non-linearity of the curve was observed. The growth per cycle of both the Lu(thd)$_3$/O$_3$ and the TMA/O$_3$ cycles could be deduced from the measured growth per supercycle and the Lu/(Lu+Al) mole fraction. The result is depicted in Fig. 3b.

The growth per cycle for TMA/O$_3$ increased slightly with the number of Lu(thd)$_3$/O$_3$ cycles but was similar to the value obtained from Al$_2$O$_3$ deposition (~0.9 Å), hence there was little influence of Lu(thd)$_3$/O$_3$ on TMA/O$_3$. The deduced growth per cycle for Lu(thd)$_3$/O$_3$ was 0.1 Å, independent of the supercycle. It should be noted that the growth per cycle for Lu$_2$O$_3$ using Lu(thd)$_3$/O$_3$ was found to be 0.17 Å consistent with the trends reported before (8). Hence, the Lu(thd)$_3$/O$_3$ deposition was inhibited on Al$_2$O$_3$-like surfaces. It should be noted that, because the growth per cycle was much lower for Lu(thd)$_3$/O$_3$ than for TMA/O$_3$, the deposition of Lu-rich films required a very large n:m ratio and led to a rather low overall growth rate and to long process times.

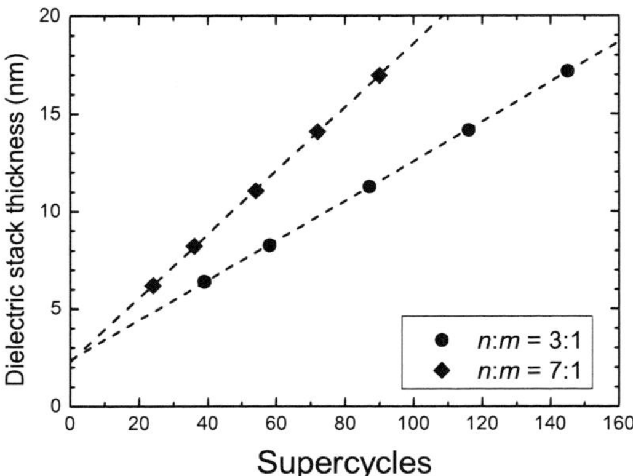

Figure 2: Growth curves for $Lu_xAl_{2-x}O_3$ deposited using both 3:1 and 7:1 supercycles. The resulting dielectric stack thicknesses increased linearly with the number of supercycles.

The film density of $Lu_xAl_{2-x}O_3$ determined by XRR as well as mass measurements is shown in Fig. 4a as a function of Lu/(Lu+Al) mole fraction x. Values were consistent with the expected bulk densities. The atomic-force micrograph of a 15 nm thick $LuAlO_3$ film in Fig. 4b was characterized by small islands on a 10 nm scale with heights of 5–10 Å. The extracted rms roughness was 1.5-2 Å. ToF-SIMS (data not show) indicated C impurities on a level typical for oxides grown by ALD using metal-organic precursors. The C content was undetectable by XPS, so the absolute level was well below 1%. XRD showed that all $Lu_xAl_{2-x}O_3$ films were amorphous as grown. For stoichiometric $LuAlO_3$ films, The IS-XRD data in Fig. 5 demonstrated crystallization at 940°C into the metastable hexagonal phase (6,9); Al-rich $Lu_xAl_{2-x}O_3$ films ($x\sim0.25$) did not show any crystallization up to 1100°C.

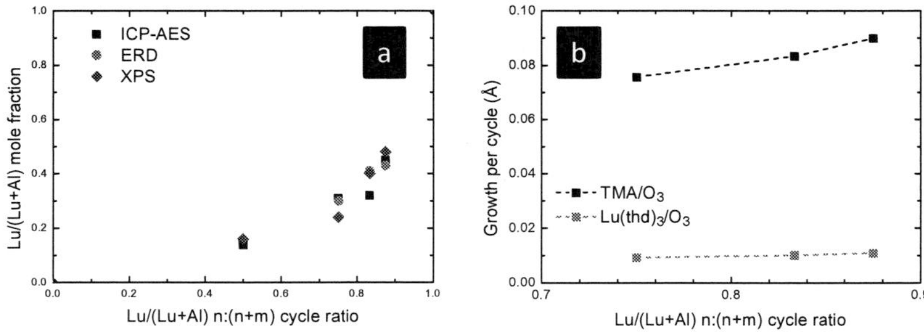

Figure 3: (a) Lu/(Lu+Al) mole fraction x of the $Lu_xAl_{2-x}O_3$ films as a function of the $n:m$ $Lu(thd)_3/O_3$: TMA/O_3 cycle ratio. (b) Growth per cycle of individual $Lu(thd)_3/O_3$ and the TMA/O_3 cycles as a function of the $n:m$ $Lu(thd)_3/O_3$: TMA/O_3 cycle ratio.

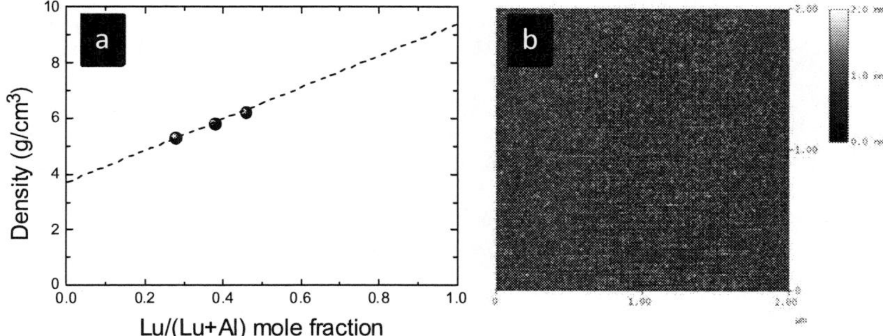

Figure 4: (a) Density of $Lu_xAl_{2-x}O_3$ as a function of the Lu/(Lu+Al) mole fraction x. The densities are consistent with a linear interpolation between bulk Lu_2O_3 and Al_2O_3 (dashed line). (b) Atomic-force micrograph of a 15 nm thick $LuAlO_3$ film.

Conclusion

In conclusion, we have demonstrated the atomic-layer deposition of $Lu_xAl_{2-x}O_3$ thin films using $Lu(thd)_3$ and TMA in combination with O_3 as an oxidizer. Good control of the deposition process as well as of the Lu/(Lu+Al) composition was obtained. While the full composition range with $0 \leq$ Lu/(Lu+Al) ≤ 1 was in principle accessible by the process, high Lu concentrations led to long process times because of the low growth per $Lu(thd)_3/O_3$ cycle. This was further aggravated by the observation of an even lower growth per $Lu(thd)_3/O_3$ cycle in combination with TMA/O_3, possibly due to growth inhibition of $Lu(thd)_3/O_3$ on Al-O-like surfaces. All films showed bulk densities and low roughness. The process was thus well suited to study the prospects of $Lu_xAl_{2-x}O_3$ as interpoly or blocking dielectrics in advanced NVM devices (6,10,11).

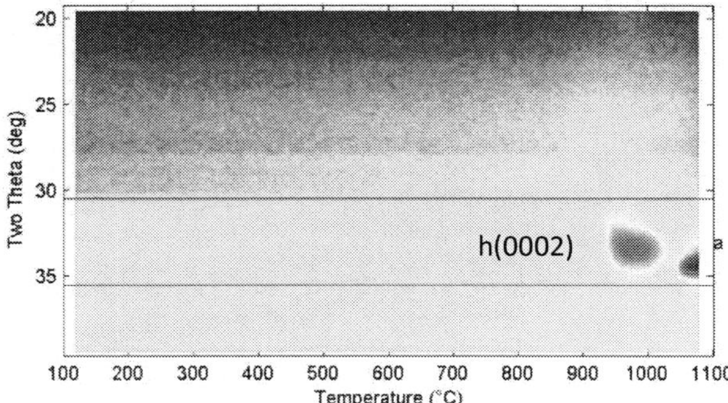

Figure 5: *In situ* XRD data of $LuAlO_3$. Crystallization was observed at 940°C.

References

1. International Technology Roadmap for Semiconductors, see www.itrs.net
2. M.T. Bohr, R.S. Chau, T. Ghani, and K. Mistry, *IEEE Spectrum*, **44**(10), 29 (2007).
3. B. Govoreanu, J.A. Kittl, J. De Vos, A. Rothschild, P. Blomme, D.Wellekens, D. Ruiz Aguado, M. Jurczak, and J. Van Houdt, *ECS Trans.*, **19**(2), 649 (2009).
4. J. R. Power, D. Shum, Y. Gong, S. Bogacz, J. Haeupel, H. Estel, R. Strenz, R. Kakoschke, K. Van Der Zanden, and R. Allinger, *Proc. Int. Conf. Memory Technology Design 2008*, G. Baccarani, E. Gnani, and M. Rudan, Editors, p. 93, University of Bologna Press, Bologna (2008).
5. C. Adelmann, D. Pierreux, J. Swerts, D. Dewulf, A. Hardy, H. Tielens, A. Franquet, B. Brijs, A. Moussa, T. Conard, M.K. Van Bael, J.W. Maes, M. Jurczak, J.A. Kittl, and S. Van Elshocht, *Chem. Vap. Depos.* **16**, 170 (2010).
6. C. Adelmann, J. Swerts, O. Richard, T. Conard, M. Popovici, V.V. Afanas'ev, L. Breuil, A. Cacciato, K. Opsomer, B. Brijs, H. Tielens, G. Pourtois, H. Bender, C. Detavernier, M. Jurczac, S. Van Elshocht, and J.A. Kittl, *ECS Trans.*, 33, 31 (2010).
7. M. Meuris, P. W. Mertens, A. Opdebeeck, H. F. Schmidt, M. Depas, G. Vereecke, M. M. Heyns, and A. Philipossian, *Solid State Technol.*, **38**, 109 (1995).
8. L. Niinistö, J. Päiväsaari, J. Niinistö, M. Putkonen, and M. Nieminen, *phys. stat. sol. (a)*, **201**, 1443 (2004); J. Päiväsaari, M. Putkonen, and L. Niinistö, L., *Thin Solid Films*, **472**, 275 (2005).
9. F. Bertaut and J. Mareschal, *C. R. Séances Acad. Sci.*, **257**, 867 (1963); G. Szabo and R.A. Pâris, *C. R. Séances Acad. Sci. Ser. C*, **267**, 1816 (1968).
10. D. Wellekens, S. Van Elshocht, C. Adelmann, J. Meersschaut, J. Swerts, J.A. Kittl, A. Cacciato, I. Debusschere, M. Jurczak, and J. Van Houdt, *Proc. IEEE Intern. Memory Workshop 2010*, p. 130, Monterey (2010).
11. L. Breuil, C. Adelmann, G. Van den bosch, A. Cacciato, M.B. Zahid, M. Toledano Luque, A. Suhane, A. Arreghini, R. Degraeve, S. Van Elshocht, I. Debusschere, J. Kittl, M. Jurczak, and J. Van Houdt, *Proc. 40th Europ. Solid-State Dev. Res. Conf.*, p. 440, Sevilla (2010).

ECS Transactions, 34 (1) 479-482 (2011)
10.1149/1.3567623 ©The Electrochemical Society

Evolution of STI Gap fill Technology

James C Chen[a], Yingjie Chen[a], Rong Gao[a], Charley Cheng[a], Xianyuan Li[a], Ganming Zhao[a], Diana Chan[b], and Terrance Lee[b]

[a] Applied Materials China, Shanghai 201203, P. R. China
[b] Applied Materials, Santa Clara, California 95054, USA

As the IC industry progresses along the ITRS roadmap, continually producing more demanding devices at each technology node, new processes are required to realize the yield requirement for success. In this paper, we present and review the evolution of STI (Shallow Trench Isolation) application, and recent developments in the High-Density Plasma Chemical Vapor Deposition (HDP CVD) process and High Aspect Ratio Process (HARP) CVD process. Aspects relevant to the HDP CVD process and using different processes to achieve gap filling and planarization are discussed.

HDP-CVD Process

The application of HDP-CVD for STI module has been proven successful. The HDP-CVD process achieves the bottom-up fill of a trench by simultaneously depositing and physical sputtering of the gap-filling oxide. Source RF power provides energy to ignite and sustain the plasma, producing ions and radicals for surface reactions. Bias RF power provides energy to the plasma and increases the ion's directional velocity towards the wafer. The ions strike the film surface, resulting in physical sputtering which keeps the trench open for bottom-up gap-fill to proceed. Fig. 1 is a concept of HDP-CVD process. The mixture of Ar/O_2 has traditionally been used as the sputtering material at technology nodes of $0.18\mu m$ and higher. For the $0.15\mu m$ technology node and below, one problem associated with sputtering is that there is an angular dependence to the sputtering material. The sputtered SiO_2 can re-deposit on the other side of the trench, causing excess buildup. In order to reduce the amount of re-deposition, the HDP-CVD process has evolved from using Ar as the sputtering material, to O_2 only and then adding lighter atoms such as He and H_2. The He process can provide a gap fill down to $0.11\mu m$ node and the H_2 process can extend HDP-CVD marginally down to the 65nm node. Fig.2 shows collision kinematics of different sputtering materials.

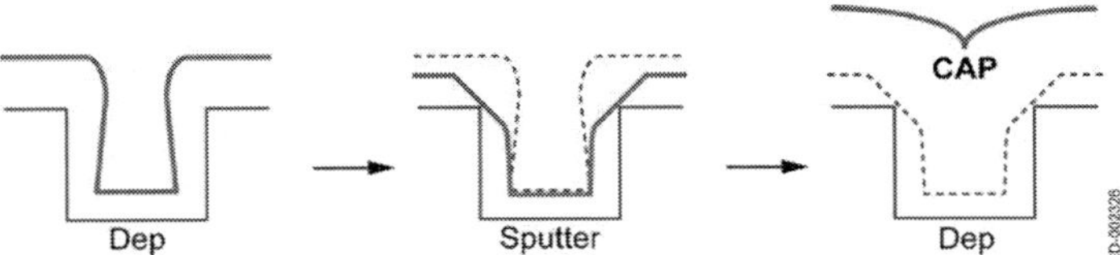

Figure 1. HDP-CVD process concept

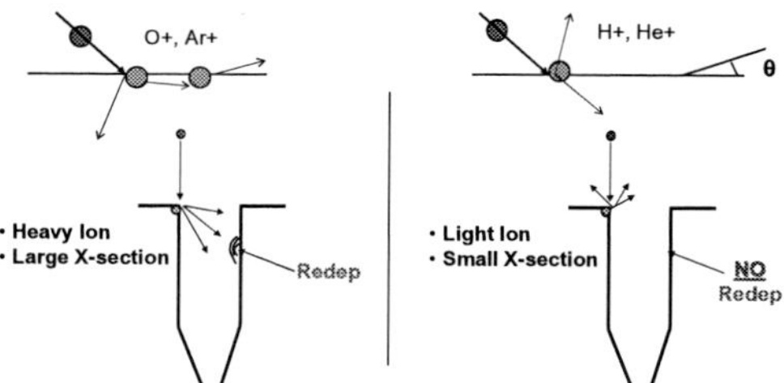

Figure 2. Collision Kinematics of Heavy and Light Atoms

HDP-CVD with Fluorine Assisted Etch Processes

For more aggressive structures, Integrated Process Modulation (IPM) has been used to fill these structures. IPM process is an extension of the Deposition - Etch - Deposition (DED) process. It is a series of deposition/etch cycles, using a NF3-based plasma dry etch preventing the trench from closing while depositing the film (1), as shown in Fig.3. Each deposition is followed by a chemical etch to keep the trench open and allows the bottom-up gap-fill to proceed to completion. The etch is able to reduce the growth on the sidewall as well as to reduce the height of the "hats" on top of the trench, reducing the aspect ratio of the trench and making gap-fill progressively easier. Fig.4 is a comparison of the traditional HDP-CVD process and IPM process. The IPM process has been extended to 65nm devices (2).

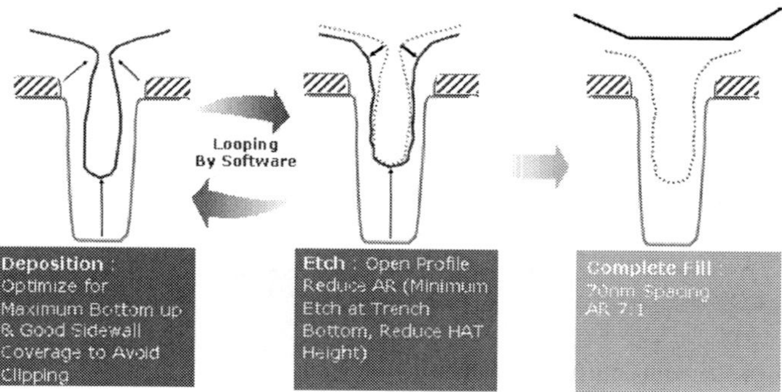

Figure 3. Integrated Process Modulation (IPM)

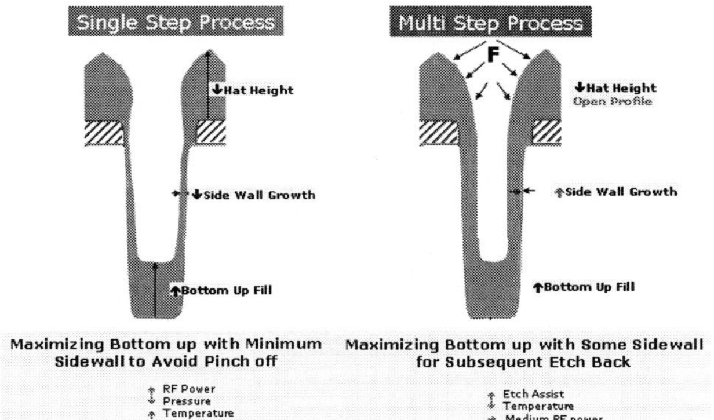

Figure 4. HDP STI Gap Fill Approach

SACVD HARP Process

Although HDP-CVD is the most commonly used gap fill technology for STI due to its good gap fill performance, ease of polishing and superior film properties. However, as technology node <65nm, gap fill has become increasingly challenging for HDP-CVD. A High Aspect Ratio Process (HARP) is developed using an Ozone/TEOS-based sub-atmospheric chemical vapor deposition process to meet deeper gap fill requirements (AR>8). In the HARP process, ozone (O_3) and tetra-ethyl-ortho silicate (TEOS) is introduced into a pressure-controlled chamber. O_3 decomposes into O_2 and O- free radicals at an elevated temperature, which react with TEOS on a hot surface to form a SiO_2 film. In addition, unlike the conventional gap-fill technology, HARP films are stress-tunable films and can induce stress in the channel and enhance carrier mobility. For critical STI device structures, a new process, eHARP, has been developed for advanced memory and logic devices. The new eHARP film improves upon the first-generation HARP film gap fill capability by introducing water vapor to the TEOS/ozone chemistry during the deposition. This results in a more tightly packed, denser film that enables seamless, void-free gap fill in features with AR >12:1. In some instances where trenches have a vertical profile, the HARP process has evolved into a "dep/etch/dep" process. The Siconi etching step acts to reshape the partially filled HARP liner allowing easier gap fill capability in the subsequent capping step as seen in Figure 5. Now HARP has been used for 45nm devices and beyond (3) and HARP/Siconi/HARP processes have been qualified for 28nm technology node.

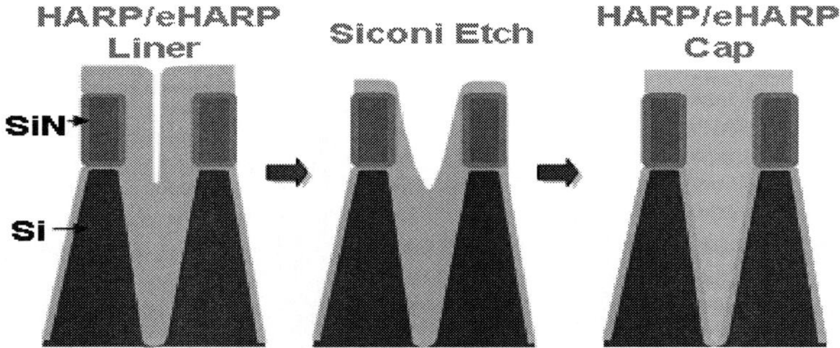

Figure 5. HARP/Siconi etch /HARP Process Scheme

Summary

The evolution of STI gap fill technology was discussed in this paper. This material can be used as a guideline for STI gap fill study. However, developing a gap fill process is not so trivial especially for high aspect ratio structure.

REFERENCES

1. D. Li, X. C. Chen, and L. Zhang, "HDP-CVD dep/etch/dep process for improved deposition into high aspect ratio features," US Patent #6,908,862 published June 21, 2005
2. Y.J Chen, H.Yuan, Z.Y Zhang, Nicola Li, "Advanced HDP STI Gap-fill Development in 65nm Logic Device" *CTSIC 2010.*
3. Tilke, A.T., Hampp, R., Stapelmann, C., Culmsee, M., Conti, R., Wille, W., Jaiswal, R., Galiano, M., Jain, A. " STI Gap-Fill Technology with High Aspect Ratio Process for 45nm CMOS and beyond " *Advanced Semiconductor Manufacturing Conference, 2006.*

ECS Transactions, 34 (1) 483-487 (2011)
10.1149/1.3567624 ©The Electrochemical Society

Annealing Effect on the Electrical Properties of La$_2$O$_3$/InGaAs MOS Capacitors

T. Kanda[a], D. Zade[a], Y. -C. Lin[c], K. Kakushima[b], P. Ahmet[a], K. Tsutsui[b],
A. Nishiyama[b], N. Sugii[b], E. Y. Chang[c], K. Natori[a], T. Hattori[a], and H. Iwai[a]

[a]Frontier Research Center, Tokyo Institute of Technology,
4259, Nagatsuta, Midori-ku, Yokohama 226-8052, Japan
[b]Interdisciplinary Graduate School of Science and Engineering,
Tokyo Institute of Technology, 4259, Nagatsuta, Midori-ku, Yokohama 226-8052, Japan
[c]National Chiao Tung University, Hsinchu 30010, Taiwan, ROC

The electrical characteristics of InGaAs MOS capacitors with 8-nm-thick La$_2$O$_3$ gate dielectrics have been measured. The effects of annealing temperature and annealing time on the interface state densities (D_{it}) have been extracted. It has been found that the low D_{it} can be achieved by lowering the annealing temperature for an extended period of time.

Introduction

MOS devices with III-V channels, especially InGaAs, have been of great interests as one of the promising candidates for future MOSFET. One of the major issues in InGaAs MOS devices is the lack of good electrical properties at the high-k/substrate interface. Among various high-k materials, Al$_2$O$_3$ (k ~8-9) has been reported to have small interface state density with reduced hysteresis in the capacitance-voltage (C-V) curve (1,2). Among higher k-value dielectrics, needed to further scaling the oxide thickness, La$_2$O$_3$ has been identified as a good candidate for Si-CMOS technology (3). In this study, we have investigated the electrical characteristics of MOS capacitors with La$_2$O$_3$ as the gate dielectric upon various annealing temperatures ranging from 300 to 450 °C with duration of 0.5, 5, or 30 minutes.

Experiment

300-nm-thick Si doped InGaAs layers with indium (In) concentration of 53% and gallium (Ga) concentration of 47% were epitaxially grown on lattice matched n-InP (100) wafers. The doping concentration Si is 1×10^{17} cm^{-3}. Fabrication process of La$_2$O$_3$/InGaAs MOS capacitors is shown in Fig. 1. Wafers were degreased by acetone and ethanol, and rinsed in deionized water, followed by etching native oxides in a 20% HF solution for 3 minutes at room temperature. An 8-nm-thick La$_2$O$_3$ layer was deposited by an e-beam evaporation (10^{-6} Pa) at room temperature. A 50-nm-thick W film was successively *in situ* deposited by an RF magnetron sputtering. After the gate lithography and SF$_6$ reactive ion etching (RIE), a 50-nm-thick Al contact layer on the backside of the substrates was deposited by a thermal evaporation. Rapid thermal post-metallization annealing (PMA) was performed in forming gas (F.G) (N$_2$:H$_2$=97%:3%) ambient in the range from 300 to 450 °C for 0.5, 5, or 30 minutes. C-V characteristics were measured at frequency in the range from 1 kHz to 1 MHz.

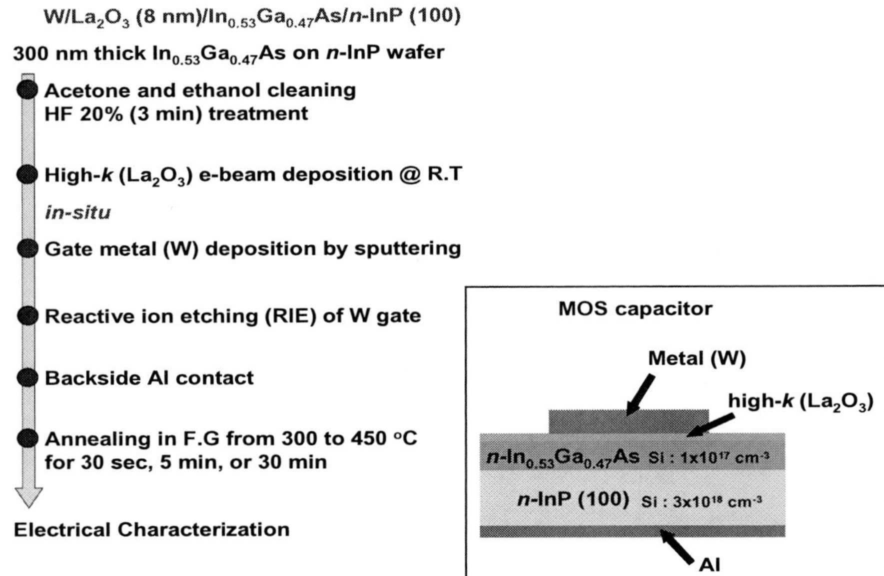

Figure 1. Fabrication process of La$_2$O$_3$/InGaAs MOS capacitors.

Result and Discussion

Figure 2 (a) and (b) show room temperature C-V characteristics with annealing at 300 °C and 450 °C for 5 minutes, respectively. The measured frequency range is from 1 kHz to 1 MHz and the measured gate voltage range is from -2 to 2 V. Frequency dispersion in accumulation and inversion is observed. The frequency dispersion in the accumulation region (positive gate voltage) is effect on the interface traps which is referred as weak Fermi-level pinning effect (4-6). The strong frequency dispersion in inversion region (negative gate voltage), is the effect on minority carrier response due to the presence of interface trap states (7). On the other hand, the C-V curves measured at low frequency (1 kHz and 10 kHz) in Fig. 2 (b) differ from those at higher frequency at V_g > 0.5 V. These behaviors were caused by the gate leakage current. Indeed, the gate leakage current after annealing at the temperature from 300 to 450 °C for 5 minutes as a function of gate voltage increased with increasing annealing temperature (Fig. 3). The same tendency was also observed with the samples annealing for 0.5 and 30 minutes. Indeed, samples with thicker La$_2$O$_3$ layer (15 and 20 nm) did not show this behavior. The gate leakage current of a 15-nm-thick La$_2$O$_3$ layer deposited sample was observed about 10^{-2} A/cm^2 which was orders less than that of the 8-nm-thick La$_2$O$_3$ layer deposited sample.

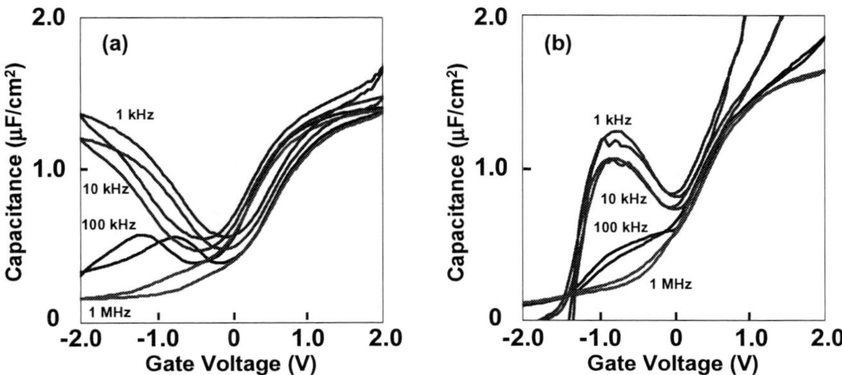

Figure 2. *C-V* curves of La$_2$O$_3$/InGaAs MOS capacitors compared various annealing condition. (a) 300 °C 5min and (b) 450 °C 5min in F.G ambient.

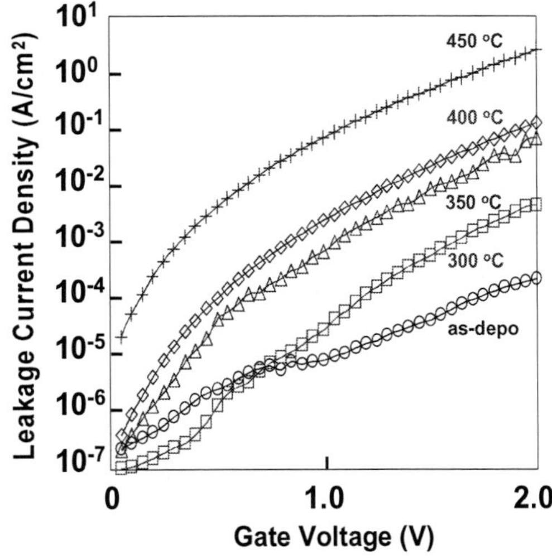

Figure 3. Gate leakage currents of the 8-nm-thick La$_2$O$_3$ layer deposited InGaAs MOS capacitors compared to various annealing condition and as-depo. The annealing were performed in F.G ambient at the range of 300 - 450 °C for 5min.

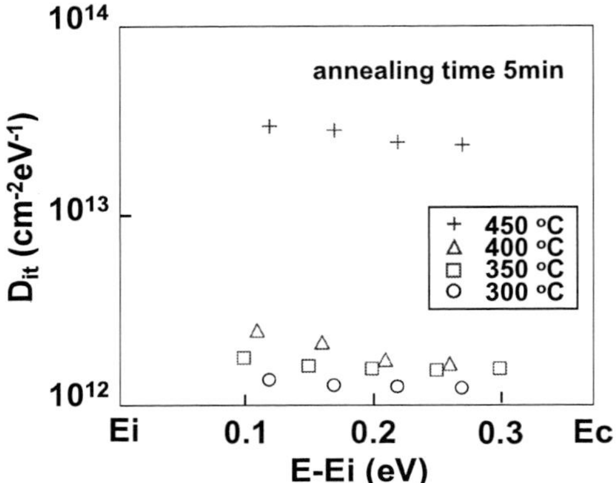

Figure 4. Interface state distribution of La$_2$O$_3$/InGaAs MOS capacitors depending on annealing temperature. The annealing were performed in F.G ambient at the range of 300 - 450 °C for 5min.

Figure 4 shows interface state distribution of the La$_2$O$_3$/InGaAs MOS capacitor in the upper half of the band gap after 5 minutes of forming gas annealing from 300 to 450 °C. The interface trap densities (D_{it}) were extracted by conductance method (8,9). The lowest D_{it} of 10^{12} cm^{-2}eV^{-1} was achieved with 300 °C annealed samples. Values of D_{it} gradually increase with increasing the annealing temperature. However, when annealed at 450 °C, the D_{it} showed a dramatic increase above 10^{13} cm^{-2}eV^{-1}.

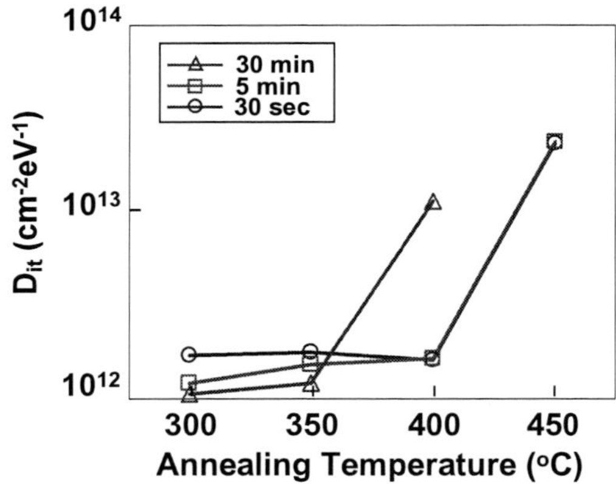

Figure 5. Interface state densities of La$_2$O$_3$/InGaAs MOS capacitors depending on annealing temperature and annealing time.

The influences of the annealing temperature and time on the D_{it} are shown in Fig 5. Lower D_{it} values can be achieved at 300 °C and 350 °C with longer annealing time. The D_{it} value at 400 °C for 30 minutes increased. However, reducing the annealing time can suppress this degradation. The result shows that lower annealing temperature (300 - 350 °C) with longer annealing time (30 minutes) is preferable to reduce the D_{it}. The lowest D_{it} value of 1.1×10^{12} cm^{-2}eV^{-1} is obtained from the 30 minutes anneal at 300 °C.

Conclusion

We have investigated the interface state densities (D_{it}) of InGaAs MOS capacitors with La$_2$O$_3$ as the gate dielectric upon annealing condition from 300 to 450 °C with various annealing time. Lower annealing temperature (300 - 350 °C) and longer annealing time (30 minutes) is preferable for reducing the D_{it}. The low D_{it} value is 1.1×10^{12} cm^{-2}eV^{-1} at 300 °C for 30 minutes.

Acknowledgement

This study was supported by Japan Science and Technology Agency (JST).

References

1. H. C. Chiu, *et al.*, *Appl. Phys. Lett.*, **93**, 202903 (2008).
2. D. Lin, *et al.*, *IEDM.*, 327-330 (2009).
3. J. –P. Maria, *et al.*, *J. Appl. Phys.*, **90**, 3476-3482 (2001).
4. K. Martens, *et al.*, *Micro-electronic. Eng.*, **84**, 2146-2149 (2007).
5. P. K. Hurley, *et al.*, *ECS trans.*, **25**, 113-127 (2009).
6. Y. Hwang, *et al.*, *J. Appl. Phys.*, **108**, 034111 (2010).
7. A. Dimoulas, *et al.*, *Appl. Phys. Lett.*, **86**, 032908 (2005).
8. Y. Hwang, *et al.*, *Appl. Phys. Lett.*, **96**, 102910 (2010).
9. H. Yang, *et al.*, *Jpn. J. Appl. Phys.*, vol. 44, No. 48 (2005).

ECS Transactions, 34 (1) 489-494 (2011)
10.1149/1.3567625 ©The Electrochemical Society

Metal Inserted Poly-Si Stacks with La$_2$O$_3$ Gate Dielectrics for Scaled EOT and V$_{FB}$ Control by Oxygen Incorporation

T. Kawanago[a], K. Kakushima[b], P. Ahmet[a], K. Tsutsui[b], A. Nishiyama[b], N. Sugii[b], K. Natori[a], T. Hattori[a], and H. Iwai[a]

[a] Frontier Research Center, Tokyo Institute of Technology
[b] Interdisciplinary Graduate School of Science and Engineering, Tokyo Institute of Technology, 4259 S2-20, Nagatsuta, Midori-ku, Yokohama, 226-8503, Japan

Metal-Inserted Poly-Si (MIPS) stacks for gate oxide scaling have been presented with La$_2$O$_3$ gate dielectrics. An equivalent oxide thickness (EOT) of 0.69nm is achieved with good interfacial property by high temperature annealing. The flatband voltage (V$_{FB}$) can be modulated by oxygen incorporation in conjunction with Si removal process with less than 1Å EOT degradation.

Introduction

Continuous scaling in the equivalent oxide thickness (EOT) with high-k/metal gate stacks is crucially important to suppress not only gate leakage current but also the severe short-channel effect for highly scaled FETs [1]. Achieving EOT below 0.8nm essentially requires a direct contact of high-k/Si structure with good interfacial property. La$_2$O$_3$ can achieve the direct contact of high-k/Si structure owing to the material nature to form La-silicate at the interface [2]. However, the high-k/Si interface is sensitive to the oxygen partial pressure during the process. Excess silicate formation results in an increase of EOT. Thus, how to control the oxygen during fabrication process is significantly important. It is reported that Metal-Inserted Poly-Si (MIPS) structure can prevent the oxygen diffusion from the atmosphere to gate stacks [3]. In this study, the effect of Si cap layer on high temperature annealed MOS devices with La$_2$O$_3$ dielectric is experimentally investigated. Moreover, oxygen incorporation for V$_{FB}$ tuning without EOT penalty was developed in the previous work [4]. It is of great interest to combine the oxygen incorporation with MIPS structure for V$_{FB}$ tuning and scaled EOT simultaneously. Therefore, oxygen incorporation after Si removal through the metal electrode is also demonstrated.

Experimental Procedure

La$_2$O$_3$ was deposited on HF-last n-Si wafers for MOS capacitors by e-beam evaporation, followed by *in-situ* W metal deposition by RF sputtering. The W thickness is 5nm. A TiN was deposited on the W metal with various thicknesses as a barrier layer. After the TiN deposition, a 100nm thick Si cap layer was deposited. The TiN and Si was also deposited by RF sputtering. The metal was patterned by reactive ion etching (RIE) with SF$_6$ chemistry to form gate electrodes. The substrate impurity concentration of MOS capacitors is 3×10^{15}cm^{-3}. The post-metallization annealing in forming gas (H$_2$:N$_2$=3:97%) was performed at 800°C for 30min. After the Si cap layer removal by wet etching using TMAH [5], oxygen ambient annealing (O$_2$:N$_2$=5:95%) was applied to supply additional oxygen into gate stacks. An Al contact layer on backside of the substrate was deposited. Finally, recovery annealing (FGA) was performed at 420°C for 30min. Process flow is summarized in Figure 1. EOT and V_{FB} were estimated by NCSU CVC program [6].

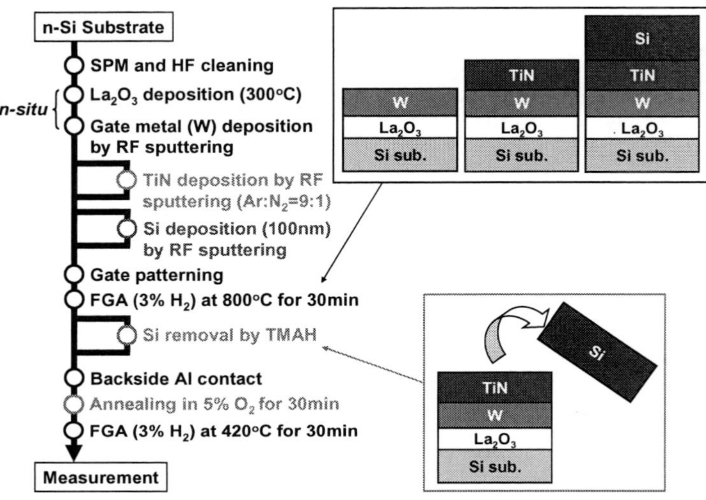

Figure 1 Fabrication process of MOS capacitors

Results and Discussion

Figure 2 shows the comparison of C-V characteristics with various gate stacks. Increase of EOT after FGA at 800°C 30min was suppressed compared with that of W single layer. The EOT with Si removal process is less than half that of W single layer structure. Control the amount of oxygen during annealing is crucially important for scaled EOT.

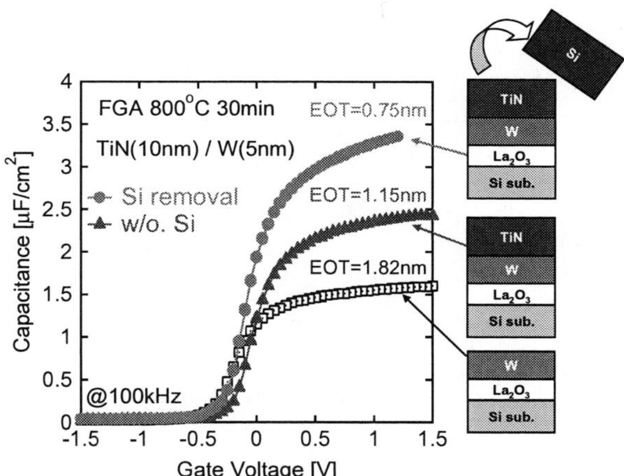

Figure 2 Comparison of C-V characteristics with various gate stacks

Figure 3 shows the comparison of C-V curves between simulation and experiment. TiN thickness is 10nm. The experimental C-V curve is in good agreement with ideal C-V curve calculated by NCSU CVC, indicating an excellent interfacial property. It was demonstrated that C-V characteristics with 0.69nm in EOT close to the ideal C-V curve was attained with a combination of Si cap layer and high temperature annealing.

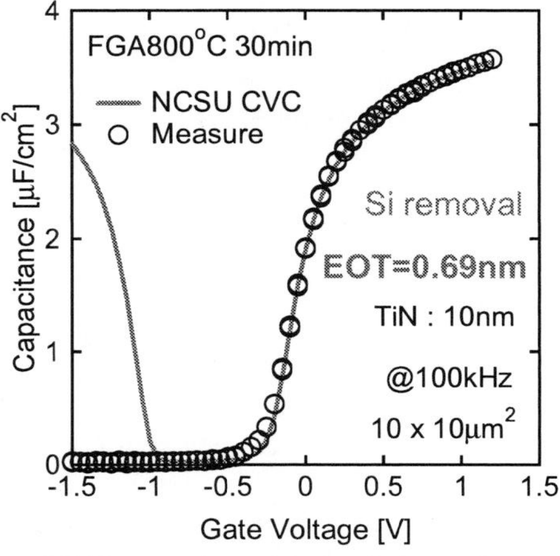

Figure 3 Comparison of C-V curves between simulation and experiment

Figure 4 shows the EOT as a function of TiN thickness with and without Si cap layer. Although TiN layer on W gate is also effective to inhibit the increase of EOT, EOT increment was dramatically suppressed by Si cap layer on TiN. On the other hand, impact of TiN thickness on EOT increase is negligibly small. Recent researches were also adopted MIPS structure because of consistency with conventional CMOS process [7]. It was found that increase of EOT due to silicate reaction at La_2O_3/Si interface can be inhibited by MIPS structure.

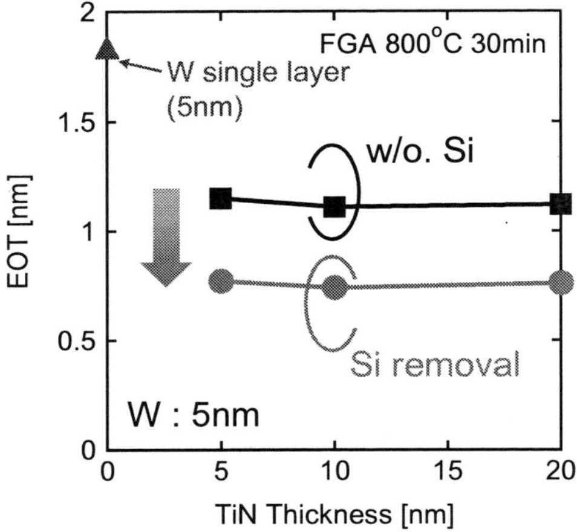

Figure 4 EOT as a function of TiN thickness with and without Si Cap layer

Next, oxygen ambient annealing was performed in TiN/W stacked structure. In previous study [4], oxygen ambient annealing was conducted to W single layer of 5nm in thickness. Thus, annealing condition was investigated to supply additional oxygen through TiN/W stacks. Figure 5 shows the effect of oxygen incorporation through TiN/W stacks on C-V characteristics as a function of TiN thickness. Oxygen ambient annealing was performed at 400°C for 30min, followed by recovery annealing. Large positive V_{FB} shift can be obtained with decreasing TiN thickness while little EOT degradation was confirmed from C-V characteristics. It was found that additional oxygen was also introduced through the TiN/W stack layer.

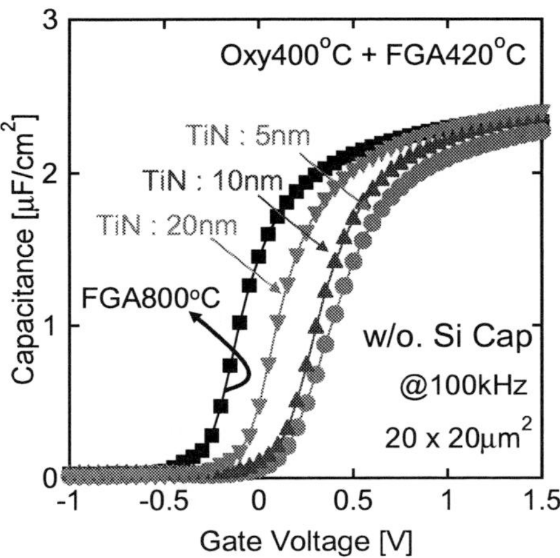

Figure 5 Effect of oxygen incorporation on C-V curves as a function of TiN thickness

Oxygen annealing through TiN/W stack structure was also performed after Si layer removal by wet process using TMAH. Figure 6 shows the demonstration of positive V_{FB} shift by oxygen incorporation after Si removal. TiN thickness is 10nm.

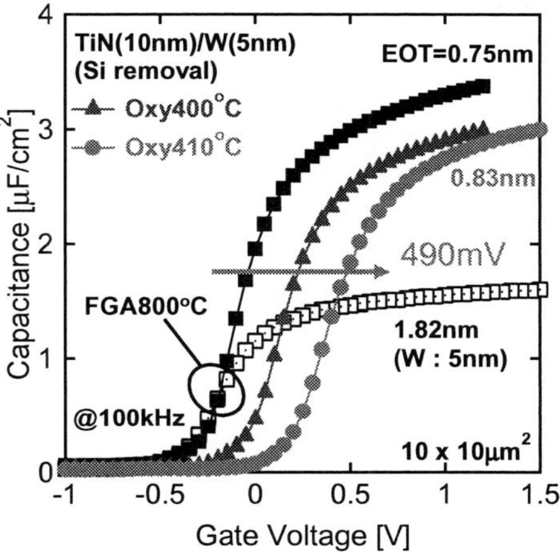

Figure 6 Demonstration of positive V_{FB} shift by oxygen incorporation after Si removal

The positive V_{FB} shift by 490mV was in good agreement with previous experimental results [4]. Although decreasing of capacitance can be observed by oxygen ambient annealing, EOT degradation is 0.8Å extracted with NUSC CVC.

Conclusion

Effect of Si cap layer in MOS devices with La_2O_3 dielectrics was investigated to extend further the EOT scaling. It was found that increment of EOT due to excess silicate reaction was dramatically suppressed by control the amount of oxygen with Si cap layer (MIPS structure). Scaled EOT and C-V characteristics close to the ideal C-V curve was achieved at the same time by combination of high temperature annealing with Si cap layer; in addition, additional oxygen can be supplied through TiN/W stack structure. Positive V_{FB} shift was obtained by oxygen incorporation at the cost of less than 1Å EOT degradation. It was demonstrated that oxygen incorporation with scaled EOT can be obtained by combining oxygen ambient annealing with Si removal process.

Acknowledgments

This study was supported by New Energy and Industrial Technology Development Organization (NEDO), and Grant-in-Aid for research fellows of Japan Society for the Promotion of Science (JSPS).

References

1. T. Skotnicki et al., *IEEE Trans. Electron Devices*, vol. 55, no. 1, pp. 96, (2008).

2. K. Kakushima et al., *Technical digest of International Workshop on Dielectric Thin Films for Future ULSI Devices*: Science and Technology, pp. 9, (2008).
3. C. Choi et al., *Microelectronic Engineering*, 86, pp. 1737, (2009).
4. T. Kawanago et al., *Proc. 40th European Solid-State Device Research Conference*, pp. 301 (2010).
5. M. Saitoh et al., *IEDM Tech. Dig.*, pp. 187, (2004).
6. J. R. Hauser et al., *Proc. AIP Conf.*, pp. 235, (1998).
7. T. Ando et al., *IEDM Tech. Dig.*, pp. 423, (2009).

ECS Transactions, 34 (1) 495-501 (2011)
10.1149/1.3567626 ©The Electrochemical Society

Characteristics of HfSiAlON Gate Dielectric Prepared by Physical Vapor Deposition

Gaobo Xu and Qiuxia Xu

Key Laboratory of Microelectronics Devices & Integrated Technology,
Institute of Microelectronics, Chinese Academy of Sciences, Beijing 100029, China

A gate stack with HfSiAlON as gate dielectric and MoAlN as gate electrode has been developed to obtain p-type band-edge metal gate stack. A corresponding gate dielectric equivalent oxide thickness (EOT) of 18Å, and effective work function (EWF) of ~5.14eV were obtained by a gate last process. HfSiAlON dielectric is formed by depositing HfSiON with a AlN_x cap at the top or bottom of interface on an ultrathin SiO_2 by physical vapor deposition (PVD), and followed by an annealing at 900°C for 30s. Further studies found that the EWF enhancement is dependent on not only the AlN_x thickness but also its location. The EWF enhancement increases with the increase of the AlN_x thickness. And the EWF enhancement is more pronounced cap for AlN_x at the bottom interface. The corresponding mechanism is discussed.

Introduction

With continuous scaling down of CMOS devices, high-k gate dielectrics and metal gate electrodes are required to replace SiO_2 or SiON gate dielectric and the conventional doped poly-Si gate electrodes due to the issues such as high gate leakage current and poly-Si depletion etc. To satisfy V_{th} requirements in high performance CMOS devices, a comprehensive understanding on new high k gate dielectric and metal gate materials are strongly required, especially on the following aspects as material compatibility, process integration with the traditional CMOS process flow, suitable EOT and EWF (4.05eV for NMOS, 5.17eV for PMOS) etc (1-3).

The EWF of a metal gate electrode depends on not only the work function of the bulk metal but also the gate dielectric material, in particularly interfacial layer properties of the gate dielectric. The impurity piled up at interface is considered to be a very effective way for the EWF modulation (4-6).

In this paper, the high-k gate dielectric of HfSiON with AlN_x interfacial layer was prepared by PVD, followed by a rapid thermal anneal, to form HfSiAlON gate dielectric. To ensure a high interface quality and isolate the high-k layer from the Si surface, an ultra-thin SiO_2 was grown before the high-k layer deposition. MoAlN was deposited on the HfSiAlON gate dielectric as the gate electrode. The physical and electrical properties of HfSiAlON gate dielectric were investigated.

Fabrication

Capacitors were fabricated on N(100) substrates with a dopant concentration of 2 $\times$ 10^{15}cm^{-3}. After a standard clean process, silicon wafers were treated with a HF+IPA+H$_2$O solution to remove the contamination of particle, metal ion and organism, as well as to suppress the growth of nature oxide (7). The SiO$_2$ interface was grown by rapid thermal oxidation. AlN$_x$/HfSiON/AlN$_x$ gate dielectric was deposited on the SiO$_2$ interface by PVD. Then a post deposition annealing (PDA), using rapid thermal annealing (RTA), was performed in N$_2$ with a trace of O$_2$ ambience at 900℃ for 30s, to form HfSiAlON gate dielectric. MoAlN metal gate electrode was deposited by PVD. Finally, all samples were metallized with Al(Si) on backside at 400ºC in an N$_2$ ambience for 40 min.

Results

Physical Characterization

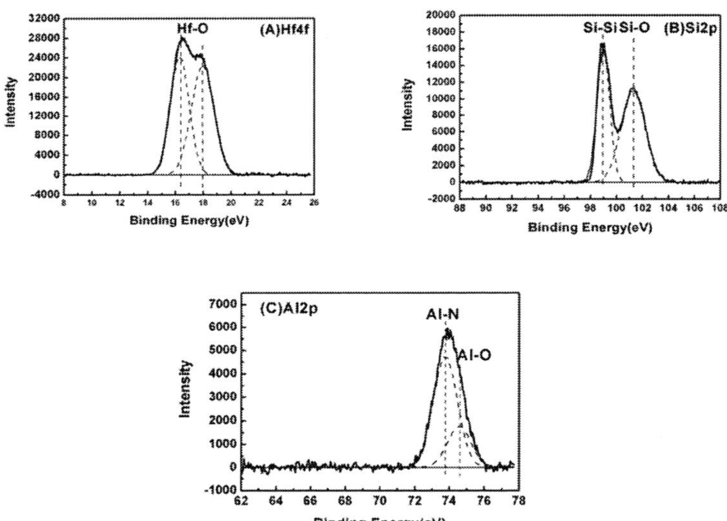

Fig.1 XPS spectra for HfSiAlON gate dielectric after 900°C PDA

In this study, the chemical composition and bonding states of HfSiAlON gate dielectric were investigated by X-ray photoelectron spectroscopy (XPS) measurement with a monochromatic Al Kα (1486.6 eV) X-ray incident. The binding energy was referenced to C$_{1S}$ at 284.65eV. Fig.1 shows the result of XPS analysis of HfSiAlON gate dielectric after 900°C PDA. The Hf composition in HfSiAlON gate dielectric is 5.12 at.%, the Al composition is 9.64 at.%, the Si composition is 24.68 at.% , the oxygen composition is 30.03 at.% and the N composition ratio is about 13.45%. As shown in Fig.1(C), Al-O bonds are formed after 900°C PDA.

<u>Electrical Characterization</u>

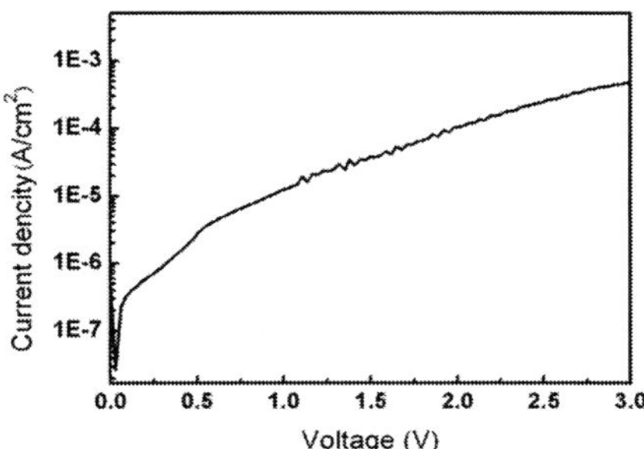

Fig.2 J_g-V_g curves of MoAlN/HfSiAlON gate dielectric.

High-frequency (1MHz) capacitance-voltage (C-V) characteristics of MoAlN/HfSiAlON gate stack were measured at room temperature by using MDC C-V analyzer. The leakage-current (J_g-V_g) characteristics were measured by using Keithley 4200 semiconductor parameter analyzer.

A thin EOT of 18 Å is extracted from the C-V curve (not shown), which is fitted by Quantum Mechanical C-V simulator developed by the Device Group at UC Berkeley. The measurements show that the gate leakage current of HfSiAlON gate dielectric with an 18Å EOT is 9.4×10^{-5}A/cm^2 at V_{fb}+1V (Fig.2). Compared to the pure SiO$_2$ with the same EOT, the HfSiAlON gate dielectric yields a gate leakage reduction of more than 10^4 times. This is because the physical thickness of HfSiAlON gate dielectric is much thicker than SiO$_2$ for the same EOT, which reduce the leakage current greatly.

There is always Fermi-level pinning effect existed at different degree between gate electrode and high-k gate dielectric, especially for PMOS device, which causes a lot of troubles for P-type work function tuning. In this study, MoAlN metal gate was used on HfSiAlON gate dielectric as the gate electrode. The calculation of the metal gate work function is according to the equations as below:

$$\Phi_m = \Phi_s + \Phi_{ms} \qquad [1]$$

$$\Phi_{ms} = V_{FB} + \frac{Q_0}{\varepsilon_0 \varepsilon_{HK}} \times EOT \qquad [2]$$

Here, Φ_m and Φ_s are the work functions with respect to the metal gate and Si substrate, V_{FB} is the flat band voltage, Q_0 is the equivalent electronic charge, ε_0 is the vacuum permittivity and ε_{HK} is the permittivity of high-k gate dielectric. The value of Φ_{ms} is equal to V_{FB} value when EOT=0 (8,9). In order to exact the work function of MoAlN metal gate, HfSiAlON films were prepared with different EOT. The V_{FB} versus EOT plot is shown in Fig.3. The effective work function of MoAlN metal gate on HfSiAlON gate dielectric exacted is 5.14eV.

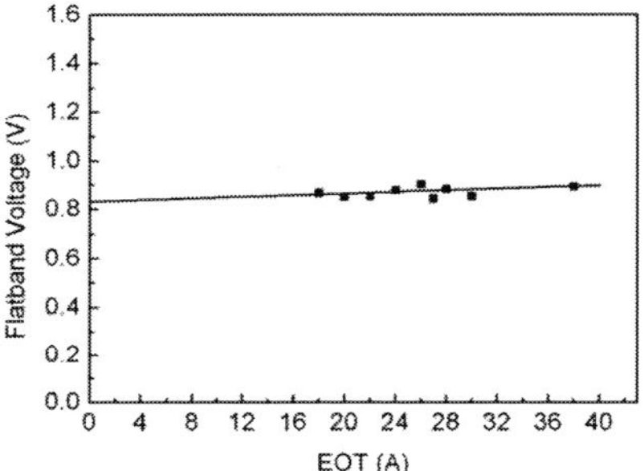

Fig.3 V_{FB} versus EOT plot of MoAlN metal gate on HfSiAlON gate dielectrics.

Effects of AlN$_x$ Layer on electrical properties of HfSiAlON Gate Dielectric

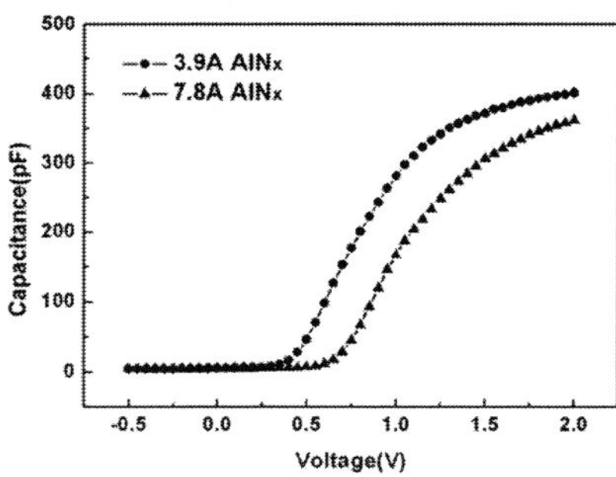

Fig.4 C-V curves of HfSiAlON gate dielectrics with different thickness of AlN$_x$.

Effect of the AlN$_x$ thickness. It is found that the EWF enhancement is dependent on the thickness of AlN$_x$ interfacial layer, as shown in Fig.4. When the thickness of AlN$_x$ interfacial layer is 3.9 Å, the EOT of HfSiAlON is 27Å and the flatband voltage (V_{FB}) is 0.46V. When the thickness of AlN$_x$ interfacial layer is 7.8Å, the EOT of HfSiAlON is 30Å and the flatband voltage (V_{FB}) is 0.71V. It is shown that with the

AlN$_x$ thickness increase, the V$_{FB}$ shift increases forward to the positive direction and EOT increases also. This is because AlN$_x$ reacts on SiO$_2$ buffer layer and Al-O$_x$ dipoles are formed, which is helpful to modulate the energy band and shift the threshold voltage (10-12). With the AlN$_x$ thickness increase, more Al-O$_x$ dipoles are formed, which induces the positive shift of flatband voltage increase. While, because AlO$_x$ is a relatively lower high-k layer, thicker AlO$_x$ interfacial layer will result in the EOT increase.

<u>Effect the location of AlN$_x$.</u> It is also found that the location of AlN$_x$ interfacial layer is important for modulating the metal gate EWF. To clarify the relationship, HfSiAlON gate dielectrics with AlN$_x$ at different location were prepared and TaN metal gate was deposited as gate electrode. The measurement results are compared with TaN/HfSiON gate stack, as shown in Fig.5.

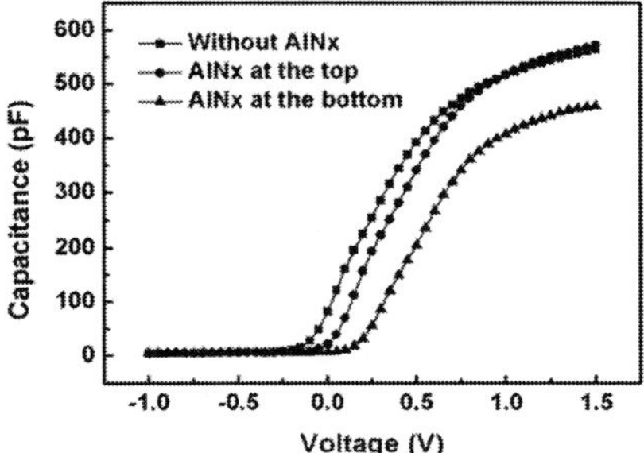

Fig.5 C-V curves of TaN/HfSiON and TaN/HfSiAlON with AlN$_x$ at different locations.

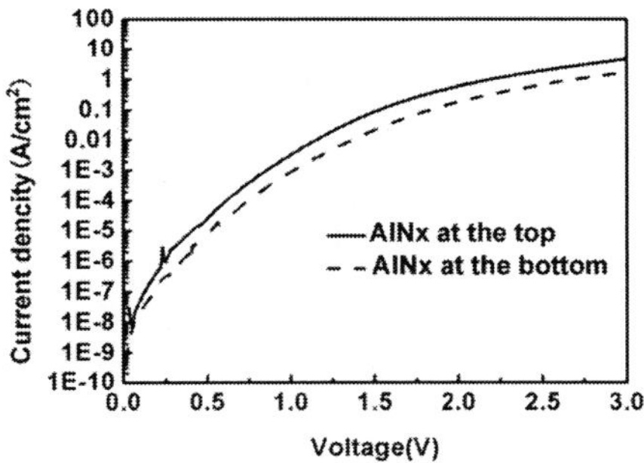

Fig.6 J$_g$-V$_g$ curves of TaN/HfSiAlON with AlN$_x$ at different locations.

Without the AlN$_x$ interfacial layer, the EOT of HfSiON gate dielectric is 15Å and the V$_{FB}$ of TaN/HfSiON gate stack is -0.11V. When it is located at the top interface, the EOT of HfSiAlON gate dielectric is 15Å and the V$_{FB}$ of TaN/HfSiAlON gate stack is 0.02V. When it is located at the bottom interface, the EOT of HfSiAlON gate dielectric is 19Å and the V$_{FB}$ of TaN/HfSiAlON gate stack is 0.21V. Compared to the

TaN/HfSiON gate stack, the flatband voltages of TaN/HfSiAlON gate stacks with different AlN_x locations both increase forward to the positive direction. And, TaN/HfSiAlON gate stack with AlN_x at the bottom interface has a largest flatband voltage shift ($\triangle V_{FB}$=0.313V). However, an EOT increase of 4Å is also observed. This is also because AlN_x reacts with SiO_2 buffer layer and $Al-O_x$ dipoles are formed, modulating the energy band and shift the threshold voltage forward to the positive direction. While, the different location of AlN_x has different effect on the threshold voltage shift. When AlN_x is located at the top interface, Al atoms have to diffuse through the whole gate dielectric and therefore less Al atoms react with SiO_2 buffer layer, resulting in fewer $Al-O_x$ dipoles formed and smaller flatband voltage shift obtained. When AlN_x is located at the bottom interface, more $Al-O_x$ dipoles are formed and larger flatband voltage shift is obtained. The different locations of AlN_x have little effect on the leakage current density, due to the same physical thicknesses. As shown in Fig.6, the leakage current density is all 3.9×10^{-3} A/cm^2 (V_g=V_{fb}+1V).

Conclusion

The HfSiAlON high-k gate dielectrics were fabricated by a PVD method, followed by a PDA using rapid thermal annealing in N_2 with a trace of O_2 ambience. The HfSiAlON gate dielectric with an 18Å EOT is obtained. The corresponding leakage current is only $9.37\times10^{-5} A/cm^2$ (V_g= V_{fb}+1V), about four orders of magnitude lower than poly-Si /SiO_2 gate stack with the same EOT. The effective work function of gate stack for MoAlN metal electrode on HfSiAlON gate dielectric is 5.14eV. It proofs that MoAlN is an ideal metal gate material for PMOSFET. The thickness and the location of AlN_x interfacial layer have strong effects on electrical properties of HfSiAlON, which has been investigated carefully. The excellent characteristics observed in HfSiAlON gate dielectric suggest that it may be a very promising gate dielectric to replace the conventional SiO_2 or SiON as gate dielectric for advanced CMOS devices in gate last process.

Acknowledgments

The authors thank the State Key Development Program for Basic Research of china (Grant No. 2006CB302704) and the National Natural Science Foundation of China (Grant No. 60776030) for the financial support, Dapeng Chen, Wenwu Wang, Huaxiang Yin, Shumin Chai, Yongliang Li, Huajie Zhou, Haiping Shang, Song Yi and all engineers at Integrated Circuit Advanced Process Center of IMECAS for their support and corporation in the sample fabrication.

References

1. J. Robertson, *Eur. Phys. J. Appl. Phys.* 28, 265 (2004)
2. G. D. Wilk et al., *J. Appl. Phys.* 89(10), p.5243 (2001)
3. M. Houssa et al. *Materials Science and Engineering R* 51, 37 (2006)
4. H. J. Li and M. I. Gardner *IEEE Electron Devices Lett.* 26(7), 441(2005)
5. H.N. Alshareef et al., *App. Phy. Lett.* 88, 112114 (2006)
6. W.W. Wang et al., *J. Appl. Phys.* 105, 064108 (2009)

7. Q. X. Xu et al., *IEEE Trans. Electron Devices* 51, 113 (2004)
8. H. J. Zhou and Q. X. Xu *Chin. J. Semiconductors* 28, 1532(2007)
9. G. B. Xu and Q. X. Xu *Chin. Phys.* 2, 768(2009)
10. X.L. Wang et al., *App. Phy. Lett.* 96, 152907 (2010)
11. Z.C. Yang et al., *App. Phy. Lett.* 94, 252905 (2009)
12. O. Sharia et al., *Phy. Rev. B* 77, 085326 (2008)

ECS Transactions, 34 (1) 503-508 (2011)
10.1149/1.3567627 ©The Electrochemical Society

Deposition of VO_X films by reactive sputtering and its properties

Xiaoying Wei, Kailiang Zhang *, Fang Wang, Lanlan Wang, Yang Zhang, and Kai Song,

School of Electronics Information Engineering, Tianjin Key Laboratory of Film Electronic & Communication Devices, Tianjin University of Technology, Tianjin, 300384 , *corresponding author , kailiang_zhang@163.com

VO_X thin films were deposited on the $Cu/Ti/SiO_2$ substrates by reactive sputtering at room temperature and annealed under vacuum at different temperatures. The crystal structure and morphology of VO_X films were characterized respectively by XRD and AFM, while the electrical switching properties were tested by Semiconductor Device Analyzer. The effects of annealing temperature on structure and morphology were discussed. Results showed that the film as-deposited was amorphous, and crystallized into different phases (e.g. V_2O_5, V_2O_3 etc.) with the increasing of annealing temperature. AFM results indicated that the crystal grain sizes of VO_X thin films were enlarged from 300 to 500℃. The in-situ morphologies have obvious changes from RT to 75℃. The device unit produced by films annealed at 400℃ can achieve a reversible metal−insulator transition (MIT), whose reset and set voltage are 0.56V and 1.12V, respectively.

Introduction

Recently, the binary oxide-based resistive random access memory (RRAM) has attracted growing interests as one of the most promising alternatives for easing challenges of flash scaling down, owing to its great potential of nonvolatility, low power consumption, high scalability, multibit storage, and CMOS compatibility (1-2) . The work on Resistive RAM (RRAM) has not been done using VO_X as the resistance change material.

Vanadium takes various valence states. There are a number of oxide forms, including V_2O_5, VO_2, V_2O_3 and V_6O_{13} etc., existing in vanadium oxide. Therefore Vanadium oxide films exhibit unique optical, electrical, electrochemical and thermal switching properties, including a phase transition from the semiconductor to metal (3-4). The interests in VOx films have increased in recent years due to their potential applications in various optoelectronic devices such as catalyst, smart windows for solar cells, electronic information displays and color memory devices, cathode coatings in high-capacity lithium batteries and optical switches (5).

The properties of films strongly rely on the source materials, deposition techniques and deposition parameters. The stoichiometry of thin films VO_X is easily deviated by small changes in synthesis parameters. A lot of techniques have been applied to deposit VO_X thin films, such as reactive sputtering, pulsed laser deposition, sol-gel process and evaporation. Among these methods, reactive sputtering is the common method in producing VOx films because it has many advantages in process stability and production cycle time (a high deposition rate) (6).

In this paper, VO_X films were deposited on the $Cu/Ti/SiO_2$ substrates by reactive sputtering at room temperature and annealed under vacuum at different temperatures. The crystal structure, surface morphology and reversible resistive switching of the VO_X films are investigated.

Experimental

A 200-nm-SiO_2 on the P-type Si wafer was used as the substrate for device fabrication. Both 5-nm-thick Ti adhesion layer and 500-nm-thick Cu bottom electrode were deposited on the SiO_2 layer. Subsequently, a 300-nm-thick VO_X thin film was deposited on Cu bottom electrode by reactive sputtering. Deposition conditions for VO_X thin films were listed in Table I.

TABLE I. Deposition conditions for VO_X thin films

Deposition conditions	Parameters
Operation pressure	1pa
Deposition temperature	RT
Deposition time	20min
$O_2/(O_2+Ar)$	20%
Sputtering power	200W
Annealing time	1h
Annealing temperature	300°C,400°C,500°C

The crystal structure of VO_X thin films were characterized by X-ray diffraction (XRD). The X-rays operating parameters are a Cu Kα of wavelength and a scan range of 10°-60°. Microstructure and morphology of VO_X films were characterized by AFM (Agilent 5600LS).The current-voltage (I-V) characteristics were measured by Agilent B1500A.

Results and Discussion

Fig.1. shows the XRD spectra for VO_X films on $Cu/Ti/SiO_2$ substrate annealed at different temperatures. In Fig.1, there were no vanadium oxide peaks shown in the film as-deposited, suggesting that the film has an amorphous structure. After thermal annealing, the films show a polycrystalline structure. When VO_X films were annealed at 300°C, there were V_2O_5, V_4O_9 and V_7O_{13} mixture phases. The films annealed at 400°C and 500°C coincide with V_4O_9, V_4O_7 phases and V_4O_7, V_2O_3 phases, respectively. It indicated that valence states of vanadium transit from high valence states to low valence states with the increasing of annealing temperature.

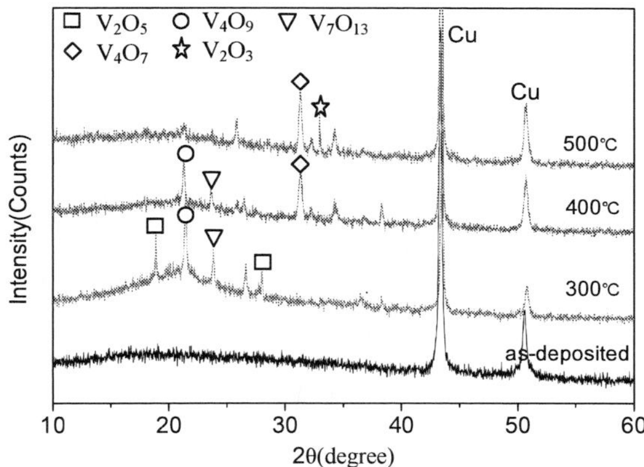

Figure 1. The XRD patterns of the VO$_X$ films annealed at different temperatures.

The AFM images of VO$_X$ films annealed at different temperatures are presented in Fig.2. The images of (a), (b), (c) and (d) depict the mirco-topography of films as-deposited, and annealed at 300℃, 400℃, 500℃ respectively. As can be seen that average surface grain size of the VO$_X$ films were enlarged, and coupling and merger phenomena could be observed with the increasing of annealing temperature. RMS roughness of films for as-deposited, annealed at 300℃, 400℃ and 500℃ were 1.91m, 1.52nm, 1.23nm and 1.22nm, respectively. The RMS roughness lightly decreased with the increase of annealing temperature. It can be explained that the higher annealing temperature will enhance the mobility of molecules or ions in the films. The movement of molecules or ions may induce the crystallization and the variation of surface morphology of the films.

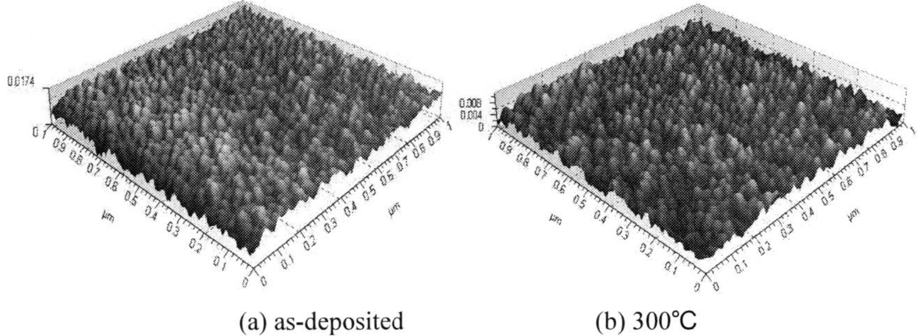

(a) as-deposited   (b) 300°C

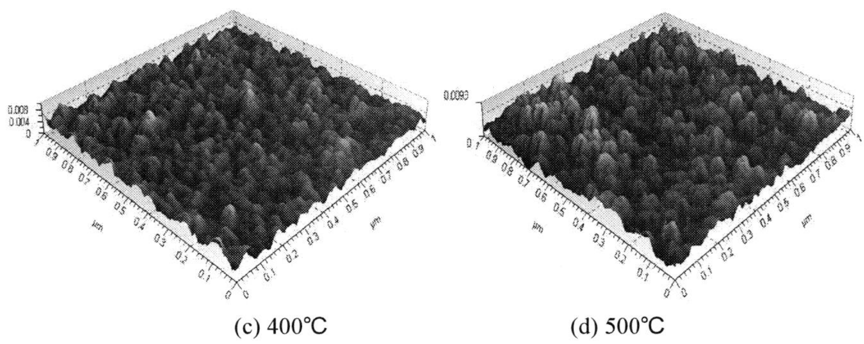

(c) 400°C　　　　　　　　(d) 500°C

Figure 2. The AFM image of the VO$_X$ films annealed at different temperatures.

Fig.3 presents the AFM morphology changes of VO$_X$ films during an in-situ heating. Fig.3 (a) and Fig.3(b) were the AFM micro-structure at RT and 75°C. The average grain sizes of VO$_X$ films have obvious changes from RT to 75°C. When the temperature increased from RT to 75°C, the average grain size increased from 60nm to 200nm, and RMS roughness increased from 10.8nm to 12.4nm. Both RMS roughness and average grain size for the VO$_X$ films at 75°C are bigger than those at RT.

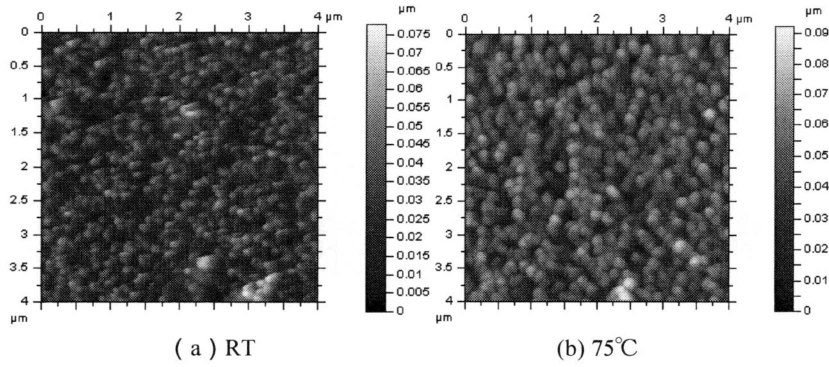

(a) RT　　　　　　　　(b) 75°C

Figure 3. The in-situ morphology changes of VO$_X$ films at different temperatures.

Fig.4 shows the schematic view of the structure used to test the reversible metal−insulator transition (MIT) of VO$_X$ films. Both bottom and top electrodes were Cu. The current-voltage (I-V) characteristics were charactered by voltage sweep.

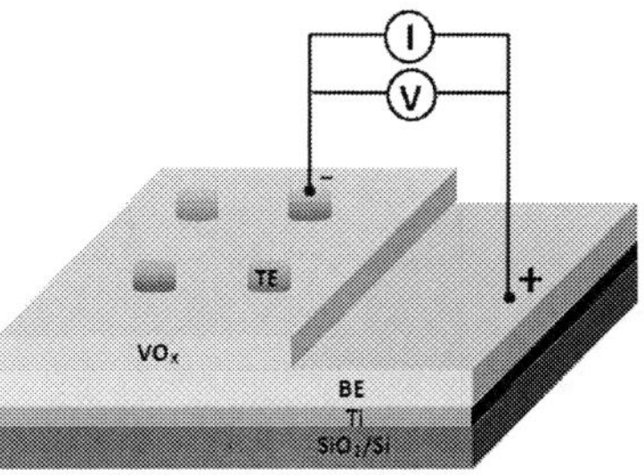

Figure 4. Schematic drawing of the test circuit used for Cu/VO$_X$/Cu memory unit.

Fig.5 presents the typical I-V curves of forming, reset and set processes of the Cu/VOx/Cu structure applied by the voltage sweep. The current compliance (CC=5mA) were used in the forming and set processes in order to avoid irreversible memory unit damage. The inset shows the R-V curve of the reset process. When a positive voltage was applied to the Cu/VO$_X$/Cu structure for the first time, the device attained ON-state at 1.63V, and the resistance switched from high state to low state, the forming process was achieved. Then when applying a positive voltage to device, an abrupt resistivity changed at 0.56V, and the device was on HRS (high resistance state), the reset process was achieved. After reset process, the set process which is similar with forming process was achieved, the set voltage was 1.12V, and the device was on LRS (low resistance state). Based on the above resistive switching characteristics, the Cu/VO$_X$/Cu structure memory unit can achieve reversible metal−insulator transition (MIT). The LRS/HRS resistance ratio of reset process is about 10^3.

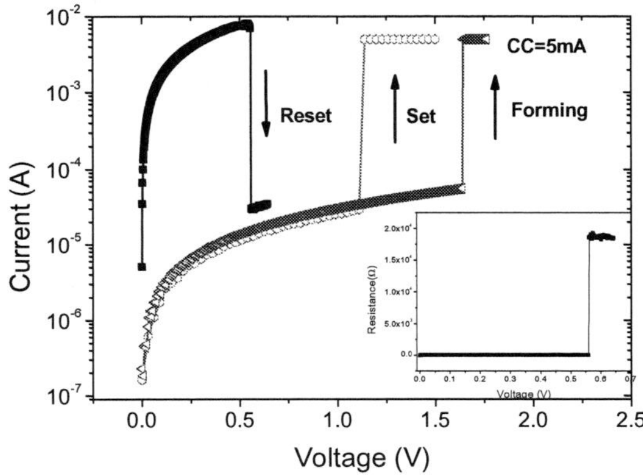

Figure 5. I-V characteristics of Cu/VO$_X$/Cu memory unit and the R-V curve of the reset process

Conclusion

The crystal structure and morphology of VO_X films annealed at different temperatures were investigated. There were no vanadium oxide peaks remaining of the film as-deposited, with the increasing of annealing temperature, V_2O_5, VO_2 and V_2O_3 exist simultaneously in the films. Average surface grain sizes of the VO_X films were enlarged with the increasing of annealing temperature. Average grain sizes for the VO_X films observed in an in-situ morphology investigation are bigger at 75°C than at RT. The $Cu/VO_X/Cu$ structure can achieve a reversible metal–insulator transition (MIT). This paper offers VO_X material as a potential candidate to be applied in the future RRAM.

Acknowledgments

We acknowledge the financial support provided by the National Natural Science Foundation of China under Grant No 60806030, and Tianjin Natural Science Foundation under Grant No 08JCYBJC14600, No 10SYSYJC27700 and Tianjin Science and Technology Developmental Funds of Universities and Colleges under Grant No ZD200709.

References

1. Weihua Guan, Shibing Long, Qi liu, Ming Liu, Wei Wang, *IEEE Electron Device Lett.*, **29**(5), 434- 436 (2008).
2. Masanori Kawai, Kimihiko Ito, Noriya Ichikawa, and Yuichi Shimakawa, *Appl. Phys. Lett.*, **96**(7), 072106 (2010).
3. Dmitry Ruzmetov, Gokul Gopalakrishnan, Jiangdong Deng, Venkatesh Narayanamurti, and Shriram Ramanathan1, *Journal of Applied Physis.*, **106**(8), 083702 (2009).
4. Xiaochun Wu, Fachun Lai, Limei Lin, Yongzeng Li, Lianghui Lin, Yan Qu, Zhigao Huang, *Applied Surface Science.*, **255**(5), 2840-2844 (2008).
5. R. G. Mani and S. Ramanathan, *Appl. Phys. Lett.*, **91**(6), 062104 (2007).
6. A. Khodin, Hak-In Hwang , Sung-Min Hong, V. Zalessky,T. Leonova, M. Belov , E. Outkina, *Materials Science and Engineering B.*, **141**(3),108-114 (2007).

ECS Transactions, 34 (1) 509-514 (2011)
10.1149/1.3567628 ©The Electrochemical Society

ALD Ru and its Application in DRAM MIM-Capacitors and Interconnect

Marc Schaekers, Johan Swerts, Laith Altimime and Zsolt Tőkei

Imec, Kapeldreef 75, 3001 Leuven, Belgium

Ruthenium is proposed as electrode for next generation DRAM metal-insulator-metal capacitors and interconnect barrier/seed. This study compares different ALD processes to deposit Ru and RuO_2 films. ALD with oxygen results in the deposition of either Ru or RuO_2, depending on deposition conditions. We also report the growth and scalability of Ru films by plasma-enhanced ALD (PEALD) with ammonia. The inorganic precursor RuO_4 was studied as an alternative chemical pathway, of which the characteristics and film properties were studied at temperature range 100-250°C using various process conditions. The quality of films deposited with thermal ALD and PEALD is compared with respect to the targeted applications. In particular, the investigation of the thermal stability of the different Ru and RuO_2 films under typical DRAM anneal conditions is presented.

Introduction

In next generation DRAM metal-insulator-metal capacitors (MIMCAP) structures, metal electrodes and a high-k dielectric allow continuous scaling of the size of DRAM cells. Perovskites like $SrTiO_3$ (STO) are introduced as the insulator in MIMCAP (1). One concern in the selection of metal electrodes for MIMCAP is the chemical stability of the interfaces between the dielectric and the electrodes. Ru and its oxide, RuO_2, are promising candidates to replace TiN as electrode in contact with STO because RuO_2 is conductive. To limit the amount of expensive ruthenium, the MIMCAP stack can include a TiN layer plus a thin Ru (or RuO_2) layer.

Advanced interconnect (IC) technology typically consists of Cu lines and vias, electrochemically plated on a Cu/Ta/TaN seed/barrier deposited with Physical Vapor Deposition (PVD). Improvements of PVD technology have extended its use for many nodes. For sub-25 nm line widths, the total thickness and resistivity restrictions form a major hurdle, not only for PVD but also for simple ALD alternatives. Modified stacks containing a mixed barrier-seed and seedless Cu electroplating are explored; among the different candidates, Ru-based layers have been identified as very promising (2).

In this work, thermal and plasma enhanced ALD processes for Ru and RuO_2 are compared. In-depth mechanistic studies have revealed that most of these so-called ALD processes might better be labeled as "pulsed CVD" because of the occurrence of decomposition or second order reactions, but conveniently we will stick to "ALD". Both MIMCAP en IC applications require conformal deposition of thin Ru films, but other process restrictions are quite different. First, ALD with (Methylcyclopentadienyl)(Pyrrolyl)Ruthenium (MCPR) and O_2 as a co-reactant is considered (3). A variation of the oxygen pulse time results in a deposition of either Ru or RuO_2. The presence of oxygen makes this process not suitable for interconnect because it would increase contact resistance. Using an ammonia plasma instead of oxygen, a PEALD process of Ru for interconnects was studied (4). The plasma impacts the conformality, and while it could maintain sufficient sidewall quality for interconnect aspect ratios, it is not expected to be useful for very high aspect ratio MIMCAP structures.

The inorganic precursor RuO_4 in combination with hydrogen is an interesting alternative, as thermal process with no added oxygen. The safety issues of RuO_4 were tackled by using Torus, a blend of RuO_4 in a fluorocarbon solvent supplied by Air Liquide (5). The process characteristics and properties of films deposited with Torus were studied at temperature range 100-250°C using various process conditions. Depending on the hydrogen dose a deposition of Ru or RuO_2 can be selected (6).

The properties of the different thermal ALD and Plasma Enhanced ALD (PEALD) have been described in detail before (3,4,6); in this study the quality of deposited films will be compared.

Experimental

ALD Ru layers were deposited with MCPR and oxygen at 310°C on TiN and STO surfaces in a 300 mm ALD chamber. Two processing cycles for the thermal ALD process have been used for the deposition of Ru and RuO_2 respectively. Ru is deposited with 4s Ru, 1s Ar purge, 1s O_2 and 1s Ar purge (4:1:1:1), and RuO_2 using the same Ru and Ar pulses, but 4s O_2 pulse (4:1:4:1). A direct RF plasma is applied for PEALD of Ru using MCPR pulsed with either NH_3 at 330°C.

Depositions using Torus were performed in an experimental ALD reactor on coupons. Remote RF plasma can be applied to the gas mixture. A spectroscopic ellipsometer is attached to the chamber to perform in-situ monitoring. The Torus was kept at room temperature in a bubbler with argon used as carrier gas. Hydrogen, 20% in argon, was used as reagent for ruthenium depositions.

Substrates used were ALD TiN, PVD TiN and $SrTiO_3$ (STO). 10nm of stoichiometric STO (Sr:Ti=1) was deposited with ALD on TiN and given a crystallization anneal at 600°C prior to Ru deposition.

X-ray reflectometry (XRR), grazing incidence X-ray diffraction (GIXRD), sheet resistance (Rs), laser scattering (haze map by KLA-Tencor SP2), Rutherford backscattering spectrometry (RBS), elastic recoil detection (ERD), X-ray photoelectron spectroscopy (XPS), Static time-of-flight secondary ion mass spectroscopy (TOFSIMS), atomic force microscopy (AFM) and scanning electron microscopy (SEM) were used to measure the thickness, density, crystallinity, resistivity, chemical composition, surface closure and roughness. The thermal stability of the deposited films was investigated with SEM after anneal in nitrogen or oxygen. The film thickness (5nm) and the anneal time (60 s) was chosen irrespective of deposition process to match typical DRAM conditions.

Results and discussion

Deposition of Ru films: surface sensitivity

Deposition on metals. All ruthenium processes show some degree of surface sensitivity, but there exist large differences between them.

The deposition of Ru from MCPR/O_2 on PVD TiN has large non-uniformity in both thickness and composition (3). On ALD TiN the deposited film was much more uniform and smooth (fig. 1a). Also on sputtered Ru the deposition occurred smoothly.

By PEALD from MCPR and N_2/NH_3 plasma a Ru growth per cycle of 0.04 nm was obtained on TiN and TaN (4). The minimal thickness of the metal nitride to enable Ru growth has been determined to be as low as 0.7-0.8 nm which is promising for scaling. Growth studies on scaled and thick ALD TiN or TaN still showed a limited growth inhibition during the first 20 cycles followed by a linear growth behavior. TOFSIMS suggests Ru layer closure for a film thickness around 2 nm. AFM revealed that the RMS roughness values were below 0.4 nm for film thicknesses up to 20 nm. Compared to ALD, PEALD shows less growth inhibition and smoother films (fig. 1).

With in-situ ellipsometry only a very minor incubation on TiN was detected (< 5 cycles) with the Torus/H_2 reaction system (6). A roughness below 0.4 nm RMS on a 5 nm film (60 cycles) on TiN was measured. The as-deposited films were inspected by SEM and appeared smooth across the deposition temperature range (similar to fig. 1b).

Deposition on dielectrics. For the top electrode applications, Ru or RuO_2 layers need to be deposited on the dielectric. It is found that the deposition using the same MCPR/O_2 processes for Ru and RuO_2 used on TiN does not give a decent film on STO unless a surface pre-treatment is done using H_2/He remote plasma. A monolayer reduction of the surface is expected during such treatment and is favorable to the nucleation of Ru on the STO surface, but might deteriorate the electrical properties of the dielectric. A linear growth is obtained then with a deposition rate around 0.58 Å/cycle. Also the PEALD process suffers from severe incubation on dielectrics; RBS showed that only 1.10^{14} Ru atoms/cm^2, i.e. less than a monolayer, were deposited on SiO_2 after 120 PEALD cycles.

With Torus/H_2, the difference between TiN and STO substrates is much smaller, although the incubation time on STO, about 10 cycles, was still longer than on TiN. Deposition on thermally grown SiO_2 was also tested but appeared to be more difficult: no deposition was detected after 100 cycles. When the deposition was preceded by an argon plasma treatment in the ALD chamber, the incubation time dropped to less than 10 cycles. This confirms a general trend that was also observed by us with MCPR and by others (7), of better nucleation after plasma treatment.

Thermal stability of Ru films

Anneal in nitrogen ambient. The surface smoothness can also be seen in figures 2 and 3, after anneal in nitrogen. From anneal temperatures of 600°C onward, the ruthenium surface begins to change. We have seen a similar behavior of films deposited with ALD or PEALD and metal-organic precursors (figure 2) as with Torus (figure 3).

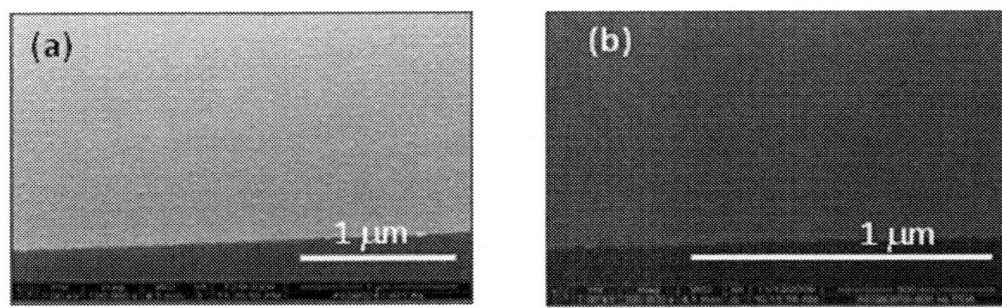

Figure 1. 5nm Ru on TiN, as deposited with MCPR ALD (a) and PEALD (b).

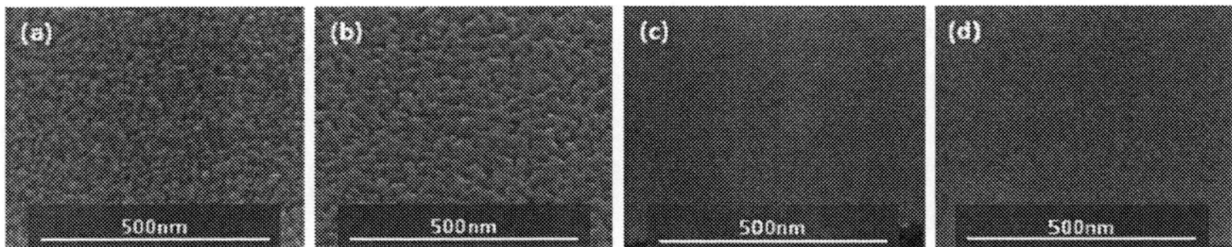

Figure 2. Ru films on TiN, deposited with MCPR ALD (a-b) or PEALD (c-d), after 60s anneal in nitrogen at 500°C (a, c) or 600°C (b, d).

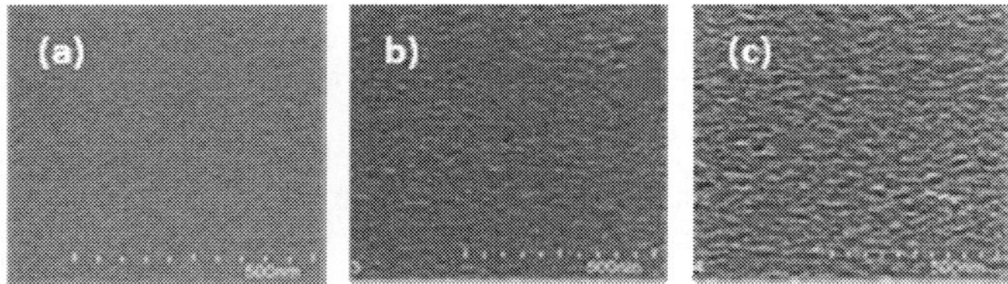

Figure 3. Effect of a 60 s anneal in nitrogen on the surface of 5 nm Ru on TiN deposited with Torus/H_2. a: 500°C, b: 600°C, c: 650°C.

<u>Anneal in oxygen ambient.</u> Anneal of Ru in oxygen atmosphere causes crystallization to occur already at very low temperature (figure 4). The behavior of films deposited with any chemistry is very similar and indicates that this might be a fundamental material behavior.

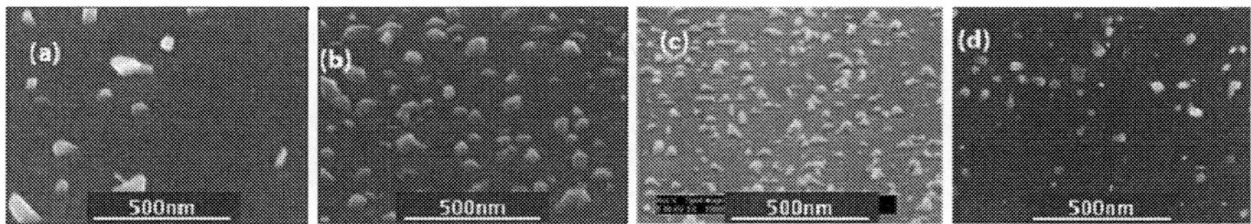

Figure 4. Effect of a 60 s anneal in oxygen on the surface of Ru films a: MCPR/O_2 at 400°C, b: MCPR PEALD at 400°C, c: MCPR PEALD at 350°C, d; Torus/H_2 at 350°C

Deposition of RuO_2 films

<u>Surface sensitivity of the deposition.</u> Longer O_2 pulse in MCPR/O_2 induces an O-rich ambient, and thus a RuO_2 layer is formed. Composition and atomic configuration of the substrates have a strong impact on the deposition. Depositions of RuO_2 on TiN occur

much more easily and uniform than on STO. With Torus, the decomposition of RuO_4 at 200°C at low hydrogen pressure results in deposition of a RuO_2 film (6). At 200°C, with 2s Torus pulses but without hydrogen, 0.08 nm RuO_2 growth per cycle was obtained on TiN.

<u>Thermal stability in nitrogen ambient.</u> RuO_2 films deposited on TiN at 200°C, without hydrogen pulse, were investigated further. The thermal stability in nitrogen was examined with SEM following anneal during 60s at 500°C to 650°C. The surface remained unchanged until 600°C.

<u>Thermal stability in oxygen ambient.</u> The thermal stability in oxygen was tested as well. ALD RuO_2 from $MCPR/O_2$ didn't change visibly upon anneal. Unexpectedly, crystal formation was observed after anneal of Torus RuO_2 in oxygen at 400°C (Fig. 5b). With RuO_2 from other precursors this had not been observed so these films are suspected to be Ru-rich. To improve the stoichiometry of the RuO_2 films, the Torus pulses were alternated with oxygen pulses (10s at 0.9mbar O_2). Smooth RuO_2 films were deposited this way; no crystal formation was observed on those films after oxygen anneal at 400°C (Fig. 5c).

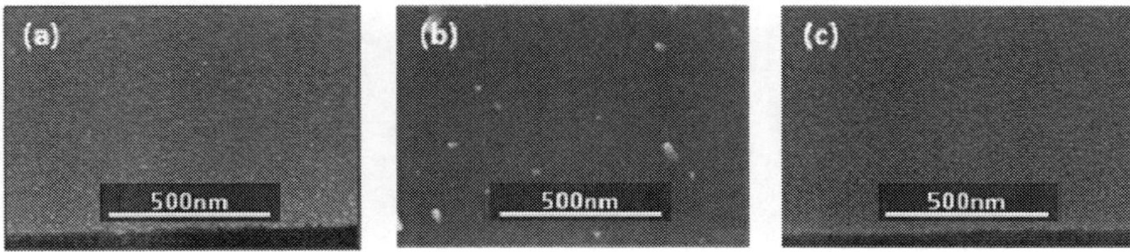

Figure 5. Surface of RuO_2 after 60s anneal at 400°C in oxygen. A: $MCPR/O_2$ ALD; b: Torus; c: $Torus/O_2$.

Conclusions

Ru and RuO_2 layers were deposited on selected metallic and dielectric substrates by an O_2-based thermal ALD process, by PEALD and by a thermal process without added oxygen. The nucleation period plays a key role in the deposition, inducing sensitivity to the substrates. The PEALD process shows less inhibition than the corresponding ALD process with the same Ru precursor, in agreement with a better performance of the ALD process on plasma treated surface. A short incubation time is important for interconnect application, where the maximum thickness is very limited. The thermal stability in nitrogen and oxygen ambient, important for MIMCAP bottom electrode application, is comparable for films deposited with a different chemistry.

References

1. J.A. Kittl, K. Opsomer, M. Popovici et al., ECS Trans. **19** (1), 29 (2009)
2. O.-K. Kwon, S.-H. Kwon, H.-S. Park and S.-W. Kang, J. Electrochem. Soc. **151** (12), C753 (2004)
3. C. Zhao, M.A. Pawlak, M. Popovici, M. Schaekers, E. Sleeckx, E. Vancoille, D.J. Wouters, Zs. Tokei and J.A. Kittl, ECS Trans. 25 (4), 377 (2009)

4. J. Swerts, S. Armini, L. Carbonell, A. Delabie, A. Franquet, S. Mertens, T. Witters, M. Schaekers, Zs. Tokei, G. Beyer, V. Gravey, A. Cockburn, K. Shah, J. Aubuchon and S. Van Elschocht, presented at AVS 57th Int. Symp., Albuquerque 2010, to be published in J. Vac. Sci. Technol.
5. J. Gatineau, K. Yanagita and C. Dussarrat, Microelectron. Eng. **83**, 2248 (2006)
6. M. Schaekers, B. Capon, C. Detavernier and N. Blasco, ECS Trans. **33** (2), 135 (2010)
7. T. Park, D. Choi and H. Jeon, 10th Int. Conf. on ALD, Seoul 2010, M22 (2010)

ECS Transactions, 34 (1) 515-521 (2011)
10.1149/1.3567629 ©The Electrochemical Society

Evaluation of Metallization Options for Advanced Cu Interconnects Application

N. Jourdan, L. Carbonell, N. Heylen, J. Swerts, S. Armini, A. Maestre Caro, S. Demuynck, K. Croes, G. Beyer, Z. Tökei, S. Van Elshocht, and E. Vancoille

imec, Kapeldreef 75, B-3001 Leuven, Belgium

The traditional Cu interconnect barrier/seed process consisting of PVD-Ta based barrier/Cu-seed will reach its limit between 20 nm and 30 nm wide trench dimension. To extend Cu interconnect technology further, possible solutions such as PVD-RuTa, PEALD-Ru-based, CVD-Co, PVD/CVD-self-formed-$MnSi_xO_y$ and self-assembled monolayers (SAMs) are studied. It is shown that both PVD-RuTa and CVD-Co possess the so-called seed enhancement capability allowing Cu filling of narrow recesses. However, they exhibit limitations in terms of Cu-diffusion barrier efficiency, electromigration reliability and scalability. Despite, the concept of SAM [NH_2-SAM(C_3)] as Cu diffusion barrier is demonstrated, it requires maturity and compatibility within the process flow (e.g. adhesion with the Cu overlayer). Finally, it is considered that PEALD-Ru-based alloys and CVD-based MnSixOy films are serious candidates for sub-30 nm wide trench technologies because of their conformal nature and ability to act as an efficient Cu diffusion barrier in the range of 2 nm thickness.

Introduction

In an Integrated Circuit (IC), the interconnect provides the wiring architecture to distribute clock and other signals to the various functional blocks; and the necessary power and ground connections. For about three decades, aluminum wires embedded in Silicon dioxide have been the interconnect of choice in silicon chips. Scaling of the ICs over the years has made the signal propagation in the circuit mainly limited by the so-called interconnect RC time delay rather than by the intrinsic gate delay. The RC time delay is proportional to the product of the total resistance and the capacitance of the line. As a result, new materials for both wires and the dielectric media had to be developed to enable faster chips. In the late nineties, copper has been introduced as an alternative to aluminum for its lower resistivity and higher electromigration resistance performance. As regards the dielectric, many materials with lower dielectric constant than SiO_2 (commonly referred as low-k materials) have been developed but their integration in the modern ICs is still under investigation. Beyond the benefits mentioned above, Cu interconnection also offers lower power consumption, manufacturing complexity and cost compared to the aluminum interconnect technology.

A major research effort has gone into making Cu a viable interconnect solution. New technologies such as chemical mechanical polishing (CMP) and electroplating of Cu have been developed along with a new patterning architecture known as Damascene. In such a process flow, after patterning of the dielectric, first a conductor barrier is deposited to isolate the Cu metallization from the rest of the IC (Cu is a fast diffuser in Silicon and several deep levels are well known). Then a thin Cu-seed layer is deposited to provide a low-resistance conduction path for the plating current that drives the subsequent electrodeposition process. It also assists the nucleation of the plated Cu film. Finally, the Cu line patterning is completed by a CMP operation to remove the excess of the Cu on top of the features.

The line width scaling with technology node results in a dramatic increase of the Cu line effective resistivity. This is partly due to the electron grain boundary/surface scattering (1) and partly because the higher resistive diffusion barrier occupies a larger portion in the Cu line metallization. As a result, a reduction in barrier layer thickness is required. The ITRS roadmap illustrates very clearly how challenging the Cu metallization will be in the coming years. In the 2009 edition, the latest forecast predicts a barrier thickness of 0.7 nm for a metal 1 half pitch of 7.5 nm wide (for MPU/ASIC application, assuming Cu interconnect technology is used) to be in production in 2024. The Cu interconnect may not be able to go so far and new materials, concepts might be needed. However, in the meantime, Cu-technology has to be extended as much as possible.
The traditional Tantalum-based diffusion barrier and commonly used Physical Vapor Deposition (PVD) technique for formation of both barrier and Cu-seed films has gone through continuous improvement from both process and hardware points of view and has led to reliable ICs for several generations. However, this scheme cannot take up the challenges mentioned above. Indeed, the optimized version of such a process flow leading to void free trenches and passing reliability criteria is limited to ~30 nm half pitch metallization. As a result, for further scalability, alternatives to the traditional scheme must be found. Below, we present an overview of new metallization options involving concepts such as seed enhancement layer, direct plating, self-formed barrier and self-assembled monolayers. In each case, the state of the art, limitations and required further investigation will be discussed.

Metallization Options to Extend the Cu Interconnect Technology

<u>1/ Physical Vapor Deposition (PVD) of RuTa</u>

The PVD-Cu-seed layer continuity can be marginal in aggressive trench dimensions, leading to voids during electroplating and eventually to device failure. One alternative to overcome this issue is to use a layer, which has the so-called ''seed enhancement'' property. Ru possesses such a capability (2). If this element is alloyed with Ta, with well established diffusion barrier aptitude, one can assume that a single RuTa alloy layer can act as a diffusion barrier as well as a Cu-seed enhancement layer. In our study, we made RuTa alloy layers of about 10% of Ta atoms by means of PVD using a target with the same composition.

For seed enhancement property evaluation, a thin RuTa layer of about 3nm followed by a thin Cu-seed layer of 7 nm was deposited in trenches as narrow as 20 nm width. For comparison the same structures were made using 1.5 nm of the standard Ta-based diffusion barrier. As it can be seen in top view SEM images (figure 1a), after CMP, voids are present in the case of Ta-based barrier, and not in the case of the RuTa layer. Also, from electrical point of view, only RuTa liners enabled successful metallization down to 20 nm width trenches as demonstrated by high line resistance yield in the range of 60 to

80% depending on the Cu-seed thickness/process used. All Ta-based barriers combined with the same Cu-seeds failed (figure 1b).

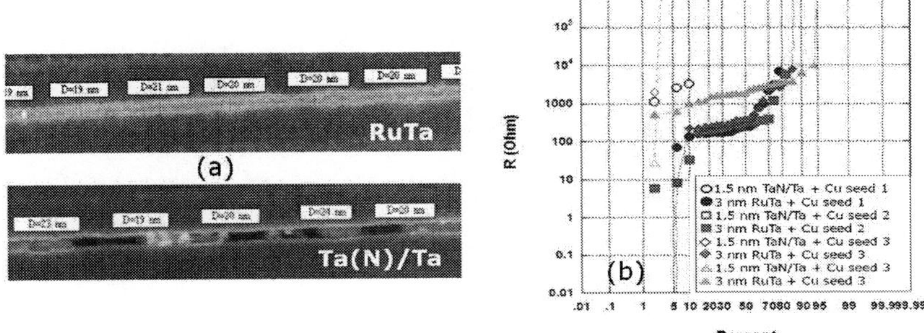

Figure 1. (a) Top view SEM images after CMP of 20 nm width lines using either 3 nm RuTa or 1.5 nm Ta(N)/Ta layers as the barrier and both combined to a 7 nm PVD-Cu seed. (b) Comparison of 20 nm width line resistance distribution between 1.5 nm TaN/Ta and 3 nm RuTa combined with the same Cu seed conditions.

These two results clearly show that RuTa layer allows repairing marginal Cu-seed during electroplating step and to extend PVD technology down to 20 nm width dimensions. However, it has been found that RuTa layers have poor Cu diffusion barrier efficiency and electromigration reliability performance.

<u>2/ Plasma Enhanced Atomic layer Deposition (PEALD) of Ru –based films</u>

As seen in the previous paragraph, Ru is a very attractive material to extend Cu interconnect technology, once its poor barrier efficiency is solved by appropriate alloying or film stacking. Beyond its seed enhancement capability, Ru can also be directly Cu-electroplated. Then, the filling is no longer impacted by the PVD-Cu seed overhang, which can result in pinch-off and void formation at top of the trenches. In order to get the maximum advantage of Ru capability in advanced technology nodes, it has to be deposited in a conformal manner and to be Dual Damascene (DD) structure compatible (i.e. no oxygen environment during liner deposition because of the presence of exposed Cu). The most advanced technology leading to a conformal layer is of course Atomic Layer Deposition (ALD). However, most of the ALD-Ru processes use Oxygen as reactant and therefore they are not DD compatible. Therefore, we developed a PEALD-Ru process using (Methylcyclopentadienyl)(Pyrrolyl) Ruthenium and an ammonia plasma (3). Another constraint with ALD-based processes is that Ru does not grow on Silicon Oxide and requires a nucleation/adhesion metal nitride layer such as TaN or TiN. It has been found that a nucleation layer of less than 1nm is sufficient to initiate the Ru growth and film closure at about 2nm thickness. Furthermore, Ru layers on 300mm wafer size exhibit a very good within wafer thickness uniformity (<5%), an excellent wafer to wafer uniformity (<1%), a density of about 95% of the bulk film, a surface roughness below 0.4 nm, a high purity and finally a very good step coverage as demonstrated on narrow trenches from 65 nm to 25 nm width (figure 2). To address the poor barrier efficiency of Ru, thin PEALD-Ru-based alloys films are under development and their compatibility with direct Cu electroplating is subject to further investigation.

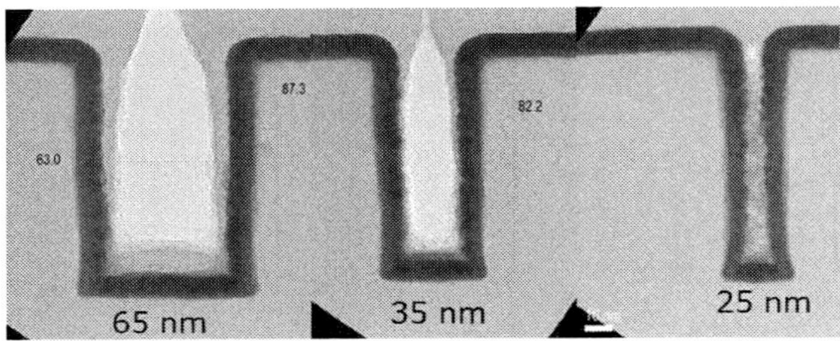

Figure 2. Cross-section TEM images of ALD-TaN (3 nm)/PEALD-Ru (5 nm) liner in 65 nm to 25 nm wide trenches.

3/ Metal Organic Chemical Vapor Deposition (MOCVD) of Co

Another material of interest as a seed enhancement layer is Co (2), which is more cost effective than Ru. For conformality purpose, thin Co layers were deposited by means of MOCVD technique using di-cobalt hexacarbonylbutylacetylene and H_2 as co-reactants. This process also requires a thin nucleation/adhesion layer such as TaN. After deposition, the films are subject to an in-situ H_2 plasma to densify the film and to reduce the amount of impurities within the layer. Such a film exhibits an excellent within wafer thickness uniformity of about 3% on 300mm wafer size, C and O contents below 6% and finally a film resistivity of about 50 $\mu\Omega$.cm for a 10 nm thick film. The integration of CVD-Co in the Cu interconnect technology was found to be very difficult, because of its incompatibility with acidic chemistries, which could be used during subsequent processes. For example, such a chemistry, after CMP operation leads to a strong Cu galvanic corrosion and Co dissolution on sidewalls as shown in figure 3a. After CMP process optimization, the seed repair capability of Co was proven electrically on 30 nm half pitch lines as illustrated in figure 3b. Indeed, when the standard Ta-based barrier is associated to a marginal Cu-seed, low line sheet resistance yield and spread distribution is obtained, while, the insertion of a 2 nm thick Co layer between the barrier and the seed enables higher yield and tighter distribution. Furthermore, it was found that the Cu-diffusion barrier efficiency of the stack is driven by the Ta-based under layer (i.e. 2 nm Co has a poor barrier performance) and a minimum thickness is needed for diffusion barrier purpose. This limits the scalability of Ta/Co stack. It was also shown that such a stack cannot meet the 10 years electromigration reliability lifetime specification.

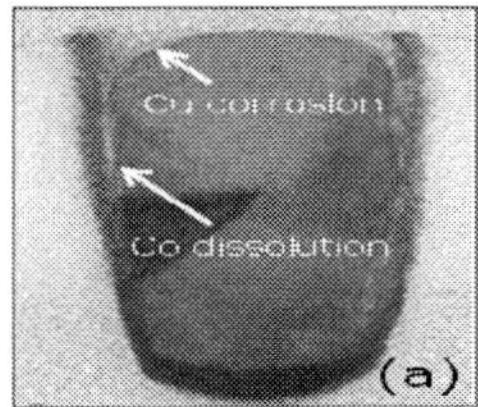

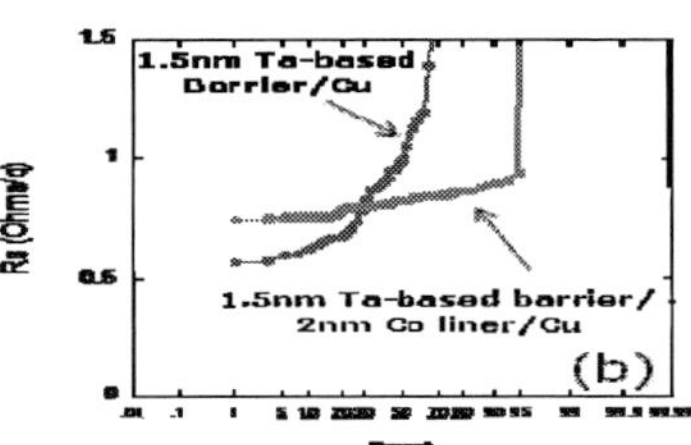

Figure 3. (a) Cross section TEM image of 50 nm Cu line after CMP with TaN/Co liner. (b) Impact of Co liner between barrier and seed on line resistance distribution.

4/ Manganese-silicate Self-formed Diffusion Barrier

Recently, Manganese-silicate self-formed barriers have received a lot of interest to extend the Cu interconnect technology for their excellent Cu diffusion barrier property. Originally, we evaluated the feasibility of such a barrier by depositing by PVD a CuMn (8 at. %) alloy layer directly onto the dielectric followed by a Cu electroplating operation and a thermal anneal. The later operation being responsible for migration of Mn atoms towards the Oxygen containing dielectric to form a thin $MnSi_xO_y$ layer (4). The formation of such a barrier has been demonstrated in 30 nm wide trenches as shown in figure 4a. Furthermore, the intrinsic barrier performance of the layer has been checked through TDDB test on planar capacitors (4, 5) and benchmarked to the standard Ta-based barrier and to a barrier-less process (i.e. direct Cu deposition onto the dielectric) (figure 4b). As can be seen, 2.2 nm of a self-formed barrier outperforms the barrier-less process in terms of Mean Time To failure (MTTF) and exhibits a comparable result to a 6 nm Ta-based barrier. This indicates that such a thin self-formed layer acts as an efficient Cu diffusion barrier and would be a serious candidate for the 16 nm technology node and beyond. However, due to the PVD approach, the scalability of such a process was found to be limited to ~40 nm wide trenches. Therefore, we developed a MOCVD-based process using Bisethylcyclopentadienyl manganese, $(EtCp)_2Mn$ as the precursor. The Mn dose incorporation on tetraethylorthosilicate, (TEOS)-SiO_2 was investigated by RBS and a linear behavior was found in the range of 100°C to 400°C. At 400°C (~ limit for Cu interconnect technology), a 4 nm thick MnSixOy with an amorphous structure (favorable for diffusion barrier purpose) was formed as shown in the cross-section TEM image in figure 4c. Currently, the barrier efficiency and electromigration reliability performance of the MOCVD-based MnSixOy layers are under investigation.

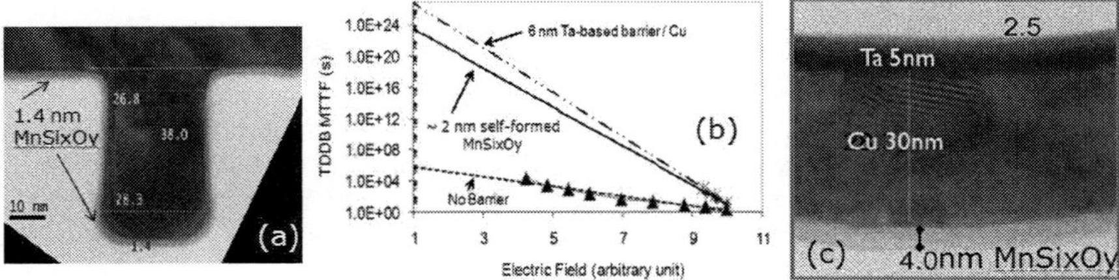

Figure 4: (a) cross section TEM image of 30 nm wide trench with 1.4 nm self-formed MnSixOy generated by PVD. (b) Comparison of ~2 nm MnSixOy (-PVD) with Ta-based barrier and barrier less structures in terms of TDDB. (c) Cross section TEM image of TEOS/MnSixOy (4 nm)/Cu/Ta stack. The MnSixOy layer is generated by MOCVD.

5/ Self-Assembled Monolayers (SAMs)

SAMs consist of self-assembled monomolecular organic films (figure 5a). And those with thickness ≤ 2 nm are of interest for diffusion barrier application in advanced Cu interconnect technologies (6). The molecules are composed of organic chains bonded by silane head group chemistry and terminated with desired functional groups (figure 5b). The organic chain length impacts the film thickness and quality (packing density, order). The head group will determine the selectivity and adhesion with respect to the substrate, packing density and thermal stability. Finally, the terminal group dominates the adhesion towards the overlayer (e.g. Cu for interconnect application). A comprehensive screening of SAMs molecules with different head group ($SiCl_3$, $Si(OCH_3)_3$, $Si(OCH_3)Cl_2$) bonding to the dielectric substrate, chain length (n=3 to 21) and terminal group (CH_3, Br, CN, NH_2, C_5H_4N and SH) bonding to the Cu overlay showed the most promising molecule is the so-called NH_2-SAM(C_n) [NH_2-$(CH_2)_n$-$Si(OCH_3)_3$] (6). Furthermore, it was found that NH_2-SAM(C_3) is compatible with Back End Of Line (BEOL) thermal budget. It exhibits a density of 1.48 g/cm^3 (comparable to SiC barrier), an adhesion to PVD-Cu of 7.2 J/m^2 and a sufficient Cu diffusion barrier performance as demonstrated by TDDB test on planar capacitors shown in figure 5c. Currently, for better barrier performance, NH_2-SAM(C_{11}) formation combined to more friendly Cu deposition technique than PVD such as Chemical vapor Deposition (CVD) and Electro Less Deposition (ELD) are under investigation. At the moment, such a stack suffers from adhesion point of view and requires further optimization for any integration purpose.

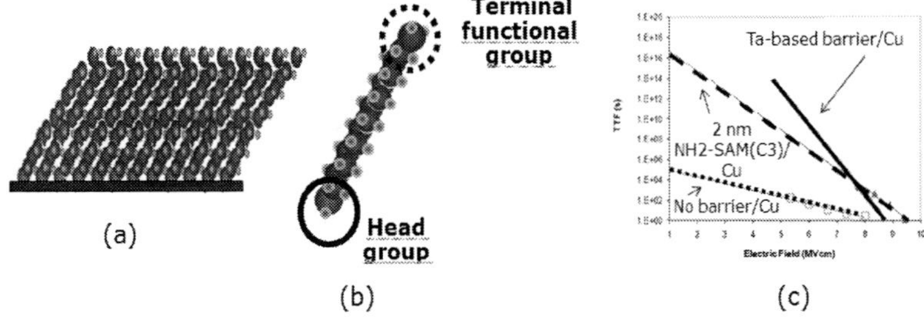

Figure 5: (a) Schematic of a monomolecular film (SAM). (b) Generic SAM molecule. (c) Comparison of a 2 nm NH_2-SAM(C_3) with Ta-based barrier and barrier less structures in terms of TDDB.

Conclusions

Possible alternatives to the traditional barrier/seed technology (PVD-Ta barrier/Cu-seed) are investigated to extend Cu interconnections in sub-30 nm wide trenches. It is shown that PVD-RuTa and CVD-Co allow repairing marginal Cu-seed continuity during the electroplating step. Such a capability has been demonstrated down to 20 nm width lines with RuTa and at least 30 nm in the case of Co. However, these liners show limitations in terms of barrier efficiency, electromogration reliability and scalability. SAMs are very attractive because they allow achieving in principle conformal organic films with thicknesses $\leq$ 2 nm. Although, the Cu diffusion barrier efficiency for 2 nm thick NH_2-SAM(C_3) film has been demonstrated, the technology requires more maturity and compatibility with subsequent operations (e.g. strong adhesion with the Cu overlayer). Finally, PEALD-Ru-based alloys and CVD-$MnSi_xO_y$ are very promising candidates for $\leq$ 20 nm wide trench technologies because of their conformality and barrier efficiency at low thickness.

References

1. G. Schindler et al., *Advanced Metallization Conference – AMC*, (2002).
2. M. Lane et al., *Appl. Phys. Lett.* 83(12), 2330 (2003).
3. J. Swerts et al., *AVS 57th International Symposium & Exhibition*, (2010).
4. H. Volders et al., *Advanced Metallization Conference - AMC*, (2008).

5. L. Zhao et al., 47th Annual IEEE International Reliability Physics Symposium – IRPS (2009).
6. A. Maestre Caro et al., *Microelectronic Engineering 85* 2047–2050, (2008).

ECS Transactions, 34 (1) 523-528 (2011)
10.1149/1.3567630 ©The Electrochemical Society

Fine Pitch Micro-bump Interconnections for Advanced 3D Chip Stacking

W. Zhang, P. Limaye, A. La Manna, E. Beyne and P. Soussan

IMEC, Kapeldreef 75, Leuven 3001, Belgium

3D integration requires the realisation of electrical interconnects between multiple vertically stacked Si devices. Solder based, fine pitch, micro-bump connections are a promising approach for making die-to-die or die-to-wafer interconnections mainly due to its lower bonding temperature as compared with Cu-to-Cu thermo-compression bonding. This paper presents thin die stacking using Cu/Sn micro-bumps with dies containing Through Silicon Via (TSV). Two different approaches named Transient Liquid Phase (TLP) bonding and Solid State Diffusion (SSB) bonding are investigated.

Introduction

The electronics industry is increasingly considering high density 3D integration using Through-Si-Vias (TSVs) in order to address the ever continuing product needs of miniaturization and performance increase for future generation of ICs (1). This results in high bandwidth interconnects which cannot be achieved using traditional wire bonded die stack assemblies. 3D integration requires a physical stacking such as die-to-die (D2D) or die-to-wafer (D2W) while forming a permanent electrical and mechanical connection between the input/output pins of the devices. Solder based micro-bump (μbump) connection techniques are promising approaches for making fine pitch D2D interconnections due to its lower bonding temperature and higher tolerance to geometrical variations as compared with Cu-to-Cu thermo-compression bonding, which typically requires bonding temperatures above 300°C for extended periods of time (1).

This paper focuses on CuSn intermetallic based μbumps as a means of die stacking. Two different approaches for forming the intermetallic bump connection are systematically studied. The first approach is referred to as Transient Liquid Phase (TLP) bonding, during which all Sn solder is molten and subsequently transformed into intermetallic compounds through reactions with the solder pad and under-bump-metallisation (UBM). These intermetallic compounds are characterized by a much higher melting point than the solder itself (2). This allows sequential stacking of multiple dies, package construction (flip-chip, overmolding) and package assembly without disturbing the intermetallic micro-joints. The typical Sn TLP bonding temperature is above 240°C, which is higher than its melting point, 232°C. The second approach is called solid state diffusion (SSD) bonding where the bonding temperature is below the melting point of Sn (3), typical below 200°C. During this bonding, the intermetallic compounds are formed by the solid state inter-diffusion instead of the liquid-solid reaction in the TLP bonding. Another difference compared to TLP bonding is that some un-reacted Sn may present in the connection after bonding. Since the SSD bonding temperature is below the melting point of Sn, it does not re-melt the solder joints at lower levels of the stack and hence also enables repeated stacking of additional layers. It is expected however that the intermetallic phases will further grow overtime, finally reacting all the available unreacted Sn in the micro joints.

Apart from Cu/Sn μbump interconnection, an alternative approach to make chip to chip interconnections is using the Inium based fine pitch solder bumps (4). Indium is chosen because it is a very soft material and has a low melting point, 156 °C. Due to the high cost of Indium, this solution is typically limeted to special applications such as high end image sensors.

Experimental

All experiments described in this paper were performed using 5x5 mm^2 imec packaging test chips. A picture showing some of the key test patterns on such a test wafer is shown in figure 1. These chips consist of a standard single damascene Cu interconnect layer in an oxide dielectric, passivated using an oxide/nitride layer. The test structures used for μbump interconnections discussed in this work are mainly the peripheral rows of daisy-chained contacts. In one version of the test chip, these connect a total of 480 25μm diameter μbump connections at 40μm pitch. Another test chip contains 8200 area array 25μm diameter μbump at 50μm pitch. For the μbump formation process, a semi-additive electroplating technique was used. First a Ti/Cu seed layer and a resist layer were deposited sequentially. Then lithography was done to define open areas for the μbumps. After developing the resist, sequential electroplating of Cu and Sn were carried out to form the μbumps. Eventually, the resist layer was stripped, and the Ti/Cu seed layer was etched away.

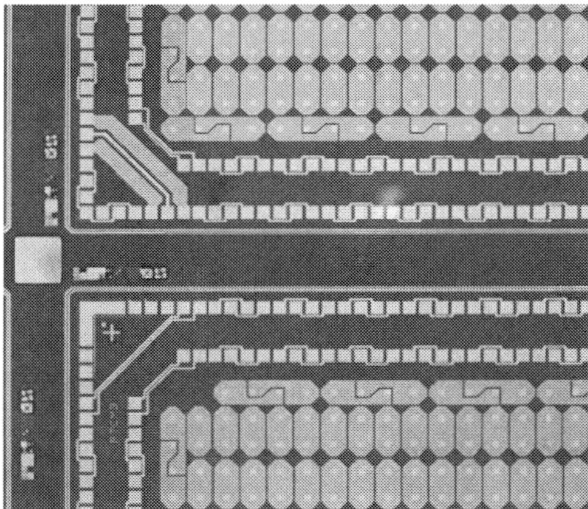

Figure 1: Photograph of a part of the imec PTCM3 test-chip used for the μbump interconnect studies

The test chip shown in figure 1 uses peripheral pad pitch of 60μm and is used in combination with TSVs. The process used for realizing these TSVs was discussed in detail in (5). The wafers are thinned to 50μm thickness prior to TSV processing. The diameter of the TSV is 25μm for a depth of 50μm. The TSV Cu-fill and the μbump formation are combined in a single process. A Ti/Cu seed layer was deposited on the 50-µm-thin TSV with aspect-ratio 2 after the TSV etch. Prior to Cu filling, a 10-µm thick negative resist mask layer was applied, which allows the simultaneous formation of the Cu-filled TSV and the Cu/Sn bump. A 3.5µm thick Sn layer was plated directly after filling of the TSVs with Cu. After resist stripping, the metallic seed layer was chemically removed.

Bonding of the test samples was done on SET FC-150 flip-chip bonder at temperatures of 150 up to 250^{0}C. Different cleaning agents, such as fluxes, were applied prior to bonding. During bonding, a pressure was applied, varying from 5 MPa to 150 MPa, The bonding time was between 3 and 20 minutes.

After bonding, the interconnects can be checked electrically by measuring the resistance and isolation of the interwoven daisy chains. Some of the samples were sent for X-section SEM investigations.

Results

A µbump consists of a solder bump and an UBM on the top die. A solderable metal "bump-pad", in this case Cu, is provided on the landing substrate. The use of a solder µbump is also an option, but not used in this work. The reaction between solder and UBM/ solderable metal results in intermetallic formation and hence the top and bottom dies are connected. In this study the UBM is Cu or a Cu/Ni bi-layer, while the solder bump comprises either pure Sn or SnAg. Figure 2 shows optical interference profile measurements of ϕ25 µm micro-bumps, where the SnAg solder bumps are formed on the Cu/Ni UBM.

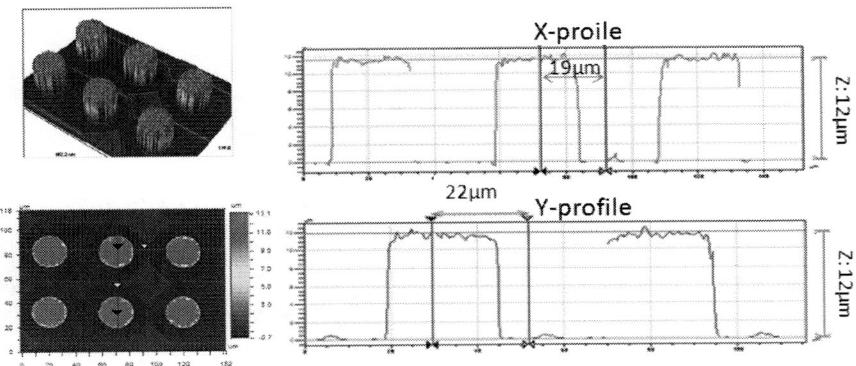

Figure 2. Veeco profilomiter images of the ϕ25 µm µbumps. The profile scans in X and Y directions indicate some roughness of the bump surface.

Both Sn and Cu are prone to be oxidized due to their low free Gibbs energy. Indeed it is found that oxides removal is critical to ensure intermatellic formation for SSD bonding. Several commercially available cleaning agents such as flux, and dilute organic acid are used to clean Cu and Sn during flip-chip bonding. The fluxing action may also be provided by the use of a so-called no-flow-underfill (NUF). This has the advantage of providing the underfill process simultanously with the assembly process. However, none of them can efficiently remove oxides at very low temperature. This limits our SSD bonding temperature a minimum of 150^0C. In fact, the combination of different cleaning agents gave the best bonding results for 150^0C. Oxides removal is less critical for TLP bonding, where liquid Sn can wet the Cu UBM to form intermetallics.

The bonding pressure is found to be another important factor to form good intermetallic joints in SSD bonding. There is a lower-limit pressure of about 20 MPa below which the solder joints contain too many defects such as voids and hence lose electrical connection. However, 150 MPa is almost the upper-limit above which the Sn is squeezed laterally and can cause electrical shorting between the adjacent bumps. For 40/15 um pitch/spacing bump connection, 50 MPa is sufficient to achieve high yielding devices.

Unlike SSD bonding, a much smaller pressure is applied to TLP bonding and 2.5-10 MPa makes no difference in electrical yield.

Considering the best trade-off between process ease and joint interface quality, bonding with NUF is chosen as the baseline process. NUF serves as an intermediate cleaning agent as well as permanent underfill. This eliminates the need for capillary underfill. Furthermore, the process throughput is increased by using the approach of an initial highly parallel die to wafer pick and place followed by a collective bonding of the dies. This approach has been demonstrated to work at full 200 mm wafer level, as shown in figure 3. More than 90% yielding devices are achieved on 40/15 μm pitch/spacing peripheral array dies for the TLP and solid state diffusion bonding approaches.

Figure 3. Optical image of the D2W assembly (top) and X-section SEM of a typical μbump interconnect chain.

Voids formation for both TLP and SSD bonding is also studied. Apart from voiding due to the entrapment of cleaning agent residues, smaller sub-micron voids known as Kirkendall voids are observed inside the Cu_3Sn-phase which is formed after Cu-Sn bonding. Ageing experiments were performed on different Cu/Sn/Cu sandwich structures with the aim of investigating the effect of the preparation method on void formation. It is found that voids are originally located at the interface between Cu and Sn but more voids are drifting away from this surface with further ageing. A large amount of voids can be observed in the centre of the intermetallic joint as well. Kirkendall voids are also found in Cu/Sn/Cu blanket film, where the sandwich film is formed by sequential electro-depositions. However, when comparing micro-bump samples with blanket film stacks which are continuously aged until full transformation to the Cu_3Sn phase, a difference in voiding density, both at the Cu-intermetallic interface and in the centre of the connection is observed, as shown in figure 4. Since the blanket Cu/Sn/Cu film is deposited by identical plating chemistry but without additional reducing agents, the presence of these reducing agent contaminants is believed to cause an increase in void nucleation and growth.

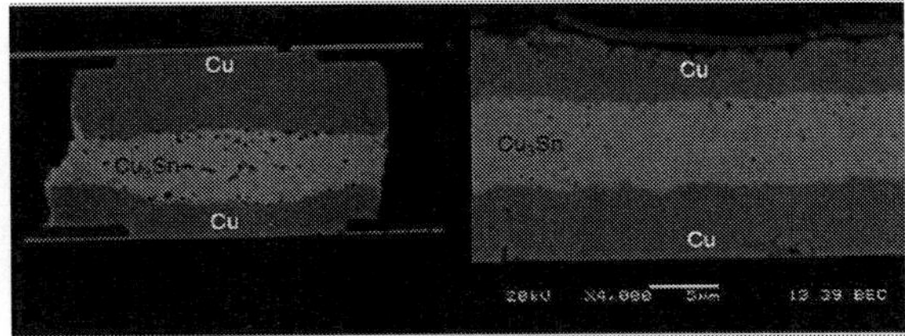

Figure 4. Comparison of a Cu/Sn/Cu µbump bonded with NUF and a blanket Cu/Sn/Cu sandwich structure after full transformation into the Cu_3Sn equilibrium phase.

Finally we show an application of the CuSn intermetallic based µbump in stacking dies formed by via last approach. 3D Wafer level Package (3D-WLP) TSV devices with 60µm pitch peripheral array TSVs are stacked on the matching Si substrate (Figure 5). These devices have daisy chain connections to monitor electrical continuity of the TSV + micro-bump connection. A high yield of the Cu-Sn stacking process is observed.

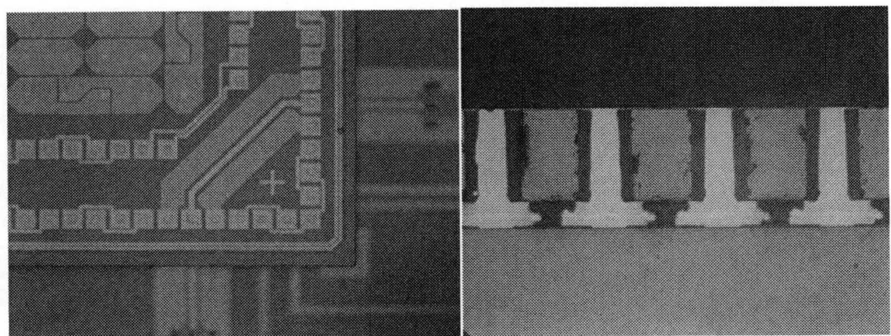

Figure 5. Top-view- optical microscope image showing a 50-µm-thin TSV top die bonded on a full thickness landing die (left) and the X-section image (right) showing the Cu TSV connected by Cu/Sn solder joint.

Conclusions

In this paper we report on thin die stacking using fine pitch Cu/Sn micro-bump. Both of transient liquid phase bonding and solid state diffusion bonding provide high electrical yield.

Acknowledgments

Contributions from Y. Civale are appreciated.

References

1. E. Beyne, P. De Moor, W. Ruythooren, R. Labie, A. Jourdain, H. Tilmans, D.S. Tezcan, P. Soussan, B. Swinnen, and R. Cartuyvels, IEEE Technical Digest of International Electron Devices Meeting, 1 (2008).
2. L. Bernstein, *J. Electrochem. Soc.,* **113**, 1282 (1966).
3. W. Zhang, P. Limaye, P. Soussan, to be published.
4. W. Zhang, W. Ruythooren, *J. Electro. Mater.,* **37**, 1095(2008).
5. Y. Civale, D.S Tezcan, H.G.G. Philipsen, P. Jaenen, R. Agarwal, F. Duval, P. Soussan, Y. Travaly, E. Beyne, IEEE International Conference on 3D System Integration, 2009, 1-4, 28-30 Sept. 2009.

ECS Transactions, 34 (1) 529-534 (2011)
10.1149/1.3567631 ©The Electrochemical Society

Temperature and Stress effects on IMC Behavior of Thin Film Cu-Al System in Wire Bond

XueFei Ming[a] and KunQuan Fan[b]

[a] R&D Department, No.58th Research Institute, CETC, China
[b] R&D Department, ASM Pacific Technology, Singapore

Wire bond is widely accepted as a predominant method for the area of electronic packaging. In this paper, authors focus on the nanoscale interfacial characteristics of thermosonic copper ball bonding on the aluminum metallization. A series of examinations and a 2.5D finite element simulation have been designed to find the IMC generation and the stress distribution. By TEM and SEM analysis, it was found at different temperatures, the IMC layers showed great difference in the thickness. By XRD and TEM analysis, the IMC also show different structures. By the finite element analysis, authors found the stress distribution when copper was bonding on the aluminum pad. By doing the research, people can have a general idea of the proper temperature and force when using copper wire bond.

Introduction

As a predominant method in making interconnection of electronic devices, wire bond is widely used in nowadays electronic packaging industry. Gold and aluminum are two main kinds of materials used as bonding wire. Copper has been a replacement material used instead of gold, as Cu has many advantages.

The copper wire can provide the same performance at smaller diameters as the gold one, however, without the high cost in the material aspect. Large-diameter copper wire can and does replace aluminum wire where a high current carrying capacity is needed or where there are problems with complex geometry. Annealing and process steps used by manufacturers enhance the ability to use large-diameter copper wire to wedge bond to silicon without damages to dies.

For the wire bond process, the intermetallic compound (IMC) between wires and pads has been widely found. To analysis the physical property of the Al-Cu system, further contributions to the knowledge about Cu/Al IMCs are surely necessary. Therefore, a study of isothermal aging of copper bump bonds on aluminum metallization pads is reported in this paper. The growth behavior of the Cu/Al IMCs and the cracks between copper ball surface and the IMC layer during aging was investigated in detail. The main IMC phases in the copper bonds were determined by micro-XRD. The experimental arrangement and corresponding results are presented, analyzed and discussed in section II. Finally, some conclusions are drawn from the study.

Experimental

The purpose is to investigate the temperature effect on the IMC formation, and the IMC cracks caused by the stress generated in the bonding process. So the authors arranged two groups of bonding samples. The heat treatment temperatures are 310°C and 200°C, respectively. All samples in this work were bonded on the silicon chips designed for power devices.

The samples were all examined by X-ray diffraction (XRD) with Cu K_α radiation ($\lambda = 1.54056$ Å) in the 2θ continuous scanning mode. The sample stage was constantly rotating during the measurement, and the initial tilt angle was set to 20° for deeper X-ray penetration into the film.

The TEM sample was first made as a sandwich comprising two thin films adhered face to face with two supporting dummy Si wafers for each condition, then subsequently ground to appropriate thickness mechanically. Perforation was made with ion milling in the center of the sandwich to generate sufficient electron transparent area for TEM observation.

Results & Analysis

A TEM picture, fig 1, shows a general distribution of the copper and aluminum distribution of the sample. It is clear that both Al and Cu layers are approximately 100nm thick as required and no reaction seems to have occurred at the interface. In figure1, layer A is aluminum, B is copper layer, while C&D layers are the silicon and glue layer. The XRD analysis shows the results matches perfect (Fig. 2). Samples both thermally treated and as-deposited were investigated by XRD, and the phases formed were identified. Table 1 summarizes all the phases that appeared at 200 and 310°C. It can be seen that annealing generates similar types of intermetallic phases and similar sequence of formation at both temperatures.

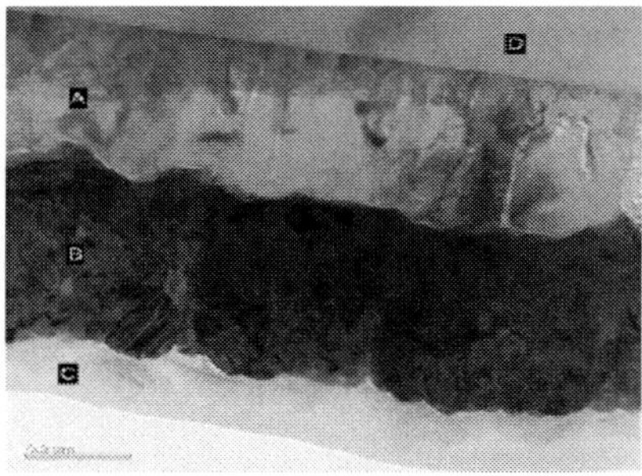

Fig. 1 TEM micrograph of the as-deposited sample (100nm Al+100nm Cu).

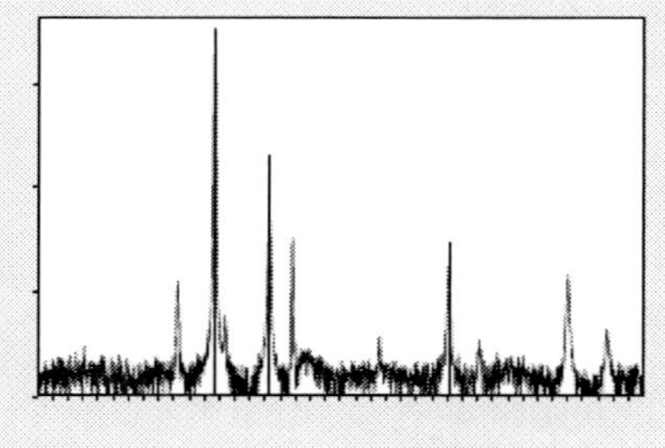

Fig. 2 XRD results of the sample as-deposited.

Cu/Al IMC phase identification

From the data of Table 1, it can be summarize that the Al/Cu layer consists of six different intermetallic compounds. The intermetallic formation is evident in the annealed samples at 200 and 300°C as indicated by a minor peak at $2\theta = 40.48°$ in the XRD pattern. However, it was difficult to identify this new phase unless it was thick enough to generate more peaks.

Table 1 List of Interface Compounds Refer to the XRD		
Sample	200°C	310°C
1#(100nmAl+100nmCu)	AlCu+CuAl2+Cu9Al4+Al2Cu3	AlCu+CuAl2+Cu9Al4+Al2Cu3

It can be seen that annealing at both temperatures generates similar types of intermetallic phases and similar sequence of formation. However, XRD results cannot be directly used for the final results of the Cu/Al IMC phase identification. Therefore, it has to be other analysis such as TEM.

And from the Table 1 it can be seen that some phases cannot be determined firmly, because the generation of intermetallic compounds is so complex. Table 1 List of interface compounds refer to the XRD. Figure 3 and 4 are the TEM pictures for same sample but annealed at different temperatures. Fig 3 is at 200°C, while Fig 4 is at 310°C. From the contrast it can be clearly found that at low temperature, the diffusion of copper and aluminum has not been completely. In fact, the growth rate of an intermediate phase during the inter-diffusion of metals follows the parabolic rate law and can be described with empirical equation:

$$\Delta = (Kt)^{1/2}$$

Where Δ is the thickness of the IMC layer; K indicates the reaction rate of IMC formation, and t represents the aging time.

In this experiment, the aging time is 30 minutes for the entire sample preparing. At different temperature, the K value changes. TEM results show that the K value increases with temperature. The principle is the high temperature provides the high energy for a faster diffusion process. However, the XRD results show that the intermetallic compounds are still the same. So the contrast suggests the intermetallic compounds generate at the beginning of annealing. From other researchers' work, in different aging time, usually at least tens of hours, the intermetallic compounds changes and some of them disappear while other new phases generate.

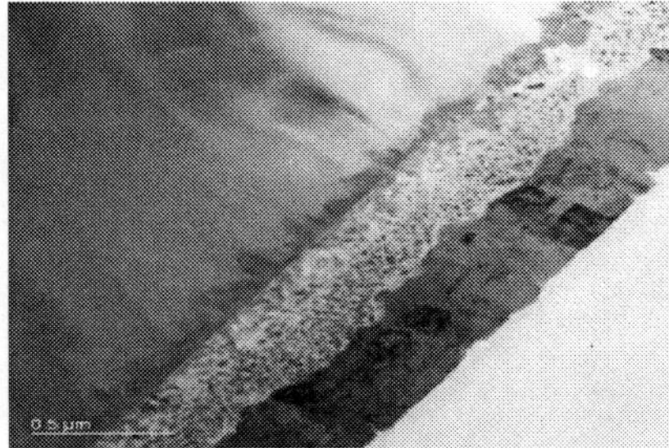

Fig. 3 TEM of the sample annealed at 200°C   Fig. 4 TEM of the sample annealed at 310°C.

In the short aging time (only 30 minutes), the beginning phases seem similar at different temperatures. Since the XRD analysis doesn't provide more details in film structures, high resolution TEM could be a great help to identify the IMCs phases.

Fig. 5 is a partial TEM image of the sample annealed at 200°C. Table 2&3 show the results of the EDX analysis of particle A & B in fig 5, respectively. It can be found that the atom ratio in particle A is approximate 1:7 (Al : Cu), so it most probably be the solid solution. For particle B, as the heating process is vacuous, the element of O might be introduced in the sample preparing process, so that might be the oxidation of Al and Cu's intermetallic compounds. Fig 5 also has a high graphics of A. Figure 6 shows a TEM image of Al-Cu diffusion of the sample annealed at 200°C, and the table 4 is the EDX data for this sample.

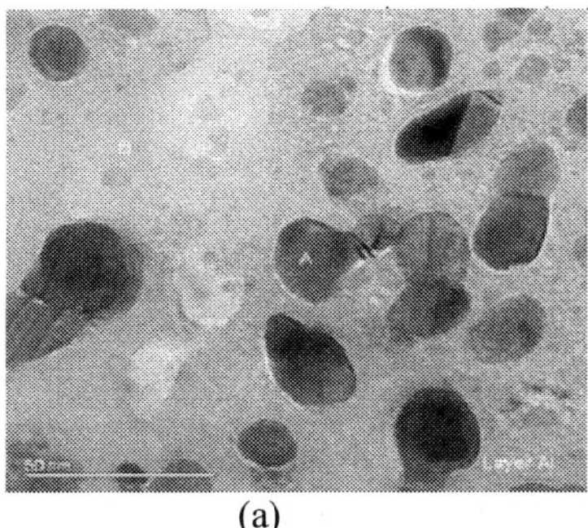
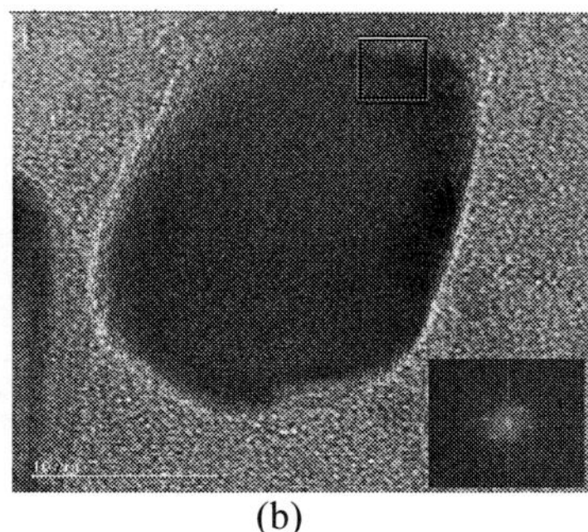

(a) (b)

Fig. 5 TEM partial image of the sample annealed at 200°C.

Table. 2 EDX results on the IMC particle A in Fig. 5 (a)				
El	AN Series	Unn.[wt.-%]	C norm.[wt.-%]	C Atom.[at.-%]
Al	13 K-series	7.42	7.42	15.87
Cu	29 K-series	92.58	92.58	84.13

Table. 3 EDX results on the IMC particle B in Fig. 5 (a)				
El	AN Series	Unn.[wt.-%]	C norm.[wt.-%]	C Atom.[at.-%]
Al	13 K-series	61.56	61.56	63.78
Cu	29 K-series	23.67	23.67	10.41
O	8 K-series	14.77	14.77	25.81

Table 4 EDX results of point 1 to 6 of the sample annealed at 200°C.

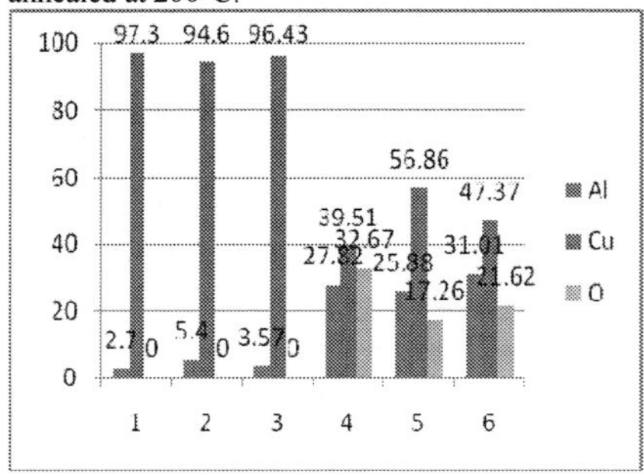

Table 5 EDX results of point 1 to 6 of the sample annealed at 310°C.

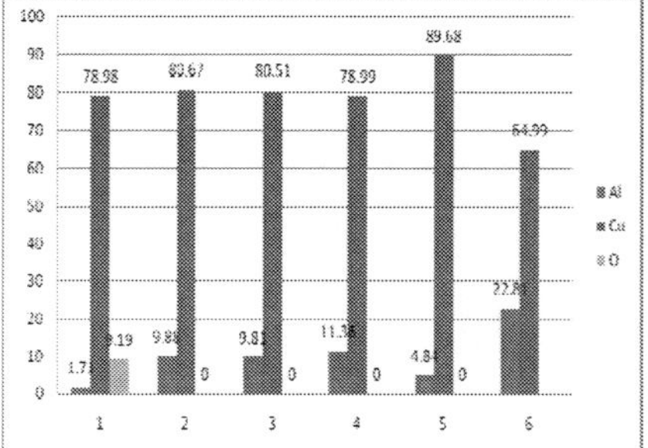

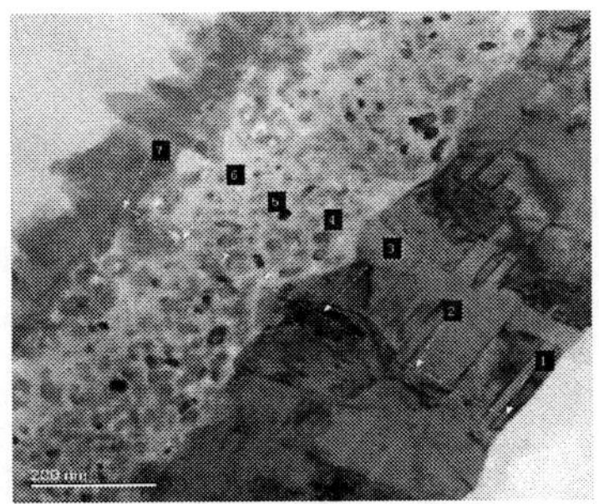

Fig. 6 Details of the Al/Cu layer diffusion of the sample annealed at 200°C.

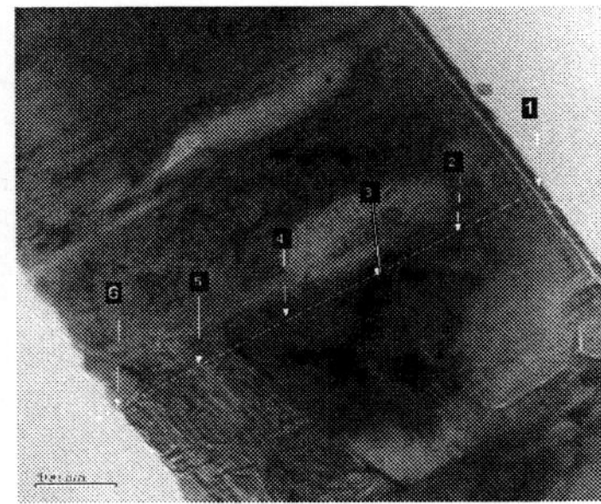

Fig. 7 Details of layer diffusion of the sample annealed at 310°C.

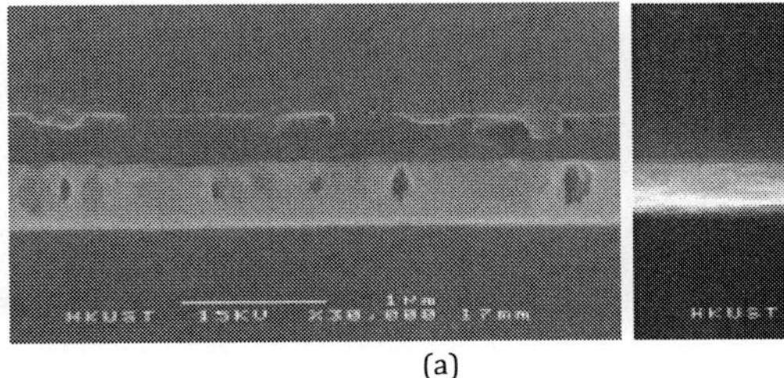

(a)

(b)

Fig. 8 (a) SEM for the annealed sample. (b) SEM for the as deposited sample.

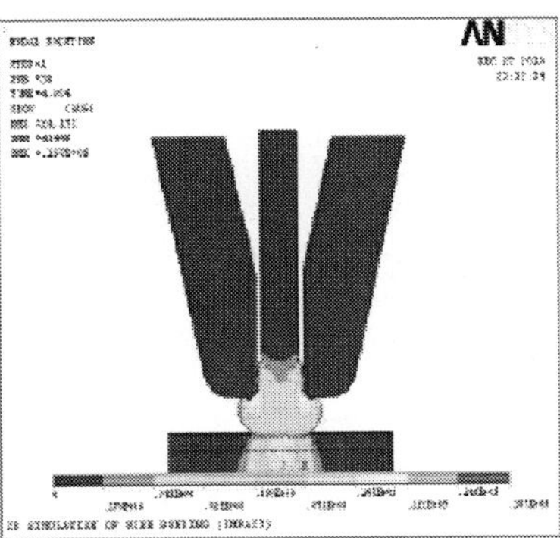

Fig. 9 FEM analysis shows the maximum stress is 157MPa.

The TEM & EDX results for the sample annealed at 310°C are shown in Fig. 7 and Table 5. At 310°C the IMCs are mainly formed by the solid solution (By SEM map data, the different distribution of Al and Cu can be found clearly. The copper is homogeneous distributed in the whole layer while aluminum is distributed inside the copper zone). But also some Cu3Al component was found in the edge of the IMC layer. For 200°C, as the temperature is not high enough, most of the IMCs are the oxide of copper and aluminum. From the TEM results, it can be found that the copper diffuses toward the aluminum and forms a layer which contains many solid solution particles.

Conclusion

The IMC growth increases with temperature. The IMC growth process was dominated by grain boundary and volume diffusion. At low temperature, since the diffusion reaction performs incompletely, it is easy to form the oxides of copper and aluminum. But at high temperature, as the diffusion reaction performs completely, the oxidation decreases and most of IMCs are solid solutions. If we compare the sample superficies between the as deposited and the annealed ones using SEM (Fig 8), it can be easily found that when aluminum is superfluous, the cracks of Al/Cu layer are much less than the others. FEM analysis indicates the stress distribution as shown in Fig 9. The maximum stress is 157MPa. The IMCs are very similar in the composition category but different in distribution. AlCu+CuAl2+Cu9Al4+Al2Cu3 are the mainly IMCs. But the different anneal conditions result in the different distributions.

Acknowledgments

The authors greatly appreciate the HKUST for helping with TEM\SEM\XRD.

References

1. S. Mori, H. Yoshida, and N. Uchiyama, *"The development of new copper ball bonding-wire"* Proc 38th Electron Comp Conf, 1988, pp. 539-545.
2. J. Kurtz, D. Cousens, and M. Dufour, *"Copper wire ball bonding"* Proc 34th Electron Comp Conf, 1984, pp. 1-5.
3. K. Toyozawa, K. Fujita, S. Minamide, and T. Maeda, *"Development of copper wire bonding application technology"* IEEE Trans Comp Hybrids Manufact Technol,Vol. CHMT-13 (1990), pp. 667-672.
4. M. G. Osborne and N. M. Murdeshwar, *"Developing wire bond interconnect solutions for copper"* Proc 3rd Annu Semicond Packag Symp, SEMICON W, 2000, pp. E1-E5.
5. Hyoung-Joon Kim, Joo Yeon Lee, Kyung-Wook Paik, Kwang-Won Koh, Jinhee Won, Sihyun Choe, Jin Lee, Jung-Tak Moon, and Yong-Jin Park, "*Effects of Cu/Al Intermetallic Compound (IMC) on Copper Wire and Aluminum Pad Bondability*", IEEE Trans on Comp And Packag Technol, Vol. 26, No. 2 (2003), pp. 367-374.
6. Harman G, *Wire bonding in microelectronics, materials, processes, reliability, and yield*, second ed., McGraw-Hill (New York, 1997), pp. 21.
7. Wul FW, Breach CD, Stephan D et al, *"Characterisation of intermetallic growth in copper and gold ball bonds on aluminum metallisation"* Proc 6th Electronics Packaging Technology Conf, Singapore, Dec. 2004, pp. 348-353.
8. Harman G. *Wire bonding in microelectronics*, McGraw-Hill (New York, 1989), pp. 135.
9. Ellis TW, Levine L, Wicen R et al, *"Copper wire bonding"* Proc of SEMICON conf, 2000.

ECS Transactions, 34 (1) 535-540 (2011)
10.1149/1.3567632 ©The Electrochemical Society

Review of Silicon Nanowire Oxidation

X. Shi, R. Kurstjens, I. Vos, J.-L. Everaert, M. Schaekers

Imec, Kapeldreef 75, 3001 Leuven, Belgium

Silicon nanowires (SNWs) have recently attracted a lot of interest due to its explosive potential for nanoelectronic applications. To materialize these applications, the key role of thermal oxidation behavior of SNWs must be thoroughly understood. In the first half of the paper we review the results on SNW oxidation published up to date. It is commonly observed that SNW oxidation shows a retarded oxidation rate and stronger temperature dependence compared to those of planar Si; in addition, the SNW oxide thickness is also a strong function of SNW diameter. In the second half, we report our results on the SNW oxidation using an in-situ steam generation (ISSG) approach. ISSG oxidation is applied to oxidize the SNWs generated by a top-down lithography-etch approach, as an important technique to reduce the SNW diameter below 10nm. It is found that SNW oxidation with ISSG also exhibits a retared oxidation compared to that of planar Si using the same ISSG process, and a less dependance on SNW sizes than furnace SNW oxidation.

Introduction

Silicon nanowires (SNWs) have over the past years attracted a lot of interest for nanoelectronic applications, such as light emitter diodes (LEDs), field effect transistors (FETs), sensors and solar cells (1, 2), due to their unique optical and electrical properties (3). For these applications, the thermal oxidation of SNWs plays an important role, for example to provide a high quality dielectric shell to protect SNWs or to reduce the SNW diameter below 5nm to obtain visible photoluminescence.

There are two basic approaches of synthesizing SNWs: top-down and bottom-up approach. In a top-down approach a large piece of material is cut down to small pieces through different means such as lithography and electrophoresis (4) whereas in a bottom-up approach the nanowire is synthesized by the combination of constituent ad-atoms. The Vapor-Liquid-Solid (VLS) growth method is the most well-known bottom-up approach (5). These two types of SNWs have been applied extensively for the oxidation studies (4, 6-10).

In the paper we first review the SNW oxidation results reported in literature, which are usually achieved using conventional furnace oxidation or rapid thermal oxidation (RTO). Then we report the preliminary results of SNW oxidation with the in-situ steam generation (ISSG) oxidation method, a radical oxidation process (11).

Review of SNW Oxidation

The oxidation of blanket Si wafers is well governed by the Deal-Grove model, which describes the oxidation of blanket Si as a one dimensional process. The oxidation can be divided into two stages: at short oxidation times, it is an interface-reaction controlled process, while at the long times it is a diffusion dominant process (12). In contrast, the oxidation of SNW is a more complicated process: it is a two-dimensional oxidation process, involving a non-uniform deformation of the oxide around SNW cores. The two-dimensional deformation causes stress, which can further impact the oxidation of SNW. In 1988 Kao *et al* developed a two-dimensional thermal oxidation model and studied the stress effect in SNW oxidation (10). The model suggests that the viscous stress associated with the oxide deformation is the fundamental force of the retardation in SNW oxidation.

Furnace Oxidation of SNW

Furnace oxidation is the first oxidation method applied in SNW oxidation. Many articles are available for the furnace oxidation of SNWs synthesized with different methods (4, 6-8). For instance, H. Liu *et al* reported the results of oxidation of SNWs formed by lithography-etch top-down approach (4), while D. Shir *et al* and C. Buttner *et al* reported the results of furnace oxidation of SNWs prepared by Vapor-Liquid-Solid (VLS) bottom-up method (7, 8). The typical temperature range of the furnace oxidation is between 650-950°C. There are a few phenomena commonly observed: 1) SNW oxidation exhibits a retarded effect compared to that of blanket Si. The retardation is strongly influenced by the oxidation time and the temperature, for a longer oxidation time and a lower temperature, a more pronounced retardation is observed; 2) the SNW oxidation also strongly depends on the SNW dimension. The smaller SNWs are more difficult to oxidize. All these can be explained by the two-dimensional oxidation causing the non-uniform deformation of the oxide around SNWs. B. Liu *et al* also reported that adding a chlorinated gas, e.g. trichloroethane (TCA), during oxidation could slightly enhance the SNW oxidation (6).

RTO Oxidation of SNW

RTO is an oxidation process which heats the Si wafer to a high temperature (up to 1200°C) on a timescale of several seconds. S. Krylyuk *et al* applied RTO in SNW oxidation and reported the results (9). The temperature range investigated is between 900 and 1000°C. Compared to the furnace oxidation of SNW, RTO provides faster oxidation rates and a weaker dependence of oxide thickness on the SNW diameter.

SNW Oxidation with ISSG

In this part we will discuss the preliminary results of SNW oxidation with in-situ steam generation (ISSG) oxidation approach. The in-situ steam is generated by adding H_2 during oxidation. The combustion reaction takes place at the hot wafer surface and generates the radical oxygen, which dominates the oxidation process. Equation 1 shows the chemical reaction formula of ISSG process. For SNW oxidation, it is expected that the radical oxygen might help to release the stress caused by deformation, and further enhance the SNW oxidation. ISSG is a high temperature process. A higher temperature results in a higher viscosity of SiO_2. In the experiment, the ISSG temperature range is selected between 1000 ~ 1150°C. The typical ISSG pressure is between 7 ~ 15Torr.

$$O_2 + H_2 \dashrightarrow H_2O + O^* \tag{1}$$

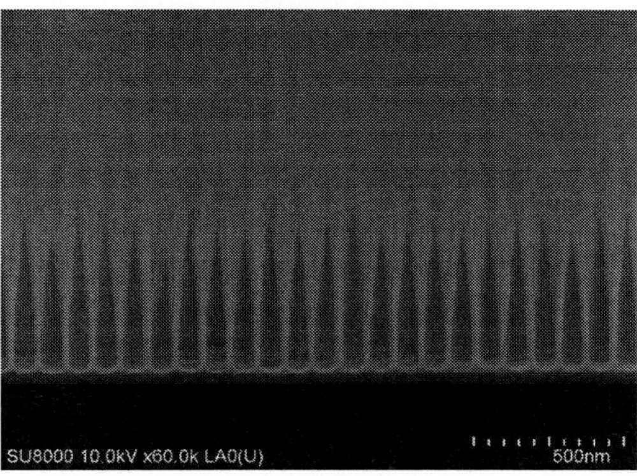

Figure 1 The SEM picture of SNW array fabricated by the lithography-etch approach.

The SNWs are fabricated by the lithography-etch approach on 300mm wafers. Figure 1 shows a scanning electron microscope (SEM) picture of the SNW array. The diameters of SNWs vary from 55nm to 20nm. Figure 2 (a) shows a cross-SEM (XSEM) picture of SNW's with a diameter ~ 20nm. Subsequently the wafers with SNWs are oxidized with ISSG.

(a) X-SEM picture before oxidation.

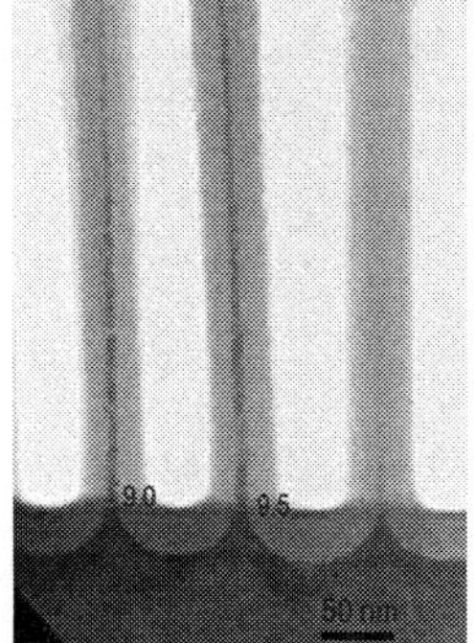

(b) TEM picture after oxidation.

Figure 2 SNW pictures before and after oxidation: (a) The X-SEM picture of SNWs with a diameter ~ 20nm, taken after etch and before oxidation; (b) The TEM picture of SNWs after ISSG oxidation at 1100°C.

The measurement of SNW oxidation is always a challenge. Usually transmission electron microscopy (TEM) is applied to measure the SNW oxide thickness. Figure 2 (b) shows a TEM picture of SNWs after ISSG oxidation at 1100°C. However, TEM measurement requires a long sample preparation and cannot provide a fast feedback.

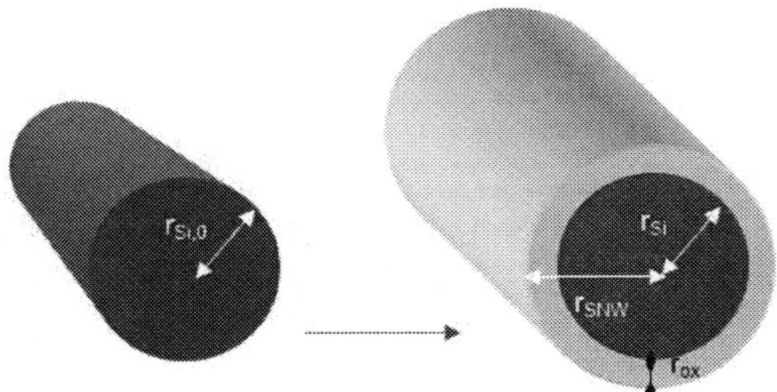

Figure 3 The schematics of SNW before and after oxidation. $r_{Si,0}$ represents the SNW radius before oxidation, while r_{SNW}, r_{Si} and r_{ox} represent the SNW radius, the remaining Si core radius and SNW oxide thickness after oxidation, respectively.

$$r_{Si}^2 = \left(r_{SNW}^2 - \left(\frac{\Omega_{SiO2}}{\Omega_{Si}} \right) r_{Si,0}^2 \right) \left(\frac{\Omega_{Si}}{\Omega_{Si} - \Omega_{SiO2}} \right) \qquad [2]$$

$$r_{ox} = r_{SNW} - r_{Si} \qquad [3]$$

A method is developed to extract the SNW oxide thickness by comparing the SNW radius change before and after oxidation (7). The model is based on the constant Si atoms before and after oxidation. The SNW radius before and after oxidation can be measured with top-view SEM. Based on the radius change, the radius of the remaining Si core in SNW can be extracted by using Equation 2. In Equation 2, Ω_{Si} is the atomic volume of Si (20 Å^3), while Ω_{SiO2} is the molecular volume of SiO_2 (45 Å^3). $r_{Si,0}$ represents the SNW radius before oxidation, while r_{SNW}, r_{Si} and r_{ox} represent the SNW radius, the remaining Si core radius, and SNW oxide thickness after oxidation, respectively. Figure 3 shows the detailed schematics. Both $r_{Si,0}$ and r_{SNW} can be measured with top-view SEM. Once the Si core radius after oxidation r_{Si} is extracted with Equation 2, the SNW oxide thickness r_{ox} can be calculated with Equation 3.

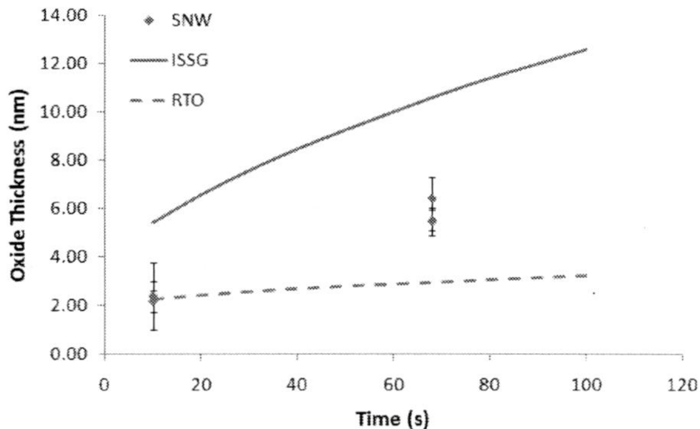

Figure 4 The SNW oxidation by ISSG oxidation at 1100°C. The ISSG and RTO results on blanket Si are listed as reference.

Figure 4 shows the SNW oxidation with ISSG at 1100°C. The oxidation curves of ISSG and RTO at the similar condition on blanket Si are listed as reference. Compared to the ISSG oxidation curve on blanket Si, the ISSG oxidation on SNW yields a thinner oxide thickness and exhibits a retarded oxidation. This retardation can be attributed to the fact that two dimensional oxidation causes the non-uniform deformation, which further causes the stress and retards the oxidation process. However, the ISSG oxidation on SNW exhibits a thicker oxide thickness than the RTO oxidation at the same oxidation condition on blanket Si. It clearly indicates that the in-situ steam caused by adding H_2 during the oxidation enhances the SNW oxidation. This enhancement could be considered from two respects: firstly, in-situ steam yields the radical oxygen, while secondly, it also causes oxide more viscous and helps to release the stress caused by the deformation of the two dimensional oxidation. The relative high ISSG temperature could also help to release the stress. Compared to the furnace SNW oxidation, the SNW oxidation with ISSG yields a faster oxidation rate.

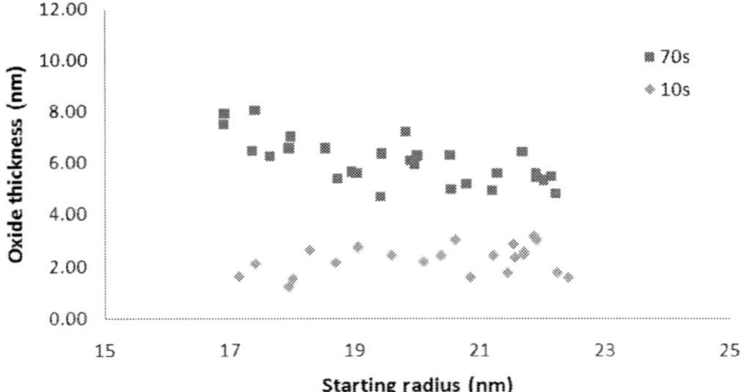

Figure 5 The weak dependence of ISSG SNW oxide on SNW size. The ISSG oxidation is at 1100°C, while the ISSG process time is 70s or 10s.

As described above, the furnace oxidation has a more pronounced retarded oxidation on small SNW's, because the small SNWs have a big surface curvature which causes more deformation during the two-dimensional oxidation. As a consequence the furnace oxidation exhibits a strong dependence on SNW size. However, unlike the results of furnace and RTO oxidation, the ISSG SNW oxidation exhibits hardly any influence from the SNW sizes. Figure 5 shows the ISSG oxidation results of SNW with SNW radius ranging from 17nm to 23nm. Within this radius range, it is hardly to see the influence of SNW radius on SNW oxide thickness. It could be due to the same two effects of ISSG mentioned previously: the in-situ steam generated by adding H_2 during oxidation and a relative high temperature of 1100°C. Both effectively cause the oxide more viscous, which further helps to release the stress. As a consequence the dependence of SNW oxidation on the SNW diameters for ISSG is not as significant as that of furnace oxidation.

Summary

In the paper we first review the SNW oxidation with furnace and RTO reported in the literature. The retarded oxidation and the strong dependence on the SNW sizes are observed. In the second part we report the preliminary results of SNW oxidation with ISSG. Compared to the furnace SNW oxidation, ISSG provides a faster SNW oxidation and a smaller dependence on SNW diameter.

References

1. Y. Cui and C. M. Lieber, *Science* **291**, 851 (2001).
2. X. Duan, C. Niu, V. Sahi, J. Chen, J. W. Parce, S. Empedocles, and J. L. Goldman, *Nature (London)* **425**, 274 (2003).
3. X. Zhao, C. M. Wei, L. Yang, and M. Y. Chou, *Phys. Rev. Lett.* **92** 236805 (2004).
4. H. I. Liu, D. K. Biegelsen, F. A. Ponce, N. M. Johnson, and R. F..W. Pease, *J. Vac. Sci. Technol.* B **11**, 2532 (1993).
5. R. S. Wagner and W. C. Ellis *Appl. Phys. Lett.* **4**, 89 (1964).
6. B. Liu, Y. Wang, T. Ho, K. K. Lew, S. M. Eichfed, J. M. Redwing, T. S. Mayer, and S. E. Mohney, *J. Vac. Sci. Technol.* **A 26**, 370 (2008)
7. C. C. Buttner and M. Zacharias, *Appl. Phys. Lett.* **89** 263106 (2006).
8. D. Shir, B. Z. Liu, A. M. Mohammad, K. K. Lew, and S. E. Mohney, *J. Vac. Sci. Technol.* **B 24** 1333 (2006).
9. S. Krylyuk, A. V. Davydov, I. Levin, A. Motayed, and M. D. Vaudin, *Appl. Phys. Lett.* **94** 063113 (2009).
10. D. B. Kao, J. P. McVittie, W. D. Nix, and K. C. Saraswat, *IEEE Trans. Electron Devices* **35**, 25 (1988).
11. T. Y. Luo, M. Laughery, G. A. Brown, H. N. Al-Shareef, V. H. C. Watt, A. Karamcheti, M. D. Jackson, and H. R. Huff, *IEEE Electron Device Lett.* **21** 9 (2000).
12. B. E. Deal and A. S. Grove, *J. Appl. Phys.* **36** 3770 (1965).

ECS Transactions, 34 (1) 541-546 (2011)
10.1149/1.3567633 ©The Electrochemical Society

Effect of film thickness on resistance switching characteristics for Cu/NiO/Pt structure

Yang Zhang, Kailiang Zhang[*], Fang Wang, Xiaoying Wei, and Jinshi Zhao

School of Electronics Information Engineering, Tianjin Key Laboratory of Film Electronic & Communication Devices, Tianjin University of Technology, Tianjin, 300384, CHINA, [*]corresponding author, kailiang_zhang@163.com

The effect of NiO film thickness on resistance switching characteristics of a Cu/NiO/Pt structure is investigated. I-V results show that forming voltage decreases with the reduction of film thickness, and finally drops down to the set voltage. NiO film thickness should be controlled under a typical thickness in order to implement lower power consumption device design. The underlying mechanism of the phenomenon is attributed to the important role that Cu atoms play in the structure. Due to redox existing in the interface between Cu and NiO film, Cu atoms are ionized and pushed to the opposite electrode when the positive voltage is applied to Cu electrode, Cu conductive filaments are formed. In opposite, Cu atoms lost their electrons and are pushed off from Pt electrode when the positive voltage is applied to the Pt electrode, resulting in the rupture of Cu filaments.

Introduction

Since the Flash memory is facing reliability issues and geometrical limitation, including several different mechanisms of charge loss and fluctuation (1), such as stress-induced leakage current (SILC) through the tunnel oxide (2,3), charge trapping in the dielectrics after cycling, charge detrapping from the tunnel oxide (4,5) and random telegraph signal (6,7), resistance random access memory(RRAM) in metal/insulator/metal (MIM) structure has been widely investigated due to their potential application in the next generation non-volatile memories (8-12).

Many materials have been found to exhibit resistance switching behaviors, including binary oxides (13-15), perovskite oxides (16) and so on. Among them, binary oxides have been considered to promising candidates, because of a large on/off ratio, long endurance, high switching speed and CMOS process compatibility (17-20). However, a forming process is needed for binary oxides to activate the pristine devices to show resistance switching character. Typically, the forming voltage is much higher than the set voltage, which is not desired for the device design and testing (10, 21, 22). Therefore, it is meaningful to lower the forming voltage and even eliminate the forming process.

In our work, a forming-free resistance switching in NiO-based device with a Cu/NiO/Pt structure has been observed by decreasing the NiO film thickness. The effect of film thickness is discussed. Finally, a possible resistance switching mechanism of the forming-free phenomenon is proposed.

Experiments

The structure of Cu/NiO/Pt/Ti was fabricated on SiO$_2$/Si substrates. The top-electrode was formed by Cu with a circle dimension of 50μm. The Ti layer serves as an adhesion layer for Pt (bottom-electrode). The NiO films with various thicknesses (20nm to 600nm) were deposited by the reaction sputtering under the 5% oxygen partial pressure. Electronic characterization was done by the Agilent B1500A semiconductor parameter analyzer and a compliance current was used to protect devices from hard breakdown in the forming and set process. The crystal structure of the NiO thin film is studied by an x-ray diffraction (XRD), in which the NiO thin film exhibits a polycrystalline structure.

Results and Discussion

The structure shown in Figure 1 is used to characterize the electronic characters of NiO films. Figure 2 shows the I-V characteristics of the NiO thin films. By applying a bias voltage to a certain value, the pristine resistance at a high value (HRS or OFF state) is transformed into a low value (LRS or ON state), which is called the forming process. The compliance current is required to prevent the device from hard breakdown. This sample exhibits typical characteristics of a bipolar resistance switching. The forming process is necessary to initialize the device. Except Cu as top electrodes, bipolar resistance switchings with forming process were not realized in NiO/Pt with Pt or Ni as top electrodes. It indicates that the Cu electrode has a large influence on resistance switching.

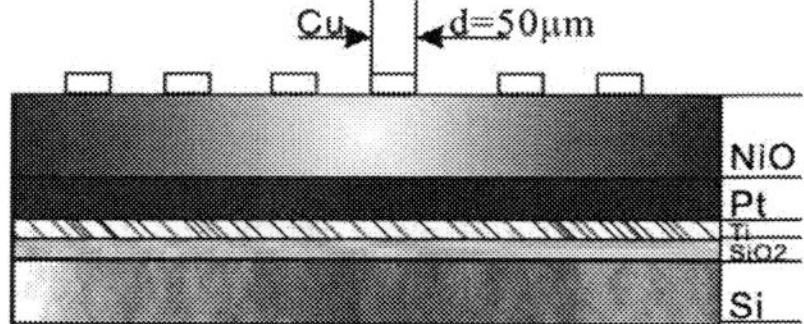

Figure 1. The schematic of the test device structure.

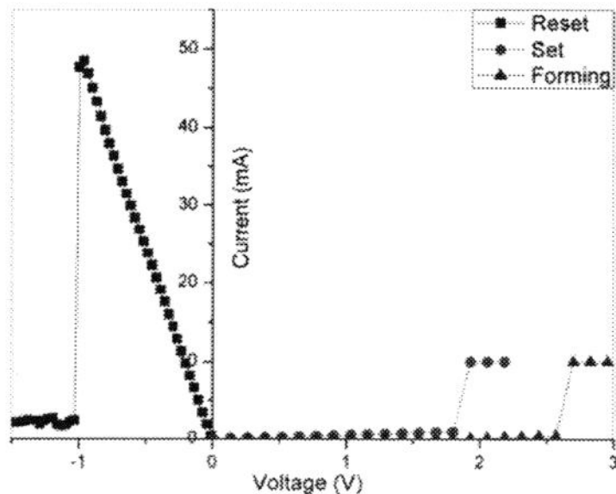

Figure 2. I-V characteristics of Cu/NiO/Pt structure device.

Figure 3 shows the pristine resistance of the NiO thin film as a function of the film thickness. The pristine resistance increases linearly with thickness. The influence of test cycles on the pristine resistance variation is studied by applying the same bias voltage, lower than the forming voltage, on the pristine device for several times. The results are shown in Figure 4. Along the direction of axis of ordinates, the pristine resistance decreases as the bias voltage increases. Along the cross axle direction, the pristine resistance reduces with the increase of testing times for the same sweeping voltage, which is measured at the same voltage (such as 2v, 10v and 19.56v). This phenomenon suggests that the Cu atoms were ionized and Cu ions were pushed to the opposite electrode because of the redox existence in the interface between Cu and NiO film. Comparing the results between the 1st and 2nd cycle, the resistance measured by 10v at the 1st cycle is lower than that measured by 2v at the 2nd cycle. It can be attributed to the unstable Cu filament. As shown in Figure 5, by sweeping the same voltage several times, the pristine resistance becomes lower and lower till the forming process occurs. It can be explained by Figure 8 (a-c). As the same bias voltage was applied, lower than the forming voltage, the electrical field is not high enough to push Cu ions to the opposite electrode immediately. Afterwards Cu ions were pushed to the position shown in Figure 8 (b). And finally Cu ions attaches Pt electrode shown in Figure 8 (c), and the forming process completes.

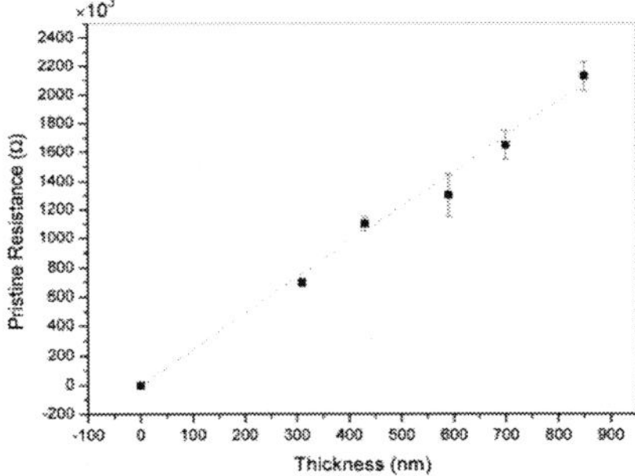

Figure 3. Pristine resistance as a function of NiO film thickness.

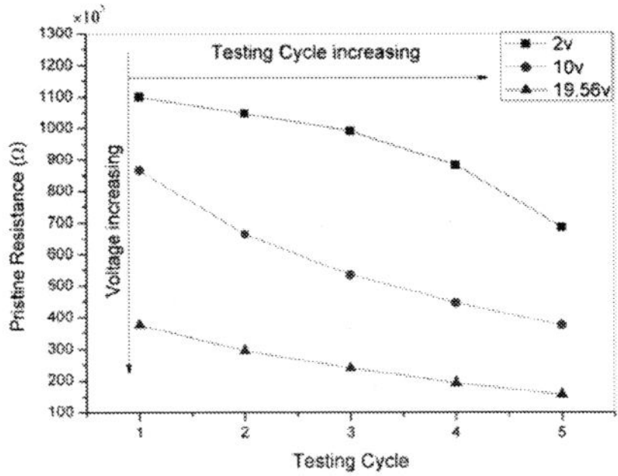

Figure 4. Pristine resistance as a function of the number of testing cycles.

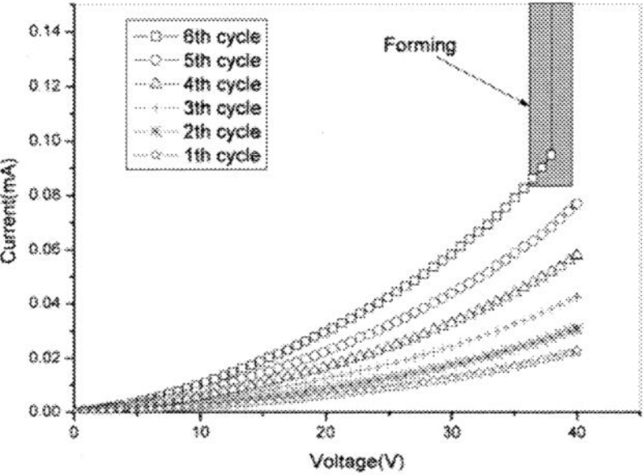

Figure 5. The trends of the pristine resistance by sweeping same voltage several times.

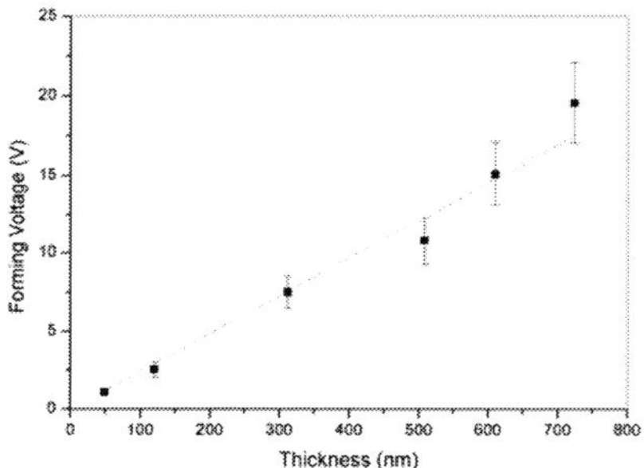

Figure 6. The forming voltage as a function of film thickness

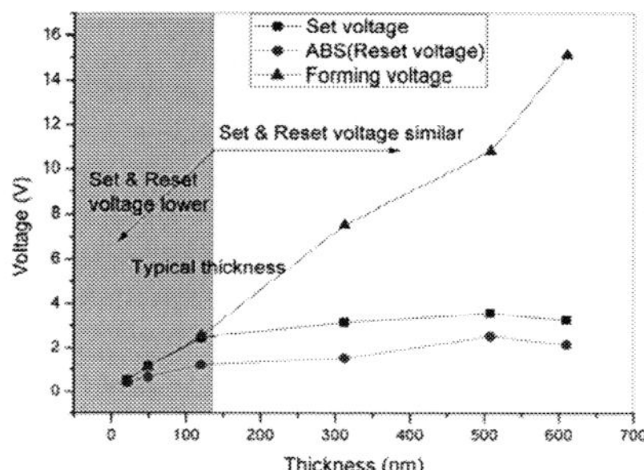

Figure 7. Forming, set and reset voltages as a function of film thickness.

The forming ($V_{forming}$), set (V_{set}) and reset voltage (V_{reset}) as a function of NiO film thickness are shown in Figure 6 and Figure 7. When the film thickness is higher than the typical thickness, the forming voltage decreases linearly with film thickness, while the set and reset voltage show a weak dependence on thickness. In contrast, as the thickness is lower than the typical thickness, all the $V_{forming}$, V_{set} and V_{reset} decrease with the thickness and the forming voltage is similar to the set voltage. The phenomenon can be explained by Figure 8.

In the forming process, a certain electrical field should be reached to push Cu ions to the opposite electrode. Hence, the forming voltage becomes higher for thicker films shown in Figure 8 (a-c, f, g). In contrast, Cu ions move only a short distance about the typical thickness in the set and reset process shown in Figure 8 (d, e, h, i). However, in Figure 8(j, k) Cu ions can be pushed easily through the bulk of NiO film (due to a thickness thinner than typical thickness) to form Cu filaments. Since the forming and set processes have a similar moving distance shown in Figure 8(j-l), their voltages are similar.

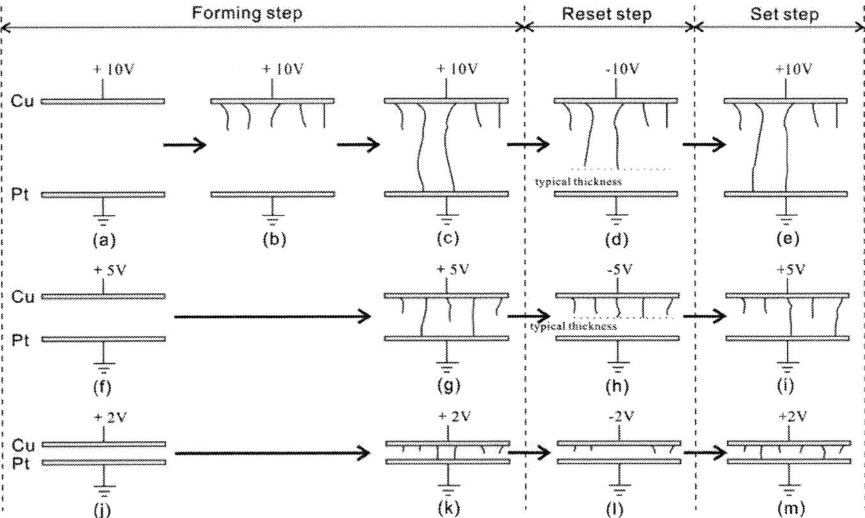

Figure 8. Schematic diagrams of resistance switching mechanism for devices with different NiO film thickness: (a-e) film thickness lager than the typical thickness; (f-i) film thickness around the typical thickness; and (j-m) film thickness thinner than the typical thickness.

Conclusions

The device with the Cu/NiO/Pt structure exhibits typical characteristics of a bipolar resistance switching. There is a typical thickness existing for the Cu/NiO/Pt structure. Under the typical thickness, the forming voltage and set voltage are similar; $V_{forming}$, V_{set} and V_{reset} descend with the thickness decrease. Because of redox existence in the interface between Cu and NiO film, Cu atoms are ionized and Cu ions are pushed to the opposite electrode when the positive voltage is applied to Cu electrode, Cu conductive filaments formed. In opposite, Cu atoms lost their electrons and were pushed off from Pt electrode (Cu filaments ruptured) when the positive voltage is applied to the Pt electrode.

Acknowledgments

We acknowledge the financial support provided by the National Natural Science Foundation of China under Grant No 60806030, and Tianjin Natural Science Foundation under Grant No 08JCYBJC14600, No 10SYSYJC27700 and Tianjin Science and Technology Developmental Funds of Universities and Colleges under Grant No ZD200709.

References

1. D. Ielmini, *Microelectronic Engineering* **86** (7-9), 1870-1875 (2009).
2. S. Satoh, G. Hemink, et al., *IEEE Transactions on Electron Devices* **45** (2), 482-486 (1998).
3. D. Ielmini, A. S. Spinelli, et al., *IEEE Transactions on Electron Devices* **49** (11), 1955-1961 (2002).
4. N. Mielke, H. Belgal, et al., *IEEE Transactions on Device and Materials Reliability* **4** (3), 335-344 (2004).
5. N. Mielke, H. P. Belgal, et al., *2006 IEEE International Reliability Physics Symposium Proceedings 44th Annual*, 29-35 (2006).
6. N. Tega, H. Miki, et al., *2006 International Electron Devices Meeting (IEDM)*, 1-4 (2006).
7. C. Monzio Compagnoni, R. Gusmeroli, et al., *IEEE Transactions on Electron Devices* **55** (1), 388-395 (2008).
8. H. Y. Jeong, Y. I. Kim, et al., *Nanotechnology* **21** (11), 115203 (2010).
9. T. W. Hickmott, *Journal of Applied Physics* **104** (10) (2008).
10. Q. N. Mao, Z. G. Ji, et al., *Journal of Physics D-Applied Physics* **43** (39) (2010).
11. D. Ielmini, F. Nardi, et al., *Electron Device Letters* **31** (4), 353-355 (2010).
12. S.-B. Long, Q. Liu, et al., *2010 10th IEEE International Conference on Solid-State and Integrated Circuit Technology (ICSICT)*, 1163-1165 (2010).
13. Y. Wu, B. Lee, et al., *Electron Device Letters* **31** (12), 1449-1451 (2010).
14. S. Kawabata, M. Nakura, et al., *2010 IEEE International Memory Workshop (IMW)*, 1-2 (2010).
15. J. Y. Son, Y. H. Shin, et al., *ACS Nano* **4** (5), 2655-2658 (2010).
16. M. P. Tendulkara, J. R. Jameson, et al., *2009 10th Annual Non-Volatile Memory Technology Symposium (NVMTS)*, 48-51 (2009).
17. F. Nardi, D. Ielmini, et al., *2010 IEEE International Memory Workshop (IMW)*, 1-4 (2010).
18. R. C. Tenent, D. T. Gillaspie, et al., *Journal of the Electrochemical Society* **157** (3), H318-H322 (2010).
19. C. Ku uml geler, R. Weng, et al., *Thin Solid Films*, 2258-2260 (2010).
20. S. B. Lee, S. Park, et al., *Applied Physics Letters* **95** (12) (2009).
21. H. B. Lv, M. Yin, et al., *2008 Joint Non-Volatile Semiconductor Memory Workshop and International Conference on Memory Technology and Design, Proceedings*, 52-53 (2008).
22. C. Yu-Sheng, W. Tai-Yuan, et al., *2009 International Symposium on VLSI Technology, Systems, and Applications (VLSI-TSA)*, 37-38 (2009).

ECS Transactions, 34 (1) 547-550 (2011)
10.1149/1.3567634 ©The Electrochemical Society

Optical constants of ZnO films

B. Huang and H.-D. Yang

Department of Electronic Engineering, Jinan University, Guangzhou 510632, China.

ZnO films have been deposited on glass substrates by an RF magnetron sputtering under various powers. The optical constants of the ZnO films are determined by employing the Kramers–Kronig theory using the spectral dependence of the real part of the refractive index.

Introduction

Zinc oxide films have received broad attention due to its remarkable properties and potential applications in solar cells, acousto-optic devices and optoelectronic devices (1–4). In view of these applications, there is a strong need for the accurate optical constants of ZnO film over a wide range of wavelength. Although optical constants of the bulk ZnO crystal have been measured by Yoshikawa and Adachi using spectroscopic ellipsometry (5), this powerful technique is extremely surface sensitive which limits its application to the ZnO film. This is because that the optical properties of ZnO film are strongly dependent upon the deposition conditions and even the preparation methods (6, 7). The other most commonly used technique, as applied in this study, involves measuring the normal incidence transmission coefficient T and/or the near normal incidence reflection coefficient R (8).

In this paper, the ZnO films were deposited on glass substrates by using an RF magnetron sputter. The optical constants of the ZnO films have been determined by employed the Kramers–Kronig (KK) theory to obtain the spectral dependence of the real part of the refractive index. The absorption coefficient was evaluated from the transmittance (T) and reflectance (R) data.

Experiment

The ZnO films were deposited on glass slides by RF magnetron sputtering using a ZnO target (99.99%) in a mixture gas of Ar (99.99%) and O_2 (99.99%) with Ar/O_2 ratio 3:2 under various power: 100 W, 120 W and 140 W, and these samples are labeled as P1, P2, P3, respectively. The total pressure was maintained at 2Pa during film deposition. The total sputtering time was 90 minutes. The transmission and reflectance spectra of the samples were measured by a Varian Cary-300 spectrophotometer at the range from 320 nm to 800 nm at room temperature in air. A Sloan Dektak3 surface profilometer was used to determine the thickness of the films.

Results and discussion

According to the Kramers–Kronig relationship, the absorption coefficient $\alpha(\lambda)$ and the refractive index $n(\lambda)$ of thin films can be related by (9)

$$n(\lambda) - 1 = \frac{1}{2\pi^2} \int_0^\infty \frac{\alpha(\lambda')d\lambda'}{1 - \lambda'^2 / \lambda^2} \qquad [1]$$

where λ' is the running variable for the wavelength in the wavelength range $[0, \infty]$. In practice, we could only obtain $\alpha(\lambda)$ in the wavelength $\lambda_1 \sim \lambda_2$ nm from the transmittance (T) and reflectance (R) data using the relation (10)

$$\alpha(\lambda) = -\frac{1}{d} \ln\left(\frac{T}{(1-R)} \right) \qquad [2]$$

where d is the film thickness. In addition, we need to have values of $\alpha(\lambda)$ at short and long wavelength. In this paper, we use the similar relation employed by Xue et al (9) and Bhattacharyya et al (11). In short, the wavelength range $[0, \infty]$ may be split up into three different parts: the low wavelength limit ranging, the estimable wavelength range as recorded by Varian Cary-300 spectrophotometer, and the higher wavelength limit. In the lower wavelength limit, the optical absorption coefficient for direct band gap semiconductor near the band edge, is given by (9,11):

$$\alpha(\lambda) = A(\frac{hc}{\lambda} - E_g)^{1/2} \qquad [3]$$

where A is a constant, h is Planck's constant, c is the speed of light, and E_g is the band gap. In the higher wavelength limit $\alpha(\lambda)$ is treated as constant, since the optical absorption is small.

When the absorption coefficient is obtained, the wavelength dependent extinction coefficient $k(\lambda)$ may be related to the absorption coefficient as:

$$k(\lambda) = \frac{\alpha(\lambda)\lambda}{4\pi} \qquad [4]$$

Correspondingly, the real and the imaginary components of the complex dielectric constant of thin films are:

$$\varepsilon_1 = n^2(\lambda) - k^2(\lambda), \qquad [5]$$
and
$$\varepsilon_2 = 2n(\lambda)k(\lambda). \qquad [6]$$

By applying the theory above, the optical parameters of ZnO films deposited under various power can now be deduced. Transmission spectra of the ZnO films deposited by different RF power are shown in Fig.1. All films exhibit a high transmittance in the visible region. It is found that the transmittance decreases with the increase of RF power, and there is a sharp absorption edge at about 390 nm. Fig. 2 shows the absorption coefficients of ZnO films as a function of the wavelength. The absorption coefficients in the ultraviolet region are larger than those in the visible region and increase with the RF power.

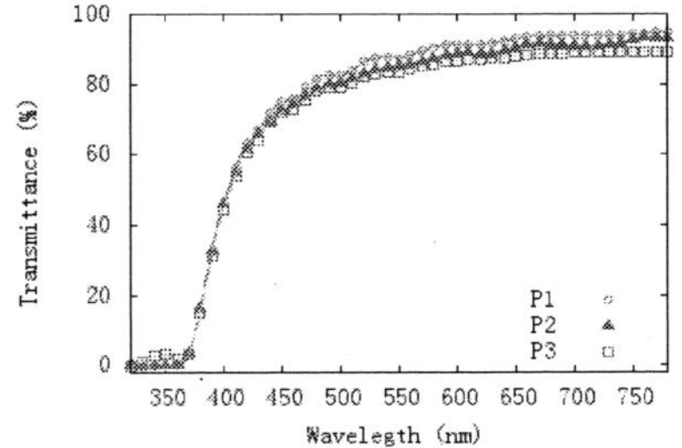

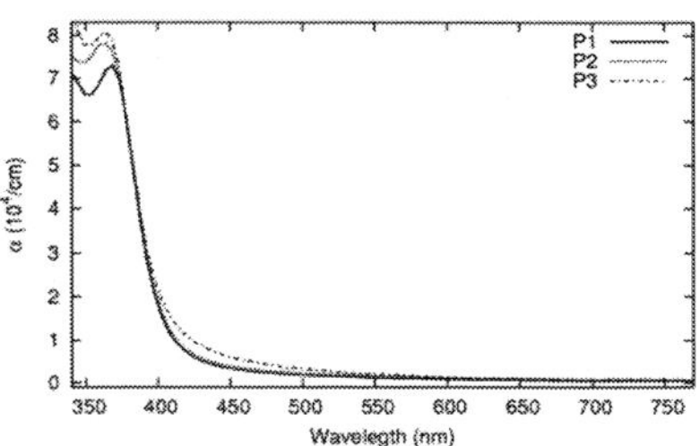

Fig.1. The transmission spectra of the samples as a function of the wavelength.

Fig.2. Absorption coefficients of ZnO thin films as a function of the wavelength.

The derived values of the film refraction indices n are displayed in Fig.3, along with the values of the ZnO single crystals measured by Yoshikawa et al. using spectroscopic ellipsometry (5). It is evident that these samples have a lower refraction indices n compared with those of the ZnO crystals. It is found that the values of n decrease with the increasing of RF power. The deposited films usually contain voids. The refractive index of the film is a function of the packing density of the film (12). The results obtained also indicate sample P1 has a higher packing density compared with those of other samples. This can be attributed to the fact that the films have higher growth rate at higher power. If the growth rate is too fast, the film is prone to have voids.

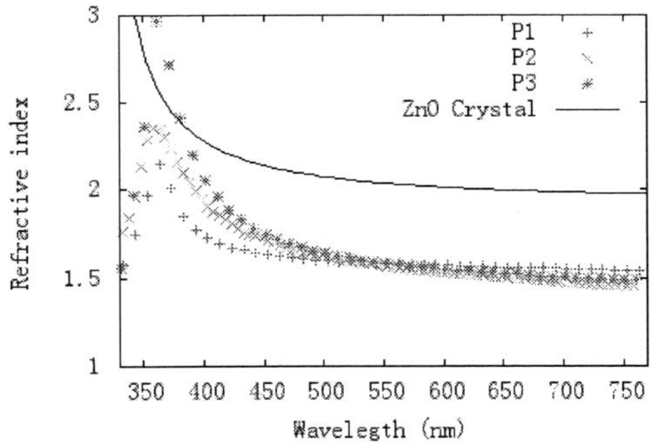

Fig.3. Refractive index of ZnO films as a function of the wavelength.

Conclusion

ZnO films have been deposited on glass substrates by RF magnetron sputtering under various power. The absorption coefficient was evaluated from the transmittance (T) and reflectance (R) data, and the KK relation has been employed to determine the optical constants of the ZnO films. The results show that the KK relation is a simple and effective method for determining the optical constants for ZnO films from transmittance and reflectance spectra.

Acknowledgments

The work was financially supported by The Science and Technology Project of Jinan University (No. 51208054); The Fundamental Research Funds for the Central Universities (11610701); The Guangdong Province and Ministry of Education of China (No. 2008A090400011); The Guangdong-Hong Kong Technology Cooperation Funding Project (No. 2008A011800005, No. 2009205115, No.20090101-4); The Natural Science Foundation of Guangdong Province (No. 06025195, No. 9151027501000076); Visiting Scholar Foundation of State Key Lab of Silicon Materials at Zhejiang University (No. SKL2009-13).

References

1. U.Ozgur, Y. I. Alivov, C. Liu, A. Teke, M. A. Reshchikov, S. Dogan, V. Avrutin, S.-J.Cho, H. Morkoc, *J. Appl. Phys.*, **98(4)**, 041301 (2005).
2. S.-H. Kim, H.-K. Kim, T.-Y. Seong, *Appl. Phys. Lett.*, **86(11)**, 112101 (2005).
3. D.C. Look, B. Claflin, Phys. *Stat. Sol. B*, **241**, 624 (2004) .
4. R. L. Hoffman, *J. Appl. Phys.*, **95(10)**, 5813, (2004).
5. H. Yoshikawa, S. Adachi, *Jpn. J. Appl. Phys.*, **36**, 6237 (1997).
6. X.Zi-qiang, D.Hong, L.Yan, and C.Hang, Mat. *Sci. Semicond. Proc.*, **9**, 132 (2006).
7. R. Ghosh, D. Basak, and S. Fujihara, *J. Appl. Phys.*, **96**, 2689 (2004).
8. D. Poelman, P. F. Smet, *J. Phys. D*, **36**, 1850 **(2003)**.
9. S.W. Xue, X.T. Zu, W.G. Zheng, H.X. Deng, Z. Xiang, *Physica B*, **381**, 209 (2006).
10. V. Srikant, University of California, Santa Barbara, 1996.
11. S.R. Bhattacharyya, R.N. Gayen, R. Paul, and A.K. Pal, *Thim. Sol. Film.*, **517**, 5530 (2009).
12. S. Bandyopadhyay, G. K. Paul, S. K. Sen, *Sol. En. Mat.*, **71**, 103 (2002).

ECS Transactions, 34 (1) 551-555 (2011)
10.1149/1.3567636 ©The Electrochemical Society

Influence of the Pressure on ZnO:Al Film Deposited by DC Magnetron Reactive Sputtering

S. Yu[a], H. D. Yang[a,b], B. Huang[a,b], J. D. Shi[a] and L. X. Zeng[a]

[a] Department of Electronic Engineering, Jinan University, Guangzhou 510632, China
[b] State Key Lab of Silicon Materials, Zhejiang University Hangzhou 310027, China.

Films were deposited on glass substrates by a DC magnetron sputtering method using a ZnO target doped with 2wt% Al_2O_3. Effects of sputtering on the structural, optical, and electrical properties of ZnO:Al films were investigated. The results showed that film's crystalline quality could be improved along the c-axis direction by increasing the sputtering pressure. Below 12.5Pa, grain size became more refined and the film's surface became more uniform. The resistivity of ZnO:Al films reached a minimum of 1.66×10^{-3} Ω·cm at 12.5Pa. The films' average transmittances were higher than 80% in the visible (400nm–800nm) region. Furthermore, the bathochromic shift phenomenon of absorption edge had been found in this process. Consequently, the properties of optical, electrical degraded with the increase of pressure above 12.5 Pa.

Introduction

Transparent conductive oxide (TCO) with low resistivity and high transmittance in visible light has broad applications in the optical fields (1,8). Under normal circumstances, TCO is constituted by wide band-gap non-degenerate semiconductor which have low resistivity and high transmittance in visible and near infrared region. Due to its good chemical, mechanical stability and high reduction resistance and rich in natural reservoir, ZnO transparent conductive film has been regarded as one the most promising TCO materials (2).

Al-doped ZnO (ZnO:Al) provided additional advantages, such as lower cost, lower depositing temperature, and a higher stability in hydrogen plasma environment (3). Among various coating methods including vacuum deposition, ion beam and ion assisted deposition method, pulsed laser deposition (PLD), sol-gel method, chemical vapor deposition (4), magnetron sputtering is one of the most commonly used.

Experimental procedure

The experiments started with a ZnO target of 2wt.% Al_2O_3. The target is bounded to a 5mm thick copper plate. The substrate, a regular glass plate, is first cleaned with acetone, ethanol, and distilled water through an ultrasonic clean and then dried with nitrogen before putting into the sputter chamber. The chamber base pressure was kept at 2.5×10^{-3} Pa and the sputtering gas used was pure argon (Ar). The distance between target and substrate is 30 mm for an optimal sputter yield at the glowing threshold; in addition, the substrate was not heated during sputtering. For all the samples, sputtering power was kept at 80 W and the sputter time 45 minutes. Films were deposited at the sputtering pressure of 10.4Pa, 11Pa, 12.5Pa, 14.5Pa and 16Pa in argon environment, respectively. Surface morphology was observed with scanning electron microscope (SEM) and thickness measurement was performed using a step apparatus. The average thickness obtained was about 700nm. The crystal structure properties of films were investigated by using the X-ray diffraction method (XRD). The effect on optical and electrical properties was investigated by using a UV spectrophotometer and a four-point probe.

Results and Discussions

<u>Microstructure and Morphology</u>

Electrical and optical property of film was affected not only by its composition, but also by the corresponding surface morphology and the microstructures. X-ray diffraction maps of films in different argon sputtering pressure and scanning angle for 20 ~ 80 ° are shown in figure 1. It is evident that strong diffraction peaks in (100), (002), (101), (110) crystal faces were observed. Through a detailed comparison and analysis of X-ray diffraction of ZnO 、 ZnAl$_2$O$_4$ 、 Al$_2$O$_3$, we found that there was no other crystal diffraction peaks except those of ZnO. This validated that the films deposited were a six-party fiber smelting structure of ZnO. The peak of (002) increased when the pressure increased from 10.4Pa to 16Pa. This implies that an increase of argon pressure rendered the grain growth along the c-axial direction and the quality of crystallization was enhanced. Furthermore, FWHM also increased with pressure implying smaller grains. Meanwhile, the peak of (100) and (101) were found to increase as well. The average of grain size can be estimated according to Xie-Le formula (5):

$$D = \frac{0.9\lambda}{\beta\cos\theta}$$

[1]

D: grain size λ :X-ray wavelength θ :Prague Angle β : FWHM

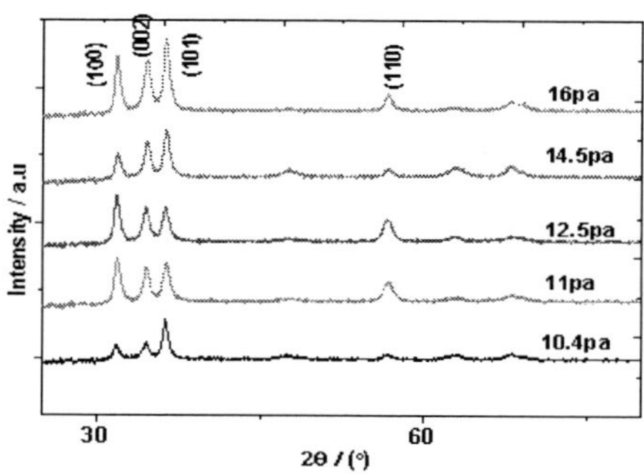

Figure 1 XRD patterns of 2wt.% Al$_2$O$_3$-doped ZnO thin films with various sputtering pressures

TABLE I. Fabrication process parameters of ZnO:Al film

Sample No.	Sputtering pressure/Pa	Deposition power/W	Distance /mm	Deposition time/ min	Thickness / nm
1	10.4	80	30	45	683
2	11	80	30	45	605
3	12.5	80	30	45	601
4	14.5	80	30	45	740
5	16	80	30	45	755

Table 1 the process parameters of the films fabricated, the distance between target and the substrate is 30mm.

Figure 2 (a) ~ (e) showed the SEM diagram in different sputtering pressure. It showed a very irregular grain sizes at 10.4Pa. Grains became larger and more uniform with increasing pressure, grain boundaries became clearer, grains appeared to be coalesced, and grain surface roughness increased.

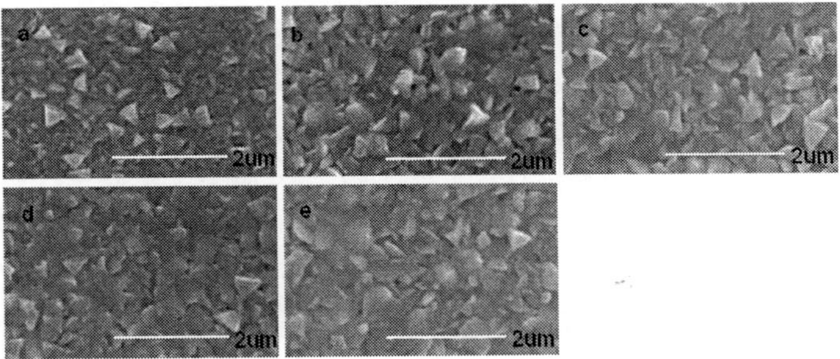

Figure 2 SEM images of ZnO:Al films. The films were fabricated in sputtering pressures of 10.4Pa (a), 11Pa (b), 12.5Pa (c),14.5Pa (d), and16Pa (e)

Electrical and Optical properties

The relationship between the resistivity of ZnO:Al films and pressure was shown in figure 3. From the graph, it is clear that the resistivity was the highest at 10.4 Pa and dropped significantly at 11Pa to about $2\times10^{-3}\Omega\cdot cm$ and remained evenly low at higher pressures. Figure 4 displayed transmittances of films under various pressures. It is evident that there was a steep absorption edge in the area of visible light and near infrared (400 ~ 900nm). The transmittances of ZnO:Al films in various sputtering pressure all exceeded 80% due to the light absorbed less by defects and impurities in the film. The transmittances decreased when the pressure was at 14.5Pa and 16Pa. It is suspected that effective forbidden band width decreased at higher pressure and absorption edges moved to the L-waves direction, i.e., an appeared bathochromic shift.

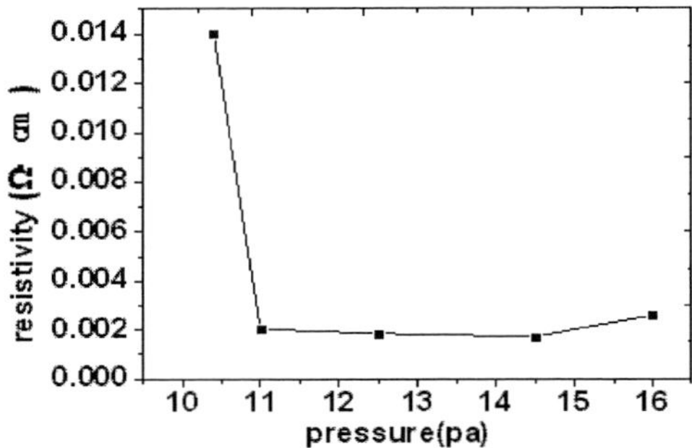

Figure 3   resistivity patterns of ZnO thin films with different sputtering pressures

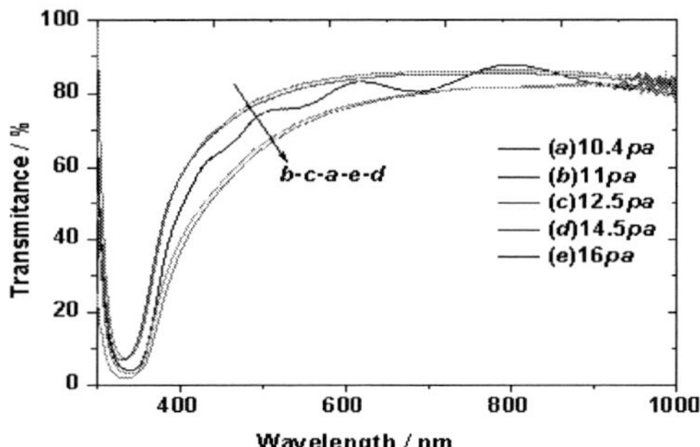

Figure 4   Transmittance curves of ZnO: Al thin film with different sputtering pressures

Conclusions

In conclusion, the influence of pressure on ZnO:Al films deposited by DC magnetron reactive sputtering has been studied. As pressure increases, the diffraction peak of (002) of film increases gradually, the best Crystallization quality could be found under the pressure of 14.5Pa, in which the surface of film was most uniform. Film's resistivity appeared to have a trend which decreases at first and then increases with increasing sputtering pressure. When the pressure was at 12.5Pa, the resistivity was at the minimum $1.66 \times 10^{-3} \Omega \cdot$ cm. The transmittances of ZnO:Al films fabricated in the sputtering pressures of 10.4Pa、 11Pa、 12.5Pa、 14.5Pa、 16Pa exceeded 80% in the range of 400~800nm.

Acknowledgements

The work was financially supported by The Guangdong Province and Ministry of Education of China (No. 2008A090400011);The Guangdong Hong Kong Technology

Cooperation Funding Project(No. 2008A011800005, No.2009205115, No.20090101-4); The Natural Science Foundation of Guangdong Province (No.06025195,No. 9151027501000076); Visiting Scholar Foundation of State Key Lab of Silicon Materials at Zhejiang University (No. SKL2009-13); The Funding Project of Shenzhen Hong Kong Innovation Circle.

References

1. T. Minami, *Semiconductor Science and Technology*, 20 (4), 35 (2005).
2. K. C. Park, D. Y. Ma and K. H. Kim, *Thin Solid Films*, 305 (1/2), 201 (1997).
3. T. Minami, T. Yamanoto and T. Miyata, *Thin Solid Films*,366 (1/2), 63268(2000).
4. Y. P. Liu, F. Chen and A. B. Guo, *Vacuum and cryogenic,* 13(1), 1 (2007).
5. G. J. Fang, D. J. Li and B. L. Yao, *Phys. Stat. Sol. A*, 418(2), 156 (2002).
6. J. W. Xu, H. Wang, M. F. Ren and J. Cheng, *Journal of materials science and engineering*, 25(5), 765 (2007).
7. D. J. Kwak, K. II Park and B. S. Kim, *Journal of the Korean Physical Society*, 45(1), 206 (2004).
8. M. H. Jiang and X. Y. Liu, *Semiconductor photoelectric*, 30(3), 417 (2004).

ECS Transactions, 34 (1) 557-561 (2011)
10.1149/1.3567637 ©The Electrochemical Society

Study of the electrical and optical properties of the silicon carbide thin film

R. Luo[a], H. D. Yang[a,b], B. Huang[a,b] and B. Y. Xu[a]

[a] Department of Electronic Engineering, Jinan University, Guangzhou 510632, China
[b] State Key Lab of Silicon Materials, Zhejiang University, Hangzhou, 310027, China

Boron-doped SiC thin films were deposited using Plasma Enhanced Chemical Vapor Deposition (PECVD) with SiH_4, CH_4, H_2, and B_2H_6. Effects of carbon concentration on the electrical and optical properties were studied. The results showed that the optical band gap of the film was above 2.2eV when CH_4/SiH_4 was tuned between 0.3 to 0.4 and B_2H_6/SiH_4 from 0 to 0.1; in addition, the electrical conductivity was above 1×10^{-7} S/cm. The characteristics obtained can easily meet the stringent electrical and optical requirements of the corresponding solar cell.

Introduction

As a promising semiconductor material, silicon carbide has a wide band gap and high thermal electronic conductivity (1). Because of these properties, it is used as a window layer (P layer) of PIN thin-film solar cell (2). Hamakawa et al. (3) reported that the optical band gap of SiC films could be as high as 2.8eV with the increase of carbon content. However, too much carbon can reduce the conductivity (3). Fortunately, this negative impact can be mitigated by boron doping (4). Therefore, the impact of carbon concentration and boron doping, i.e., by varying the flow of CH_4 and B_2H_6, to the electrical and optical properties of the film should be investigated with the goal to identify the optimal processing window for boron-doped a-SiC:H thin films.

Experimental Methods

a-SiC:H films were deposited in a 13.56MHz PECVD chamber system using SiH_4, CH_4, H_2 and B_2H_6. The carbon content was controlled by varying the CH_4/SiH_4 flow ratio. Similarly, the change of the boron content was controlled by varying the flow ratio of B_2H_6/SiH_4. In order to facilitate the analysis, we defined R_x as the flow ratio of CH_4/SiH_4 and R_y as the flow ratio of B_2H_6/SiH_4. The value of R_x was from 0.1 to 0.4, and the value of R_y from 0 to 0.4. The flow of SiH_4 (hydrogen dilution 20%) was fixed to 20sccm. All relevant parameters were shown in Table 1.

TABLE 1 Technological parameters of the thin films

Parameters	Value
SiH_4 (sccm)	20 (20%)
CH_4 (sccm)	2~8 (99%)
H_2 (sccm)	50
B_2H_6 (sccm)	0~8 (1 %)

substrate temperature(^{o}C)	150
RF power (W)	36
reaction pressure (Pa)	100

     Shimadzu UV-2550 UV-Vis Spectrophotometer was used to obtain the absorption spectrum at room temperature, which was later used to calculate the optical band gap using the Tauc formula. To measure the conductivity, the aluminum metallization was required. The GTP-S ellipsometer was used to measure the thickness and the Keithley 2635 SYSTEM Sourcemeter was used to obtain the electric current and the voltage between the electrodes.

Results and Discussions

<u>Optical band gap</u>

     To measure the optical band gap, the absorption spectra of thin film is needed. For example, when R_x is 0.15 and R_y is 0, absorption spectrum obtained is shown in Fig.1.

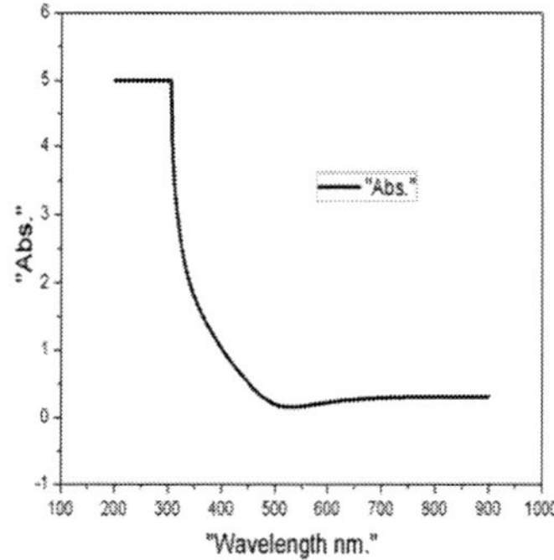

Figure 1 Absorption spectrum of the film when R_x=0.15,R_y=0

Figure 2 Relationship between $(\alpha h v)^{1/2}$ and hv when R_x=0.15,R_y=0

The Tauc formula is shown below,

$$(\alpha h v)^{1/2} = B (h v - E_g) \qquad [1]$$

where *hv* represents sub-energy of the incident light, α is the absorption coefficient of wavelength, and B is a constant. The value of the optical band gap is the linear part of the curve $(\alpha h v)^{1/2}$ at the intersect of hv (5, 6). Fig. 2 shows the corresponding optical band-gap of 2.02eV from Fig. 1.

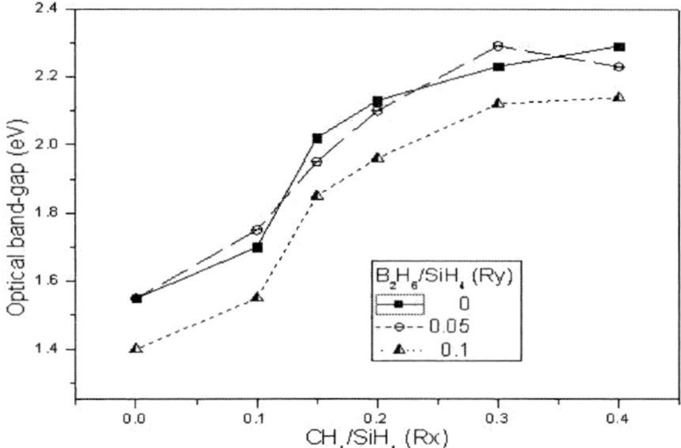

Figure 3 Plot of optical band gap of the films as a funciton of R_x density when R_y is fixed

Figure 3 shows that the optical band-gap becomes larger with increasing R_y. As shown by Zhang *et al.* more CH_n are decomposed with increasing R_x (7). In addition, the introduction of the carbon inhibits grain growth and thus widens the optical band-gap (8). But it is worth noting that the rate of increase becomes smaller when R_x is larger than 0.15 since Si−C bonds are close to saturation resulting in a smaller the rate of increase.

Figure 3 also shows that the curve of which R_y=0 nearly coincides with the curve of which R_y=0.05, but it varies a lot when R_y changes from 0.05 to 0.1.This is consistent with the research of D. L. Xia (10), in which boron doping tends to reduce the desorption energy of H, further leads to reduce the hydrogen content in the compounds, and ultimately renders the optical band gap narrowed. Additionally, boron doping will change the structure of thin film, inhibit the crystallization, and increases the amorphous component in the film. All of these will narrow the optical band gap (9, 10).

It is evident that optical band gap can be above 2.2eV when R_y ranges from 0 to 0.1 and R_x from 0.3 to 0.4.

Conductivity

The extent of aluminum metallization must be known in order to obtain the correct values of conductivity. Furthermore, the thickness of the film and the electric current should also be measured. The conductivity can thus be calculated with the equation below.

$$\sigma = I\omega \,/\, LdU \qquad\qquad [2]$$

where U is the voltage between two electrodes of aluminum, I is the electric current measured, ω is the distance between two aluminum electrodes, L is the length of electrode, and d is the thickness of the film. In our experiment, ω=3mm, L=2cm.

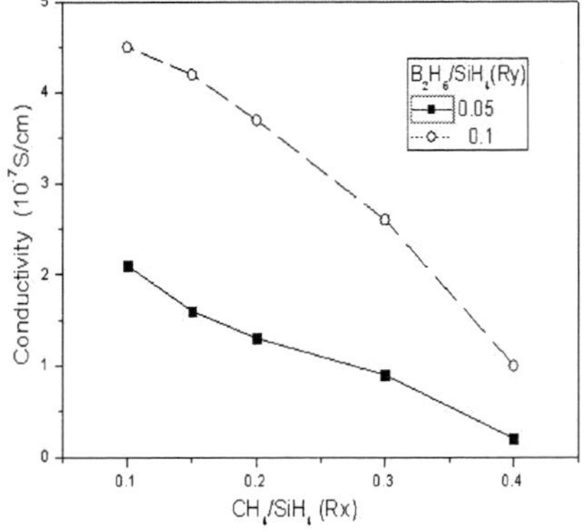

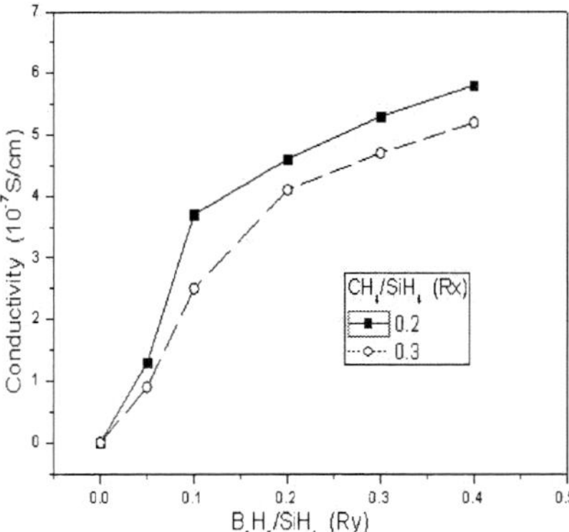

Figure 4 Plot of conductivity of the films as a function of R_x density when R_y is fixed to 0.05 or 0.1

Figure 5 Plot of conductivity of the films as a function of R_y density when R_x is fixed to 0.2 or 0.3

Figure 4 shows that conductivity decreases with the increase of R_x. This is caused by the formation of Si—C bonds, or caused by the replacement of Si—H bonds by Si—CH$_n$ bonds. The amount of these bonds increases with increasing R_x. The downward trend is more obvious when $R_x > 0.3$. W. Q. Han suggested it was caused by the percolation of the film. When the crystalline state rate is higher than the Kan value of percolation, nanocrystals in the film form a conductance pathway. The decrease of crystalline components results in the decrease of optical band gap of the film (11). The crystalline state rate is less than the Kan value of percolation when $R_x > 0.3$. As a result, the band gap decreases sharply.

Figure 5 shows that the conductivity of the film increases with the increase of R_y when R_x is fixed to 0.2 or 0.3. Boron doping overcomes the effects of electronic conductive and converts n-type films into p-type films (12) and significantly improves the conductivity of films. The growth trend slows down when $R_y > 0.1$. Too much boron is counterproductive since boron atoms can replace hydrogen atoms in Si—H bonds, forming Si—B bonds without electrical activity (13). The results obtained also shows that the conductivity is above 1×10^{-7}S/cm only when $R_y > 0$ and $R_x < 0.4$.

Conclusion

The boron-doped a-SiC: H films were deposited on glass substrates with PECVD method. Under the current experimental conditions, an optical band gap above 2.2eV and the conductivity above 1×10^{-7}S/cm can be obtained when the flow ratio of CH_4/SiH_4 is from 0.3 to 0.4 and B_2H_6/SiH_4 from 0 to 0.1. The films fabricated by this manner met the optical and electrical requirements of window layer for the solar cell.

Acknowledgments

The work was financially supported by The Guangdong Province and Ministry of Education of China (No. 2008A090400011); The Guangdong-Hong Kong Technology

Cooperation Funding Project (No.2008A011800005, No.2009205115, No.20090101-4); The Natural Science Foundation of Guangdong Province (No. 06025195, No. 9151027501000076); Visiting Scholar Foundation of State Key Lab of Silicon Materials at Zhejiang University (No.SKL2009-13); The Funding Project of Shenzhen Hong Kong Innovation Circle.

References

1. K. W. Yang, J. Pan and Y. T. Yang, *Semiconductor Information*, 37(2), 13(2000).
2. X. Y. Han, G. J. Li, G. F. Hou, et al , *Journal of semiconductors*, 1548, 29(2008).
3. M. Ma, S. Aoyama, H. Okamoto and Hamakawa, *Solar Energy Materials and Solar Cells*, 41, 453(1996).
4. Z. Li, X. W. Zhang, G.R. Han, *Chinese journal of vacuum science and technology*, 29(2), 119(2009)
5. R. Tsu, P. Menna and A. H. Mahan, *Solar Cells* 189, 21(1987).
6. J. P. Conde, V. Chu, M. F. d. Silva, A. Kling, Z. Dai, J.C. Soares, S. Arekat, A. Fedorov, M. N. Berberan-Santos, F. Giorgis and C.F. Pirri, *J.Appl. Phys.*, 3327, 85 (1999).
7. Y. Y. Zhang, P. Y. Du, R. Zhang, G. R. Han and W. J. Weng, *Journal of Non-Crystalline Solids,* 1435, 354(2008).
8. J. Robertson, *Philos. Mag.* 615, B66 (1992).
9. S. Yamasaki, A. Matsuda, K. Tanaka, *Jpn. J.Appl. Phys.*, 21(12), 789(1982).
10. D. L. Xia, H. F. Wang, Z. Z. Shi, X. L. Zhang, J. Liu, *Imaging Science and Photochemistry,*29(4), 296(2010).
11. W. Q. Han, G. R. Han, D. L. Nie and Z. S. Ding, *acta energiae.solaris sinica.*, 17(1), 87(1996).
12. X. W. Zhang and G. R. Han, *journal of functional materials and devices,* 9(2), 150(2003).
13. K. Wine, R. A. Street, N. M. Johnson, et al. *Phys Rev.*, 3120, B42 (1990).

ECS Transactions, 34 (1) 563-566 (2011)
10.1149/1.3567638 ©The Electrochemical Society

Influence of Vacuum-annealing Temperature on the Properties of Direct Current (DC) Magnetron Sputtered ZAO Thin Films

J. D. Shi[a], H. D. Yang[a,b], B. Huang[a,b], B. Y. Xu[a] and S. Yu[a]

[a] Department of Electronic Engineering, Jinan University, Guangzhou 510632, China
[b] State Key Lab of Silicon Materials, Zhejiang University, Hangzhou 310027, China

This study focused on the thermal annealing effects on the optical and electrical properties as well as the microstructure and morphology of the ZAO thin films. Film quality is significantly improved after vacuum-annealed at 160°C or above. The growth rate along (002) direction, before reached a maximum at 330°C, was found to be faster than those not annealed; in addition, the corresponding FWHM decreased to 0.74° at 330°C and grain size grew to 12nm at 400°C. Furthermore, the resistivity of ZAO thin films decreased and reached a minimum of $9.36 \times 10^{-4} \, \Omega \cdot cm$, while the average transmittance increased to 82% in the visible region.

Introduction

Solar battery is a photovoltaic device which uses solar photovoltaic effect and therefore must be manufactured with transparent conductive oxide (TCO) thin films[1,2]. ITO ($In_2O_3+SnO_2$) transparent conductive film is currently widely used in industry[3]. Howerer, its use is limited because of the poor thermal stability and the relatively weak radiation resistance. On the other hand, aluminum-doped zinc oxide (ZAO) thin film is attractive due to its non-toxicity, low cost, and high thermal stability with electrical and optical properties comparable to ITO[4-6]. It has been shown extensively that annealing is an effective measure to improve the properties of ZAO thin film [7-9].

In this paper, all ZAO thin films have been fabricated on glass slide substrates by direct current (DC) magnetron sputtering and annealed at various temperatures. A study of the effects of annealing temperatures on the corresponding crystal structural, surface topography, electrical and optical properties was carried out systematically.

Experimental

All ZAO thin films have been fabricated on glass substrates by a Shenyang Tiancheng vacuum 500W DC sputtering system with a ceramic $ZnO:Al_2O_3$ target (98:2wt%) in a base pressure of 1.5×10^{-3}Pa. The distance between glass substrate and ceramic target surface was about 3mm. The corresponding sputter has been carried out in a pressrue of 12.5Pa using Ar. The sputtering power was maintained at 80W for 45 minutes. After deposition, films were selectively annealed at various temperatures of 160°C, 250°C, 330°C, 400°C, 460°C in vacuum.

X'Pert Pro MPD XRD was used to measure the crystal structure of the films. The surface morphology was investigated by SEM. The sheet resistance were investigated by a SOY-5 FPP. Optical transmittance measurements were carried out using a UV-2400 UV-VIS spectrophotometer. All measurements were carried out at room temperature.

Result and discussion

Structure and surface morphology

Figure 1 (a) displays X-ray patterns of the ZAO films annealed at various temperatures in vacuum. It shows that as the annealing temperature increased, the (100), (101) and (110) peaks decreased initially and then increased. When the annealing temperature increased from room temperature to 330℃, the intensity of (002) peak increased gradually to the highest . During the process of crystallization, ZAO thin film grew along (002) crystal surface preferentially with a tendency to be epitaxially cylindrical. This implies that the microstructure of samples was transformed into a high c-axis orientated polycrystal with hexagonal wurtzite structure; in addition, the growth direction of crystalline grains was perpendicular to glass substrates.

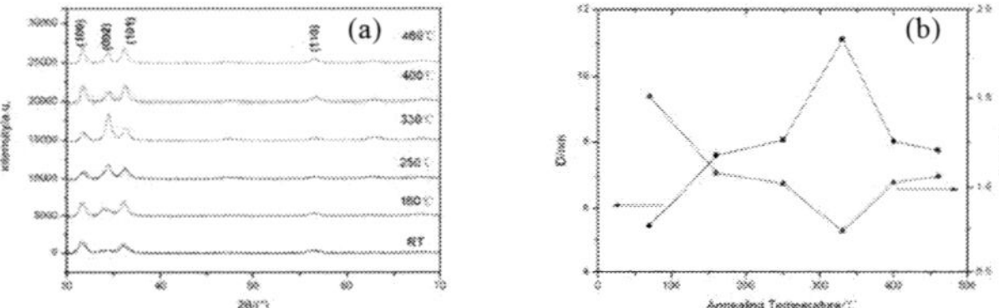

Fig.1 (a)X-ray patterns; (b) FWHM for peak (002) and diameter of crystalline grain of the ZAO films annealed at different temperatures in vacuum.

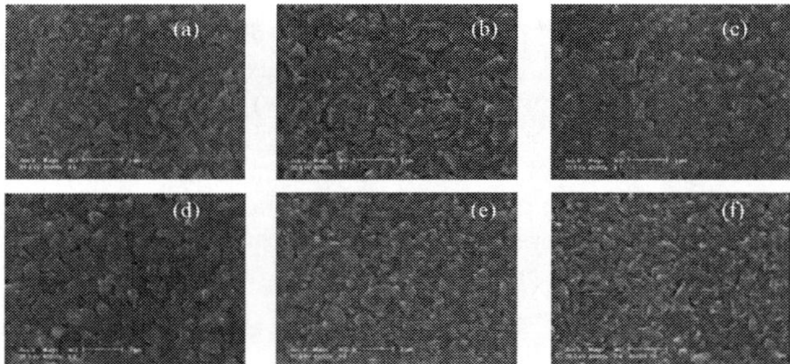

Fig.2 SEM images of ZAO thin films at different annealing temperature: (a)not annealed; (b) 160℃;(c)250℃;(d)330℃;(e)400;(f)480℃.

XRD results, together with the corresponding full-width-half-maximum (FWHM) for peak (002) and diameter of crystalline grain of the ZAO films annealed at various temperatures are shown on Fig.1 (b). The grain size was calculated using Scherrer formula [10]

$$D = 0.9 * \lambda / (FWHM * \cos\theta) \quad (1)$$ [1]

Where λ is the wavelength of X-ray (about 0.154 06nm), θ is the the half-diffraction angle of diffraction peak. It could be seen that the FWHM of (002) peak firstly decreased to a minimum at 330 °C and then increased at above 330°C. The grain size increased as the result of the annealing. When the annealing temperature exceeded 400 °C, the grain size decreased sharply which was inversely proportional to the FWHM of (002) peak.

As seen from the Fig.2 (SEM images: magnification 40,000 times), the grains grew irregularly during deposition. After annealing at more than 160°C, the edges of grains started to round-off as driven thermodynamically by the minimization from surface tension. The crystalline quality of ZAO thin films showed an improvement at temperatures between 160 and 400 °C. However, when the annealing temperature exceeded 400°C, the surface morphology became rough.

<u>Electrical and optical properties</u>

High electrical and optical performance is crucial for ZAO films as the electrodes for Solar battery. Annealing improved the electrical properties, as demonstrated in Fig.3 (a). Between room temperature and 330°C, the resistivity of ZAO thin films significantly reduced from $3.28\times10^{-3}\,\Omega\cdot$ cm to $9.36\times10^{-4}\,\Omega\cdot$ cm. However with the continue increase of annealing temperature, the resistivity began to increase. The increase of resistivity is probably due to the degradation of film crystalline quality, which is consistent with the degradation of optical properties shown in Fig. (b) and (c).

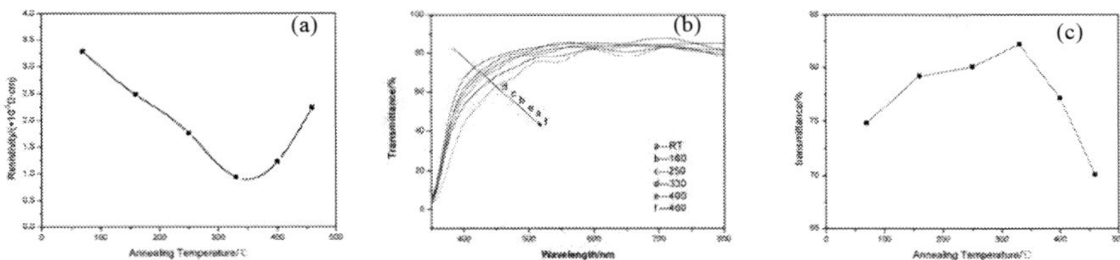

Fig.3 The (a) resistivity, (b) transmittance and (c) average transmittance of ZAO thin films at different annealing temperature.

Figure 3(c) displays the normalized UV-VIS transmittance of ZAO thin films after annealing. The figures show that the average transmittance of ZAO films increased with annealing temperature increasing and reached a maximum of 82% at 330°C. But when the annealing temperature increased above 330°C, the average transmittance reduced drastically and droped as low as 70% at 480°C. In the annealing temperature range of room temperature to 330°C, the figure exhibits a blue shift of the absorption edge. The carrier concentration increased with annealing temperature increasing, resulting in the brodening of effective crystal band gap, so the absorption edge moved towards shortwave direction. When annealing temperature exceeded 330°C near the softening temperature of the substrate, the films' crystal quality deteriorated dramatically. The degree of the metallization of films also increased while film gradually darkened, which reduced the transmittance sharply.

Conclusions

In conclusion, the properties of ZAO films have been shown capable of improvement by thermal annealing. It is found that the intensity of (002) peak increased significantly with FWHM decreased to 0.74° at 330°C and grain size grew to 12nm at 400°C. At the same time, the resistivity of ZAO thin films decreased after anneal and reached a minimum of $9.36 \times 10^{-4} \Omega \cdot$ cm while the average transmittance increased to 82% in visible region. A blue shift of the absorption edge was observed. The overall results showed that the optimal annealing temperature is 330°C for ZAO thin films.

Acknowledgments

The work was financially supported by the Guangdong Province and Ministry of Education of China (No. 2008A090400011); The Guangdong-Hong Kong Technology Cooperation Funding Project (No. 2008A011800005, No.2009205115, No.20090101-4); The Natural Science Foundation of Guangdong Province (No.0602519, No.9151027501000076); Visiting Scholar Foundation of State Key Lab of Silicon Materials at Zhejiang University (No. SKL2009-13); The Funding Project of Shenzhen Hong Kong Innovation Circle.

References

1. X.Y. Li, H.J. Li, Z. J. Wang, H. Xia, Z. Y. Xiong, J. X. Wang and B. C. Yang, *Optics Communications*, **282**(2), 247 (2009).
2. H. L. Hartnagel, A. K. Jain and C. Jagadish, *Semiconducting Transparent Thin Films*, p.306, Paston Press, UK (1995).
3. H. Zhu, E. Bunte, J. Hüpkes, H. Siekmann and S.M. Huang, *Thin Solid Films*, **517**(10), 3161 (2006).
4. S. J. Pearton, D. P. Norton, K. Ip, Y. W. Heo and T. Steiner, *Prog. Mater. Sci.*, **50**, 293 (2005).
5. E. G. Fu, D. M. Zhuang, G. Zhang, W. F. Yang and M. Zhao, *Applied Surface Science*, **217**, 88 (2003).
6. T. Minami, H. Nanto and S. Takata, *Thin Solid Films*, **124**, 43 (1985).
7. G.J. Fang, D. J. Li and B. L. Yao, *Thin Solid Films*, **418**(2), 156 (2002).
8. R. P. Wang, H. Muto, Y. Yamada and T. Kusumori, *Thin Solid Films*, **411**, 69 (2002).
9. K. H. Bang, D. K. Hwang and J. M. Myoung, *Applied Surface Science*, **207**, 359 (2003).
10. R. J. Hong, J. D. Shao, H. B. He and Z. X. Fan, *Journal of Crystal Growth.*, **284**(3/4), 347 (2005).

ECS Transactions, 34 (1) 567-570 (2011)
10.1149/1.3567639 ©The Electrochemical Society

Study of Phosphorus Out-Diffusion from High Density Plasma CVD Phosphosilicate Glass Process

Li Min[a], Zhang Zi Ying[b] and Qiang Xu[a]

[a] Semiconductor Manufacturing International Corporation, No.18 Wen Chang Rd., BDA, Beijing, 100176, P.R.China
[b] Semiconductor Manufacturing International Corporation, No.18 Zhang Jiang Rd., Pudong New Area, Shanghai, 201203, P.R.China

High Density Plasma (HDP) CVD phosphosilicate glass (PSG) process applied on pre-metal dielectric (PMD) layer has plasma damage concern which can impact the device performance and reliability. A PETEOS cap layer is used for the plasma damage reduction. In this paper, the diffusion of phosphorus through PSG to PETEOS cap layer is investigated. The PMD film stack consists of 2000Å PSG/6000Å PETEOS. Secondary Ion Mass Spectroscopy (SIMS) was used to measure the P distribution in the PMD film stack. Transmission Electron Microscope (TEM) was used to confirm the contact glue layer and W encroachment along with the PSG and PETEOS interface. We found that O_2 plasma treatment or DI-water clean of PSG surface could effectively reduce the P accumulation at the interface. It is also found that HDP USG transition layer between PSG and PETEOS could inhibit P out-diffusion. The fundamental of P out-diffusion also has been discussed.

Introduction

HDP-CVD is currently the most commonly used gap-fill technology for sub-65nm logic PMD process because of its excellent gap-fill capability and good film quality. HDP-CVD phosphosilicate glass is used to stabilize semiconductor devices as it traps impurity ions such as Na which would otherwise drift to the oxide/silicon interface and affect the device electrical properties. However, a main concern with HDP-CVD process is the electrical damage to thin gate oxide caused by the charges in the plasma. It has been reported that PECVD process creates less plasma damages than HDP-CVD process. To reduce the plasma damage, HDP-PSG and PETEOS bi-layer film stack is applied successfully at the pre-metal dielectric application. But the contact glue layer Ti/TiN and W encroachment has always been observed along with the interface between HDP-PSG and PETEOS (Fig1). In this work, we discussed the formation fundamental of the interface and have tackled this problem by HDP-PSG surface optimization.

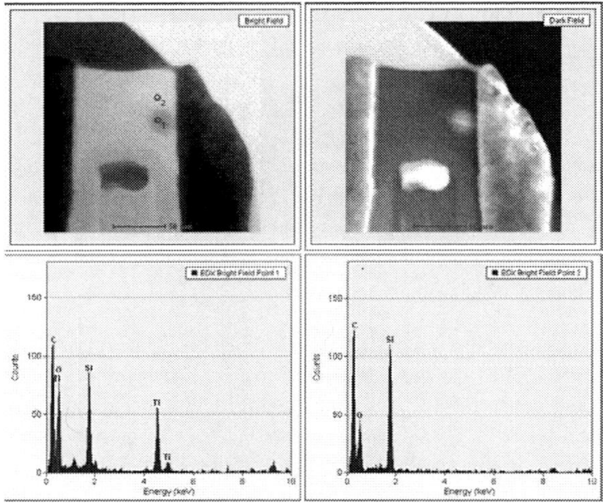

Fig1. Ti/TiN extrusion between HDPPSG and PETEOS interface at the middle of contact hole

Experimental

HDPPSG and PETEOS films were deposited using the conventional HDP-CVD and PECVD tool, respectively. The precursor gases of HDPPSG consisted of He, O_2, SiH_4 and PH_3. The precursor gases for PETEOS formation consisted of TEOS, O_2 and He. The PMD film stack was made of 2000Å PSG/6000Å PETEOS. The P% concentration is about 4%. SMIS was used to obtain the P depth profile.

SMIS analysis was performed using a CAMECA ims-6f system. The experimental conditions were O^{2+} primary ion source with 1kV impact energy. The raster area was 200μm*200μm square with a 66 μm diameter analysis area. An electron flood gun was used for charge neutralization; the current was set at about 10mA.

Results and Discussion

Impact of HDPPSG surface treatment

After HDPPSG deposition, distilled water wet clean or O_2 plasma treatment was performed, respectively, then capped by PETEOS layer. Fig 2 shows the SMIS results on samples with the two different surface treatment methods. It is found that applying wet clean or plasma treatment to the PSG surface could effectively reducing P out-diffusion from PSG. The Phosphorus peak can be obviously reduced compared to the baseline condition.

<u>Impact of transition layer</u>

After HDPPSG deposition, in-situ deposit HDPUSG film as transition layer for HDPPSG and PETEOS film. Figure3 shows the SIMS depth profile on this sandwich structure. We overlaid the four SIMS depth profiles into one figure for comparison. We found that the ability to inhibit P out-diffusion was different with different USG thickness—the higher USG thickness, the better the ability to block P out-diffusion.

<u>Fundamental discussion</u>

We firstly focus on the origin of the weak interface. An obvious Phosphorus peak was observed at the interface of HDPPSG and PETEOS for the baseline condition. Adding surface treatment as well as transition layer both can reduce the P concentration dramatically. One mechanism could be responsible for this phenomenon: Indeed, after HDPPSG deposition, some unfixed P atoms would like to out-diffusion along with PSG crystal lattice and accumulate at the film surface, and then diffuse to PETEOS cap layer, which caused bad quality of interface PETEOS film. Since the interface P-doped PETEOS film has a higher wet etch rate (WER) than HDP PSG and bulk PETEOS films, as a consequence, during the following wet clean steps, the acid could easily attack the interface to form a seam. Barrier layer always deposited along with the seam and induced serious contact short.

Distilled water clean can remove the accumulated Phosphorus at the interface; O_2 plasma treatment can fix the active P atoms as well as sputter away some surface atoms; while USG transition layer can block the P out-diffuse and enhance the adhesion between HDPPSG and PETEOS cap layer. All of these confirmed the hypothesis of P out-diffusion impact on interface.

Conclusion

In this paper we reported the airy ILD gaps on W lines, discussed the formation fundamental of TiN/Ti and W encroachment along with the HDPPSG and PETEOS interface. From the above experiments, we conclude that: 1) Surface treatments including distilled water clean or O_2 plasma treatment could improve the interface quality. 2) HDPUSG is a good material as the capping and transition material for PSG. Phosphorus out-diffusion could be effectively blocked.

Acknowledgments

Authors would like to thank Dai HaiBo of SMIC Failure Lab and Process Integration Engineers for the SIMS and TEM analysis and technical discussion.

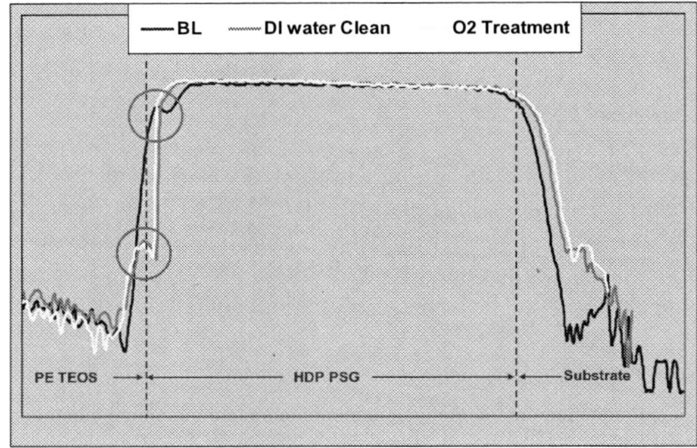

Fig2. SMIS depth profiles of HDPPSG with DI water clean or plasma treatment. P peaks were observed after PETEOS capping

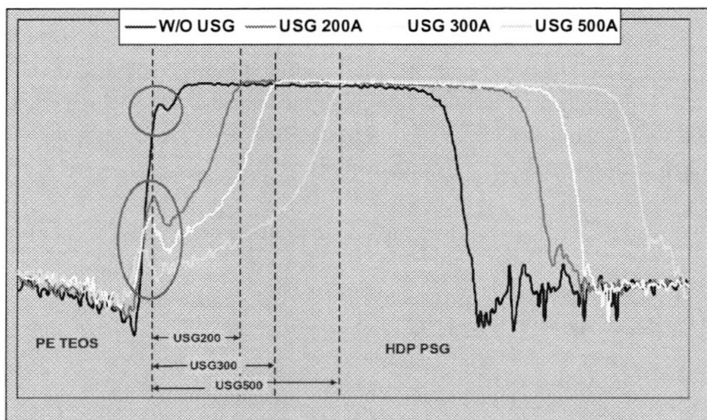

Fig3. SIMS depth profiles of HDPPSG film with HDPUSG transition layer. Higher HDPUSG thickness, less P diffused into PETEOS interface

References

1. Donggun Park, Int. Symp. p.15 (1997).
2. Y.W. The, J. Sudijono, C. Ching, S. Venkataraman and A. Jain, MRS, (2006).
3. H.Liu et al., Advanced Metallization Conf., p.623 (2006).
4. S. Bothra, C.T. Gabriel, Electrochem. Soc, 42, p.67 (1996).
5. C. Chen, et.al., TED, V.47,no.7, p.1355 (2000)

ECS Transactions, 34 (1) 571-576 (2011)
10.1149/1.3567640 ©The Electrochemical Society

Effects of substrate temperature on resistive switching of TiO$_X$ thin film

Lanlan Wang, Kailiang Zhang *, Fang Wang, and Kai Song

School of Electronics Information Engineering, Tianjin Key Laboratory of Film
Electronic & Communication Devices, Tianjin University of Technology, Tianjin, China,
300384, *corresponding author, kailiang_zhang@163.com

Titanium oxide thin films (TiO$_x$) were deposited on Cu
(111)/Ti/SiO$_2$/Si substrates by reactive sputtering with various
substrate temperatures, ranging from room temperature (RT) to
350°C. X-ray diffraction (XRD) analysis showed that the TiO$_x$ film
at RT was amorphous. In contrast, some sub-stoichiometric oxide
films (Ti$_2$O$_3$ and Ti$_3$O$_5$) were formed with the temperatures
ranging from 200 to 350°C. In addition, the peak of (104) Ti$_2$O$_3$
phase became higher before disappeared at 350°C with substrate
temperatures increasing from 200 to 300°C. Electrical test results
showed that both films deposited at 350°C and RT exhibited stable
resistance switching behaviors, while the sample at 350°C yields a
set and reset voltage as low as 0.58V and 0.45V, respectively.
However, the fluctuation of V$_{reset}$ for the samples became too large
for resistive switching to be reversible when the temperature
increased from 200°C to 300°C.

Introduction

Recently, various binary transition metal oxides, such as TiO$_2$ (1) and NiO (2), have
attracted considerable attention due to its potential applications for nonvolatile resistance
random access memory (RRAM). They offer a possibility of high density integration,
lower power operation, easier down-scaling and CMOS (complementary metal oxide
semiconductor) compatibility. Among them, TiO$_2$ has been actively studied for its high
on/off resistance ratio, low on/off voltage, and long retention time compared to the other
binary transition metal oxides.

TiO$_2$ films have been deposited by pulse laser deposition (PLD) (3) or by the thermal
oxidation of TiN or Ti layers to fabricated MIM structures (4-6). However, the thermal
oxidation technique is usually not easy to obtain a smooth and dense film due to oxygen
diffusion. The quality of surface could affect the electric properties of the film. Recently,
it is reported that TiO$_2$ thin films can be easily formed by reactive sputtering (7).
Compared to other fabrication techniques, the reactive sputtering method is very cost-
effective and more compatible with current industry manufacturing processes.

In this report, TiO$_x$ thin films were fabricated on Cu (111)/Ti/SiO$_2$/Si substrates by
reactive sputtering. Moreover, the effects of substrate temperature on the structure

properties of TiO_x films were systematically investigated, and stable resistive switching behavior (RSB) was observed at the substrate temperature of 350°C with a narrow dispersion of the resistance states and switching voltages.

Experimental Details

TiO_x thin films were deposited with reactive sputtering system on $Cu(111)/Ti/SiO_2/Si$ substrates. Ti metal target (60 mm, 99.995%) and a gas mixture of Ar/O_2 of 10:1 were used to deposit TiO_x thin films. The base vacuum was evacuated to 3.0×10^{-4} Pa, and growth pressure was 1Pa. During the deposition of TiO_x thin films, the DC power was maintained at 120W and the substrate temperature was set at values ranging from RT to 350°C. 500nm Cu thin film as the bottom electrode was deposited by DC magnetron sputtering.

The thickness of TiO_x thin films were measured by step profiler (Vecco, Dektak 150). The surface morphology of thin films was characterized by an atomic force microscope (AFM). The crystalline phase of TiO_x thin films were identified by an X-ray diffraction method. The current-voltage characteristics were measured at room temperature with a semiconductor parameter analyzer (Agilent B1500A) in sweep mode. A current compliance (CC) was used to protect samples from damages due to high currents (8).

Results and Discussion

<u>Surface Morphology and Crystalline Phase Analysis</u>

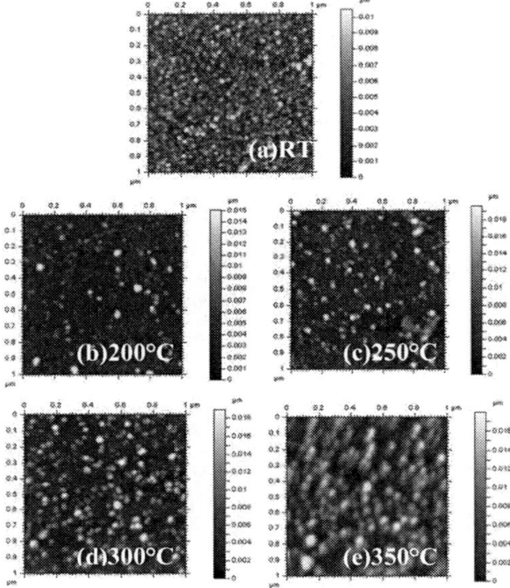

Figure 1 AFM images of TiO_x thin film deposited at different temperatures.

Figure 1 illustrates the morphology of the films deposited at various temperatures. Topography of the surface (1μm×1μm) was scanned. From the AFM image, it is evident that TiO_x film presented in Fig. 1(a) is mostly smooth and dense. The surface becomes rougher with increasing temperature. Fig. 1(a)–(d) also showed that samples at RT are relatively homogeneous in comparison to those of 200°C and 250°C where some large grains are evident. When the substrate temperature is at 350°C, TiO_x grains shown in Fig.1 (e) are uniform and dense grains with size larger than that of the sample growth at RT. The results of profiler measurement show that the film thickness growth at RT, 200°C, 250°C, 300°C and 350°C are 105nm, 78nm, and 154nm, 100nm, and 85nm, respectively.

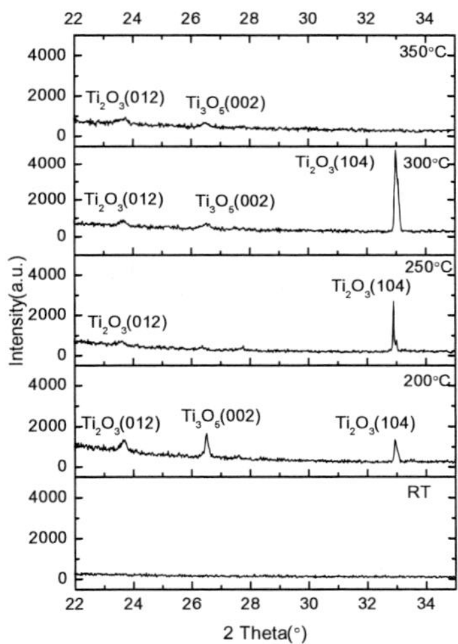

Figure 2 XRD spectra of TiO_x thin films deposited under different substrate temperatures.

XRD spectra of TiO_x films grown at various temperatures are shown in Figure 2. It is clear that XRD peaks depend strongly on the substrate temperature. TiO_x film grown at room temperature is usually amorphous. However, it is possible to see electrically insulating Ti_2O_3 (9) or Ti_3O_5 with the increasing substrate temperature. The oxide peaks at $(012)Ti_2O_3$, $(002)Ti_3O_5$, and (104) Ti_2O_3 (33.001°) are observed in samples 200°C, 250°C and 300°C. In addition, the largest peak, i.e., $(104)Ti_2O_3$, almost completely vanished at 350°C. This is consistent with the literature (10) that Ti_2O_3 starts to transform to TiO_2 at about 250°C-300°C. Therefore, raising the substrate temperature can reduce the low valence oxide of the film deposited. Figure 2 also confirms that the TiO_2 film deposited at 350°C is amorphous. Furthermore, AFM and XRD results show that the substrate temperature affects the film microstructure and the stoichiometry of the film.

<u>Resistance switching analysis</u>

Typical current–voltage (I–V) curves of TiO_x film deposited at 350°C obtained were shown in Figure 3. The current-voltage characteristic of the symmetric $Cu/TiO_x/Cu$ system showed a distinct unipolar resistive switching. A high electrical stress with a CC of 10 mA is necessary to initiate the switching property of this structure, which is known as the forming process, as illustrated in the inset. After the forming process, the device was in the LRS (low resistance state). Then sweeping the voltage from zero to a certain value (the reset voltage), the device is switched to the HRS (high resistance state) with a sudden drop of current. When sweeping is reversed, an abrupt jump of current appeared at a specific voltage (the set voltage), which is always higher than the reset voltage with the resistance switched back to the LRS again.

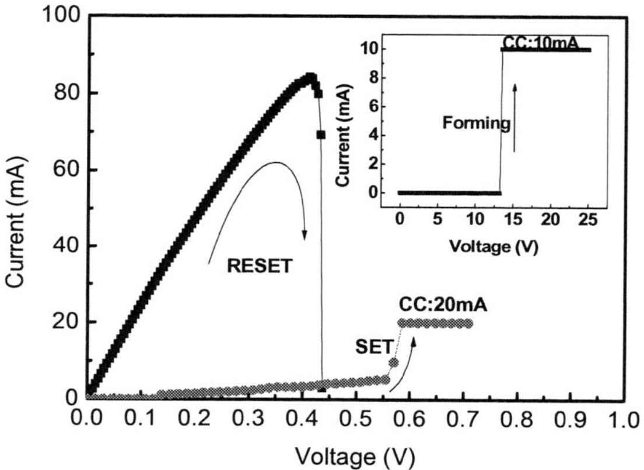

Figure 3 I-V characteristics of $Cu/350°C$-TiO_x/Cu structure

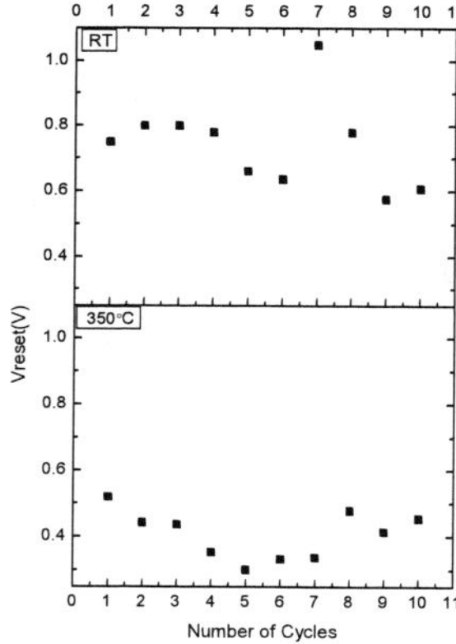

Figure 4 Reset voltages of $Cu/TiO_2/Cu$ structures with TiO_2 deposited at RT and 350°C

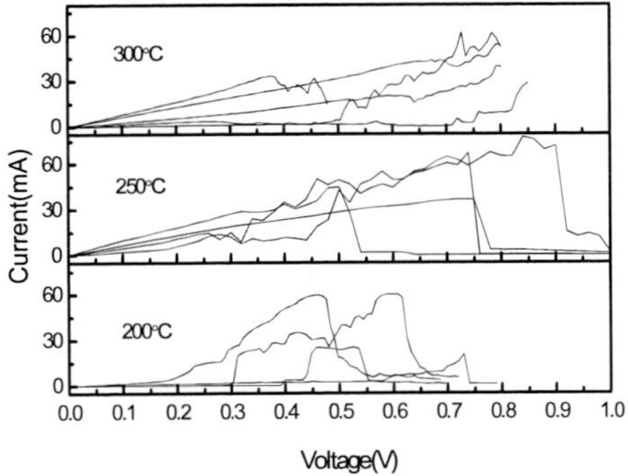

Figure 5 The reset processes of the sample deposited under 200°C, 250°C and 300°C

Figure 4 showed the switching characteristics of the samples deposited at RT and 350°C. The forming voltage values for the sample RT and 350°C are 2.7V and 13.5V. However, the fluctuation of V_{reset} for the sample 350°C is smaller in comparison to that of RT sample while the average reset voltage for the sample 350°C is 0.45V, which is lower than 0.75V of the sample RT.

The resistance switching has been reported to be closely related to the formation and rupture of conducting filaments (11) formed by percolation of defects in TiO_2 films. The initial voltage for the sample at 350°C is much higher than that of the sample at RT, indicating that fewer defects appear at high substrate temperature. Moreover, because TiO_2 film is known as n-type wide-band-gap semiconductor (12), the grain boundaries may also make a considerable contribution to the formation of the conductive filament and the stable resistive switching. The crystalline grains become larger with increasing substrate temperatures resulting in a decrease of grain boundaries, which implies a rise of forming voltage as observed. Especially, the film deposited at 350°Cexhibits more stable resistance switching behaviors than the film at RT, and is more favorable to practical application.

Figure 5 shows the reset processes of the sample deposited at 200°C, 250°C and 300°C. The strong peak of (104)Ti_2O_3 for the sample at 300°C indicates that the film has large amount of (104)Ti_2O_3. With the peak of (104)Ti_2O_3 becoming stronger, the RSB for the sample 200°C, 250°C and 300°C seems to be unstable. (104)Ti_2O_3 crystallization has just started to form at 200°C and the resistance state can still return from high resistance state to low resistance. The fluctuation of V_{reset} becomes larger though because of the increase of (104) Ti_2O_3 crystallization at 250°C. For the sample at 300°C, however, (104) Ti_2O_3 is abundant and reversible resistance change behavior disappears.

Conclusions

In summary, we studied the effect of the substrate temperature on the resistive switching of TiO_x film. Results show that before switching to set process the film deposited at 350°C keeps stable reset processes with the V_{reset} as low as 0.45V. For the samples at 200°C, 250°C and 300°C, the fluctuation of V_{reset} become larger gradually, and finally reversible resistive switching disappear at 300°C due to the existence of (104) Ti_2O_3.

Acknowledgments

We acknowledge the financial support provided by the National Natural Science Foundation of China under Grant No 60806030, and Tianjin Natural Science Foundation under Grant No 08JCYBJC14600, No 10SYSYJC27700 and Tianjin Science and Technology Developmental Funds of Universities and Colleges under Grant No ZD200709.

References

1. H.Schroeder, D.S.Jeong, *Microelectronic Engineering*, 84, 1982 (2007).
2. U.Russo, D.Ielmini, C.Cagli, A.L.Lacaita, *IEEE Trans. Electron Dev*, 56, 186 (2009).
3. X.Cao, X.M.Li, W.D.Yu, Y.W.Zhang, *J.Inorg.Mater*, 24, 50 (2009).
4. H.Shima, F.Takano, H. Muramatsu, H.Akinaga, I.H.Inoue, H.Takagi, *Appl. Phys. Lett*, 92, 043510 (2008).
5. C.P. Hsiung, J.Y. Gan, S.H. Tseng, N.H.Tai, P.J.Tzeng, C.H.Lin, F.Chen, M.J.Tsai, *Electrochem. Solid-State Lett*, 12, G31 (2009).
6. K.M. Kim, B.J. Choi, C.S. Hwang, *Appl. Phys. Lett*, 90, 242906 (2007).
7. C.K.Chung, M.W. Liao,C.W. Lai, *Thin Solid Films*, 518, 1415 (2009).
8. X.Cao, X.M.Li, X.D.Gao, Y.W.Zhang, X.J.Liu, Q.Wang, L.D.Chen, *Appl.Phys.A*, 97, 883 (2009).
9. J.Jo, H.Choi, S.Kang, *Thin Solid Films*, 516, 8693 (2008).
10. X.H.Wang, Y.Y.Xue, Ph.D.Thesis, Wuhan University of Technology, China (2003).
11. L.Yang, C.Kuegeler, K.Szot, A.Ruediger, R.Waser, *Appl.Phys.Lett*, 95, 013109 (2009).
12. Wan-Gee Kim, Shi-Woo Rhee, *Microelectronic Engineering*, 86, 2153 (2009).

ECS Transactions, 34 (1) 577-582 (2011)
10.1149/1.3567641 ©The Electrochemical Society

Electrical and Optical Properties of Zinc Oxide Thin Films Deposited by Magnetron Sputtering

Xiaoqin Ding, Yunfeng Lai

Institute of Micro/Nano Electronic Devices and Solar Cells, School of Physics & Information Engineering, Fuzhou University, Fuzhou 350108, PR China

The Al doped ZnO(ZnO:Al) films were deposited on glass substrates with direct current (DC) or radio frequency (RF) magnetron sputtering by using an alloy or ceramic target. The influence of the substrate temperature on the crystal structure, surface morphology, optical and electronic properties of the ZnO:Al films was investigated. Highly transparent ZnO:Al films with optical band gap of 3.25-3.29eV were deposited by sputtering a ceramic target. The X-ray diffraction results show that all ZnO:Al films grow in a hexagonal-wurtzite phase with highly c-axis preferred orientation and the average crystal size increases upon the promotion of substrate temperatures. The resistivity of ZnO:Al films decreases with an increasing substrate temperature. The resistivity further decreases to 3.566×10^{-3} $\Omega \cdot$cm with annealing in nitrogen.

1. Introduction

Zinc oxide (ZnO), one of the most important binary II–VI compounds, is a direct-band gap semiconductor with a wurtzite structure (1). Due to its unique wide band gap, piezoelectric properties and high transparency, ZnO has attracted considerable attention and has been widely used in solar cells, blue and ultraviolet light emitting devices, pressure transducers and optoacoustic devices (2). Pure ZnO usually contains various intrinsic defects such as Zn vacancies, interstitial Zn, O vacancies and interstitial O (3-4), which make the pure ZnO films have poor electrical properties. However, the electrical properties of ZnO films can be improved by doping Al, because the Al atom substitutes the zinc atom or occupies the interstitial site, which changes the defect environment and improves the photoelectric properties (5-8). There are various techniques for thin film deposition (9–14). Magnetron sputtering seems to be one of the best schemes for the deposition of ZnO thin film, which is contributed to its low cost and good repeatability.

In this study, the Al doped ZnO (ZnO:Al) thin films were deposited on glass substrate with direct current (DC) or radio frequency (RF) sputtering mode. We changed the substrate temperature but kept the other parameters to study the effects of deposition parameters on the micro-structure and photoelectric properties of ZnO:Al thin films.

2. Experimental

ZnO:Al films were deposited on glass substrates in a magnetron sputtering system by using a Zn-Al (Al=2 wt%) alloy target or a ZnO-Al$_2$O$_3$ (Al$_2$O$_3$=2 wt %) ceramic target. In order to get a clean surface, the glass substrates were ultrasonically cleaned in acetone then rinsed in alcohol, followed by drying in an oven. During the deposition, the distance between target and substrate was 75mm, the sputtering power was 110W, the deposition time were the same for all samples and the deposition pressure was kept at 0.5 Pa. Other deposition parameters are shown in table 1.

The crystal structures of thin films were characterized by using X-ray diffraction (XRD) patterns, the surface morphology was observed by using AFM (Park Science, AP2000L), the optical properties were measured by using a spectrophotometer (Lambda35UV/VIS) and the electrical properties were acquired by using Hall effect measuring system (Ecopia HMS-300).

TABLE 1: Deposition parameters of ZnO:Al films

Parameters	Group I	Group II	Group III	Group IV
Sputtering mode	RF	RF	DC	RF
Target	Alloy	Alloy	Alloy	Ceramic
$O_2/(Ar+O_2)$ /%	4.2	4.2	4.2	0
Deposition time/h	0.5	1	0.5	1
Substrate temperature/°C	RT to 250	RT to 250	200 to 250	100 to 250

3. Results and Discussions

3.1. Crystal structural and surface morphology

Fig. 1(a) shows the XRD patterns of ZnO:Al films deposited at different substrate temperatures by RF reactive sputtering with an alloy target (deposition parameters are shown in Group I of Table 1). Strong (002) diffraction peaks are observed on most curves, indicating ZnO preferential c-axis orientation. The intensity of (002) peak increases when the substrate temperature is raised, which provides energy to enhance the atom mobility and results in a crystallinity improvement of ZnO:Al films (15). Table 2 shows the parameters of the (002) diffraction peak of ZnO:Al films. It can be found that FWHM of the (002) peaks decreases as substrate temperatures increase and reaches a minimum of ~0.204° (substrate temperature=250°C), which indicates that 250°C is a suitable substrate temperature to deposit ZnO:Al films with high crystal quality.

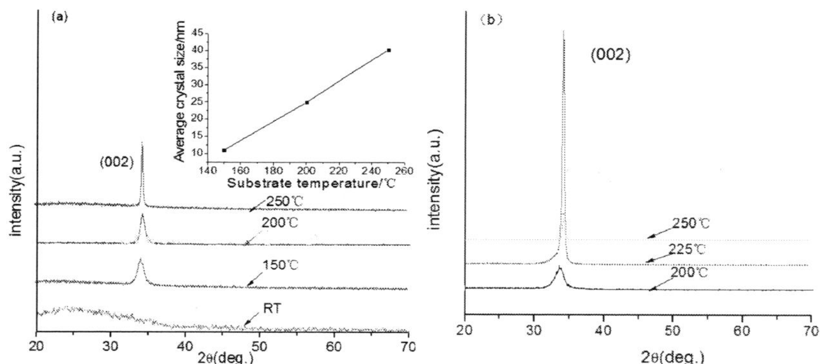

Fig.1 XRD patterns of the ZnO:Al films deposited by RF and DC sputtering mode as respectively shown in (a) and (b). The inset shows the average crystal size (D) of the thin films.

According to the Scherrer formula, the average crystal size (D) could be estimated by:

$$D = \frac{0.89\lambda}{\beta\cos\theta} \tag{1}$$

where λ, β and θ respectively represents wavelength (1.5406 Å), FWHM and diffraction angle of the (002) peak. The inset in Fig.1 (a) shows the dependence of D on substrate temperatures. When the substrate temperature is raised, the D increases. However, at the DC sputtering mode (shown in Fig.1(b)), the intensity of the (002) peak increases with temperature increasing and reaches a maximum at 225℃. As the temperature increases over 225℃, the intensity of (002) peak decreases, which means that over-heating causes a negative influence on the crystallographic properties of ZnO (16).

TABLE2: Parameters of the (002) peak in XRD of Fig.1 (a)

Substrate temperature /℃	RT	100	150	200	250
FWHM /°	-	0.787	0.734	0.326	0.204
D /nm	-	10.49	11.03	24.92	40.31
2θ /°	-	34.467	34.151	33.804	34.381

Fig. 2 shows the surface morphologies of the ZnO:Al films deposited with the parameters shown in Group IV of Table 1. At 100℃, the film has a large root mean square (RMS) roughness of 13.8nm and exhibits columnar structure due to a limited atomic mobility but it decreases to 4.84nm at 150 ℃, which is due to an accelerative surface diffusion in higher substrate temperatures, accompanying with an improved crystallization. With the further increasing of substrate temperature, the roughness slightly changes and the conjoint growth of grains could be observed at 250℃ as shown in Fig 2(d), which is also consistent with Hong's results (19).

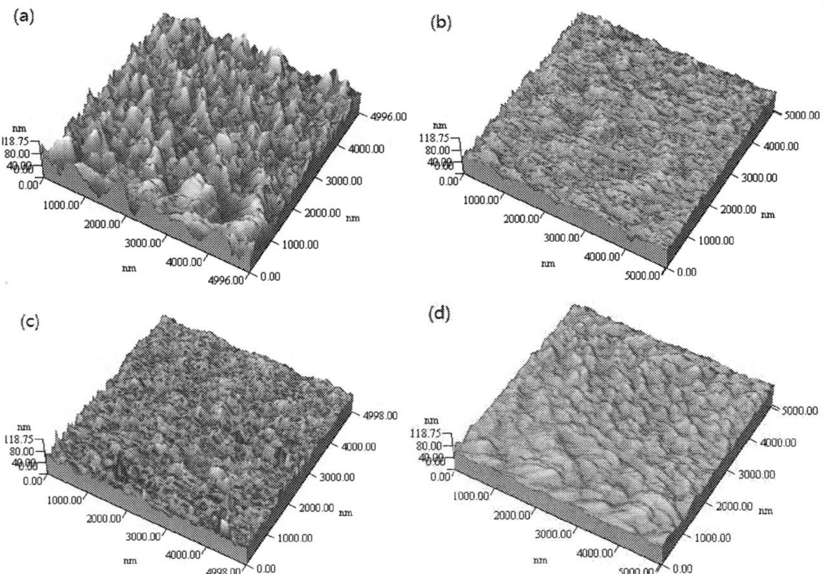

Fig.2 Morphologies of the ZnO:Al films deposited at different substrate temperatures (Group IV) (a)100°C; (b)150°C; (c) 200°C; (d) 250°C.

3.2. Optical properties

The transmittance spectra of the ZnO:Al thin films are also studied to investigate the optical properties of the films. As shown in Fig 3(a), a sharp absorption edge in the transmittance spectra is observed in the wavelength range of 370–380 nm whose corresponding photon energy is equal to the band gap of bulk ZnO (16). In Fig 3(b), the transmittance varies from 80-90% in the visible light region. Absorption coefficient (α) has been estimated from the transmissions spectra and given by:

$$\alpha = \frac{1}{d}\ln(\frac{\sqrt{(1-R)^4 + 4T^2R^2} - (1-R)^2}{2TR^2})$$

[2]

where d is the thickness of film, R and T respectively indicate the reflectance and the transmittance [16]. The following equation could be used to estimate the optical band gap, E_g:

$$\alpha(v)hv = A(hv - E_g)^{1/N}$$

[3]

where hv is the photon energy, N depends on the type of electron transition and it is set to be 2 for a direct-band gap semiconductor (16, 17). The optical band gap is determined by extrapolating the linear part of equation [3] to intercept the energy axis (hv) at $\alpha hv=0$. In this study, the corresponding optical band gap of ZnO:Al films by an alloy target slightly increases from 3.30eV to 3.40eV with a raising substrate temperature. When the target is changed to be a ceramic one, the optical band gap also increases from 3.25eV to 3.29eV (shown in Fig. 3(b)) with an increasing substrate temperature and the absorption edge shifts to the shorter wavelength region. The carrier concentration (n) also increases from $2.28\times10^{20}/cm^3$ to $9.193\times10^{20}/cm^3$ (shown in table 3), with the substrate temperature increasing from room temperature to 250°C. According to the Burstein-Moss theory (18, 19), the band gap widening is proportional to $n^{2/3}$, which is confirmed by our experimental results.

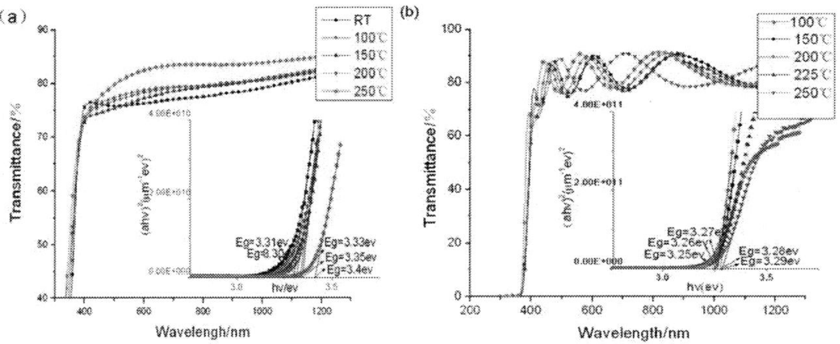

Fig.3 Optical transmittance of ZnO:Al films deposited with parameters shown in Group I (a) and Group IV (b). The inset shows the corresponding dependence of $(\alpha hv)^2$ on hv.

TABLE 3: Electrical properties of ZnO:Al thin films (Group II) with annealing at 400°C for 1 hour

Substrate temperature /°C	Carrier concentration /cm^{-3}	Mobility /cm^2·v^{-1}·s^{-1}	Resistivity /Ω·cm
RT	-2.280×10^{20}	0.910	3.008×10^{-2}
100	-3.223×10^{20}	1.732	1.118×10^{-2}
150	-6.580×10^{20}	2.107	4.502×10^{-3}
200	-7.853×10^{20}	2.100	3.790×10^{-3}
250	-9.193×10^{20}	1.904	3.566×10^{-3}

3.3 Electrical properties

The dependence of electrical resistivity on substrate temperature is shown in Fig. 4. It can be seen that the resistivity of films decreases rapidly with an increasing substrate temperature. It is probably attributed to the increasing of atomic mobile energy in high temperature, which results in a crystallinity improvement and makes crystallines bigger in size and defects less in number. The XRD results in Fig. 1 could be evidence.

Table 3 shows the electrical properties of the ZnO:Al films (Group II) with annealing in nitrogen at 400°C for 1 hour. It can be seen that all films are n-type semiconductor. Carrier concentration (n) increases with a raising substrate temperature and reaches a highest value of 9.193×10^{20}/cm^3 at 250°C. The film's resistivity (ρ) decreases with an increase of substrate temperature from RT to 250°C. A minimal ρ of 3.566×10^{-3} Ω·cm is obtained at 250°C. The ρ is proportional to the reciprocal of product of n and the carrier mobility (μ), which is determined by the following formula:

$$\rho = \frac{1}{nq\mu} \qquad [4]$$

The ZnO:Al film deposited at 250°C shows a lowest resistivity, which probably results from a largest product of n and μ. In addition, compared with the resistivity of the as-deposited films, the resistivity of annealed films drastically declines. It is probably attributed to better crystallization and less grain boundaries, which could also be proved by an increasing μ as shown in Table 3.

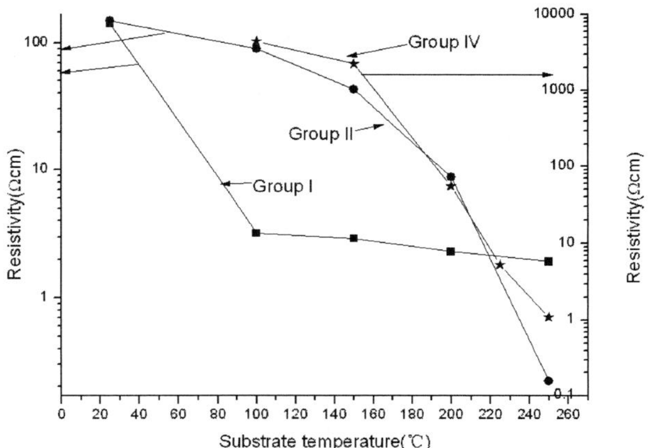

Fig.4. Influence of substrate temperatures on the electrical resistivity of ZnO:Al films.

4. Conclusions

ZnO:Al thin films were deposited on glass substrates by DC or RF magnetron sputtering at different substrate temperatures. The crystallinity of ZnO:Al films is improved with an increasing substrate temperature. High-quality ZnO:Al thin films with a preferential c-axis orientation and smoothest surface could be obtained at the substrate temperature of 250°C. In visible light region, the transmittance of the ZnO:Al thin films is in the range of 80-90% with the band gap of ~3.3eV. The electrical properties of the ZnO:Al films could be improved by annealing in high temperature with a lowest electrical resistivity of $3.566 \times 10^{-3} \Omega \cdot cm$ at substrate temperature 250°C.

Acknowledgments

The project was financially supported by the National Natural Science Foundation of China (Grant No. 61006003), the Natural Science Foundation of Fujian Province (Grant No. 2010J05134), the Scientific Research Foundation for the Introduced Talents of Fuzhou University and the Science and Technology Development Foundation of Fuzhou University (Grant No. 2009-XQ-32).

References

1. W. Gao and Z.W. Li *Ceram Int* 30, 1155 (2004).
2. S. S and R. S. Srinivasa, *Thin Solid Films* 515, 8718 (2007).
3. X. C. Wang, X.M. Chen and B.H. Yang, *J. Alloys Compd.* 488,232 (2009).
4. Z. Q. Xu, H. Deng and Y. Li, *Mater. Res. Bull.* 41,354 (2006).
5. H. J. Jin, M. J. Song and C. B. Park, *Physical B* 404, 1097 (2009).
6. L. J. Li, H. Deng, L. P. Dai, J. J. Chen, Q. L. Yuan and Y. Li, *Mater. Res. Bull.* 43, 1456 (2008).
7. Y. Z. Yoo and T. Sekiguchi, *Appl. Phys. Lett.* 84,502 (2004).
8. Y.G. Wang, S.P. Lau, X. H. Zhang and H. H. Hng, *J. Appl. Phys.* 94,354(2003).
9. B. S. Li, Y.C. Liu, Z. S. Chu, D. Z. Shen, Y. M. Lu, J. Y. Zhang and X.W. Fan, *J. Appl. Phys.* 91, 501(2002).
10. M.J. Alam and D.C. Cameron, *J. Vac. Sci. Technol.* A19, 1642 (2001).
11. A. J. C. Fiddes, K. Durose, A. W. Brinkman, J. Woods, P. D. Coates and A. J. Banister, *J. Cryst. Growth* 159, 210 (1996).
12. T. Ohgaki, N. Ohashi, H. Kakemoto, S. Wada, Y. Adachi, H. Haneda and T. Tsurumi, *J. Appl. Phys.* 93, 1961(2003).
13. Y. R. Ryu, S. Zhu, J. D. Budai, H. R. Chandrasekhar and H. W. White, *J. Appl. Phys.* 88,201(2000).
14. T. Minami, S. Ida and T. Miyata, *Thin Solid Films* 416, 92 (2002).
15. S. J. Kang and Y. H. Joung, *Appl. Surf. Sci.* 253, 7330 (2007).
16. Z. B. Ayadi, L. E. Mir, K. Djessas and S. Alaya, *Thin Solid Films* 517, 6305 (2009).
17. A. T and V. G, *Appl. Surf. Sci.* 256, 5056 (2010).
18. Y. H. Kim and K. S. Lee, *Appl. Surf. Sci.*255,7251 (2009)
19. R. J. Hong and X. Jing, *Appl. Surf. Sci.* 207, 341 (2003).

ECS Transactions, 34 (1) 1173-1182 (2011)
©The Electrochemical Society

Author Index

Van Bael, M. K.	473
Van Elshocht, S.	473
Vancoille, E.	515
Vanhellemont, J.	1151
Vanherle, W.	455
Vath III, C. J.	843
Vérove, C.	389
Vinet, M.	37
Vogel, M.	1059
Vos, I.	535
Vos, R.	671
Wada, M.	671
Waldfried, C.	433
Waldron, N.	647
Wan, X.	1053
Wang, A.	399
Wang, B.	103
Wang, C.	603
Wang, D.	1117
Wang, F.	25, 195
Wang, G.	725
Wang, H.	107
Wang, J.	731
Wang, J.	973
Wang, J.	81
Wang, K.	355
Wang, L.	25, 195, 503, 571
Wang, L.	277
Wang, P.	1109
Wang, P.	17, 155
Wang, Q.	1129
Wang, Q.	195
Wang, Q.	149
Wang, S.	683
Wang, S.	383
Wang, S.	793
Wang, W.	1047
Wang, W.	1017, 1065
Wang, W.	719
Wang, W.	103
Wang, X.	405
Wang, X.	445
Wang, X.	683
Wang, Y.	209
Wang, Y.	149
Wang, Y.	1123
Wang, Y.	173
Wang, Z.	949
Warren, P.	799
Washburn, C.	249
Wasisto, H.	979
Watanabe, S.	615
Wei, J. C.	343
Wei, M.	61
Wei, X.	195, 503, 541, 705
Wei, X.	819
Weiwei, L.	691
Wensheng, Q.	143
Westerman, R.	421
Wimplinger, M.	1023
Witters, L.	455
Witters, L.	647
Wu, B.	49
Wu, B.	997
Wu, G.	1053
Wu, H.	49
Wu, J. Y.	1165
Wu, J.	719, 731, 737
Wu, L.	699
Wu, Q.	269, 303
Wu, W.	931
Xi, Y.	985
Xia, J.	439
Xiao, S.	277
Xiao, S.	173
XIAO, W.	49
Xie, C.	215
Xie, D.	997
Xie, J.	1053
Xing, C.	677, 743, 787
Xing, W.	189
Xu, B. Y.	557
Xu, B.	563
Xu, C.	737